# Cryptography 101

## From Theory to Practice

Rolf Oppliger

ARTECH HOUSE
BOSTON | LONDON
artechhouse.com

**Library of Congress Cataloging-in-Publication Data**
A catalog record for this book is available from the U.S. Library of Congress.

**British Library Cataloguing in Publication Data**
A catalogue record for this book is available from the British Library.

Cover design by Charlene Stevens

ISBN 13: 978-1-63081-846-3


**685 Canton Street**
**Norwood, MA 02062**



10 9 8 7 6 5 4 3 2 1

*To my family*

# Contents

# III PUBLIC KEY CRYPTOSYSTEMS 325

# Foreword

Fifty years ago, there was no Internet. Not everyone had a telephone, and they were all analog. There was no WWW, no "cloud." Our cars were all mechanical, microwave ovens were still too big and expensive for most homes, and there were only a handful of universities offering classes in the new field of "computer science." If we wanted to find some facts about coconut exports from Malaysia or a chronological list of Viceroys of Peru, there was no Google: Instead, we needed to visit a library and consult paper books and encyclopedia.

The world has changed considerably in a matter of decades—less than an average human lifespan. Many children growing up today may never have seen a landline telephone or an analog clock, referenced a bound encyclopedia, or know what a "long-distance" call means other than in a history class or by watching old movies on their tablets. The pace of innovation seems to be increasing, outpacing the predictions of technologists and science fiction authors alike.

Some basic concepts thread themselves through these changes in technology and have been present for millennia. How do we secure information from being read by the unauthorized? How do we prevent forgeries and theft? How do we verify identity at a distance? And how can we be sure that the measures we take to safeguard our information, property, and identities are proof against malicious attack? Whether our information is on a paper ledger or a spreadsheet in a cloud container, we want to have confidence that it is protected.

Cryptography has been—and continues to be—a linchpin of our stored information and communications' trustworthiness. We can encrypt economic transactions ranging from online shopping to cryptocurrency to international banking so as to prevent theft and fraud. People use encryption to hide political and religious communications from oppressive governments. We all employ digital signatures to sign contracts. Companies use cryptographic hash codes to protect the software in electronic voting machines and critical security patches. Cryptography is all around us; although we don't see it, we depend on cryptography's correct functioning in dozens of different ways every day.

Rolf Oppliger wrote his first book on cryptography in 2005, and I wrote the forward to it. We both knew the history of cryptography: For much of the past, it was the domain of government agents, clandestine lovers, and smugglers. We also saw how it had become a critical component of network communications, international finance, and everyday commerce in a matter of a few decades. I don't think either of us envisioned how much the world would embrace fast-changing digital technologies and thus have an even greater reliance on strong cryptography.

What we (and others) have also learned—repeatedly—over the last few decades is that knowing powerful cryptographic methods is not sufficient on its own. The implementations of those algorithms and supporting mechanisms (e.g., random number generation, key sharing) must be correct as well. That is why this current book delves deeper into considerations of how algorithms are actually constructed and realized. As such, it presents enhanced tutelage in the basics of how and why to use cryptography suitable for its increasing importance.

In the next 15–20 years, we are likely to see much more innovation. Many of the items around us will have embedded, communicating processors in the Internet of Things. Self-driving vehicles will be commonplace, quantum computation may be widely used, and direct human-computer interfaces might provide greater bandwidth from system to brain than our biological eyes can sustain. Whatever technology may be deployed, I am confident that cryptography will still be a crucial element to protect our privacy and our digital property, attest to our identities, and safeguard our systems' integrity. To function in that world, people will need to understand cryptographic mechanisms and their applications. You hold in your possession a valuable guide to how to do just that. As such, this is more than a primer—it is a prospectus for future success.

*Eugene H. Spafford*
*Professor, Purdue University*
*June 2021*

# Preface

*Necessity is the mother of invention,*
*and computer networks are the mother of modern cryptography.*

— Ronald L. Rivest[1]

*Any sufficiently advanced technology is indistinguishable from magic.*

— Arthur C. Clarke[2]

With the current ubiquity of computer networks and distributed systems in general and the Internet in particular, cryptography has become a key enabling technology to secure the information infrastructure(s) we are building, using, and counting on in daily life. This is particularly true for modern cryptography.[3] The important role of (modern) cryptography is, for example, pointed out by the first quote given above. As explained later in the book, the quoted cryptographer—Ronald L. Rivest—is one of the leading pioneers and experts of modern cryptography, and he coinvented the Rivest, Shamir, Adleman (RSA) public key cryptosystem that is omnipresent and in widespread use today.

According to the second quote given above, cryptography sometimes looks magical and seems to solve problems one would consider impossible to solve at first sight, such as exchanging a secret key in public, tossing a coin over a public network, or proving knowledge of some fact without leaking any information about it. Unfortunately, most of these solutions don't come with a mathematically rigorous proof, but only with a reduction proof that is relative to some other problem and relies on specific assumptions. Consequently, we don't really know whether the

1 In "Cryptography as Duct Tape," a short note written to the Senate Commerce and Judiciary Committees in opposition to mandatory key recovery proposals on June 12, 1997 (the note is available electronically at http://theory.lcs.mit.edu/∼rivest/ducttape.txt).

2 This quote is also known as Clarke's third law.

3 In Chapter 1, we explain what modern cryptography means and how it actually differs from classical cryptography.

claimed solution really or only seemingly works and is therefore "indistinguishable from magic." The fundamental question whether cryptography is science or magic was first asked by James L. Massey in a 2001 eyeopening talk[4] and later reopened by Dieter Gollmann in 2011.[5] Even today it remains a key question and a particularly interesting and challenging research topic. It boils down to the question of whether one-way functions really exist (given the fact that we "only" have candidates for such functions at hand). At the end of the book, we'll revisit this point and argue that telling cryptography apart from magic and illusion is sometimes difficult if not impossible, and we'll coin the term *cryptollusion* for this purpose.

Due to its important role, computer scientists, electrical engineers, and applied mathematicians should be educated in both the theoretical principles and practical applications of cryptography. Cryptography is a tool, and as such it can provide security only if used properly. If not used properly, it may fail to provide security, and the result may even be worse than not using cryptography in the first place. This is because users think they are protected, whereas in reality this is not the case, and this, in turn, may lead to inappropriate or incorrect behavior. We know from daily life that incorrect user behavior may lead to security breaches.

There are many books that can be used for educational purposes (e.g., [1–33] itemized in alphabetical order).[6] Among these books, I particularly recommend [2, 3, 13, 24, 26, 28] to teach classes, [20] to serve as a handy reference for cryptographic algorithms and protocols (also available electronically on the Internet[7]), [12] to provide a historical perspective, and [4] to overview practically relevant cryptographic standards.

In the early 2000s (when most of the books mentioned above were not yet available), I spent a considerable amount of time writing a manuscript that I could use to lecture and teach classes on cryptography. I then decided to turn the manuscript into a book that would provide a comprehensive overview about contemporary cryptography. In 2005, the first edition of the resulting book was published in Artech House's Information Security and Privacy Series[8] [34], and the second edition was released in 2011 [35]. The target audience of either edition included mathematicians, computer scientists, and electrical engineers, both in research and practice. It goes without saying that this also included computer practitioners, consultants, and information security officers who wanted to gain insight into this fascinating and quickly evolving field.

4 https://techtv.mit.edu/videos/16442-cryptography-science-or-magic.

5 https://www.youtube.com/watch?v=KOlOqb8q2Gs.

6 In addition to these books, Dan Boneh and Victor Shoup are drafting "A Graduate Course in Applied Cryptography" that is available from https://toc.cryptobook.us.

7 http://www.cacr.math.uwaterloo.ca/hac.

8 https://www.esecurity.ch/serieseditor.html.

Many things have happened since then, and two trends have made it particularly difficult to come up with a third edition of the book: On the one hand, the entire field has expanded in many ways, making it more and more difficult to remain comprehensive. On the other hand, the gap between theory and practice has become very large (if not too large). To put it in Albert Einstein's words: "In theory, theory and practice are the same. In practice, they are not." Today, there are theoretical cryptographers working on proofs and new constructs that hardly have real-world consequences, whereas there are practical cryptographers working on trying to securely implement cryptographic primitives and combine them in security protocols and applications in some ad hoc manner. Working in one of these worlds does not necessarily mean that one also understands the other world. It has therefore become important to let practitioners know about theoretical results and breakthroughs and theorists know about what can actually go wrong in the practice. So theorists and practitioners can learn from each other, but to do so they have to understand each other first.

Against this background, I have changed both the title and the focus of the new book: The new title is *Cryptography 101: From Theory to Practice*, and the new focus is to bring theory and practice in line and to bridge the currently existing gap between them. Honestly speaking, another subtitle would have been *From Practice to Theory*, because it better mirrors my own course of action, but this order is somehow unusual and therefore not used here. Anyway, the resulting book is intended to be accessible to both theorists and practitioners, and to provide them with sufficient information to understand the other world. Its structure still follows [34] and [35], but most parts have been written entirely from scratch.

*Cryptography 101: From Theory to Practice* is still tutorial in nature. It starts with two chapters that introduce the topic and briefly overview the cryptographic systems (or cryptosystems) in use today. The book then addresses unkeyed cryptosystems (Part I), secret key cryptosystems (Part II), and public key cryptosystems (Part III). Part III also includes cryptographic protocols that make use of public key cryptography, such as key exchange and zero-knowledge protocols. Finally, the book finishes with some conclusions (Part IV) and several appendixes about discrete mathematics, probability theory, information theory, and complexity theory.

Each chapter is intended to be comprehensive on its own and almost all include a list of references that can be used for further study. Where necessary and appropriate, I have added uniform resource locators (URLs) as footnotes to the text. The URLs point to corresponding information pages on the Web. While care has been taken to ensure that the URLs are valid now, unfortunately—due to the dynamic nature of the Internet and the Web—I cannot guarantee that these URLs and their contents remain valid forever. In regard to the URLs, I apologize for any information page that may have been removed or replaced since the time of writing and publishing of the book. To make the problem less severe, I have not included URLs that can be expected to be removed or replaced soon.

Readers who like to experiment with cryptographic systems are invited to download, install, and play around with some of the many software packages that have been written and are available for demonstrational and educational purposes. Among these packages, I particularly recommend CrypTool (with its many variants) that is publicly and freely available,[9] and that provides insight into the basic working principles of the cryptographic algorithms and protocols in use today. If you are particularly interested and savy in mathematics, then you may use the open source software system Sage[10] that yields a viable alternative to commercial software packages, like Maple, Mathematica, or MATLAB.

If you want to implement and market some of the cryptographic techniques or systems addressed in this book, then you must be cautious and note that the entire field of cryptography is tied up in patents and corresponding patent claims. Consequently, the situation is highly involved, and you must make sure that you have appropriate licenses or a good lawyer (and preferably both). This is particularly important if you intend to distribute and commercialize the implementation.

To make things worse, regulations for both the use and the import and export of cryptographic products differ from country to country.[11] Most importantly, with regard to the export of cryptographic products, the situation is involved. In the United States, for example, the Bureau of Industry and Security (BIS) of the Department of Commerce (DoC) has been in charge of export controls since 1996. The rules governing exports and reexports of cryptographic products are found in the Export Administration Regulations (EAR). If a U.S. company wants to sell a cryptographic product overseas, it must have export approval according to the EAR.

In January 2000, the DoC published a regulation implementing the White House's announcement of a new framework for U.S. export controls on encryption items.[12] The policy was in response to the changing global market, advances in technology, and the need to give U.S. industry better access to these markets, while continuing to provide essential protections for national security. The regulation enlarged the use of license exceptions, implemented the changes agreed to at the Wassenaar Arrangement[13] on export controls for conventional arms and dual-use

9 https://www.cryptool.org.

10 https://www.sagemath.org.

11 There are usually no regulations for the import of cryptographic products. A respective survey is, for example, available at http://www.cryptolaw.org. As of this writing, the latest version is 27.0 from 2013.

12 The announcement was made on September 16, 1999.

13 The Wassenaar Arrangement (https://www.wassenaar.org) is a treaty originally negotiated in July 1996 and signed by 31 countries to restrict the export of dual-use goods and technologies to specific countries considered to be dangerous.

goods and technologies in December 1998, and eliminated the deemed export rule for encryption technology. In addition, new license exception provisions were created for certain types of encryption, such as source code and toolkits. Some countries, such as Cuba, Iran, Iraq, Libya, North Korea, Sudan, and Syria, are exempted from the regulation. We are not going to address legal issues surrounding the use and export of cryptographic products in this book, but note again that you may want to talk to a lawyer before you either use or export cryptographic products.

Last, but not least, it is important to note that *Cryptography 101: From Theory to Practice* addresses only the materials that are published and available in the open literature. These materials are, for example, presented and discussed at the conferences held by the International Association for Cryptologic Research (IACR[14]). There may (or may not) be additional and complementary materials available in the backyards of secret services and intelligence agencies. These materials are subject to speculations and rumors; sometimes they provide a starting point for best-selling books and movie plots, and sometimes they even turn out to be true in the aftermath. In this book, we do not speculate about these materials. It is, however, important to note and keep in mind that these materials still exist and that their mere existence may make this book or parts of it obsolete (once their existence becomes known). For example, the notion of public key cryptography was invented by employees of a British intelligence agency a few years before it was published in the open literature (Section 1.3). Also, the data encryption standard (DES) was designed to be resistant against differential cryptanalysis—a cryptanalytical attack against block ciphers that was discussed in the public literature only two decades after the standardization of the DES (Section 9.6.1.4). There are certainly many other (documented or undocumented) examples that may illustrate the point.

I hope that *Cryptography 101: From Theory to Practice* serves your needs. Also, I would like to take the opportunity to invite you as a reader to let me know your opinions and thoughts. If you have something to correct or add, please let me know. If I have not expressed myself clearly, please let me know, too. I appreciate and sincerely welcome any comment or suggestion in order to update the book in future editions and to turn it into an appropriate reference book that can be used for educational purposes. The best way to reach me is to send a message to `rolf.oppliger@esecurity.ch`. You may also visit my homepage at `rolf-oppliger.com` or `rolf-oppliger.ch`, the book's homepage at `https://books.esecurity.ch/crypto101.html` (to find post errata lists, additional information, and complementary material), or my blogs at `https://blog.esecurity.ch` for information security and privacy in general, and `https://cryptolog.esecurity.ch` for cryptology in particular. I'm looking forward to hearing from you in one way or another.

14 https://www.iacr.org.

## References


[1] Aumasson, J.P., *Serious Cryptography: A Practical Introduction to Modern Encryption*. No Starch Press, San Francisco, CA, 2017.

[2] Buchmann, J.A., *Introduction to Cryptography*, 2nd edition. Springer-Verlag, New York, 2004.

[3] Delfs, H., and H. Knebl, *Introduction to Cryptography: Principles and Applications*, 3rd edition. Springer-Verlag, New York, 2015.

[4] Dent, A.W., and C.J. Mitchell, *User's Guide to Cryptography and Standards*. Artech House Publishers, Norwood, MA, 2004.

[5] Easttom, C., *Modern Cryptography: Applied Mathematics for Encryption and Information Security*. McGraw-Hill Education, 2015.

[6] Ferguson, N., and B. Schneier, *Practical Cryptography*. John Wiley & Sons, New York, 2003.

[7] Ferguson, N., B. Schneier, and T. Kohno, *Cryptography Engineering: Design Principles and Practical Applications*. John Wiley & Sons, New York, 2010.

[8] Garrett, P.B., *Making, Breaking Codes: Introduction to Cryptology*. Prentice Hall PTR, Upper Saddle River, NJ, 2001.

[9] Goldreich, O., *Foundations of Cryptography: Volume 1, Basic Tools*. Cambridge University Press, Cambridge, UK, 2007.

[10] Goldreich, O., *Foundations of Cryptography: Volume 2, Basic Applications*. Cambridge University Press, Cambridge, UK, 2009.

[11] Hoffstein, J., J. Pipher, and J.H. Silverman, *An Introduction to Mathematical Cryptography*, 2nd edition. Springer-Verlag, New York, 2014.

[12] Kahn, D., *The Codebreakers: The Comprehensive History of Secret Communication from Ancient Times to the Internet*. Scribner, New York, 1996.

[13] Katz, J., and Y. Lindell, *Introduction to Modern Cryptography*, 2nd edition. Chapman & Hall/CRC, Boca Raton, FL, 2014.

[14] Koblitz, N., *A Course in Number Theory and Cryptography*, 2nd edition. Springer-Verlag, New York, 1994.

[15] Koblitz, N., *Algebraic Aspects of Cryptography*. Springer-Verlag, New York, 2004.

[16] Konheim, A.G., *Computer Security and Cryptography*. Wiley-Interscience, 2007.

[17] Luby, M., *Pseudorandomness and Cryptographic Applications*. Princeton University Press, Princeton, NJ, 2019.

[18] Mao, W., *Modern Cryptography: Theory and Practice*. Prentice Hall PTR, Upper Saddle River, NJ, 2003.

[19] Martin, K.M., *Everyday Cryptography: Fundamental Principles & Applications*, 2nd edition. Oxford University Press, New York, 2017.

[20] Menezes, A., P. van Oorschot, and S. Vanstone, *Handbook of Applied Cryptography*. CRC Press, Boca Raton, FL, 1996.

[21] Mollin, R.A., *RSA and Public-Key Cryptography*. Chapman & Hall/CRC, Boca Raton, FL, 2002.

[22] Mollin, R.A., *Codes: The Guide to Secrecy From Ancient to Modern Times*. Chapman & Hall/CRC, Boca Raton, FL, 2005.

[23] Mollin, R.A., *An Introduction to Cryptography*, 2nd edition. Chapman & Hall/CRC, Boca Raton, FL, 2006.

[24] Paar, C., and J. Pelzl, *Understanding Cryptography: A Textbook for Students and Practitioners*, Springer-Verlag, New York, 2010.

[25] Schneier, B., *Applied Cryptography: Protocols, Algorithms, and Source Code in C*, 20th Anniversary Edition. John Wiley & Sons, New York, 2015.

[26] Smart, N., *Cryptography Made Simple*. Springer-Verlag, New York, 2016.

[27] Stanoyevitch, A., *Introduction to Cryptography with Mathematical Foundations and Computer Implementations*. Chapman & Hall/CRC, Boca Raton, FL, 2010.

[28] Stinson, D., and M. Paterson, *Cryptography: Theory and Practice*, 4th edition. Chapman & Hall/CRC, Boca Raton, FL, 2018.

[29] Talbot, J., and D. Welsh, *Complexity and Cryptography: An Introduction*. Cambridge University Press, Cambridge, UK, 2006.

[30] Vaudenay, S., *A Classical Introduction to Cryptography: Applications for Communications Security*. Springer-Verlag, New York, 2006.

[31] Von zur Gathen, J., *CryptoSchool*. Springer-Verlag, New York, 2015.

[32] Wang, et al., *Mathematical Foundations of Public Key Cryptography*. CRC Press, Boca Raton, FL, 2015.

[33] Yan, S.Y., *Computational Number Theory and Modern Cryptography*. John Wiley & Sons, New York, 2013.

[34] Oppliger, R, *Contemporary Cryptography*. Artech House Publishers, Norwood, MA, 2005.

[35] Oppliger, R, *Contemporary Cryptography*, 2nd edition. Artech House Publishers, Norwood, MA, 2011.

# Acknowledgments

There are many people involved in the writing and publication of a book about cryptography. This already applied to the two editions of *Contemporary Cryptography*, but it also applies to this book. In particular, I thank my students and course participants who have found numerous errors in my previous books and who are likely to find errors in this one, François Weissbaum for reading large parts of the manuscript, pointing me to some embarrassing errors, and discussing interesting questions with me; Ilias Cherkaoui for reviewing the entire manuscript; and Gene Spafford for providing the foreword. Once again, the staff at Artech House was enormously helpful in producing and promoting the book. Among these people, I am particularly grateful to David Michelson, Merlin Fox, and Soraya Nair. My most important thanks go to my family—my wife Isabelle and our beloved youngsters Lara and Marc. Without their encouragement, support, patience, and love, this book would not exist.

# Chapter 1

# Introduction

In this chapter, we pitch the field and introduce the topic of the book, namely cryptography, at a high operating altitude and level of abstraction. More specifically, we elaborate on cryptology (including cryptography) in Section 1.1, address cryptographic systems (or cryptosystems for short) in Section 1.2, provide some historical background information in Section 1.3, and outline the rest of the book in Section 1.4. The aim is to lay the basics to understand and put into proper perspective the contents of the book.

## 1.1 CRYPTOLOGY

The term *cryptology* is derived from the Greek words "kryptós," meaning "hidden," and "lógos," meaning "word." Consequently, the term cryptology can be paraphrased as "hidden word." This refers to the original intent of cryptology, namely to hide the meaning of words and to protect the confidentiality and secrecy of the respective data accordingly. As will (hopefully) become clear throughout the book, this viewpoint is too narrow, and the term cryptology is currently used for many other security-related purposes and applications in addition to the protection of the confidentiality and secrecy of data.

More specifically, cryptology refers to the mathematical science and field of study that comprises cryptography and cryptanalysis.

- The term *cryptography* is derived from the Greek words "kryptós" (see above) and "gráphein," meaning "to write." Consequently, the meaning of the term cryptography can be paraphrased as "hidden writing." According to the Internet security glossary provided in Request for Comments (RFC) 4949 [1], cryptography also refers to the "mathematical science that deals with transforming data to render its meaning unintelligible (i.e., to hide its semantic content), prevent its undetected alteration, or prevent its unauthorized use. If the transformation is reversible, cryptography also deals with restoring encrypted data to intelligible form." Consequently, cryptography refers to the process of protecting data in a very broad sense.

- The term *cryptanalysis* is derived from the Greek words "kryptós" (see above) and "analýein," meaning "to loosen." Consequently, the meaning of the term can be paraphrased as "to loosen the hidden word." This paraphrase refers to the process of destroying the cryptographic protection, or—more generally—to study the security properties and possibilities to break cryptographic techniques and systems. Again referring to [1], the term cryptanalysis refers to the "mathematical science that deals with analysis of a cryptographic system in order to gain knowledge needed to break or circumvent[1] the protection that the system is designed to provide." As such, the cryptanalyst is the antagonist of the cryptographer, meaning that his or her job is to break or—more likely—circumvent the protection that the cryptographer has designed and implemented in the first place. Quite naturally, there is an arms race going on between the cryptographers and the cryptanalysts (but note that an individual person may have both skills, cryptographic and cryptanalytical ones).

Many other definitions for the terms cryptology, cryptography, and cryptanalysis exist and can be found in the literature (or on the Internet, respectively). For example, the term cryptography is sometimes said to more broadly refer to the study of mathematical techniques related to all aspects of information security (e.g., [2]). These aspects include (but are not restricted to) data confidentiality, data integrity, entity authentication, data origin authentication, nonrepudiation, and/or many more. Again, this definition is broad and comprises anything that is directly or indirectly related to information security.

In some literature, the term cryptology is said to also include steganography (in addition to cryptography and cryptanalysis).

- The term *steganography* is derived from the Greek words "steganos," meaning "impenetrable," and "gráphein" (see above). Consequently, the meaning of the term can be paraphrased as "impenetrable writing." According to [1], the

1 In practice, circumventing (bypassing) the protection is much more common than breaking it. In his 2002 ACM Turing Award Lecture (https://amturing.acm.org/vp/shamir_2327856.cfm or https://www.youtube.com/watch?v=KUHaLQFJ6Cc), for example, Adi Shamir—a coinventor of the RSA public key cryptosystem—made the point that "cryptography is typically bypassed, not penetrated," and this point was so important to him that he put it as a third law of security (in addition to "absolutely secure systems do not exist" and "to halve your vulnerability you have to double your expenditure").

term refers to "methods of hiding the existence of a message or other data. This is different than cryptography, which hides the meaning of a message but does not hide the message itself." Let us consider an analogy to clarify the difference between steganography and cryptography: if we have money to protect or safeguard, then we can either hide its existence (by putting it, for example, under a mattress), or we can put it in a safe that is as burglarproof as possible. In the first case, we are referring to steganographic methods, whereas in the second case, we are referring to cryptographic ones. An example of a formerly widely used steganographic method is invisible ink. A message remains invisible, unless the ink is subject to some chemical reaction that causes the message to reappear and become visible again. Currently deployed steganographic methods are much more sophisticated, and can, for example, be used to hide information in electronic files. In general, this information is arbitrary, but it is typically used to identify the owner or the recipient of a file. In the first case, one refers to *digital watermarking*, whereas in the second case one refers to *digital fingerprinting*. Both are active areas of research.

The relationship between cryptology, cryptography, cryptanalysis, and steganography is illustrated in Figure 1.1. In this book, we only focus on cryptography in a narrow sense (this is symbolized with the shaded box in Figure 1.1). This can also be stated as a disclaimer: we elaborate on cryptanalysis only where necessary and appropriate, and we do not address steganography at all. There are other books that address cryptanalysis (e.g., [3–5]) or provide useful information about steganography in general (e.g., [6, 7]) and digital watermarking and fingerprinting in particular (e.g., [8, 9]).

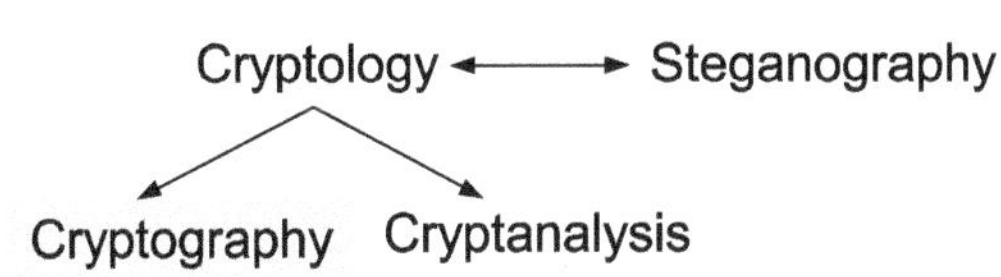


**Figure 1.1** The relationship among cryptology, cryptography, cryptanalysis, and steganography.

Interestingly, cryptographic and steganographic technologies and techniques are not mutually exclusive, and it may make a lot of sense to combine them. For example, the open source disk encryption software VeraCrypt[2] employs steganographic techniques to hide the existence of an encrypted disk volume (known as a "hidden volume"). It is possible and likely that we will see more combinations of cryptographic and steganographic technologies and techniques in future products.

## 1.2 CRYPTOGRAPHIC SYSTEMS

According to [1], the term *cryptographic system*[3] (or *cryptosystem*) refers to "a set of cryptographic algorithms together with the key management processes that support use of the algorithms in some application context." Again, this definition is fairly broad and comprises all kinds of cryptographic algorithms and—as introduced below—protocols. Hence, the notion of an algorithm[4] captured in Definition 1.1 is key for a cryptographic system.

**Definition 1.1 (Algorithm)** *An* algorithm *is a well-defined computational procedure that takes a value as input and turns it into another value that represents the output.*

In addition to being well-defined, it is sometimes required that the algorithm halts within a reasonable amount of time (for any meaningful definition of "reasonable"). Also, Definition 1.1 is rather vague and not mathematically precise. It neither states the computational model for the algorithm, nor does it say anything about the problem the algorithm is supposed to solve, such as, for example, computing a mathematical function. Consequently, from a theoretical viewpoint, an algorithm can be more precisely defined as a well-defined computational procedure for a well-defined computational model for solving a well-defined problem. This definition, however, is a little bit clumsy and therefore not widely used in field. Instead, people usually prefer the simpler (and more intuitive) definition stated above.

In practice, a major distinction is made between algorithms that are deterministic and algorithms that are not (in which case they are probabilistic or randomized).

2 VeraCrypt (https://www.veracrypt.fr) is a fork of the formerly very popular but discontinued TrueCrypt project.

3 In some literature, the term *cryptographic scheme* is used to refer to a cryptographic system. Unfortunately, it is seldom explained what the difference is between a (cryptographic) scheme and a system. So for the purpose of this book, we don't make a distinction, and we use the term cryptographic system to refer to either of them. We hope that this simplification is not too confusing. In the realm of digital signatures, for example, people often use the term digital signature scheme that is not used in this book. Instead, we consistently use the term digital signature system to refer to the same construct.

4 The term *algorithm* is derived from the name of the mathematician Mohammed ibn-Musa al-Khwarizmi, who was part of the royal court in Baghdad and lived from about 780 to 850.

- An algorithm is *deterministic* if its behavior is completely determined by the input. This also means that the algorithm always generates the same output for the same input (if executed multiple times).
- An algorithm is *probabilistic* (or *randomized*) if its behavior is not completely determined by the input, meaning that the algorithm internally employs some (pseudo)random values.[5] Consequently, a probabilistic algorithm may generate different outputs each time it is executed with the same input.

Today, probabilistic algorithms play a much more important role in cryptography than they used to play in former times. Anyway, an algorithm may be implemented by a computer program that is written in a specific programming language, such as Pascal, C, or Java. Whenever we describe algorithms in this book, we don't use a specific programming language, but we use a more formal and simpler notation that looks as follows:

| (input parameters) |
|---|
| computational step |
| ... |
| computational step |
| (output parameters) |

The input and output parameters are written in brackets at the beginning and at the end of the algorithm description, whereas the body of the algorithm consists of a sequence of computational steps that are executed in the specified order. Throughout the book, we sometimes introduce cryptosystems as sets of algorithms that are each written in this notation.

If more than one entity is involved in the execution of an algorithm (or the computational procedure it defines, respectively), then one is in the realm of protocols—a term that originates from diplomacy. Definition 1.2 captures the notion of a protocol.

**Definition 1.2 (Protocol)** *A* protocol *is a distributed algorithm in which two or more entities take part.*

Alternatively, one can define a protocol as a distributed algorithm in which a set of more than one entity (instead of two or more entities) takes part. In this case, it is immediately clear that an algorithm also represents a protocol, namely one that is degenerated in the sense that the set consists of just one entity. Hence,

5 A value is random (pseudorandom) if it is randomly (pseudorandomly) generated.

an algorithm can also be seen as a special case of a protocol. The major distinction between an algorithm and a protocol is that only one entity is involved in the former, whereas two or more entities are involved in the latter. This distinguishing fact is important and must be kept in mind when one talks about algorithms and protocols—not only cryptographic ones. It means that in a protocol the different entities may have to send messages to each other, and hence that a protocol may also comprise communication steps (in addition to computational steps). As such, protocols tend to be more involved than algorithms and this also affects their analysis. Similar to an algorithm, a protocol may be deterministic or probabilistic, depending on whether the protocol internally employs random values. For the purpose of this book, we are mainly interested in cryptographic algorithms and protocols as suggested in Definitions 1.3 and 1.4.

**Definition 1.3 (Cryptographic algorithm)** *A* cryptographic algorithm *is an algorithm that employs and makes use of cryptographic techniques and mechanisms.*

**Definition 1.4 (Cryptographic protocol)** *A* cryptographic protocol *is a protocol that employs and makes use of cryptographic techniques and mechanisms.*

Remember the definition for a cryptographic system (or cryptosystem) given at the beginning of this section. According to this definition, a cryptosystem may comprise more than one algorithm, and the algorithms need not be executed by the same entity, that is, they may be executed by multiple entities in a distributed way. Consequently, this notion of a cryptosystem also captures the notion of a cryptographic protocol as suggested in Definition 1.4. Hence, another way to look at cryptographic algorithms and protocols is to say that a cryptographic algorithm is a *single-entity cryptosystem*, whereas a cryptographic protocol is a *multi-entity* or *multiple entities cryptosystem*. These terms, however, are not used in the literature, and hence we don't use them in this book either.

At this point in time, it is important to note that a typical cryptographic application may consist of multiple (cryptographic) protocols, that these protocols and their concurrent execution may interact in subtle ways, and that the respective interdependencies may be susceptible to *multi-protocol attacks*.[6] As its name suggests, more than one protocol is involved in such an attack, and the adversary may employ messages from one protocol execution to construct valid-looking messages for other protocols or executions thereof. If, for example, one protocol uses digital signatures for random-looking data and another protocol is an authentication protocol in which an entity must digitally sign a nonce to authenticate itself, then an adversary can use the first protocol as an oracle and has it digitally sign a nonce from the second

6 The notion of a *chosen protocol* or *multi-protocol attack* first appeared in a 1997 paper [10], but the problem had certainly preexisted before that.

protocol. This is a simple and straightforward attack that can be mitigated easily (for example, by using two distinct keys). However, more involved interactions and interdependencies are possible and likely exist, and hence multi-protocol attacks tend to be powerful and difficult to mitigate. Fortunately, many such attacks have been described in scientific papers, but only few have been mounted in the field so far—at least as far as we know today.

In the cryptographic literature, it is common to use human names for entities that participate in cryptographic protocols, such as a Diffie-Hellman key exchange. For example, in a two-party protocol the participating entities are usually called *Alice* and *Bob*. This is a convenient way of making things unambiguous with relatively few words, since the pronoun *she* can be used for Alice, and *he* can be used for Bob. The disadvantage of this naming scheme is that people automatically assume that the entities refer to human beings This need not be the case, and Alice, Bob, and all other entities are rather computer systems, cryptographic devices, hardware modules, smartcards, or anything along these lines. In this book, we don't follow the tradition of using Alice, Bob, and the rest of the gang. Instead, we use single-letter characters, such as A, B, and so on, to refer to the entities that take part and participate in a cryptographic protocol. This is admittedly less fun, but more appropriate (see, for example, [11] for a more comprehensive reasoning about this issue). In reality, the entities refer to social-technical systems that may have a user interface, and the question of how to properly design and implement such an interface is key to the overall security of the system. If this interface is not appropriate, then phishing and many other types of social engineering attacks become trivial to mount.

The cryptographic literature provides many examples of more or less useful cryptographic protocols. Some of these protocols are overviewed, discussed, and put into perspective in this book. To formally describe a (cryptographic) protocol in which A and B take part, we use the following notation:

| **A**<br>(input parameters) | | **B**<br>(input parameters) |
|---|---|---|
| ... | | ... |
| computational step | | computational step |
| ... | | ... |
| | $\longrightarrow$ | |
| | ... | |
| | $\longleftarrow$ | |
| ... | | ... |
| computational step | | computational step |
| ... | | ... |
| (output parameters) | | (output parameters) |

Some input parameters may be required on either side of the protocol (note that the input parameters need not be the same). The protocol then includes a sequence of computational and communication steps. Each computational step may occur only on one side of the protocol, whereas each communication step requires data to be transferred from one side to the other. In this case, the direction of the data flow is indicated by an arrow. The set of all data that is communicated this way refers to the *protocol transcript*. Finally, some parameters may be output on either side of the protocol. These output parameters actually represent the result of the protocol execution. Similar to the input parameters, the output parameters need not be the same on either side. In many cases, however, the output parameters are the same. In the case of the Diffie-Hellman key exchange, for example, the output is the session key that can subsequently be used to secure communications.

### 1.2.1 Classes of Cryptographic Systems

Cryptographic systems may or may not use secret parameters (e.g., cryptographic keys). If secret parameters are used, then they may or may not be shared among the participating entities. Consequently, there are three classes of cryptographic systems that can be distinguished.[7] They are captured in Definitions 1.5–1.7.

**Definition 1.5 (Unkeyed cryptosystem)** *An* unkeyed cryptosystem *is a cryptographic system that uses no secret parameter.*

**Definition 1.6 (Secret key cryptosystem)** *A* secret key cryptosystem *is a cryptographic system that uses secret parameters that are shared among the participating entities.*

**Definition 1.7 (Public key cryptosystem)** *A* public key cryptosystem *is a cryptographic system that uses secret parameters that are not shared among the participating entities.*

In Chapter 2, we informally introduce and briefly overview the most important representatives of these classes. These representatives are then formally addressed in Part I (unkeyed cryptosystems), Part II (secret key cryptosystems), and Part III (public key cryptosystems) of the book. In these parts, we provide more formal definitions of both the cryptosystems and their security properties. In the rest of this section, we continue to argue informally about the security of cryptographic systems and the different perspectives one may take.

7 The classification scheme was created by Ueli M. Maurer.

### 1.2.2 Secure Cryptographic Systems

The goal of cryptography is to design, implement, and employ cryptographic systems that are "secure." But to make precise statements about the security of a particular cryptosystem, one must formally define the term security. Unfortunately, reality looks different, and the literature is full of cryptographic systems that are claimed to be secure without providing an appropriate definition for it. This is dissatisfactory, mainly because anything can be claimed to be secure, unless its meaning is precisely nailed down.

Instead of properly defining the term security and analyzing whether a cryptographic system meets this definition, people often like to argue about key lengths. This is because the key length is a simple and very intuitive security parameter. So people frequently use it to characterize the cryptographic strength of a system. This is clearly an oversimplification, because the key length is a suitable (and meaningful) measure of security if and only if an exhaustive key search is the most efficient way to break it. In practice, however, this is seldom the case, and there are often simpler ways to break the security of a system (e.g., simply by reading out some keying material from the memory). In this book, we avoid discussions about key lengths. Instead, we refer to the recommendations[8] compiled and hosted by BlueKrypt.[9] They provide practical advice to decide what key lengths are appropriate for any given cryptosystem (even using different methods).

In order to discuss the security of a cryptosystem, there are two perspectives one may take: a theoretical one and a practical one. Unfortunately, the two perspectives are fundamentally different, and one may have a cryptosystem that is theoretically secure but practically insecure (e.g., due to a poor implementation), or—vice versa—a cryptosystem that provides a sufficient level of security in practice but is not very sophisticated from a theoretical viewpoint. Let us have a closer look at both perspectives before we focus on some general principles for the design of cryptographic systems.

#### 1.2.2.1 Theoretical Perspective

According to what has been said above, one has to start with a precise definition for the term "security" before one can argue about the security of a particular cryptosystem. What does it mean for such a system to be "secure"? What properties does it have to fulfill? In general, there are two questions that need to be answered here:

8 https://www.keylength.com.

9 https://www.bluekrypt.com.

1. Who is the adversary; that is, what are his or her capabilities and how powerful is he or she?
2. What is the task the adversary has to solve in order to be successful; that is, to break the security of the system?

An answer to the first question comprises the specification of several parameters related to the adversary, such as his or her computing power, available memory, available time, types of feasible attacks, and access to a priori information. For some of these parameters, the statements can be coarse, such as the computing power and the available time are finite or infinite. The result is a *threats model* (i.e., a model of the adversary one has in mind and against whom one wants to protect oneself).

An answer to the second question is even more tricky. In general, the adversary's task is to find (i.e., compute, guess, or otherwise determine) one or several pieces of information that he or she should not be able to know. If, for example, the adversary is able to determine the cryptographic key used to encrypt a message, then he or she must clearly be considered to be successful. But what if he or she is able to determine only half of the key, or—maybe even more controversial—a single bit of the key? Similar difficulties occur in other cryptosystems that are used for other purposes than confidentiality protection.

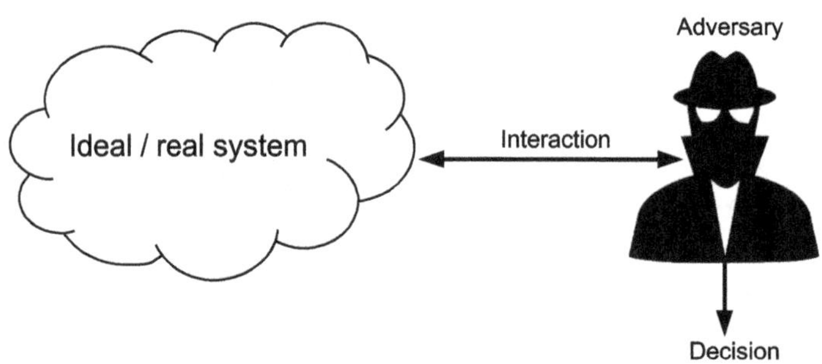


**Figure 1.2** Security game in the ideal/real simulation paradigm.

The preferred way to deal with these difficulties is to define a *security game* in the *ideal/real simulation paradigm* illustrated in Figure 1.2. On the left side, there is a system that is either *ideal* or *real*, meaning that it is either a theoretically perfect system for the task one has in mind (ideal system) or it is a real-world implementation thereof (real system). If, for example, the real system is a block cipher used for symmetric encryption, then the ideal system can be thought of as a pseudorandom permutation. The adversary on the right side can interact with the (ideal or real) system in some predefined way, and his or her task is to tell the two cases apart. Again referring to a block cipher, it may be the case that the adversary can have arbitrary plaintext messages be encrypted or arbitrary ciphertext messages be decrypted. The type of interaction matters and must therefore be part of the specification of the security game. In either case, the adversary wins the game if he or she is able to tell whether he or she is interacting with an ideal or a real system with a probability that is better than guessing. If the adversary is not able to do so, then the real system is indistinguishable from an ideal one, and hence it has all the relevant properties of it, including its security. It is therefore as secure as the ideal system. Most security proofs in cryptography follow this line of argumentation and define the task the adversary needs to solve (in order to be successful) in this game-theoretic way. We will see many examples throughout the book.

As captured in Definition 1.8, a definition for a secure cryptographic system must start with the two questions itemized above and their respective answers.

**Definition 1.8 (Secure cryptographic system)** *A cryptographic system is* secure *if a well-defined adversary cannot break it, meaning that he or she cannot solve a well-defined task.*

This definition gives room for several notions of security. In principle, there is a distinct notion for every possible adversary combined with every possible task. As a general rule of thumb, we say that strong security definitions assume an adversary that is as powerful as possible and a task to solve that is as simple as possible. If a system can be shown to be secure in this setting, then there is a security margin. In reality, the adversary is likely less powerful and the task he or she must solve is likely more difficult, and this, in turn, means that it is very unlikely that the security of the system gets broken.

Let us consider a provocative question (or mind experiment) to clarify this point: If we have two safes $S_1$ and $S_2$, where $S_1$ cannot be cracked by a schoolboy within one minute and $S_2$ cannot be cracked by a group of world-leading safe crackers within one year, and we ask the question whether $S_1$ or $S_2$ can be considered to be more secure, then we would unanimously opt for $S_2$. The argument would go as follows: If even a group of world-leading safe crackers cannot crack $S_2$ within one year, then it very unlikely that a real-world criminal is able to crack it in a shorter but more realistic amount of time, such as one hour. In contrast, the security level $S_1$ provides is much less convincing. The fact that even a schoolboy can crack $S_1$ within one minute makes us believe that the criminal is likely to be successful in cracking the safe within one hour. This gives us a much better feeling about the security of $S_2$. In cryptography, we use a similar line of argumentation when we talk about the strength of a security definition.

If we say that an adversary cannot solve a task, then we can still distinguish the two cases in which he or she is simply not able solve it (independent from his or her computational resources) or he or she can in principle solve it but doesn't have the computational resources to do so, meaning that it is computationally too expensive. This distinction brings us to the following two notions of security.

**Unconditional security:** If an adversary with infinite computing power is not able to solve the task within a finite amount of time, then we are talking about *unconditional* or *information-theoretic security*. The mathematical theories behind this notion of security are probability theory and information theory, as briefly introduced in Appendixes B and C.

**Conditional security:** If an adversary is in principle able to solve the task within a finite amount of time but doesn't have the computational resources to do so,[10] then we are talking about *conditional* or *computational security*. The mathematical theory behind this notion of security is computational complexity theory, as briefly introduced in Appendix D.

The distinction between unconditional and conditional security is at the core of modern cryptography. Interestingly, there are cryptosystems known to be secure in the strong sense (i.e., unconditionally or information-theoretically secure), whereas there are no cryptosystems known to be secure in the weak sense (i.e., conditionally or computationally secure). There are many cryptosystems that are assumed to be computationally secure, but no proof is available for any of these systems. In fact, not even the existence of a conditionally or computationally secure cryptosystem has been proven so far. The underlying problem is that it is generally impossible to prove a lower bound for the computational complexity of solving a problem. To some extent, this is an inherent weakness or limitation of complexity theory as it stands today.

In addition to the unconditional and conditional notions of security, people often use the term *provable security* to refer to another—arguably strong—notion of security. This goes back to the early days of public key cryptography, when Whitfield Diffie and Martin E. Hellman proposed a key exchange protocol [12][11] (Section 12.3) and a respective security proof. In fact, they showed that their protocol is secure unless somebody is able to solve a hard mathematical problem (that is strongly related to the discrete logarithm problem). The security of the protocol is thus reduced to a hard mathematical problem (i.e., if somebody is able to solve

10 It is usually assumed that the adversary can run algorithms that have a polynomial running time.

11 This paper is the one that officially gave birth to public key cryptography. There is a companion paper entitled "Multiuser Cryptographic Techniques" that was presented by the same authors at the National Computer Conference on June 7–10, 1976.

the mathematical problem, then he or she is also able to break the security of the protocol). This is conceptually similar to proving that squaring a circle with compass and straightedge is impossible. This well-known fact from geometry can be proven by reducing the problem of squaring a circle to the problem of finding a non-zero polynomial $f(x) = a_n x^n + \ldots + a_1 x + a_0$ with rational coefficients $a_i$ for $i = 0, 1, \ldots, n$, such that $\pi$ is a root; that is, $f(\pi) = 0$. Because we know that $\pi$ is not algebraic (it is transcendental), we know that such a polynomial does not exist and cannot be found in the first place. Conversely, this implies that a circle cannot be squared with a compass and straightedge.

More specifically, the basic concept of a reduction proof is to show that breaking a cryptosystem is computationally equivalent to solving a hard mathematical problem. This means that one must prove the following two directions:

- If the hard problem can be solved, then the cryptosystem can be broken.
- If the cryptosystem can be broken, then the hard problem can be solved.

Diffie and Hellman only proved the first direction, and they did not prove the second direction.[12] This is unfortunate, because the second direction is important for security. If we can prove that an adversary who is able to break a cryptosystem is also able to solve the hard problem, then we can reasonably argue that it is unlikely that such an adversary exists, and hence that the cryptosystem in question is likely to be secure. Michael O. Rabin[13] was the first researcher who proposed a cryptosystem [14] that can be proven to be computationally equivalent to a mathematically hard problem (Section 13.3.2).

The notion of (provable) security has fueled a lot of research since the late 1970s. In fact, there are many (public key) cryptosystems proven secure in this sense. It is, however, important to note that a complexity-based proof is not absolute, and that it is only relative to the (assumed) intractability of the underlying mathematical problem(s). This is a similar situation to proving that a problem is $\mathbf{NP}$-hard. It proves that the problem is at least as difficult as other problems in $\mathbf{NP}$, but it does not provide an absolute proof of its computational difficulty.[14]

There are situations in which a security proof requires an additional assumption, namely that a cryptographic primitive–typically a cryptographic hash function—behaves like a random function. This leads to a new paradigm and

12 This was changed in [13] and some follow-up publications.

13 In 1976, Michael O. Rabin and Dana S. Scott jointly received the ACM Turing Award for their work on nondeterministic machines that is key to theoretical computer science and complexity theory. This work was not yet directly related to cryptography.

14 Refer to Appendix D to get a more detailed overview about $\mathbf{NP}$and $\mathbf{NP}$-hard problems.

methodology to design cryptographic systems that are "provably secure" in the *random oracle model* (in contrast to the *standard model*) [15]. The methodology employs the idea of an ideal system (as introduced above) and consists of the following three steps:

- First, one designs an ideal system that uses random functions[15] (also known as random oracles), most notably cryptographic hash functions.
- Second, one proves the security of the ideal system.
- Third, one replaces the random functions with real ones.

As a result, one obtains an implementation of the ideal system in the real world (where random functions do not exist). Due to the use of random oracles, this methodology is known as *random oracle methodology*, and—as mentioned above—it yields cryptosystems that are secure in the random oracle model. As further addressed in Section 8.4, such cryptosystems and their respective "security proofs" are widely used in the field, but they must be taken with a grain of salt. In fact, it has been shown that it is possible to construct cryptographic systems that are provably secure in the random oracle model, but become totally insecure whenever the cryptographic hash function used in the protocol (to replace the random oracle) is instantiated. This theoretical result is worrisome, and since its publication many researchers have started to think controversially about the random oracle methodology and the usefulness of the random oracle model per se. At least it must be noted that formal analyses in the random oracle model are not security proofs in purely mathematical parlance. The problem is the underlying ideal assumptions about the randomness properties of the cryptographic hash functions. This is not something that is otherwise used in mathematically rigorous proofs.

In this book, we don't consider provable security (with or without the random oracle model) as a security notion of its own. Instead, we see it as a special case of conditional security, namely one where the intractability assumption represents the condition.

#### 1.2.2.2 Practical Perspective

So far, we have argued about the security of a cryptosystem from a purely theoretical viewpoint. In practice, however, any (theoretically secure) cryptosystem must be implemented, and there are many things that can go wrong here (e.g., [16]). For example, the cryptographic key in use may be kept in memory and extracted from

15 The notion of a random function is briefly introduced in Section 2.1.2 and more thoroughly addressed in Chapter 8.

there, for example, using a cold boot attack[16] [17], or the user of a cryptosystem may be subject to all kinds of phishing and social engineering attacks.

Historically, the first such attacks tried to exploit the compromising emanations that occur in all information-processing physical systems. These are unintentional intelligence-bearing signals that, if intercepted and properly analyzed, may disclose the information transmitted, received, handled, or otherwise processed by a piece of equipment. In the late 1960s and early 1970s, the U.S. National Security Agency (NSA) coined the term *TEMPEST* to refer to this field of study (i.e., to secure electronic communications equipment from potential eavesdroppers), and vice versa, the ability to intercept and interpret those signals from other sources.[17] Hence, the term TEMPEST is a codename (not an acronym[18]) that is used broadly to refer to the entire field of emission security or emanations security (EMSEC). There are several U.S. and NATO standards that basically define three levels of TEMPEST requirements: NATO SDIP-27 Levels A, B, and C. This is beyond the scope of this book.

In addition to cold boot attacks and exploiting compromising emanations, people have been very innovative in finding possibilities to mount attacks against presumably tamper-resistant hardware devices that employ invasive measuring techniques (e.g., [18, 19]). Most importantly, there are attacks that exploit side channel information an implementation may leak when a computation is performed. Side channel information is neither input nor output, but refers to some other information that may be related to the computation, such as timing information or power consumption. Attacks that try to exploit such information are commonly referred to as

16 This attack exploits the fact that many dynamic random access memory (DRAM) chips don't lose their contents when a system is switched off immediately, but rather lose their contents gradually over a period of seconds, even at standard operating temperatures and even if the chips are removed from the motherboard. If kept at low temperatures, the data on these chips persist for minutes or even hours. In fact, the researchers showed that residual data can be recovered using simple techniques that require only temporary physical access to a machine, and that several popular disk encryption software packages, such as Microsoft's BitLocker, Apple's FileVault, and TrueCrypt (the predecessor of VeraCrypt) were susceptible to cold boot attacks. The feasibility and simplicity of such attacks has seriously challenged the security of many disk encryption software solutions.

17 https://www.nsa.gov/news-features/declassified-documents/cryptologic-spectrum/assets/files/tempest.pdf.

18 The U.S. government has stated that the term TEMPEST is not an acronym and does not have any particular meaning (it is therefore not included in this book's list of abbreviations and acronyms). However, in spite of this disclaimer, multiple acronyms have been suggested, such as "Transmitted Electro-Magnetic Pulse / Energy Standards & Testing," "Telecommunications ElectroMagnetic Protection, Equipment, Standards & Techniques," "Transient ElectroMagnetic Pulse Emanation STandard," "Telecommunications Electronics Material Protected from Emanating Spurious Transmissions," and—more jokingly—"Tiny ElectroMagnetic Particles Emitting Secret Things."

*side channel attacks*. Let us start with two mind experiments to illustrate the notion of a side channel attack.[19]

- Assume somebody has written a secret note on a pad and has torn off the paper sheet. Is there a possibility to reconstruct the note? An obvious possibility is to go for a surveillance camera and examine the respective recordings. A less obvious possibility is to exploit the fact that pressing the pen on the paper sheet may have caused the underlying paper sheet to experience some pressure, and this, in turn, may have caused the underlying paper sheet to show the same groove-like depressions (representing the actual writing). Equipped with the appropriate tools, an expert may be able to reconstruct the note. Pressing the pen on a paper sheet may have caused a side channel to exist, even if the original paper sheet is destroyed.
- Consider a house with two rooms. In one room are three light switches and in the other room are three lightbulbs, but the wiring of the light switches and bulbs is unknown. In this setting, somebody's task is to find out the wiring, but he or she can enter each room only once. From a mathematical viewpoint, one can argue (and prove) that this task is impossible to solve. But from a physical viewpoint (and taking into account side channel information), the task can be solved: One can enter the room with the light switches, permanently turn on one bulb, and turn on another bulb for some time (e.g., a few seconds). One then enters the room with the lightbulbs. The bulb that is lit is easily identified and refers to the switch that has been permanently switched on. But the other two bulbs are not lit, and hence one cannot easily assign them to the respective switches. But one can measure the temperature of the lightbulbs. The one that is warmer more likely refers to the switch that has been switched on for some time. This information can be used to distinguish the two cases and to solve the task accordingly. Obviously, the trick is to measure the temperature of the lightbulbs and to use this information as a side channel.

In analogy to these mind experiments, there are many side channel attacks that have been proposed to defeat the security of cryptosystems, some of which have turned out to be very powerful. The first side channel attack that opened people's eyes and the field in the 1990s was a timing attack against a vulnerable implementation of the RSA cryptosystem [20]. The attack exploited the correlation between a cryptographic key and the running time of the algorithm that employed the key. Since then, many implementations of cryptosystems have been shown to be vulnerable against timing attacks and some variants, such as cache timing attacks or branch prediction analysis. In 2003, it was shown that remotely mounting timing

19 The second mind experiment was proposed by Artur Ekert.

attacks over computer networks is feasible [21], and since 2018 we know that almost all modern processors that support speculative and out-of-order command execution are susceptible to sophisticated timing attacks.[20] Other side channel attacks exploit the power consumption of an implementation of an algorithm that is being executed (usually called power consumption or power analysis attacks [22]), faults that are induced (usually named differential fault analysis [23, 24]), protocol failures [25], the sounds that are generated during a computation [26, 27], and many more.

Side channel attacks exploit side channel information. Hence, a reasonable strategy to mitigate a specific side channel attack is to avoid the respective side channel in the first place. If, for example, one wants to protect an implementation against timing attacks, then timing information must not leak. At first sight, one may be tempted to add a random delay to every computation, but this simple mechanism does not work (because the effect of random delays can be compensated by having an adversary repeat the measurement many times). But there may be other mechanisms that work. If, for example, one ensures that all operations take an equal amount of time (i.e., the timing behavior is largely independent from the input), then one can mitigate such attacks. But constant-time programming has turned out to be difficult, certainly more difficult than it looks at first sight. Also, it is sometimes possible to blind the input and to prevent the adversary from knowing the true value. Both mechanisms have the disadvantage of slowing down the computations. There are fewer possibilities to protect an implementation against power consumption attacks. For example, dummy registers and gates can be added on which useless operations are performed to balance power consumption into a constant value. Whenever an operation is performed, a complementary operation is also performed on a dummy element to assure that the total power consumption remains balanced according to some higher value. Protection against differential fault analysis is less general and even more involved. In [23], for example, the authors suggest a solution that requires a cryptographic computation to be performed twice and to output the result only if they are the same. The main problem with this approach is that it roughly doubles the execution time. Also, the probability that the fault will not occur twice is not sufficiently small (and this makes the attack harder to implement, but not impossible). The bottom line is that the development of adequate and sustainable protection mechanisms to mitigate differential fault analysis attacks remains a timely research topic. The same is true for failure analysis and acoustic cryptanalysis, and it may even be true for other side channel attacks that will be found and published in the future. Once a side channel is known, one can usually do something to avoid the respective attacks. But keep in mind that many side channels may exist that are still unknown today.

20 The first such attacks have been named *Meltdown* and *Spectre*. They are documented at https://www.spectreattack.com.

The existence and difficulty of mitigating side channel attacks have inspired theoreticians to come up with a model for defining and delivering cryptographic security against an adversary who has access to information leaked from the physical execution of a cryptographic algorithm [28]. The original term used to refer to this type of cryptography is *physically observable cryptography*. More recently, however, researchers have coined the term *leakage-resilient cryptography* to refer to essentially the same idea. Even after many years of research, it is still questionable whether physically observable or leakage-resilient cryptography can be achieved in the first place (e.g., [29]). It is certainly a legitimate and reasonable design goal, but it may not be a very realistic one.

#### 1.2.2.3 Design Principles

In the past, we have seen many examples in which people have tried to improve the security of a cryptographic system by keeping secret its design and internal working principles. This approach is sometimes referred to as "security through obscurity." Many of these systems do not work and can be broken trivially.[21] This insight has a long tradition in cryptography, and there is a well-known cryptographic principle—*Kerckhoffs's principle*[22]—that basically states that a cryptographic system should be designed so as to remain secure, even if the adversary knows all the details of the system, except for the values explicitly declared to be secret, such as secret keys [30]. We follow this principle in this book, and we only address cryptosystems for which we can assume that the adversary knows the details. This assumption is in line with our requirement that the adversaries should be assumed to be as powerful as possible (to obtain the strong security definitions detailed in Section 1.2.2).

In spite of Kerckhoffs's principle, the design of a secure cryptographic system remains a difficult and challenging task. One has to make assumptions, and it is not clear whether these assumptions really hold in reality. For example, one usually assumes a certain set of countermeasures to protect against specific attacks. If the adversary attacks the system in another way, then there is hardly anything that can be done about it. Similarly, one has to assume the system to operate in a "typical" environment. If the adversary can manipulate the environment, then he or she may be able to change the operational behavior of the system, and hence to open up new vulnerabilities. The bottom line is that cryptographic systems that are based on make-believe, ad hoc approaches, and heuristics are often broken in the field in some new and ingenious ways. Instead, the design of a secure cryptographic system should be based on firm foundations. Ideally, it consists of the following two steps:

21 Note that "security through obscurity" may work outside the realm of cryptography.

22 The principle is named after Auguste Kerckhoffs, who lived from 1835 to 1903.

1. In a *definitional step*, the problem the cryptographic system is intended to solve is identified, precisely defined, and formally specified.
2. In a *constructive step*, a cryptographic system that satisfies the definition distilled in step one, possibly while relying on intractability assumptions, is designed.

Again, it is important to note that most parts of modern cryptography rely on intractability assumptions and that relying on such assumptions seems to be unavoidable. But there is still a huge difference between relying on an explicitly stated intractability assumption or just assuming (or rather hoping) that an ad hoc construction satisfies some unspecified and vaguely specified goals. This basically distinguishes cryptography as a science from cryptography as an art.

## 1.3 HISTORICAL BACKGROUND INFORMATION

Cryptography has a long and thrilling history that is addressed in many books (e.g., [31–33]). In fact, probably since the very beginning of the spoken and—even more importantly—written word, people have tried to transform "data to render its meaning unintelligible (i.e., to hide its semantic content), prevent its undetected alteration, or prevent its unauthorized use" [1]. According to this definition, these people have always employed cryptography and cryptographic techniques. The mathematics behind these early systems may not have been very advanced, but they still employed cryptography and cryptographic techniques. For example, Gaius Julius Caesar[23] used an encryption system in which every letter in the Latin alphabet was substituted with the letter that is found three positions afterward in the lexical order (i.e., "A" is substituted with "D," "B" is substituted with "E," and so on). This simple additive cipher is known as *Caesar cipher* (Section 9.2). Later on, people employed encryption systems that used more advanced and involved mathematical transformations. Many books on cryptography contain numerous examples of historically relevant encryption systems—they are not repeated in this book; the encryption systems in use today are simply too different.

Until World War II, cryptography was considered to be an art (rather than a science) and was primarily used in military and diplomacy. The following two developments and scientific achievements turned cryptography from an art into a science:

23 Gaius Julius Caesar was a Roman emperor who lived from 102 BC to 44 BC.

- During World War II, Claude E. Shannon[24] developed a mathematical theory of communication [34] and a related communication theory of secrecy systems [35] when he was working at AT&T Laboratories.[25] After their publication, the two theories started a new branch of research that is commonly referred to as *information theory* (refer to Appendix C for a brief introduction to information theory). It is used to prove the unconditional security of cryptographic systems.
- As mentioned earlier, Diffie and Hellman developed and proposed the idea of public key cryptography at Stanford University in the 1970s.[26] Their vision was to employ trapdoor functions to encrypt and digitally sign electronic documents. As introduced in Section 2.1.3 and further addressed in Chapter 5, a trapdoor function is a function that is easy to compute but hard to invert—unless one knows and has access to some trapdoor information. This information represents the private key held by a particular entity.

Diffie and Hellman's work culminated in a key agreement protocol (i.e., the Diffie-Hellman key exchange protocol described in Section 12.3) that allows two parties that share no secret to exchange a few messages over a public channel and to establish a shared (secret) key. This key can, for example, then be used to encrypt and decrypt data. After Diffie and Hellman published their discovery, a number of public key cryptosystems were developed and proposed. Some of these systems are still in use, such as RSA [37] and Elgamal [38]. Other systems, such as the ones based on the knapsack problem,[27] have been broken and are no longer in use. Some practically important public key cryptosystems are overviewed and discussed in Part III of this book.

24 Claude E. Shannon was a mathematician who lived from 1916 to 2001.

25 Similar studies were done by Norbert Wiener, who lived from 1894 to 1964.

26 Similar ideas were pursued by Ralph C. Merkle at the University of California at Berkeley [36]. More than a decade ago, the British government revealed that public key cryptography, including the Diffie-Hellman key agreement protocol and the RSA public key cryptosystem, was invented at the Government Communications Headquarters (GCHQ) in Cheltenham in the early 1970s by James H. Ellis, Clifford Cocks, and Malcolm J. Williamson under the name *nonsecret encryption* (NSE). You may refer to the note "The Story of Non-Secret Encryption" written by Ellis in 1997 to get the story (use a search engine to find the note in an Internet archive). Being part of the world of secret services and intelligence agencies, Ellis, Cocks, and Williamson were not allowed to openly talk about their discovery.

27 The *knapsack problem* is a well-known problem in computational complexity theory and applied mathematics. Given a set of items, each with a cost and a value, determine the number of each item to include in a collection so that the total cost is less than some given cost and the total value is as large as possible. The name derives from the scenario of choosing treasures to stuff into your knapsack when you can only carry so much weight.

Since the early 1990s, we have seen a wide deployment and massive commercialization of cryptography. Today, many companies develop, market, and sell cryptographic techniques, mechanisms, services, and products (implemented in hardware or software) on a global scale. There are cryptography-related conferences and trade shows[28] one can attend to learn more about products that implement cryptographic techniques, mechanisms, and services. One must be cautious here, because the term *crypto* is sometimes also used in a narrow sense that is limited to blockchain-based distributed ledger technologies (DLTs). This is especially true since the rise of Bitcoin and other cryptocurrencies.

The major goal of this book is to provide a basic understanding of what is actually going on in the field. If you want to learn more about the practical use of cryptography to secure Internet and Web applications, you may also refer to [39–42] or any other book about Internet and Web security in general, and Internet security protocols in particular. These practical applications of cryptography are not addressed (or repeated) in this book.

In Section D.5, we briefly introduce the notion of a quantum computer. As of this writing, the issue of whether it is possible to build and operate a sufficiently large and stable quantum computer is still open and controversially discussed in the community. But if such a computer can be built, then we know that many cryptosystems in use today can be broken efficiently. This applies to almost all public key cryptosystems (because these systems are typically based on the integer factorization problem or discrete logarithm problem that can both be solved on a quantum computer in polynomial time), and it partly applies to secret key cryptosystems (because it is known how to reduce the steps required to perform an exhaustive key search for an $n$-bit cipher from $2^n$ to $2^{n/2}$). Against this background, people have started to look for cryptographic primitives that remain secure even if sufficiently large and stable quantum computers can be built and operated. The resulting area of research is known as *post-quantum cryptography* (PQC). In the last couple of years, PQC has attracted a lot of interest and funding, and many researchers have come up with proposals for PQC. In the case of secret key cryptography, resistance against quantum computers can be provided by doubling the key length. This is simple and straightforward. In the case of public key cryptography, however, things are more involved and new design paradigms are needed here. This is where code-based, hash-based, lattice-based, isogeny-based, and multivariate-based cryptosystems come into play. As further addressed in Section 18.3, there is a NIST competition going on to evaluate and standardize one or more post-quantum public key cryptographic algorithms. However, unless major advances in building large quantum computers are made, it is unlikely that PQC will be widely used in the field. The currently known

28 The most important trade show is the RSA Conference held annually in the United States, Europe, and Asia. Refer to https://www.rsaconference.com for more information.

post-quantum cryptosystems are simply too inefficient or require significantly longer keys than the traditional (non-PQC) ones.

## 1.4 OUTLINE OF THE BOOK

The rest of this book is organized as follows:

- In Chapter 2, "Cryptographic Systems," we introduce, briefly overview, and put into perspective the three classes of cryptographic systems (i.e., unkeyed cryptosystems, secret key cryptosystems, and public key cryptosystems) with their major representatives.
- In Chapter 3, "Random Generators," we begin Part I on unkeyed cryptosystems by elaborating on how to generate random values or bits.
- In Chapter 4, "Random Functions," we introduce the notion of a random function that plays a pivotal role in contemporary cryptography. Many cryptosystems used in the field can, at least in principle, be seen as a random function.
- In Chapter 5, "One-Way Functions," we focus on functions that are one-way. Such functions are heavily used in public key cryptography.
- In Chapter 6, "Cryptographic Hash Functions," we conclude Part I by outlining cryptographic hash functions and their use in contemporary cryptography.
- In Chapter 7, "Pseudorandom Generators," we begin Part II on secret key cryptosystems by elaborating on how to use a pseudorandom generator seeded with a secret key to generate pseudorandom values (instead of truly random ones).
- In Chapter 8, "Pseudorandom Functions," we do something similar for random functions: We explain how to construct pseudorandom functions that use a secret key as input and generate an output that looks as if it were generated by a random function.
- In Chapter 9, "Symmetric Encryption," we overview and discuss the symmetric encryption systems that are most frequently used in the field.
- In Chapter 10, "Message Authentication," we address message authentication and explain how secret key cryptography can be used to generate and verify message authentication codes.

- In Chapter 11, "Authenticated Encryption," we conclude Part II by combining symmetric encryption and message authentication in authenticated encryption. In many regards, AE represents the state of the art of cryptography as it stands today.
- In Chapter 12, "Key Establishment," we begin Part III on public key cryptosystems by elaborating on the practically relevant problem of how to establish a secret key that is shared between two or more entities.
- In Chapter 13, "Asymmetric Encryption," we overview and discuss the asymmetric encryption systems that are most frequently used in the field.
- In Chapter 14, "Digital Signatures," we elaborate on digital signatures and respective digital signature systems (DSSs).
- In Chapter 15, "Zero-Knowledge Proofs of Knowledge," we explore the notion of an interactive proof system that may have the zero-knowledge property, and discuss its application in zero-knowledge proofs of knowledge for entity authentication.
- In Chapter 16, "Key Management," we begin Part IV by discussing some aspects related to key management that represents the Achilles' heel of applied cryptography.
- In Chapter 17, "Summary," we conclude with some remarks about the current state of the art in cryptography.
- In Chapter 18, "Outlook," we predict possible and likely future developments and trends in the field, including PQC.

This book includes several appendixes. Appendixes A to D summarize the mathematical foundations and principles, including discrete mathematics (Appendix A), probability theory (Appendix B), information theory (Appendix C), and complexity theory (Appendix D). Finally, a list of mathematical symbols and a list of abbreviations and acronyms are provided at the end of the book.

Note that cryptography is a field of study that is far too broad to be addressed in a single book, and that one has to refer to additional and more in-depth material, such as the literature referenced at the end of each chapter, to learn more about a particular topic. Another possibility to learn more is to experiment and play around with cryptographic systems. As mentioned in the Preface, there are several software packages that can be used for this purpose, among which CrypTool is a particularly good choice.

The aims of this book are threefold: (1) to provide an overview, (2) to give an introduction into each of the previously mentioned topics and areas of research and development, and (3) to put everything into perspective. Instead of trying to be comprehensive and complete, this book tries to ensure that you still see the forest for the trees. In the next chapter, we delve more deeply into the various representatives of the three classes of cryptosystems introduced above.

## References

[1] Shirey, R., *Internet Security Glossary, Version 2*, RFC 4949 (FYI 36), August 2007.

[2] Menezes, A., P. van Oorschot, and S. Vanstone, *Handbook of Applied Cryptography*. CRC Press, Boca Raton, FL, 1996.

[3] Stamp, M., and R.M. Low, *Applied Cryptanalysis: Breaking Ciphers in the Real World*. John Wiley & Sons, New York, 2007.

[4] Swenson, C., *Modern Cryptanalysis: Techniques for Advanced Code Breaking*. Wiley Publishing, Indianapolis, IN, 2008.

[5] Joux, A., *Algorithmic Cryptanalysis*. Chapman & Hall/CRC, Boca Raton, FL, 2009.

[6] Cole, E., *Hiding in Plain Sight: Steganography and the Art of Covert Communication*. Wiley Publishing, Indianapolis, IN, 2003.

[7] Wayner, P., *Disappearing Cryptography: Information Hiding — Steganography and Watermarking*, 3rd edition. Morgan Kaufmann Publishers, Burlington, MA, 2009.

[8] Arnold, M., M. Schmucker, and S.D. Wolthusen, *Techniques and Applications of Digital Watermarking and Content Protection*. Artech House Publishers, Norwood, MA, 2003.

[9] Katzenbeisser, S., and F. Petitcolas, *Information Hiding*. Artech House Publishers, Norwood, MA, 2015.

[10] Kelsey, J., B. Schneier, and D. Wagner, "Protocol Interactions and the Chosen Protocol Attack," *Proceedings of the 5th International Workshop on Security Protocols*, Springer-Verlag, 1997, pp. 91–104.

[11] Oppliger, R., "Disillusioning Alice and Bob," *IEEE Security & Privacy*, Vol. 15, No. 5, September/October 2017, pp. 82–84.

[12] Diffie, W., and M.E. Hellman, "New Directions in Cryptography," *IEEE Transactions on Information Theory*, IT-22(6), 1976, pp. 644–654.

[13] Maurer, U.M., "Towards the Equivalence of Breaking the Diffie-Hellman Protocol and Computing Discrete Logarithms," *Proceedings of CRYPTO '94*, Springer-Verlag, LNCS 839, 1994, pp. 271–281.

[14] Rabin, M.O., "Digitalized Signatures and Public-Key Functions as Intractable as Factorization," MIT Laboratory for Computer Science, MIT/LCS/TR-212, 1979.

[15] Bellare, M., and P. Rogaway, "Random Oracles are Practical: A Paradigm for Designing Efficient Protocols," *Proceedings of the 1st ACM Conference on Computer and Communications Security*, 1993, pp. 62-73.

[16] Anderson, R., "Why Cryptosystems Fail," *Communications of the ACM*, Vol. 37, No. 11, November 1994, pp. 32–40.

[17] Halderman, J.A., et al., "Lest We Remember: Cold Boot Attacks on Encryption Keys," *Communications of the ACM*, Vol. 52, No. 5, May 2009, pp. 91–98.

[18] Anderson, R., and M. Kuhn, "Tamper Resistance—A Cautionary Note," *Proceedings of the 2nd USENIX Workshop on Electronic Commerce*, November 1996, pp. 1–11.

[19] Anderson, R., and M. Kuhn, "Low Cost Attacks on Tamper Resistant Devices," *Proceedings of the 5th International Workshop on Security Protocols*, Springer-Verlag, LNCS 1361, 1997, pp. 125–136.

[20] Kocher, P., "Timing Attacks on Implementations of Diffie-Hellman, RSA, DSS, and Other Systems," *Proceedings of CRYPTO '96*, Springer-Verlag, LNCS 1109, 1996, pp. 104–113.

[21] Brumley, D., and D. Boneh, "Remote timing attacks are practical," *Proceedings of the 12th Usenix Security Symposium*, USENIX Association, 2003.

[22] Kocher, P., J. Jaffe, and B. Jun, "Differential Power Analysis," *Proceedings of CRYPTO '99*, Springer-Verlag, LNCS 1666, 1999, pp. 388–397.

[23] Boneh, D., R. DeMillo, and R. Lipton, "On the Importance of Checking Cryptographic Protocols for Faults," *Proceedings of EUROCRYPT '97*, Springer-Verlag, LNCS 1233, 1997, pp. 37–51.

[24] Biham, E., and A. Shamir, "Differential Fault Analysis of Secret Key Cryptosystems," *Proceedings of CRYPTO '97*, Springer-Verlag, LNCS 1294, 1997, pp. 513–525.

[25] Bleichenbacher, D., "Chosen Ciphertext Attacks Against Protocols Based on the RSA Encryption Standard PKCS #1," *Proceedings of CRYPTO '98*, Springer-Verlag, LNCS 1462, 1998, pp. 1–12.

[26] Asonov, D., and R. Agrawal, "Keyboard Acoustic Emanations," *Proceedings of IEEE Symposium on Security and Privacy*, 2004, pp. 3–11.

[27] Zhuang, L., Zhou, F., and J.D. Tygar, "Keyboard Acoustic Emanations Revisited," *Proceedings of ACM Conference on Computer and Communications Security*, November 2005, pp. 373–382.

[28] Micali, S., and L. Reyzin, "Physically Observable Cryptography," *Proceedings of Theory of Cryptography Conference (TCC 2004)*, Springer-Verlag, LNCS 2951, 2004, pp. 278–296.

[29] Renauld, M., et al., "A Formal Study of Power Variability Issues and Side-Channel Attacks for Nanoscale Devices," *Proceedings of EUROCRYPT 2011*, Springer-Verlag, LNCS 6632, 2011, pp. 109–128.

[30] Kerckhoffs, A., "La Cryptographie Militaire," *Journal des Sciences Militaires*, Vol. IX, January 1883, pp. 5–38, February 1883, pp. 161-191.

[31] Kahn, D., *The Codebreakers: The Comprehensive History of Secret Communication from Ancient Times to the Internet*. Scribner, New York, 1996.

[32] Bauer, F.L., *Decrypted Secrets: Methods and Maxims of Cryptology*, 2nd edition. Springer-Verlag, New York, 2000.

[33] Levy, S., *Crypto: How the Code Rebels Beat the Government—Saving Privacy in the Digital Age*. Viking Penguin, New York, 2001.

[34] Shannon, C.E., "A Mathematical Theory of Communication," *Bell System Technical Journal*, Vol. 27, No. 3/4, July/October 1948, pp. 379–423/623–656.

[35] Shannon, C.E., "Communication Theory of Secrecy Systems," *Bell System Technical Journal*, Vol. 28, No. 4, October 1949, pp. 656–715.

[36] Merkle, R.C., "Secure Communication over Insecure Channels," *Communications of the ACM*, Vol. 21, No. 4, April 1978, pp. 294–299.

[37] Rivest, R.L., A. Shamir, and L. Adleman, "A Method for Obtaining Digital Signatures and Public-Key Cryptosystems," *Communications of the ACM*, Vol. 21, No. 2, February 1978, pp. 120–126.

[38] Elgamal, T., "A Public Key Cryptosystem and a Signature Scheme Based on Discrete Logarithm," *IEEE Transactions on Information Theory*, Vol. 31, No. 4, 1985, pp. 469–472.

[39] Oppliger, R., *Internet and Intranet Security*, 2nd edition. Artech House Publishers, Norwood, MA, 2002.

[40] Oppliger, R., *Security Technologies for the World Wide Web*, 2nd edition. Artech House Publishers, Norwood, MA, 2003.

[41] Oppliger, R., *SSL and TLS: Theory and Practice*, 2nd edition. Artech House Publishers, Norwood, MA, 2016.

[42] Oppliger, R., *End-to-End Encrypted Messaging*. Artech House Publishers, Norwood, MA, 2020.

# Chapter 2

# Cryptographic Systems

As mentioned in Section 1.2.1, there are three classes of cryptographic systems: unkeyed cryptosystems, secret key cryptosystems, and public key cryptosystems. In this chapter, we introduce, overview, and provide some preliminary definitions for the most important representatives of these classes in Sections 2.1–2.3, and we conclude with some final remarks in Section 2.4. All cryptosystems mentioned in this chapter will be revisited and more thoroughly addressed in later chapters of the book.

## 2.1 UNKEYED CRYPTOSYSTEMS

According to Definition 1.5, unkeyed cryptosystems use no secret parameters. The most important representatives are random generators, random functions, one-way functions, and cryptographic hash functions. We address them in this order.

### 2.1.1 Random Generators

Randomness is the most important ingredient for cryptography, and most cryptographic systems in use today depend on some form of randomness. This is certainly true for key generation and probabilistic encryption, but it is also true for many other cryptographic algorithms and systems. In Section 9.3, we will see that a perfectly secret encryption system (i.e., the one-time pad), requires a random bit for the encryption of every single plaintext message bit. This means that the one-time pad (in order to provide perfect secrecy) requires huge quantities of random bits. But also in many other cases do we need ways to generate sequences of random values or bits. This is where the notions of a *random generator* or a *random bit generator* as captured in Definition 2.1 come into play.

**Definition 2.1 (Random generator)** *A* random generator *is a device that outputs a sequence of statistically independent and unbiased values. If the output values are bits, then the random generator is also called a* random bit generator.

A random bit generator is depicted in Figure 2.1 as a gray box. It is important to note that such a generator has no input, and that it only generates an output. Also, because the output is a sequence of statistically independent and unbiased bits, all bits occur with the same probability; that is, $\Pr[0] = \Pr[1] = 1/2$, or—more generally—all $2^k$ different $k$-tuples of bits occur with the same probability $1/2^k$ for all integers $k \geq 1$. Luckily, there are statistical tests that can be used to verify these properties. Passing these tests is a necessary but usually insufficient prerequisite for the output of a random generator to be suitable for cryptographic purposes and applications.

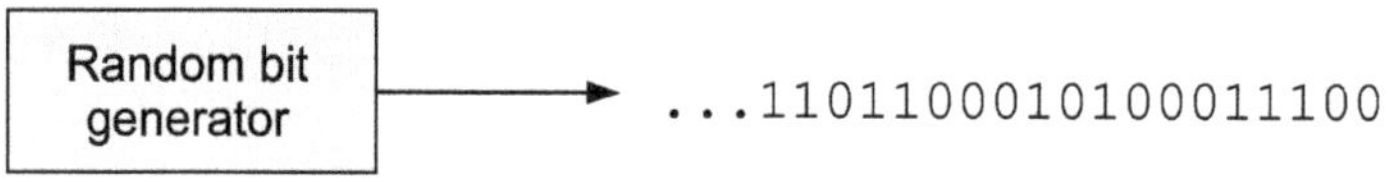


**Figure 2.1** A random bit generator.

It is also important to note that a random (bit) generator cannot be implemented in a deterministic way. Instead, it must be inherently nondeterministic, meaning that an implementation must use some physical events or phenomena (for which the outcomes are impossible to predict). Alternatively speaking, every (true) random generator requires a naturally occurring source of randomness. The engineering task of finding such a source and exploiting it in a device that may serve as a generator of binary sequences that are free of biases and correlations is a very challenging one. As mentioned above, the output of such a generator must fulfill statistical tests, but it must also withstand various types of sophisticated attacks.

Random generators and their design principles and security properties are further addressed in Chapter 3.

### 2.1.2 Random Functions

As mentioned above, a random generator is to output some random-looking values. In contrast, a *random function* (sometimes also called a *random oracle*) is not characterized by its output, but rather by the way it is chosen. This idea is captured in Definition 2.2.

**Definition 2.2 (Random function)** *A* random function *is a function* $f : X \rightarrow Y$ *that is chosen randomly from* $\text{Funcs}[X, Y]$ *(i.e., the set of all functions that map elements of the domain* $X$ *to elements of the codomain* $Y$*).*

For input value $x \in X$, a random function can output any value $y = f(x) \in f(X) \subseteq Y$. The only requirement is that the same input $x$ always maps to the same output $y$. Except for that, everything is possible and does not really matter (for the function to be random).

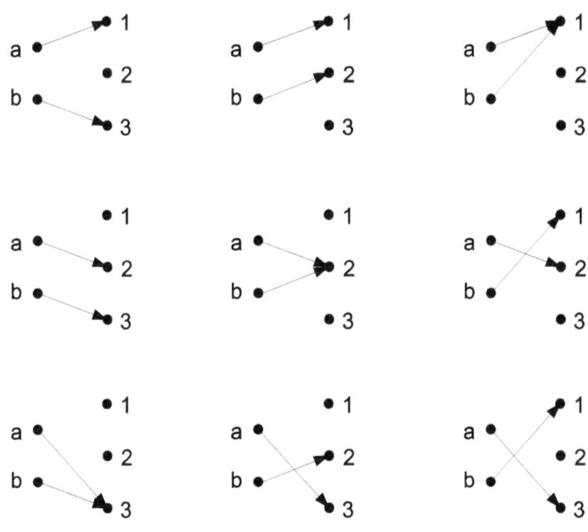

**Figure 2.2** Elements of $\text{Funcs}[\{a, b\}, \{1, 2, 3\}]$.

Note that there are $|Y|^{|X|}$ functions in $\text{Funcs}[X, Y]$, and this number is incredibly large—even for moderately sized $X$ and $Y$. If $X = \{a, b\}$ and $Y = \{1, 2, 3\}$, then $\text{Funcs}[X, Y]$ comprises $3^2 = 9$ functions, and this number is sufficiently small that its elements can be illustrated in Figure 2.2. If we moderately increase the sets and let $X$ be the set of all 2-bit strings and $Y$ the set of all 3-bit strings, then $|X| = 2^2$ and $|Y| = 2^3$, and hence $\text{Funcs}[X, Y]$ already comprises $(2^3)^{2^2} = (2^3)^4 = 2^{12} = 4,096$ functions. What this basically means is that there are $4,096$ possible functions to map 2-bit strings to 3-bit strings. This can no longer be illustrated. Let us consider an even more realistic setting, in which $X$ and $Y$ are both 128 bits long. Here, $\text{Funcs}[X, Y]$ comprises

$$(2^{128})^{2^{128}} = 2^{128 \cdot 2^{128}} = 2^{2^7 \cdot 2^{128}} = 2^{2^{135}}$$

functions, and this number is so large that it would require $2^{135}$ bits, if one wanted to number the functions and use an index to refer to a particular function from $\text{Funcs}[X, Y]$. This is clearly impractical.

Also note that random functions are not meant to be implemented. Instead, they represent conceptual constructs that are mainly used in security proofs. Referring to the security game of Figure 1.2 in Section 1.2.2.1, the random function yields the ideal system and it is shown that no adversary can tell the real system (for which the security needs to be proven) apart from it. If this can in fact be shown, then the real system behaves like a random function, and this, in turn, means that the adversary must try out all possibilities. Since there are so many possibilities, this task can be assumed to be computationally infeasible, and hence one can conclude that the real system is secure for all practical purposes.

If $X = Y$ and $\text{Funcs}[X, X]$ is restricted to the set of all possible permutations of $X$ (i.e., $\text{Perms}[X]$), then we can say that a *random permutation* is a randomly chosen permutation from $\text{Perms}[X]$. Everything said above is also true for random permutations.

We revisit the notions of random functions and random permutations, as well as the way they are used in cryptography in Chapter 4.

### 2.1.3 One-Way Functions

Informally speaking, a function $f : X \rightarrow Y$ is one way if it is easy to compute but hard to invert. Referring to complexity theory, *easy* means that the computation can be done efficiently (i.e., in polynomial time), whereas *hard* means that it is not known how to do the computation efficiently, that is, no efficient algorithm is known to exist.[1] More formally, the notion of a *one-way function* is captured in Definition 2.3 and illustrated in Figure 2.3.

**Definition 2.3 (One-way function)** *A function $f : X \rightarrow Y$ is* one way *if $f(x)$ can be computed efficiently for all $x \in X$, but $f^{-1}(f(x))$ cannot be computed efficiently, meaning that $f^{-1}(y)$ cannot be computed efficiently for $y \in_R Y$.*

In this definition, $X$ represents the domain of $f$, $Y$ represents the range, and the expression $y \in_R Y$ reads as "an element $y$ that is randomly chosen from $Y$." Consequently, it must be possible to efficiently compute $f(x)$ for all $x \in X$,

1 Note that it is not impossible that such an algorithm exists; it is just not known.

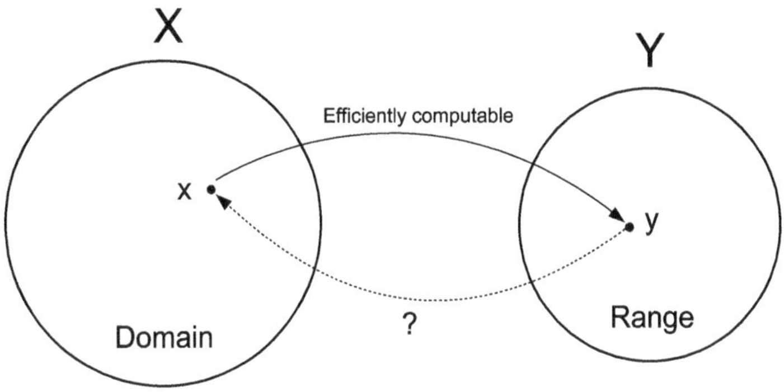


**Figure 2.3** A one-way function.

whereas it must not—or only with a negligible probability—be possible to compute $f^{-1}(y)$ for a $y$ that is randomly chosen from $Y$. To be more precise, one must state that it may be possible to compute $f^{-1}(y)$, but that the entity that wants to do the computation does not know how to do it. In either case, Definition 2.3 is not precise in a mathematically strong sense, because we have not yet defined what an efficient computation is. In essence, a computation is said to be efficient if the (expected) running time of the algorithm that does the computation is bounded by a polynomial in the length of the input. Otherwise (i.e., if the expected running time is not bounded by a polynomial in the length of the input), the algorithm requires super-polynomial time and is said to be inefficient. For example, an algorithm that requires exponential time is clearly superpolynomial. This notion of efficiency (and the distinction between polynomial and superpolynomial running time algorithms) is yet coarse, but still the best we have to work with.

There are many real-world examples of one-way functions. If, for example, we have a telephone book, then the function that assigns a telephone number to each name is easy to compute (because the names are sorted alphabetically) but hard to invert (because the telephone numbers are not sorted numerically). Also, many physical processes are inherently one way. If, for example, we smash a bottle into pieces, then it is generally infeasible to put the pieces together and reassemble the bottle. Similarly, if we drop a bottle from a bridge, then it falls down, whereas the reverse process never occurs by itself. Last but not least, time is inherently one way, and it is (currently) not known how to travel back in time. As a consequence of this fact, we continuously age and have no possibility to make ourselves young again.

In contrast to the real world, there are only a few mathematical functions conjectured to be one way. The most important examples are centered around *modular exponentiation*: Either $f(x) = g^x \pmod{m}$, $f(x) = x^e \pmod{m}$, or $f(x) = x^2 \pmod{m}$ for a properly chosen modulus $m$. While the argument $x$ is in the exponent in the first function, $x$ represents the base of the exponentiation function in the other two functions. Inverting the first function requires computing discrete logarithms, whereas inverting the second (third) function requires computing e-th (square) roots. The three functions are used in many public key cryptosystems: The first function is, for example, used in the Diffie-Hellman key exchange protocol (Section 12.3), the second function is used in the RSA public key cryptosystem (Section 13.3.1), and the third function is used in the Rabin encryption system (Section 13.3.2). We will discuss all of these systems later in the book. It is, however, important to note that none of these functions has been shown to be one way in a mathematically strong sense, and that it is theoretically not even known whether one-way functions exist at all. This means that the one-way property of these functions is just an assumption that may turn out to be wrong (or illusory) in the future—we don't think so, but it may still be the case.

In the general case, a one-way function cannot be inverted efficiently. But there may still be some one-way functions that can be inverted efficiently, if and—as it is hoped—only if some extra information is known. This brings in the notion of a *trapdoor (one-way) function* as captured in Definition 2.4.

**Definition 2.4 (Trapdoor function)** *A one-way function $f : X \rightarrow Y$ is a* trapdoor function *(or a* trapdoor one-way function, *respectively), if there is some extra information (i.e., the* trapdoor*) with which $f$ can be inverted efficiently (i.e., $f^{-1}(f(x))$ can be computed efficiently for all $x \in X$ or $f^{-1}(y)$ can be computed efficiently for $y \in_R Y$).*

Among the functions mentioned above, $f(x) = x^e \pmod{m}$ and $f(x) = x^2 \pmod{m}$ have a trapdoor, namely the prime factorization of $m$. Somebody who knows the prime factors of $m$ can also efficiently invert these functions. In contrast, the function $f(x) = g^x \pmod{m}$ is not known to have a trapdoor if $m$ is prime.

The mechanical analog of a trapdoor (one-way) function is a padlock. It can be closed by everybody (if it is in unlocked state), but it can be opened only by somebody who holds the proper key. In this analogy, a padlock without a keyhole represents a one-way function. In the real world, this is not a particularly useful artifact, but in the digital world, as we will see, there are many applications that make use of it.

If $X$ and $Y$ are the same, then a one-way function $f : X \rightarrow X$ that is a permutation (i.e., $f \in Perms[X]$), is called a *one-way permutation*. Similar to Definition 2.4, a one-way permutation $f : X \rightarrow X$ is a *trapdoor permutation* (or a *trapdoor one-way permutation*, respectively), if it has a trapdoor. Consequently, one-way permutations and trapdoor permutations are special cases of one-way functions and trapdoor functions, namely ones in which the domain and the range are the same and the functions themselves are permutations.

One-way functions including trapdoor functions, one-way permutations, and trapdoor permutations are further addressed in Chapter 5. In this chapter, we will also explain why one has to consider families of such functions to be mathematically correct. In Part III of the book, we then elaborate on the use one-way and trapdoor functions in public key cryptography.

### 2.1.4 Cryptographic Hash Functions

Hash functions are widely used and have many applications in computer science. Informally speaking, a hash function is an efficiently computable function that takes an arbitrarily large input and generates an output of a usually much smaller size. This idea is captured in Definition 2.5 and illustrated in Figure 2.4.

**Definition 2.5 (Hash function)** *A function $h : X \rightarrow Y$ is a* hash function, *if $h(x)$ can be computed efficiently for all $x \in X$ and $|X| \gg |Y|$.*

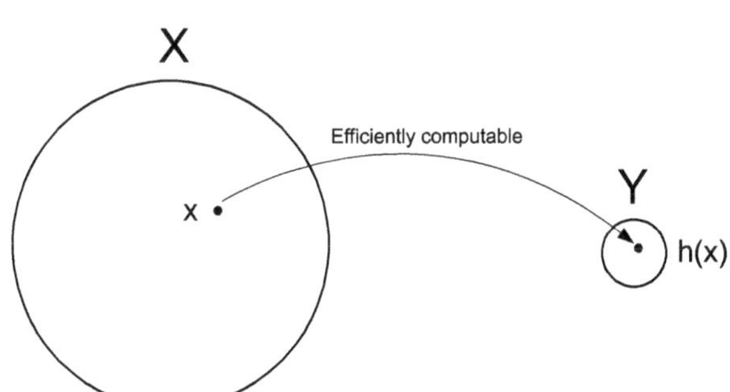


**Figure 2.4** A hash function.

The elements of $X$ and $Y$ are typically strings of characters from a given alphabet. If $\Sigma_{in}$ is the input alphabet and $\Sigma_{out}$ is the output alphabet, then a hash function $h$ can be written as $h : \Sigma_{in}^* \rightarrow \Sigma_{out}^n$, or $h : \Sigma_{in}^{n_{max}} \rightarrow \Sigma_{out}^n$ if the input size is restricted to $n_{max}$ for some technical reasons.[2] In either case, the output is $n$ characters long. In many practical settings, $\Sigma_{in}$ and $\Sigma_{out}$ are identical and refer to the binary alphabet $\Sigma = \{0, 1\}$. In such a setting, the hash function $h$ takes as input an arbitrarily long bitstring and generates as output a bitstring of fixed size $n$.

In cryptography, we are talking about strings that are a few hundred bits long. Also, we are talking about hash functions that have specific (security) properties, such as one-wayness (or preimage resistance, respectively), second-preimage resistance, and/or collision resistance. These properties are introduced and fully explained in Chapter 6. In the meantime, it suffices to know that a hash function used in cryptography (i.e., a cryptographic hash function), must fulfill two basic requirements:

- On the one hand, it must be hard to invert the function; that is, the function is one-way or preimage resistant.[3]
- On the other hand, it must be hard to find a collision, meaning that it is either hard to find a second preimage for a given hash value; that is, the function is second-preimage resistant, or it is hard to find two preimages that hash to the same value (i.e., the function is collision resistant).

According to Definition 2.6, the first requirement can be combined with any of the two possibilities from the second requirement.

**Definition 2.6 (Cryptographic hash function)** *A hash function h is* cryptographic, *if it is either one-way and second-preimage resistant or one-way and collision resistant.*

Cryptographic hash functions have many applications in cryptography. Most importantly, a cryptographic hash function $h$ can be used to hash arbitrarily sized messages to bitstrings of fixed size. This is illustrated in Figure 2.5, where the ASCII-encoded message "This is a file that includes some important but long statements. Consequently, we may need a short representation of this file." is hashed to `0x492165102a1a9e3179a2139d429e32ce4f7fd837` (in hexadecimal notation). This value represents the *fingerprint* or *digest* of the message and—in some sense—stands for it. The second-preimage resistance property implies that it is difficult—or even computationally infeasible—to find another message that hashes to the same value. It also implies that a minor modification of the message leads to a completely different hash value that looks random. If, for example, a second point were added to the message given above, then the resulting hash value would be

2 One such reason may be that the input length must be encoded in a fixed-length field in the padding.

3 The two terms are used synonymously here.

`0x2049a3fe86abcb824d9f9bc957f00cfa7c1cae16` (that is completely independent from `0x492165102a1a9e3179a2139d429e32ce4f7fd837`). If the collision resistance property is required, then it is even computationally infeasible to find two arbitrary messages that hash to the same value.[4]

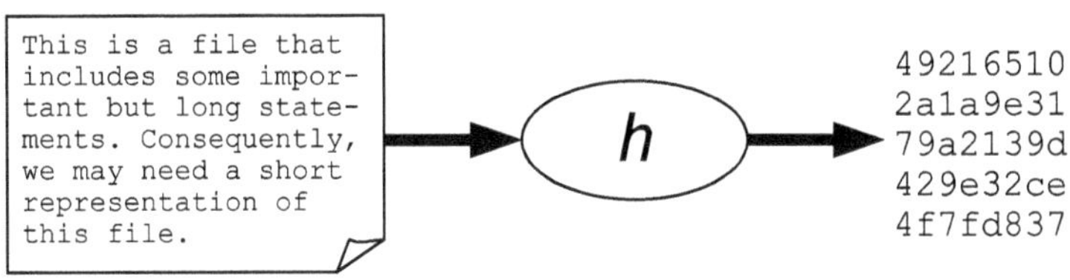


**Figure 2.5** A cryptographic hash function.

Examples of cryptographic hash functions that are used in the field are MD5, SHA-1 (depicted in Figure 2.5), the representatives from the SHA-2 family, and SHA-3 or KECCAK. These functions generate hash values of different sizes.[5]

Cryptographic hash functions, their design principles, and their security properties are further addressed in Chapter 6.

Random generators, random functions, one-way functions, and cryptographic hash functions are unkeyed cryptosystems that are omnipresent in cryptography, and that are used as building blocks in more sophisticated cryptographic systems and applications. The next class of cryptosystems we look at are secret key cryptosystems. Referring to Definition 1.6, these cryptosystems use secret parameters that are shared among all participating entities.

## 2.2 SECRET KEY CRYPTOSYSTEMS

The most important representatives of secret key cryptosystems are pseudorandom generators, pseudorandom functions, systems for symmetric encryption and message authentication, as well as authenticated encryption. For all of these systems we briefly explain what they are all about and what it means by saying that they are

4 Because the task of finding two arbitrary messages that hash to the same value is simpler than finding a message that hashes to a given value, collision resistance is a stronger property than second-preimage resistance.

5 The output of MD5 is 128 bits long. The output of SHA-1 is 160 bits long. Many other hash functions generate an output of variable size.

secure. Note, however, that the notion of security will be explained in more detail in the respective chapters in Part II of the book.

### 2.2.1 Pseudorandom Generators

In Section 2.1.1, we introduced the notion of a random generator that can be used to generate random values. If a large number of such values is needed, then it may be more appropriate to use a *pseudorandom generator* (PRG) instead of—or rather in combination with—a true random generator. More specifically, one can use a random generator to randomly generate a short value (i.e., a seed), and a PRG to stretch this short value into a much longer sequence of values that appear to be random. The notions of a PRG and a *pseudorandom bit generator* (PRBG) are captured in Definition 2.7.

**Definition 2.7 (PRG and PRBG)** *A* PRG *is an efficiently computable function that takes as input a relatively short value of length $n$, called the* seed, *and generates as output a value of length $l(n)$ with $l(n) \gg n$ that appears to be random (and is therefore called* pseudorandom*). If the input and output values are bit sequences, then the PRG is a PRBG.*

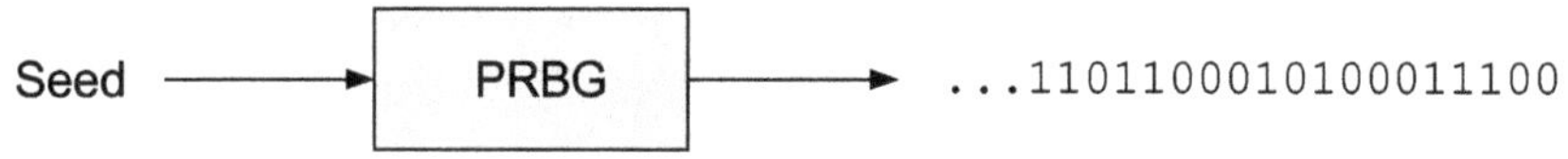


**Figure 2.6** A PRBG.

A PRBG is illustrated in Figure 2.6.[6] Formally speaking, a PRBG $G$ is a mapping from $\mathcal{K} = \{0,1\}^n$ to $\{0,1\}^{l(n)}$, where $l(n)$ represents a stretch function (i.e., a function that stretches an $n$-bit input value into a longer $l(n)$-bit output value with $n < l(n) \leq \infty$):

$$G : \mathcal{K} \longrightarrow \{0,1\}^{l(n)}$$

Note that Definition 2.7 is not precise in a mathematically strong sense, because we have not yet defined what we mean by saying that a bit sequence "appears to be random." Unlike a true random generator, a PRG operates deterministically,

6 Note the subtle difference between Figures 2.1 and 2.6. Both generators output bit sequences. But while the random bit generator has no input, the PRBG has a seed that represents the input.

and this, in turn, means that a PRG always outputs the same values if seeded with the same input value. A PRG thus represents a *finite state machine* (FSM), and hence the sequence of the generated values needs to be cyclic (with a potentially very large cycle). This is why we cannot require that the output of a PRG is truly random, but only that it appears to be so (for a computationally bounded adversary).

This is the starting point for a security definition: We say that a PRG is secure, if its output is indistinguishable from the output of a true random generator. Again referring to the security game of Section 1.2.2.1, we consider an adversary who can have a generator output arbitrarily many values, and his or her task is to decide whether these values are generated by a true random generator or a PRG (i.e., whether they are randomly or pseudorandomly generated). If he or she can do so (with a probability that is better than guessing), then the PRG does not appear to behave like a random generator and is therefore not assumed to be secure. Needless to say that the adversary can employ all statistical tests mentioned in Section 2.1.1 and further explored in Section 3.3 to make his or her decision.

We will explore PRGs and (cryptographically) secure PRGs in more detail in Chapter 7. As suggested by the title of [1], pseudorandomness and PRGs are key ingredients and have many applications in cryptography, such as key generation and additive stream ciphers.

### 2.2.2 Pseudorandom Functions

We have just seen how to use a PRG to "simulate" a true random generator (and this is why we call the respective generator *pseudorandom* instead of *random*). Following a similar line of argumentation, we may try to "simulate" a random function as introduced in Section 2.1.2 with a *pseudorandom function* (PRF). Remember that a random function $f : X \rightarrow Y$ is randomly chosen from $\text{Funcs}[X,Y]$, and that the size of this set; that is $|\text{Funcs}[X,Y]| = |Y|^{|X|}$, is so incredibly large that we cannot number its elements and use an index to refer to a particular function. Instead, we can better use a subset of $\text{Funcs}[X,Y]$ that is sufficiently small so that we can number its elements and use a moderately sized index to refer to a particular function from the subset. If we use a secret key as an index into the subset, then we can have something like a random function without its disadvantages. This is the plan, and a respective definition for a PRF is given in Definition 2.8.[7]

**Definition 2.8 (PRF)** *A* PRF *is a family $F : \mathcal{K} \times X \rightarrow Y$ of (efficiently computable) functions, where each $k \in \mathcal{K}$ determines a function $f_k : X \rightarrow Y$ that is*

7 The notion of a function family (or family of functions, respectively) is formally introduced in Section A.1.1. Here, it is sufficient to have an intuitive understanding for the term.

*indistinguishable from a random function; that is, a function randomly chosen from* $\text{Funcs}[X, Y]$.

Note that there are "only" $|\mathcal{K}|$ elements in $F$, whereas there are $|Y|^{|X|}$ elements in $\text{Funcs}[X, Y]$. This means that we can use a relatively small key to determine a particular function $f_k \in F$, and this function still behaves like a random function, meaning that it is computationally indistinguishable from a truly random function. In our security game this means that, when interacting[8] with either a random function or a PRF, an adversary cannot tell the two cases apart. In other words, he or she cannot or can only tell with a negligible probability whether he or she is interacting with a random function or "only" a pseudorandom one. Such a PRF (that is indistinguishable from a random function) then behaves like a random function and is considered to be "secure," meaning that it can be used for cryptographic purposes and applications.

A *pseudorandom permutation* (PRP) is defined similarly: A PRP is a family $P : \mathcal{K} \times X \rightarrow Y$ of (efficiently computable) permutations, where each $k \in \mathcal{K}$ determines a permutation $p_k : X \rightarrow X$ that is indistinguishable from a random permutation; that is, a permutation randomly chosen from $\text{Perms}[X]$.

PRFs and PRPs are important in modern cryptography mainly because many cryptographic constructions that are relevant in practice can be seen this way: A cryptographic hash function is a PRF (with no key); a key derivation function (KDF) is a PRF with a seed acting as key; a block cipher is a PRP; a PRG can be built from a PRF and vice versa, and so on. PRFs and PRPs and their security properties will be further addressed in Chapter 8.

### 2.2.3 Symmetric Encryption

When people talk about cryptography, they often refer to confidentiality protection using a *symmetric encryption system* that, in turn, can be used to encrypt and decrypt data. *Encryption* refers to the process that maps a plaintext message to a ciphertext, whereas *decryption* refers to the reverse process (i.e., the process that maps a ciphertext back to the plaintext message). Formally speaking, a symmetric encryption system (or *cipher*) can be defined as suggested in Definition 2.9.

**Definition 2.9 (Symmetric encryption system)** *Let $\mathcal{M}$ be a plaintext message space,[9] $\mathcal{C}$ a ciphertext space, and $\mathcal{K}$ a key space. A* symmetric encryption system *or* cipher *refers to a pair $(E, D)$ of families of efficiently computable functions:*

8 Interacting means that the adversary can have arbitrarily many input values of his or her choice be mapped to respective output values.

9 In some other literature, the plaintext message space is denoted by $\mathcal{P}$.

- *$E : \mathcal{K} \times \mathcal{M} \rightarrow \mathcal{C}$ denotes a family $\{E_k : k \in \mathcal{K}\}$ of* encryption functions *$E_k : \mathcal{M} \rightarrow \mathcal{C}$;*
- *$D : \mathcal{K} \times \mathcal{C} \rightarrow \mathcal{M}$ denotes a family $\{D_k : k \in \mathcal{K}\}$ of respective* decryption functions *$D_k : \mathcal{C} \rightarrow \mathcal{M}$.*

*For every message $m \in \mathcal{M}$ and key $k \in \mathcal{K}$, the functions $D_k$ and $E_k$ must be inverse to each other; that is, $D_k(E_k(m)) = m$.*[10]

In a typical setting, $\mathcal{M} = \mathcal{C} = \{0,1\}^*$ refers to the set of all arbitrarily long binary strings, whereas $\mathcal{K} = \{0,1\}^l$ refers to the set of all $l$ bits long keys. Hence, $l$ stands for the key length of the symmetric encryption system (e.g., $l = 128$).

While the decryption functions need to be deterministic, the encryption functions can be deterministic or probabilistic. In the second case, they usually take some random data as additional input (this is not formalized in Definition 2.9). As will become clear later in the book, probabilistic encryption functions tend to be more secure than deterministic ones.

In the description of a (symmetric or asymmetric) encryption system, we often use an algorithmic notation: Generate then stands for a key generation algorithm (that can be omitted in the symmetric case because the key is just randomly selected from $\mathcal{K}$), Encrypt for an algorithm that implements the encryption function, and Decrypt for an algorithm that implements the decryption function. Using this notation, the working principle of a symmetric encryption system is illustrated in Figure 2.7. First, the Generate algorithm generates a key $k$ by randomly selecting an element from $\mathcal{K}$. This key is distributed to either side of the communication channel (this is why the encryption system is called "symmetric" in the first place), namely to the sender (or the sending device, respectively) on the left side and the recipient (or the receiving device, respectively) on the right side. The sender can encrypt a plaintext message $m \in \mathcal{M}$ with its implementation of the encryption function or Encrypt algorithm and $k$. The resulting ciphertext $c = E_k(m) \in \mathcal{C}$ is sent to the recipient over the communication channel that can be insecure (drawn as a dotted line in Figure 2.7). On the right side, the recipient can decrypt $c$ with its implementation of the decryption function or Decrypt algorithm and the same key $k$. If the decryption is successful, then the recipient is able to retrieve and continue to use the original message $m$.

The characteristic feature of a symmetric encryption system is in fact that $k$ is the same on either side of the communication channel, meaning that $k$ is a secret shared by the sender and the recipient. Another characteristic feature is that the system can operate on individual bits and bytes (typically representing a *stream*

10 In some symmetric encryption systems, it does not matter whether one encrypts first and then decrypts or one decrypts first and then encrypts: that is, $D_k(E_k(m)) = E_k(D_k(m)) = m$.

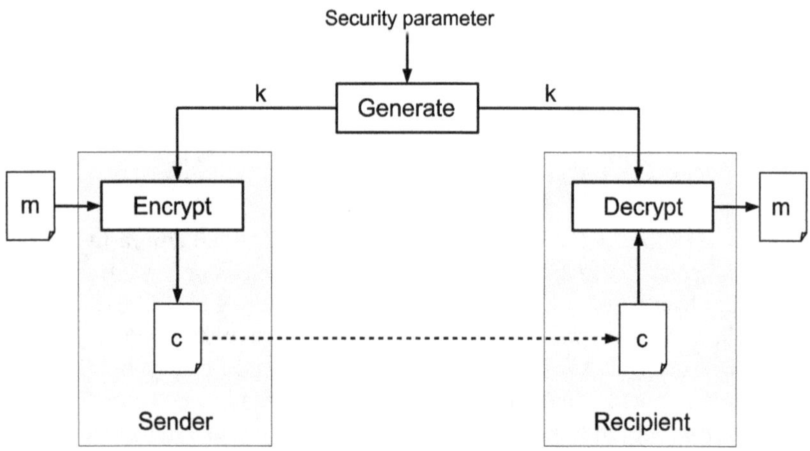


**Figure 2.7** The working principle of a symmetric encryption system.

*cipher*) or on larger blocks (typically representing a *block cipher*). While there are modes of operation that turn a block cipher into a stream cipher, the opposite is not known to be true, meaning that there is no mode of operation that effectively turns a stream cipher into a block cipher.

To make meaningful statements about the security of a symmetric encryption system or cipher, one must define the adversary and the task he or she needs to solve to be successful (Definition 1.8). With regard to the adversary, one must specify his or her computing power and the types of attacks he or she is able to mount, such as ciphertext-only attacks, chosen-plaintext attacks, or even chosen-ciphertext attacks. With regard to the task, one must specify whether he or she must decrypt a ciphertext, determine a key, determine a few bits from either the plaintext or the key, or do something else. Consequently, there are several notions of security one may come up with. If, for example, an adversary has infinite computing power but is still not able to decrypt a ciphertext within a finite amount of time, then the respective cipher is unconditionally or information-theoretically secure. We already mentioned that the one-time pad yields an example of such an information-theoretically secure cipher (that provides perfect secrecy). If the adversary is theoretically able to decrypt a ciphertext within a finite amount of time, but the computing power required to do so is beyond his or her capabilities, then the respective cipher is "only" conditionally or computationally secure. This means that the system can be broken in theory (e.g., by an exhaustive key search), but the respective attacks are assumed to be computationally infeasible to mount by the adversaries one has in mind.

In Chapter 9, we will introduce, formalize, discuss, and put into perspective several notions of security, including semantic security. If a cipher is semantically secure, then it is computationally infeasible to retrieve any meaningful information about a plaintext message from a given ciphertext, even if the adversary can mount a chosen-plaintext attack, and hence has access to an encryption oracle. All symmetric encryption systems in use today are at least semantically secure. In this chapter, we will also outline ciphers that are either historically relevant, such as the Data Encryption Standard (DES) and RC4, or practically important, such as the Advanced Encryption Standard (AES) and Salsa20/ChaCha20. Note that there are many other ciphers proposed in the literature. The majority of them have been broken, but some still remain secure.

### 2.2.4 Message Authentication

While encryption systems are to protect the confidentiality of data, there are applications that require rather the authenticity and integrity of data to be protected—either in addition or instead of the confidentiality. Consider, for example, a financial transaction. It is nice to have the confidentiality of this transaction be protected, but it is somehow more important to protect its authenticity and integrity. The typical way to achieve this is to have the sender add an *authentication tag* to the message and to have the recipient verify the tag before he or she accepts the message as being genuine. This is conceptually similar to an error correction code. But in addition to protect a message against transmission errors, an authentication tag also protects a message against tampering and deliberate fraud. This means that the tag itself needs to be protected against an adversary who may try to modify the message and/or the tag.

From a bird's eye perspective, there are two possibilities to construct an authentication tag: Either through the use of public key cryptography and a *digital signature* (as explained later in this book) or through the use of secret key cryptography and a *message authentication code* (MAC[11]). The second possibility is captured in Definition 2.10.

**Definition 2.10 (MAC)** *A* MAC *is an authentication tag that can be computed and verified with a secret parameter (e.g., a secret key).*

In the case of a message sent from one sender to one recipient, the secret parameter must be shared between the two entities. If, however, the message is sent

11 In some literature, the term *message integrity code* (MIC) is used synonymously and interchangeably. However, this term is not used in this book.

to multiple recipients, then the secret parameter must be shared among the sender and all receiving entities. In this case, the distribution and management of the secret parameter yields a major challenge (and is probably one of the Achilles' heels of the entire system).

Similar to a symmetric encryption system, one can introduce and formally define a system to compute and verify MACs. In this book, we sometimes use the term *message authentication system* to refer to such a system (contrary to many other terms used in this book, this term is not widely used in the literature). It is formalized in Definition 2.11.

**Definition 2.11 (Message authentication system)** *Let $\mathcal{M}$ be a message space, $\mathcal{T}$ a tag space, and $\mathcal{K}$ a key space. A* message authentication system *then refers to a pair $(A, V)$ of families of efficiently computable functions:*

- $A : \mathcal{K} \times \mathcal{M} \rightarrow \mathcal{T}$ *denotes a family* $\{A_k : k \in \mathcal{K}\}$ *of* authentication functions $A_k : \mathcal{M} \rightarrow \mathcal{T}$;
- $V : \mathcal{K} \times \mathcal{M} \times \mathcal{T} \rightarrow \{valid, invalid\}$ *denotes a family* $\{V_k : k \in \mathcal{K}\}$ *of* verification functions $V_k : \mathcal{M} \times \mathcal{T} \rightarrow \{valid, invalid\}$.

*For every message $m \in \mathcal{M}$ and key $k \in \mathcal{K}$, $V_k(m, t)$ must yield $valid$ if and only if $t$ is a valid authentication tag for $m$ and $k$; that is, $t = A_k(m)$, and hence $V_k(m, A_k(m))$ must yield $valid$.*

Typically, $\mathcal{M} = \{0,1\}^*$, $\mathcal{T} = \{0,1\}^{l_{tag}}$ for some fixed tag length $l_{tag}$, and $\mathcal{K} = \{0,1\}^{l_{key}}$ for some fixed key length $l_{key}$. It is often the case that $l_{tag} = l_{key} = 128$, meaning that the tags and keys are 128 bits long each.

The working principle of a message authentication system is depicted in Figure 2.8. Again, the Generate algorithm randomly selects a key $k$ from $\mathcal{K}$ that is sent to the sender (on the left side) and the recipient (on the right side). The sender uses the authentication function or Authenticate algorithm to compute an authentication tag $t$ from $m$ and $k$. Both $m$ and $t$ are sent to the recipient. The recipient, in turn, uses the verification function or Verify algorithm to check whether $t$ is a valid tag with respect to $m$ and $k$. The resulting Boolean value yields the output of the Verify algorithm; it can either be $valid$ or $invalid$.

To argue about the security of a message authentication system, we must define the adversary and the task he or she must solve to be successful (i.e., to break the security of the system). Similar to symmetric encryption systems, we may consider adversaries with infinite computing power to come up with systems that are unconditionally or information-theoretically secure, or—more realistically—adversaries with finite computing power to come up with systems that are "only" conditionally or computationally secure.

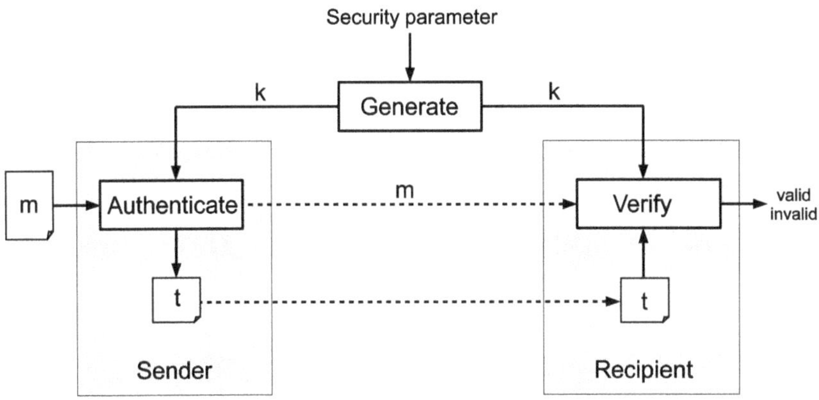


**Figure 2.8** The working principle of a message authentication system.

As we will see in Chapter 10, there are message authentication systems that are unconditionally or information-theoretically secure, but require a unique key to authenticate a single message, and there are systems that are "only" conditionally or computationally secure, but can use a single and relatively short key to authenticate multiple messages. Needless to say, this is more appropriate and useful, and hence most systems used in the field are conditionally or computationally secure. Furthermore, the notion of (computational) security we are heading for is unforgeability, meaning that it must be computationally infeasible to generate a valid tag for a new message. This requirement will be clarified and more precisely defined in Chapter 10, when we elaborate on message authentication, MACs, and respective message authentication systems.

### 2.2.5 Authenticated Encryption

For a long time, people used symmetric encryption systems to encrypt messages and message authentication systems to generate MACs that were then appended to the messages. But it was not clear how (i.e., in what order), the two cryptographic primitives had to be applied and combined to achieve the best level of security. In general, there are three approaches or generic composition methods. Let $m \in \mathcal{M}$ be a message, $k_e$ an encryption key, and $k_a$ an authentication key from respective key spaces. The generic composition methods can then be described as follows:

- In *Encrypt-then-MAC* (EtM) the message $m$ is first encrypted, and the resulting ciphertext is then authenticated, before the ciphertext and the MAC are sent together to the recipient. Mathematically speaking, the data that is sent to the recipient is $E_{k_e}(m) \parallel A_{k_a}(E_{k_e}(m))$. EtM is used, for example, in IP security (IPsec).
- In *Encrypt-and-MAC* (E&M) the message $m$ is encrypted and authenticated independently, meaning that—in contrast to EtM—a MAC is generated for the plaintext and sent together with the ciphertext to the recipient. In this case, the data that is sent to the recipient is $E_{k_e}(m) \parallel A_{k_a}(m)$. E&M is used, for example, in Secure Shell (SSH).
- In *MAC-then-Encrypt* (MtE) a MAC is first generated for the plaintext message, and the message with the appended MAC is then encrypted. The resulting ciphertext (that comprises both the message and the MAC) is then sent to the recipient. In this case, the data that is sent to the recipient is $E_{k_e}(m \parallel A_{k_a}(m))$. MtA was used, for example, in former versions of the SSL/TLS protocols [2].

Since 2001 it is known that encrypting a message and subsequently applying a MAC to the ciphertext (i.e., the EtM method), provides the best level of security [3], and most of today's security protocols follow this approach, i.e., they apply message encryption prior to authentication. More specifically, most people combine message encryption and authentication in what is called *authenticated encryption* (AE) or *authenticated encryption with associated data* (AEAD). AEAD basically refers to AE where all data are authenticated but not all data are encrypted. The exemplary use case is an IP packet, where the payload is encrypted and authenticated but the header is only authenticated (because it must be accessible to the intermediate routers in the clear). AE is further addressed in Chapter 11. Most importantly, there are several modes of operation for block ciphers that provide AE (or AEAD, respectively).

The next class of cryptosystems we look at are public key cryptosystems. According to Definition 1.7, these are cryptosystems that use secret parameters that are not shared among all participating entities, meaning that they are held in private.

## 2.3 PUBLIC KEY CRYPTOSYSTEMS

Instead of sharing all secret parameters, the entities that participate in a public key cryptosystem hold two distinct sets of parameters: One that is private (collectively referred to as the *private* or *secret key* and abbreviated as $sk$), and one that is published (collectively referred to as the *public key* and abbreviated as $pk$).[12] A necessary but usually not sufficient prerequisite for a public key cryptosystem to be secure is that both keys—the private key and the public key—are yet mathematically related, but it is still computationally infeasible to compute one from the other. Another prerequisite is that the public keys are available in a form that provides authenticity and integrity. If somebody is able to introduce faked public keys, then he or she is usually able to mount very powerful attacks. This is why we usually require public keys to be published in some certified form, and hence the notions of (public key) certificates and public key infrastructures (PKI) come into play. This topic will be further addressed in Section 16.4.

The fact that public key cryptosystems use secret parameters that are not shared among all participating entities suggests that the respective algorithms are executed by different entities, and hence that such cryptosystems are best defined as sets of algorithms (that are then executed by these different entities). We adopt this viewpoint in this book and define public key cryptosystems as sets of algorithms. In the case of an asymmetric encryption system, for example, there is a key generation algorithm Generate, an encryption algorithm Encrypt, and a decryption algorithm Decrypt. The Generate and Encrypt algorithms are usually executed by the sender of a message, whereas the Decrypt algorithm is executed by the recipient(s). As discussed later, other public key cryptosystems may employ other sets of algorithms.

In the following sections, we briefly overview the most important public key cryptosystems used in the field, such as key establishment, asymmetric encryption, and digital signatures. The overview is superficial here, and the technical details are provided in Part II of the book (this also applies to zero-knowledge proofs of knowledge that are not addressed here).

Because public key cryptography is computationally less efficient than secret key cryptography, it usually makes a lot of sense to combine both types of cryptography in *hybrid cryptosystems*. In such a system, public key cryptography is mainly used for authentication and key establishment, whereas secret key cryptography is used for everything else (most notably bulk data encryption). Hybrid cryptosystems are frequently used and very widely deployed in the field. In fact, almost every non-trivial application of cryptography employs some form of hybrid cryptography.

### 2.3.1 Key Establishment

If two or more entities want to employ and make use of secret key cryptography, then they must share a secret parameter that represents a cryptographic key. Consequently, in a large system many secret keys must be generated, stored, managed,

12 It depends on the cryptosystem whether it matters which set of parameters is used to represent the private key and which set of parameters is used to represent the public key.

used, and destroyed (at the end of their life cycle) in a secure way. If, for example, $n$ entities want to securely communicate with each other, then there are

$$\binom{n}{2} = \frac{n(n-1)}{1 \cdot 2} = \frac{n^2 - n}{2}$$

such keys. This number grows in the order of $n^2$, and hence the establishment of secret keys is a major practical problem—sometimes called the $n^2$-problem—and probably the Achilles' heel of the large-scale deployment of secret key cryptography. For example, if $n = 1,000$ entities want to securely communicate with each other, then there are

$$\binom{1,000}{2} = \frac{1,000^2 - 1,000}{2} = 499,500$$

keys. Even for moderately large $n$, the generation, storage, management, usage, and destruction of all such keys is prohibitively expensive and the antecedent distribution of them is next to impossible. Things even get worse in dynamic systems, where entities may join and leave at will. In such a system, the distribution of keys is impossible, because it is not even known in advance who may want to join. This means that one has to establish keys when needed, and there are basically two approaches to achieve this:

- The use of a *key distribution center* (KDC) that provides the entities with the keys needed to securely communicate with each other;
- The use of a *key establishment protocol* that allows the entities to establish the keys themselves.

A prominent example of a KDC is the Kerberos authentication and key distribution system [4]. KDCs in general and Kerberos in particular have many disadvantages. The most important disadvantage is that each entity must unconditionally trust the KDC and share a master key with it. There are situations in which this level of trust is neither justified nor can it be accepted by the participating entities. Consequently, the use of a key establishment protocol that employs public key cryptography yields a viable alternative that is advantageous in most situations and application settings.

In a simple key establishment protocol, an entity randomly generates a key and uses a secure channel to transmit it to the peer entity (or peer entities, respectively). The secure channel can be implemented with any asymmetric encryption system: The entity that randomly generates the key encrypts the key with the public key of the peer entity. This protocol is simple and straightforward, but it has the problem that the security depends on the quality and security of the key generation process—which yields a PRG. Consequently, it is advantageous to have a mechanism in place that allows two (or even more) entities to establish and agree on a commonly shared key. This is where the notion of a key agreement or key exchange protocol comes into play (as opposed to a key distribution protocol).

Even today, the most important key exchange protocol was proposed by Diffie and Hellman in their landmark paper [5] that opened the field of public key cryptography. It solves a problem that looks impossible to solve: How can two entities that have no prior relationship and do not share any secret use a public channel to agree on a shared secret? Imagine a room in which people can only shout messages at each other. How can two persons in this room agree on a secret? As we will see in Chapter 12, the *Diffie-Hellman key exchange protocol* solves this problem in a simple and ingenious way. In this chapter, we will also discuss a few variants and their security, and say a few words about the use of quantum cryptography as an alternative way of exchanging keying material.

### 2.3.2 Asymmetric Encryption Systems

Similar to a symmetric encryption system, an asymmetric encryption system can be used to encrypt and decrypt plaintext messages. The major difference between a symmetric and an asymmetric encryption system is that the former employs secret key cryptography and respective techniques, whereas the latter employs public key cryptography and respective techniques.

As already emphasized in Section 2.1.3, an asymmetric encryption system can be built from a trapdoor function—or, more specifically—from a family of trapdoor functions. Each public key pair comprises a public key $pk$ that yields a one-way function and a private key $sk$ that yields a trapdoor (needed to efficiently compute the inverse of the one-way function). To send a secret message to the recipient, the sender uses the recipient's public key and applies the respective one-way function to the plaintext message. The resulting ciphertext is then sent to the recipient. The recipient, in turn, is the only entity that supposedly holds the trapdoor (information) needed to invert the one-way function and to decrypt the ciphertext accordingly. More formally, an asymmetric encryption system can be defined as in Definition 2.12.

**Definition 2.12 (Asymmetric encryption system)** *An* asymmetric encryption system *consists of the following three efficient algorithms:*

- $\mathsf{Generate}(1^k)$ *is a probabilistic key generation algorithm that takes as input a security parameter* $1^k$ *(in unary notation*[13]*), and generates as output a public key pair* $(pk, sk)$ *that is in line with the security parameter.*
- $\mathsf{Encrypt}(pk, m)$ *is a deterministic or probabilistic encryption algorithm that takes as input a public key* $pk$ *and a plaintext message* $m$*, and generates as output a ciphertext* $c = \mathsf{Encrypt}(pk, m)$.
- $\mathsf{Decrypt}(sk, c)$ *is a deterministic decryption algorithm that takes as input a private key* $sk$ *and a ciphertext* $c$*, and generates as output a plaintext message* $m = \mathsf{Decrypt}(sk, c)$.

*For every plaintext message* $m$ *and public key pair* $(pk, sk)$*, the* Encrypt *and* Decrypt *algorithms must be inverse to each other; that is,* $\mathsf{Decrypt}(sk, \mathsf{Encrypt}(pk, m)) = m$.

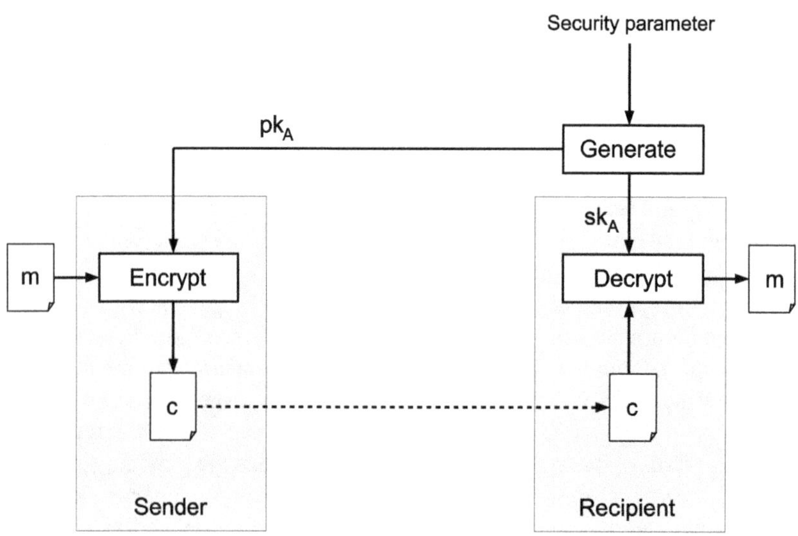


**Figure 2.9** The working principle of an asymmetric encryption system.

13 This means that a 1 is repeated $k$ times.

The working principle of an asymmetric encryption system is illustrated in Figure 2.9. At the top of the figure, the Generate algorithm is to generate a public key pair for entity A that is the recipient of a message. In preparation for the encryption, A's public key $pk_A$ is provided to the sender on the left side. The sender then subjects the message $m$ to the one-way function represented by $pk_A$, and sends the respective ciphertext $c = \mathsf{Encrypt}_{pk_A}(m)$ to A. On the right side, A knows its secret key $sk_A$ that represents a trapdoor to the one-way function and can be used to decrypt $c$ and retrieve the plaintext message $m = \mathsf{Decrypt}_{sk_A}(c)$ accordingly. Hence, the output of the Decrypt algorithm is the original message $m$.

There are many asymmetric encryption systems that have been proposed in the literature, such as Elgamal, RSA, and Rabin. These systems are based on the three exemplary one-way functions mentioned in Section 2.1.3 (in this order). Because it is computationally infeasible to invert these functions, the systems provide a reasonable level of security—even in their basic forms (that are sometimes called textbook versions).

In Chapter 13, we will elaborate on these asymmetric encryption systems, but we will also address notions of security that cannot be achieved with them. The strongest notion of security is again defined in the game-theoretical setting: An adversary can select two equally long plaintext messages and has one of them encrypted. If he or she cannot tell whether the respective ciphertext is the encryption of the first or the second plaintext message with a probability that is better than guessing, then the asymmetric encryption system leaks no information and is therefore assumed to be (semantically) secure. This may hold even if the adversary has access to a decryption oracle, meaning that he or she can have any ciphertext of his or her choice be decrypted—except, of course, the ciphertext the adversary is challenged with. We will more thoroughly explain this setting and present variants of the basic asymmetric encryption systems that remain secure even in this setting.

### 2.3.3 Digital Signatures

Digital signatures can be used to protect the authenticity and integrity of messages, or—more generally—data objects. According to [6], a *digital signature* refers to "a value computed with a cryptographic algorithm and appended to a data object in such a way that any recipient of the data can use the signature to verify the data's origin and integrity." Similarly, the term digital signature is defined as "data appended to, or a cryptographic transformation of, a data unit that allows a recipient of the data unit to prove the source and integrity of the data unit and protect against forgery, e.g. by the recipient" in ISO/IEC 7498-2 [7]. Following the second definition, there are two classes of digital signatures:

- If data representing the digital signature is appended to a data unit (or message), then one refers to a *digital signature with appendix*.
- If a data unit is cryptographically transformed in a way that it represents both the data unit (or message) that is signed and the digital signature, then one refers to a *digital signature giving message recovery*. In this case, the data unit is recovered if and only if the signature is successfully verified.

In either case, the entity that digitally signs a data unit or message is called the *signer* or *signatory*, whereas the entity that verifies the digital signature is called the *verifier*. In a typical setting, both the signer and the verifier are computing devices operating on a user's behalf. The formal definitions of the two respective digital signature systems (DSS) are given in Definitions 2.13 and 2.14.

**Definition 2.13 (DSS with appendix)** *A* DSS with appendix *consists of the following three efficiently computable algorithms:*

- $\mathsf{Generate}(1^k)$ *is a probabilistic key generation algorithm that takes as input a security parameter* $1^k$*, and generates as output a public key pair* $(pk, sk)$ *that is in line with the security parameter.*
- $\mathsf{Sign}(sk, m)$ *is a deterministic or probabilistic signature generation algorithm that takes as input a signing key* $sk$ *and a message* $m$*, and generates as output a digital signature* $s$ *for* $m$*.*
- $\mathsf{Verify}(pk, m, s)$ *is a deterministic signature verification algorithm that takes as input a verification key* $pk$*, a message* $m$*, and a purported digital signature* $s$ *for* $m$*, and generates as output a binary decision whether the signature is valid.*

$\mathsf{Verify}(pk, m, s)$ *must yield* $\mathtt{valid}$ *if and only if* $s$ *is a valid digital signature for* $m$ *and* $pk$*. This means that for every message* $m$ *and every public key pair* $(pk, sk)$*,* $\mathsf{Verify}(pk, m, \mathsf{Sign}(sk, m))$ *must yield* $\mathtt{valid}$*.*

**Definition 2.14 (DSS giving message recovery)** *A* DSS giving message recovery *consists of the following three efficiently computable algorithms:*

- $\mathsf{Generate}(1^k)$ *is a probabilistic key generation algorithm that takes as input a security parameter* $1^k$*, and generates as output a public key pair* $(pk, sk)$ *that is in line with the security parameter.*
- $\mathsf{Sign}(sk, m)$ *is a deterministic or probabilistic signature generation algorithm that takes as input a signing key* $sk$ *and a message* $m$*, and generates as output a digital signature* $s$ *giving message recovery.*

- $\mathsf{Recover}(pk, s)$ *is a deterministic message recovery algorithm that takes as input a verification key* $pk$ *and a digital signature* $s$*, and generates as output either the message that is digitally signed or a notification indicating that the digital signature is invalid.*

$\mathsf{Recover}(pk, s)$ *must yield* $m$ *if and only if* $s$ *is a valid digital signature for* $m$ *and* $pk$*. This means that for every message* $m$ *and every public key pair* $(pk, sk)$*,* $\mathsf{Recover}(pk, \mathsf{Sign}(sk, m))$ *must yield* $m$*.*

Note that the Generate and Sign algorithms are identical in this algorithmic notation, and that the only difference refers to the Verify and Recover algorithms. While the Verify algorithm takes the message $m$ as input, this value is not needed by the Recover algorithm. Instead, the message $m$ is automatically recovered if the signature turns out to be valid.

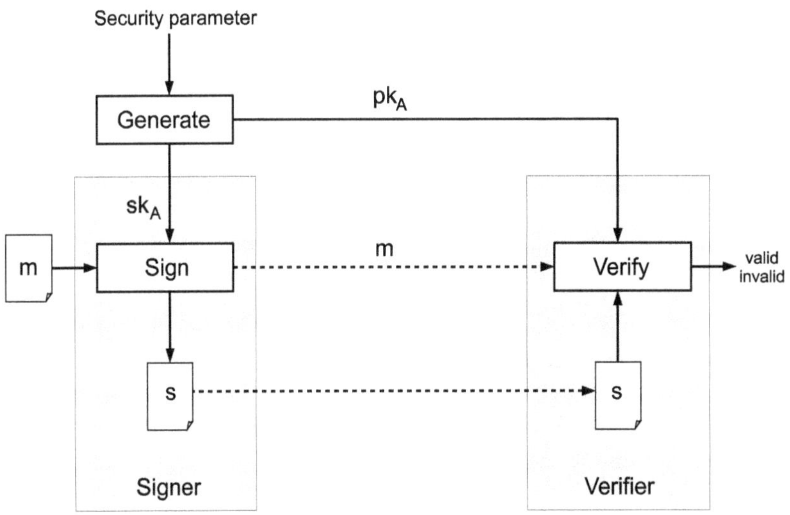


**Figure 2.10** The working principle of a DSS with appendix.

The working principle of a DSS with appendix is illustrated in Figure 2.10. This time, the Generate algorithm is applied on the left side (i.e., the signer' side). The signer uses the secret key $sk_A$ (representing the trapdoor) to sign message $m$; that is, $s = \mathsf{Sign}(sk_A, m)$. This message $m$ and the respective signature $s$ are then sent to the verifier. This verifier, in turn, uses the Verify algorithm to verify $s$. More specifically, it takes $m$, $s$, and $pk_A$ as input values and ouputs either $valid$ or $invalid$ (obviously depending on the validity of the signature).

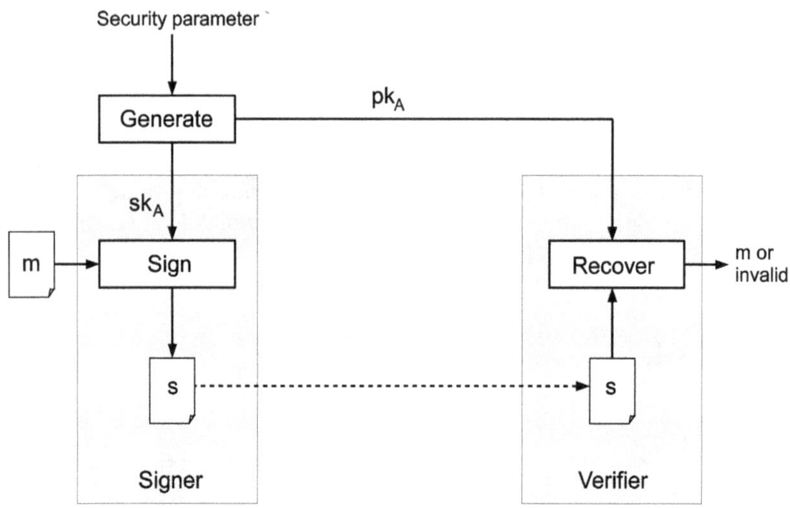


**Figure 2.11** The working principle of a DSS giving message recovery.

If the DSS is giving message recovery, then the situation is slightly different. As illustrated in Figure 2.11, the beginning is the same. But instead of sending $m$ and $s$ to the recipient, the signatory only sends $s$. The signature encodes the message. So when the recipient subjects $s$ to the Recover algorithm, the output is either $m$ (if the signature is valid) or $invalid$.

With the proliferation of the Internet in general and Internet-based electronic commerce in particular, digital signatures and the legislation thereof have become important and timely topics. In fact, many DSS with specific and unique properties have been developed, proposed, and published in the literature. Again, the most important DSSs are overviewed, discussed, and put into perspective in Chapter 14. These are RSA, Rabin, Elgamal, and some variations thereof, such as the Digital Signature Algorithm (DSA) and its elliptic curve version (ECDSA).

Similar to asymmetric encryption systems, the security discussion for digital signatures is nontrivial and subtle, and there are several notions of security discussed in the literature. The general theme is that it must be computationally infeasible for an adversary to generate a new valid-looking signature, even if he or she has access to an oracle that provides him or her with arbitrarily many signatures for messages of his or her choice. In technical parlance, this means that the DSS must resist existential forgery, even if the adversary can mount an adaptive chosen-message attack. Again, we will revisit this topic and explain the various notions of security for DSSs, as well as some constructions to achieve these levels of security in Chapter 14.

## 2.4 FINAL REMARKS

In this chapter, we briefly introduced and provided some preliminary definitions for the most important representatives of the three main classes of cryptosystems distinguished in this book, namely unkeyed cryptosystems, secret key cryptosystems, and public key cryptosystems. We want to note (again) that this classification scheme is somewhat arbitrary, and that other classification schemes may be possible as well.

The cryptosystems that are preliminarily defined in this chapter are revisited, more precisely defined (in a mathematically strong sense), discussed, and put into perspective in the remaining chapters of the book. For all of these systems, we also dive more deeply into the question of what it means for such a system to be secure. This leads us to various notions of security that can be found in the literature. Some of these notions are equivalent, whereas others are fundamentally different. In this case, it is interesting to know the exact relationship of the notions, and what notion is actually the strongest one. We then also provide cryptosystems that conform to this (strongest) notion of security.

Following this line of argumentation, it is a major theme in cryptography to better understand and formally define notions of security, and to prove that particular cryptosystems are secure in this sense. It is another theme to start with cryptographic building blocks that are known to be secure, and to ask how these building blocks can be composed or combined in more advanced cryptographic protocols or systems so that their security properties still apply. Alternatively speaking, how can secure cryptographic building blocks be composed or combined in a modular fashion so that the result remains secure? This is an interesting and practically relevant question addressed in cryptographic research areas and frameworks like *universal composability* [8] or *constructive cryptography* [9]. These topics are beyond the scope of this book and not further addressed here. Instead, we "only" provide an overview of the cryptographic building blocks that are available and that are assumed to be secure when considered in isolation. Just keep in mind that a cryptosystem that is secure in isolation does not need to remain secure when combined or composed with others. So, research areas and frameworks like universal composability and constructive cryptography are crucial for the overall security of a cryptographic application.

## References


[1] Luby, M., *Pseudorandomness and Cryptographic Applications*. Princeton Computer Science Notes, Princeton, NJ, 1996.

[2] Oppliger, R., *SSL and TLS: Theory and Practice*, 2nd edition. Artech House Publishers, Norwood, MA, 2016.

[3] Krawczyk, H., "The Order of Encryption and Authentication for Protecting Communications (or: How Secure Is SSL?)," *Proceedings of CRYPTO 2001*, Springer-Verlag, LNCS 2139, 2001, pp. 310–331.

[4] Oppliger, R., *Authentication Systems for Secure Networks*. Artech House Publishers, Norwood, MA, 1996.

[5] Diffie, W., and M.E. Hellman, "New Directions in Cryptography," *IEEE Transactions on Information Theory*, Vol. 22, No. 6, 1976, pp. 644–654.

[6] Shirey, R., *Internet Security Glossary, Version 2*, Informational RFC 4949 (FYI 36), August 2007.

[7] ISO/IEC 7498-2, *Information Processing Systems—Open Systems Interconnection Reference Model—Part 2: Security Architecture*, 1989.

[8] Canetti, R., "Universally Composable Security: A New Paradigm for Cryptographic Protocols," Cryptology ePrint Archive, *Report 2000/067*, 2000, https://eprint.iacr.org/2000/067.

[9] Maurer, U.M., "Constructive Cryptography – A New Paradigm for Security Definitions and Proofs," *Proceedings of the 2011 International Conference on Theory of Security and Applications (TOSCA 2011)*, Springer-Verlag, LNCS 6993, 2012, pp. 33–56.

# Part I

# UNKEYED CRYPTOSYSTEMS

# Chapter 3

# Random Generators

*Random numbers should not be generated with a method chosen at random.*

— Donald E. Knuth[1]

As mentioned in Section 2.1.1, randomness is the most important ingredient for cryptography, and most cryptographic systems in use today require some form of randomness (as we will see throughout the book). As suggested by Knuth's leading quote, random numbers must be generated carefully and should not be generated with a method chosen at random. This is the theme of this chapter. We give an introduction in Section 3.1, overview and discuss some possible realizations and implementations of random generators in Section 3.2, address statistical randomness testing in Section 3.3, and conclude with some final remarks in Section 3.4. Unlike many other chapters that follow, this chapter is kept relatively short.

## 3.1 INTRODUCTION

The term *randomness* is commonly used to refer to nondeterminism. If we say that something is *random*, we mean that we cannot determine its outcome, or—equivalently—that its outcome is nondeterministic. Whether randomness really exists is also a philosophical question. Somebody who believes that everything is determined and behaves in a deterministic way would typically argue that randomness does not exist. If, for example, we consider a coin toss, then a physicist knowing

1 Donald Ervin Knuth is an American computer scientist who was born in 1938.

precisely the weight of the coin, the initial position, the forces applied during the toss, the wind speed, as well as many other parameters that are relevant, should in principle be able to predict the outcome of a toss (i.e., head or tail). If this is not possible, then this is due to the fact that coin tossing is a chaotic process. Chaos is a type of behavior that can be observed in systems that are highly sensitive to initial conditions. The evolution of such systems is so sensitive to initial conditions that it is simply not possible to determine them precisely enough to make reliable predictions about the future behavior. One could say that any form of determinism is then hidden behind the complexity of the chaotic system. For the purpose of this book, we don't address the philosophical question and simply assume that systems sometimes are so chaotic that they cannot be predicated, and hence that randomness exists. This is also in line with the present knowledge in quantum physics, where we have to assume that randomness exists in the first place.

If we assume the existence of randomness, then we may ask whether it is possible to measure it in one way or another. For example, one may ask for a given value whether it is random or not. Is, for example, 13 random? Or is 27 random? Is 13 more random than 27, or—vice versa—is 27 more random than 13? Unfortunately, questions like these don't make a lot of sense unless they are considered in a specific context.

In theory, there is a measure of randomness for a finite sequence of values. In fact, the *Kolmogorov complexity* measures the minimal length of a program for a Turing machine[2] that is able to generate the sequence. Unfortunately, the Kolmogorov complexity is inherently noncomputable; that is, it is not known how to compute the Kolmogorov complexity for a given sequence of values, and hence it is not particularly useful.

If we know that a bit sequence is generated with a *linear feedback shift register* (LFSR) as introduced in Section 9.5.1, then we can use the *linear complexity* to measure its randomness. In fact, the linear complexity stands for the size of the shortest LFSR that can generate the sequence. This measure peaks to the difficulty of generating—and perhaps analyzing—the bit sequence. There is an algorithm created by Elwyn R. Berlekamp and James L. Massey [1] that can be used to compute the linear complexity of a bit sequence. Note, however, that the linear complexity (and hence also the *Berlekamp-Massey algorithm*) assumes that the bit sequence is generated with an LFSR. Consequently, it is possible that a bit sequence has a large linear complexity but can still be generated easily with other means. This possibility cannot be excluded.

For the rest of this chapter we leave aside the questions whether randomness exists and how to measure it. Instead, we elaborate on the question of how to generate random values. In Definition 2.1, we introduced the notions of a random

2 Refer to Section D.5 for an introduction to Turing machines.

generator and a random bit generator—the latter was also depicted in Figure 2.1. In short, it is a device or algorithm that outputs a sequence of statistically independent and unbiased bits. This means that the bits occur with the same probability; that is, $\Pr[0] = \Pr[1] = 1/2$, and that all $2^k$ possible $k$-tuples occur approximately equally often; that is, with probability $1/2^k$ for all $k \in \mathbb{N}$.

If we can generate random bits, then we can also generate (uniformly distributed) random numbers of any size. If, for example, we want to construct an $n$-bit random number $a$, then we set $b_n = 1$, use the random generator to generate $n-1$ random bits $b_{n-1}, \ldots, b_1$, and set

$$a = \sum_{i=1}^{n} b_i 2^{i-1} \tag{3.1}$$

Similarly, if we want to construct a number that is randomly selected from the interval $[0, m]$ for $m \in \mathbb{N}$, then we set $n$ to the length of $m$, i.e., $n = \lfloor \log m \rfloor + 1$, and use the random generator to generate $n$ random bits $b_1, \ldots, b_n$. If $a$ constructed according to (3.1) is smaller or equal to $m$, then we use it. If, however, $a$ is bigger than $m$, then we don't use it and generate another number instead. Consequently, in what follows we only elaborate on the generation of random bits, and we assume that the construction of random numbers from random bits can always be done this way.

## 3.2 REALIZATIONS AND IMPLEMENTATIONS

The relevant RFC 4086 (BCP 106) [2] recommends the use of special hardware to generate truly random bits. There are, however, situations in which special hardware is not available, and software must be used instead. Consequently, there is room for both hardware-based and software-based random generators. Some general ideas about how to realize and implement such generators are overviewed next. Afterward, we also introduce and outline some deskewing techniques.

### 3.2.1 Hardware-Based Random Generators

Hardware-based random generators exploit the randomness that occurs in physical processes and phenomena. According to [3], examples of such processes and phenomena include:

- The elapsed time between emission of particles during radioactive decay;
- The thermal noise from a semiconductor diode or resistor;

- The frequency instability of a free-running oscillator (e.g., [4]);
- The amount a metal-insulator-semiconductor capacitor is charged during a fixed period of time (e.g., [5]);
- The air turbulence within a sealed disk drive that causes random fluctuations in disk drive sector read latency times (e.g., [6, 7]);
- The sound from a microphone or video input from a camera.

It goes without saying that any other physical process or phenomena may be employed by a hardware-based random generator. When designing such a generator, it may also be a natural choice to take advantage of the inherent randomness of quantum physics and to resort to the use of a quantum process as a source of randomness. This generally allows selecting a very simple process as a source of randomness. The first example itemized above falls into this category. But due to the use of radioactive materials, the respective generators are involved and may cause health concerns for its users.

A simpler technique to build a hardware-based random generator based on quantum physics is to employ an optical quantum process as a source of randomness. From a quantum physics viewpoint, light consists of elementary particles known as photons. In many situations, photons exhibit a random behavior. If, for example, a photon is sent to a semitransparent mirror, then the photon is reflected with a probability of 0.5 and it continues its transmission with the remaining probability of 0.5. Using two single-photon detectors, the two cases can be distinguished and the result can be turned into one bit. There are commercially available random generators that work this way and can generate millions of (random) bits per second.

Hardware-based random generators could easily be integrated into contemporary computer systems. However, this is not always the case, and hence hardware-based random generators are neither readily available nor widely deployed in the field.

### 3.2.2 Software-Based Random Generators

First of all, it is important to note that designing a random generator in software is even more difficult than doing so in hardware. Again, a process is needed to serve as a source of randomness. According to [3], the following processes may be used to serve as a source of randomness for a software-based random generator:

- The system clock (e.g., [8]);
- The elapsed time between keystrokes or mouse movements;

- The content of input/output buffers;
- The input provided by the user;
- The values of operating system variables, such as system load or network statistics.

Again, this list is not comprehensive, and many other processes may be used as a source of randomness for a software-based random generator. Anyway, the behavior of a process may vary considerably depending on various factors, such as the computer platform, operating system, and software in use, but it may still be difficult to prevent an adversary from observing (or even manipulating) a process. For example, if an adversary has a rough idea of when a random bit sequence was generated, he or she can guess the content of the system clock at that time with a high degree of accuracy. Consequently, care must be taken when the system clock and the identification numbers of running processes are used to generate random bit sequences. This type of problem gained a lot of publicity in 1995, when it was found that the encryption in Netscape browsers could easily be broken due to the limited range of its randomly generated values. Because the values used to determine session keys could be established without too much difficulty, even U.S. domestic browsers with 128-bit session keys carried at most 47 bits of entropy in their session keys [9]. Shortly after this revelation, it was found that the MIT implementation of Kerberos version 4 [10] and the magic cookie key generation mechanism of the X windows system both suffered from a similar weakness.

Sometimes, it is possible to use external (i.e., external to the computer system that needs the randomness) sources of randomness. For example, a potential source of randomness is the unpredictable behavior of the stock market. This source, however, has some disadvantages of its own. For example, it is sometimes predictable (e.g., during a crash), it can be manipulated (e.g., by spreading rumors or by placing a large stock transaction), and it is never secret, but the search for external sources of randomness remains an interesting research area. The U.S. NIST has a respective project that aims at implementing interoperable randomness beacons.[3] They provide a source of randomness, but—due to their public nature—their output should never be used directly as cryptographic keys.

In [2], it is argued that the best overall strategy for meeting the requirement for unguessable random bits in the absence of a single reliable source is to obtain random input from a large number of uncorrelated sources and to mix them with a strong mixing function. A strong mixing function, in turn, is one that combines two or more inputs and produces an output where each output bit is a different complex nonlinear function of all input bits. On average, changing an input bit will change

3 https://csrc.nist.gov/projects/interoperable-randomness-beacons.

about half of the output bits. However, because the relationship is complex and nonlinear, no particular output bit is guaranteed to change when any particular input bit is changed. A trivial example for such a function is addition modulo $2^{32}$. More general strong mixing functions (for more than two inputs) can be constructed using other cryptographic systems, such as cryptographic hash functions or symmetric encryption systems.

### 3.2.3 Deskewing Techniques

Any source of random bits may be defective in the sense that the output bits are biased (i.e., the probability of the source emitting a 1 is not equal to $1/2$) or correlated (i.e., the probability of the source emitting a 1 depends on previously emitted bits). There are several *deskewing techniques* that can be used to generate a random bit sequence from the output of such a defective random bit generator:

- If, for example, a defective random generator outputs biased (but uncorrelated) bits, then a simple deskewing technique created by John von Neumann can be used to suppress the bias [11]. Therefore, the output bit sequence is grouped into a sequence of pairs of bits, where each pair of bits is transformed into a single bit:

  - A `10` pair is transformed to `1`;
  - A `01` pair is transformed to `0`;
  - `00` and `11` pairs are discarded.

  For example, the biased bit sequence 11010011010010 is transformed to 001, and this binary sequence is still uncorrelated and now also unbiased. It goes without saying that von Neumann's technique is far from being optimal, and so the technique was later generalized and optimized to achieve an output rate near the source entropy [12–14].

- If a defective random generator outputs correlated bits, then the situation is generally more involved. A simple deskewing technique is to decrease correlation by combining two (or even more) sequences.

The cost of applying any deskewing technique is that the original bit sequence is shortened. In the case of von Neumann's technique, for example, the length of the resulting sequence is at most one quarter of the length of the original sequence. This illustrates the fact that a deskewing technique can also be seen as a compression technique. After its application, any bias is silently removed and no further compression should be possible.

In practice, a simple and straightforward deskewing technique is to pass a bit sequence whose bits are biased and/or correlated through a cryptographic hash function or a block cipher (with a randomly chosen key). The result can be expected to have good randomness properties, otherwise the cryptogrphic primitives are revealed to be weak.

## 3.3 STATISTICAL RANDOMNESS TESTING

While it is impossible to give a mathematical proof that a generator is a random (bit) generator, statistical randomness testing may help detecting certain kinds of defects or weaknesses. This is accomplished by taking a sample output sequence of the generator and subjecting it to some statistical randomness tests. Each test determines whether the sequence that is generated possesses a certain attribute that a truly random sequence would be likely to exhibit.

The first statistical tests for random numbers were published by Maurice George Kendall and Bernard Babington-Smith in 1938 [15]. The tests were built on statistical hypothesis tests, such as the chi-square test that had originally been developed to distinguish whether or not experimental phenomena match up with their theoretical probabilities. Kendall and Babington-Smith's original four tests took as their null hypothesis the idea that each number in a given random sequence has an equal chance of occurring, and that various other patterns in the data should also be distributed equiprobably. These tests are:

- The *frequency test* checks that all possible numbers (e.g., 0, 1, 2, ...) occur with roughly the same frequency (i.e., the same number of times);
- The *serial test* checks the same thing for sequences of two numbers (e.g., 00, 01, 02, ...);
- The *poker test* checks for certain sequences of five numbers at a time based on hands in a poker game;
- The *gap test* looks at the distances between 0s (in the binary case, for example, 00 would be a distance of 0, 010 would be a distance of 1, 0110 would be a distance of 2, ...).

The tests do not directly attest randomness. Instead, each test may provide some probabilistic evidence that a generator produces sequences that have "good" randomness properties. In other words, if a sequence fails any of these tests, then the generator can be rejected as being nonrandom. Only if the sequence passes all tests (within a given degree of significance, such as 5%) can the generator be accepted as being random. This is conceptually similar to primality testing outlined in Appendix A.2.4.3.

As random numbers became more and more common, more tests of increasing sophistication were developed and are currently being used:

- In 1995, George Marsaglia published a suite (or battery) of statistical randomness tests and respective software, collectively referred to as the *Diehard tests*.[4] They are widely used in practice.
- In 2007, Pierre L'Ecuyer and Richard Simard published another test suite (together with respective software) that is known as *TestU01* [16].
- In the recent past, the U.S. NIST has become active in the field and published a statistical test suite for random and pseudorandom number generators suitable for cryptographic applications. The suite comprises, among other tests, the universal statistical test that was created by Ueli M. Maurer [17]. The basic idea behind Maurer's universal statistical test is that it should not be possible to significantly compress (without a loss of information) the output sequence of a random generator. Alternatively speaking, if a sample output sequence can be significantly compressed, then the respective generator should be rejected as being defective. Instead of actually compressing the sequence, Maurer's universal statistical test computes a quantity that is related to the length of the compressed sequence. The NIST test suite has been revised several times, and the most recent revision was done in April 2010 [18].

The Diehard tests, TestU01, and the NIST test suite represent the current state of the art in statistical randomness testing. If you have to test an output sequence, then you should go for any or several of these tests.

## 3.4 FINAL REMARKS

Random generators are at the core of most systems that employ cryptographic techniques in one way or another. If, for example, a secret key cryptosystem is being used, then a random generator should generate a shared secret to be established between the communicating peers. If a public key cryptosystem is used, then a random generator should be used to generate the respective public key pairs. Furthermore, if the cryptosystem is probabilistic, then a random generator should be used for every cryptographic operation, such as encryption or digital signature generation.

4 http://www.stat.fsu.edu/pub/diehard.

In this chapter, we elaborated on random generators and overviewed and discussed some possible realizations and implementations thereof. There are hardware-based and software-based random generators. In either case, deskewing techniques may be used to improve the defectiveness of a random generator, and statistical randomness testing may be used to evaluate the quality of the output. In practice, it is often required that random bit generators conform to a security level specified in FIPS PUB 140-2 [19]. Hence, there is room for conformance testing as well as evaluation and certification here.

From an application viewpoint, it is important to be able to generate truly random bits (using a random generator) and to use them as a seed for a PRG. The PRG can then be used to generate a potentially infinite sequence of pseudorandom bits. It depends on a secret key (i.e., the seed), and therefore represents a secret key cryptosystem. PRGs are addressed in Chapter 7.

## References


[1] Massey, J., "Shift-Register Synthesis and BCH Decoding," *IEEE Transactions on Information Theory*, Vol. 15, No. 1, 1969, pp. 122–127.

[2] Eastlake, D., J. Schiller, and S. Crocker, "Randomness Requirements for Security," Request for Comments 4086, Best Current Practice 106, June 2005.

[3] Menezes, A., P. van Oorschot, and S. Vanstone, *Handbook of Applied Cryptography*. CRC Press, Boca Raton, FL, 1996.

[4] Fairfield, R.C., R.L. Mortenson, and K.B. Koulhart, "An LSI Random Number Generator (RNG)," *Proceedings of CRYPTO '84*, 1984, pp. 203–230.

[5] Agnew, G.B., "Random Sources for Cryptographic Systems," *Proceedings of EUROCRYPT '87*, Springer-Verlag, LNCS 304, 1988, pp. 77–81.

[6] Davis, D., R. Ihaka, and P. Fenstermacher, "Cryptographic Randomness from Air Turbulance in Disk Drives," *Proceedings of CRYPTO '94*, Springer-Verlag, LNCS 839, 1994, pp. 114–120.

[7] Jakobsson, M., et al., "A Practical Secure Physical Random Bit Generator," *Proceedings of the ACM Conference on Computer and Communications Security*, 1998, pp. 103–111.

[8] Lacy, J.B., D.P. Mitchell, and W.M. Schell, "CryptoLib: Cryptography in Software," *Proceedings of the USENIX Security Symposium IV*, USENIX Association, October 1993, pp. 1–17.

[9] Goldberg, I., and D. Wagner, "Randomness and the Netscape Browser—How Secure Is the World Wide Web?" *Dr. Dobb's Journal*, January 1996.

[10] Dole, B., S. Lodin, and E.H. Spafford, "Misplaced Trust: Kerberos 4 Session Keys," *Proceedings of the ISOC Network and Distributed System Security Symposium*, 1997, pp. 60–70.

[11] Von Neumann, J., "Various Techniques Used in Connection with Random Digits," *Journal of Research of the National Bureau of Standards*, Vol. 12, 1951, pp. 36–38.

[12] Elias, P., "The Efficient Construction of an Unbiased Random Sequence," *The Annals of Mathematical Statistics*, Vol. 43, No. 3, 1972, pp. 865–870.

[13] Blum, M., "Independent Unbiased Coin Flips from a Correlated Biased Source—A Finite State Markov Chain," *Combinatorica*, Vol. 6, No. 2, 1986, pp. 97–108.

[14] Peres, Y., "Iterating Von Neumann's Procedure for Extracting Random Bits," *The Annals of Statistics*, Vol. 20, No. 1, 1992, pp. 590–597.

[15] Kendall, M.G., and B. Babington-Smith, "Randomness and Random Sampling Numbers," *Journal of the Royal Statistical Society*, Vol. 101, No. 1, 1938, pp. 147–166.

[16] L'Ecuyer, P., and R. Simard, "TestU01: A C Library for Empirical Testing of Random Number Generators," *ACM Transactions on Mathematical Software*, Vol. 33, No. 4, Article 22, August 2007.

[17] Maurer, U.M., "A Universal Statistical Test for Random Bit Generators," *Journal of Cryptology*, Vol. 5, 1992, pp. 89–105.

[18] U.S. Department of Commerce, National Institute of Standards and Technology, *A Statistical Test Suite for Random and Pseudorandom Number Generators for Cryptographic Applications*, Special Publication 800-22, Revision 1a, April 2010.

[19] U.S. Department of Commerce, National Institute of Standards and Technology, *Security Requirements for Cryptographic Modules*, FIPS PUB 140-2, May 2001, http://csrc.nist.gov/publications/fips/fips140-2/fips1402.pdf.

# Chapter 4

# Random Functions

This chapter is very short and has a clear mission: It introduces, briefly explains, and puts into perspective the notion of a random function that is key to theoretical considerations in many security analyses and proofs. In some literature, random functions are called random oracles, and we use the terms synonymously and interchangeably here. We introduce the topic in Section 4.1, elaborate on how one would implement a random function (if one wanted to) in Section 4.2, and conclude with some final remarks in Section 4.3. Contrary to all other chapters and appendixes in this book, this chapter does not come with a list of references.

## 4.1 INTRODUCTION

In the previous chapter, we elaborated on random generators, and we argued that a "good" random generator must output values (or bits in the case of a random bit generator) that have "good" randomness properties—for whatever one may call "good." We also said that these properties cannot be measured directly, but can only be measured indirectly by showing that no statistical randomness test is able to find a defect. The point to remember is that a random generator outputs values that look random, and hence that it is mainly characterized by its output.

When we talk about random functions (or random oracles) in this chapter, the focus is not the output of such a function, but rather the way it is chosen. We already captured in Definition 2.2 that it is a function $f : X \rightarrow Y$ that is chosen randomly from $\mathrm{Funcs}[X, Y]$; that is, the set of all functions that map elements of a domain $X$ to elements of a range $Y$. We also said that there are $|Y|^{|X|}$ such functions, and that this number is incredibly large—even for moderately sized $X$ and $Y$. Remember that in a realistic setting, in which $X$ and $Y$ are sequences of 128 bits each, $\mathrm{Funcs}[X, Y]$ comprises $2^{2^{135}}$ functions, and that $2^{135}$ bits are needed to refer to a particular function in $\text{Funcs}[X,Y]$. This is clearly infeasible to be used in the field, and yet this huge quantity is what makes random functions attractive in security considerations: If one can show that a function behaves like a random function, then one can be pretty sure that an adversary is not going to guess or otherwise determine the map of a given argument. The set of possibilities is simply too large.

A special class of random functions occurs if $X = Y$ and one only considers permutations of $X$; that is, $\text{Perms}[X]$. Instead of $|X|^{|X|}$, there are $|X|!$ permutations in $\text{Perms}[X]$, so the mathematical formula is different, but the resulting number of possibilities is again incredibly large. In our example with 128-bit strings, for example, it is $|2^{128}|!$—again a number beyond imagination. Following the same line of argumentation as above, a randomly chosen permutation from $\text{Perms}[X]$ yields a *random permutation*.

Note that the term random function is somehow misleading, because it may lead one to believe that some functions are more random than others or—more generally—that randomness is a property that can be attributed to a function per se. This is not true, and the attribute random in the term random function only refers to the way it is chosen from the set of all possibilities. The same reasoning also applies to the term random permutation.

## 4.2 IMPLEMENTATION

In contrast to the header of this section, random functions (and random permutations) are purely theoretical and conceptual constructs that are not meant to be implemented. Referring to the security game introduced in Section 1.2.2.1 (and illustrated in Figure 1.2), a random function yields an ideal system an adversary has to tell apart from a real system (i.e., a given cryptosystem for which the security needs to be shown). If the adversary cannot tell the two systems apart, then the cryptosystem under consideration behaves like a random function, and this, in turn, means that the adversary must try all possible functions. Since there are so many possibilities (as there are elements in $\text{Funcs}[X,Y]$ or $\text{Perms}[X]$), this task can be assumed to be computationally intractable for the adversary one has in mind, and hence the cryptosystem can be assumed to be secure.

According to the way it is defined, a random function can output any value $y = f(x) \in f(X) \subseteq Y$ for input value $x \in X$. The only requirement is that the same input value $x$ always maps to the same output value $y$. Except for that, everything is possible and does not really matter (for the function to be random). In an extreme case, for example, it is even possible that the function that maps all $x \in X$ to zero is chosen randomly from $\text{Funcs}[X, Y]$. This means that it is a perfectly valid random function in spite of the fact that this may contradict intuition.

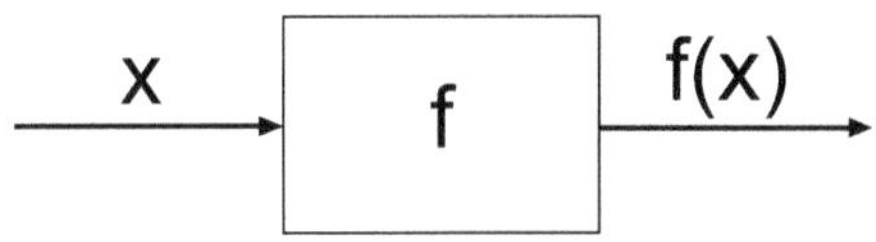


**Figure 4.1** A random function $f$.

Let us assume that we still want to implement a random function (in spite of the concerns expressed above). How would we do this? A random function is best thought of as a black box that has a particular input-output behavior that can be observed by everybody, meaning that anybody can feed input values $x \in X$ into the box and observe the respective output values $f(x) \in Y$. This is illustrated in Figure 4.1. Again, the only requirement is that if a specific input value $x$ is fed multiple times into the box, then the output value $f(x)$ must always be the same. This can be achieved easily with some internal state.

Another way to think about a random function $f$ is as a large random table $T$ with entries of the form $T[x] = (x, f(x))$ for all $x \in X$. The table can either be statically determined or dynamically generated.

- If $T$ is statically determined, then somebody must have flipped coins (or used any other source of randomness) to determine $f(x)$ for all $x \in X$ and put these values into the table.
- If $T$ is dynamically generated, then there must be an algorithm and a respective program that initializes the table with empty entries and that proceeds as follows for every input value $x \in X$: It checks whether $T[x]$ is empty. If it is empty, then it randomly chooses $f(x)$ from $Y$, writes this value into the appropriate place of $T[x]$, and returns it as a result. If $T[x]$ is not empty, then it returns the corresponding value as a result.

In either case, implementing a random function (if one has a random generator to start with) seems trivial, and it is a simple programming exercise to implement one. However, one should not get too excited about the result, because—due to the fact that it is a purely theoretical and conceptual construct—it won't serve any useful purpose in the field, meaning that one cannot solve any real-world problem with a random function.

## 4.3 FINAL REMARKS

The sole purpose of this chapter was to introduce the notions of a random function and a random permutation. Now that we have done so, we are ready to use them in various settings. In fact, many cryptographic primitives and cryptosystems used in the field can be seen as either a random function or a random permutation. More precisely, they are not truly random but show a similar behavior and are thus indistinguishable from them. To emphasize this subtle difference, such functions and permutations are called pseudorandom, and we are going to address pseudorandom functions and permutations in Chapter 8. For example, a cryptographic hash function yields a pseudorandom function, and a block cipher yields a pseudorandom permutation. We will come across random and pseudorandom functions and permutations at many places throughout the book, and it is therefore key to understand their notion and the intuition behind their design.

# Chapter 5

# One-Way Functions

As mentioned before, one-way functions (and trapdoor functions) play a pivotal role in contemporary cryptography, especially in the realm of public key cryptography.[1] In this chapter, we elaborate on such functions. More specifically, we introduce the topic in Section 5.1, overview and discuss a few candidate one-way functions in Section 5.2, elaborate on integer factorization algorithms and algorithms for computing discrete logarithms in Sections 5.3 and 5.4, briefly introduce elliptic curve cryptography in Section 5.5, and conclude with some final remarks in Section 5.6.

## 5.1 INTRODUCTION

In Section 2.1.3, we introduced the notion of a one-way function in Definition 2.3 and depicted it in Figure 2.1. We said that a function $f : X \rightarrow Y$ is *one way*, if $f(x)$ can be computed efficiently for all $x \in X$, but $f^{-1}(f(x))$ cannot be computed efficiently, meaning that $f^{-1}(y)$ cannot be computed efficiently for $y \in_R Y$. We also said that the definition is not precise in a mathematically strong sense and that one must first introduce some complexity-theoretic basics (mainly to define more accurately what is meant by saying that one can or one cannot "compute efficiently").

Using complexity theory as summarized in Appendix D, the notion of a one-way function is now more precisely captured in Definition 5.1.

1 In many textbooks, one-way functions are introduced together with public key cryptography. In this book, however, we follow a different approach and introduce one-way functions independently from public key cryptography as unkeyed cryptosystems.

**Definition 5.1 (One-way function)** *A function $f : X \rightarrow Y$ is* one way *if the following two conditions are fulfilled:*

- *The function $f$ is easy to compute, meaning that it is known how to efficiently compute $f(x)$ for all $x \in X$; that is, there is a probabilistic polynomial-time (PPT) algorithm*[2] *$A$ that outputs $A(x) = f(x)$ for all $x \in X$.*
- *The function $f$ is hard to invert, meaning that it is not known how to efficiently compute $f^{-1}(f(x))$ for $x \in_R X$ (or $f^{-1}(y)$ for $y \in_R Y$); that is, there is no known PPT algorithm $A$ that outputs $A(f(x)) = f^{-1}(f(x))$ for $x \in_R X$ (or $A(y) = f^{-1}(y)$ for $y \in_R Y$).*[3] *In either case, it is assumed that the values of $X$ (and $Y$) are uniformly distributed, and hence that $x$ (or $y$) is sampled uniformly at random from $X$ (or $Y$).*

Another way to express the second condition is to say that any PPT algorithm $A$ that tries to invert $f$ on a randomly chosen element of its range[4] only succeeds with a probability that is negligible.[5] More formally, this means that there is a positive integer $n_0 \in \mathbb{N}$, such that for every PPT algorithm $A$, every $x \in X$, every polynomial $p(\cdot)$, and all $n_0 \leq n \in \mathbb{N}$ the following relation holds:

$$\Pr[A(f(x), 1^b) \in f^{-1}(f(x))] \leq \frac{1}{p(n)}$$

In this notation, $b$ refers to the bit length of $x$, and hence the PPT algorithm $A$ is actually given two arguments: $f(x)$ and $b$ in unary notation. The sole purpose of the second argument is to allow $A$ to run in time polynomial in the length of $x$, even if $f(x)$ is much shorter than $x$. In the typical case, $f$ is length-preserving, and hence the second argument $1^b$ is redundant and represents just a technical subtlety. Using the two arguments, the algorithm $A$ is to find a preimage of $f(x)$. The formula basically says that the probability of $A$ being successful is negligible, meaning that it is smaller than the reciprocal of any polynomial in $n$. This captures the notion of a one-way function in a mathematically precise way.

2 In Section 1.2, we introduced the notion of a probabilistic algorithm. As its name suggests, a PPT algorithm is probabilistic and runs in polynomial time.

3 Note that $A$ is not required to find the correct value of $x$. It is only required to find some inverse of $f(x)$ or $y$. If, however, the function $f$ is injective, then the only inverse of $f(x)$ or $y$ is $x$.

4 See Section A.1.1 for the subtle difference between the codomain and the range of a function.

5 The notion of a *negligible function* is captured in Definition D.1. In essence, the success probability of a PPT algorithm $A$ is negligible if it is bound by a polynomial fraction. It follows that repeating $A$ polynomially (in the input length) many times yields a new algorithm that also has a success probability that is negligible. Put in other words, events that occur with negligible probability remain negligible even if the experiment is repeated polynomially many times. This property is important for complexity-theoretic considerations.

In addition to this notation, there are other notations that can be used to express the same idea. For example, the following notation is sometimes also used in the literature:

$$\Pr[(f(z) = y : x \xleftarrow{r} \{0,1\}^b; y \leftarrow f(x); z \leftarrow A(y, 1^b)] \leq \frac{1}{p(n)}$$

It says that if $x$ is sampled uniformly at random from $\{0,1\}^b$ (i.e., all $x \in \{0,1\}^b$ are equally probable), $y$ is assigned $f(x)$, and $z$ is assigned $A(y, 1^b)$, then the probability that $f(z)$ equals $y$—and hence $A$ is successful in inverting $f$—is negligible. The two notations are equivalent, and either one can be used.

Besides the notion of a one-way function, we also introduced the notion of a trapdoor (one-way) function in Definition 2.4. We said that a one-way function $f : X \rightarrow Y$ is a *trapdoor function* (or a *trapdoor one-way function*, respectively), if there is some extra information, i.e., a *trapdoor*, with which $f$ can be inverted efficiently, i.e., $f^{-1}(f(x))$ can be computed efficiently for $x \in_R X$ (or $f^{-1}(y)$ for $y \in_R Y$). Consequently, the notion of a trapdoor function can be defined by simply prepending the words "unless some extra information (i.e., a trapdoor) is known" in front of the second condition in Definition 5.1. This is not done here. Instead, we refer to an alternative way of defining a trapdoor function in Definition 5.2. Again, the two ways of defining a trapdoor function are equivalent.

**Definition 5.2 (Trapdoor function)** *A one-way function $f : X \rightarrow Y$ is a* trapdoor function *if there is a trapdoor information t and a PPT algorithm I that can be used to efficiently compute $x' = I(f(x), t)$ with $f(x') = f(x)$.*

Referring to Section 2.1.3, a *one-way permutation* is a one-way function function $f : X \rightarrow X$, where $X$ and $Y$ are the same and $f$ is a permutation; that is, $f \in Perms[X]$. Similarly, a one-way permutation $f : X \rightarrow X$ is a *trapdoor permutation* (or a *trapdoor one-way permutation*, respectively), if it has a trapdoor. We already stressed the fact that one-way permutations and trapdoor permutations are special cases of one-way functions and trapdoor functions, namely ones in which the domain and the range are the same and the functions are permutations.

Instead of talking about one-way functions, trapdoor functions, one-way permutations, and trapdoor permutations, people often refer to families of such functions and permutations. To understand why this is the case, one has to consider a complexity-theoretic argument: Many cryptographic functions required to be one way output bit strings of fixed length. For example, cryptographic hash functions are required to be one way (among other things) and output strings of 160 or more—but always a fixed number of—bits (Chapter 6). Given this fact, one may ask how computationally expensive it is to find a preimage of a given value, and hence to invert such a function. If the function outputs $n$-bit values, then $2^n$ tries are usually sufficient to invert the function and find a preimage for a given hash value.[6] Because $2^n$ is constant for a fixed $n \in \mathbb{N}$, the computational complexity to invert the function—using the big-O notation introduced in Section D.3—is $O(1)$ and hence trivial. So the complexity-theoretic answer to the question stated above is not particularly useful, and one cannot say that inverting such a function is intractable. If one wants to use complexity-theoretic arguments, then one cannot have a constant $n$. Instead, one must make $n$ variable, and it must be possible to let $n$ grow arbitrarily large. Consequently, one has to work with a potentially infinite *family*[7] of functions, and there must be at least one function for every possible value of $n$. Following this line of argumentation, the notion of a family of one-way functions is captured in Definition 5.3.

**Definition 5.3 (Family of one-way functions)** *A family of functions* $F = \{f_i : X_i \rightarrow Y_i\}_{i \in I}$ *is a* family of one-way functions *if the following two conditions are fulfilled:*

- *$I$ is an index set that is infinite;*
- *For every $i \in I$ there is a function $f_i : X_i \rightarrow Y_i$ that is one-way according to Definition 5.1.*

The notion of a family also applies to trapdoor functions, one-way permutations, and trapdoor permutations. For example, a family of one-way functions $\{f_i : X_i \rightarrow Y_i\}_{i \in I}$ is a *family of one-way permutations*, if for every $i \in I$ $f_i$ is a permutation over $X_i$ (i.e., $Y_i = X_i$), and it is a *family of trapdoor functions*, if every $f_i$ has a trapdoor $t_i$. In this book, we sometimes use the terms one-way functions, trapdoor functions, one-way permutations, and trapdoor permutations when we should use the terms families of such functions and permutations (to be mathematically correct). We make this simplification, because we think that it is more appropriate and sometimes simpler to understand—but we are well aware of the fact that we lack formal correctness.

The notion of a one-way function suggests that $x$—in its entirety—-cannot be computed efficiently from $f(x)$. This does not exclude the case that some parts of $x$ can be determined, whereas other parts cannot. In 1982, it was shown by Andrew C.

6 In a more thorough analysis, the probability of successfully inverting the function and finding a preimage in $t$ tries is equal to $1 - (1 - 1/2^n)^t$. If $t = 2^n$, then this value is very close to 1. After $2^n$ tries, it is therefore almost certain that a preimage is found.

7 In some literature, the terms "classes," "collections," or "ensembles" are used instead of "families." These terms refer to the same idea.

Yao[8] that every one-way function $f$ must have at least one *hard-core predicate*; that is, a predicate[9] that can be computed efficiently from $x$ but not from $f(x)$ [1]. The respective notion of a hard-core predicate is rendered more precise in Definition 5.4 and is illustrated in Figure 5.1.

**Definition 5.4 (Hard-core predicate)** *Let $f : X \rightarrow Y$ be a one-way function. A* hard-core predicate *of $f$ is a predicate $B : X \rightarrow \{0, 1\}$ that fulfills the following two conditions:*

- *$B(x)$ can be computed efficiently for all $x \in X$; that is, there is a PPT algorithm $A$ that can output $B(x)$ for all $x \in X$.*
- *$B(x)$ cannot be computed efficiently from $y = f(x) \in Y$ for $x \in_R X$; that is, there is no known PPT algorithm $A$ that can output $B(x)$ from $y = f(x)$ for $x \in_R X$.*

The point is that PPT algorithm $A$ can compute $B(x)$ from $x$ but not from $y = f(x)$. Again, there are multiple possibilities to formally express this idea; they are all equivalent and not repeated here.

After Yao had introduced the notion of a hard-core predicate, Manuel Blum[10] and Silvio Micali[11] used it to design and come up with a PRG that is—as they called it—cryptographically secure [2]. The notion of a cryptographically secure PRG is further addressed in Section 7.3. In short, such a PRG applies a one-way function and extracts a hard-core predicate from the result in each step. This is not very efficient, but it allows one to mathematically argue about the cryptographic strength of such a PRG.

## 5.2 CANDIDATE ONE-WAY FUNCTIONS

As mentioned in Section 2.1.3, there are only a few functions conjectured to be one way, and almost all practically relevant functions are centered around modular exponentiation: The discrete exponentiation function $f(x) = g^x \pmod{m}$, the

8 In 2000, Andrew Chi-Chih Yao received the Turing Award in recognition of his contributions to the theory of computation, pseudorandom number generation, cryptography, and communication complexity.

9 A *predicate* is a function whose output is a single bit.

10 In 1995, Manuel Blum received the Turing Award for his contributions to the foundations of computational complexity theory and its application to cryptography and program checking.

11 In 2012, Silvio Micali, along with Shafi Goldwasser, received the ACM Turing Award for his transformative work that laid the complexity-theoretic foundations for the science of cryptography, and in the process pioneered new methods for efficient verification of mathematical proofs in complexity theory.

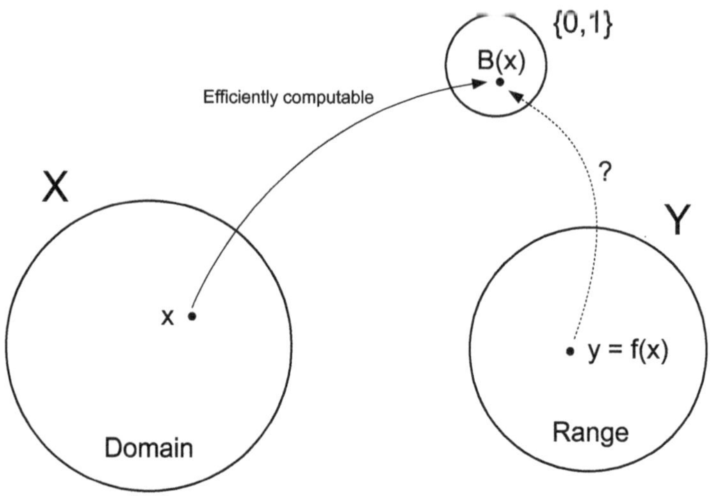


**Figure 5.1** A hard-core predicate of a one-way function $f$.

RSA function $f(x) = x^e \pmod m$, and the modular square function $f(x) = x^2 \pmod m$ for some properly chosen moduli $m$. These functions are at the core of public key cryptography as it stands today (Part III of the book). Unless somebody is able to build a sufficiently large quantum computer, they cannot be inverted efficiently, meaning that the best known algorithms to invert them on currently available hardware are superpolynomial (i.e., they have an exponential or subexponential time complexity). These algorithms are briefly addressed in Sections 5.3 and 5.4.

In the following subsections, we use the discrete exponentiation, RSA, and modular square functions to construct three families of one-way functions: **Exp**, **RSA**, and **Square**. Note that we use boldface characters to refer to a function family, whereas we use normal characters to refer to distinct functions within these families. It is known that the most significant bit (MSB) yields a hard-core predicate for **Exp**, **RSA**, and **Square**, whereas the least significant bit (LSB) yields another hard-core predicate for **RSA** and **Square** [3, 4].

### 5.2.1 Discrete Exponentiation Function

In the real numbers $\mathbb{R}$, the exponentiation function maps arbitrary elements $x \in \mathbb{R}$ to other elements $y = \exp(x) = e^x \in \mathbb{R}$, whereas the logarithm function does the opposite; that is, it maps $x$ to $\ln(x)$. This is true for the base $e$, but it is equally true for any other base $a \in \mathbb{R}$. Formally, the two functions can be expressed as follows:

$$\begin{aligned} \text{Exp}: \mathbb{R} &\longrightarrow \mathbb{R} \\ x &\longmapsto a^x \end{aligned} \qquad \begin{aligned} \text{Log}: \mathbb{R} &\longrightarrow \mathbb{R} \\ x &\longmapsto \log_a x \end{aligned}$$

In $\mathbb{R}$, both the exponentiation function and the logarithm function are continuous and can be computed efficiently, using approximation procedures. However, in a discrete algebraic structure, it is usually not possible to use the notion of continuity or to approximate a solution. In fact, there are cyclic groups (that need to be discrete and finite) in which the exponentiation function—that is called the *discrete exponentiation function*—can be computed efficiently (using, for example, the square-and-multiply algorithm outlined in Algorithm A.3), but the inverse function—that is called the *discrete logarithm function*—cannot be computed efficiently (i.e., no efficient algorithm is known to exist and all known algorithms have a superpolynomial time complexity). If we write such a group $G$ multiplicatively, then we refer to the multiplication as the group operation. Also, we know from discrete mathematics that such a group has many elements that may serve as a generator (or primitive root), and we fix a particular one. Using such a group $G$ with generator $g$, we can express the discrete exponentiation and logarithm functions as follows:

$$\begin{aligned} \text{Exp}: \mathbb{N} &\longrightarrow G \\ x &\longmapsto g^x \end{aligned} \qquad \begin{aligned} \text{Log}: G &\longrightarrow \mathbb{N} \\ x &\longmapsto \log_g x \end{aligned}$$

The discrete exponentiation function Exp maps a positive[12] integer $x$ to the element $g^x$ of $G$ by multiplying the generator $g$ $x$ times to itself. The discrete logarithm function Log does the opposite: It maps an element $x$ of $G$ to the number of times $g$ must be multiplied to itself to yield $x$. In either case, the finite set $\{1, \ldots, |G|\}$ can be used instead of $\mathbb{N}$.

Depending on the nature of $G$, no efficient algorithm may be known to exist to compute Log. The most prominent example of a group that is widely used in practice is a properly crafted subgroup of $\langle \mathbb{Z}_p^*, \cdot \rangle$; that is, the set of all integers between 1 and $p-1$ (where $p$ is prime) together with the multiplication operation. We sometimes

12 Note that the restriction to positive integers only is not mandatory, and that Exp can also be defined for negative integers. However, it simplifies things considerably if one only considers positive integers here.

use $\mathbb{Z}_p^*$ to refer to this group. Because the order of $\mathbb{Z}_p^*$ is $p-1$ and this number is even, $\mathbb{Z}_p^*$ has at least two nontrivial subgroups; that is, one of order 2 and one of order $q = (p-1)/2$. If $q$ is a composite number, then the second subgroup may have further subgroups. We are mainly interested in the other case, where $q$ is prime. In this case, $q$ is called a *Sophie Germain prime* and $p$ is called *safe* (Section A.2.4.4). The $q$-order subgroup of $\mathbb{Z}_p^*$ is the one we use as $G$. It is isomorphic to $\mathbb{Z}_q$ (Definition A.20), and it is generated by an element $g$ of order $q$ that is called a *generator*. If we iteratively apply the multiplication operation to $g$, all $q$ elements of $G$ are generated sooner or later. Needless to say there are groups in which the algorithms to compute discrete logarithms are less efficient than in this group, such as groups of points on elliptic curves (Section 5.5).

The discrete exponentiation function is well understood in mathematics, and there are many properties that can be exploited in interesting ways. One such property is that it yields a group isomorphism (Section A.1.3) from the additive group $\langle \mathbb{Z}_{p-1}, + \rangle$ to the multiplicative group $\langle \mathbb{Z}_p^*, \cdot \rangle$. This means that

$$\text{Exp}(x+y) = \text{Exp}(x) \cdot \text{Exp}(y)$$

and this, in turn, implies the fundamental exponent law $g^{x+y} = g^x g^y$.

Earlier in this chapter, we argued that we have to consider families of one-way functions to use complexity-theoretic arguments. In the case of the discrete exponentiation function in $\mathbb{Z}_p^*$, $p$ and $g$ may define an index set $I$ as follows:

$$I := \{(p,g) \mid p \in \mathbb{P}^*;\ g \text{ generates } G \subset \mathbb{Z}_p^* \text{ with } |G| = q = (p-1)/2\}$$

In this notation, $\mathbb{P}^*$ refers to the set of all safe primes (Section A.2.4), and $G$ denotes a $q$-element sugroup of $\mathbb{Z}_p^*$. Using $I$, one can define the **Exp** family of discrete exponentiation functions

$$\textbf{Exp} := \{\text{Exp}: \mathbb{N} \longrightarrow G,\ x \longmapsto g^x\}_{(p,g)\in I}$$

and the **Log** family of discrete logarithm functions

$$\textbf{Log} := \{\text{Log}: G \longrightarrow \mathbb{N},\ x \longmapsto \log_g x\}_{(p,g)\in I}$$

If one wants to use **Exp** as a family of one-way functions, then one has to make sure that discrete logarithms cannot be computed efficiently in the group that is considered. This is where the *discrete logarithm assumption* (DLA) comes into play. It suggests that a PPT algorithm $A$ that is to compute a discrete logarithm can succeed only with a probability that is negligible.[13]

There are several problems phrased around the DLA and the conjectured one-way property of the discrete exponential function: The *discrete logarithm problem* (DLP), the (computational) *Diffie-Hellman problem* (DHP), and the *decisional Diffie-Hellman problem* (DDHP) that are rendered more precise in Definitions 5.5 to 5.7. In these definitions, the problems are specified in an abstract notation using a cyclic group $G$ and a generator $g$, whereas the numerical examples are given in a specific group, namely $\mathbb{Z}_7^* = \{1, 2, 3, 4, 5, 6\}$ with generator $g = 5$.[14] We don't care about the Pohlig-Hellman algorithm here, and hence we don't use a prime-order subgroup of $\mathbb{Z}_7^*$.

**Definition 5.5 (DLP)** *Let $G$ be a cyclic group with generator $g$. The DLP is to determine $x \in \mathbb{N}$ from $G$, $g$, and $g^x \in G$.*

In our numerical example with $\mathbb{Z}_7^*$ and $g = 5$, the DLP for $g^x = 4$ yields $x = 2$, because $g^x = 5^2 \pmod 7 = 4$. Here, the group is so small that all possible values of $x$ can be tried out. It goes without saying that this does not work for large groups (i.e., groups with many elements). Also, the cyclic nature of $G$ makes it impossible to find an efficient algorithm to solve the DLP by approximation. The situation is fundamentally different from the continuous case of using the exponentiation and logarithm functions in $\mathbb{R}$.

**Definition 5.6 (DHP)** *Let $G$ be a cyclic group, $g$ a generator of $G$, and $x$ and $y$ two positive integers that are smaller than the order of $G$; that is, $0 < x, y < |G|$. The DHP is to determine $g^{xy}$ from $G$, $g$, $g^x$, and $g^y$.*

In our example, $x = 3$ and $y = 6$ yield $g^x = 5^3 \pmod 7 = 6$ and $g^y = 5^6 \pmod 7 = 1$. Here, the DHP is to determine $g^{xy} = 5^{18} \pmod 7 = 1$ from $g^x = 6$ and $g^y = 1$. As its name suggests, the DHP is at the core of the Diffie-Hellman key exchange protocol that is addressed in Section 12.3.

13 It is important to note that computing a discrete logarithm may be possible and efficient for some input values. For example, we see in Section 5.4.1.3 that there is an efficient algorithm due to Stephen Pohlig and Martin Hellman [5] that can be used to compute discrete logarithms in $\mathbb{Z}_p^*$ if $p - 1$ has only small prime factors (Section 5.4.1). To avoid this situation, one requires that $|\mathbb{Z}_p^*| = p - 1$ is only divided by 2 and a prime $q$. This, in turn, means that $\mathbb{Z}_p^*$ has a $q$-element subgroup, and all computations are done in this group. This defeats the Pohlig-Hellman algorithm to efficiently compute discrete logarithms.

14 Note that the element 5 is a generator, as it generates all 6 elements of $\mathbb{Z}_7^*$: $5^0 = 1 \equiv 1 \pmod 7 \rightarrow \underline{1}, 5^1 = 5 \equiv 5 \pmod 7 \rightarrow \underline{5}, 5^2 = 25 \equiv 4 \pmod 7 \rightarrow \underline{4}, 5^3 = 125 \equiv 6 \pmod 7 \rightarrow \underline{6}, 5^4 = 625 \equiv 2 \pmod 7 \rightarrow \underline{2}$, and $5^5 = 3125 \equiv 3 \pmod 7 \rightarrow \underline{3}$. The resulting elements 1, 5, 4, 6, 2, and 3 (that are underlined) form the entire group. Another generator would be $g = 4$. In general, there are $\phi(\phi(p))$ generators in $\mathbb{Z}_p^*$. In this example, $p = 7$ and hence there are $\phi(\phi(p)) = \phi(6) =$ generators, namely 4 and 5.

**Definition 5.7 (DDHP)** *Let $G$ be a cyclic group, $g$ a generator of $G$, and $x$, $y$, and $z$ three positive integers that are smaller than the order of $G$; that is, $0 < x, y, z < |G|$. The DDHP is to decide whether $g^{xy}$ or $g^z$ solves the DHP for $g^x$ and $g^y$, if one is given $G$, $g$, $g^x$, $g^y$, $g^{xy}$, and $g^z$.*

In our example, $x = 3$, $y = 6$, and $z = 2$ yield $g^x = 5^3 \pmod 7 = 6$, $g^y = 5^6 \pmod 7 = 1$, and $g^z = 5^2 \pmod 7 = 4$. In this case, the DDHP is to determine whether $g^{xy} = 1$ (see above) or $g^z = 4$ solves the DHP for $g^x = 6$ and $g^y = 1$. Referring to the previous example, the correct solution here is 1.

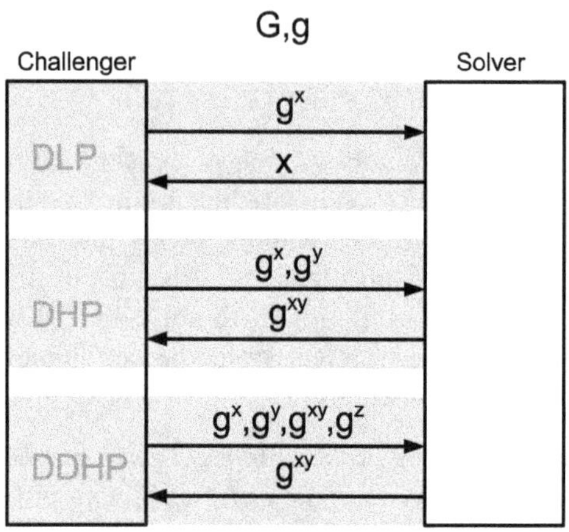


**Figure 5.2** The DLP, DHP, and DDHP.

All DLA-based problems, i.e., the DLP, DHP, and DDHP, are illustrated in Figure 5.2. An interesting question is how they relate to each other. This is typically done by giving complexity-theoretic reductions from one problem to another (Definition D.4). In fact, it can be shown that DHP $\leq_P$ DLP (i.e., the DHP polytime reduces to the DLP), and that DDHP $\leq_P$ DHP (i.e., the DDHP polytime reduces to the DHP) in a finite group. We can therefore give an ordering with regard to the computational complexities of the three DLA-related problems:

$$\text{DDHP} \leq_P \text{DHP} \leq_P \text{DLP}$$

This basically means that the DLP is the hardest problem, and that one can trivially solve the DHP and the DDHP if one is able to solve the DLP. In many groups, the DLP and the DHP are known to be computationally equivalent [6, 7], but this needs to be the case in all groups. If one has a group in which the DHP is as hard as the DLP, then one can safely build a cryptosystem on top of the DHP. Otherwise, one better relies on cryptosystems that are based on the DLP. As a side remark we note that if one found a group in which the DHP is easy but the DLP is hard, then one would also have a good starting point to build a fully homomorphic encryption system (Section 13.5). This would be good news and could possibly enable many interesting applications.

Interestingly, there are also groups in which the DDHP can be solved in polynomial time, whereas the fastest known algorithm to solve the DHP still requires subexponential time. This means that in such a group, the DDHP is in fact simpler to solve than the DHP, and such groups are sometimes called *gap Diffie-Hellman groups*, or *GDH groups* for short. Such groups are used, for example, in the BLS DSS briefly mentioned in Section 14.2.7.4.

In either case, it is important to better understand the DLP and to know the algorithms (together with their computational complexities) that can be used to solve it and compute discrete logarithms accordingly. Some of these algorithms are overviewed in Section 5.4.

### 5.2.2 RSA Function

The RSA function refers to the second type of modular exponentiation mentioned in Section 2.1.3, namely $f(x) = x^e \pmod{m}$, where $m$ represents a composite integer usually written as $n$. More specifically, $n$ is the product of two distinct primes $p$ and $q$; that is, $n = pq$, and $e$ is relatively prime to $\phi(n)$, where $\phi(n)$ refers to Euler's totient function (Section A.2.6). Using $n$ and $e$, the *RSA function* can be defined as follows:

$$\begin{array}{rccc} \text{RSA}_{n,e} : & \mathbb{Z}_n & \longrightarrow & \mathbb{Z}_n \\ & x & \longmapsto & x^e \end{array}$$

The function operates on $\mathbb{Z}_n$ and basically computes the e-th power of $x \in \mathbb{Z}_n$. Since $e$ is relatively prime to $\phi(n)$, the function is bijective for all $x$ that are invertible modulo $n$; that is, $gcd(x, n) = 1$. But, in this context, it can be shown that the function is also bijective if $gcd(x, n) = p$ or $gcd(x, n) = q$. This means that $\text{RSA}_{n,e}$ yields a permutation on the elements of $\mathbb{Z}_n$, and that it has an inverse function that computes e-th roots. To be able to compute the inverse function, one must know the multiplicative inverse element $d$ of $e$ modulo $\phi(n)$. The same RSA function parametrized with $n$ and $d$ (i.e., $\text{RSA}_{n,d}$), can then be used to compute the inverse of $\text{RSA}_{n,e}$ as follows:

$$\begin{aligned}\text{RSA}_{n,d} : \mathbb{Z}_n &\longrightarrow \mathbb{Z}_n \\ x &\longmapsto x^d\end{aligned}$$

Both $x^e$ and $x^d$ can be written modulo $n$, and in either case it is clear that the result yields an element of $\mathbb{Z}_n$. $\text{RSA}_{n,e}$ can be efficiently computed using modular exponentiation. In order to compute $\text{RSA}_{n,d}$, however, one must know either $d$, one prime factor of $n$ (i.e., $p$ or $q$), or $\phi(n)$. Any of these values yields a trapdoor to the one-wayness of $\text{RSA}_{n,e}$. As of this writing, no polynomial-time algorithm is known to compute any of these values from $n$ and $e$ (unless one has a sufficiently large quantum computer at hand).

If we want to turn the RSA function into a family of one-way functions, then we must define an index set $I$. This can be done as follows:

$$I := \{(n,e) \mid n = pq; p,q \in \mathbb{P}; p \neq q; 1 < e < \phi(n); (e, \phi(n)) = 1\}$$

It means that $p$ and $q$ are distinct primes, $n$ is their product, and $1 < e < \phi(n)$ is randomly chosen and coprime to $\phi(n)$. Using $I$, the family of RSA functions can be defined as follows:

$$\textbf{RSA} := \{\text{RSA}_{n,e} : \mathbb{Z}_n \longrightarrow \mathbb{Z}_n,\ x \longmapsto x^e\}_{(n,e)\in I}$$

This family of RSA functions is called the *RSA family*. Because e and d are somehow symmetric in the sense that they are multiplicative inverse to each other modulo $\phi(n)$, the family comprises $\text{RSA}_{n,e}$ and $\text{RSA}_{n,d}$. Also, because each RSA function $\text{RSA}_{n,e}$ has trapdoors (as mentioned above) and yields a permutation over $\mathbb{Z}_n$, the RSA family actually represents a family of trapdoor permutations. This suggests that $\text{RSA}_{n,e}$ is hard to invert (for sufficiently large $n$ and without knowing a trapdoor). This fact has not be proven so far, meaning that it is only assumed that $\text{RSA}_{n,e}$ is hard to invert. In fact, the *RSA assumption* suggests that any PPT algorithm can invert $\text{RSA}_{n,e}$ with a success probability that is negligible. There is a slightly stronger version of the RSA assumption known as the *strong RSA assumption*. It differs from the RSA assumption in that the success probability for the PPT algorithm remains negligible even if it can select the value of e.

An obvious way to invert $\text{RSA}_{n,e}$ is to determine a trapdoor, most notably the prime factorization of $n$. This means that the *integer factoring problem* (IFP) captured in Definition 5.8 needs to be solved.

**Definition 5.8 (IFP)** *Let $n \in \mathbb{N}$ be positive integer. The IFP is to determine the distinct values $p_1, \ldots, p_k \in \mathbb{P}$ and $e_1, \ldots, e_k \in \mathbb{N}$ with $n = p_1^{e_1} \cdots p_k^{e_k}$.*

The IFP is well defined, because every positive integer can be factored uniquely up to a permutation of its prime factors (Theorem A.7). Note that the IFP need not always be intractable, but that there are instances of the IFP that are assumed to be so. In fact, the *integer factoring assumption* (IFA) suggests that any PPT algorithm that is to factorize $n$ succeeds with only a negligible probability. Again, this is just an assumption for which we don't know whether it really holds. To better understand the IFP and the plausibility of the IFA, we have a look at the currently available integer factorization algorithms in Section 5.3.

Under the IFA and RSA assumption, the *RSA problem* (RSAP) captured in Definition 5.9 is computationally intractable.

**Definition 5.9 (RSAP)** *Let $(n, e)$ be a public key with $n = pq$ and $c \equiv m^e \pmod n$ a ciphertext. The RSAP is to determine $m$; that is, computing the $e^{th}$ root of $c$ modulo $n$, if $d$, the factorization of $n$, i.e., $p$ and $q$, and $\phi(n)$ are unknown.*

It is obvious that RSAP $\leq_P$ IFP (i.e., the RSAP polytime reduces to the IFP). This means that one can invert the RSA function if one can solve the IFP. The converse, however, is not known to be true, meaning that it is not known whether a simpler way exists to invert the RSA function than to solve the IFP.

According to the strong RSA assumption, the value of $e$ may also be considered as a parameter. In this case, the resulting problem is called the *flexible RSAP*: For any given $n$ and $c$, find values for $e$ and $m$ such that $c \equiv m^e \pmod n$. Obviously, the flexible RSAP is not harder to solve than the RSAP, meaning that one can solve the flexible RSAP if one can solve the RSAP (simply fix an arbitrary value for $e$ and solve the respective RSAP).

The RSAP is at the core of many public key cryptosystems, including, for example, the RSA public key cryptosystem used for asymmetric encryption (Section 13.3.1) and digital signatures (Section 14.2.1).

### 5.2.3 Modular Square Function

If we start with the "normal" RSA function in $\mathbb{Z}_n^*$ (where $n$ is a composite integer and the product of two primes), but we replace $e$ that needs to be relatively prime to $\phi(n)$ with the fixed value 2,[15] then the resulting function represents the *modular square function*. It is defined as follows:

$$\begin{aligned} \text{Square}_n : \mathbb{Z}_n^* &\longrightarrow QR_n \\ x &\longmapsto x^2 \end{aligned}$$

15 Note that $e = 2$ is not a valid value for the "normal" RSA function because 2 cannot be relatively prime to $\phi(n)$. This value is equal to $p - 1$ times $q - 1$ that are both even, so $\phi(n)$ is also even. This, in turn, means that 2 and $\phi(n)$ have 2 as a common divisor and hence cannot be coprime.

Note that 2 is not relatively prime to $\phi(n)$, and hence $\text{Square}_n$ is not bijective and does not yield a permutation over $\mathbb{Z}_n^*$. In fact, the range of the modular square function is $QR_n$; that is, the set of quadratic residues or squares modulo $n$ (Section A.3.7), and this is a proper subgroup of $\mathbb{Z}_n^*$; that is, $QR_n \subset \mathbb{Z}_n^*$. This means that there are values $x_1, x_2, \ldots$ in $\mathbb{Z}_n^*$ that are mapped to the same value $x^2$ in $QR_n$,[16] and hence $\text{Square}_n$ is not injective and the inverse *modular square root function*

$$\begin{aligned} \text{Sqrt}_n : QR_n &\longrightarrow \mathbb{Z}_n^* \\ x &\longmapsto x^{1/2} \end{aligned}$$

is not defined (since it is not injective). To properly define it, one has to make sure that $\text{Square}_n$ is injective and surjective, and hence bijective. This can be achieved by restricting its range to $QR_n$, where $n$ is a Blum integer (Definition A.32); that is, $n$ is the product of two primes $p$ and $q$ that are both equivalent to 3 modulo 4; that is, $p \equiv q \equiv 3 \pmod 4$.[17] In this case, the modular square function is bijective and yields a permutation over $QR_n$, and hence the modular square root function always has a solution. In fact, every $x \in QR_n$ has four square roots modulo $n$, of which one is again an element of $QR_n$. This unique square root of $x$ is called the *principal square root* of $x$ modulo $n$.

To turn the modular square function into a family of trapdoor functions, one can define an index set $I$ as follows:

$$I := \{n \mid n = pq; p, q \in \mathbb{P}; p \neq q; |p| = |q|; p, q \equiv 3 \pmod 4\}$$

Using $I$, one can define the family of modular square functions as follows:

$$\textbf{Square} := \{\text{Square}_n : QR_n \longrightarrow QR_n,\ x \longmapsto x^2\}_{n \in I}$$

This family is called the *Square family*, and the family of inverse functions is defined as follows:

$$\textbf{Sqrt} := \{\text{Sqrt}_n : QR_n \longrightarrow QR_n,\ x \longmapsto x^{1/2}\}_{n \in I}$$

This family is called the *Sqrt family*. The Square family is used by some public key cryptosystems, including the Rabin public key cryptosystem (Section 13.3.2). In the case of the "normal" RSA function, we said that the problems of computing e-th

16 To be precise, there are always two values $x_1$ and $x_2$ that are mapped to an element in $QR_n$.

17 Note that it is not sufficient to restrict the range to $QR_n$. If, for example, $n = 15$ (that is the product of two primes but is not a Blum integer), then $QR_n = \{1, 4\}$, but the list of squares of $QR_n$ only comprises 1 (note taht $4^2 \pmod{15} = 1$). In contrast, if $n = 21$ (that is a Blum integer), then $QR_n = \{1, 4, 16\}$ and the list of squares is the same set.

roots in $\mathbb{Z}_n$ and factoring $n$ are not known to be computationally equivalent. This is different here: Modular squares can always be computed efficiently, but modular square roots (if they exist) can be computed efficiently if and only if the prime factorization of $n$ is known. This means that the problems of computing square roots in $QR_n$ and factoring $n$ are computationally equivalent. This fact distinguishes the security properties of the Rabin public key cryptosystem from those of RSA. But it also has some practical disadvantages that makes it less attractive to be used in the field.

For every $n \in I$, the prime factors $p$ and $q$ or $\phi(n)$ yield trapdoors. Consequently, if one can solve the IFP, then one can also invert the modular square function for $n$ and break the respective public key cryptosystem. We look at the algorithms to solve the IFP next, and we therefore use the *L-notation* introduced in Section D.4. It allows us to specify where the time complexity of an algorithm really is in the range between a fully polynomial-time algorithm and a fully exponential-time algorithm. It really helps to get yourself familiar with the L-notation before delving into the details of the next two sections.

## 5.3 INTEGER FACTORIZATION ALGORITHMS

The IFP has attracted many mathematicians in the past, and there are several integer factorization algorithms to choose from. Some of these algorithms are special-purpose, whereas others are general-purpose.

- *Special-purpose algorithms* depend upon and take advantage of special properties of the composite integer $n$ that needs to be factorized, such as its size, the size of its smallest prime factor $p$, or the prime factorization of $p - 1$. Hence, these algorithms can only be used if certain conditions are fulfilled.
- In contrast, *general-purpose algorithms* depend upon nothing and work equally well for all $n$.

In practice, algorithms of both categories are usually combined and used one after another. If one is given an integer $n$ with no further information, then one first tests its primality before one applies integer factorization algorithms. In doing so, one employs special-purpose algorithms that are optimized to find small prime factors first before one turns to the less efficient general-purpose ones. Hence, the invocation of the various algorithms during the factorization of $n$ gives room for optimization—in addition to the optimization of the algorithms themselves.

Let us overview and briefly discuss the most important (special-purpose and general-purpose) algorithms next. We use $b$ to refer to the bit length of $n$; that is,

$b = log_2(n) = ln(n)/ln(2) = ln(n)/0.693\ldots \approx ln(n)$. For the sake of simplicity, we ignore the constant factor $ln(2) \approx 0.693\ldots$ in the denominator of the fraction, and we approximate the bit length of $n$ with the natural logarithm here.

### 5.3.1 Special-Purpose Algorithms

We start with trial division before we briefly touch on more sophisticated algorithms, such as $P \pm 1$ and elliptic curve method (ECM), and Pollard Rho.

#### 5.3.1.1 Trial Division

If $n$ is a composite integer, then it must have at least one prime factor that is less or equal to $\sqrt{n}$. Consequently, one can factorize $n$ by trying to divide it by all prime numbers up to $\lfloor\sqrt{n}\rfloor$. This simple and straightforward algorithm is called *trial division*. If, for example, $n$ is a 1024-bit integer, then it requires $\sqrt{2^{1024}} = (2^{1024})^{1/2} = 2^{1024/2} = 2^{512}$ divisions in the worst case. The algorithm thus has a time complexity of $O(\sqrt{n})$, and this value grows exponentially with the bit length $b$:

$$O(\sqrt{n}) = O(e^{ln\sqrt{n}}) = O(e^{ln(n^{1/2})}) = O(e^{\frac{1}{2}ln(n)}) = O(e^{\frac{b}{2}})$$

Exponential time complexity is beyond what is feasible today, and hence the trial division algorithm can only be used for the factorization of integers that are sufficiently small (e.g., smaller than $10^{12}$) or smooth (Appendix A.2.5). This turns trial division into a special-purpose algorithm. In contrast to the time complexity, the space complexity of the algorithm is negligible (because there are no intermediate results that need to be stored and processed).

#### 5.3.1.2 $P \pm 1$ and ECM

In the early 1970s John M. Pollard[18] developed and proposed a special-purpose integer factorization algorithm known as $p-1$ [8]. The algorithm got its name from the fact that it can be used to factorize an integer $n$ and find a prime factor $p$, if and only if $p-1$ is $B$-smooth (Definition A.27). If, for example, $p = 13$, then $p-1 = 12$ is 3-smooth and 3 is the respective smoothness bound (because $12 = 3 \cdot 2^2$, and hence all prime factors are less or equal than 3). Generally speaking, if $p = q_1^{k_1} \ldots q_r^{k_r}$ for $q_1, \ldots, q_r \in \mathbb{P}$ and $k_1, \ldots, k_r \in \mathbb{Z}$, then $p-1$ is $B$-smooth, if all $q_i \leq B$ for $i = 1, \ldots, r$ (and $B$-powersmooth, if all $q_i^{k_i} \leq B$).

When one wants to apply Pollard's $p-1$ integer factorization algorithm, one typically neither knows the prime factors of $p-1$ (i.e., $q_1, \ldots, q_r$), nor the

18 John Michael Pollard is a British mathematician who was born in 1941.

smoothness bound $B$. Instead, one has to start with a value for $B$ that looks reasonable, such as $B = 128$. If one later finds out that it doesn't work with this value, then one may multiply $B$ with $s = 2, 3, \ldots$ and work with the respective values $B = 128 \cdot s$. Since one also doesn't know the $q_i$ (and $q_i^{k_i}$) for $i = 1, \ldots, r$, one computes the auxiliary value

$$M = \prod_{q \in \mathbb{P}; q^\alpha \leq B} q^\alpha$$

as the product of all primes $q \in \mathbb{P}$ that are smaller or equal than $B$ with some exponent $\alpha$. This value depends on $q$ and $B$, and could therefore also be written as $\alpha(q, B)$. It refers to the largest integer such that $q^\alpha \leq B$; that is, $\alpha = \lfloor log(B)/log(q) \rfloor$.

According to the assumption that $p - 1$ is $B$-smooth and the way $M$ is generated (as the product of all primes less or equal than $B$), $p - 1$ divides $M$ (i.e., $p - 1 | M$), and this means—according to Fermat's little theorem (Theorem A.9)—that for all $a \in \mathbb{Z}$ with $gcd(a, n) = 1$

$$a^M \equiv a^{k(p-1)} \equiv 1 \pmod{p}$$

must hold for some $k \in \mathbb{N}$, and hence $p$ divides $a^M - 1$:

$$p | a^M - 1$$

Because $p$ also divides $n$, we may compute $gcd(a^M - 1, n)$. The result may be trivial (i.e., 1 or $n$), but it may also refer to a nontrivial divisor $p$. Otherwise, $B$ can be multiplied with the next-bigger $s$ or an entirely new value for $B$ can be used. In either case, the algorithm can be repeated until a valid prime factor $1 < p < n$ is found.

**Algorithm 5.1** Pollard's $p - 1$ integer factorization algorithm.

$(n)$

---

repeat
  select $B$ and $a$
  $M = \prod_{q \in \mathbb{P}; q^\alpha \leq B} q^\alpha$
  $p = gcd(a^M - 1, n)$
  if $p = 1$ or $p = n$ then restart the algorithm or exit
until $1 < p < n$

---

$(p)$

The resulting $p-1$ integer factorization algorithm is summarized in Algorithm 5.1. It takes as input $n$, and it tries to output a prime factor $p$ of $n$. The algorithm is probabilistic in the sense that it selects a smoothness bound $B$ and a small integer $a > 1$, such as $a = 2$. In the general case, $a$ is likely to be coprime with $n$ (otherwise $gcd(a, n)$ already reveals a prime factor of $n$). It then computes $M$ as described above, and determines the greatest common divisor of $a^M - 1$ and $n$.[19] If the result is 1 or $n$, then only a trivial factor is found, and the algorithm can be repeated (with another value for $B$[20]). If the result is different from 1 or $n$, then a prime factor $p$ is found and the algorithm terminates.

If, for example, we want to factorize $n = 299$, then we may fix $a = 2$ and start with $B = 3$. In this case, $M = 2 \cdot 3 = 6$ and $gcd(a^M - 1, n) = gcd(2^6 - 1, 299) = gcd(63, 299) = 1$. This divisor is trivial, and hence the algorithm must be repeated with a larger value for $B$. If $B = 4$, then $M = 2^2 \cdot 3 = 12$ and $gcd(a^M - 1, n) = gcd(2^{12} - 1, 299) = gcd(2^{12} \bmod 299 - 1, 299) = gcd(208, 299) = 13$. This value is nontrivial (i.e., $1 < 13 < 299$), and hence it is a divisor that yields the output of the algorithm. If we chose a larger value for $B$, such as $B = 10$, then $M = 2^3 \cdot 3^2 \cdot 5 \cdot 7 = 2520$ and $gcd(2^{2520} \bmod 299 - 1, 299) = 13$ would yields the same divisor of 299.

Note again that one knows neither the prime factorization of $p-1$ nor $B$ before the algorithm begins. So one has to start with an initially chosen $B$ and perhaps increase the value during the execution of the algorithm. Consequently, the algorithm is practical only if $B$ is sufficiently small. For the typical size of prime numbers in use today, the probability that the algorithm succeeds is rather small. But the existence of the algorithm is still the main reason why some cryptographic standards require that RSA moduli are the product of strong primes (Section A.2.4.4).[21]

After the publication of Pollard's $p - 1$ algorithm, several researchers came up with modifications and improvements. For example, in the early 1980s, Hugh C. Williams[22] proposed the $p + 1$ algorithm [9] that is similar to Pollard's $p - 1$ algorithm, but requires $p + 1$ to be smooth (instead of $p - 1$). The resulting algorithm is mathematically more involved than Pollard's $p - 1$ algorithm, as it uses Lucas sequences to perform exponentiation in a quadratic field. This is not further addressed here.

In the late 1980s, Hendrik W. Lenstra[23] proposed yet another modification of Pollard's $p - 1$ algorithm that is even more efficient [10]. Due to the fact that it uses

19 To compute this value, it is not necessary to compute the huge integer $a^M$. Instead, the algorithm can compute $a^M \bmod n$, an integer that is smaller than $n$.

20 More precisely, if $gcd(a^M - 1, n) = 1$, then $B$ is too small, and if $gcd(a^M - 1, n) = n$, then $B$ is too big.

21 A prime $p$ is strong if $p - 1$ has at least one large prime factor.

22 Hugh Cowie Williams is a Canadian mathematician who was born in 1943.

23 Hendrik Willem Lenstra is a Dutch mathematician who was born in 1949.

elliptic curves over $GF(p)$, it is called ECM. It is still a special-purpose algorithm with a subexponential time complexity of $L_p[1/2, \sqrt{p}]$, where $p$ refers to the smallest prime factor of $n$. In a typical setting, all prime factors of $n$ are equally large, and hence $p$ is roughly of size $\sqrt{n}$. In this case, $p$ is sufficiently large so that the ECM cannot be considered a threat against the standard RSA public key cryptosystem that uses two primes. However, it must still be taken into account, especially when one implements multiprime RSA, where $n$ may have more than two prime factors. This is nonstandard but sometimes still used in the field. For example, the tenth and eleventh Fermat numbers, $F_{10} = 2^{2^{10}} + 1$ and $F_{11} = 2^{2^{11}} + 1$, were factorized with the ECM in 1995 and 1988, respectively. Due to the fact that $F_{10}$ and $F_{11}$ consist of 4 and 5 prime factors, these results went largely unnoticed in public (in contrast to $F_8$ and $F_9$ mentioned below).

#### 5.3.1.3 Pollard Rho

In 1975 (and hence only one year after the release of his $p-1$ algorithm), Pollard proposed another probabilistic integer factorization algorithm that has become known as *Pollard Rho* (or *Pollard* $\rho$) [11]. Again, it is well suited to factorize integers that have a small prime factor (i.e., $n = p_1^{k_1} \ldots p_r^{k_r}$ for $p_1, \ldots, p_r \in \mathbb{P}$, $k_1, \ldots, k_r \in \mathbb{Z}$, and at least one of the prime factors $p_1, \ldots, p_r$ is small). A slightly modified version of this algorithm was used, for example, to factorize the 78-digit eighth Fermat number $F_8 = 2^{2^8} + 1$ in 1980, as this number unexpectedly turned out to have a small prime factor.

At its core, the Pollard Rho algorithm uses a simple function $f$ and a starting value $x_0$ to recursively compute a sequence $(x_i)_{i \geq 0}$ of pseudorandom values according to $x_i = f(x_{i-1})$ for $i = 1, 2, \ldots$ Typically, a quadratic congruential generator (QCG) is used to serve as $f$, meaning that the recursive formula becomes

$$x_{i+1} = (x_i^2 + a) \bmod n \tag{5.1}$$

for some $a, n \in Z$ and $a < n$.[24] This generates a sequence of values $x_0, x_1, x_2, \ldots$ that eventually falls into a cycle, meaning that the $x$-values start repeating themselves (this needs to be the case because there only $n$ values between 0 and $n - 1$). The expected time until this happens and the expected length of the cycle are both proportional to $\sqrt{n}$.

If, for example, we start with $x_0 = 2$ and subject this starting value to $x_{i+1} = (x_i^2 + 1) \bmod 209$ (meaning that $n = 209$ and $a = 1$), then the sequence that

24 Note that a linear congruential generator (LCG) as introduced in Chapter 7 could also be used here (instead of a QCG). Compared to a QCG, however, a LCG generates sequences of numbers with pseudorandomness properties that are not as good as those generated with a QCG.

is generated starts with the value $x_1 = 5$ before it falls into a cycle with $x_2 = 26$, $x_3 = 49$, $x_4 = 102$, and $x_5 = 163$. The next value to be generated (i.e., $x_6$) is equal to $x_2$, and hence the length of the cycle is $6-2=4$. The name rho ($\rho$) is taken from the fact that there is a starting sequence (also known as *tail*) that finally leads to a *cycle*. This can be well visualized with the Greek letter rho.

To find a cycle, one can proceed naïvely as just described (and store all intermediate values), or one can use a cycle detection method created by Robert Floyd that has small memory requirements. The idea is to check for each $k = 1, 2, \ldots$ whether $x_k$ equals $x_{2k}$. Technically, this means that one uses two variables $x$ and $y$ (with the same starting value $x_0$) and has one of them proceed twice as fast; that is, $x_{i+1} = f(x_i)$ and $y_{i+1} = f(f(y_i))$ for $i > 0$. This means that each possible cycle length is tried out. If the value is correct, then the respective $x$ and $y$ values are equal. In our example, we start with $k = 1$ and compute $x_1 = 5$ and $y_1 = x_2 = 26$. For $k = 2$, we have $x_2 = 26$ and compute $y_2 = x_4 = 102$. For $k = 3$, we have $x_3 = 49$ and $y_3 = x_6 = 26$. Finally, for $k = 4$, we have $x_4 = 102$ and $y_4 = x_8 = 102$. This means that $x_4$ and $x_8$ are equal and that all intermediate values are part of a cycle with length four. This confirms the previous result in a way that doesn't require to store intermediate values.

The Pollard Rho integer factorization algorithm uses Floyd's cycle-finding method to probabilistically factorize $n$. Let us assume that we have found a cycle and hence two values $x$ and $y$ for which $x \equiv y \pmod{p}$ but $x \not\equiv y \pmod{n}$. The first equivalence suggests that $p$ divides $x - y$; that is, $p \mid (x-y)$, whereas the second equivalence suggests that $n$ doesn't divide $x - y$; that is, $n \nmid (x-y)$. In combination, the result is that a prime factor $p$ for $n$ may be derived from such a pair $(x, y)$ by computing the greatest common divisor from $|x - y|$ and $n$; that is, $p = gcd(|x-y|, n)$. If $p = n$, then the algorithm fails and must be restarted from scratch. The resulting algorithm is summarized in Algorithm 5.2.[25] There is even an improved version created by Richard P. Brent [12] that is not addressed here.

Let us consider a simple example with $n = 8051$ and $f(x) = x^2 + 1 \pmod{8051}$. After having initialized $x = y = 2$ and $p = 1$, the algorithm enters the while-loop. In the first iteration, the new value for $x$ is $2^2 + 1 = 5$ (the modulus $n = 8051$ does not matter here), and the new value for $y$—after having applied $f$ twice—is $5^2 + 1 = 26$. The algorithm thus computes $gcd(|5-26|, 8051) = gcd(21, 8051) = 1$, and this yields the trivial prime factor 1. This means that the algorithm remains in the while-loop. In the second iteration, the new value for $x$ is again $f(5) = 26$, and the new value for $y$ is $f(f(26)) = f(26^2+1) = f(677) = 677^2+1 = 458{,}330 \equiv 7474 \pmod{8051}$. So after the second iteration, $x = 26$ and $y = 7474$. This time, $gcd(|26 - 7474|, 8051) = gcd(7448, 8051) = 1$, and hence

25 Note that the algorithm can be optimized in a way that the gcd is not computed in each step of the loop.

**Algorithm 5.2** Pollard Rho integer factorization algorithm.

$(n)$

---

$x = 2$
$y = 2$
$p = 1$
while $p = 1$
    $x = f(x)$
    $y = f(f(y))$
    $p = gcd(|x - y|, n)$
if $p = n$ then exit and return failure

---

$(p)$

the while-loop must be iterated again. In the third iteration, $x$ is set to $f(26) = 677$, and $y$ is set to $f(f(7474))$. $f(7474) = 55,860,677 \equiv 2839 \pmod{8051}$ and $f(2839) = 2839^2 + 1 = 8,059,922 \equiv 871 \pmod{8051}$, meaning that $f(f(7474))$ is equal to 871. After the third iteration, one has $x = 677$ and $y = 871$, and hence $gcd(|677-871|, 8051) = gcd(194, 8051) = 97$. This terminates the while-loop, and because $97 \neq 8051$, 97 is indeed a nontrivial prime factor of 8051. It is simple to find the other prime factor by computing $8051/97 = 83$. Note that another function $f$ may have found 83 first, and that there is nothing special about 97 or 83. If $p$ were equal to $n$ in the end, then the algorithm would fail and need to be started from scratch (using another function $f$).

Obviously, the efficiency of the algorithm depends on how fast an $(x, y)$-pair with $1 < gcd(|x - y|, n) < n$ is found. Due to the birthday paradox (Section 6.1), such a pair is found after approximately $1.2\sqrt{p}$ numbers have been chosen and tried out. This means that the algorithm has a time complexity of $O(\sqrt{p})$, where $p$ is the smallest prime factor of $n$. As $p$ is at most $\sqrt{n}$, this means that the time complexity of the Pollard Rho integer factorization algorithm is

$$O(\sqrt{\sqrt{n}}) = O(\sqrt[4]{n}) = O(n^{1/4}) = O(e^{\ln(n^{1/4})}) = O(e^{\frac{1}{4}\ln(n)}) = O(e^{\frac{b}{4}})$$

This complexity is again exponential in the length of $n$, and hence the algorithm can only be used if $p$ is small compared to $n$. For the sizes of integers in use today, the algorithm is still impractical (this also applies to Brent's improved version). Contrary to the time complexity, the space complexity of the algorithm (and most of its variants) is negligible. Again, this is mainly due to Floyd's cycle-finding method.

### 5.3.2 General-Purpose Algorithms

In contrast to special-purpose algorithms, general-purpose integer factorization algorithms work equally well for all integers $n$, meaning that there is no special requirement regarding the structure of $n$ or its prime factors. Most importantly, these algorithms are used if $n$ is the product of two equally sized prime factors.

Most general-purpose integer factorization algorithms in use today exploit an old idea that is due to Fermat. The idea starts from the fact that every odd integer $n \geq 3$ can be written as the difference of two squares:

$$n = x^2 - y^2$$

for some integers $x, y \in \mathbb{N}$, where $y$ may also be zero. According to the third binomial formula, $x^2 - y^2$ is equal to $(x+y)(x-y)$, and this suggests that $p = (x+y)$ and $q = (x - y)$ are the two (nontrivial) factors of $n$ (in fact, it can be shown that $x = (p+q)/2$ and $y = (p-q)/2$). For example, if one wants to factorize $n = 91$, then one has to find two integers for which the difference of the squares is equal to this value. In this example, $x = 10^2 = 100$ and $y = 3^2 = 9$ satisfy this property, and hence $p = 10 + 3 = 13$ and $q = 10 - 3 = 7$ yield the two (prime) factors of 91. Note that $13 \cdot 7 = 91$, as well as $10 = (13+7)/2$ and $3 = (13-7)/2$.

Fermat also proposed a simple method to find a valid $(x, y)$-pair: Start with $x = \lceil \sqrt{n} \rceil$ and compute $z = x^2 - n$. If $z$ is a square, meaning there is a $y$ with $y^2 = z$, then $(x, y)$ is a valid pair. Otherwise, one increments $x$ and repeats the method for this value. In the end, one has $x$ and $y$ and can determine the two factors $(x + y)$ and $(x - y)$.

If, for example, $n = 10033$, then one starts with $x = \lceil \sqrt{10033} \rceil = 101$, computes $101^2 - 10033 = 10201 - 10033 = 168$, and recognizes that 168 is not a square. Hence, one repeats the method with $x = 102$: $102^2 - 10033 = 10404 - 10033 = 371$. This is not a square either, and one repeats the method with $x = 103$: $103^2 - 10033 = 10609 - 10033 = 576$. This is a square (since $576 = 24^2$), and hence $(103, 24)$ is a valid pair and can be used to compute the (nontrivial) factors $103 + 24 = 127$ and $103 - 24 = 79$ (note that $127 \cdot 79 = 10033$).

Fermat's factorization method is efficient (and hence practical) if $x$ and $y$ are similarly sized and not too far away from $\sqrt{n}$. Otherwise, the method gets inefficient, because $x$ is always incremented by only one. In general, one cannot make the assumption that $x$ and $y$ are near, and hence one follows a more general approach. In fact, one looks for $(x, y)$-pairs that satisfy

$$x^2 \equiv y^2 \pmod{n}$$

and

$$x \not\equiv \pm y \pmod{n}$$

If such a pair is found, then a nontrivial factor of $n$ can be found with a success probability of $1/2$ by computing $gcd(x - y, n)$ (with another probability of $1/2$ a trivial factor 1 or $n$ is found). So factorizing $n$ can be reduced to finding such $(x, y)$-pairs. There are several algorithms that can be used for this purpose, such as *continued fraction*[26] and some sieving methods based on [15], such as the *quadratic sieve* (QS[27]) [16] and the *number field sieve* (NFS) [17]. All of these algorithms have a subexponential time complexity. Continued fraction, for example, has a time complexity of

$$O(e^{\sqrt{2ln(n)ln(ln(n))}}) = L_n[1/2, \sqrt{2}]$$

For integers under 120 decimal digits or so, the QS is the most efficient general-purpose integer factorization algorithm with a time complexity of $L_n[1/2, 1]$. For several years it was believed that this complexity is as good as it can possibly be. The QS was used, for example, to factorize the 129-digit integer RSA-129 (see below). To factorize integers beyond 120 digits, however, the NFS is the factorization algorithm of choice today. There are actually two variants of the NFS: the *special number field sieve* (SNFS) and the *general number field sieve* (GNFS). As its name suggests, the GNFS is still a general-purpose algorithm, whereas the SNFS is a special-purpose one. It applies to integers of the form $n = r^e - s$ for small $r$ and $|s|$. Both algorithms have a time complexity of $L_n[1/3, c]$, where $c = \sqrt[3]{32/9} \approx 1.526$ for the SNFS and $c = \sqrt[3]{64/9} \approx 1.923$ for the GNFS. The SNFS was used, for example, in 1990 to factorize the 155-digit ninth Fermat number $F_9 = 2^{2^9} + 1$ [18].[28] While the SNFS is asymptotically the most efficient variant of the NFS, the GNFS works for all integers and is simpler to implement in a distributed environment. It is therefore often used in distributed integer factorization projects launched on the Internet.

Similar to the QS algorithm, the NFS algorithm (and its variants) consists of two steps, of which one—the so-called relation collection step—can be parallelized and optimized by the use of special hardware devices. In fact, there have been many

26 This method was first described in the 1930 [13] and later adapted for implementation on computer systems in the 1970s [14].

27 The QS algorithm was proposed at EUROCRYPT '84. This fact illustrates the importance of number theory in general, and integer factorization in particular, for modern cryptography.

28 Note $F_9$ consists of 3 prime factors, whereas all previous Fermant numbers are either prime themselves ($F_1 - F_4$) or are the product of only 2 prime factors ($F_5 - F_8$).

such devices proposed in the literature. Examples include the The Weizmann Institute Key-Locating Engine (TWINKLE) [19, 20], The Weizmann Institute Relation Locator (TWIRL) [21], SHARK [22], and Yet Another Sieving Device (YASD) [23]. The design and implementation of these devices is a topic of its own that is beyond the scope of this book and therefore not addressed here.

### 5.3.3 State of the Art

After the publication of the RSA public key cryptosystem (Section 13.3.1), a challenge was posted in the August 1977 issue of *Scientific American* [24]. In fact, an amount of USD 100 was offered to anyone who could decrypt a message that was encrypted using a 129-digit integer acting as modulus. The number became known as RSA-129, and it was not factored until 1994 (with a distributed version of the QS algorithm [25]).

$$\begin{aligned} \text{RSA-129} &= 1143816257578888676692357799761466120102182967212 \\ &\quad 4236256256184293570693524573389783059712356395870 \\ &\quad 5058989075147599290026879543541 \\ &= 3490529510847650949147849619903898133417764638493 \\ &\quad 387843990820577 \\ &\quad * \\ &\quad 3276913299326670954996198819083446141317764296799 \\ &\quad 2942539798288533 \end{aligned}$$

Until 2007, RSA Security[29] had sponsored a few cryptographic challenges, including the RSA Factoring Challenge, to learn more about the actual difficulty of factoring large integers of the type used in the RSA public key cryptosystem. Most importantly, the following RSA numbers have been factorized so far (among many others):

- RSA-576 (2003)
- RSA-640 (2005)
- RSA-704 (2012)

29 As its name suggests, RSA Security was a company founded in 1986 to market the RSA public key cryptosystem and some related cryptographic techniques. In 2006, it was acquired by EMC Corporation, and in 2016, EMC Corporation was acquired by Dell Technologies. More recently, on September 2, 2020, RSA announced its new status as an independent company (that is independent from either EMC or Dell).

- RSA-768 (2009[30])

The factorization of RSA-576 and RSA-640 was awarded with a cash prize of USD 10,000 and USD 20,000. All subsequent RSA challenge numbers were factorized after the challenge became inactive (i.e., after the acquisition of RSA Security), so that the originally promised cash prize was not paid out. But people still continue with the challenge, and the latest achievements include RSA-240 referring to a 795-bit number (December 2019) and RSA-250 referring to a 829-bit number (February 2020). Furthermore, the next big RSA challenge numbers (in the sense that a cash prize was originally promised) are 896, 1,024, 1,536, and 2,048 bits long.

The bottom line is that the current state of the art in factorizing large integers is still below 1,024 bits. So an RSA key of that size should still be sufficient. But due to the fact that such keys tend to have a long lifetime, it is safer and highly recommended to switch to longer keys.

As mentioned earlier, the NFS (including the SNFS and the GNFS) is the best known algorithm to factorize integers with more than 120 digits. However, this is only true as long as nobody is able to build a quantum computer (Section D.5). If somebody had a quantum computer at hand, then he or she could use a polynomial-time algorithm to solve the IFP that is due to Peter W. Shor [26, 27].[31] More specifically, Shor's algorithm has a cubic time complexity (i.e., $O((\ln n)^3)$) and a linear space complexity (i.e., $O(\ln n)$). This means that a quantum computer needs $c \cdot \ln n$ qubits to factorize an integer $n$ (for a small constant factor $c$). This translates to a few thousands of qubits, and this is far beyond what is technically feasible today, as people are currently able to build quantum computers with "only" about 50–70 qubits.

## 5.4 ALGORITHMS FOR COMPUTING DISCRETE LOGARITHMS

There are several public key cryptosystems whose security is based on the computational intractability of the DLP (Definition 5.5) in a cyclic group or some related problem. If somebody were able to efficiently compute discrete logarithms, then he or she would be able to break these systems. It is therefore important to know the best (i.e., most efficient) algorithms that can be used to actually compute discrete logarithms. Again, there are two classes of such algorithms:

30 Surprisingly, RSA-768 was factored three years before RSA-704, even though it refers to a larger integer.

31 In 1999, Shor won the prestigious Gödel Prize for this work and the development of algorithms to solve the IFP and the DLP on a quantum computer.

- *Generic algorithms* work in any cyclic group, meaning that they do not attempt to exploit the special properties of the group in which the discrete logarithms need to be computed.
- *Nongeneric* or *special-purpose algorithms* attempt to exploit special properties of the cyclic group in which the discrete logarithms need to be computed, meaning that these algorithms are specifically designed to work in a specific group.

It goes without saying that nongeneric (or special-purpose) algorithms are typically more efficient than generic ones. But we start with generic ones first.

### 5.4.1 Generic Algorithms

There are a few generic algorithms that can be used to solve the DLP in a cyclic group $G$ with generator $g$. Due to a result of Victor Shoup [28], we know that $O(\sqrt{|G|})$ is a lower bound for the time complexity of a generic algorithm, and that improvements are only possible if the prime factorization of $|G|$ is known. In this case (and if the prime factors of $|G|$ are sufficiently small), then the Pohlig-Hellman algorithm [5] based on the Chinese remainder theorem (CRT, Theorem A.8) can be used to more efficiently solve the DLP.[32] Let

$$|G| = p_1^{e_1} p_2^{e_2} \dots p_l^{e_l}$$

be the prime factorization of $|G|$. To compute the discrete logarithm $x = \log_g h$, the Pohlig-Hellman algorithm follows a divide-and-conquer approach: Rather than dealing with the large group $G$, it computes smaller discrete logarithms $x_i$ in the subgroups of order $p_i^{e_i}$ (e.g., using Shanks' baby-step giant-step algorithm or Pollard Rho) and then uses the CRT to compile the desired value $x$ from all $x_i$ $(i = 1, \dots, l)$. The time complexity of the resulting algorithm depends on the prime factors of $|G|$, and to mitigate the attack, $|G|$ must have a prime factor that is at least $2^{160}$.

Before we address the aforementioned algorithms of Shanks and Pollard, we say a few words about brute-force search.

#### 5.4.1.1 Brute-Force Search

The simplest and most straightforward generic algorithm to solve the DLP is *brute-force search*, meaning that one successively computes powers of $g$ (i.e., $g^1$, $g^2$, $g^3$,...), until one reaches $h$. For a randomly chosen $h$, one can expect to find the

32 According to [5], the algorithm was independently discovered by Roland Silver (a few years earlier) and Richard Schroeppel and H. Block, none of whom published the result.

correct exponent after checking half of the possible values (i.e., $|G|/2$). This means that brute-force search has a time complexity of

$$O(|G|)$$

To make a brute-force search prohibitively expensive, one must use a group $G$ that is sufficiently large. For example, in the case of the cyclic group $\mathbb{Z}_p^*$ for prime $p$, $(p-1)/2$ checks are required on the average to compute a discrete logarithm. So $|G| = p - 1$ should be at least in the order of $2^{80}$. It goes without saying that this only holds if a brute-force attack is the only feasible attack, and that this is seldom the case (as there are almost always more efficient algorithms to compute discrete logarithms). Two examples are outlined next.

#### 5.4.1.2 Baby-Step Giant-Step Algorithm

One algorithm that is more efficient than brute-force search is generally credited to Daniel Shanks.[33] It is basically a time-memory trade-off, meaning that it runs faster than brute-force search, but it also uses some extra memory.

The goal of the algorithm is to determine the value $x = \log_g h$. The algorithm therefore requires a parameter $t$ that is typically set to the ceiling of $\sqrt{|G|}$; that is, $t = \lceil \sqrt{|G|} \rceil$. Using this parameter, $x$ can be written as

$$x = x_g t - x_b$$

with $0 \leq x_g, x_b < t$.[34] The algorithm is named *baby-step giant-step*, because it consists of two respectively named steps:

- In the *giant-step*, the algorithm computes the pairs $(j, (g^t)^j = g^{tj})$ for all $0 \leq j < |G|/t \approx t$, sorts these pairs according to the second component $g^{jt}$, and stores them in a (now sorted) table.
- In the *baby-step*, the algorithm computes $hg^i$ for $i = 0, 1, 2, \ldots$ until it yields a value that is contained in the table from the giant-step. In this case, $hg^i = g^x g^i$ is equal to $g^{tj}$, and this, in turn, suggests that

$$\frac{g^{tx_g}}{g^{x_b}} g^i = g^{tj}$$

33 Daniel Shanks was an American mathematician who lived from 1917 to 1996. The baby-step giant-step algorithm appeared in a 1971 publication by him, but it is sometimes rumored that the Russian mathematician Alexander Gelfond, who lived from 1906 to 1968, already knew the algorithm in 1962.

34 The subscripts stand for "giant" and "baby."

This is because $g^x g^i = g^{x_g t - x_b} g^i = g^{t x_g} g^{-x_b} g^i = g^{t x_g}/g^{x_b} \cdot g^i$. The equation holds for $x_b = i$ and $x_g = j$, and hence $x = x_g t - x_b = jt - i$.

The baby-step giant-step algorithm outputs $x$ that solves the DLP for $h$ in $G$ with generator $g$.

If, for example, $G$ is $\mathbb{Z}_{29}^*$ and the generator is $g = 11$, then we may want to solve the DLP for $h = 3$ (meaning that we are looking for $x \in \mathbb{N}$ with $11^x \pmod{29} = 3$). Using the baby-step giant-step algorithm, one first sets $t = \lceil\sqrt{28}\rceil = 6$, and then computes the giant-step table as follows:

| $j$ | $g^{tj}$ |
|---|---|
| 0 | $11^{6\cdot 0} = 11^0 \equiv \underline{1} \pmod{29}$ |
| 1 | $11^{6\cdot 1} = 11^6 \equiv \underline{9} \pmod{29}$ |
| 2 | $11^{6\cdot 2} = 11^{12} \equiv \underline{23} \pmod{29}$ |
| 3 | $11^{6\cdot 3} = 11^{18} \equiv \underline{4} \pmod{29}$ |
| 4 | $11^{6\cdot 4} = 11^{24} \equiv \underline{7} \pmod{29}$ |
| 5 | $11^{6\cdot 5} = 11^{30} \equiv \underline{5} \pmod{29}$ |

Strictly following the algorithm, this table would have to be sorted according to the second component $g^{tj}$ (i.e., the value that is underlined). Because the numbers are small, this sorting step is not performed here. Instead, we jump directly into the baby-step: In the first iteration, $i$ is set to zero, and $hg^i$ equals $3 \cdot 11^0 = 3 \cdot 1 = 3$. This value does not match any of the underlined values in the giant-step table. In the second iteration, $i$ is set to one, and $hg^i$ equals $3 \cdot 11^1 = 3 \cdot 11 = 33 \equiv 4 \pmod{29}$. This value matches the value for $j = 3$ in the giant-step table. This means that $i = 1$ and $j = 3$, and hence $x = 3 \cdot 6 - 1 = 17$. The correctness of this value follows from $11^{17} \pmod{29} = 3$.

The time and space complexities of the baby-step giant-step algorithm are both $O(\sqrt{|G|})$ (for $t = \lceil\sqrt{|G|}\rceil$). In a group of order $2^{80}$, this complexity is $O(\sqrt{2^{80}}) = O(2^{80/2}) = O(2^{40})$. In other words, in order to obtain a complexity of $2^{80}$, one must employ a group with $2^{160}$ elements. In the case of $\mathbb{Z}_p^*$, for example, this suggests that $p$ must be at least 160 bits long. Unfortunately, there are even more powerful (special-purpose) algorithms to solve the DLP in $\mathbb{Z}_p^*$, and hence even larger bit lengths for $p$ are usually required.

#### 5.4.1.3 Pollard Rho

As pointed out in 1978 by Pollard [29], a similar idea as used in his Rho integer factorization algorithm can be used to solve the DLP with the same time complexity as the baby-step giant-step algorithm but only little memory. In 1999, it was shown that the algorithm can be parallelized [30], and this makes *Pollard Rho* ($\rho$) the most efficient generic algorithm to solve the DLP known to date.

The Pollard Rho algorithm exploits the birthday paradox and uses a pseudorandom (iterating) function $f : G \rightarrow G$ to find a sequence of group elements $(y_i)_{i \geq 0} = y_0, y_1, y_2, \ldots$ that represents a random walk (through all elements of the group). The walk is computed according to $y_0 = g^{\alpha_0} h^{\beta_0}$ for some $\alpha_0$ and $\beta_0$ randomly chosen from $\{0, 1, \ldots, |G|-1\}$ and $y_{k+1} = f(y_k)$ for $k \in \mathbb{N}$. This means that it must be possible to efficiently compute $\alpha_{k+1}$ and $\beta_{k+1}$ from $\alpha_k$ and $\beta_k$ such that $f(y_k) = y_{k+1} = g^{\alpha_{k+1}} h^{\beta_{k+1}}$. This implies that while computing $y_k$ ($k \in \mathbb{N}$), one can keep track of the corresponding sequences of exponents; that is, $(\alpha_k)$ and $(\beta_k)$ with $y_k = g^{\alpha_k} h^{\beta_k}$. Sooner or later, one finds a pair of matching values $(y_i, y_j)$, and in this case, $y_i = y_j$ suggests that $g^{\alpha_i} h^{\beta_i} = g^{\alpha_j} h^{\beta_j}$. Since $h = g^x$, this implies

$$g^{\alpha_i} g^{\beta_i x} \equiv g^{\alpha_j} g^{\beta_j x}$$

If $gcd(\beta_i - \beta_j, |G|) = 1$, then $x$ can be computed as

$$x = (\alpha_j - \alpha_i)(\beta_i - \beta_j)^{-1} \pmod{|G|}$$

Note that the iterating function $f$ can be arbitrary and that Pollard made some proposals here. These proposals have been optimized to compute sequences of group elements that more closely resemble a random walk.

The Pollard Rho algorithm is simple and straightforward, but—according to Shoup's result mentioned above—it is as efficient as a generic algorithm can possibly be (with time complexity $O(\sqrt{|G|})$). To get more efficient algorithms, one has to consider nongeneric (special-purpose) algorithms as addressed next.

### 5.4.2 Nongeneric (Special-Purpose) Algorithms

All algorithms addressed so far (except the Pohlig-Hellman algorithm) are completely independent from the group in which the discrete logarithms need to be computed, meaning that they work in all cyclic groups. As mentioned earlier, this is not true for nongeneric (special-purpose) algorithms. These algorithms exploit special properties of the groups in use and are therefore more powerful.

Most importantly, the *index calculus method* (ICM) yields a probabilistic algorithm to compute discrete logarithms in $\mathbb{Z}_p^*$ and some other groups (but it does not work in all cyclic groups).[35] Without going into details, we note that the time

35 The ICM was first described by a French mathematician named Maurice Kraitchik in a book on number theory published in 1922. After the discovery of the Diffie-Hellman key exchange, the method was revisited and many cryptographers working in the early days of public key cryptography shaped it and presented it in the form we know it today.

complexity of the ICM is $L_p[1/2, c]$ for some small constant $c$, such as $\sqrt{2}$. Also, the ICM inspired many researchers to come up with new algorithms or apply existing algorithms to the DLP. For example, the NFS algorithm that has been invented to solve the IFP can also be used to solve the DLP. Remember that it has a running time complexity of $L_p[1/3, c]$, and hence it is one of the algorithms of choice to be used in $\mathbb{Z}_p^*$.

If applied to an extension field $\mathbb{F}_q$ with $q = p^n$ for some prime $p \in \mathbb{P}$, even more efficient versions of the ICM exist. Most of them have a time complexity of $L_q[1/3, c]$ for some small constant $c$ (this is comparable to the NFS algorithm). In 2013, however, Antoine Joux significantly improved the state of the art in solving the DLP in an extension field with small characteristic. The time complexity of his algorithm is almost $L_q[1/4, c]$ [31, 32]. Although this is not yet polynomial (remember that the first parameter in the L-notation must be zero for an algorithm to run in polynomial time), it is pretty close and people sometimes call it quasi-polynomial. In some extension fields, this is by far the most efficient algorithm that can be used today to solve the DLP.

### 5.4.3 State of the Art

If we are working in $\mathbb{Z}_p^*$, then state of the art in computing discrete logarithms is directly comparable to the state of the art in factoring integers. There are only a few exceptional cases in which computing discrete logarithms is more efficient than factoring integers. This suggests that the bit length of $p$ (if we work in $\mathbb{Z}_p^*$) should be comparable to the bit length of the modulus $n$ used in a cryptosystem whose security depends on the IFA. In either case, 1,024 bits provide a security level that is comparable to 80 bits in the secret key case, whereas 2,048 bits provide a security level that is comparable to 112 bits. It goes without saying that this security level is more appropriate today. If we work in $\mathbb{Z}_p^*$, then special care must be taken that $p-1$ does not have only small prime factors. Otherwise, the Pohlig-Hellman algorithm can be used to efficiently compute discrete logarithms.

If we work in a group in which the nongeneric (special-purpose) algorithms do not work, then the state of the art in computing discrete logarithms is worse than the state of the art in factoring integers. In this case, we have to use generic algorithms (that do not have subexponential running times). The best algorithms we can use in this case have a time complexity of $O(\sqrt{|G|})$. This fact is, for example, exploited by the XTR public key cryptosystem[36] and elliptic curve cryptography as addressed next.

36 The term XTR stands for ECSTR, which is an abbreviation for "Efficient and Compact Subgroup Trace Representation" [33].

## 5.5 ELLIPTIC CURVE CRYPTOGRAPHY

Public key cryptosystems get their security from the assumed intractability of inverting a one-way function, but inverting such a function may not be equally difficult in all algebraic structures and groups. For example, we have seen in the previous section that there are nongeneric algorithms to invert the discrete exponentiation function in the multiplicative group of $\mathbb{Z}_p^*$ with subexponential running time, but that these algorithms do not work in all cyclic groups. In particular, the algorithms do not work in groups of points on an elliptic curve over a finite field. In such a group, one has to use a generic algorithm to compute a discrete logarithm, and these algorithms have an exponential time complexity of $O(\sqrt{|G|})$.

If an instance of the DLP is harder to solve in a particular cyclic group, then the cryptosystems that are based on this instance may employ shorter keys to achieve the same level of security. This is the major advantage of *elliptic curve cryptography* (ECC): It works with shorter keys and can therefore be implemented more efficiently. Note, however, that it is still possible that nongeneric algorithms with subexponential running time to solve the DLP exist for such groups—we simply don't know any of them. The lower bound $O(\sqrt{|G|})$ only applies to generic algorithms, so it cannot be taken as an argument to exclude their existence.

ECC employs groups of points on an elliptic curve over a finite field $\mathbb{F}_q$, where $q$ is either an odd prime (in the case of a prime field) or a power of a prime. In the second case, only the prime 2 is used in standard ECC, meaning that only extension fields of characteristic 2 ($q = 2^m$ for some $m \in \mathbb{N}$) are considered. Such fields are also called binary extension fields, and $m$ refers to the degree of such a field.[37] To keep things as simple as possible, we make the following two restrictions regarding the elliptic curves we consider:

- First, we only consider elliptic curves over a prime field for some odd prime $p \in \mathbb{P}$, denoted as $\mathbb{Z}_p$.
- Second, we only consider elliptic curves over $\mathbb{Z}_p$ defined by the Weierstrass equation

$$y^2 \equiv x^3 + ax + b \pmod{p} \tag{5.2}$$

for $a, b \in \mathbb{Z}_p$ and $4a^3 + 27b^2 \not\equiv 0 \pmod{p}$.

37 An odd prime is typically used for software implementations, whereas a power of 2 is typically used for hardware implementations.

For any given $a$ and $b$ in $\mathbb{Z}_p$, (5.2) yields pairs of solutions $x, y \in \mathbb{Z}_p$ that can be formally expressed as follows:

$$\begin{aligned} E(\mathbb{Z}_p) = \{(x,y) \quad | \quad & x, y \in \mathbb{Z}_p \ \wedge \\ & y^2 \equiv x^3 + ax + b \pmod{p} \ \wedge \\ & 4a^3 + 27b^2 \not\equiv 0 \pmod{p}\} \end{aligned}$$

The resulting set $E(\mathbb{Z}_p)$ consists of all $(x, y) \in \mathbb{Z}_p \times \mathbb{Z}_p = \mathbb{Z}_p^2$ that yield a solution to (5.2). We can graphically interpret $(x, y)$ as a point in the $(x, y)$-plane, where $x$ represents the coordinate on the horizontal axis and $y$ represents the coordinate on the vertical axis. Such an $(x, y)$ refers to a point on the elliptic curve $E(\mathbb{Z}_p)$. The respective plot looks chaotic and does not resemble the elliptic curves one can define over the real numbers $\mathbb{R}$ that are sometimes used to visualize the notion of an elliptic curve.

In addition to the points on the curve, one also considers a *point at infinity* (typically denoted $\mathcal{O}$). This point yields the identity element that is required to form a group. If we use $E(\mathbb{Z}_p)$ to refer to an elliptic curve defined over $\mathbb{Z}_p$, then $\mathcal{O}$ is usually included implicitly.

Let us consider an exemplary elliptic curve that we are going to use in the book. For $p = 23$ and $a = b = 1$ (note that $4 \cdot 1^3 + 27 \cdot 1^2 \not\equiv 0 \pmod{23}$), the elliptic curve $y^2 \equiv x^3 + x + 1$ is defined over $\mathbb{Z}_{23}$. Besides $\mathcal{O}$, the elliptic curve $E(\mathbb{Z}_{23})$ comprises the following 27 elements or points in the $(x, y)$-plane:

$$\begin{array}{ccccccc} (0,1) & (0,22) & (1,7) & (1,16) & (3,10) & (3,13) & (4,0) \\ (5,4) & (5,19) & (6,4) & (6,19) & (7,11) & (7,12) & (9,7) \\ (9,16) & (11,3) & (11,20) & (12,4) & (12,19) & (13,7) & (13,16) \\ (17,3) & (17,20) & (18,3) & (18,20) & (19,5) & (19,18) & \end{array}$$

Together with $\mathcal{O}$, this sums up to 28 points or elements. In general, let $n$ be the number of points on an elliptic curve over a finite field $\mathbb{F}_q$ with $q$ elements. We then know that $n$ is of the order of $q$. In fact, there is a theorem due to Helmut Hasse[38] that bounds $n$ as

$$q + 1 - 2\sqrt{q} \leq n \leq q + 1 + 2\sqrt{q}$$

In our example, Hasse's theorem suggests that $E(\mathbb{Z}_{23})$ has $23 + 1 - 2\sqrt{23} = 14.44\ldots$ and $23 + 1 + 2\sqrt{23} = 35.56\ldots$ elements. 28 is in this range.

In addition to a set of elements, a group must also have an associative operation. In ECC, this operation is usually called addition (mainly for historical

38 Helmut Hasse was a German mathematician who lived from 1898 to 1979.

reasons), meaning that two points on an elliptic curve are added.[39] In general, the addition operation can be explained geometrically or algebraically. The geometric explanation is particularly useful if two points on an elliptic curve over $\mathbb{R}$ are added. Let $P = (x_1, y_1)$ and $Q = (x_2, y_2)$ be two distinct points on an elliptic curve $E(\mathbb{R})$. The sum of $P$ and $Q$, denoted $R = (x_3, y_3)$, is constructed in three steps:

1. Draw a line through $P$ and $Q$;
2. This line intersects $E(\mathbb{R})$ in a third point;
3. $R$ is the reflection of this point on the $x$-axis.

If $P = (x_1, y_1)$, then the double of $P$, denoted $R = (x_3, y_3)$, can be constructed in a similar way:

1. Draw the tangent line to $E(\mathbb{R})$ at $P$;
2. This line intersects $E(\mathbb{R})$ in a second point;
3. $R$ is the reflection of this point on the $x$-axis.

The following algebraic formulas for the sum of two points and the double of a point can be derived from the respective geometric interpretation:

1. $P + \mathcal{O} = \mathcal{O} + P = P$ for all $P \in E(\mathbb{Z}_q)$.
2. If $P = (x, y) \in E(\mathbb{Z}_q)$, then $(x, y) + (x, -y) = \mathcal{O}$. The point $(x, -y)$ is denoted $-P$ and called the negative of $P$. Note that $-P$ is indeed a point on the elliptic curve (e.g., $(3, 10) + (3, 13) = \mathcal{O}$).
3. Let $P = (x_1, y_1) \in E(\mathbb{Z}_q)$ and $Q = (x_2, y_2) \in E(\mathbb{Z}_q)$ with $P \neq -Q$, then $P + Q = (x_3, y_3)$ where

$$\begin{aligned} x_3 &= \lambda^2 - x_1 - x_2 \\ y_3 &= \lambda(x_1 - x_3) - y_1 \end{aligned}$$

and

$$\lambda = \begin{cases} \frac{y_2 - y_1}{x_2 - x_1} & if\ P \neq Q \\ \frac{3x_1^2 + a}{2y_1} & if\ P = Q \end{cases}$$

39 By contrast, the group operation in $\mathbb{Z}_p^*$ is multiplication. The differences in the resulting additive notation and multiplicative notation can sometimes be confusing.

Consider the elliptic curve $E(\mathbb{Z}_{23})$ defined above. Let $P = (3, 10)$ and $Q = (9, 7)$ be two elements from this group. Then $P + Q = (x_3, y_3)$ is computed as follows:

$$\begin{aligned}
\lambda &= \frac{7-10}{9-3} = \frac{-3}{6} = \frac{-1}{2} = 11 \in \mathbb{Z}_{23} \\
x_3 &= 11^2 - 3 - 9 = 6 - 3 - 9 = -6 \equiv 17 \pmod{23} \\
y_3 &= 11(3-(-6)) - 10 = 11(9) - 10 = 89 \equiv 20 \pmod{23}
\end{aligned}$$

Consequently, $P + Q = (17, 20) \in E(\mathbb{Z}_{23})$.

On the other hand, if one wants to add $P = (3, 10)$ to itself, then one has $P + P = 2P = (x_3, y_3)$, and this point is computed as follows:

$$\begin{aligned}
\lambda &= \frac{3(3^2)+1}{20} = \frac{5}{20} = \frac{1}{4} = 6 \in \mathbb{Z}_{23} \\
x_3 &= 6^2 - 6 = 30 \equiv 7 \pmod{23} \\
y_3 &= 6(3-7) - 10 = -24 - 10 = -11 \equiv 12 \pmod{23}
\end{aligned}$$

Consequently, $2P = (7, 12)$, and the procedure can be iterated to compute arbitrary multiples of $P$: $3P = (19, 5)$, $4P = (17, 3)$, $5P = (9, 16)$, $6P = (12, 4)$, $7P = (11, 3)$, $8P = (13, 16)$, $9P = (0, 1)$, $10P = (6, 4)$, $11P = (18, 20)$, $12P = (5, 4)$, $13P = (1, 7)$, $14P = (4, 0)$, $15P = (1, 16)$, $16P = (5, 19)$, $17P = (18, 3)$, $18P = (6, 19)$, $19P = (0, 22)$, $20P = (13, 7)$, $21P = (11, 20)$, $22P = (12, 19)$, $23P = (9, 7)$, $24P = (17, 20)$, $25P = (19, 18)$, $26P = (7, 11)$, $27P = (3, 13)$, and $28P = \mathcal{O}$. After having reached $nP = \mathcal{O}$, a full cycle is finished and everything starts from scratch. So $29P = P = (3, 10)$, $30P = 2P = (7, 12)$, and so on. In this example, the order of the group $n$ is 28, and—according to Lagrange's theorem—the order of an element must divide $n$. For example, the point $(4, 0)$ has order 2 (that divides 28); that is, $2(4, 0) = 28P = \mathcal{O}$. In ECC, all standard curves are chosen so that $n$ is prime. In such a group, every element has order $n$ and may serve as a generator. This is different from other cyclic groups, where a generator needs to be found in the first place.

In 1901, Henri Poincaré[40] proved that for every elliptic curve $E(\mathbb{Z}_p)$, the group of points on that curve (together with the point at infinity) and the addition operation as explained above form a cyclic group. ECC uses such a group and exploits the fact that a DLP can be defined in it. As captured in Definition 5.10, the resulting DLP is called the *elliptic curve discrete logarithm problem* (ECDLP) and it is structurally identical to the DLP from Definition 5.5.

40 Henri Poincaré was a French mathematician who lived from 1854 to 1912.

**Definition 5.10 (ECDLP)** *Let $E(\mathbb{F}_q)$ be an elliptic curve over $\mathbb{F}_q$, $P$ a point on $E(\mathbb{F}_q)$ of order $n = |E(\mathbb{F}_q)|$, and $Q$ another point on $E(\mathbb{F}_q)$. The ECDLP is to determine an $x \in \mathbb{Z}_n$ with $Q = xP$.*

The ECDLP thus asks for the number of times a point $P$ on an elliptic curve needs to be added to itself so that the total sum hits another point $Q$ on the curve. Compared to the DLP in a cyclic group $G$, $P$ plays the role of the generator $g$ and $Q$ refers to $h$ (with $h = g^x$).

As mentioned above, subexponential algorithms are not known to exist to solve the ECDLP. This has the positive side effect (from the cryptographer's viewpoint) that the resulting elliptic curve cryptosystems are equally secure with smaller key sizes than their conventional counterparts, meaning that the strength-per-key-bit is substantially bigger in ECC than it is in conventional DLP-based cryptosystems. Thus, smaller parameters can be used in ECC than with DLP-based systems for the same level of security. The advantages that can be gained from smaller parameters include faster computations in some cases[41] as well as shorter keys and certificates. These advantages are particularly important in environments where resources like computing power, storage space, bandwidth, and power consumption are constrained (e.g., smartcards). For example, to reach the security level of 2,048 (3,072) bits in a conventional public key cryptosystem like RSA, it is estimated that 224 (256) bits are sufficient in ECC [34]. This is more or less the order of magnitude people work with today.

Based on the intractability assumption of the ECDLP, Neal Koblitz [35] and Victor Miller [36] independently proposed elliptic curve cryptosystems in the mid-1980s. The cryptosystems are best viewed as elliptic curve versions of DLP-based cryptosystems, in which the group $\mathbb{Z}_p^*$ (or a subgroup thereof) is replaced by a group of points on an elliptic curve over a finite field. Consequently, there are elliptic curve variants of cryptosystems that only need the mathematical structure of a group, such as Diffie-Hellman, Elgamal, DSA, and many more.[42] In fact, almost all DLP-based cryptosystems have an elliptic curve variant (some of them are addressed in Part III of this book). As mentioned above, these cryptosystems have the advantage that they can be used with shorter keys. This is different when it comes to IFP-based cryptosystems like RSA or Rabin. The respective cryptosystems employ elliptic curves over a ring $\mathbb{Z}_n$ (instead of a finite field) and are mathematically more involved (e.g., [37–40]). They are mainly of academic interest, because they offer almost no practical advantage over RSA or Rabin. In fact, they rely on the same mathematical

41 Most importantly, 256-bit ECC is faster than 3072-bit RSA when computing with private keys. When computing with public keys, 3072-bit RSA is still faster than 256-bit ECC, especially when the standard RSA public exponent $e = 2^{16} + 1$ is used.

42 There is even an ECC version for the Diffie-Hellman integrated encryption scheme (DHIES) known as the elliptic curve integrated encryption scheme (ECIES).

problem, so an adversary being able to factorize $n$ can either break RSA and Rabin or any elliptic curve variant thereof.

The efficiency advantages of ECC are not free and come with some disadvantages. Most importantly, elliptic curve cryptosystems tend to be more involved and less well understood than their DLP-based counterparts. This means that care must be taken to implement the systems properly and to avoid pitfalls that are well known. For example, it has been shown that the ECDLP sometimes reduces to the DLP in an extension field, where the ICM can be used to compute discrete logarithms [41]. Because the reduction is only efficient for a special class of elliptic curves—so-called *supersingular curves*—such curves should be avoided in the first place. Luckily, there is a simple test to ensure that an elliptic curve is not supersingular. Some other vulnerabilities and potential attacks are known and discussed in the literature (e.g., [42–46]). A related disadvantage of ECC is that it may be possible to hide backdoors. This became particularly obvious when it was publicly disclosed that the *Dual Elliptic Curve Deterministic Random Bit Generator* (Dual_EC_DRBG) standardized in NIST SP 800-90A (2006) contained a backdoor.[43] The standard was withdrawn in 2014, but since then people have remained worried about the possibility of backdoors being placed in standardized and not sufficiently well understood groups based on elliptic curves.

A distinguishing feature of ECC is that every user may select a different elliptic curve $E(\mathbb{F}_q)$—even if all users employ the same finite field $\mathbb{F}_q$. From a security perspective, this flexibility has advantages (because it provides agility), but it also has disadvantages (because it makes interoperability difficult and—as mentioned above—it may raise concerns about backdoors). Furthermore, implementing an elliptic curve cryptosystem is neither simple nor straightforward, and there are usually many possibilities to do so, some of them even covered by patent claims.

Against this background, several standardization bodies have become active and are trying to specify various aspects of ECC and its use in the field. Most importantly, the *elliptic curve digital signature algorithm* (ECDSA) is the elliptic curve analog of the DSA that was originally proposed in 1992 by Scott Vanstone[44] in response to NIST's request for public comments on their first proposal for the DSA [47]. Starting in 1998, it was first accepted as an ISO/IEC standard; that is, ISO/IEC 14888-3 [48] that was later complemented by ISO/IEC 15946-1 [49], the ANSI (X9.62) [50], and the IEEE in their standard specifications for public-key cryptography (IEEE Std 1363-2000). Finally, NIST has incorporated the ECDSA in FIPS 186 (since version 2) that is now (in its fourth version) the primary reference

43 The possibility that the Dual_EC_DRBG may comprise a backdoor was first reported by Dan Shumov and Niels Ferguson at the rump session of the CRYPTO 2007 conference. It was later confirmed by the revelations of Edward Snowden.

44 Scott Vanstone was a Canadian mathematician and cryptographer who lived from 1947 to 2014.

for the ECDSA and the curves that can be used [51]. It recommends various elliptic groups over a prime field $\mathbb{F}_p$ (or $GF(p)$) or a characteristic two (or binary) finite field $\mathbb{F}_{2^m}$ (or $GF(2^m)$) for some $m \in N$ (standing for the degree of the field). The curves are summarized in Table 5.1, where the details can be found in [51]. In the left column, five curves over prime fields, denoted as P-XXX (where XXX stands for the bit length of the field size), are itemized that are frequently used in the field. This is particularly true for P-256 that is by far the most widely deployed elliptic curve. In the right column, five curves over binary fields, denoted as B-XXX, and five Koblitz curves—sometimes also called anomalous binary curves—denoted as K-XXX, are itemized. These curves are less frequently used in the field.

**Table 5.1**
Elliptic Curves Specified for ECDSA

| $GF(p)$ | $GF(2^m)$ |
|---|---|
| P-192 | K-163, B-163 |
| P-224 | K-233, B-233 |
| P-256 | K-283, B-283 |
| P-384 | K-409, B-409 |
| P-512 | K-571, B-571 |

Mainly due to the the Dual_EC_DRBG incident, people are worried about elliptic curves suggested by U.S. governmental bodies like NIST. This also applies to the curves promoted by the Standards for Efficient Cryptography Group (SECG[45]) that are in line with the NIST recommendations. The elliptic curves specified by the SECG (as of 2010) are summarized in Table 5.2. Most importantly, secp256k1 is the elliptic curve that is used for ECDSA signatures in Bitcoin.

In addition to the elliptic curves promoted by NIST and SECG, there are only a few alternatives. In 2005, for example, a German working group called ECC-Brainpool specified a suite of elliptic curves that are collectively referred to as the *Brainpool curves*. They are supported by the IETF [52] and used in many Internet security protocols. But their security is not without question,[46] and hence researchers have launched the *SafeCurves* project[47] to evaluate the strength of currently deployed elliptic curves and specify new curves in full transparency (also with regard to their design criteria). The most important curves that have come out

45 The SECG is an industry consortium founded in 1998 to develop commercial standards to facilitate the adoption of efficient cryptography and interoperability across a wide range of computing platforms.

46 https://bada55.cr.yp.to/bada55-20150927.pdf.

47 https://safecurves.cr.yp.to.

**Table 5.2**
Elliptic Curves Specified by SECG

| $GF(p)$ | $GF(2^m)$ |
|---|---|
| secp192k1, secp192r1 | sect163k1, sect163r1, sect163r2 |
| secp224k1, secp224r1 | sect233k1, sect233r1 |
| secp256k1, secp256r1 | sect239k1 |
| secp384r1 | sect283k1, sect283r1 |
| secp521r1 | sect409k1, sect409r1 |
| | sect571k1, sect571r1 |

of this project are *Curve25519* (*Ed25519*),[48] *Ed448-Goldilocks*, and *E-521*. These curves are increasingly used on the Internet, for example, in the realm of end-to-end encrypted (E2EE) messaging [53].

## 5.6 FINAL REMARKS

In this chapter, we elaborated on one-way functions and trapdoor functions. We also defined the notion of a family of such functions, and we overviewed and discussed some functions that are conjectured to be one way or trapdoor. More specifically, we looked at the discrete exponentiation function, the RSA function, and the modular square function. We further looked at hard-core predicates and algorithms for factoring integers and computing discrete logarithms. Having some basic knowledge about these algorithms is important to understand the state of the art in public key cryptography.

Most public key cryptosystems in use today are based on one (or several) of the conjectured one-way functions. This is also true for ECC that operates in cyclic groups in which known special-purpose algorithms to compute discrete logarithms do not work. From a practical viewpoint, ECC is interesting because it allows us to use smaller keys while maintaining the same level of security (compared to other public key cryptosystems). This is advantageous especially when it comes to implementing cryptographic systems and applications in environments that are somehow restricted, such as smartcards. For the purpose of this book, however, we don't make a major distinction between public key cryptosystems that are based on the DLP and systems that are based on the ECDLP. They both use the

48 Because Curve25519 cannot be used for digital signatures (including ECDSA) natively, people have come up with a DSS called Ed25519 (http://ed25519.cr.yp.to).

mathematical group structure, but the underlying operations between group elements are significantly different.

It is sometimes recommended to use cryptosystems that combine different types of one-way functions in one way or another. If one of these functions turns out not to be one-way, then the other functions may still prevail and keep on securing the cryptosystem. Obviously, this strategy becomes useless if all functions simultaneously turn out not to be one-way or a hardware device can be built that allows an adversary to efficiently invert them, such as a quantum computer. As this is not likely to be the case anytime soon, the strategy still remains reasonable.

## References


[1] Yao, A.C., "Theory and Application of Trapdoor Functions," *Proceedings of 23rd IEEE Symposium on Foundations of Computer Science*, IEEE Press, Chicago, 1982, pp. 80–91.

[2] Blum, M., and S. Micali, "How to Generate Cryptographically Strong Sequences of Pseudo-Random Bits," *SIAM Journal of Computing*, Vol. 13, No. 4, November 1984, pp. 850–863.

[3] Alexi, W., et al., "RSA and Rabin Functions: Certain Parts Are as Hard as the Whole," *SIAM Journal of Computing*, Vol. 17, No. 2, 1988, pp. 194–209.

[4] Goldwasser, S., and S. Micali, "Probabilistic Encryption," *Journal of Computer and System Sciences*, Vol. 28, No. 2, April 1984, pp. 270–299.

[5] Pohlig, S., and M.E. Hellman, "An Improved Algorithm for Computing Logarithms over GF(p)," *IEEE Transactions on Information Theory*, Vol. 24, No. 1, January 1978, pp. 106–110.

[6] Maurer, U.M., "Towards the Equivalence of Breaking the Diffie-Hellman Protocol and Computing Discrete Logarithms," *Proceedings of CRYPTO '94*, Springer-Verlag, LNCS 839, 1994, 271–281.

[7] Maurer, U.M., and S. Wolf, "The Diffie-Hellman Protocol," *Designs, Codes, and Cryptography*, Special Issue on Public Key Cryptography, Vol. 19, No. 2–3, 2000, pp. 147–171.

[8] Pollard, J.M., "Theorems of Factorization and Primality Testing," *Proceedings of the Cambridge Philosophical Society*, Vol. 76, No. 3, 1974, pp. 521–528.

[9] Williams, H.C., "A $p + 1$ Method of Factoring," *Mathematics of Computation*, Vol. 39, No. 159, July 1982, pp. 225–234.

[10] Lenstra, H.W., "Factoring Integers with Elliptic Curves," *Annals of Mathematics*, Vol. 126, 1987, pp. 649–673.

[11] Pollard, J.M., "A Monte Carlo Method for Factorization," *BIT Numerical Mathematics*, Vol. 15, No. 3, 1975, pp. 331–334.

[12] Brent, R.P., "An Improved Monte Carlo Factorization Algorithm," *BIT Numerical Mathematics*, Vol. 20, 1980, pp. 176–184.

[13] Lehmer, D.H., and R.E. Powers, "On Factoring Large Numbers," *Bulletin of the American Mathematical Society*, Vol. 37, 1931, pp. 770–776.

[14] Morrison, M.A., and J. Brillhart, "Method of Factoring and the Factorization of $\mathbb{F}_7$," *Mathematics of Computation*, Vol. 29, 1975, pp. 183–205.

[15] Dixon, J.D., "Asymptotically Fast Factorization of Integers," *Mathematics of Computation*, Vol. 36, No. 153, 1981, pp. 255–260.

[16] Pomerance, C., "The Quadratic Sieve Factoring Algorithm," *Proceedings of EUROCRYPT '84*, Springer-Verlag, 1984, pp. 169–182.

[17] Lenstra, A.K., and H.W. Lenstra, *The Development of the Number Field Sieve*. Springer-Verlag, LNCS 1554, New York, 1993.

[18] Lenstra, H.W., et al., "The Factorization of the ninth Fermat Number," *Mathematics of Computation*, Vol. 61, 1993, pp. 319–349.

[19] Shamir, A., "Factoring Large Numbers with the TWINKLE Device," *Proceedings of Cryptographic Hardware and Embedded Systems (CHES '99)*, Springer-Verlag, LNCS 1717, 1999, pp. 2–12.

[20] Lenstra, A.K., and A. Shamir, "Analysis and Optimization of the TWINKLE Factoring Device," *Proceedings of EUROCRYPT 2000*, Springer-Verlag, LNCS 1807, 2000, pp. 35–52.

[21] Shamir, A., and E. Tromer, "Factoring Large Numbers with the TWIRL Device," *Proceedings of CRYPTO 2003*, Springer-Verlag, LNCS 2729, 2003, pp. 1–26.

[22] Franke, J., et al., "SHARK: A Realizable Special Hardware Sieving Device for Factoring 1024-Bit Integers," *Proceedings of Cryptographic Hardware and Embedded Systems (CHES 2005)*, Springer-Verlag, LNCS 3659, 2005, pp. 119–130.

[23] Geiselmann, W., and R. Steinwandt, "Yet Another Sieving Device," *Proceedings of CT-RSA 2004*, Springer-Verlag, LNCS 2964, 2004, pp. 278–291.

[24] Gardner, M., "A New Kind of Cipher That Would Take Millions of Years to Break," *Scientific American*, Vol. 237, pp. 120–124.

[25] Atkins, D., "The Magic Words Are Squeamish Ossifrage," *Proceedings of ASIACRYPT '94*, Springer-Verlag, LNCS 917, 1995, pp. 263–277.

[26] Shor, P.W., "Algorithms for Quantum Computation: Discrete Logarithms and Factoring," *Proceedings of the IEEE 35th Annual Symposium on Foundations of Computer Science (FOCS)*, Santa Fe, NM, November 1994, pp. 124–134.

[27] Shor, P.W., "Polynomial-Time Algorithms for Prime Factorization and Discrete Logarithms on a Quantum Computer," *SIAM Journal of Computing*, October 1997, pp. 1484–1509.

[28] Shoup, V., "Lower Bounds for Discrete Logarithms and Related Problems," *Proceedings of EUROCRYPT '97*, Springer-Verlag, LNCS 1233, 1997, pp. 256–266.

[29] Pollard, J.M., "Monte Carlo Methods for Index Computation (mod p)," *Mathematics of Computation*, Vol. 32, No. 143, 1978, pp. 918–924.

[30] van Oorschot, P.C., and M.J. Wiener, "Parallel Collision Search with Cryptanalytic Applications," *Journal of Cryptology*, Vol. 12, 1999, pp. 1–28.

[31] Joux, A., "A New Index Calculus Algorithm with Complexity $L(1/4 + o(1))$ in Very Small Characteristic," Cryptology ePrint Archive, Report 2013/095, 2013.

[32] Barbulescu, R., et al., "A Quasi-Polynomial Algorithm for Discrete Logarithm in Finite Fields of Small Characteristic," Cryptology ePrint Archive, Report 2013/400, 2013.

[33] Lenstra, A.K., "The XTR Public Key System," *Proceedings of CRYPTO 2000*, Springer-Verlag, LNCS 1880, 2000, pp. 1–19.

[34] Lenstra, A.K., and E.R. Verheul, "Selecting Cryptographic Key Sizes," *Journal of Cryptology*, Vol. 14, No. 4, 2001, pp. 255–293.

[35] Koblitz, N., "Elliptic Curve Cryptosystems," *Mathematics of Computation*, Vol. 48, No. 177, 1987, pp. 203–209.

[36] Miller, V., "Use of Elliptic Curves in Cryptography," *Proceedings of CRYPTO '85*, LNCS 218, Springer-Verlag, 1986, pp. 417–426.

[37] Koyama, K., et al., "New Public-Key Schemes Based on Elliptic Curves over the Ring $\mathbb{Z}_n$," *Proceedings of CRYPTO '91*, LNCS 576, Springer-Verlag, 1992, pp. 252–266.

[38] Demytko, N., "A New Elliptic Curve Based Analogue of RSA," *Proceedings of EUROCRYPT '93*, LNCS 765, Springer-Verlag, 1994, pp. 40–49.

[39] Koyama, K., "Fast RSA-Type Schemes Based on Singular Cubic Curves $y^2 + axy = x^3 \pmod{n}$," *Proceedings of EUROCRYPT '95*, LNCS 921, Springer-Verlag, 1995, pp. 329–340.

[40] Kuwakado, H., and K. Koyama, "Security of RSA-Type Cryptosystems Over Elliptic Curves Against Håstad Attack," *Electronic Letters*, Vol. 30, No. 22, October 1994, pp. 1309–1318.

[41] Menezes, A., T. Okamoto, and S.A. Vanstone, "Reducing Elliptic Curve Logarithms to Logarithms in a Finite Field," *IEEE Transactions on Information Theory*, Vol. 39, 1993, pp. 1639–1646.

[42] Koblitz, N.I., *A Course in Number Theory and Cryptography*, 2nd edition. Springer-Verlag, New York, 1994.

[43] Blake, I., G. Seroussi, and N. Smart, *Elliptic Curves in Cryptography*, Cambridge University Press, Cambridge, U.K., 2000.

[44] Hankerson, D., A. Menezes, and S.A. Vanstone, *Guide to Elliptic Curve Cryptography.* Springer-Verlag, New York, 2004.

[45] Washington, L.C., *Elliptic Curves: Number Theory and Cryptography*, 2nd edition. Chapman & Hall/CRC, Boca Raton, FL, 2008.

[46] Shemanske, T.R., *Modern Cryptography and Elliptic Curves: A Beginner's Guide*, American Mathematical Society, 2017.

[47] Vanstone, S., "Responses to NIST's Proposal," *Communications of the ACM*, Vol. 35, No. 7, July 1992, pp. 50–52.

[48] ISO/IEC 14888-3, *IT Security techniques—Digital signatures with appendix—Part 3: Discrete logarithm based mechanisms*, 2018.

[49] ISO/IEC 15946-1, *Information technology—Security techniques—Cryptographic techniques based on elliptic curves—Part 1: General*, 2016.

[50] ANSI X9.62, *Public Key Cryptography for the Financial Services Industry: the Elliptic Curve Digital Signature Algorithm (ECDSA)*, 2005.

[51] U.S. NIST FIPS 186-4, *Digital Signature Standard (DSS)*, July 2013.

[52] Lochter, M., and J. Merkle, *Elliptic Curve Cryptography (ECC) Brainpool Standard Curves and Curve Generation*, RFC 5639, March 2010.

[53] Oppliger, R., *End-to-End Encrypted Messaging*. Artech House Publishers, Norwood, MA, 2020.

# Chapter 6

# Cryptographic Hash Functions

In this chapter, we elaborate on cryptographic hash functions. More specifically, we introduce the topic in Section 6.1, address a basic construction—the Merkle-Damgård construction—in Section 6.2, elaborate on the historical perspective and the development of the cryptographic hash functions used in the field in Section 6.3, overview in detail some exemplary hash functions in Section 6.4, and conclude with some final remarks in Section 6.5.

## 6.1 INTRODUCTION

As mentioned in Section 2.1.4 and captured in Defintion 2.5, a *hash function* is an efficiently computable function $h : \Sigma_{in}^* \rightarrow \Sigma_{out}^n$ that takes an arbitrarily long[1] input word $x \in \Sigma_{in}^*$ (with $\Sigma_{in}$ representing the input alphabet and $\Sigma_{in}^*$ the domain of $h$) and generates an output word $y \in \Sigma_{out}^n$ (with $\Sigma_{out}$ representing the output alphabet and $\Sigma_{out}^n$ the range of $h$) of fixed size $n$. According to Definition 2.6, such a hash function is *cryptographic* if has some of the following three properties.

- A hash function $h$ is *preimage resistant* or *one-way* if it is computationally infeasible to find an input word $x \in \Sigma_{in}^*$ with $h(x) = y$ for an output word $y$ that is sampled uniformly at random from $\Sigma_{out}^n$, i.e., $y \in_R \Sigma_{out}^n$.
- A hash function $h$ is *second-preimage resistant* or *weak collision resistant* if it is computationally infeasible to find a second input word $x' \in \Sigma_{in}^*$ with $x' \neq x$ and $h(x') = h(x)$ for an input word $x$ that is sampled uniformly at random from $\Sigma_{in}^*$ (i.e., $x \in_R \Sigma_{in}^*$).

1 Remember from Section 2.1.4 that for technical reasons one usually has to assume a maximum length $n_{max}$ for input words. In this case, the hash function is formally expressed as $h : \Sigma_{in}^{n_{max}} \rightarrow \Sigma_{out}^n$.

- A hash function $h$ is *collision-resistant* or *strong collision resistant* if it is computationally infeasible to find two input words $x, x' \in \Sigma_{in}^*$ with $x' \neq x$ and $h(x') = h(x)$.

Again referring to Definition 2.6, a cryptographic hash function must be one-way and either second-preimage resistant or collision-resistant. There are a few additional comments to make at this point:

- In some literature, collision-resistant hash functions are called *collision free*. This term is wrong, because—due to the pigeonhole principle—collisions must always occur if one uses a hash function that compresses arbitrarily long input words to output words of a fixed (and shorter) length.
- In a complexity-theoretic setting, one cannot say that finding a collision for a given hash function is a difficult problem. In fact, finding a collision (for a given hash function) is only a problem instance (refer to Appendix D.2 for a discussion about the difference between a problem and a problem instance). This is because there is always an efficient algorithm that finds a collision, namely one that simply outputs two words that hash to the same value. Thus, the concept of collision resistance only makes sense if one considers a sufficiently large family (or class) of hash functions from which one is chosen at random.[2] An algorithm to find collisions must then work for all hash functions of the family, including the one that is chosen at random. This also means that complexity theory is an inappropriate tool to argue about the collision resistance of a particular hash function (e.g., SHA-1).
- A collision-resistant hash function is always second-preimage resistant because it is then computationally infeasible to find any collision, including one for a particular preimage. The converse, however, is not true; that is, a second-preimage resistant hash function need not be collision resistant (this is why the terms *weak collision resistant* and *strong collision resistant* are sometimes used in the first place). Consequently, collision resistance implies second-preimage resistance, but not vice versa.
- A (strong or weak) collision-resistant hash function need not be preimage resistant. Consider the following pathological example to illustrate this point:[3] If $g$ is a collision resistant hash function that generates an $n$-bit output, then

2 This line of argumentation is similar to the one that has led us to talk about families of one-way functions (instead of one-way functions) in the previous chapter.

3 The example was created by Ueli M. Maurer.

one can define a $(n+1)$-bit hash function $h$ as follows:

$$h(x) = \begin{cases} 1 \mid\mid x & \text{if} \mid x \mid = n \\ 0 \mid\mid g(x) & \text{otherwise} \end{cases}$$

The hash function $h$ is still collision-resistant: If $h(x)$ begins with 1, then there is no collision, and if $h(x)$ begins with 0, then finding a collision means finding a collision for $g$ (which is assumed to be computationally intractable due to the collision resistance property of $g$). But $h$ is not preimage-resistant. For all $h(x)$ that begin with a one, it is trivial to find a preimage (just drop the one) and to invert $h$ accordingly. Consequently, $h$ is a hash function that is collision-resistant but not preimage-resistant. The bottom line is that preimage resistance and collision resistance are inherently different properties that must be distinguished accordingly.

- Last but not least, we note that the notion of a collision can be generalized to a multicollision where more than two input words are hashed to the same value. More specifically, an $r$-collision is an $r$-tuple of input values $(x_1, \ldots, x_r)$, such that $h(x_1) = \ldots = h(x_r)$. For $r = 2$, an $r$-collision refers to the standard notion of a collision. So the more interesting cases occur for $r > 2$. We already mentioned here that finding multicollisions is not substantially more difficult than finding "normal" collisions, but we more thoroughly address the issue and its implications at the end of the chapter.

In practice, $\Sigma_{in}$ and $\Sigma_{out}$ are often set to $\{0,1\}$, and hence a respective hash function can be seen as a well-defined mapping from $\{0,1\}^*$ to $\{0,1\}^n$, where $n$ refers to the output length of the hash function.

A practically relevant question is how large the parameter $n$ should be. If it is large, then the cryptographic hash function is not so efficient (because the hash values require a lot of resources to process, store, and transmit). On the other hand, it cannot be too short either, because otherwise finding collisions becomes simple. So there is a trade-off to make: $n$ should be as short as possible, but as long as needed.

Against this background, a lower bound for $n$ is usually obtained by the *birthday attack*. This attack is based on the *birthday paradox* that is well known in probability theory. It basically says that the probability of two persons in a group sharing the same birthday is greater than $1/2$, if the group (chosen at random) has more than 23 members. This number is surprisingly low, and this is why it is called a paradox. To obtain the result, one employs a sample space $\Sigma$ that consists of all $n$-tuples over the 365 days of the year (i.e., $\mid \Sigma \mid = 365^n$). Let Pr$[\mathcal{A}]$ be the probability that at least two out of $n$ persons have the same birthday, meaning that a birthday collision occurs. This value is difficult to compute directly. It is much simpler to compute $\Pr[\overline{\mathcal{A}}]$, which is the probability that all $n$ persons have distinct (i.e., different) birthdays, meaning that no birthday collision occurs, and to derive $\Pr[\mathcal{A}]$ from there. Following this line of argumentation, $\Pr[\mathcal{A}]$ can be computed for $0 \leq n \leq 365$ as follows:

$$\begin{aligned}
\Pr[\mathcal{A}] &= 1 - \Pr[\overline{\mathcal{A}}] \\
&= 1 - \frac{|\overline{\mathcal{A}}|}{|\Sigma|} \\
&= 1 - \frac{365}{365} \cdot \frac{364}{365} \cdot \ldots \cdot \frac{365-n+1}{365} \\
&= 1 - 365 \cdot 364 \cdot \ldots \cdot (365-n+1) \cdot \frac{1}{365^n} \\
&= 1 - \frac{365!}{(365-n)!} \cdot \frac{1}{365^n} \\
&= 1 - \frac{365!}{(365-n)!365^n}
\end{aligned}$$

On line 3, the first person has 365 possible days for which he or she does not produce a collision, the second person has $365 - 1 = 364$ possible days for which he or she does not produce a collision, and so on, until the n-th person has $365-n+1$ possible days for which he or she does not produce a collision. The overall probability $\Pr[\overline{\mathcal{A}}]$ is therefore

$$\frac{365}{365} \cdot \frac{364}{365} \cdot \ldots \cdot \frac{365-n+1}{365}$$

and this value can be used in the formula given above. Obviously (and again due to to the pigeonhole principle), $\Pr[\mathcal{A}]$ is equal to 1 for $n > 365$. In this case, it is no longer possible that all $n$ persons have different birthdays.

The surprising fact is that $\Pr[\mathcal{A}]$ grows very rapidly, and that $n$ must only be 23 to reach a probability greater or equal than $1/2$.[4] More specifically, if $n = 23$, then

$$\begin{aligned}
\Pr[\mathcal{A}] &= 1 - \frac{365!}{(365-23)!365^{23}} \\
&= 1 - \frac{365!}{(342)!365^{23}} \\
&= 1 - \frac{365 \cdot 364 \cdots 343}{365^{23}} \approx 0.508.
\end{aligned}$$

4 For $n = 40$, the probability is already about 0.9 or 90%.

In contrast, if we fix the date and ask for the number of persons that are required to make the probability that at least one person has exactly this date as his or her birthday, then $n$ must be much larger. In this case, $\Pr[\mathcal{A}]$ refers to the probability that at least one of $n$ persons has the given date as his or her birthday, and $\Pr[\overline{\mathcal{A}}]$ refers to the opposite (i.e., the probability that no person has his or her birthday on that particular date). We can compute

$$\begin{aligned}\Pr[\mathcal{A}] &= 1 - \Pr[\overline{\mathcal{A}}] \\ &= 1 - \left(\frac{364}{365}\right)^n\end{aligned}$$

because each person has a probability of $364/365$ to have a birthday that is distinct from the given date. If we want $\Pr[\mathcal{A}]$ to be larger than $1/2$, then we must solve

$$1 - \left(\frac{364}{365}\right)^n \geq \frac{1}{2}$$

for $n$. This means that

$$\left(\frac{364}{365}\right)^n \geq \frac{1}{2}$$

and hence

$$n \cdot \log\left(\frac{364}{365}\right) \geq \log\left(\frac{1}{2}\right)$$

This, in turn, means that

$$n \geq \frac{\log\left(\frac{1}{2}\right)}{\log\left(\frac{364}{365}\right)} \approx 251$$

and hence $n$ must be equal or larger than about 251 (the exact value largely depends on the accuracy of computing the logarithm function). Compare this value to 23 that has been the answer to the previous question. The difference exceeds intuition and is therefore called a paradox.

Applying the line of argumentation to hash functions means that finding two persons with the same birthday reveals a collision (i.e., a collision in the strong sense), whereas finding a person with a given birthday reveals a second preimage (i.e., a collision in the weak sense). Hence, due to the birthday paradox, one can argue that collision resistance is inherently more difficult to achieve than second-preimage resistance, and this, in turn, suggests that collision resistance is a stronger property than second-preimage resistance.

Mathematically speaking, finding collisions for a hash function $h$ is the same problem as finding birthday collisions. For a hash function that produces $n$-bit output values, there are not 365 possible values but $2^n$. The question is how many messages $x_1, x_2, \ldots, x_t$ one has to hash until one has a reasonable probability that $h(x_i) = h(x_j)$ for two distinct messages $x_i \neq x_j$. Following the line of argumentation given above, one can compute the probability that no collision occurs—let's again call it $\Pr[\overline{\mathcal{A}}]$—as follows:

$$\begin{aligned}
\Pr[\overline{\mathcal{A}}] &= \left(\frac{2^n - 0}{2^n}\right)\left(\frac{2^n - 1}{2^n}\right)\left(\frac{2^n - 2}{2^n}\right)\cdots\left(\frac{2^n - (t-1)}{2^n}\right) \\
&= \prod_{i=0}^{t-1} \frac{2^n - i}{2^n} \\
&= \prod_{i=0}^{t-1} \left(1 - \frac{i}{2^n}\right)
\end{aligned}$$

Remember that the Taylor series expansions of the exponential function is defined as follows:

$$\begin{aligned}
e^x &= \sum_{i=0}^{\infty} \frac{x^i}{i!} \\
&= \frac{x^0}{0!} + \frac{x^1}{1!} + \frac{x^2}{2!} + \frac{x^3}{3!} + \ldots \\
&= 1 + x + \frac{x^2}{2} + \frac{x^3}{6} + \ldots
\end{aligned}$$

This series converges very rapidly for very small values $x \ll 1$, and hence only the first two summands really matter. Therefore, the following approximation can be used:

$$e^{-x} \approx 1 - x$$

Because $1/2^n \ll 1$, we can use this approximation to rewrite $\Pr[\overline{\mathcal{A}}]$ as follows:

$$
\begin{aligned}
\Pr[\overline{\mathcal{A}}] &= \prod_{i=0}^{t-1}\left(1-\frac{i}{2^n}\right) \\
&\approx \prod_{i=0}^{t-1} e^{-\frac{i}{2^n}} \\
&= e^{-\frac{0}{2^n}} \cdot e^{-\frac{1}{2^n}} \cdot e^{-\frac{2}{2^n}} \cdot e^{-\frac{3}{2^n}} \cdot \ldots \cdot e^{-\frac{(t-1)}{2^n}} \\
&= e^{-\frac{0+1+2+3+\ldots+(t-1)}{2^n}} \\
&= e^{-\frac{t(t-1)}{2\cdot 2^n}} \\
&= e^{-\frac{t(t-1)}{2^{n+1}}}
\end{aligned}
$$

Recall that $\Pr[\overline{\mathcal{A}}]$ refers to the probability that no collision occurs, but that we are interested in the complementary probability $\Pr[\mathcal{A}] = 1 - \Pr[\overline{\mathcal{A}}]$; that is, the probability that a collision is found in $t$ messages $x_1, x_2, \ldots, x_t$. More specifically, we wonder how many messages need to be hashed until a collision occurs with probability $\Pr[\mathcal{A}]$. This, in turn, means that we have to express $t$ in relation to $\Pr[\mathcal{A}]$. We start with the approximation

$$\Pr[\mathcal{A}] \approx 1 - e^{-\frac{t(t-1)}{2^{n+1}}}$$

and solve it for $t$. We therefore rewrite the approximation as

$$1 - \Pr[\mathcal{A}] \approx e^{-\frac{t(t-1)}{2^{n+1}}}$$

and compute the logarithm on either side:

$$\ln(1 - \Pr[\mathcal{A}]) \approx -\frac{t(t-1)}{2^{n+1}} \cdot \ln e = -\frac{t(t-1)}{2^{n+1}}$$

The right equation holds because $\ln e = 1$. The approximation can be written as

$$-t(t-1) \approx 2^{n+1} \cdot \ln(1 - \Pr[\mathcal{A}])$$

and

$$t(t-1) \approx 2^{n+1} \cdot (-\ln(1 - \Pr[\mathcal{A}]))$$

Because $-\ln x = \ln(1/x)$, this can be written as

$$t(t-1) \approx 2^{n+1} \cdot \ln\left(\frac{1}{1 - \Pr[\mathcal{A}]}\right)$$

For $t \gg 1$, it holds that $t^2 \approx t(t-1)$. This means that we can approximate $t$ as follows:

$$t \approx \sqrt{2^{n+1} \cdot \ln\left(\frac{1}{1-\Pr[\mathcal{A}]}\right)} = 2^{(n+1)/2}\sqrt{\ln\left(\frac{1}{1-\Pr[\mathcal{A}]}\right)}$$

This approximation stands for the relationship between the number of hashed messages $t$ needed for a collision as compared to the output length $n$ of the hash function and the collision probability $\Pr[\mathcal{A}]$. The most important consequence is that the number of messages needed to find a collision is roughly equal to the square root of the number of possible output values (i.e., about $\sqrt{2^n} = 2^{n/2}$), and this is why the birthday attack is sometimes also called the *square root attack*. If, for example, we want to find a collision for a hash function with an 80-bit output and success probability 0.5 (50%), then we have to hash about

$$t \approx 2^{(81)/2}\sqrt{\ln\left(\frac{1}{1-0.5}\right)} \approx 2^{40.2}$$

messages. This is perfectly feasible today, and hence we have to use hash functions that generate significantly longer output values. The smallest length in use today is 128 bits. Due to the birthday attack, the respective security level is equivalent to 64 bits, and hence people currently prefer hash functions that generate longer output values. In fact, it is commonly agreed that 160 bits is the minimum output length for a cryptographic hash function in use today, and people prefer even longer output values (as we will see later in this chapter).[5]

In addition to preimage, second-preimage, and collision resistance, there are a few other properties of hash functions sometimes mentioned (and discussed) in the literature. We don't use them in this book, but we still want to mention them for the sake of completeness.

- A hash function $h$ is *noncorrelated* if its input bits and output bits are not correlated in one way or another. Needless to say, this property must be fulfilled by all cryptographic hash functions used in the field.

5 One reason why people may prefer longer hash values is the impact a quantum computer may have. Using such a computer, the security level of an $n$-bit hash function decreases to the cube root of $n$ (instead of the square root of $n$). This means that a cryptographic hash function that generates 256-bit hash values has a security level of $\sqrt{2^{256}} = 2^{256/2} = 2^{128}$ against conventional computers, but only $\sqrt[3]{2^{256}} = 2^{256/3} \approx 2^{85}$ against quantum computers. This clearly speaks in favor of using longer hash values.

- A hash function $h$ is *generalized collision resistant* if it is computationally infeasible to find two input values $x$ and $x'$ with $x' \neq x$ such that $h(x)$ and $h(x')$ are similar in some specific way (e.g., they are equal in some bits).
- A hash function $h$ is *weakened collision resistant* if it is computationally infeasible to find two input values $x$ and $x'$ with $x' \neq x$ and $h(x) = h(x')$ such that $x$ and $x'$ are similar in some specific way (e.g., they are equal in some bits). Hence, the notion of weakened collision resistance is conceptually similar to the notion of generalized collision resistance, but the similarity applies to the preimages.

According to Definition 2.6, there are cryptographic hash functions that are one-way and second-preimage resistant (type I) and somewhat stronger functions that are one-way and collision-resistant (type II). In many applications, the use of cryptographic hash functions of type I would be sufficient, but people still prefer to use cryptographic hash functions of type II, mainly to achieve a security margin. This is reasonable from a security perspective, but it may lead to a deprecation of a function that would still work perfectly fine in a particular context.

In general, there are many ways to construct a cryptographic hash function. Referring to ISO/IEC 10118-1 [1], there are hash functions that employ block ciphers (ISO/IEC 10118-2 [2]), dedicated hash functions (ISO/IEC 10118-3 [3]), and hash functions based on modular arithmetic (ISO/IEC 10118-4 [4]). Mainly due to their performance advantages, dedicated hash functions are usually the preferred choice.

## 6.2 MERKLE-DAMGÅRD CONSTRUCTION

Most dedicated cryptographic hash functions in use today follow a construction that was independently proposed by Ralph C. Merkle and Ivan B. Damgård in the late 1980s [5, 6].[6] According to this construction, an *iterated hash function* $h$ is computed by repeated application of a collision-resistant compression function $f : \Sigma^{b+l} \longrightarrow \Sigma^l$ with $b, l \in \mathbb{N}$ to successive blocks $x_1, \ldots, x_n$ of a message $x$.[7] As illustrated in Figure 6.1, the compression function $f$ takes two input values:

1. A $b$-bit message block $x_i$ for $i = 1, \ldots, n$;
2. An $l$-bit chaining value $H_i$ for $i = 1, \ldots, n$.

6 Both papers were presented at CRYPTO '89.

7 Note that the input alphabet $\Sigma_{in}$ and the output alphabet $\Sigma_{out}$ are assumed to be the same here (denoted as $\Sigma$).

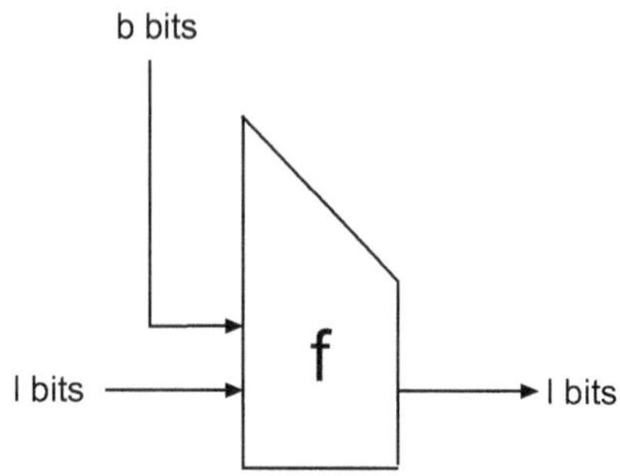


**Figure 6.1** A compression function $f$.

In a typical setting, $l$ is 160 or 256 bits[8] and $b$ is 512 bits. The output of the compression function can be used as a new $l$-bit chaining value, which is input to the next iteration of the compression function. This is continued until all message blocks are exhausted.

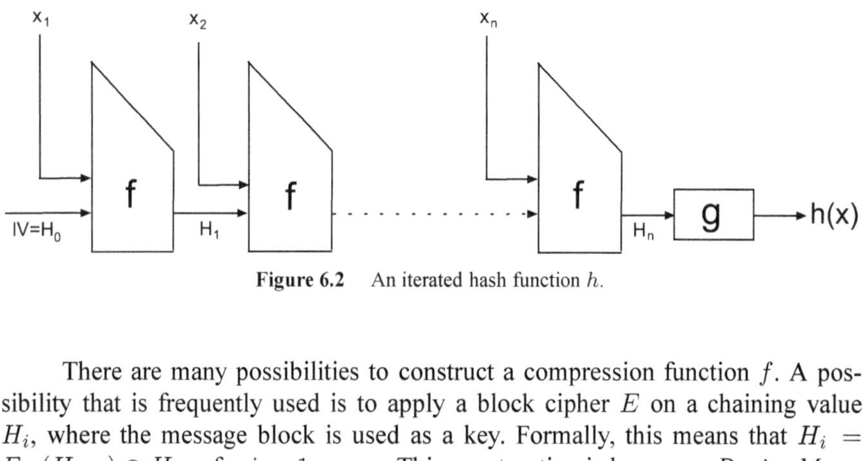


**Figure 6.2** An iterated hash function $h$.

There are many possibilities to construct a compression function $f$. A possibility that is frequently used is to apply a block cipher $E$ on a chaining value $H_i$, where the message block is used as a key. Formally, this means that $H_i = E_{x_i}(H_{i-1}) \oplus H_{i-1}$ for $i = 1, \ldots, n$. This construction is known as *Davies-Meyer compression function*, and it is used in most cryptographic hash functions outlined in Section 6.4.

Using an arbitrary compression function, an iterated hash function $h$ can be constructed as illustrated in Figure 6.2. In this figure, $f$ represents the compression

8 In the past, $l$ used to be 128 bits. However, all cryptographic hash functions that use this bitlength for $l$ have become susceptible to some birthday attacks. Most of these attacks are not particularly devastating, but their mere existence suggests that one should move to longer values for $l$, such as 160 or 256 bits.

function and $g$ represents an output function (that can also be the identity function). The message $x$ is padded to a multiple of $b$ bits and divided into a sequence of $n$ $b$-bit blocks $x_1, \ldots, x_n$. The compression function $f$ is then repeatedly applied, starting with an initial value ($IV = H_0$) and the first message block $x_1$, and continuing with each new chaining value $H_i$ and successive message block $x_{i+1}$ for $i = 1, \ldots, n$. After the last message block $x_n$ is processed, the final chaining value $H_n$ is subject to the output function $g$,[9] and the output of this function yields the output of the hash function $h$ for $x$; that is, $h(x)$. The bottom line is that an iterative hash function $h$ for $x = x_1x_2 \ldots x_n$ can be recursively defined as follows:

$$\begin{aligned} H_0 &= IV \\ H_i &= f(H_{i-1}, x_i) \text{ for } i = 1, \ldots, n \\ h(x) &= g(H_n) \end{aligned}$$

As mentioned above, the message to be hashed must be padded to a multiple of $b$ bits. One possibility is to pad $x$ with zeros. Padding with zeros, however, may also introduce some ambiguity about $x$. For example, the message 101110 padded to 8 bits would be 10111000 and it is then unclear how many trailing zeros were present in the original message. Several methods are available to resolve this problem. Merkle proposed to append the bitlength of message $x$ at the end of $x$. Following this proposal, the padding method of choice in currently deployed hash functions is to append a one, a variable number of zeros, and the binary encoding of the length of the original message to the message. We revisit this simple padding scheme when we discuss the hash function MD4 in the subsequent section.

Merkle and Damgård showed that finding a collision for $h$ in their construction; that is, finding two input values $x$ and $x'$ with $x \neq x'$ and $h(x) = h(x')$, is at least as hard as finding a collision for the underlying compression function $f$. This means that if $f$ is a collision-resistant compression function and $h$ is an iterated hash function that employs $f$ in the proposed way, then $h$ is a cryptographic hash function that is also collision-resistant. Put in other words: The iterated hash function inherits the collision resistance property from the underlying compression function. This can be turned into Theorem 6.1.

**Theorem 6.1** *If the compression function $f$ is collision-resistant, then the iterated hash function $h$ that is built according to the Merkle-Damgård construction is also collision-resistant.*

*Proof.* Suppose we can construct a collision for $h$; that is, $h(x) = h(x')$ for two different $x = x_1x_2 \ldots x_n$ and $x' = x'_1x'_2 \ldots x'_{n'}$. It then follows that $H_n = H_{n'}$,

9 In some literature, the output function is also called the *finalization function*.

and hence $f(H_{n-1}, x_n) = f(H_{n'-1}, x'_{n'})$. Either we have found a collision for the compression function $f$ or the inputs for $f$ are the same; that is, $(H_{n-1}, x_n) = (H_{n'-1}, x'_{n'})$. This means that $H_{n-1}$ and $H_{n'-1}$ are equal, and $x_n$ and $x'_{n'}$ are equal. Since $x_n$ and $x'_{n'}$ refer to the last block of the respective messages, they also include the lengths. It thus follows that the lengths are equal, as well; that is, $n = n'$. Combining $H_{n-1} = H_{n'-1}$ with $H_{n-1} = f(H_{n-2}, x_{n-1})$ and $H_{n'-1} = f(H_{n'-2}, x'_{n'-1})$, we can recursively continue the proof. Because the messages $x$ and $x'$ are different, $x_i$ and $x'_i$ must be different at some point $i$, and this, in turn, means that we get a collision for $f$. This contradicts our assumption and completes the proof.

□

In the literature, there are many collision-resistant compression functions that can be turned into collision-resistant cryptographic hash functions with the Merkle-Damgård construction. Some examples are given in this chapter. More recently, researchers have come up with alternative designs for cryptographic hash functions that are also collision-resistant. Most importantly, SHA-3 no longer follows the Merkle-Damgård, which is outlined toward the end of the chapter.

## 6.3 HISTORICAL PERSPECTIVE

The driving force behind the development of cryptographic hash functions was public key cryptography in general and digital signatures in particular. Consequently, the former company RSA Security (Section 5.3.3) played a crucial role in the development and deployment of many practically relevant cryptographic hash functions. The first such function developed by RSA Security was acronymed MD—standing for *message digest*. It was proprietary and never published. MD2 specified in RFC 1319 [7] was the first published cryptographic hash function that was used in the field.[10] When Merkle proposed a cryptographic hash function called SNEFRU that was several times faster than MD2,[11] RSA Security responded with MD4[12] specified in RFC 1320 [8] and addressed in Section 6.4.1. MD4 took advantage of the fact that newer processors could do 32-bit operations, and it was therefore able to run

10 MD2 was, for example, used in the secure messaging products of RSA Security.

11 The function was proposed in 1990 in a Xerox PARC technical report entitled *A Software One Way Function.*

12 There was an MD3 cryptographic hash function, but it was superseded by MD4 before it was ever published or used.

faster than SNEFRU. In 1991, SNEFRU and some other cryptographic hash functions were successfully attacked[13] using differential cryptanalysis [9]. Furthermore, some weaknesses were found in a version of MD4 with only two rounds instead of three [10]. This did not break MD4, but it made RSA Security sufficiently nervous that it was decided to strengthen MD4. The result was MD5 specified in RFC 1321 [11] and addressed in Section 6.4.2. MD5 is more secure than MD4, but it is also a little bit slower.

During the 1990s, a series of results showed that MD4 was insecure [12] and MD5 was partially broken [13, 14].[14] The situation changed fundamentally when a group of Chinese researchers led by Xiaoyun Wang published collisions for MD4, MD5, and a few other cryptographic hash functions in 2004 and 2005 [15].[15] The collisions referred to pairs of two-block (i.e., 1,024-bit) messages and two such message pairs were presented. Since this presentation, many colliding message pairs have been found and researchers have been able to improve the collision attacks against MD5 or apply them in more realistic settings. In 2008, for example, a group of researchers was able to create a rogue CA certificate due to an MD5 collision.[16] The bottom line is that MD4, MD5, and the other cryptographic hash functions attacked by Wang et al. should no longer be used. Nevertheless, MD4 and MD5 may still serve as study objects for the design principles of iterated hash functions. As such, they are also addressed in this book.

In 1993, the U.S. NIST proposed the *Secure Hash Algorithm* (SHA), which is similar to MD5, but more strengthened and also a little bit slower. Probably after discovering a never-published weakness in the original SHA proposal,[17] the U.S. NIST revised it and called the new version SHA-1. As such, SHA-1 is specified in

13 The attack was considered successful because it was shown how to systematically find a collision (i.e., two messages with the same hash value).

14 In 1993, Bert den Boer and Antoon Bosselaers found collisions for the compression function of MD5 [13]. In fact, they found pairs of different message blocks and chaining values that compress to the same value. In 1996, Hans Dobbertin improved this result by finding collisions for different message blocks that employ the same chaining value [14]. Because this chaining value was still different from the IV employed by MD5, this result couldn't be turned into a real attack against the collision resistance property of MD5.

15 The original paper is available at http://eprint.iacr.org/2004/199.pdf. It was submitted for the CRYPTO '04 Conference. Due to a translation error in the Chinese version of Bruce Schneier's book entitled *Applied Cryptography*, however, the paper had a subtle flaw and was rejected for the conference. Nevertheless, Wang still attended the conference and presented a corrected version of her results during the rump session. A corrected version of the paper was later submitted to EUROCRYPT '05 and published in the respective conference proceedings [15].

16 http://www.win.tue.nl/hashclash/rogue-ca.

17 At CRYPTO '98, Florent Chabaud and Antoine Joux published a weakness of SHA-0 [16]. This weakness was fixed by SHA-1, so it is reasonable to assume that they found the original weakness.

the Federal Information Processing Standards Publication (FIPS PUB) 180,[18] also known as *Secure Hash Standard* (SHS). FIPS PUB 180 was first released in 1995.

The second revision of FIPS PUB 180, FIPS PUB 180-2, was released in August 2002 and became effective in February 2003. In addition to superseding FIPS 180-1, FIPS 180-2 added three new algorithms that produce and output larger hash values: SHA-256, SHA-384, and SHA-512. The SHA-1 algorithm specified in FIPS 180-2 is the same algorithm as the one specified in FIPS 180-1, although some of the notation has been modified to be consistent with the notation used in SHA-256, SHA-384, and SHA-512. In February 2004, NIST published a change notice for FIPS 180-2 to additionally include SHA-224.[19] SHA-224 is identical to SHA-256, but uses different initialization values and truncates the final hash value to 224 bits. SHA-224, SHA-256, SHA-384, and SHA-512 generate hash values of 224, 256, 384, and 512 bits. The complexities of collision attacks against these hash functions is roughly $\sqrt{2^{224}} = 2^{\frac{224}{2}} = 2^{112}$ (SHA-224), $2^{128}$ (SHA-256), $2^{192}$ (SHA-384), and $2^{256}$ (SHA-512). This means that the respective security levels are comparable to those that can be achieved with 3DES, AES-128, AES-192, and AES-256 in this order. So the hash lengths are not randomly chosen but intentionally crafted.

The third revision of FIPS PUB 180, FIPS PUB 180-3, was released in October 2008 and officially introduced SHA-224 as part of the SHS.

In March 2012, a fourth revision of FIPS PUB 180, FIPS PUB 180-4, was released [19].[20] In addition to SHA-1, SHA-224, SHA-256, SHA-384, and SHA-512, this document also specifies SHA-512/224 and SHA-512/256, two versions of SHA-512 that generate 224-bit or 256-bit output values. All algorithms specified in FIPS PUB 180-4 together with their length parameters (i.e., message, block, word, and hash value sizes), are summarized in Table 6.1. The major distinction is between SHA-1, SHA-224, and SHA-256 on the one hand, and SHA-384, SHA-512, SHA-512/224, and SHA-512/256 on the other hand. While the first group operates on 32-bit words and 512-bit message blocks, the second group operates on 64-bit words and 1024-bit message blocks. In either case, a message block consists of 16 words. Except from this distinction, the inner working principles of all cryptographic hash functions specified in FIPS PUB 180-4 are very similar. This will become clear when we step through the functions in Section 6.4.4.

The cryptographic hash functions specified in FIPS PUB 180-4 can be used for the processing of sensitive unclassified information. Only the use of SHA-1 must be considered with care and should be avoided whenever possible. According to NIST Special Publication 800-131A, for example, SHA-1 should no longer be used for the

18 SHA-1 is also specified in informational RFC 4634 [17].

19 SHA-224 is also specified in informational RFC 3874 [18].

20 The applicability clause of FIPS PUB 180-4 was slightly revised in August 2015. The publication is currently complemented by FIPS PUB 202 (as addressed below).

**Table 6.1**
Secure Hash Algorithms as Specified in FIPS 180-4 [19]

| Algorithm | Message Size | Block Size | Word Size | Hash Value Size |
|---|---|---|---|---|
| SHA-1 | $< 2^{64}$ bits | 512 bits | 32 bits | 160 bits |
| SHA-224 | $< 2^{64}$ bits | 512 bits | 32 bits | 224 bits |
| SHA-256 | $< 2^{64}$ bits | 512 bits | 32 bits | 256 bits |
| SHA-384 | $< 2^{128}$ bits | 1,024 bits | 64 bits | 384 bits |
| SHA-512 | $< 2^{128}$ bits | 1,024 bits | 64 bits | 512 bits |
| SHA-512/224 | $< 2^{128}$ bits | 1,024 bits | 64 bits | 224 bits |
| SHA-512/256 | $< 2^{128}$ bits | 1,024 bits | 64 bits | 256 bits |

generation of digital signatures (for other applications, the use of SHA-1 may still be appropriate).

One year after the publication of collisions for MD4, MD5, and other cryptographic hash functions, Wang et al. presented collisions for SHA-1 [20]. Their attack requires only $2^{69}$ (instead of $2^{80}$ in a brute-force attack) hash operations to find a collision in SHA-1, but the attack can be improved theoretically to $2^{63}$ operations. The attack was widely discussed in the media and made people nervous about the security of all deployed cryptographic hash functions. In November 2005, NIST held a workshop on the topic. The recommendations that came out of this workshop were that one should:

1. Transition rapidly to the stronger SHA-2 family of hash functions—especially for digital signatures;
2. Encourage cryptographic hash function research to better understand hash function design and attacks;
3. Run a competition for one (or several) standardized cryptographic hash function(s).

The third recommendation was rapidly put in place and a respective SHA-3 competition was initiated in 2007.[21] Until the end of October 2008, NIST received 64 submissions from the international cryptographic research community. In December 2008, NIST selected 51 algorithms for round 1 of the SHA-3 competition. In July 2009, this field was narrowed to 14 algorithms for round 2, and in December

21 http://www.nist.gov/hash-competition.

2010, NIST announced five finalists for round 3: BLAKE,[22] Grøstl,[23] KECCAK,[24] JH,[25] and Skein[26] (in alphabetical order). BLAKE, Grøstl, and KECCAK were European proposals, whereas JH was from Singapore, and Skein was the only proposal from the United States.

On October 2, 2012, NIST officially announced the selection of KECCAK as the winner of the SHA-3 competition and the new SHA-3 hash function (the names KECCAK and SHA-3 are sometimes used synonymously and interchangeably). KECCAK was designed by a team of cryptographers from Belgium and Italy. According to NIST, KECCAK was chosen "for its elegant design, large security margin, good general performance, excellent efficiency in hardware implementations, and for its flexibility." It is intended to complement the existing SHA-2 family of cryptographic hash functions (rather than to replace it), and hence SHA-2 and SHA-3 are likely to coexist in the future (and they really do as one can observe today).

In addition to the cryptographic hash functions proposed by RSA Security and NIST, there are also a few other proposals, such as RIPEMD-128 and RIPEMD-160 [21, 22],[27] HAVAL [23], Tiger and Tiger2,[28] as well as Whirlpool.[29] The security of these hash functions is not as thoroughly examined as the security of the other hash functions mentioned above. Therefore, their security should be taken with a grain of salt, and they should be used defensively. Anyway, they are not addressed in this book.

## 6.4 EXEMPLARY HASH FUNCTIONS

In this section, we delve more deeply into some exemplary hash functions that are used in the field. This includes MD4, MD5, SHA-1, the SHA-2 family, and KECCAK/SHA-3. MD4, MD5, and SHA-1 are addressed mainly because they are important milestones in the development of cryptographic hash functions. From today's perspective, they should no longer be used (because people have been able to find collisions). Instead, the cryptographic hash functions of choice are the representatives from the SHA-2 family, the finalists of the SHA-3 competition,

22 http://131002.net/blake.

23 http://www.groestl.info.

24 http://keccak.noekeon.org.

25 http://www3.ntu.edu.sg/home/wuhj/research/jh.

26 http://skein-hash.info.

27 There are also 256- und 320-bit versions of RIPEM, known as RIPEM-256 and RIPEM-320. But these versions of RIPEM are hardly used in practice.

28 http://www.cs.technion.ac.il/∼biham/Reports/Tiger.

29 Whirlpool is a predecessor of KECCAK. Further information about Whirlpool is available at http://www.larc.usp.br/∼pbarreto/WhirlpoolPage.html.

and—most importantly—KECCAK/SHA-3. If you are not particularly interested in the technical details of these hash functions, then you may skip this section without difficulty.

### 6.4.1 MD4

As mentioned above, MD4 was proposed in 1990 and specified in RFC 1320 [8].[30] It follows the Merkle-Damgård construction and uses a Davies-Meyer compression function with $b = 512$ and $l = 128$, so the length of an MD4 hash value is 128 bits. MD4 was designed to be efficiently executed on 32-bit processors with a little-endian architecture.[31]

Let $m = m_0 m_1 \ldots m_{s-1}$ be an $s$-bit message that is to be hashed. MD4 first generates an array $w$ of $n$ 32-bit words

$$w = w[0] \parallel [1] \parallel \ldots \parallel w[n-1]$$

where $n$ is a multiple of 16; that is, $n \equiv 0 \pmod{16}$. Hence, the bitlength of $w$ is a multiple of $32 \cdot 16 = 512$ bits. The $s$ bits of $m$ are distributed as follows:

$$\begin{aligned} w[0] &= m_0 m_1 \ldots m_{31} \\ w[1] &= m_{32} m_{33} \ldots m_{63} \\ &\ldots \\ w[n-1] &= m_{s-32} m_{s-31} \ldots m_{s-1} \end{aligned}$$

More specifically, the array $w$ is constructed in two steps:

- First, $m$ is padded so that the bitlength is congruent to 448 modulo 512. The padding consists of a one and a variable number of zeros. Note that padding is always performed, even if the length of the message is already congruent to 448 modulo 512. Also note that the padded message is 64 bits short of being a multiple of 512 bits.

30 The original version of MD4 was published in October 1990 in RFC 1196. A slightly revised version was published soon after in April 1992 in RFC 1320. This was the same time when the RFC documents specifying MD2 [7] and MD5 [11] were also published.

31 A *little-endian architecture* means that a 4-byte word $a_1a_2a_3a_4$ is stored as $a_4a_3a_2a_1$ and represents the integer $a_4 2^{24} + a_3 2^{16} + a_2 2^8 + a_1$. In a *big-endian architecture*, the same 4-byte word would be stored as $a_1a_2a_3a_4$ and represent the integer $a_1 2^{24} + a_2 2^{16} + a_3 2^8 + a_4$. It is rumored that Rivest designed MD4 for a little-endian architecture mainly because he observed that big-endian architectures are generally faster and could therefore better afford the processing penalty (of reversing each word before processing it).

- Second, a 64-bit binary representation of $s$ (i.e., the bitlength of the original message before padding), is appended to the result of the first step. In the unlikely case of $s$ being greater than $2^{64}$, only the low-order 64 bits of $s$ are used (i.e., $s$ is computed modulo $2^{64}$). In either case, the 64 bits yield the last two words of $w$; that is, $w[n-2]$ and $w[n-1]$.

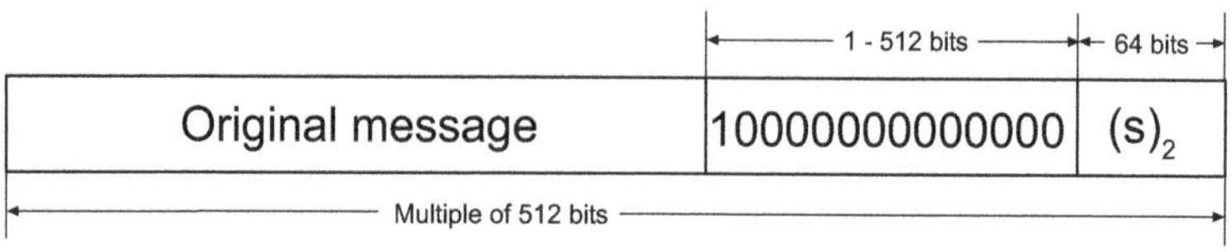


**Figure 6.3** The structure of a message preprocessed to be hashed using MD4.

The structure of a message preprocessed to be hashed using MD4 is illustrated in Figure 6.3. The bitlength is a multiple of 512 bits, and hence the number of 32-words is a multiple of 16. This is the starting point of the MD4 algorithm overviewed in Algorithm 6.1. It takes as input an $s$-bit message $m$, and it generates as output a 128-bit hash value $h(m)$. Internally, the algorithm uses four 32-bit registers $A$, $B$, $C$, and $D$.

The algorithm starts with the construction of an $n$-word array $w$ as described above and the initialization of the four registers with constant values. These values are constructed as follows:

| | | |
|---|---|---|
| 0111 0110 0101 0100 0011 0010 0001 0000 | = | 0x76543210 = 0x67452301 |
| 1111 1110 1101 1100 1011 1010 1001 1000 | = | 0xFEDCBA98 = 0xEFCDAB89 |
| 1000 1001 1010 1011 1100 1101 1110 1111 | = | 0x89ABCDEF = 0x98BADCFE |
| 0000 0001 0010 0011 0100 0101 0110 0111 | = | 0x01234567 = 0x10325476 |

On the left side, there are four lines of eight 4-bit counters. The counter starts on the right sides of the first line with 0000 and goes up to 1111 on the left side of the second line. The same is done on line three and four, but this time it starts with 0000 on the left side of line four and goes up to 1111 on the right side of line three. In each line, the eight 4-bit values represent eight hexadecimal digits that are written in big-endian format. The eight digits refer to four bytes, and if each of these bytes is written in little-endian format, then one ends with the values that stand in the rightmost column. These are the constant values that are used to initialize the four registers $A$, $B$, $C$, and $D$ (in this order).

After this initialization, the array $w$ is processed in $n/16$ iterations. In each iteration, the next 16 words (512 bits) of $w$ are stored in array $X$ and the values of

**Algorithm 6.1** The MD4 hash function (overview).

$(m)$

---

Construct $w = w[0] \parallel w[1] \parallel \ldots \parallel w[n-1]$
$A =$ `0x67452301`
$B =$ `0xEFCDAB89`
$C =$ `0x98BADCFE`
$D =$ `0x10325476`
for $i = 0$ to $n/16 - 1$ do
  for $j = 0$ to 15 do $X[j] = w[i \cdot 16 + j]$
  $A' = A$
  $B' = B$
  $C' = C$
  $D' = D$
  Round 1 (Algorithm 6.2)
  Round 2 (Algorithm 6.3)
  Round 3 (Algorithm 6.4)
  $A = A + A'$
  $B = B + B'$
  $C = C + C'$
  $D = D + D'$

---

$(h(m) = A \parallel B \parallel C \parallel D)$

the four registers $A$, $B$, $C$, and $D$ are stored in $A'$, $B'$, $C'$, and $D'$ for later reuse. In the main part of the algorithm, the compression function is applied in three rounds (i.e., Round 1, Round 2, and Round 3). These rounds are summarized in Algorithms 6.2, 6.3, and 6.4. In each round, the registers are updated in a specific way using the 16 words of $X$. Finally, the four registers are updated by adding back the original values that have been stored in $A'$, $B'$, $C'$, and $D'$. After the $n/16$ iterations, the actual contents of the four registers are concatenated, and the resulting $4 \cdot 32 = 128$ bits yield the output $h(m)$ of the hash function.

In the three round functions of Algorithms 6.2, 6.3, and 6.4, the "normal" Boolean operators $\wedge$ (AND), $\vee$ (OR), $\oplus$ (XOR), and $\neg$ (NOT) are applied bitwise on words to form the following three logical functions $f$, $g$, and $h$:

$$\begin{aligned} f(X,Y,Z) &= (X \wedge Y) \vee ((\neg X) \wedge Z) \\ g(X,Y,Z) &= (X \wedge Y) \vee (X \wedge Z) \vee (Y \wedge Z) \\ h(X,Y,Z) &= X \oplus Y \oplus Z \end{aligned}$$

Each of these functions is used in one particular round: $f$ in round 1, $g$ in round 2, and $h$ in round 3. Each function takes as input three 32-bit words and produces as output a 32-bit word. The truth table of the functions is illustrated in Table 6.2.

**Algorithm 6.2** Round 1 of the MD4 hash function.

1. $A = (A + f(B,C,D) + X[0]) \overset{\curvearrowright}{\hookleftarrow} 3$
2. $D = (D + f(A,B,C) + X[1]) \overset{\curvearrowright}{\hookleftarrow} 7$
3. $C = (C + f(D,A,B) + X[2]) \overset{\curvearrowright}{\hookleftarrow} 11$
4. $B = (B + f(C,D,A) + X[3]) \overset{\curvearrowright}{\hookleftarrow} 19$
5. $A = (A + f(B,C,D) + X[4]) \overset{\curvearrowright}{\hookleftarrow} 3$
6. $D = (D + f(A,B,C) + X[5]) \overset{\curvearrowright}{\hookleftarrow} 7$
7. $C = (C + f(D,A,B) + X[6]) \overset{\curvearrowright}{\hookleftarrow} 11$
8. $B = (B + f(C,D,A) + X[7]) \overset{\curvearrowright}{\hookleftarrow} 19$
9. $A = (A + f(B,C,D) + X[8]) \overset{\curvearrowright}{\hookleftarrow} 3$
10. $D = (D + f(A,B,C) + X[9]) \overset{\curvearrowright}{\hookleftarrow} 7$
11. $C = (C + f(D,A,B) + X[10]) \overset{\curvearrowright}{\hookleftarrow} 11$
12. $B = (B + f(C,D,A) + X[11]) \overset{\curvearrowright}{\hookleftarrow} 19$
13. $A = (A + f(B,C,D) + X[12]) \overset{\curvearrowright}{\hookleftarrow} 3$
14. $D = (D + f(A,B,C) + X[13]) \overset{\curvearrowright}{\hookleftarrow} 7$
15. $C = (C + f(D,A,B) + X[14]) \overset{\curvearrowright}{\hookleftarrow} 11$
16. $B = (B + f(C,D,A) + X[15]) \overset{\curvearrowright}{\hookleftarrow} 19$

- The function $f$ is known as the *selection function*, because if the n-th bit of $X$ is 1, then it selects the n-th bit of $Y$ for the n-th bit of the output. Otherwise (i.e., if the n-th bit of $X$ is 0), it selects the n-th bit of $Z$ for the n-th bit of the output.
- The function $g$ is known as the *majority function*, because the n-th bit of the output is 1 if and only if at least two of the three input words' n-th bits are 1.
- Last but not least, the function $h$ simply adds all input words modulo 2.

In addition to these logical functions, MD4 also employs the integer addition modulo $2^{32}$ operation (+) and the circular left shift (rotate) operation for words. More specifically, $X \overset{\curvearrowright}{\hookleftarrow} c$ refers to the $c$-bit left rotation (circular left shift) of word $w$ (with $0 \leq c \leq 31$).

Note that all operations employed by MD4 are fast and can be implemented efficiently in hardware and software. Also note that rounds 2 and 3 take into account two constants $c_1$ (in round 2) and $c_2$ (in round 3). They are defined as follows:

- $c_1 = \lfloor 2^{30}\sqrt{2} \rfloor = 1,518,500,249$ (decimal) $=$ 10110101000001001111001 10011001 (binary) $=$ 0x5A827999 (hexadecimal);

**Algorithm 6.3** Round 2 of the MD4 hash function.

1. $A = (A + g(B, C, D) + X[0] + c_1) \overset{\curvearrowleft}{\hookleftarrow} 3$
2. $D = (D + g(A, B, C) + X[4] + c_1) \overset{\curvearrowleft}{\hookleftarrow} 5$
3. $C = (C + g(D, A, B) + X[8] + c_1) \overset{\curvearrowleft}{\hookleftarrow} 9$
4. $B = (B + g(C, D, A) + X[12] + c_1) \overset{\curvearrowleft}{\hookleftarrow} 13$
5. $A = (A + g(B, C, D) + X[1] + c_1) \overset{\curvearrowleft}{\hookleftarrow} 3$
6. $D = (D + g(A, B, C) + X[5] + c_1) \overset{\curvearrowleft}{\hookleftarrow} 5$
7. $C = (C + g(D, A, B) + X[9] + c_1) \overset{\curvearrowleft}{\hookleftarrow} 9$
8. $B = (B + g(C, D, A) + X[13] + c_1) \overset{\curvearrowleft}{\hookleftarrow} 13$
9. $A = (A + g(B, C, D) + X[2] + c_1) \overset{\curvearrowleft}{\hookleftarrow} 3$
10. $D = (D + g(A, B, C) + X[6] + c_1) \overset{\curvearrowleft}{\hookleftarrow} 5$
11. $C = (C + g(D, A, B) + X[10] + c_1) \overset{\curvearrowleft}{\hookleftarrow} 9$
12. $B = (B + g(C, D, A) + X[14] + c_1) \overset{\curvearrowleft}{\hookleftarrow} 13$
13. $A = (A + g(B, C, D) + X[3] + c_1) \overset{\curvearrowleft}{\hookleftarrow} 3$
14. $D = (D + g(A, B, C) + X[7] + c_1) \overset{\curvearrowleft}{\hookleftarrow} 5$
15. $C = (C + g(D, A, B) + X[11] + c_1) \overset{\curvearrowleft}{\hookleftarrow} 9$
16. $B = (B + g(C, D, A) + X[15] + c_1) \overset{\curvearrowleft}{\hookleftarrow} 13$

- $c_2 = \lfloor 2^{30}\sqrt{3} \rfloor = 1,859,775,393 = 1101110110110011110101110100001$ = 0x6ED9EBA1.

A reference implementation of MD4 in the C programming language is given in Appendix A of [8]. Again, note that the security of MD4 was breached a long time ago, and that it must not be used anymore. We only use it as a starting point to explain MD5 and SHA-1.

### 6.4.2 MD5

As mentioned above, MD5 is a strengthened version of MD4 that was proposed in 1991 and specified in RFC 1321 [11]. It is conceptually and structurally very similar to MD4, and hence Algorithm 6.5 that overviews the MD5 hash function looks like Algorithm 6.1 that overviews MD4. The main difference is that MD5 invokes four rounds (instead of three). This is advantageous from a security viewpoint, but it is disadvantageous from a performance viewpoint. In fact, the additional round in MD5 decreases its performance in proportion (i.e., for about one third or 30%). The four rounds of MD5 are specified in Algorithms 6.6 to 6.9. There are a few differences in the rounds and operations they consist of:

**Algorithm 6.4** Round 3 of the MD4 hash function.

1. $A = (A + h(B,C,D) + X[0] + c_2) \hookleftarrow 3$
2. $D = (D + h(A,B,C) + X[8] + c_2) \hookleftarrow 9$
3. $C = (C + h(D,A,B) + X[4] + c_2) \hookleftarrow 11$
4. $B = (B + h(C,D,A) + X[12] + c_2) \hookleftarrow 15$
5. $A = (A + h(B,C,D) + X[2] + c_2) \hookleftarrow 3$
6. $D = (D + h(A,B,C) + X[10] + c_2) \hookleftarrow 9$
7. $C = (C + h(D,A,B) + X[6] + c_2) \hookleftarrow 11$
8. $B = (B + h(C,D,A) + X[14] + c_2) \hookleftarrow 15$
9. $A = (A + h(B,C,D) + X[1] + c_2) \hookleftarrow 3$
10. $D = (D + h(A,B,C) + X[9] + c_2) \hookleftarrow 9$
11. $C = (C + h(D,A,B) + X[5] + c_2) \hookleftarrow 11$
12. $B = (B + h(C,D,A) + X[13] + c_2) \hookleftarrow 15$
13. $A = (A + h(B,C,D) + X[3] + c_2) \hookleftarrow 3$
14. $D = (D + h(A,B,C) + X[11] + c_2) \hookleftarrow 9$
15. $C = (C + h(D,A,B) + X[7] + c_2) \hookleftarrow 11$
16. $B = (B + h(C,D,A) + X[15] + c_2) \hookleftarrow 15$

- The majority function $g$ of MD5 was changed from

  $$g(X,Y,Z) = (X \wedge Y) \vee (X \wedge Z) \vee (Y \wedge Z)$$

  to

  $$g(X,Y,Z) = ((X \wedge Z) \vee (Y \wedge (\neg Z))$$

  to make it less symmetric. Also, a new logical function $i$ was introduced to be used in the fourth round. This function is defined as follows:

  $$i(X,Y,Z) \quad = \quad Y \oplus (X \vee (\neg Z))$$

  The other two logical functions, $f$ and $h$, remain unchanged. The truth table of all logical functions employed by MD5 is illustrated in Table 6.3. Note that the columns for $f$ and $h$ are the same as in Table 6.2, the column for $g$ is slightly different, and the column for $i$ is entirely new.

- The MD5 hash function employs a 64-word table $T$ (instead of $c_1$ and $c_2$) that is constructed from the sine function. Let $T[i]$ be the i-th element of the table $T$, then

  $$T[i] = \lfloor 4,294,967,296 \cdot |\sin(i)| \rfloor$$

**Table 6.2**
Truth Table of the Logical Functions Employed by MD4

| X | Y | Z | f | g | h |
|---|---|---|---|---|---|
| 0 | 0 | 0 | 0 | 0 | 0 |
| 0 | 0 | 1 | 1 | 0 | 1 |
| 0 | 1 | 0 | 0 | 0 | 1 |
| 0 | 1 | 1 | 1 | 1 | 0 |
| 1 | 0 | 0 | 0 | 0 | 1 |
| 1 | 0 | 1 | 0 | 1 | 0 |
| 1 | 1 | 0 | 1 | 1 | 0 |
| 1 | 1 | 1 | 1 | 1 | 1 |

where $i$ is in radians. Because $4,294,967,296$ is equal to $2^{32}$ and $|\sin(i)|$ is a number between 0 and 1, each element of $T$ is an integer that represents 32 bits. Consequently, the table $T$ provides a "randomized" set of 32-bit patterns, which should eliminate any regularities in the input. The elements of $T$ are listed in Table 6.4. As an example, let's look at $T[1]$: The sine of 1 in radian is $0.8414709848\ldots$, and this value multiplied with $4,294,967,296$ is equal to a value whose integer part is 3614090360 or 0xD76AA478. This is the first entry in $T$. All other entries are generated the same way.

Again, a reference implementation for MD5 in the C programming language is given in Appendix A of [11]. It is known that MD5 is susceptible to collision attacks, and such attacks have been successfully mounted in the past. While a "normal" collision search attack requires $2^{64}$ messages to be hashed, the collision attack of Wang et al. [15] requires about $2^{39}$ messages to be hashed and the best-known attack only $2^{32}$. This value is sufficiently small to turn the attack into a practical threat. The bottom line is that MD5 (like MD4) must not be used anymore.

### 6.4.3 SHA-1

SHA-1 is conceptually and structurally similar to MD5 (and hence also to MD4). As mentioned in Section 6.3, it was the first cryptographic hash function standardized by the U.S. NIST [19] as part of the SHS (that currently also comprises a few other cryptographic hash functions from the SHA-2 family). The two most important differences between SHA-1 and MD5 are that SHA-1 was designed to run optimally on computer systems with a big-endian architecture (instead of a little-endian architecture), and that it employs five registers $A$, $B$, $C$, $D$, and $E$ (instead of only

**Algorithm 6.5** The MD5 hash function (overview).

$(m)$

---

Construct $w = w[0] \parallel w[1] \parallel \ldots \parallel w[n-1]$
$A =$ `0x67452301`
$B =$ `0xEFCDAB89`
$C =$ `0x98BADCFE`
$D =$ `0x10325476`
for $i = 0$ to $n/16 - 1$ do
    for $j = 0$ to 15 do $X[j] = w[i \cdot 16 + j]$
    $A' = A$
    $B' = B$
    $C' = C$
    $D' = D$
    Round 1 (Algorithm 4.6)
    Round 2 (Algorithm 4.7)
    Round 3 (Algorithm 4.8)
    Round 4 (Algorithm 4.9)
    $A = A + A'$
    $B = B + B'$
    $C = C + C'$
    $D = D + D'$

---

$(h(m) = A \parallel B \parallel C \parallel D)$

four).[32] Five times 32 equals 160, and hence the output length of a SHA-1 hash value is 160 bits (instead of 128 bits as with MD4 and MD5).

Instead of $f$, $g$, $h$, and $i$, SHA-1 uses a sequence of 80 logical functions $f_0$, $f_1, \ldots, f_{79}$ that are defined as follows:

$$f_t(X,Y,Z) = \begin{cases} Ch(X,Y,Z) = (X \wedge Y) \oplus ((\neg X) \wedge Z) & 0 \leq t \leq 19 \\ Parity(X,Y,Z) = X \oplus Y \oplus Z & 20 \leq t \leq 39 \\ Maj(X,Y,Z) = (X \wedge Y) \oplus (X \wedge Z) \oplus (Y \wedge Z) & 40 \leq t \leq 59 \\ Parity(X,Y,Z) = X \oplus Y \oplus Z & 60 \leq t \leq 79 \end{cases}$$

Note that the $Parity$ function occurs twice (for $20 \leq t \leq 39$ and $60 \leq t \leq 79$). Also note that $\oplus$ can be replaced by $\vee$ in the formulas to compute $Ch$ and $Maj$ without changing the result. The truth table of these functions is illustrated in Table 6.5 (where for $Parity$ function occurs twice for the sake of completeness).

Instead of $c_1$ and $c_2$ (as with MD4) or the 64 words of the $T$ table (as with MD5), SHA-1 uses 4 constant 32-bit words that are used to build a sequence of 80

32 Needless to say, the register $E$ requires a new initialization value that is 0xC3D2E1F0 (see Algorithm 6.10).

**Table 6.3**
Truth Table of the Logical Functions Employed by MD5

| X | Y | Z | f | g | h | i |
|---|---|---|---|---|---|---|
| 0 | 0 | 0 | 0 | 0 | 0 | 1 |
| 0 | 0 | 1 | 1 | 0 | 1 | 0 |
| 0 | 1 | 0 | 0 | 1 | 1 | 0 |
| 0 | 1 | 1 | 1 | 0 | 0 | 1 |
| 1 | 0 | 0 | 0 | 0 | 1 | 1 |
| 1 | 0 | 1 | 0 | 1 | 0 | 1 |
| 1 | 1 | 0 | 1 | 1 | 0 | 0 |
| 1 | 1 | 1 | 1 | 1 | 1 | 0 |

words $K_0, K_1, \ldots, K_{79}$. This sequence is defined as follows:

$$K_t = \begin{cases} \lfloor 2^{30}\sqrt{2} \rfloor = \texttt{0x5A827999} & 0 \leq t \leq 19 \\ \lfloor 2^{30}\sqrt{3} \rfloor = \texttt{0x6ED9EBA1} & 20 \leq t \leq 39 \\ \lfloor 2^{30}\sqrt{5} \rfloor = \texttt{0x8F1BBCDC} & 40 \leq t \leq 59 \\ \lfloor 2^{30}\sqrt{10} \rfloor = \texttt{0xCA62C1D6} & 60 \leq t \leq 79 \end{cases}$$

Note that the first two words are equal to $c_1$ and $c_2$ from MD4, and that the second two words are generated in a similar way.

The SHA-1 hash function is overviewed in Algorithm 6.10. There are three main differences to MD4 (Algorithm 6.1) and MD5 (Algorithm 6.5):

- First, the preprocessing of a message is similar to MD4 and MD5, but there are two subtle differences:

  1. As mentioned above, SHA-1 assumes a big-endian architecture (i.e., the leftmost byte of the binary representation of $s$ also appears leftmost).
  2. While $w$ is an array of 32-bit words in MD4 and MD5, SHA-1 uses an array $b$ of 16-word blocks instead. Hence, $b[i]$ $(i = 0, 1, \ldots, n-1)$ refers to a 16-word block that is $16 \cdot 32 = 512$ bits long.

- Second, SHA-1 uses each 16-word block from $b$ to recursively derive an 80-word message schedule $W$ as follows:

$$W_t = \begin{cases} b & 0 \leq t \leq 15 \\ (W_{t-3} \oplus W_{t-8} \oplus W_{t-14} \oplus W_{t-16}) \hookleftarrow 1 & 16 \leq t \leq 79 \end{cases}$$

**Algorithm 6.6** Round 1 of the MD5 hash function.

1. $A = (A + f(B,C,D) + X[0] + T[1]) \hookleftarrow 7$
2. $D = (D + f(A,B,C) + X[1] + T[2]) \hookleftarrow 12$
3. $C = (C + f(D,A,B) + X[2] + T[3]) \hookleftarrow 17$
4. $B = (B + f(C,D,A) + X[3] + T[4]) \hookleftarrow 22$
5. $A = (A + f(B,C,D) + X[4] + T[5]) \hookleftarrow 7$
6. $D = (D + f(A,B,C) + X[5] + T[6]) \hookleftarrow 12$
7. $C = (C + f(D,A,B) + X[6] + T[7]) \hookleftarrow 17$
8. $B = (B + f(C,D,A) + X[7] + T[8]) \hookleftarrow 22$
9. $A = (A + f(B,C,D) + X[8] + T[9]) \hookleftarrow 7$
10. $D = (D + f(A,B,C) + X[9] + T[10]) \hookleftarrow 12$
11. $C = (C + f(D,A,B) + X[10] + T[11]) \hookleftarrow 17$
12. $B = (B + f(C,D,A) + X[11] + T[12]) \hookleftarrow 22$
13. $A = (A + f(B,C,D) + X[12] + T[13]) \hookleftarrow 7$
14. $D = (D + f(A,B,C) + X[13] + T[14]) \hookleftarrow 12$
15. $C = (C + f(D,A,B) + X[14] + T[15]) \hookleftarrow 17$
16. $B = (B + f(C,D,A) + X[15] + T[16]) \hookleftarrow 22$

The 16 words of $b$ become the first 16 words of $W$. The remaining $80 - 16 = 64$ words of $W$ are generated according to the formula. Note that neither MD4 nor MD5 uses a message schedule.

- Third, SHA-1 does not operate in "official" rounds. Instead, the algorithm has an inner loop that is iterated 80 times. Since the iterations use four different $f$-functions, the resulting algorithm can still be seen as one that internally operates in rounds (sometimes called *stages*).

Because SHA-1 uses five registers $A$, $B$, $C$, $D$, and $E$ (instead of only four) and the operations are more involved, SHA-1 is a little bit less efficient than its predecessors. On the other hand, SHA-1 hash values are longer than MD4 and MD5 hash values (160 instead of 128 bits), and hence SHA-1 is potentially more resistant against collision attacks. So people had been advocating the use of SHA-1 (instead of MD4 or MD5) for quite some time. But as mentioned above, SHA-1 was theoretically broken in 2005 [20]. While a "normal" collision attack against SHA-1 requires $2^{80}$ messages to be hashed, this attack requires only $2^{69}$ messages. This is clearly more efficient than brute force, but the respective attack is still difficult to mount in practice. Since 2005, the attack has been improved (e.g., [24]), both in terms of finding collisions for reduced-round versions of SHA-1, as well as finding collisions for the full version of SHA-1.

**Algorithm 6.7** Round 2 of the MD5 hash function.

1. $A = (A + g(B, C, D) + X[1] + T[17]) \hookleftarrow 5$
2. $D = (D + g(A, B, C) + X[6] + T[18]) \hookleftarrow 9$
3. $C = (C + g(D, A, B) + X[11] + T[19]) \hookleftarrow 14$
4. $B = (B + g(C, D, A) + X[0] + T[20]) \hookleftarrow 20$
5. $A = (A + g(B, C, D) + X[5] + T[21]) \hookleftarrow 5$
6. $D = (D + g(A, B, C) + X[10] + T[22]) \hookleftarrow 9$
7. $C = (C + g(D, A, B) + X[15] + T[23]) \hookleftarrow 14$
8. $B = (B + g(C, D, A) + X[4] + T[24]) \hookleftarrow 20$
9. $A = (A + g(B, C, D) + X[9] + T[25]) \hookleftarrow 5$
10. $D = (D + g(A, B, C) + X[14] + T[26]) \hookleftarrow 9$
11. $C = (C + g(D, A, B) + X[3] + T[27]) \hookleftarrow 14$
12. $B = (B + g(C, D, A) + X[8] + T[28]) \hookleftarrow 20$
13. $A = (A + g(B, C, D) + X[13] + T[29]) \hookleftarrow 5$
14. $D = (D + g(A, B, C) + X[2] + T[30]) \hookleftarrow 9$
15. $C = (C + g(D, A, B) + X[7] + T[31]) \hookleftarrow 14$
16. $B = (B + g(C, D, A) + X[12] + T[32]) \hookleftarrow 20$

In 2011, the U.S. NIST finally deprecated SHA-1, and disallowed its use for digital signatures by the end of 2013. This was a wise decision, because a few years later two devastating attacks brought SHA-1 to the end of its life cycle: An attack called SHAttered[33] in 2017 and a chosen-prefix attack called "SHA-1 is a Shambles" in 2019.[34] In light of these attacks, SHA-1 should really be replaced by another cryptographic hash function as soon as possible, preferably by a representative of the SHA-2 family or KECCAK/SHA-3.

### 6.4.4 SHA-2 Family

As mentioned above, FIPS PUB 180-4 [19] specifies multiple cryptographic hash functions in addition to SHA-1 that are collectively referred to as the *SHA-2 family*. The output length of these hash functions is part of their name, so SHA-224 refers to a function that outputs 224-bit hash values. The same is true for SHA-256, SHA-384, and SHA-512. SHA-512/224 and SHA-512/256 are similar to SHA-512, but their output is truncated to 224 and 256 bits, respectively.

The cryptographic hash functions from the SHA-2 family employ the same $Ch$ and $Maj$ functions that are used in SHA-1. They can be applied to 32-bit or 64-bit

33 https://shattered.io.

34 https://sha-mbles.github.io.

**Algorithm 6.8** Round 3 of the MD5 hash function.

1. $A = (A + h(B,C,D) + X[5] + T[33]) \stackrel{\curvearrowleft}{\hookleftarrow} 4$
2. $D = (D + h(A,B,C) + X[8] + T[34]) \stackrel{\curvearrowleft}{\hookleftarrow} 11$
3. $C = (C + h(D,A,B) + X[11] + T[35]) \stackrel{\curvearrowleft}{\hookleftarrow} 16$
4. $B = (B + h(C,D,A) + X[14] + T[36]) \stackrel{\curvearrowleft}{\hookleftarrow} 23$
5. $A = (A + h(B,C,D) + X[1] + T[37]) \stackrel{\curvearrowleft}{\hookleftarrow} 4$
6. $D = (D + h(A,B,C) + X[4] + T[38]) \stackrel{\curvearrowleft}{\hookleftarrow} 11$
7. $C = (C + h(D,A,B) + X[7] + T[39]) \stackrel{\curvearrowleft}{\hookleftarrow} 16$
8. $B = (B + h(C,D,A) + X[10] + T[40]) \stackrel{\curvearrowleft}{\hookleftarrow} 23$
9. $A = (A + h(B,C,D) + X[13] + T[41]) \stackrel{\curvearrowleft}{\hookleftarrow} 4$
10. $D = (D + h(A,B,C) + X[0] + T[42]) \stackrel{\curvearrowleft}{\hookleftarrow} 11$
11. $C = (C + h(D,A,B) + X[3] + T[43]) \stackrel{\curvearrowleft}{\hookleftarrow} 16$
12. $B = (B + h(C,D,A) + X[6] + T[44]) \stackrel{\curvearrowleft}{\hookleftarrow} 23$
13. $A = (A + h(B,C,D) + X[9] + T[45]) \stackrel{\curvearrowleft}{\hookleftarrow} 4$
14. $D = (D + h(A,B,C) + X[12] + T[46]) \stackrel{\curvearrowleft}{\hookleftarrow} 11$
15. $C = (C + h(D,A,B) + X[15] + T[47]) \stackrel{\curvearrowleft}{\hookleftarrow} 16$
16. $B = (B + h(C,D,A) + X[2] + T[48]) \stackrel{\curvearrowleft}{\hookleftarrow} 23$

words. In the case of SHA-224 and SHA-256, these two functions are complemented by the following four functions that take a 32-bit input word $X$ and generate a 32-bit output word:

$$
\begin{aligned}
\Sigma_0^{\{256\}}(X) &= (X \stackrel{\curvearrowright}{\hookrightarrow} 2) \oplus (X \stackrel{\curvearrowright}{\hookrightarrow} 13) \oplus (X \stackrel{\curvearrowright}{\hookrightarrow} 22) \\
\Sigma_1^{\{256\}}(X) &= (X \stackrel{\curvearrowright}{\hookrightarrow} 6) \oplus (X \stackrel{\curvearrowright}{\hookrightarrow} 11) \oplus (X \stackrel{\curvearrowright}{\hookrightarrow} 25) \\
\sigma_0^{\{256\}}(X) &= (X \stackrel{\curvearrowright}{\hookrightarrow} 7) \oplus (X \stackrel{\curvearrowright}{\hookrightarrow} 18) \oplus (X \hookrightarrow 3) \\
\sigma_1^{\{256\}}(X) &= (X \stackrel{\curvearrowright}{\hookrightarrow} 17) \oplus (X \stackrel{\curvearrowright}{\hookrightarrow} 19) \oplus (X \hookrightarrow 10)
\end{aligned}
$$

Note that the last terms in the computation of $\sigma_0^{\{256\}}$ and $\sigma_1^{\{256\}}$ comprise the $c$-bit right shift operator ($\hookrightarrow$) instead of the $c$-bit circular right shift (right rotation) operator ($\stackrel{\curvearrowright}{\hookrightarrow}$) used elsewhere.

All other hash functions from the SHA-2 family use similar functions that take a 64-bit input word $X$ and generate a 64-bit output. These functions are defined as

**Algorithm 6.9** Round 4 of the MD5 hash function.

1. $A = (A + i(B,C,D) + X[0] + T[49]) \stackrel{\curvearrowleft}{\hookleftarrow} 6$
2. $D = (D + i(A,B,C) + X[7] + T[50]) \stackrel{\curvearrowleft}{\hookleftarrow} 10$
3. $C = (C + i(D,A,B) + X[14] + T[51]) \stackrel{\curvearrowleft}{\hookleftarrow} 15$
4. $B = (B + i(C,D,A) + X[5] + T[52]) \stackrel{\curvearrowleft}{\hookleftarrow} 21$
5. $A = (A + i(B,C,D) + X[12] + T[53]) \stackrel{\curvearrowleft}{\hookleftarrow} 6$
6. $D = (D + i(A,B,C) + X[3] + T[54]) \stackrel{\curvearrowleft}{\hookleftarrow} 10$
7. $C = (C + i(D,A,B) + X[10] + T[55]) \stackrel{\curvearrowleft}{\hookleftarrow} 15$
8. $B = (B + i(C,D,A) + X[1] + T[56]) \stackrel{\curvearrowleft}{\hookleftarrow} 21$
9. $A = (A + i(B,C,D) + X[8] + T[57]) \stackrel{\curvearrowleft}{\hookleftarrow} 6$
10. $D = (D + i(A,B,C) + X[15] + T[58]) \stackrel{\curvearrowleft}{\hookleftarrow} 10$
11. $C = (C + i(D,A,B) + X[6] + T[59]) \stackrel{\curvearrowleft}{\hookleftarrow} 15$
12. $B = (B + i(C,D,A) + X[13] + T[60]) \stackrel{\curvearrowleft}{\hookleftarrow} 21$
13. $A = (A + i(B,C,D) + X[4] + T[61]) \stackrel{\curvearrowleft}{\hookleftarrow} 6$
14. $D = (D + i(A,B,C) + X[11] + T[62]) \stackrel{\curvearrowleft}{\hookleftarrow} 10$
15. $C = (C + i(D,A,B) + X[2] + T[63]) \stackrel{\curvearrowleft}{\hookleftarrow} 15$
16. $B = (B + i(C,D,A) + X[9] + T[64]) \stackrel{\curvearrowleft}{\hookleftarrow} 21$

follows:

$$\begin{aligned}
\Sigma_0^{\{512\}}(X) &= (X \stackrel{\curvearrowleft}{\hookrightarrow} 28) \oplus (X \stackrel{\curvearrowleft}{\hookrightarrow} 34) \oplus (X \stackrel{\curvearrowleft}{\hookrightarrow} 39) \\
\Sigma_1^{\{512\}}(X) &= (X \stackrel{\curvearrowleft}{\hookrightarrow} 14) \oplus (X \stackrel{\curvearrowleft}{\hookrightarrow} 18) \oplus (X \stackrel{\curvearrowleft}{\hookrightarrow} 41) \\
\sigma_0^{\{512\}}(X) &= (X \stackrel{\curvearrowleft}{\hookrightarrow} 1) \oplus (X \stackrel{\curvearrowleft}{\hookrightarrow} 8) \oplus (X \hookrightarrow 7) \\
\sigma_1^{\{512\}}(X) &= (X \stackrel{\curvearrowleft}{\hookrightarrow} 19) \oplus (X \stackrel{\curvearrowleft}{\hookrightarrow} 61) \oplus (X \hookrightarrow 6)
\end{aligned}$$

While SHA-1 uses four 32-bit words to represent the constants $K_0, K_1, \ldots, K_{79}$, SHA-224 and SHA-256 use the same sequence of 64 distinct 32-bit words to represent the constants

$$K_0^{\{256\}}, K_1^{\{256\}}, \ldots, K_{63}^{\{256\}}.$$

These 64 words are generated by taking the first 32 bits of the fractional parts of the cube roots of the first 64 prime numbers. In hexadecimal notation, these words are as follows:

```
428a2f98 71374491 b5c0fbcf e9b5dba5 3956c25b 59f111f1 923f82a4 ab1c5ed5
d807aa98 12835b01 243185be 550c7dc3 72be5d74 80deb1fe 9bdc06a7 c19bf174
e49b69c1 efbe4786 0fc19dc6 240ca1cc 2de92c6f 4a7484aa 5cb0a9dc 76f988da
```

**Table 6.4**
The Elements of Table $T$ Employed by the MD5 Hash Function

| | | | |
|---|---|---|---|
| $T$[1]=0xD76AA478 | $T$[17]=0xF61E2562 | $T$[33]=0xFFFA3942 | $T$[49]=0xF4292244 |
| $T$[2]=0xE8C7B756 | $T$[18]=0xC040B340 | $T$[34]=0x8771F681 | $T$[50]=0x432AFF97 |
| $T$[3]=0x242070DB | $T$[19]=0x265E5A51 | $T$[35]=0x6D9D6122 | $T$[51]=0xAB9423A7 |
| $T$[4]=0xC1BDCEEE | $T$[20]=0xE9B6C7AA | $T$[36]=0xFDE5380C | $T$[52]=0xFC93A039 |
| $T$[5]=0xF57C0FAF | $T$[21]=0xD62F105D | $T$[37]=0xA4BEEA44 | $T$[53]=0x655B59C3 |
| $T$[6]=0x4787C62A | $T$[22]=0x02441453 | $T$[38]=0x4BDECFA9 | $T$[54]=0x8F0CCC92 |
| $T$[7]=0xA8304613 | $T$[23]=0xD8A1E681 | $T$[39]=0xF6BB4B60 | $T$[55]=0xFFEFF47D |
| $T$[8]=0xFD469501 | $T$[24]=0xE7D3FBC8 | $T$[40]=0xBEBFBC70 | $T$[56]=0x85845DD1 |
| $T$[9]=0x698098D8 | $T$[25]=0x21E1CDE6 | $T$[41]=0x289B7EC6 | $T$[57]=0x6FA87E4F |
| $T$[10]=0x8B44F7AF | $T$[26]=0xC33707D6 | $T$[42]=0xEAA127FA | $T$[58]=0xFE2CE6E0 |
| $T$[11]=0xFFFF5BB1 | $T$[27]=0xF4D50D87 | $T$[43]=0xD4EF3085 | $T$[59]=0xA3014314 |
| $T$[12]=0x895CD7BE | $T$[28]=0x455A14ED | $T$[44]=0x04881D05 | $T$[60]=0x4E0811A1 |
| $T$[13]=0x6B901122 | $T$[29]=0xA9E3E905 | $T$[45]=0xD9D4D039 | $T$[61]=0xF7537E82 |
| $T$[14]=0xFD987193 | $T$[30]=0xFCEFA3F8 | $T$[46]=0xE6DB99E5 | $T$[62]=0xBD3AF235 |
| $T$[15]=0xA679438E | $T$[31]=0x676F02D9 | $T$[47]=0x1FA27CF8 | $T$[63]=0x2AD7D2BB |
| $T$[16]=0x49B40821 | $T$[32]=0x8D2A4C8A | $T$[48]=0xC4AC5665 | $T$[64]=0xEB86D391 |

```
983e5152 a831c66d b00327c8 bf597fc7 c6e00bf3 d5a79147 06ca6351 14292967
27b70a85 2e1b2138 4d2c6dfc 53380d13 650a7354 766a0abb 81c2c92e 92722c85
a2bfe8a1 a81a664b c24b8b70 c76c51a3 d192e819 d6990624 f40e3585 106aa070
19a4c116 1e376c08 2748774c 34b0bcb5 391c0cb3 4ed8aa4a 5b9cca4f 682e6ff3
748f82ee 78a5636f 84c87814 8cc70208 90befffa a4506ceb bef9a3f7 c67178f2
```

Consider the first word as an example: According to its definition, it refers to the first 32 bits of the fractional part of the cube root of the first prime number, which is 2. $\sqrt[3]{2} = 1.25992104\ldots$ and hence its fractional part is $0.25992104\ldots$ In binary notation, this value can be written as 0.01000010100010100010111110011000 that translates to 0x428a2f98. The validity of this word can also be verified the other way round: 428a2f98 when interpreted as a hexadecimal fraction represents

$$\frac{4}{16}+\frac{2}{16^2}+\frac{8}{16^3}+\frac{10}{16^4}+\frac{2}{16^5}+\frac{15}{16^6}+\frac{9}{16^7}+\frac{8}{16^8}$$

and this value is approximately equal to 0.2599210496991873. So when one adds back the nonfractional part (which is 1), one gets $1.25992104\ldots$ which is the cube-root of 2 and the original starting point of our considerations.

Similarly, SHA-384, SHA-512, SHA-512/224, and SHA-512/256 use a sequence of 80 distinct 64-bit words to represent the constants

$$K_0^{\{512\}}, K_1^{\{512\}}, \ldots, K_{79}^{\{512\}}$$

**Algorithm 6.10** The SHA-1 hash function (overview).

$(m)$

---

Construct $b = b[0] \parallel b[1] \parallel \ldots \parallel b[n-1]$
$A =$ `0x67452301`
$B =$ `0xEFCDAB89`
$C =$ `0x98BADCFE`
$D =$ `0x10325476`
$E =$ `0xC3D2E1F0`
for $i = 0$ to $n-1$ do
  Derive message schedule $W$ from $b[i]$
  $A' = A$
  $B' = B$
  $C' = C$
  $D' = D$
  $E' = E$
  for $t = 0$ to 79 do
    $T = (A \overset{\curvearrowright}{\hookleftarrow} 5) + f_t(B, C, D) + E + K_t + W_t$
    $E = D$
    $D = C$
    $C = B \overset{\curvearrowright}{\hookleftarrow} 30$
    $B = A$
    $A = T$
  $A = A + A'$
  $B = B + B'$
  $C = C + C'$
  $D = D + D'$
  $E = E + E'$

---

$(h(m) = A \parallel B \parallel C \parallel D \parallel E)$

These 80 words represent the first 64 bits of the fractional parts of the cube roots of the first 80 prime numbers (so the first 32 bits of the first 64 values are the same of before). In hexadecimal notation, these 64-bit words are as follows:

```
428a2f98d728ae22 7137449123ef65cd b5c0fbcfec4d3b2f e9b5dba58189dbbc
3956c25bf348b538 59f111f1b605d019 923f82a4af194f9b ab1c5ed5da6d8118
d807aa98a3030242 12835b0145706fbe 243185be4ee4b28c 550c7dc3d5ffb4e2
72be5d74f27b896f 80deb1fe3b1696b1 9bdc06a725c71235 c19bf174cf692694
e49b69c19ef14ad2 efbe4786384f25e3 0fc19dc68b8cd5b5 240ca1cc77ac9c65
2de92c6f592b0275 4a7484aa6ea6e483 5cb0a9dcbd41fbd4 76f988da831153b5
983e5152ee66dfab a831c66d2db43210 b00327c898fb213f bf597fc7beef0ee4
c6e00bf33da88fc2 d5a79147930aa725 06ca6351e003826f 142929670a0e6e70
27b70a8546d22ffc 2e1b21385c26c926 4d2c6dfc5ac42aed 53380d139d95b3df
650a73548baf63de 766a0abb3c77b2a8 81c2c92e47edaee6 92722c851482353b
a2bfe8a14cf10364 a81a664bbc423001 c24b8b70d0f89791 c76c51a30654be30
```

**Table 6.5**
Truth Table of the Logical Functions Employed by SHA-1

| X | Y | Z | $Ch = f_{0...19}$ | $Parity = f_{20...39}$ | $Maj = f_{40...59}$ | $Parity = f_{60...79}$ |
|---|---|---|---|---|---|---|
| 0 | 0 | 0 | 0 | 0 | 0 | 0 |
| 0 | 0 | 1 | 1 | 1 | 0 | 1 |
| 0 | 1 | 0 | 0 | 1 | 0 | 1 |
| 0 | 1 | 1 | 1 | 0 | 1 | 0 |
| 1 | 0 | 0 | 0 | 1 | 0 | 1 |
| 1 | 0 | 1 | 0 | 0 | 1 | 0 |
| 1 | 1 | 0 | 1 | 0 | 1 | 0 |
| 1 | 1 | 1 | 1 | 1 | 1 | 1 |

```
d192e819d6ef5218 d69906245565a910 f40e35855771202a 106aa07032bbd1b8
19a4c116b8d2d0c8 1e376c085141ab53 2748774cdf8eeb99 34b0bcb5e19b48a8
391c0cb3c5c95a63 4ed8aa4ae3418acb 5b9cca4f7763e373 682e6ff3d6b2b8a3
748f82ee5defb2fc 78a5636f43172f60 84c87814a1f0ab72 8cc702081a6439ec
90befffa23631e28 a4506cebde82bde9 bef9a3f7b2c67915 c67178f2e372532b
ca273eceea26619c d186b8c721c0c207 eada7dd6cde0eb1e f57d4f7fee6ed178
06f067aa72176fba 0a637dc5a2c898a6 113f9804bef90dae 1b710b35131c471b
28db77f523047d84 32caab7b40c72493 3c9ebe0a15c9bebc 431d67c49c100d4c
4cc5d4becb3e42b6 597f299cfc657e2a 5fcb6fab3ad6faec 6c44198c4a475817
```

While SHA-1, SHA-224, and SHA-256 require messages to be padded to be a multiple of 512 bits, all other hash functions from the SHA-2 family require messages to be padded to be a multiple of 1024 bits. In this case, the length of the original message is encoded in the final two 64-bit words (instead of the final two 32-bit words). Everything else remains the same.

All functions from the SHA-2 family operate on eight registers[35] $A$, $B$, $C$, $D$, $E$, $F$, $G$, and $H$. They are initialized as follows:

- For SHA-256, the 8 initialization values refer to the first 32 bits of the fractional parts of the square roots of the first 8 prime numbers, i.e., 2, 3, 5, 7, 11, 13, 17, and 19.
- For SHA-384, the 8 initialization values refer to the first 64 bits of the fractional parts of the square roots of the ninth through sixteenth prime numbers, i.e., 23, 29, 31, 37, 41, 43, 51, and 53.
- For SHA-512, the 8 initialization values are generated in exactly the same way as for SHA-256. But instead of taking the first 32 bits, the first 64 bits are

35 In the original specification, these registers are named $H_0$ to $H_7$.

taken from the respective values. So the first halves of the initialization values for SHA-512 match the initialization values for SHA-256.

- For any truncated version of SHA-512 (i.e., SHA-512/224 and SHA-512/256), FIPS PUB 180-4 specifies an algorithm to compute the respective initialization values (not repeated here).

Finally, FIPS PUB 180-4 does not specify the generation of the 8 initialization values for SHA-224. These values refer to the second 32 bits of the 64-bit values that are computed for SHA-384. If, for example, the first initialization value for SHA-384 is 0xcbbb9d5dc1059ed8, then the respective initialization value for SHA-224 is just 0xc1059ed8.

Let us briefly elaborate on the various hash functions that constitute the SHA-2 family. We address SHA-256 first, because SHA-224 is a slight deviation from it. This also applies to SHA-512 and the other hash functions from the SHA-2 family.

**Algorithm 6.11** The SHA-256 hash function (overview).

$(m)$

---

Construct $b = b[0] \parallel b[1] \parallel \ldots \parallel b[n-1]$
$A =$ `0x6A09E667` $B =$ `0xBB67AE85`
$C =$ `0x3C6EF372` $D =$ `0xA54FF53A`
$E =$ `0x510E527F` $F =$ `0x9B05688C`
$G =$ `0x1F83D9AB` $H =$ `0x5BE0CD19`
for $i = 0$ to $n-1$ do
  Derive message schedule $W$ from $b[i]$
  $A' = A \quad B' = B \quad C' = C \quad D' = D$
  $E' = E \quad F' = F \quad G' = G \quad H' = H$
  for $t = 0$ to 63 do
    $T_1 = H + \Sigma_1^{\{256\}}(E) + Ch(E,F,G) + K_t^{\{256\}} + W_t$
    $T_2 = \Sigma_0^{\{256\}}(A) + Maj(A,B,C)$
    $H = G$
    $G = F$
    $E = D + T_1$
    $D = C$
    $C = B$
    $B = A$
    $A = T_1 + T_2$
  $A = A + A' \quad B = B + B'$
  $C = C + C' \quad D = D + D'$
  $E = E + E' \quad F = F + F'$
  $G = G + G' \quad H = H + H'$

---

$(h(m) = A \parallel B \parallel C \parallel D \parallel E \parallel F \parallel G \parallel H)$

#### 6.4.4.1 SHA-256

The SHA-256 hash function is overviewed in Algorithm 6.11. It is structurally very similar to SHA-1. But instead of five registers, it uses eight registers that are initialized differently (as mentioned above). Also, the message schedule $W$ comprises only 64 words (instead of 80) and is prepared in a slightly different way. In fact, it is recursively derived from $b$ as follows:

$$W_t = \begin{cases} b & 0 \leq t \leq 15 \\ \sigma_1^{\{256\}}(W_{t-2}) + W_{t-7} + \sigma_0^{\{256\}}(W_{t-15}) + W_{t-16} & 16 \leq t \leq 63 \end{cases}$$

It goes without saying that the compression function at the heart of SHA-256 also deviates from SHA-1 and employs two temporary variables $T_1$ and $T_2$ (instead of just one).

#### 6.4.4.2 SHA-224

The SHA-224 hash function is the same as SHA-256 with different initialization values (as mentioned above) and the the final hash value $h(m)$ truncated to the leftmost 224 bits. This means that only 7 of the 8 registers are used to form the output, and hence that register $H$ does not influence the output. Everything else remains the same.

#### 6.4.4.3 SHA-512

The SHA-512 hash function is overviewed in Algorithm 6.12. Again, it is conceptually and structurally very similar to SHA-1 and SHA-256, but its word size is now 64 bits. Again, the message schedule $W$ is recursively derived from $b$ according to the same formula used for SHA-256. The only difference is that the message schedule comprises 80 words, and hence that $t$ runs from 0 to 79 and has 80 possible values (instead of 64). The respective formula is as follows:

$$W_t = \begin{cases} b & 0 \leq t \leq 15 \\ \sigma_1^{\{512\}}(W_{t-2}) + W_{t-7} + \sigma_0^{\{512\}}(W_{t-15}) + W_{t-16} & 16 \leq t \leq 79 \end{cases}$$

The final output of the SHA-512 hash function is the concatenation of the 8 registers $A$ to $H$. Since $8 \cdot 64 = 512$, the output value is 512 bits long.

#### 6.4.4.4 SHA-384

The SHA-384 hash function is the same as SHA-512 with different initialization values (as mentioned above) and the final hash value $h(m)$ truncated to the leftmost

**Algorithm 6.12** The SHA-512 hash function (overview).

$(m)$

---

Construct $b = b[0] \parallel b[1] \parallel \ldots \parallel b[n-1]$
$A =$ `0x6A09E667F3BCC908` $B =$ `0xBB67AE8584CAA73B`
$C =$ `0x3C6EF372FE94F82B` $D =$ `0xA54FF53A5F1D36F1`
$E =$ `0x510E527FADE682D1` $F =$ `0x9B05688C2B3E6C1F`
$G =$ `0x1F83D9ABFB41BD6B` $H =$ `0x5BE0CD19137E2179`
for $i = 0$ to $n-1$ do
  Derive message schedule $W$ from $b[i]$
  $A' = A \quad B' = B \quad C' = C \quad D' = D$
  $E' = E \quad F' = F \quad G' = G \quad H' = H$
  for $t = 0$ to 79 do
    $T_1 = H + \Sigma_1^{\{512\}}(E) + Ch(E,F,G) + K_t^{\{512\}} + W_t$
    $T_2 = \Sigma_0^{\{512\}}(A) + Maj(A,B,C)$
    $H = G$
    $G = F$
    $E = D + T_1$
    $D = C$
    $C = B$
    $B = A$
    $A = T_1 + T_2$
  $A = A + A' \quad B = B + B'$
  $C = C + C' \quad D = D + D'$
  $E = E + E' \quad F = F + F'$
  $G = G + G' \quad H = H + H'$

---

$(h(m) = A \parallel B \parallel C \parallel D \parallel E \parallel F \parallel G \parallel H)$

384 bits. This means that only 6 of the 8 registers are used to form the output $(6 \cdot 64 = 384)$, and that registers $G$ and $H$ are not part of the final output.

#### 6.4.4.5 SHA-512/224 and SHA-512/256

The SHA-512/224 and SHA-512/256 hash functions are the same as SHA-512 with different initialization values (as mentioned above) and the final hash value $h(m)$ truncated to the leftmost 224 or 256 bits. We mention these hash functions just for the sake of completeness here; they are not widely used in the field.

### 6.4.5 KECCAK and the SHA-3 Family

As mentioned in Section 6.3, KECCAK[36] is the algorithm selected by the U.S. NIST as the winner of the public SHA-3 competition in 2012.[37] It is the basis for FIPS PUB 202 [25] that complements FIPS PUB 180-4 [19], and it specifies the SHA-3 family that comprises four cryptographic hash functions and two extendable-output functions (XOFs). While a cryptographic hash function outputs a value of fixed length, an XOF has a variable-length output, meaning that its output may have any desired length. XOFs may have many potential applications and use cases, ranging from pseudorandomness generation and key derivation, to message authentication, authenticated encryption, and stream ciphers. Except from their different output lengths, cryptographic hash functions and XOFs look very similar and may even be based on the same construction (as exemplified here).

- The four SHA-3 cryptographic hash functions are named SHA3-224, SHA3-256, SHA3-384, and SHA3-512. As in the case of SHA-2, the numerical suffixes indicate the lengths of the respective hash values.[38]
- The two SHA-3 XOFs are named SHAKE128 and SHAKE256,[39] where the numerical suffixes refer to the security levels (in terms of key length equivalence). SHAKE128 and SHAKE256 are the first XOFs that have been standardized by NIST or any other standardization body.

The SHA-3 hash functions and XOFs employ different padding schemes (as addressed below). In December 2016, NIST released SP 800-185[40] that specifies complementary functions derived from SHA-3. In particular, it specifies four types of SHA-3-derived functions: cSHAKE, KMAC, TupleHash, and ParallelHash. The acronym cSHAKE stands for "customizable SHAKE," and it refers to a SHAKE XOF that can be customized using a function name and a customization bit string. KMAC is a keyed MAC construction that is based on KECCAK or cSHAKE, respectively (Section 10.3.2). Finally, and as their names suggest, TupleHash is a SHA-3-derived function that can be used to hash a tuple of input strings, and ParallelHash can be used to take advantage of the parallelism available in modern processors (using a particular block size). The details of these SHA-3-derived functions are not further addressed here; the details can be found in the NIST SP referenced above.

36 http://keccak.team.

37 http://csrc.nist.gov/groups/ST/hash/sha-3/sha-3_standardization.html.

38 The SHA-2 hash functions are named SHA-224, SHA-256, SHA-384, and SHA-512.

39 The acronym SHAKE stands for "Secure Hash Algorithm with Keccak."

40 https://nvlpubs.nist.gov/nistpubs/SpecialPublications/NIST.SP.800-185.pdf

Unlike all cryptographic hash functions addressed so far, KECCAK and the SHA-3 hash functions do not rely on the Merkle-Damgård construction, but on a so-called *sponge construction*[41] that is based on a permutation operating on a data structure known as the *state*. The state, in turn, can either be seen as a (one-dimensional) bitstring $S$ of length $b$ or a three-dimensional array $\mathbf{A}[x, y, z]$ of bits with appropriate values for $x$, $y$, and $z$ (i.e., $xyz \leq b$).

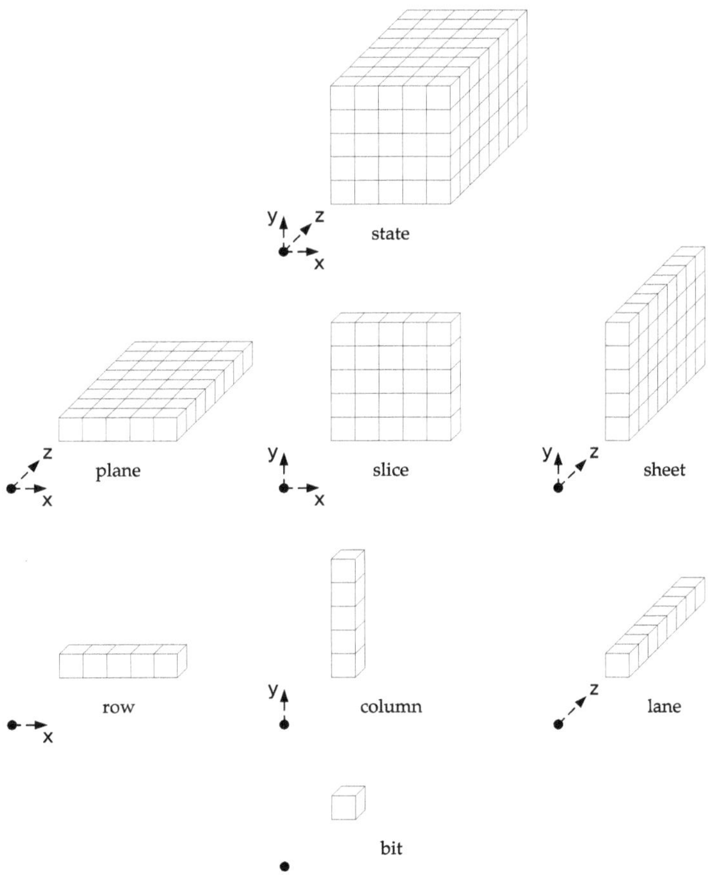


**Figure 6.4** The KECCAK state and its decomposition (© keccak.team).

41 http://sponge.noekeon.org.

The KECCAK state represented as an array **A** and its decomposition (as addressed below) are illustrated in Figure 6.4. This figure is used in this book with kind permission from the developers of KECCAK under a Creative Commons Attribution 4.0 International License (together with Figures 6.8–6.11).[42]

In the case of SHA-3, $b = 1600$, $0 \leq x, y < 5$, and $0 \leq z < w$ (with $w = 2^l = 64$ for $l = 6$ as addressed below). Consequently, the state is either a string $S$ or a $(5 \times 5 \times 64)$-array **A** of 1600 bits (as depicted in Figure 6.4). For all $0 \leq x, y < 5$ and $0 \leq z < w$, the relationship between $S$ and **A** is as follows:

$$\mathbf{A}[x, y, z] = S[w(5y + x) + z]$$

Following this equation, the first element $\mathbf{A}[0, 0, 0]$ translates to $S[0]$, whereas the last element $\mathbf{A}[4, 4, 63]$ translates to $S[64(5 \cdot 4) + 4) + 63] = S[64 \cdot 24 + 63] = S[1599]$.

Referring to Figure 6.4, there are several possibilities to decompose **A**. If one fixes all values on the $x$-, $y$-, and $z$-axes, then one refers to a single *bit*; that is, $bit[x, y, z] = \mathbf{A}[x, y, z]$. If one fixes the values on the $y$- and $z$-axes and only considers a variable value for the $x$-axis, then one refers to a *row*; that is, $row[y, z] = \mathbf{A}[\cdot, y, z]$. If one does something similar and consider a variable value for the $y$-axis ($z$-axis), then one refers to a *column* (*lane*); that is, $column[x, z] = \mathbf{A}[x, \cdot, z]$ ($lane[x, y] = \mathbf{A}[x, y, \cdot]$). Lanes are important in the design of KECCAK because they can be stored in a word and a 64-bit register of a modern processor.

If one fixes the values on the $y$-axis and consider variables value for the $x$- and $z$-axes, then one refers to a *plane*, $plane[y] = \mathbf{A}[\cdot, y, \cdot]$. Again, one can do something similar and consider a fixed value for the $z$-axis ($x$-axis) to refer to a *slice* (*sheet*), $slice[z] = \mathbf{A}[\cdot, \cdot, z]$ ($sheet[x] = \mathbf{A}[x, \cdot, \cdot]$). Some of these terms are used to describe the working principles of KECCAK and its step mappings.

If one wants to convert **A** into a string $S$, then the bits of **A** are concatenated as follows to form $S$:

$$\begin{aligned}
S &= \mathbf{A} = plane[0] \parallel plane[1] \parallel \ldots \parallel plane[4] \\
&= lane[0,0] \parallel lane[1,0] \parallel \ldots \parallel lane[4,0] \parallel \\
&\quad\; lane[0,1] \parallel lane[1,1] \parallel \ldots \parallel lane[4,1] \parallel \\
&\quad\; lane[0,2] \parallel lane[1,2] \parallel \ldots \parallel lane[4,2] \parallel \\
&\quad\; lane[0,3] \parallel lane[1,3] \parallel \ldots \parallel lane[4,3] \parallel \\
&\quad\; lane[0,4] \parallel lane[1,4] \parallel \ldots \parallel lane[4,4]
\end{aligned}$$

42 https://keccak.team/figures.html.

$$
\begin{aligned}
= \; & bit[0,0,0] \; || \; bit[0,0,1] \; || \; bit[0,0,2] \; || \ldots || \; bit[0,0,63] \; || \\
& bit[1,0,0] \; || \; bit[1,0,1] \; || \; bit[1,0,2] \; || \ldots || \; bit[1,0,63] \; || \\
& bit[2,0,0] \; || \; bit[2,0,1] \; || \; bit[2,0,2] \; || \ldots || \; bit[2,0,63] \; || \\
& \ldots \\
& bit[3,4,0] \; || \; bit[3,4,1] \; || \; bit[3,4,2] \; || \ldots || \; bit[3,4,63] \; || \\
& bit[4,4,0] \; || \; bit[4,4,1] \; || \; bit[4,4,2] \; || \ldots || \; bit[4,4,63]
\end{aligned}
$$

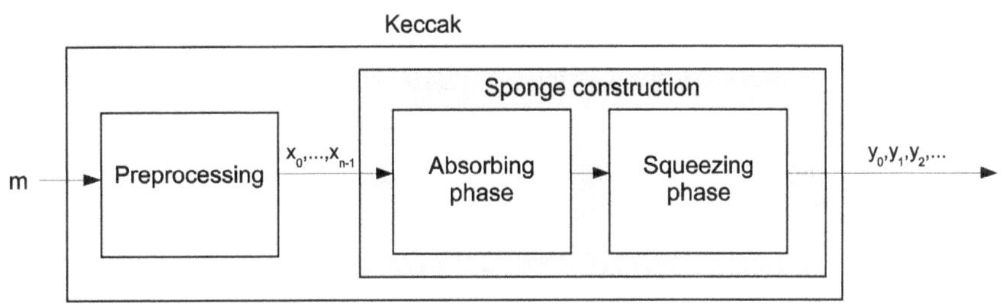


**Figure 6.5** KECCAK and the sponge construction.

The working principles of KECCAK and the sponge construction (used by KECCAK) are overviewed in Figure 6.5. A message $m$ that is input on the left side is preprocessed and properly padded (as explained below) to form a series of $n$ message blocks $x_0, x_1, \ldots, x_{n-1}$. These blocks are then subject to the sponge construction that culminates in a series of output blocks $y_0, y_1, y_2, \ldots$ on the right side. One of the specific features of KECCAK is that the number of output blocks is arbitrary and can be configured at will. In the case of a SHA-3 hash function, for example, only the first output block $y_0$ is required and from this block only the least significant bits are used (the remaining bits are discarded). But in the case of a SHA-3 XOF (i.e., SHAKE128 or SHAKE256), any number of output blocks may be used.

As its name suggests, a sponge construction can be used to absorb and squeeze bits. Referring to Figure 6.5, it consists of two phases:

1. In the *absorbing phase*, the $n$ message blocks $x_0, x_1, \ldots, x_{n-1}$ are consumed and read into the state.

2. In the *squeezing phase*, an output $y_0, y_1, y_2, \ldots$ of configurable length is generated from the state.

The same function $f$ (known as KECCAK $f$-function or $f$-permutation) is used in either of the two phases. The following parameters are used to configure the input and output sizes as well as the security of KECCAK:

- The parameter $b$ refers to the state width (i.e., the bitlength of the state). For KECCAK, it can take any value $b = 5 \cdot 5 \cdot 2^l = 25 \cdot 2^l$ for $l = 0, 1, \ldots, 6$ (i.e., 25, 50, 100, 200, 400, 800, or 1600 bits), but the first two values are only toy values that should not be used in practice. For SHA-3, it is required that $l = 6$, and, hence, $b = 25 \cdot 2^6 = 25 \cdot 64 = 1600$ bits. Since $b = 5 \cdot 5 \cdot 2^l$, the state can be viewed as a cuboid with width 5 (representing the $x$-axis), height 5 (representing the $y$-axis), and length $w = 2^l$ or—as in the case of SHA-3—$2^6 = 64$ (representing the $z$-axis). Anyway, $b$ is the sum of $r$ and $c$ (i.e., $b = r + c$).
- The parameter $r$ is called the *bit rate* (or *rate* in short). Its value is equal to the length of the message blocks, and hence it determines the number of input bits that are processed simultaneously. This also means that it stands for the speed of the construction.
- The parameter $c$ is called the *capacity*. Its value is equal to the state width minus the rate, and it refers to the double security level of the construction (so a construction with capacity 256 has a security level of 128).

**Table 6.6**
The KECCAK Parameter Values for the SHA-3 Hash Functions

| Hash Function | $n$ | $b$ | $r$ | $c$ | $w$ |
|---|---|---|---|---|---|
| SHA3-224 | 224 | 1600 | 1152 | 448 | 64 |
| SHA3-256 | 256 | 1600 | 1088 | 512 | 64 |
| SHA3-384 | 384 | 1600 | 832 | 768 | 64 |
| SHA3-512 | 512 | 1600 | 576 | 1024 | 64 |

The KECCAK parameter values for the SHA-3 hash functions are summarized in Table 6.6. Note that $b$ and $w$ are equal to 1600 and 64 in all versions of SHA-3. Also note that there is a trade-off between the rate $r$ and the capacity $c$. They must sum up to $b$. But whether $r$ or $c$ is made large depends on the application. For any security level it makes sense to select a $c$ that is twice as large and to use the remaining bits for $r$. If, for example, one wants to achieve a security level of 256, then $c$ should be 512 and the remaining $1600 - 512 = 1088$ bits determine $r$.

Before a message $m$ can be processed, it must be padded properly (to make sure that the input has a bitlength that is a multiple of $r$). KECCAK uses a relatively simple padding scheme known as *multirate padding*. It works by appending to $m$ a predetermined bit string $p$, a one, a variable number of zeros, and a terminating one. The number of zeros is chosen so that the total length of the resulting bit string is a multiple of $r$. This can be expressed as follows:

$$\text{Padding}(m) = \underbrace{m \parallel p \parallel 10^{*}1}_{\text{multiple of } r}$$

Note that the string $0^* = 0 \ldots 0$ can also be empty, meaning that it may comprise no zeros at all. Also note that the value of $p$ depends on the mode in which KECCAK is used. When using it as a hash function (and hence as a SHA-2 replacement), $p$ refers to the 2-bit string 01. Contrary to that, when using it to generate a variable-length output, $p$ refers to the 4-bit string 1111. The subtleties of these choices are provided in [25]. Anyway, the minimum number of bits appended to $m$ when used as a hash function is 4 (i.e., 0111), whereas the maximum number of bits appended is $r + 3$ (if the last message block consists of $r - 1$ bits). In the other case (i.e., when used to generate a variable-length output), at least 6 bits and at most $r + 5$ bits are appended. In either case, the result of the padding process is a series of message blocks $x_i$ ($i = 0, 1, \ldots$) each of which has a length of $r$ bits.

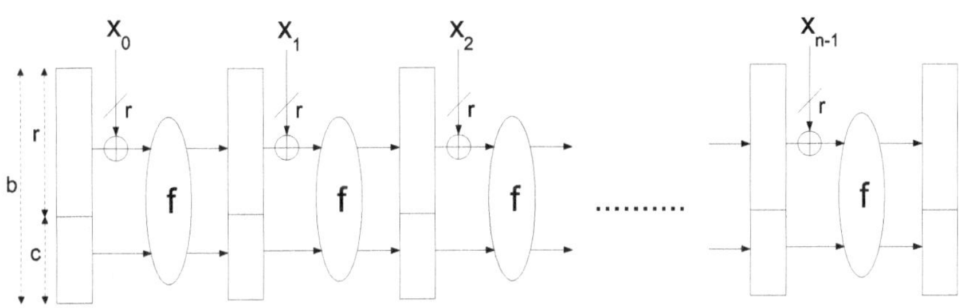


**Figure 6.6** The KECCAK absorbing phase.

As mentioned above, the sponge construction used by KECCAK is based on a permutation of the state. This permutation is called $f$-function or $f$-permutation, and it permutes the $2^b$ possible values of the state. As illustrated in Figures 6.6 and

6.7, the same $f$-function is used in both the absorbing and squeezing phase. It takes $b = r + c$ bits as input and generates an output of the same length. Internally, the $f$-function consists of $n_r$ round functions with the same input and output behavior, meaning that they all take $b$ bits as input and generate $b$ bits of output. Remember that $l$ determines the state width according to $b = 25 \cdot 2^l$ (and that SHA-3 uses the fixed values $l = 6$ and hence $b = 1600$). The value $l$ also determines the number of rounds according to the following formula:

$$n_r = 12 + 2l$$

So the possible state widths 25, 50, 100, 200, 400, 800, and 1600 come along with respective numbers of rounds: 12, 14, 16, 18, 20, 22, and 24. The longer the state width, the larger the number of rounds (to increase the security level). As SHA-3 fixes the state width to 1600 bits, the number of rounds is also fixed to 24 (i.e., $n_r = 24$).

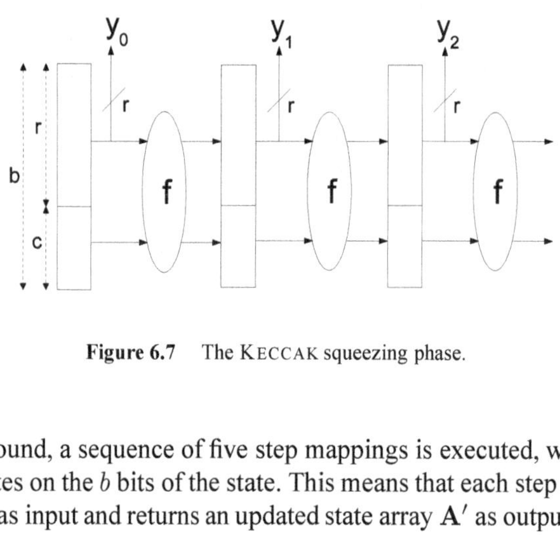


**Figure 6.7** The KECCAK squeezing phase.

In each round, a sequence of five step mappings is executed, where each step mapping operates on the $b$ bits of the state. This means that each step mapping takes a state array $\mathbf{A}$ as input and returns an updated state array $\mathbf{A}'$ as output. The five step mappings are denoted by Greek letters: theta ($\theta$), rho ($\rho$), pi ($\pi$), chi ($\chi$), and iota ($\iota$). While the first step mapping $\theta$ must be applied first, the order of the other step mappings is arbitrary and does not matter. If graphically interpreted in $\mathbf{A}$, the step mappings are relatively simple and straightforward.[43] But if defined mathematically, the respective formulas look clumsy and involved. For $0 \leq x, y \leq 4$ and $0 \leq z \leq w$, the definitions look as follows:

43 An animation is available at http://celan.informatik.uni-oldenburg.de/kryptos/info/keccak/overview.

$$\begin{aligned}
\theta &: \mathbf{A}'[x,y,z] = \mathbf{A}[x,y,z] + \sum_{y'=0}^{4} \mathbf{A}[x-1,y',z] + \sum_{y'=0}^{4} \mathbf{A}[x+1,y',z-1] \\
\rho &: \mathbf{A}'[x,y,z] = \mathbf{A}[x,y,z-(t+1)(t+2)/2] \\
&\quad \text{with } 0 \leq t < 24 \text{ and } \begin{pmatrix} 0 & 1 \\ 2 & 3 \end{pmatrix}^t \begin{pmatrix} 1 \\ 0 \end{pmatrix} = \begin{pmatrix} x \\ y \end{pmatrix} \text{ in GF}(5)^{2\times 2} \\
&\quad \text{or } t = -1 \text{ if } x = y = 0 \\
\pi &: \mathbf{A}'[x,y] = \mathbf{A}[x',y'] \text{ with } \begin{pmatrix} x \\ y \end{pmatrix} = \begin{pmatrix} 0 & 1 \\ 2 & 3 \end{pmatrix} \begin{pmatrix} x' \\ y' \end{pmatrix} \\
\chi &: \mathbf{A}'[x] = \mathbf{A}[x] + (\mathbf{A}[x+1]+1) \cdot \mathbf{A}[x+2] \\
\iota &: \mathbf{A}' = \mathbf{A} + \text{RC}[i_r]
\end{aligned}$$

The definitions are explained in the text that follows. The addition and multiplication operations are always performed bitwise in $GF(2)$. This suggests that the addition is equal to the Boolean XOR operation ($\oplus$) and the multiplication is equal to the Boolean AND operation ($\wedge$). With the exception of the round constants $\text{RC}[i_r]$ used in step mapping $\iota$, the step mappings are the same in all rounds. We now look at each of the step mappings individually.

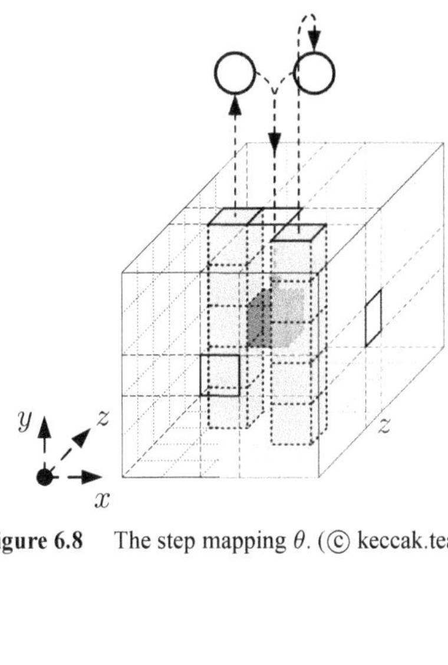


**Figure 6.8** The step mapping $\theta$. (© keccak.team)

#### 6.4.5.1 Step Mapping $\theta$

The step mapping $\theta$ is formally defined as

$$\mathbf{A}'[x,y,z] = \mathbf{A}[x,y,z] + \sum_{y'=0}^{4} \mathbf{A}[x-1,y',z] + \sum_{y'=0}^{4} \mathbf{A}[x+1,y',z-1]$$

This formula means that each bit of the state is replaced with the modulo 2 sum of itself and the bits of two adjacent columns, namely that of $column[x-1,z]$ and $column[x+1,z-1] = \mathbf{A}[x+1,\cdot,z-1]$. An algorithm that can be used to compute $\theta$ and turn $\mathbf{A}$ into $\mathbf{A}'$ is sketched in Algorithm 6.13 and illustrated in Figure 6.8. Note that the algorithm can be made more efficient if the bits of an entire lanes are processed simultaneously. The resulting algorithm deviates from Algorithm 6.13 (and this may be confusing for the reader).

**Algorithm 6.13** Step mapping $\theta$.

($\mathbf{A}$)

---

for $x = 0$ to 4 do
    for $z = 0$ to $w-1$ do
        $\mathrm{C}[x,z] = \mathbf{A}[x,0,z] \oplus \mathbf{A}[x,1,z] \oplus \mathbf{A}[x,2,z] \oplus \mathbf{A}[x,3,z] \oplus \mathbf{A}[x,4,z]$
for $x = 0$ to 4 do
    for $z = 0$ to $w-1$ do
        $\mathrm{D}[x,z] = \mathrm{C}[(x-1) \bmod 5, z] \oplus \mathrm{C}[(x+1) \bmod 5, (z-1) \bmod w]$
for $x = 0$ to 4 do
    for $y = 0$ to 4 do
        for $z = 0$ to $w-1$ do
            $\mathbf{A}'[x,y,z] = \mathbf{A}[x,y,z] \oplus \mathrm{D}[x,z]$

---

($\mathbf{A}'$)

#### 6.4.5.2 Step Mapping $\rho$

The step mapping $\rho$ is formally defined as

$$\mathbf{A}'[x,y,z] = \mathbf{A}[x,y,z-(t+1)(t+2)/2]$$

with appropriately chosen values for $t$ (as specified above). According to this definition, $\rho$ processes each lane $lane[x,y] = \mathbf{A}[x,y,\cdot]$ of the state individually by rotating its bits by a value—called the *offset*—that depends on the $x$ and $y$ coordinates of the lane. This means that for each bit of the lane, the $z$ coordinate is

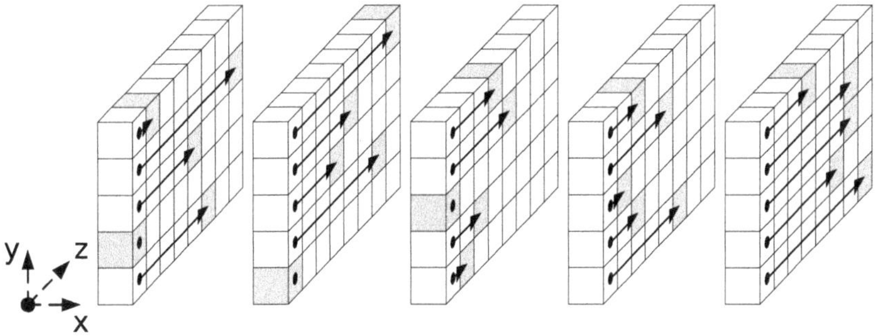


**Figure 6.9** The step mapping $\rho$. (© keccak.team)

modified by adding the offset modulo $w$. The algorithm that computes the $\rho$ mapping by turning $\mathbf{A}$ into $\mathbf{A}'$ is sketched in Algorithm 6.14 and illustrated in Figure 6.9.

**Algorithm 6.14** Step mapping $\rho$.

$(\mathbf{A})$

---

for $z = 0$ to $w - 1$ do $\mathbf{A}'[0, 0, z] = \mathbf{A}[0, 0, z]$
$(x, y) = (1, 0)$
for $t = 0$ to 23 do
    for $z = 0$ to $w$ do $\mathbf{A}'[x, y, z] = \mathbf{A}[x, y, (z - (t + 1)(t + 2)/2) \bmod w]$
    $(x, y) = (y, (2x + 3y) \bmod 5)$

---

$(\mathbf{A}')$

The function to compute the offset can be expressed mathematically (as given above) or in a table (as given in Table 6.7). Note that the offset for $lane[0, 0]$ is equal to zero, so this lane is left as is and is not subject to a rotation.

#### 6.4.5.3 Step Mapping $\pi$

The step mapping $\pi$ is formally defined as

$$\mathbf{A}'[x, y] = \mathbf{A}[x', y'] \text{ with } \begin{pmatrix} x \\ y \end{pmatrix} = \begin{pmatrix} 0 & 1 \\ 2 & 3 \end{pmatrix} \begin{pmatrix} x' \\ y' \end{pmatrix}$$

Again, it operates on lanes and rearranges them at some different position. In fact, each lane $lane[x', y']$ goes to a new position (with coordinates $x$ and $y$) according

**Table 6.7**
The Offset Values Used by the Step Mapping $\rho$

| | $x=3$ | $x=4$ | $x=0$ | $x=1$ | $x=2$ |
|---|---|---|---|---|---|
| $y=2$ | 153 | 231 | 3 | 10 | 171 |
| $y=1$ | 55 | 276 | 36 | 300 | 6 |
| $y=0$ | 28 | 91 | 0 | 1 | 190 |
| $y=4$ | 120 | 78 | 210 | 66 | 253 |
| $y=3$ | 21 | 136 | 105 | 45 | 15 |

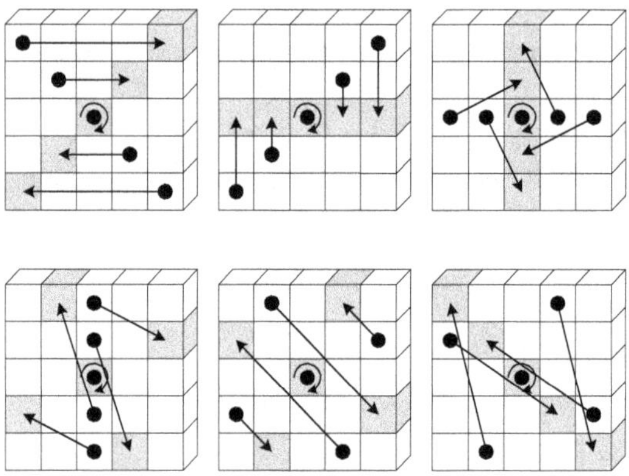

**Figure 6.10** The step mapping $\pi$. (© keccak.team)

to a multiplication of $x'$ and $y'$ with the (2x2)-matrix given above (note that the elements of the matrix are from $\mathbb{Z}_5$ or $GF(5)$). Again, the only exception is the $lane[0, 0]$ that remains unchanged and is not repositioned.

If one applies the matrix multiplication, then the following equations must hold in $\mathbb{Z}_5$:

$$\begin{aligned} x &= y' \\ y &= 2x' + 3y' \end{aligned}$$

Solving these equations for $y'$ and $x'$ yields $y' = x$ on the one hand and $2x' = y - 3y' = y - 3x$ on the other hand. To isolate $x'$ in the second equation, one can multiply either side with $2^{-1} = 3 \in \mathbb{Z}_5$ results in

$$\begin{aligned} 2x' &= y - 3x \\ 3(2x') &= 3(y - 3x) \\ 6x' &= 3y - 9x \\ x' &= 3y + x = x + 3y \end{aligned}$$

(because $6x' \equiv x' \pmod 5$ on the left side and $-9x \equiv x \pmod 5$ on the right side). These are the equations that are used to determine the new position of a lane from its old position. The respective algorithm is sketched in Algorithm 6.15 and illustrated in Figure 6.10.

**Algorithm 6.15** Step mapping $\pi$.

$(\mathbf{A}[x, y])$

---

for $x = 0$ to 4 do
  for $y = 1$ to 4 do
    for $z = 0$ to $w - 1$ do
      $\mathbf{A}'[x, y, z] = \mathbf{A}[(x + 3y) \pmod 5), x, z]$

---

$(\mathbf{A}')$

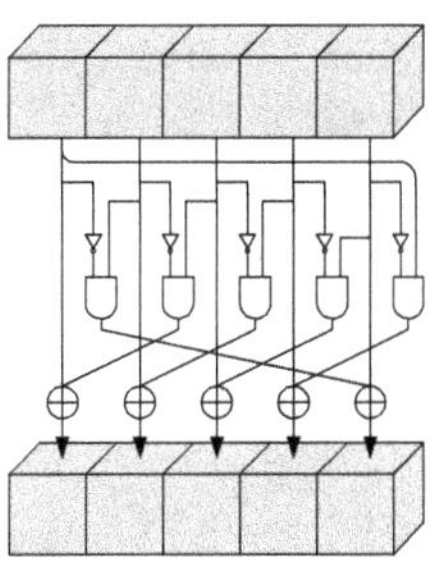

**Figure 6.11** The step mapping $\chi$. (© keccak.team)

#### 6.4.5.4 Step Mapping $\chi$

The step mapping $\chi$ is formally defined as

$$\mathbf{A}'[x] = \mathbf{A}[x] + (\mathbf{A}[x + 1] + 1) \cdot \mathbf{A}[x + 2]$$

It is the only nonlinear mapping in the $f$-function of KECCAK. Again, it operates on lanes and adds each lane $lane[x, y]$ modulo 2 (XOR) with the product (logical AND) of the inverse of $lane[x + 1, y]$ and $lane[x + 2, y]$. The respective algorithm is sketched in Algorithm 6.16 and illustrated in Figure 6.11 (for a single row).

**Algorithm 6.16** Step mapping $\chi$.

($\mathbf{A}$)

---

for $x = 0$ to 4 do
  for $y = 1$ to 4 do
    for $z = 0$ to $w - 1$ do
      $\mathbf{A}'[x, y, z] = \mathbf{A}[x, y, z] \oplus ((\mathbf{A}[(x + 1) \bmod 5, y, z] \oplus 1) \cdot \mathbf{A}[(x + 2) \bmod 5, y, z])$

---

($\mathbf{A}'$)

#### 6.4.5.5 Step Mapping $\iota$

Finally, the step mapping $\iota$ is formally defined as

$$\mathbf{A}' = \mathbf{A} + \mathrm{RC}[i_r]$$

The idea of this mapping is to add modulo 2 a predefined $w$-bit constant $\mathrm{RC}[i_r]$ to $lane[0, 0]$, and to leave all other 24 lanes as they are. The constant $\mathrm{RC}[i_r]$ is parameterized with the round index $0 \leq i_r \leq n_r - 1 = 23$, meaning that it yields a different value in every round.

**Algorithm 6.17** Step mapping $\iota$.

($\mathbf{A}, i_r$)

---

for $x = 0$ to 4 do
  for $y = 1$ to 4 do
    for $z = 0$ to $w - 1$ do
      $\mathbf{A}'[x, y, z] = \mathbf{A}[x, y, z]$
$RC = 0^w$
for $j = 0, \ldots, l$ do $RC[2^j - 1] = rc(j + 7i_r)$
for $z = 0$ to $w - 1$ do $\mathbf{A}'[0, 0, z] = \mathbf{A}'[0, 0, z] \oplus RC[z]$

---

($\mathbf{A}'$)

The algorithm to compute $\iota$ is shown in Algorithm 6.17. First, the content of $\mathbf{A}'$ is copied to $\mathbf{A}$. Then $lane[0, 0]$ is updated as described above.

Each round constant mainly consists of zeros and is generated with an auxiliary function $rc$. This function takes as input parameter a byte $t$ and returns as output parameter a bit $rc(t)$. It is defined as the output of an LFSR that is defined as follows:

$$rc[t] \quad = \quad (x^t \mod x^8 + x^6 + x^5 + x^4 + 1) \mod x \text{ in GF}(2)[x]$$

The algorithm that can be used to implement the LFSR and to compute the auxiliary function $rc$ is shown in Algorithm 6.18. The resulting round constants are listed in Table 6.8, where $RC[i_r]$ refers to the round constant used in round $i_r$.

**Algorithm 6.18** Auxiliary function $rc$.

$(t)$

---

if $t \bmod 255 = 0$ then return 1
$R = 10000000$
for $i = 1$ to $t \bmod 255$ do
    $R = 0 \parallel R$
    $R[0] = R[0] \oplus R[8]$
    $R[4] = R[4] \oplus R[8]$
    $R[5] = R[5] \oplus R[8]$
    $R[6] = R[6] \oplus R[8]$
    $R = \text{Trunc}_8[R]$

---

$(R[0])$

**Table 6.8**
The 24 Round Constants Employed by SHA-3

| | | | |
|---|---|---|---|
| $RC[0]$ | 0x0000000000000001 | $RC[12]$ | 0x000000008000808B |
| $RC[1]$ | 0x0000000000008082 | $RC[13]$ | 0x800000000000008B |
| $RC[2]$ | 0x800000000000808A | $RC[14]$ | 0x8000000000008089 |
| $RC[3]$ | 0x8000000080008000 | $RC[15]$ | 0x8000000000008003 |
| $RC[4]$ | 0x000000000000808B | $RC[16]$ | 0x8000000000008002 |
| $RC[5]$ | 0x0000000080000001 | $RC[17]$ | 0x8000000000000080 |
| $RC[6]$ | 0x8000000080008081 | $RC[18]$ | 0x000000000000800A |
| $RC[7]$ | 0x8000000000008009 | $RC[19]$ | 0x800000008000000A |
| $RC[8]$ | 0x000000000000008A | $RC[20]$ | 0x8000000080008081 |
| $RC[9]$ | 0x0000000000000088 | $RC[21]$ | 0x8000000000008080 |
| $RC[10]$ | 0x0000000080008009 | $RC[22]$ | 0x0000000080000001 |
| $RC[11]$ | 0x000000008000000A | $RC[23]$ | 0x8000000080008008 |

Given a state **A** and a round index $i_r$, the round function Rnd refers to the transformation that results from applying the step mappings $\theta$, $\rho$, $\pi$, $\chi$, and $\iota$ in a

**Algorithm 6.19** KECCAK-$p[b, n_r]$.

$(S, n_r)$

---

convert $S$ into state $\mathbf{A}$
for $i_r = 2l + 12 - n_r, \ldots, 2l + 12 - 1$ do $\mathbf{A} = \text{Rnd}(\mathbf{A}, i_r)$
convert $\mathbf{A}$ into $b$-bit string $S'$

---

$(S')$

particular order:[44]

$$\text{Rnd}(\mathbf{A}, i_r) = \iota(\chi(\pi(\rho(\theta(\mathbf{A})))), i_r)$$

In general, the KECCAK-$p[b, n_r]$ permutation consists of $n_r$ iterations of the round function Rnd as specified in Algorithm 6.19. The algorithm takes a $b$-bit string $S$ and a number of rounds ($n_r$) as input parameters and computes another $b$-bit string $S'$ as output parameter. The algorithm is fairly simple: $S$ is converted to the state $\mathbf{A}$, $n_r$ round functions Rnd are applied to the state, and the resulting state is finally converted back to the output string $S'$.

The KECCAK-$f$ family of permutations refers to the specialization of the KECCAK-$p$ family to the case where $n_r = 12 + 12l$. This means:

$$\text{KECCAK-}f[b] = \text{KECCAK-}p[b, 12 + 2l]$$

Alternatively speaking, the KECCAK-$p[1600, 24]$ permutation, which underlies the six SHA-3 functions, is equivalent to KECCAK-$f[1600]$.

During the SHA-3 competition, the security of KECCAK was challenged rigorously. Nobody found a possibility to mount a collision attack that is more efficient than brute-force. This has not changed since then, and hence people feel confident about the security of SHA-3. But people also feel confident about the security of the cryptographic hash functions from the SHA-2 family.[45] So whether SHA-3 will be successfully deployed in the field is not only a matter of security. There may be other reasons to stay with SHA-2 or move to SHA-3 (or even to any other cryptographic hash function). The effect of these reasons is difficult to predict, and hence it is hard to tell whether SHA-3 will be successful in the long term and how long this may take.

44 As mentioned above, the step mapping $\theta$ must be applied first, whereas the order of the other step mappings is arbitrary.

45 A summary of the security of SHA-1, SHA-2, and SHA-3 is given in appendix A.1 of [25].

## 6.5 FINAL REMARKS

As outlined in this chapter, most cryptographic hash functions in use today follow the Merkle-Damgård construction. This means that a collision-resistant compression function is iterated multiple times (one iteration per block of the message). There are two remarks to make:

- First, each iteration can only start if the preceding iteration has finished. This suggests that the hash function may become a performance bottleneck. This point was already made in 1995 [26], when it was argued that the hash rates of MD5 may be insufficient to keep up with high-speed networks. The problem is in fact the iterative nature of MD5 and its block chaining structure, which prevent parallelism. It is possible to modify the MD5 algorithm to accommodate a higher throughput, but it is certainly simpler and more appropriate to design and come up with cryptographic hash functions that natively support parallelism.
- Second, the design of compression functions that are collision-resistant looks more like an art than a science, meaning that there is no specific theory on how to make compression functions maximum resistant to collisions (or multicollisions).

Following the second remark, there are hardly any design criteria that can be used to design and come up with new compression functions (for cryptographic hash functions that follow the Merkle-Damgård construction) or entirely new cryptographic hash functions (such as KECCAK). This lack of design criteria contradicts the importance of cryptographic hash functions in almost all cryptographic systems and applications in use today.

If people lack proper design criteria and come up with ad hoc designs, then they may not be sure about the collision resistance of the respective cryptographic hash functions. Sometimes, they try to improve collision resistance by concatenating two (or even more) such functions. For example, instead of using MD5 or SHA-1 alone, they may apply one function after the other and concatenate the respective hash values.[46] Intuition suggests that the resulting (concatenated) hash function is more collision-resistant than each function applied individually. Unfortunately, intuition is wrong and illusive here. In 2004, it was shown by Joux that finding multicollisions is not much harder than finding "normal" collisions in this setting, and that this implies that concatenating the results of several iterated hash functions in order to build a new one does not yield a secure construction [27]. Since then, we

46 Such a construction was used, for example, in SSL 3.0 to derive keying material from a premaster secret.

know that concatenating two (or even more) iterated hash functions does not make the resulting hash function significantly more collision-resistant.

We use cryptographic hash functions a lot throughout the book. When combined with a key, for example, they are widely used for key derivation (Section 7.4 and message authentication (Section 10.3.2). An alternative design for cryptographic hash functions that are particularly well suited for message authentication was originally proposed by J. Larry Carter and Mark N. Wegman in the late 1970s [28, 29], and later refined by other researchers (e.g., [30]). Instead of using a single hash function, they consider families of such functions from which a specific function is (pseudo)randomly selected. Such a family consists of all hash functions $h : X \rightarrow Y$ that map values from $X$ to values from $Y$. More specifically, a family $H$ of hash functions $h$ is called *two-universal*, if for every $x, y \in X$ with $x \neq y$ it holds that

$$\Pr_{h \xleftarrow{r} H}[h(x) = h(y)] \leq \frac{1}{|Y|}$$

This suggests that the images are uniformly distributed in $Y$, and that the probability of having two images collide is as small as possible (given the size of $Y$). This notion of universality can be generalized beyond two, but two-universal families of hash functions are the most important use case. Using universal families of hash functions is called *universal hashing*, and universal hashing is the basic ingredient for Carter-Wegman MACs addressed in Section 10.3.3. At this point in time, we just want to introduce the term and put it into perspective. MACs require a secret key, and hence message authentication is one of the topics that is addressed in Part II of the book.

## References

[1] ISO/IEC 10118-1, *Information technology — Security techniques — Hash-functions — Part 1: General*, 2016.

[2] ISO/IEC 10118-2, *Information technology — Security techniques — Hash-functions — Part 2: Hash-functions using an n-bit block cipher*, 2010.

[3] ISO/IEC 10118-3, *Information technology — Security techniques — Hash-functions — Part 3: Dedicated hash-functions*, 2004.

[4] ISO/IEC 10118-4, *Information technology — Security techniques — Hash-functions — Part 4: Hash-functions using modular arithmetic*, 1998.

[5] Merkle, R.C., "One Way Hash Functions and DES," *Proceedings of CRYPTO '89*, Springer-Verlag, LNCS 435, 1989, pp. 428–446.

[6] Damgård, I.B., "A Design Principle for Hash Functions," *Proceedings of CRYPTO '89*, Springer-Verlag, LNCS 435, 1989, pp. 416–427.

[7] Kaliski, B., *The MD2 Message-Digest Algorithm*, Request for Comments 1319, April 1992.

[8] Rivest, R.L., *The MD4 Message-Digest Algorithm*, Request for Comments 1320, April 1992.

[9] Biham, E., and A. Shamir, "Differential Cryptanalysis of Snefru, Khafre, REDOC-II, LOKI, and Lucifer," *Proceedings of CRYPTO '91*, Springer-Verlag, LNCS 576, 1991, pp. 156–171.

[10] den Boer, B., and A. Bosselaers, "An Attack on the Last Two Rounds of MD4," *Proceedings of CRYPTO '91*, Springer-Verlag, LNCS 576, 1991, pp. 194–203.

[11] Rivest, R.L., *The MD5 Message-Digest Algorithm*, Request for Comments 1321, April 1992.

[12] Dobbertin, H., "Cryptanalysis of MD4," *Journal of Cryptology*, Vol. 11, No. 4, 1998, pp. 253–271.

[13] den Boer, B., and A. Bosselaers, "Collisions for the Compression Function of MD5," *Proceedings of EUROCRYPT '93*, Springer-Verlag, LNCS 765, 1993, pp. 293–304.

[14] Dobbertin, H., "The Status of MD5 After a Recent Attack," *CryptoBytes*, Vol. 2, No. 2, Summer 1996.

[15] Wang, X., and H. Yu, "How to Break MD5 and Other Hash Functions," *Proceedings of EUROCRYPT '05*, Springer-Verlag, LNCS 3494, 2005, pp. 19–35.

[16] Chabaud, F., and A. Joux, "Differential Collisions in SHA-0," *Proceedings of CRYPTO '98*, Springer-Verlag, LNCS 1462, 1998, pp. 56–71.

[17] Eastlake, 3rd, D., and T. Hansen, *US Secure Hash Algorithms (SHA and HMAC-SHA)*, Request for Comments 4634, July 2006.

[18] Housley, R., *A 224-Bit One-Way Hash Function: SHA-224*, Request for Comments 3874, September 2004.

[19] U.S. Department of Commerce, National Institute of Standards and Technology, *Secure Hash Standard*, FIPS PUB 180-4, March 2012.

[20] Wang, X., Y. Yin, and R. Chen, "Finding Collisions in the Full SHA-1," *Proceedings of CRYPTO 2005*, Springer-Verlag, LNCS, 2005.

[21] Dobbertin, H., A. Bosselaers, and B. Preneel, "RIPEMD-160: A Strengthened Version of RIPEMD," *Proceedings of the 3rd International Workshop on Fast Software Encryption*, Springer-Verlag, LNCS 1039, 1996, pp. 71–82.

[22] Preneel, B., A. Bosselaers, and H. Dobbertin, "The Cryptographic Hash Function RIPEMD-160," *CryptoBytes*, Vol. 3, No. 2, 1997, pp. 9–14.

[23] Zheng, Y., J. Pieprzyk, and J. Seberry, "HAVAL — A One-Way Hashing Algorithm with Variable Length of Output," *Proceedings of AUSCRYPT' 92*, Springer-Verlag, LNCS 718, 2003, pp. 83–104.

[24] Stevens, M., P. Karpman, and T. Peyrin, "Freestart collision for full SHA-1," IACR ePrint Archive, 2015, https://eprint.iacr.org/2015/967.pdf.

[25] U.S. Department of Commerce, National Institute of Standards and Technology, *SHA-3 Standard: Permutation-Based Hash and Extendable-Output Functions*, FIPS PUB 202, August 2015.

[26] Touch, J., *Report on MD5 Performance*, RFC 1810, June 1995.

[27] Joux, A., "Multicollisions in Iterated Hash Functions, Application to Cascaded Constructions," *Proceedings of CRYPTO 2004*, Springer-Verlag, LNCS 3152, 2004, pp. 306–316.

[28] Carter, J.L., and M.N. Wegman, "Universal Classes of Hash Functions," *Journal of Computer and System Sciences*, Vol. 18, 1979, pp. 143–154.

[29] Wegman, M.N., and J.L. Carter, "New Hash Functions and Their Use in Authentication and Set Equality," *Journal of Computer and System Sciences*, Vol. 22, 1981, pp. 265–279.

[30] Naor, M., and M. Yung, "Universal One-Way Hash Functions and their Cryptographic Applications ," *Proceedings of the Twenty First Annual ACM Symposium on Theory of Computing (STOC '89)*, ACM Press, pp. 33–43.

# Part II

# SECRET KEY CRYPTOSYSTEMS

# Chapter 7

# Pseudorandom Generators

*Anyone who considers arithmetical methods of producing random digits is, of course, in a state of sin.*

— John von Neumann[1]

In Chapter 3, we introduced random generators and argued that it is inherently difficult to implement them. We also concluded that it is sometimes appropriate to combine them with a way of stretching a truly random bit sequence into a much longer bit sequence that appears to be random. This is where pseudorandom generators and pseudorandom bit generators come into play. They are the main topic of this chapter: We provide an introduction in Section 7.1, overview some exemplary (ad hoc) constructions in Section 7.2, elaborate on cryptographically secure pseudorandom generators in Section 7.3, and conclude with some final remarks (especially focusing on key derivation) in Section 7.4. In the end, we want to be able to argue against John von Neumann's quote and to show that random—or at least random-looking—digits can in fact be produced with arithmetical methods.

## 7.1 INTRODUCTION

According to Section 2.2.1 and Definition 2.7, a *pseudorandom generator* (PRG) is an efficiently computable function that takes as input a relatively short value of length $n$, called the *seed*, and generates as output a value of length $l(n)$ with $l(n) \gg n$ that appears to be random (and is therefore called pseudorandom). If the

1 John von Neumann was a Hungarian-American mathematician who lived from 1903 to 1957.

input and output values are bit sequences (as depicted in Figure 2.6), then the PRG represents a *pseudorandom bit generator* (PRBG). In this chapter, we sometimes use the acronym PRG to refer to both a PRG and a PRBG.

The seed of a PRG can be seen as a secret key that is the input to the "efficiently computable function" to generate a particular output. As such, a PRG is a secret key cryptosystem, and is therefore qualified to introduce Part II. The key question is how to construct a function whose output looks random and cannot be distinguished from the output of a true random generator. This is what this chapter is all about.

In Section 2.2.1, we also said that a PRBG $G$ is a mapping from the key space $\mathcal{K} = \{0,1\}^n$ to $\{0,1\}^{l(n)}$, where $l(n)$ represents a stretch function; that is, a function that stretches an $n$-bit value into a much longer $l(n)$-bit value with $n < l(n) \leq \infty$:

$$G : \mathcal{K} \longrightarrow \{0,1\}^{l(n)}$$

Combining Definition 2.7 with this formalism suggests that a PRG is an efficiently computable function that defines the mapping $G$. With regard to this mapping, we said in Definition 2.7 that the output must "appear to be random." This must be defined more precisely. What does it mean that an output sequence appears to be random, given the fact that we cannot measure randomness is a meaningful way? This is particularly challenging, because—unlike a true random generator—a PRG operates deterministically, and hence it always outputs the same values if seeded with the same input values.

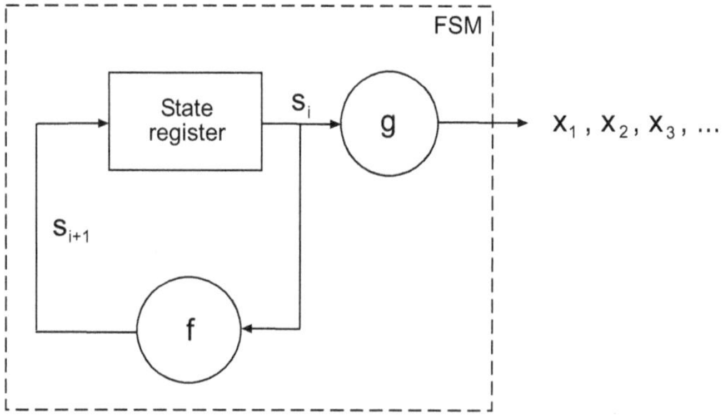


**Figure 7.1** An idealized model of a PRG representing an FSM.

The fact that a PRG operates deterministically means that it represents a *finite state machine* (FSM). An idealized model of a PRG representing an FSM is illustrated in Figure 7.1. The model comprises the following components:

- A state register;
- A next-state or *state-transition function* $f$;
- An output function $g$.

The state register is initialized with an $n$-bit seed $s_0$. In each cycle $i \geq 0$, the next-state function $f$ computes $s_{i+1}$ from $s_i$; that is, $s_{i+1} = f(s_i)$, and $s_i$ is subject to the output function $g$. The result is $x_i = g(s_i)$, and the bit sequence

$$(x_i)_{i \geq 1} = x_1, x_2, x_3, \ldots$$

yields the output of the PRG. In this idealized model, the function $f$ operates recursively on the state register, and there is no other input to the PRG than the seed the PRG actually begins with. Some PRGs used in practice slightly deviate from this idealized model by allowing the state register to be reseeded periodically. This may be modeled by having a function $f$ take into account additional sources of randomness (this possibility is not illustrated in Figure 7.1). In this case, however, the distinction between a PRG and a true random bit generator gets fuzzy.

In an FSM, the number of states is finite and depends on the length of the state register. If the seed comprises $n$ bits, then there are at most $2^n$ possible states (there are fewer states if the FSM is not well designed). This means that after at most $2^n$ cycles, the FSM will be in the same state it originally started with, and hence the sequence of output values starts repeating itself. Alternatively speaking, the sequence of output values is cyclic (with a potentially very large cycle). This is why we cannot require that the output of a PRG is truly random, but only that it appears to be so. Somebody who is able to wait for a very long time and has access to huge amounts of memory will eventually recognize that the output values start to repeat themselves. This suggests that the output values are not truly randomly generated, but generated deterministically with a PRG.

This insight may help us to clearly define what is meant by saying that the output values "appear to be random," and hence to more precisely define and nail down the security of a PRG. Remember the security game from Section 1.2.2.1 and Figure 1.2: In this game, an adversary is interacting (i.e., observing the output values) with either a true random generator (representing an ideal system) or a PRG (representing a real system). If he or she can tell the two cases apart, then the output values of the PRG do not appear to be random, and hence it cannot be considered to be secure or even cryptographically secure. We formalize this idea and line of argumentation later in this chapter (Section 7.3).

Anyway, it is obvious that a minimal security requirement for a PRG is that the length of the seed, $n = |s_0|$, is sufficiently large so that an exhaustive search over all $2^n$ possible seeds is computationally infeasible. Also, the output bit sequence generated by the PRG must pass all relevant statistical randomness tests (Section 3.3). Note, however, that these results must be taken with a grain of salt, because passing statistical randomness tests is a necessary but usually insufficient requirement for a PRG to be (cryptographically) secure. For example, the following two PRGs pass most statistical randomness tests but are still insecure for cryptographic purposes:

- PRGs that employ the binary expansion of algebraic numbers, such as $\sqrt{3}$ or $\sqrt{5}$.
- Linear congruential generators (LCG) that take as input a seed $x_0 = s_0$ and three integer parameters $a, b, n \in \mathbb{N}$ with $a, b < n$, and that use the linear recurrence

$$x_i = ax_{i-1} + b \pmod{n}$$

  to recursively generate an output sequence $(x_i)_{i \geq 1}$ with $x_i < n$. Linear congruential generators are frequently used for simulation purposes and probabilistic algorithms (see, for example, Chapter 3 of [1]), but they are highly predictable (i.e., it is possible to infer the parameters $a$, $b$, and $n$ given just a few output values $x_i$ [2, 3]), and this makes them useless for most cryptographic applications. To some extent, this also applies to QCGs introduced in Section 5.3.1.3.

Contrary to these examples, there are PRGs that can be used for cryptographic purposes. Some exemplary (ad hoc) constructions of PRGs are overviewed and discussed in the following section. Furthermore, we will see in Section 9.5 that most additive stream ciphers in use today, such as RC4/ARCFOUR and Salsa20, yield PRGs, meaning that additive stream ciphers and PRGs are conceptually very similar if not identical constructs.

## 7.2 EXEMPLARY CONSTRUCTIONS

Before we overview some exemplary constructions for PRGs, we want to debunk a popular fallacy: People often argue that a simple PRG can be built from a one-way function $f$ by randomly selecting a seed $s_0$, initializing the state register with this value, incrementing the state register in each cycle (i.e.,$s_{i+1} = s_i + 1$ for every $i > 0$), and subjecting the content of the state register to $f$. This generates the output values

$$\begin{aligned} x_1 &= f(s_0) \\ x_2 &= f(s_1) = f(s_0+1) \\ x_3 &= f(s_2) = f(s_0+2) \\ &\ldots \end{aligned}$$

that form the output sequence

$$(x_i)_{i\geq 1} = f(s_0), f(s_0+1), f(s_0+2), f(s_0+3), \ldots$$

Unfortunately, the resulting PRG need not be secure, and there are many situations in which the output values do not appear to be random. If, for example, $g$ is a one-way function, and $f$ extends $g$ by appending a one to the function result of $g$; that is, $f(x) = g(x)\|1$, then $f$ is still a one-way function. However, the output values that are generated with this function do not appear to be random, because each value ends with a one. The bottom line is that more involved constructions are usually required to build a PRG from a one-way function, and these constructions take advantage of hard-core predicates. They are addressed in the following section.

If we want to construct a more secure PRG, then we may start with an LFSR. In Section 9.5.1, we will argue that a single LFSR is not sufficient to construct a secure PRG, but that multiple LFSRs with irregular clocking may be used instead. We will mention A5/1 and A5/2, CSS, and E0 that follow this design paradigm. Furthermore, there are at least two LFSR-based PRG constructions that are known to have good cryptographic properties: the shrinking generator and the self-shrinking generator.

- The *shrinking generator* [4] employs two LFSRs $A$ and $S$ to generate two sequences $(a_i)_{i\geq 0}$ and $(s_i)_{i\geq 0}$. In each clock cycle $i$, the generator outputs $a_i$ if and only if $s_i = 1$. Otherwise, $a_i$ is discarded. The mode of operation of the shrinking generator is illustrated in Figure 7.2. The output sequence in this example is 0101010...

- The *self-shrinking generator* [5] employs only one LFSR $A$ to generate the sequence $(a_i)_{i\geq 0}$. In each clock cycle $i$, the generator outputs $a_{2i+1}$ if and only if $a_{2i} = 1$. Otherwise, $a_{2i+1}$ is discarded. The mode of operation of the self-shrinking generator is illustrated in Figure 7.3. The output sequence in this example is 0110...

From an implementation viewpoint, the self-shrinking generator is advantageous because it only requires one LFSR.

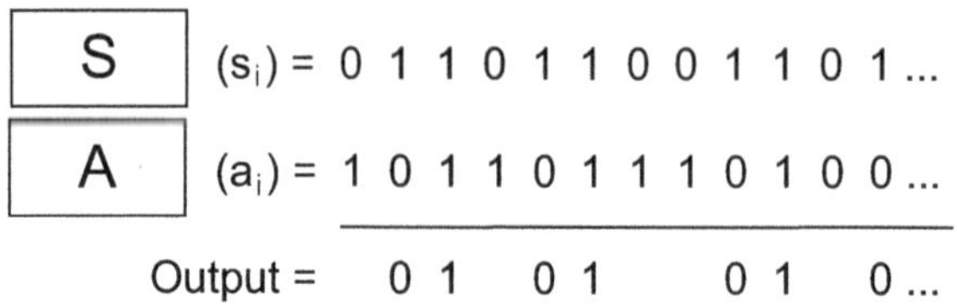


**Figure 7.2** The mode of operation of the shrinking generator.

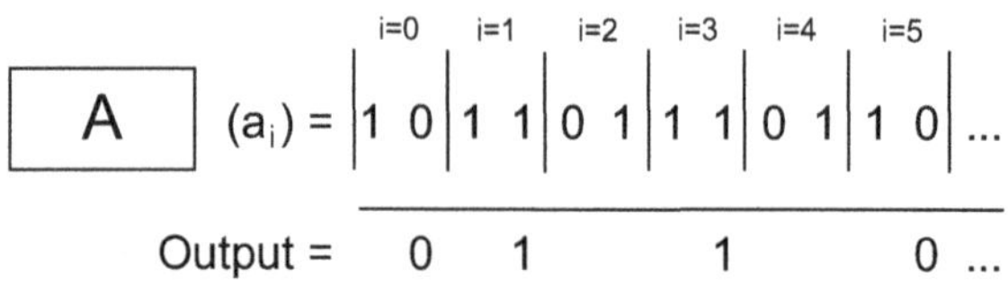


**Figure 7.3** The mode of operation of the self-shrinking generator.

LFSRs and LFSR-based PRGs are not so popular anymore, mainly because they are best implemented in hardware, and most people prefer software implementations today. A practically relevant PRG that is particularly well suited for software implementations is specified in ANSI X9.17 [6] and outlined in Algorithm 7.1. It is based on the block cipher DES (Section 9.6.1)—or Triple DES (3DES, also known as TDEA) with keying option 2.[2] The PRG takes as input a seed $s_0$, a 3DES key $k$, and an integer $n$. Here, $n$ does not refer to the length of the seed, but rather to the number of 64-bit strings that are generated to form the output; that is, $x_1, x_2, \ldots, x_n$. $D$ is an internally used 64-bit representation of the date and time, and $I$ is just an internal intermediate value that is initialized at the beginning of the algorithm (as the 3DES encryption of $D$ with key $k$). The for-loop is iterated $n$ times, where in each iteration a 64-bit output value $x_i$ is generated (the length is 64 bits, because the block size of DES and 3DES is 64 bits).

Besides ANSI X9.17, there are many other PRGs in practical use (e.g., [7–9]). For example, *Yarrow* [8] is a PRG—or rather a family of PRGs—that employs a particular cryptographic hash function and a particular block cipher, such as *Yarrow-160* with SHA-1 and 3DES in CTR mode. The design of Yarrow was later modified and the resulting family of PRGs was renamed *Fortuna* [9]. While PRGs like Yarrow and Fortuna are widely deployed in practice, there are only a few security analyses of them (e.g., [10]). The bottom line is that most PRGs used in the field are resistant against known attacks and appear to be sufficiently secure for most (cryptographic)

2 In keying option 2 (Section 9.6.1.6), only two DES keys are used, and $k$ in Algorithm 7.1 refers to both of them.

**Algorithm 7.1** ANSI X9.17 PRG.

$(s_0, k, n)$

---

$I = E_k(D)$
$s = s_0$
for $i = 1$ to $n$ do
  $x_i = E_k(I \oplus s)$
  $s = E_k(x_i \oplus I)$
  output $x_i$

---

$(x_1, x_2, \ldots, x_n)$

applications, but they have not been proven secure in a cryptographically strong sense. In the literature, they are sometimes called *practically strong*. Consequently, a PRG is practically strong if there are no known attacks against it and it is designed in an arbitrary and ad hoc way. This is fundamentally different from the cryptographically secure PRGs addressed next.

## 7.3 CRYPTOGRAPHICALLY SECURE PRGs

There are several possibilities to formally define the cryptographical strength and security of a PRG. Historically, the first definition was proposed by Manuel Blum and Silvio Micali in the early 1980s [11]. They argued that a PRG is *cryptographically secure*, if an adversary—after having seen a sequence of output bits—is not able to predict the next bit that is going to be generated with a success probability that is substantially better than guessing. Blum and Micali also proposed a PRG based on the DLA that complies with this definition (Section 7.3.1). Shortly after this seminal work, Manuel Blum—together with Leonore Blum and Michael Shub—proposed another PRG, called the *BBS PRG* or *squaring generator*,[3] that is simpler to implement but still cryptographically secure assuming the intractability of the quadratic residuosity problem (QRP, Definition A.31) [12] (Section 7.3.3), and other researchers have shown the same for the IFP and the resulting RSA PRG (Section 7.3.2). As of this writing, the BBS PRG refers to the yardstick for cryptographically secure PRGs.

As shown by Yao [14], a cryptographically secure PRG is *perfect* in the sense that no PPT algorithm can tell whether an $n$-bit string has been sampled uniformly at random from $\{0,1\}^n$ or generated with the PRG (using a proper seed) with a success probability that is substantially better than guessing. Note that this notion of

3 The acronym BBS is compiled from the first letters of the authors' names.

a perfect PRG is conceptually similar to a Turing Test[4] [15] that is used in artificial intelligence (AI). One can rephrase Yao's result by saying that a PRG that passes the Blum-Micali next-bit test is perfect in the sense that it also passes all polynomial-time (statistical) tests to distinguish it from a true random generator.

The Blum-Micali and BBS PRGs, together with the proof of Yao, represent a major achievement in the development of cryptographically secure PRGs. It has also been shown that the existence of a one-way function (with a hard-core predicate) is equivalent to the existence of a cryptographically secure PRG, i.e., one-way functions exist if and only if cryptographically secure PRGs exist [16].

To more formally address the notion of a cryptographically secure and hence perfect PRG, we revisit the security game mentioned earlier in this chapter. In this game, the adversary observes the output values of either a true random generator or a PRG, and his or her task is to tell the two cases apart, i.e., decide whether the values (he or she observes) are generated by a true random generator and a PRG. If he or she cannot do substantially better than guessing, then the PRG behaves like a true random generator and its output values can therefore be used as if they were generated by a true random generator. Put in other words: A PRG is cryptographically secure and hence perfect, if the output bit sequence cannot be distinguished (by any PPT algorithm) from an equally long output bit sequence generated by a true random generator, meaning that the two sequences are *computationally indistinguishable* or *effectively similar* (as termed in [14]). In the end, this leads to the same notion of cryptographical strength, but the tool to argue about it is *computational indistinguishability*

To follow this line of argumentation, we have to introduce the notion of a *probability ensemble* (in addition to Appendix B). Roughly speaking, this is a family of probability distributions or random variables $X = \{X_i\}_{i \in I}$, where $I$ is an index set, and each $\{X_i\}$ is a probability distribution or random variable of its own. Often $I = \mathbb{N}$ and each $X_n$ must have a certain property for sufficiently large $n$. Against this background, let

$$X = \{X_n\} = \{X_n\}_{n \in \mathbb{N}} = \{X_1, X_2, X_3, \ldots\}$$

4 The Turing Test is meant to determine if a computer program has intelligence. According to Alan M. Turing, the test can be devised in terms of a game (a so-called imitation game). It is played with three people, a man (A), a woman (B), and an interrogator (C), who may be of either sex. The interrogator stays in a room apart from the other two. The object of the game for the interrogator is to determine which of the other two is the man and which is the woman. He knows them by labels X and Y, and at the end of the game he says either "X is A and Y is B" or "X is B and Y is A." The interrogator is allowed to put questions to A and B. When talking about the Turing Test today, what is generally understood is the following: the interrogator is connected to one person and one machine via a terminal, and therefore can't see her counterparts. Her task is to find out which of the two candidates is the machine and which is the human only by asking them questions. If the machine can "fool" the interrogator, it is intelligent.

and

$$Y = \{Y_n\} = \{Y_n\}_{n \in \mathbb{N}} = \{Y_1, Y_2, Y_3, \ldots\}$$

be two probability ensembles; that is, for all $n \in \mathbb{N}^+$ $X_n$ and $Y_n$ refer to probability distributions on $\{0,1\}^n$. By saying $t \leftarrow X_n$ ($t \leftarrow Y_n$), we suggest that $t$ is sampled according to the probability distribution $X_n$ ($Y_n$). Furthermore, we say that $X$ is *polytime indistinguishable*[5] from $Y$, if for every PPT algorithm $A$ and every polynomial $p$, there exists a $n_0 \in \mathbb{N}^+$ such that for all $n > n_0$

$$\left| \Pr_{t \leftarrow X_n}[A(t) = 1] - \Pr_{t \leftarrow Y_n}[A(t) = 1] \right| \leq \frac{1}{p(n)} \tag{7.1}$$

This means that for sufficiently large $t$, no PPT algorithm $A$ can distinguish whether it was sampled according to $X_n$ or $Y_n$. In some literature, the PPT algorithm $A$ is called *polynomial-time statistical test* or *distinguisher*, and it is therefore denoted as $D$.

In computational complexity theory, it is assumed and widely believed that computationally indistinguishable probability ensembles exist, and this assumption is referred to as the general indistinguishability assumption. We say that $\{X_n\}$ is *pseudorandom* if it is polytime indistinguishable from $\{U_n\}$, where $U_n$ denotes the uniform probability distribution on $\{0,1\}^n$. More specifically, this means that for every PPT algorithm $A$ and every polynomial $p$, there exists a $n_0 \in \mathbb{N}^+$ such that for all $n > n_0$

$$\left| \Pr_{t \leftarrow X_n}[A(t) = 1] - \Pr_{t \leftarrow U_n}[A(t) = 1] \right| \leq \frac{1}{p(n)} \tag{7.2}$$

Note that (7.2) is almost identical to (7.1). The only difference is that $t$ is sampled according to the uniform probability distribution $U_n$ in the second probability of (7.2), whereas it is sampled according to $Y_n$ in (7.1). We are now ready to define the notion of a cryptographically secure PRG. This is done in Definition 7.1.

**Definition 7.1 (Cryptographically secure PRG)** *Let $G$ be a PRG with stretch function $l : \mathbb{N} \rightarrow \mathbb{N}$; that is, $l(n) > n$ for $n \in \mathbb{N}$. $G$ is* cryptographically secure *if it satisfies the following two conditions:*

- $|G(s)| = l(|s|)$ *for every* $s \in \{0,1\}^*$;
- $\{G(U_n)\}$ *is pseudorandom; that is, it is polytime indistinguishable from* $\{U_{l(n)}\}$.

5 Alternatively speaking, $X$ is indistinguishable from $Y$ in polynomial time.

The first condition captures the stretching property of the PRG (i.e., the fact that the output of the PRG is larger than its input), whereas the second condition captures the property that the generated pseudorandom bit sequence is computationally indistinguishable from (and hence practically the same as) a random bit sequence. Combining the two conditions yields a PRG that is secure in a cryptographic sense and can hence be used for cryptographic applications.

Alternatively, the notion of a cryptographically secure PRG can also be defined by first introducing the PRG advantage of $A$ with respect to PRG $G$, denoted $\mathrm{Adv}_{\mathrm{PRG}}[A, G]$. This value is a probability in the range $[0, 1]$, and it is formally defined as follows:

$$\mathrm{Adv}_{\mathrm{PRG}}[A, G] = \left| \Pr_{t \xleftarrow{r} \{0,1\}^n}[A(G(t)) = 1] \quad - \Pr_{t \xleftarrow{r} \{0,1\}^{l(n)}}[A(t) = 1] \right|$$

The PRG advantage of $A$ with respect to PRG $G$ refers to the absolute value of the difference of two probabilities, namely the probability that $A$ outputs 1 if the input is pseudorandomly generated and the probability that $A$ outputs 1 if the input is randomly generated. In the first case, $t$ is sampled uniformly at random from $\{0, 1\}^n$ and the resulting value is stretched by $G$ to $l(n)$, whereas in the second case, $t$ is sampled uniformly at random from $\{0, 1\}^{l(n)}$ and not stretched at all. Again, $A$ is a good distinguisher if its PRG advantage with respect to $G$ is close to 1, and it is a bad distinguisher if it is close to 0.

To argue about the security of PRG $G$, we are interested in the PPT algorithm $A$ with maximal PRG advantage. This yields the PRG advantage of $G$:

$$\mathrm{Adv}_{\mathrm{PRG}}[G] = \max_A \{\mathrm{Adv}_{\mathrm{PRG}}[A, G]\}$$

We say that PRG $G$ is secure, if $\mathrm{Adv}_{\mathrm{PRG}}[G]$ is negligible; that is, for every polynomial $p$, there exists a $n_0 \in \mathbb{N}^+$ such that for all $n > n_0$

$$\mathrm{Adv}_{\mathrm{PRG}}[G] \leq \frac{1}{p(n)}$$

If we have a one-way function $f$ with hard-core predicate $B$, then the following PRG $G$ with seed $s_0$ is cryptographically secure:

$$G(s_0) = B(f(s_0)), B(f^2(s_0)), \ldots, B(f^{l(n)}(s_0))$$

Talking in terms of an FSM-based PRG, the state register is initialized with $s_0$, the next-state function $f$ is the one-way function (that is iterated), and the output function $g$ computes the hard-core predicate $B$ from the state register. This idea is used in many cryptographically secure PRGs, including, for example, the Blum-Micali, RSA, and BBS PRGs mentioned above and outlined next.

### 7.3.1 Blum-Micali PRG

The Blum-Micali PRG [11] specified in Algorithm 7.2 employs the fact that the discrete exponentiation function is a (conjectured) one-way function (Section 5.2.1) with the hard-core predicate MSB. It takes as input a large prime $p$ and a generator $g$ of $\mathbb{Z}_p^*$, and it generates as output a sequence $(b_i)_{i\geq 1}$ of pseudoranom bits.

**Algorithm 7.2** The Blum-Micali PRG.

$(p, g)$

---

$x_0 \xleftarrow{r} \mathbb{Z}_p^*$
for $i = 1$ to $\infty$ do
  $x_i \equiv g^{x_{i-1}} \pmod{p}$
  $b_i = msb(x_i)$
  output $b_i$

---

$(b_i)_{i\geq 1}$

The algorithm starts with the initialization of the seed $x_0 = s_0$ with an integer that is randomly chosen from $\mathbb{Z}_p^* = \{1, \ldots, p-1\}$. Starting with this value, the discrete exponentiation function is recursively applied according to $x_i = g^{x_{i-1}} \pmod{p}$ for $i = 1, 2, \ldots$ From each $x_i$, a pseudoranom bit $b_i$ is extracted as follows: If $x_i > (p-1)/2$ then $b_i$ is set to one, otherwise it is set to zero. Hence, the hard-core predicate that is exploited in the Blum-Micali PRG is the MSB; that is, whether the result of the discrete exponentiation function is smaller or bigger than $(p-1)/2$, or—equivalently—whether it belongs to the first or second half of values.

A more thorough analysis of the Blum-Micali PRG reveals the fact that more than one bit can be extracted from $x_i$ in each round [17]. This result can be used to improve the efficiency of the Blum-Micali PRG considerably.

### 7.3.2 RSA PRG

The RSA PRG specified in Algorithm 7.3 employs the fact that the RSA function is a (conjectured) one-way function (Section 5.2.2) with the hard-core predicate LSB. Like the RSA public key cryptosystem, the RSA PRG takes as input a large integer $n$ (that is the product of two large primes $p$ and $q$) and $e$ (that is a randomly chosen integer between 2 and $\phi(n) - 1$ with $gcd(e, \phi(n)) = 1$). But unlike the RSA public key cryptosystem, the RSA PRG generates as output a sequence $(b_i)_{i\geq 1}$ of pseudorandomly generated bits (instead of a ciphertext).

**Algorithm 7.3** The RSA PRG.

$(n, e)$

---

$x_0 \xleftarrow{r} \mathbb{Z}_n^*$
for $i = 1$ to $\infty$ do
  $x_i \equiv x_{i-1}^e \pmod{n}$
  $b_i = lsb(x_i)$
  output $b_i$

---

$(b_i)_{i\geq 1}$

The RSA PRG starts with the initialization of a seed $x_0 = s_0$ with a value from $\mathbb{Z}_n^*$. It then recursively generates an output bit $b_i$ by first computing $x_i \equiv x_{i-1}^e \pmod{n}$ and then setting $b_i = lsb(x_i)$ for $i \geq 1$. Again, the output bits are compiled into bit sequence $(b_i)_{i\geq 1}$.

The RSA PRG is more efficient than the Blum-Micali PRG, but its true efficiency depends on the value for $e$. If, for example, $e = 3$, then generating an output bit requires only one modular multiplication and one modular squaring. Also, it was shown in [13] that instead of extracting a single bit from each $x_i$, a sequence of $\log n$ bits can be extracted simultaneously. This further improves the efficiency of the PRG, and current research tries to enlarge the number of bits that can be extracted simultaneously.

### 7.3.3 BBS PRG

The BBS PRG [12] specified in Algorithm 7.4 employs the fact that the modular square function is a (conjectured) one-way function (Section 5.2.2), and that—similar to the RSA function—the LSB yields a hard-core predicate. The BBS PRG takes as input a Blum integer $n$ (i.e., an integer $n$ that is the product of two primes $p$ and $q$, each of them congruent to 3 modulo 4), and it again generates as output a sequence $(b_i)_{i\geq 1}$ of pseurdorandomly generated bits.

Again, the BBS PRG starts with the initialization of a seed $x_0 = s_0$. Here, $x_0$ needs to be an integer that is coprime with $n$; that is, $gcd(x_0, n) = 1$. In each step, it then generates an output bit $b_i$ by first computing $x_i \equiv x_{i-1}^2 \pmod{n}$ and then setting $b_i = lsb(x_i)$ for $i \geq 1$. Similar to the RSA PRG, $\log n$ bits can be extracted in each step (instead of a single bit) to improve the efficiency of the PRG.

The BBS PRG is the preferred choice for most practitioners. On the one hand, it is highly efficient because it requires only one modular squaring for the generation

**Algorithm 7.4** The BBS PRG.

$(n)$

---

$x_0 \stackrel{r}{\leftarrow} \mathbb{Z}_n^*$
for $i = 1$ to $\infty$ do
  $x_i \equiv x_{i-1}^2 \pmod{n}$
  $b_i = lsb(x_i)$
  output $b_i$

---

$(b_i)_{i \geq 1}$

of an output bit. On the other hand, it has the practically relevant property that $x_i$ can be computed directly for $i \geq 1$ if one knows the prime factors $p$ and $q$ of $n$:

$$x_i = x_0^{(2^i) \bmod ((p-1)(q-1))}$$

This is a property that may be useful in some situations.

## 7.4 FINAL REMARKS

In this chapter, we introduced the notion of a PRG and elaborated on the requirements for a PRG to be practically or cryptographically secure. All PRGs in use today—independent of whether they are only practically secure or even cryptographically secure—must make the assumption that their internal state can be kept secret, and hence cannot be acquired by an adversary (otherwise, the adversary can forever after predict the bits that are generated). But in practice, it may still happen that the adversary can acquire the internal state. Maybe there is a bug in the implementation; maybe a computer has just booted for the first time and doesn't have a seed to start with; or maybe the adversary has been able to read the seed file from disk. In either case, an adversary may have been able to acquire the internal state and this may make it necessary to periodically reseed the state. Some practically secure PRGs take this into account. In addition to the generator and a seed file, for example, Fortuna [9] has an accumulator that collects and pools entropy from various sources to occasionally reseed the generator. Needless to say, such a feature is useful in the field.

There are many applications of PRGs. If a lot of keying material is required, then they can replace or at least complement true random bit generators. This is advantageous because PRGs use much less randomness than true random bit generators (they still need some randomness to start with). If a PRG is used to derive keying material from a single master key or a password, then it is often called *key derivation function* (KDF).[6] Technically speaking, a KDF can be implemented with a PRF $f_k$ (e.g., a keyed cryptographic hash function) using a construction that is conceptually similar to the one provided at the beginning of Section 7.2 (and revisited in Section 8.3.1): It takes as input a source key $k$ that must be uniform in a key space $\mathcal{K}$, a context string $c$, and a number of bytes $l$ that are required, and it outputs $n$ blocks that are generated as follows:

$$KDF(k, c, l) = f_k(c \parallel 0) \parallel f_k(c \parallel 1) \parallel \ldots \parallel f_k(c \parallel n-1) \tag{7.3}$$

If $l$ is the number of required bytes and $b$ is the block length of $f_k$, then $n = \lceil l/b \rceil$ and the respective block counter is running from 0 to $n-1$ (typically written hexadecimally, i.e., 0x00 for 0, 0x01 for 1, and so on). Remember from above that the security of the construction requires $k$ to be uniform in $\mathcal{K}$. This requirement is crucial and may not be the case in a real-world setting. If, for example, $k$ is the outcome of a key agreement, then it may happen that $k$ is biased or originate from a relatively small subset of $\mathcal{K}$. This means that some preprocessing is required to extract a uniform and pseudorandom key from the source key. There are several standards that serve this purpose, such as KDF1 to KDF4 [19], and—most importantly—the *HMAC-based extract-and-expand key derivation function* (HKDF) [20] that uses the HMAC construction (Section 10.3.2.4) and follows the extract-then-expand paradigm, meaning that it employs the HMAC construction for both the extraction of the uniform key from the source key and the expansion of this key according to (7.3). More specifically, the HKDF function consists of an HKDP extract function $KDF_{extract}$ defined as

$$HKDF_{extract}(s, k) = HMAC_s(k) = k'$$

for salt $s$ and source key $k$,[7] and an HKDF expand function $KDF_{expand}$ that uses $k'$ to recursively compute values for $T$. $T_0$ is initialized with the empty string, and

$$T_i = HMAC_{k'}(T_{i-1} \parallel c \parallel i)$$

for $i = 1, \ldots, n$ (again written hexadecimally).[8] Finally, the output of the HKDF function refers to $l$ first bytes of a string that is constructed as follows:

$$HKDF_{expand}(k', c, l) = T_1 \parallel T_2 \parallel \ldots \parallel T_n$$

6 A KDF can be seen as a PRG because it stretches a relatively short key (that represents a seed) into a much longer key or even multiple keys. Equivalently, it can also be seen as a PRF because it should look like it is being chosen from the set of all possible such functions. This viewpoint is supported, for example, by the U.S. NIST [18].

7 If no salt is provided, then the default value is an appropriately sized zero string. In HKDF terminology, $k$ refers to the input keying material (IKM).

8 In HKDF terminology, the output of the HKDF expand function refers to output keying material (OKM).

The HKDF function is used in almost all Internet security protocols to derivate arbitrary long keying material from a relatively short key.

In the realm of password hashing (i.e., where one or several keys need to be derived from a password or passphrase), there are a few complementary *password-based key derivation functions* (PBKDF), such as PBKDF1 and PBKDF2 [21, 22]. The specialties of this use case are that a password does not provide a lot of entropy and that the function needs to be slowed down to defeat (offline) password guessing attacks. This is a very old idea that has its origin in the crypt function that was originally used in the UNIX operating system to hash passwords. Another approach to defeat password guessing attacks, especially if additional hardware is available, is to make the PBKDF *memory-hard*, meaning that the execution of the function does not only require a lot of processing power, but also a lot of memory. Examples of memory-hard PBKDFs are scrypt [23] and Balloon.[9] Furthermore, there was a Password Hashing Competition (PHC[10]) going on between 2013 and 2015. Instead of official standardization bodies like NIST, it was organized by cryptographers and security practitioners. *Argon2*[11] was selected as the final PHC winner, with special recognition given to *Catena*,[12] *Lyra2*,[13] *yescrypt*, and *Makwa*. We close this outline on KDFs with the note that in some literature such functions are called *mask generation functions* (MGF). So a MGF also refers to KDF, most likely implemented with a cryptographic hash function.

## References

[1] Knuth, D.E., *The Art of Computer Programming—Volume 2: Seminumerical Algorithms*, 3rd edition. Addison-Wesley, Reading, MA, 1997.

[2] Plumstead, J., "Inferring a Sequence Generated by a Linear Congruence," *Proceedings of the 23rd IEEE Symposium on Foundations of Computer Science*, IEEE, Chicago, 1982, pp. 153–159.

[3] Krawczyk, H., "How to Predict Congruential Generators," *Journal of Algorithms*, Vol. 13, No. 4, 1992, pp. 527–545.

[4] Coppersmith, D., H. Krawczyk, and Y. Mansour, "The Shrinking Generator," *Proceedings of CRYPTO '93*, Springer-Verlag, LNCS 773, 1993, pp. 22–39.

[5] Meier, W., and O. Staffelbach, "The Self-Shrinking Generator," *Proceedings of EUROCRYPT '94*, Springer-Verlag, LNCS 950, 1994, pp. 205–214.

[6] American National Standards Institute, *American National Standard X9.17: Financial Institution Key Management*, Washington, DC, 1985.

9 https://crypto.stanford.edu/balloon.

10 https://password-hashing.net.

11 https://github.com/p-h-c/phc-winner-argon2.

12 https://github.com/medsec/catena-variants.

13 http://lyra-2.net.

[7] Gutmann, P., "Software Generation of Practically Strong Random Numbers," *Proceedings of the Seventh USENIX Security Symposium*, June 1998, pp. 243–255.

[8] Kelsey, J., B. Schneier, and N. Ferguson, "Yarrow-160: Notes on the Design and Analysis of the Yarrow Cryptographic Pseudorandom Number Generator," *Proceedings of the 6th Annual Workshop on Selected Areas in Cryptography*, Springer-Verlag, August 1999.

[9] Ferguson, N., and B. Schneier, *Practical Cryptography*. John Wiley & Sons, New York, 2003.

[10] Kelsey, J., et al., "Cryptanalytic Attacks on Pseudorandom Number Generators," *Proceedings of the Fifth International Workshop on Fast Software Encryption*, Springer-Verlag, March 1998, pp. 168–188.

[11] Blum, M., and S. Micali, "How to Generate Cryptographically Strong Sequences of Pseudo-Random Bits," *SIAM Journal of Computing*, Vol. 13, No. 4, November 1984, pp. 850–863.

[12] Blum, L., M. Blum, and M. Shub, "A Simple Unpredictable Pseudo-Random Number Generator," *SIAM Journal of Computing*, Vol. 15, May 1986, pp. 364–383.

[13] Alexi, W., et al., "RSA/Rabin Functions: Certain Parts Are as Hard as the Whole," *SIAM Journal of Computing*, Vol. 17, No. 2, April 1988, pp. 194–209.

[14] Yao, A.C., "Theory and Application of Trapdoor Functions," *Proceedings of the 23rd IEEE Symposium on Foundations of Computer Science*, 1982, pp. 80–91.

[15] Turing, A.M., "Computing Machinery and Intelligence," *Mind*, Vol. 59, No. 236, 1950, pp. 433–460.

[16] Hastad, J., et al., "A Pseudorandom Generator from Any One-Way Function," *SIAM Journal of Computing*, Vol. 28, No. 4, 1999, pp. 1364–1396.

[17] Gennaro, R., et al., "An Improved Pseudo-Random Generator Based on the Discrete Logarithm Problem," *Journal of Cryptology*, Vol. 18, No. 2, April 2005, pp. 91–110.

[18] U.S. NIST Special Publication 800-108, "Recommendation for Key Derivation Using Pseudorandom Functions," October 2009.

[19] ISO/IEC 18033-2, Information technology—Security techniques—Encryption algorithms—Part 2: Asymmetric ciphers, 2006.

[20] Krawczyk, H., and P. Eronen, *HMAC-based Extract-and-Expand Key Derivation Function (HKDF)*, RFC 5869, May 2010.

[21] U.S. NIST Special Publication 800-132, "Recommendation for Password-Based Key Derivation Part 1: Storage Applications," December 2010.

[22] Moriarty, K. (Ed.), Kaliski, B., and A. Rusch, *PKCS #5: Password-Based Cryptography Specification Version 2.1*, RFC 8018, January 2017.

[23] Percival, C., and S. Josefsson, *The scrypt Password-Based Key Derivation Function*, RFC 7914, August 2016.

# Chapter 8

# Pseudorandom Functions

In this chapter, we highlight PRFs (including PRPs) as a central theme. More specifically, we introduce the topic in Section 8.1, discuss possibilities to argue about the security of a PRF in Section 8.2, show that PRGs and PRFs are closely related in the sense that one construct one from the other in Section 8.3,[1] elaborate on the random oracle model in Section 8.4, and conclude with some final remarks in Section 8.5. In contrast to many other chapters, this chapter is theoretically motivated and sets the foundations many practical cryptosystems are based on.

## 8.1 INTRODUCTION

In Section 2.2.2 we briefly introduced the notion of a PRF, and in Definition 2.8 we gave a first definition. We said that a PRF is a family $F : \mathcal{K} \times X \rightarrow Y$ of (efficiently computable) functions, where each key $k \in \mathcal{K}$ determines a function $f_k : X \rightarrow Y$ that is indistinguishable from a random function; that is, a function randomly chosen from $\mathrm{Funcs}[X, Y]$. Because there is a function $f_k$ for every $k \in \mathcal{K}$, there are "only" $|\mathcal{K}|$ functions in $F$, whereas there are $|Y|^{|X|}$ functions in $\mathrm{Funcs}[X, Y]$ (and this number tends to be overwhelmingly large). This means that we can use a relatively small key to determine a particular function $f_k \in F$, but this function behaves like a truly random function. More specifically, it is supposed to be computationally indistinguishable from a random function.

The same line of argumentation applies to PRPs: A PRP is a family $P : \mathcal{K} \times X \rightarrow Y$ of (efficiently computable) permutations, where each $p \in \mathcal{K}$ determines a permutation $p_k : X \rightarrow X$ that is indistinguishable from a random permutation; that is, a permutation randomly chosen from $\mathrm{Perms}[X]$. So the logic of a PRP is

1 At the end of the previous chapter, we argued that a KDF can be seen as either a PRG or a PRF. This argument also provides evidence to the fact that PRGs and PRFs are closely related.

essentially the same as the logic of a PRF, and many things we say about PRFs also apply to PRPs.

In the rest of this book, we will encounter PRFs and PRPs again and again. For example, we already saw that cryptographic hash functions behave like PRFs, whereas KDFs are typically implemented with keyed hash functions and therefore yield true PRFs. In Chapter 9, we will see that block ciphers represent PRPs, and later in this chapter we will learn how to construct a PRG from a PRF and vice versa (Section 8.3). So PRFs and PRPs are in fact omnipresent in cryptography, and they are at the core of many cryptosystems in use today.

## 8.2 SECURITY OF A PRF

One of the first questions that appear in the realm of PRFs is what it means to say that a PRF is "secure." As is usually the case in security discussions, we start from the security game introduced in Section 1.2.2.1 and illustrated in Figure 1.2. In this game-theoretic setting, the adversary is interacting with either a random function (representing the ideal system) or a PRF (representing the real system), and his or her task is to tell what case applies. If he or she is able to tell the two cases apart (with a probability that is substantially greater than guessing), then he or she wins the game, and hence he or she can break the security of the PRF. The adversary can then distinguish the PRF from a truly random function, and this defeats the original purpose of the PRF (remember that a PRF is intended to be indistinguishable from a random function).

In the sequel, we want to make this intuition more concrete. We fix a function $g : X \rightarrow Y$ that can either be random (i.e., an element from $\text{Funcs}[X, Y]$, meaning that $g \xleftarrow{r} \text{Funs}[X, Y]$) or pseudorandom (i.e., an element from a PRF family $F : \mathcal{K} \times X \rightarrow Y$, meaning that $k \xleftarrow{r} \mathcal{K}$ and this key fixes a function $f_k$ from $F$), and we consider two worlds:

- In world 0, $g$ is a random function;
- In world 1, $g$ is a pseudorandom function.

Furthermore, we assume a PPT algorithm representing an adversary $A$ located in a closed room. $A$'s only possibility to interact with the outside world is to run an experiment to explore the input-output-behavior of $g$. The experiment consists of $q$ input values $x_i \in X$ $(i = 1, \ldots, q)$ that $A$ can choose and send to the outside world. There, these input values are subject to $g$, and the resulting output values $g(x_i) \in Y$ are sent back to $A$. This experiment is the only way $A$ can interact with the outside world.

In this experimental setting, $A$'s task is to decide whether it is in world 0 or 1 (i.e., whether it interacts with a random function or a PRF). After having performed the experiment with the oracle access to $g$, $A$ must make a decision (decide whether it is in world 0 or 1), and we use the term $A(g)$ to refer to this decision. Each experiment has a certain probability of returning 1, where the probability is taken over all random choices made in the experiment. The two probabilities (for the two experiments) should be evaluated separately because they are completely different.

To see how good $A$ is at determining which world it is in, one may look at the absolute difference in the probabilities that either experiment returns 1. This value is again a probability in the range $[0, 1]$. It is called PRF advantage of $A$ with respect to $F$, and it is denoted as $\text{Adv}_{\text{PRF}}[A, F]$. Obviously, $A$ is good if the PRF advantage is close to 1, and it is bad if it close to 0. There are several possibilities to formally define $\text{Adv}_{\text{PRF}}[A, F]$, and we use the following (simplified) notation here:

$$\text{Adv}_{\text{PRF}}[A, F] = \left| \Pr_{g \xleftarrow{r} F}[A(g) = 1] \quad - \Pr_{g \xleftarrow{r} \text{Funcs}[X,Y]}[A(g) = 1] \right|$$

To argue about the security of PRF $F$, we are mainly interested in the PPT algorithm $A$ with maximal PRF advantage. This yields the PRF advantage of $F$:

$$\text{Adv}_{\text{PRF}}[F] = \max_{A}\{\text{Adv}_{\text{PRF}}[A, F]\}$$

Quite naturally, we say that PRF $F$ is secure, if $\text{Adv}_{\text{PRF}}[F]$ is negligible; that is, for every polynomial $p$, there exists a $n_0 \in \mathbb{N}^+$ such that for all $n > n_0$

$$\text{Adv}_{\text{PRF}}[F] \leq \frac{1}{p(n)}$$

The bottom line is that for a secure PRF $F$, there is no PPT algorithm that can distinguish an element from $F$ from a truly random function. This means that $F$ behaves like a random function, and hence that it can be used in place of a random function.

## 8.3 RELATIONSHIP BETWEEN PRGs AND PRFs

As mentioned above, PRGs and PRFs are closely related to each other in the sense that one can construct one from the other. In this section, we outline the two respective constructions—a PRG based on a PRF and a PRF based on a PRG—but we don't prove the security claims.

### 8.3.1 PRF-Based PRG

As already mentioned in the realm of KDFs (Section 7.4), it is relatively simple and straightforward to construct a PRG $G$ with a PRF $F$: One randomly selects a key $k \in \mathcal{K}$ that fixes a particular function $f_k$ from $F$, and one then iteratively applies $f_k$ to an incrementing counter value (that may start with zero). If the resulting values are taken as the output values of $G$ seeded with $k$, then the PRF-based PRG is defined as follows:

$$G(k) = (f_k(i))_{i \geq 0} = f_k(0), f_k(1), f_k(2), f_k(3), \ldots$$

If $F$ is a secure PRF, then the PRG $G$ constructed this way can be shown to be cryptographically secure as well. Also, its efficiency depends on the efficiency of the underlying PRF.

### 8.3.2 PRG-Based PRF

More surprisingly, the close relationship also works in the opposite direction, meaning that it is possible to construct a PRF with a PRG [1]. Let $G(s)$ be a PRG with stretching function $l(n) = 2n$, meaning that its output is twice as long as its input. $G_0(s)$ refers to the first $n$ bits of $G(s)$, whereas $G_1(s)$ refers to the last $n$ bits of $G(s)$ for $s \in \{0,1\}^n$. Furthermore, $X = Y = \{0,1\}^n$, and $x = \sigma_n \cdots \sigma_2\sigma_1$ is the bitwise representation of $x$. Against this background, a PRG-based PRF $f_s : X \rightarrow Y$ can be defined as follows:

$$f_s(x) = f_s(\sigma_n \cdots \sigma_2\sigma_1) = G_{\sigma_n}(\cdots G_{\sigma_2}(G_{\sigma_1}(s)) \cdots)$$

The definition is simple, but the construction is not very intuitive. Let use therefore use a toy example to explain it. For $n = 2$, we can use a PRG $G(s)$ that is defined as follows:

$$\begin{aligned} G(00) &= 1001 \\ G(01) &= 0011 \\ G(10) &= 1110 \\ G(11) &= 0100 \end{aligned}$$

For $s = 10$ and $x = 01$ (i.e., $\sigma_2 = 0$ and $\sigma_1 = 1$), the resulting PRF looks as follows:

$$f_s(x) = f_s(\sigma_2\sigma_1) = f_{10}(01) = G_0(G_1(10)) = 11$$

To compute this value, we must first compute $G_1(10) = 10$ (i.e., the last two bits of $G(10) = 1110$) and then $G_0(10) = 11$ (i.e., the first two bits of $G(10) = 1110$). Hence, the output of $f_s$ for $x = 01$ is 11, and every value of $s$ defines a particular function $f_s$. Consequently, the family of all functions defined this way yields a PRF.

If $G$ is a cryptographically secure PRG, then the PRF $F$ constructed this way can again be shown to be secure. In contrast to the previous case, however, the efficiency of this construction is rather poor. Note that $G$ must be iterated $n$ times for an $n$-bit PRF. Also, if one wants to turn this PRF into a PRP (e.g., to construct a block cipher), then—according to Section 9.6.1.1—the PRF must be input to a three-round Feistel cipher, and this decreases the performance for another factor of three. The bottom line is that the construction is theoretically interesting, but not very practical.

## 8.4 RANDOM ORACLE MODEL

In Section 1.2.2, we introduced the notion of provable security and we mentioned that the random oracle methodology is frequently used to design cryptographic systems—most likely cryptographic protocols—that are provably secure in the so-called random oracle model. The methodology was proposed by Mihir Bellare and Philip Rogaway in the early 1990s to provide "a bridge between cryptographic theory and cryptographic practice" [2]. As such, they formalized a heuristic argument that had already been vaguely expressed in [1, 3, 4].

The random oracle methodology consists of the following three steps:[2]

1. One designs an ideal system in which all parties—including the adversary—have access to a random function (or random oracle, respectively).
2. One proves the security of the ideal system.
3. One replaces the random function with a PRF (e.g., a cryptographic hash function) and provides all parties—again, including the adversary—with a specification of this function.

As a result, one obtains an implementation of the ideal system in the real world. Bellare and Rogaway showed the usefulness of this methodology to design and analyze the security properties of some asymmetric encryption, digital signature, and zero-knowledge proof systems. Meanwhile, many researchers have used the random oracle model to analyze the security properties of many other cryptographic systems and protocols used in the field. Note, however, that a formal analysis in the

2 In some literature, steps 1 and 2 are collectively referred to as step 1.

random oracle model is not yet a security proof (because of the underlying ideal assumption), but that it still provides some useful evidence for the security of the system. In fact, Bellare and Rogaway claimed that the random oracle model can serve as the basis for efficient cryptographic systems with security properties that can at least be analyzed to some extent. Because people do not often want to pay more than a negligible price for security, such an argument for practical systems seems to be more useful than formal security proofs for inefficient systems.

There are basically two problems when it comes to an implementation of the ideal system in step 3 of the random oracle methodology:

- First, it is impossible to implement a random function by a (single) cryptographic hash function. In a random function $f$, the preimages and images are not related to each other, meaning that $x$ does not reveal any information about $f(x)$, and—vice versa—$f(x)$ does not reveal any information about $x$. If the random function is implemented with a (single) cryptographic hash function $h$, then the preimage $x$ leaks a lot of information about the image $h(x)$. In fact, $h(x)$ is then completely determined by $x$. This problem can be solved by using a (large) family of cryptographic hash functions and choosing one at random [5].
- Second, it was shown that random oracles cannot be implemented cryptographically. More specifically, it was shown in [6] that an (artificially crafted) digital signature system exists that is secure in the random oracle model but that gets insecure when the random oracle is implemented by a (family of) cryptographic hash functions.

The second problem is particularly worrisome, and since its publication many researchers have started to think controversially about the usefulness of the random oracle methodology in general and the random oracle model in particular. In fact, there is an ongoing controversy between Neal Koblitz and Alfred Menezes on one side, and the authors of the random oracle model on the other side (e.g., [7–9]).

Anyway, a proof in the random oracle model can still be regarded as evidence of security when the random oracle is replaced by a particular cryptographic hash function (according to the original claim of Bellare and Rogaway). Note that no practical protocol proven secure in the random oracle model has ever been broken when used with a cryptographic hash function such as SHA-1. The protocol used in the proof of [6] is not a natural protocol for a "reasonable" cryptographic application. Instead, it was designed specifically for the proof, and must therefore be taken with a grain of salt.

## 8.5 FINAL REMARKS

In this chapter, we elaborated on PRFs and their close relationship to PRGs. In particular, we showed that it is possible to construct a PRG if one has a PRF, and—vice versa—that it is possible to construct a PRF if one has a PRG. The constructions are conceptually simple and straightforward, but they are not thought to be implemented to serve any practical needs.

Having introduced the notion of a PRF, we then introduced, overviewed, and put into perspective the random oracle methodology that is frequently used to design cryptographic systems and protocols, and to analyze their security properties in the random oracle model. Mainly due to a negative result [6], people have started to think controversially about the random oracle model and to look for alternative approaches to analyze the security properties of cryptographic systems and protocols. In fact, security proofs avoiding the random oracle model are preferred and have started to appear in cryptographic publications. They are sometimes referred to as proofs in the standard model. Unfortunately, there also areas in which we don't have proofs in the standard model and where proofs in the random oracle model is the best we have.

## References


[1] Goldreich, O., S. Goldwasser, and S. Micali, "How to Construct Random Functions," *Journal of the ACM*, Vol. 33, No. 4, October 1986, pp. 792–807.

[2] Bellare, M., and P. Rogaway, "Random Oracles Are Practical: A Paradigm for Designing Efficient Protocols," *Proceedings of First Annual Conference on Computer and Communications Security*, ACM Press, New York, 1993, pp. 62–73.

[3] Fiat, A., and A. Shamir, "How to Prove Yourself: Practical Solutions to Identification and Signature Problems," *Proceedings of CRYPTO '86*, Springer-Verlag, LNCS 263, 1987, pp. 186–194.

[4] Goldreich, O., S. Goldwasser, and S. Micali, "On the Cryptographic Applications of Random Functions," *Proceedings of CRYPTO '84*, Springer-Verlag, LNCS 196, 1984, pp. 276–288.

[5] Canetti, R., "Towards Realizing Random Oracles: Hash Functions That Hide All Partial Information," *Proceedings of CRYPTO '97*, Springer-Verlag, LNCS 1294, 1997, pp. 455–469.

[6] Canetti, R., O. Goldreich, and S. Halevi, "The Random Oracles Methodology, Revisited," *Proceedings of 30th STOC*, ACM Press, New York, 1998, pp. 209–218.

[7] Koblitz, N., and A. Menezes, "Another Look at Provable Security," *Journal of Cryptology*, Vol. 20, January 2007, pp. 3–37.

[8] Koblitz, N., and A. Menezes, "Another Look at 'Provable Security'. II," Cryptology ePrint Archive, Report 2006/229, 2006.

[9] Koblitz, N., and A. Menezes, "The Random Oracle Model: A Twenty-Year Retrospective," Cryptology ePrint Archive: Report 2015/140, 2015.

# Chapter 9

# Symmetric Encryption

In this chapter, we elaborate on symmetric encryption in its full length. More specifically, we introduce the topic in Section 9.1, provide a brief historical perspective in Section 9.2, elaborate on perfectly secure and computationally secure encryption in Sections 9.3 and 9.4, dive deeply into stream ciphers, block ciphers, and modes of operation (for block ciphers) in Sections 9.5–9.7, and conclude with some final remarks in Section 9.8. Note that symmetric encryption systems are the most widely deployed cryptographic systems in use today, and that many books on cryptography elaborate only on these systems. Consequently, this chapter is an important and extensive one.

## 9.1 INTRODUCTION

If $\mathcal{M}$ is a plaintext message space, $\mathcal{C}$ a ciphertext space, and $\mathcal{K}$ a key space, then—according to Definition 2.9—a *symmetric encryption system* or *cipher* is a pair $(E, D)$ of families of efficiently computable functions that are defined as follows:

- $E : \mathcal{K} \times \mathcal{M} \rightarrow \mathcal{C}$ denotes a family $\{E_k : k \in \mathcal{K}\}$ of *encryption functions* $E_k : \mathcal{M} \rightarrow \mathcal{C}$;
- $D : \mathcal{K} \times \mathcal{C} \rightarrow \mathcal{M}$ denotes a family $\{D_k : k \in \mathcal{K}\}$ of respective *decryption functions* $D_k : \mathcal{C} \rightarrow \mathcal{M}$.

For every message $m \in \mathcal{M}$ and key $k \in \mathcal{K}$, the functions $D_k$ and $E_k$ must be inverse to each other (i.e., $D_k(E_k(m)) = m$). Otherwise, a ciphertext may not be decryptable, and hence the encryption system may not be very useful in the first place.[1] To illustrate the working principles of a symmetric encryption system you may revisit Figure 2.7. Note that a symmetric encryption system can be randomized in the sense that the encryption function takes additional random input. As we will see later in this chapter, randomized symmetric encryption systems have security advantages and are therefore preferred most of the time.

In some symmetric encryption systems, it doesn't matter whether one encrypts first and then decrypts or decrypts first and then encrypts. In formulas, this means

$$D_k(E_k(m)) = E_k(D_k(m)) = m$$

Taking this idea one step further, a symmetric encryption system is *commutative* if a message that is encrypted multiple times can be decrypted in arbitrary order. If a message $m$ is encrypted twice with $k_1$ and $k_2$; that is, $c = E_{k_2}(E_{k_1}(m))$, then in a commutative encryption system

$$D_{k_2}(D_{k_1}(c)) = D_{k_1}(D_{k_2}(c)) = m$$

must hold. Similarly, if $m$ is encrypted three times with $k_1$, $k_2$, and $k_3$; that is, $c = E_{k_3}(E_{k_2}(E_{k_1}(m)))$, then in a commutative encryption system

$$\begin{aligned}
D_{k_3}(D_{k_2}(D_{k_1}(c))) &= D_{k_3}(D_{k_1}(D_{k_2}(c))) = \\
D_{k_2}(D_{k_3}(D_{k_1}(c))) &= D_{k_2}(D_{k_1}(D_{k_3}(c))) = \\
D_{k_1}(D_{k_3}(D_{k_2}(c))) &= D_{k_1}(D_{k_2}(D_{k_3}(c))) = m
\end{aligned}$$

must all hold. It goes without saying that the notion of a commutative encryption can be generalized to many encryption steps. If, for example, the addition modulo 2 is used to encrypt and decrypt messages, then the order of the encryption and decryption operations does not matter, and hence the respective encryption system is commutative (for any number of encryption steps).

In order to evaluate a particular (symmetric) encryption system, one needs well-defined criteria. Referring to Shannon [1, 2],[2] the following five evaluation criteria may serve as a starting point.

**Amount of secrecy:** The ultimate goal of an encryption system is to keep plaintext messages secret. Consequently, the amount of secrecy provided by an encryption system is an important evaluation criterion. It is particularly interesting to

1 This condition is specific for symmetric encryption systems. In asymmetric encryption systems, the keys that select an encryption function and a decryption function from the corresponding families are not equal and may not be efficiently computable from one another. This point is further addressed in Chapter 13.

2 Refer to Section 1.3 for references to Shannon's original work.

be able to measure and somehow quantify the amount of secrecy a encryption system is able to provide. We are going to define perfect and computational secrecy as a possible measure to be used in the field. Except for that, we don't have the tools to more differentially argue about the amount of secrecy provided by a (symmetric) encryption system.

**Size of key:** Symmetric encryption systems employ secret keys that must be securely generated, distributed, managed, and memorized. It is therefore desirable (from an implementation and performance viewpoint) to have keys that are as small as possible. The size of the key must be at least as large as to make a brute-force attack or exhaustive key search computationally intractable. Beyond this threshold, however, the size of key is often overrated in security discussions and analyses. Long keys are not particularly more difficult to handle than short keys.

**Complexity of enciphering and deciphering operations:** To allow an efficient implementation, the enciphering and deciphering operations should not be too complex (i.e., they should be as simple as possible). This criterion used to be important in the past. Due to the power of today's computing devices, the complexity of the enciphering and deciphering operations is no longer a key issue. Currently deployed encryption systems can be efficiently implemented even on small end-user devices.

**Propagation of errors:** Different symmetric encryption systems and modes of operation have different characteristics with regard to the propagation of errors. Sometimes propagation of errors is desirable, but sometimes it is not. Hence, the nature and the characteristics of the application that needs to be secured determines the requirements with regard to error propagation.

**Expansion of messages:** In some symmetric encryption systems, the size of a message is increased by the encryption, meaning that the ciphertext is larger than the underlying plaintext message. This is not always desirable, and sometimes symmetric encryption systems are designed to minimize message expansion. If, for example, encrypted data must be fed into a fixed-length field of a communication protocol, then the symmetric encryption system must not expand the plaintext message at all.

This list is not comprehensive, and many other and (complementary) evaluation criteria may be important in a specific environment or application setting. Furthermore, not all criteria introduced by Shannon are still equally important today. For example, the "size of key" and the "complexity of enciphering and deciphering operations" criteria are not so important anymore, mainly because computer systems manage keys and run the enciphering and deciphering operations in a way that is independent from the user.

### 9.1.1 Block and Stream Ciphers

Every practically relevant symmetric encryption system processes plaintext messages unit by unit. A unit, in turn, may be either a bit or a block of bits (e.g., one or several bytes). Furthermore, the symmetric encryption system may be implemented as an FSM, meaning that the i-th ciphertext unit depends on the i-th plaintext unit, the secret key, and possibly some internal state. Depending on the existence and use of internal state, block ciphers and stream ciphers are usually distinguished.

- In a *block cipher*, the encrypting and decrypting devices have no internal state (i.e., the i-th ciphertext unit only depends on the i-th plaintext unit and the secret key). There is no memory involved, except for the internal memory that is used by the implementation of the cipher. Block ciphers are further addressed in Section 9.6.
- In a *stream cipher*, the encrypting and decrypting devices have internal state (i.e., the i-th ciphertext unit depends on the i-th plaintext unit, the secret key, and some internal state). Consequently, stream ciphers represent theoretically more advanced and more powerful symmetric encryption systems than block ciphers (in practice, things are not so clear and the question of whether block ciphers or stream ciphers are more advanced is discussed controversially). There are two major classes of stream ciphers that differ in their state transition function (i.e., the way the internal state is manipulated and the next state is computed):
    - In a *synchronous* stream cipher, the next state does not depend on the previously generated ciphertext units.
    - In a *nonsynchronous* stream cipher, the next state also depends on some (or all) of the previously generated ciphertext units.

  Synchronous stream ciphers are also called *additive stream ciphers*, and nonsynchronous stream ciphers are also called *self-synchronizing stream ciphers*. In this book, we use these terms synonymously and interchangeably. Stream ciphers are further addressed in Section 9.5.

The distinction between block ciphers and stream ciphers is less precise than one might expect. In fact, there are modes of operation that turn a block cipher into a stream cipher—be it synchronous or nonsynchronous. Some of these modes are overviewed and briefly discussed in Section 9.7—after we have introduced a few block ciphers that are most frequently used in the field.

### 9.1.2 Attacks

In Section 1.2.2, we said that we must formally define the term *security* before we can make precise statements about the security of a cryptographic system, such as a symmetric encryption system. More specifically, we must specify and nail down the adversary's capabilities and the task he or she is required to solve in order to be successful (i.e., to break the security of the system). With regard to the adversary's capabilities, there are several attacks that can be distinguished, and they are introduced here. The task to be solved and the resulting notions of security are discussed later in the chapter. Let $m_1, m_2, \ldots, m_l \in \mathcal{M}$ be a set of $l$ plaintext message units and $c_1, c_2, \ldots, c_l \in \mathcal{C}$ the set of respective ciphertext units encrypted with a particular key $k \in \mathcal{K}$.

**Ciphertext-only attacks:** In a *ciphertext-only attack*, the adversary only gets to know one or several ciphertext units $c_1, c_2, \ldots, c_l$ for some $l \geq 1$ to solve the task he or she is required to solve. Because an adversary always gets to know messages in encrypted form (otherwise, the messages would not have to be encrypted in the first place), ciphertext-only attacks are always feasible and need to be mitigated in one form or another. In fact, an encryption system that is susceptible to such attacks is totally insecure and useless, and should never be used in the field.

**Known-plaintext attacks:** In a *known-plaintext attack*, the adversary gets to know one or several plaintext message and ciphertext unit pairs $(m_1, c_1), (m_2, c_2), \ldots, (m_l, c_l)$ for some $l \geq 1$ to solve the task he or she is required to solve. In contrast to chosen-plaintext and chosen-ciphertext attacks (see below), the adversary cannot choose the plaintext message and ciphertext unit pairs. Known-plaintext attacks are possible and more likely to occur than one might expect. Note, for example, that many communication protocols have specific fields whose values are either known or can be easily guessed (for example, if they are padded with zero bytes).

**Chosen-plaintext attacks:** In a *chosen-plaintext attack* (CPA), the adversary has access to the encryption function (or the device that implements the function, respectively) and can therefore encrypt one or several plaintext message units $m_1, m_2, \ldots, m_l$ of his or her choice for some $l \geq 1$. For each $m_i$, the adversary gets to know the respective ciphertext unit $c_i$ ($1 \leq i \leq l$). In the simplest case, the adversary must choose the plaintext message units

$m_1, m_2, \ldots, m_l$ before the attack begins. In an *adaptive CPA*, however, the adversary can dynamically choose plaintext message units while the attack is going on. Needless to say, adaptive CPAs are more powerful than their nonadaptive counterparts.

**Chosen-ciphertext attacks:** In a *chosen-ciphertext attack* (CCA), the adversary has access to the decryption function (or the device that implements the function, respectively) and can therefore decrypt one or several ciphertext units $c_1, c_2, \ldots, c_l$ of his or her choice for some $l \geq 1$. For each $c_i$, the adversary gets to know the respective plaintext message unit $m_i$ ($1 \leq i \leq l$). Again, we distinguish whether the adversary must choose the ciphertext units $c_1, c_2, \ldots, c_l$ before the attack begins, or whether he or she can dynamically choose them while the attack is going on. In the second case, we call the CCA *adaptive* and use the acronym CCA2 (so CCA2 stands for an adaptive CCA). As we will see in Section 13.1, CCAs and CCA2s are particularly important in the realm of asymmetric encryption.

In general, there are many possibilities to mount such attacks, and we are going to see many examples throughout the book. Because a ciphertext-only attack is always possible, an adversary who knows the symmetric encryption system in use can mount such an attack by trying every possible key. This attack can even be parallelized if multiple processors are available. Let $|\mathcal{K}|$ be the size of the key space (i.e., the number of possible keys), $t$ the time it takes to test a key, and $p$ the number of processors performing the key search. Then each processor is responsible for approximately $|\mathcal{K}|/p$ keys, and hence it takes time $|\mathcal{K}|t/p$ to test all possibilities. On the average, one can expect to find the correct key about halfway through the search, making the expected time approximately

$$\frac{|\mathcal{K}|t}{2p} \tag{9.1}$$

This attack is known as *brute-force attack* or *exhaustive key search.* It can be mounted whenever the adversary is able to decide whether he or she has found the correct key. For example, it may be the case that the decrypted plaintext message is written in a specific language or that it otherwise contains enough redundancy to tell it apart from gibberish. Suppose, for example, that the adversary does not know the plaintext message (for a given ciphertext), but he or she knows that the plaintext message is encoded with one ASCII character per byte. This means that each byte must have a leading zero bit, and this is often enough redundancy to tell legitimate plaintext messages apart from illegitimate ones.

Unfortunately (from the adversary's perspective), a brute-force attack may also generate false positive results (i.e., keys that look like they are correct but are not the one actually used for encryption). The likelihood of this occurring depends on the relative sizes of the key and plaintext message spaces. If, for example, a block cipher has a block length of 64 bits and a key length of 80 bits, then the plaintext message and ciphertext spaces have $2^{64}$ elements, whereas the key space has $2^{80}$ elements. This means that there are on the average $2^{80}/2^{64} = 2^{80-64} = 2^{16} = 65,536$ keys that map a given plaintext message to a given ciphertext. To find the correct key—the so-called target key—among all possible keys, the adversary can consider a second plaintext message-ciphertext-pair (to further reduce the number of possible keys). If the target key is not yet unique, then he or she can use a third plaintext message-ciphertext-pair, and so on. In each step, the likelihood of finding the target key increases significantly, and normally only a few iterations are required to uniquely determine the target key.

Let us now provide a historical perspective and outline some simple ciphers one may think of at first sight. These ciphers are just examples and do not satisfy the security requirements of cryptography as it stands today.

## 9.2 HISTORICAL PERSPECTIVE

Every cipher employs one or several alphabet(s) to form plaintext message, ciphertext, and key spaces. If, for example, a single alphabet $\Sigma = \{A, \ldots, Z\}$ is used, then all spaces consist of words that can be constructed with the capital letters from $A$ to $Z$. These letters can be associated with the 26 elements of $\mathbb{Z}_{26} = \{0, 1, \ldots, 25\}$. In fact, there is a bijective mapping from $\{A, \ldots, Z\}$ into $\mathbb{Z}_{26}$, and this means that $\Sigma$ and $\mathbb{Z}_{26}$ are isomorphic (written as $\Sigma \cong \mathbb{Z}_{26}$), and hence one can work either with $\Sigma = \{A, \ldots, Z\}$ or $\mathbb{Z}_{26} = \{0, \ldots, 25\}$.

If $\Sigma \cong \mathbb{Z}_{26} = \{0, \ldots, 25\}$ and $\mathcal{M} = \mathcal{C} = \mathcal{K} = \mathbb{Z}_{26}$, then an *additive cipher* can be defined with the following encryption and decryption functions:

$$E_k : \mathcal{M} \longrightarrow \mathcal{C} \qquad\qquad D_k : \mathcal{C} \longrightarrow \mathcal{M}$$
$$m \longmapsto m + k \pmod{26} = c \qquad\qquad c \longmapsto c - k \pmod{26} = m$$

In this cipher, the decryption key is the additive inverse of the encryption key. Consequently, it is simple for anybody knowing the encryption key to determine the decryption key (that's why the encryption system is called symmetric in the first place). In Section 1.3, we mentioned the Caesar cipher that is an example of an additive cipher with the fixed key $k = 3$.

Similar to the additive cipher, one can define a *multiplicative cipher*[3] or combine an additive and a multiplicative cipher in an *affine cipher*. In this case, the key space $\mathcal{K}$ consists of all pairs $(a,b) \in \mathbb{Z}_{26}^2$ with $gcd(a,26) = 1$. As such, the key space has $\phi(26) \cdot 26 = 312$ elements and is far too small for practical use. It can, however, be used for demonstrational purposes. In fact, the encryption and decryption functions—$E_{(a,b)}$ and $D_{(a,b)}$—of an affine cipher are defined as follows:

$$E_{(a,b)} : \begin{array}{l} \mathcal{M} \longrightarrow \mathcal{C} \\ m \longmapsto am + b \pmod{26} \end{array} \qquad D_{(a,b)} : \begin{array}{l} \mathcal{C} \longrightarrow \mathcal{M} \\ c \longmapsto a^{-1}(c-b) \pmod{26} \end{array}$$

Obviously, the multiplicative inverse element of $a$ (i.e., $a^{-1}$) in $\mathbb{Z}_{26}$ is needed to properly decrypt $c$. As explained in Appendix A, the extended Euclidean algorithm (Algorithm A.2) can be used to efficiently compute this element.

An affine cipher can be broken with two known plaintext-ciphertext pairs. If, for example, the adversary knows $(F,Q) = (5,16)$ and $(T,G) = (19,6)$,[4] then he or she can set up the following system of two equivalences:

$$\begin{aligned} a5 + b &\equiv 16 \pmod{26} \\ a19 + b &\equiv 6 \pmod{26} \end{aligned}$$

The first equivalence can be rewritten as $b \equiv 16 - 5a \pmod{26}$ and used in the second equivalence: $19a + b \equiv 19a + 16 - 5a \equiv 14a + 16 \equiv 6 \pmod{26}$. Consequently, $14a \equiv -10 \equiv 16 \pmod{26}$, or $7a \equiv 8 \pmod{13}$, respectively. By multiplying either side with the multiplicative inverse element of 7 modulo 26 (which is 2), one gets $a \equiv 16 \equiv 3 \pmod{13}$, and hence $a = 3$ and $b = 1$. The adversary can now efficiently compute $D_{(a,b)}$ similar to the legitimate recipient of the encrypted message.

$\Sigma = \{A,\ldots,Z\} \cong \mathbb{Z}_{26}$ is a good choice for human beings. If, however, computer systems are used for encryption and decryption, then it is advantageous and more appropriate to use $\Sigma = \mathbb{Z}_2 = \{0,1\} \cong \mathbb{F}_2$ and to set the plaintext message, ciphertext, and key spaces to $\{0,1\}^*$. More often than not, the key space is set to $\{0,1\}^l$ (instead of $\{0,1\}^*$) for a reasonably sized key length $l$, such as 128 or 256 bits.

Additive, multiplicative, and affine ciphers are the simplest examples of *monoalphabetic substitution ciphers*. In a monoalphabetic substitution cipher, each

3 The multiplicative cipher works similar to the additive cipher. It uses multiplication instead of addition. Also, to make sure that one can decrypt all the time, one must work with $\{1,2,\ldots,26\}$ instead of $\{0,1,\ldots,25\}$.

4 This means that the letter "F" is mapped to the letter "Q" and the letter "T" is mapped to the letter "G."

letter of the plaintext alphabet is replaced by another letter of the ciphertext alphabet. The replacement is fixed, meaning that a plaintext letter is always replaced by the same ciphertext letter. In the most general case, a monoalphabetic substitution cipher can be thought of a permutation of the letters that form the (plaintext and ciphertext) alphabet. For example, A may be mapped to S, B may be mapped to M, C may be mapped to T, and so on. There are

$$|\Sigma|! = 26! = 403,291,461,126,605,635,584,000,000 > 4 \cdot 10^{26}$$

possible permutations of the 26 letters of the Latin alphabet. Each permutation represents a key, and hence the key space of a monoalphabetic substitution cipher is huge. In fact, it turns out that the key space of a monoalphabetic substitution cipher is not the problem. The problem is more related to the fact that monoalphabetic substitution ciphers cannot disguise the frequency distributions of individual letters and groups of letters. For example, the letter E is the most frequently occurring letter in English texts. So if we have a ciphertext and we notice by counting that the letter X occurs most frequently, then we have some evidence that E has been mapped to the letter X. This line of reasoning can be applied to all letters of the ciphertext, and hence it may be possible to decrypt the ciphertext using statistical arguments only (i.e., without trying out each and every possible permutation).

An early attempt to increase the difficulty of frequency analysis attacks on substitution ciphers was to disguise plaintext letter frequencies by homophony. In a *homophonic substitution cipher*, plaintext letters can be replaced by more than one ciphertext letter. Usually, the highest frequency plaintext letters are given more equivalents than lower frequency letters. In this way, the frequency distribution of ciphertext letters is flattened, making analysis more difficult.

Alternatively, *polyalphabetic substitution ciphers* flatten the frequency distribution of ciphertext letters by using multiple ciphertext alphabets in some cyclic way. All of these substitution ciphers are overviewed and discussed in the literature. Most of them, including, for example, the famous *Vigenère cipher*,[5] are easy to break today (but keep in mind that some of these ciphers had been believed to be secure for centuries until their breaking became public). In the case of the Vigenère cipher, Friedrich Kasiski and William F. Friedman published the first successful attacks in 1863 (the so-called *Kasiski test*) and 1925 (the so-called *Friedman test* or *index of coincidence*). Both statistical tests can be used to determine the number of ciphertext alphabets that are concurrently used. Once this number is known, one can break down the ciphertext into shorter trunks that are encrypted monoalphabetically. Hence frequency analysis can again be used to break each monoalphabetic substitution

5 The Vigenère cipher is a polyalphabetic substitution cipher that was published in 1585 (and considered unbreakable until 1863) and was widely deployed in previous centuries.

cipher individually, and hence the cryptanalysis of the polyalphabetic substitution cipher is reduced to the cryptanalysis of the monoalphabetic substitution cipher. Again, refer to any book about (classical) cryptography to get more information about these historically relevant ciphers and their cryptanalysis (some books are itemized in the Preface). This also applies to *Enigma*, a portable cipher machine used by the Germans during World War II to encrypt and decrypt secret messages. For the purpose of this book, we don't examine these ciphers. Instead, we focus on ciphers that are considered to be secure and hence are practically relevant today. Before that, we want to clearly distinguish between perfectly secure and computationally secure encryptions. This is the topic of the next two sections.

## 9.3 PERFECTLY SECURE ENCRYPTION

As mentioned before, the field of perfectly or information-theoretically secure encryption was pioneered by Shannon in the late 1940s [1, 2].[6] The aim was to come up with an encryption system that is perfectly secure (or secret) in the sense that it is impossible for an adversary to derive any information about a plaintext message from a given ciphertext.[7] This must be true even if the adversary has the best available computer technology at hand, and even if he or she is not limited in terms of computational resources, such as time and memory. Having such an absolute notion of security in mind, it is not obvious that perfect security (or secrecy) exists at all. But there is good and bad news: The good news is that perfect security is in fact possible and technically feasible. The bad news is that it is usually too expensive (in terms of keying material) for almost all practical purposes.

As illustrated in Figure 9.1, Shannon had a particular model of a symmetric encryption system in mind when he developed the notion of perfect or information-theoretical security. In this model, a source (left side) wants to transmit a plaintext message $m$ to the destination (right side) over an unsecure communications channel (dotted line). To secure $m$ during its transmission, the source has an encryption device and the destination has a decryption device. The devices implement an encryption and decryption algorithm[8] and are both fed with the same secret key $k$ generated by a key source. It is assumed that a secure channel exists between the key source and the encryption and decryption devices. The encryption device turns the plaintext message $m$ into a ciphertext $c$, and the decryption device does the opposite. It is assumed that the adversary has only access to the ciphertext $c$ and that he or she

6 Refer to Appendix C for an introduction to information theory.

7 This means that one has a ciphertext-only attack in mind when one talks about perfect secrecy and information-theoretically secure encryption.

8 More specifically, they implement the families $E$ and $D$ of encryption and decryption functions.

has no information about the secret key $k$ other than that obtained by observing $c$. In this situation, the adversary tries to retrieve some useful information either about the plaintext message $m$ or about the actual key $k$ in use.

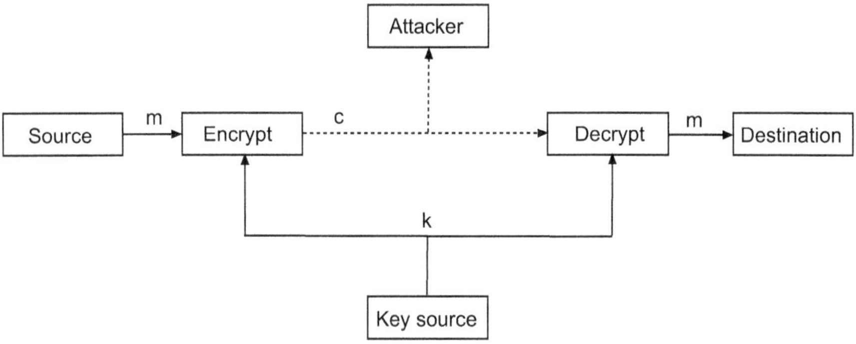


**Figure 9.1** Shannon's model of a symmetric encryption system.

A cryptographic technique originally not envisioned by Shannon is probabilistic or randomized encryption. Figure 9.2 shows such a model. In addition to the components of the original Shannon model, this model includes a random source that generates a random input $s$ for the encryption process. The random input may either be used as an additional nonsecret "key" that is transmitted to the destination and multiplexed with the ciphertext, or it may be used to randomize the plaintext, in which case the adversary does not obtain the randomizer in the clear. In either case, it is important to note that the decryption process cannot be randomized and hence that the decryption process need not be fed with $s$.

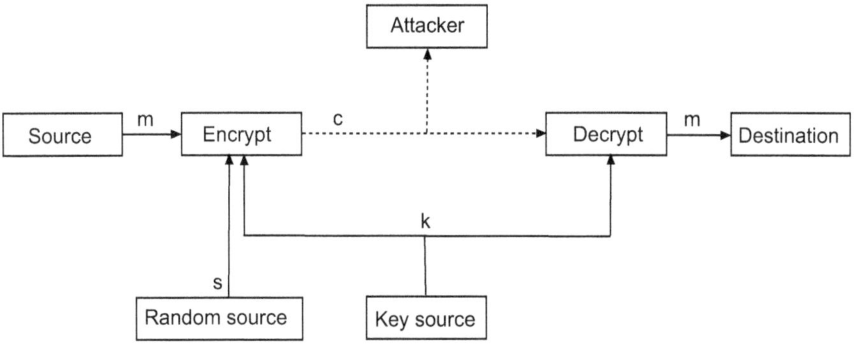


**Figure 9.2** A model of a randomized symmetric encryption system.

Having Shannon's model of a symmetric encryption system in mind, we want to explore the notion of perfectly secure encryption using probability theory first. Let $m_0, m_1 \in \mathcal{M}$ be two equally long[9] plaintext messages, and $c \in \mathcal{C}$ a ciphertext that is the encryption of either one of these messages. If the encryption is perfectly secure, then the probability that $c$ is the encryption of $m_0$ must be equal to the probability that $c$ is the encryption of $m_1$. This can be formalized as follows:

$$\Pr_{k \xleftarrow{r} \mathcal{K}}[E_k(m_0) = c] = \Pr_{k \xleftarrow{r} \mathcal{K}}[E_k(m_1) = c] \tag{9.2}$$

If (9.2) holds for all $m_0, m_1 \in \mathcal{M}$, $c \in \mathcal{C}$, and $k$ sampled uniformly at random from $\mathcal{K}$, then $c$ yields arguably no information about the plaintext message that is encrypted ($m_0$ or $m_1$), and hence the encryption is perfectly secure.

Alternatively, one can say that a symmetric encryption system is perfectly secure if for all possible pairs of plaintext messages $m_0$ and $m_1$ from $\mathcal{M}$ (again, with $|m_0| = |m_1|$) and keys $k$ sampled uniformly at random from $\mathcal{K}$ the probability distributions of the respective ciphertexts are exactly the same. Formally, this can be expressed as follows:

$$\{E_k(m_0)\} = \{E_k(m_1)\} \tag{9.3}$$

(9.2) and (9.3) nicely capture the idea. An alternative way of defining perfect security may use random variables (Definition B.2): An encryption (process) that takes place in a symmetric encryption system $(E, D)$ over $\mathcal{M}$, $\mathcal{C}$, and $\mathcal{K}$ can be seen as a discrete random experiment, where $M$ and $K$ are real-valued random variables that are distributed according to the probability distributions $P_M : \mathcal{M} \rightarrow \mathbb{R}^+$ and $P_K : \mathcal{K} \rightarrow \mathbb{R}^+$. Note that $P_M$ typically depends on the language in use, whereas $P_K$ is often uniformly distributed over all possible keys (i.e., all keys are equally probable). In either case, it is reasonable to assume that $M$ and $K$ are independent from each other. In addition to $M$ and $K$, there is a third random variable $C$ that is distributed according to $P_C : \mathcal{C} \rightarrow \mathbb{R}^+$. This random variable models the ciphertext, and hence its probability distribution $P_C$ is completely determined by $P_M$ and $P_K$. The random variable $C$ is the one that can be observed by the adversary and from which he or she may try to retrieve information about $M$ or $K$.

If the two random variables $M$ and $C$ are independent from each other, then the respective probability distributions $P_M$ and $P_{M|C}$ must be equal for all $m \in \mathcal{M}$ and $c \in \mathcal{C}$.

- $P_M$ stands for the *a priori* probability distribution that measures the probability with which each plaintext message $m$ occurs in the message space $\mathcal{M}$ (independent from a ciphertext $c$ that is observed);

9 Equally long means that $|m_0| = |m_1|$.

- $P_{M|C}$ stands for the *a posteriori* probability distribution that measures the probability with which each message $m$ is likely to be encrypted to ciphertext $c$ that is observed.

Using this terminology, a symmetric encryption system provides perfect security, if the a priori and a posteriori probability distributions are the same, meaning that the observation of a ciphertext gives the adversary no information about the plaintext message that is transmitted. This idea is captured in Definition 9.1.

**Definition 9.1 (Perfectly secure symmetric encryption system)** *A symmetric encryption system $(E, D)$ over $\mathcal{M}$, $\mathcal{C}$, and $\mathcal{K}$ is* perfectly secure *if* $P_M = P_{M|C}$.

Let us consider a simple example to illustrate the notion of a perfectly secure encryption system. Let $\mathcal{M} = \{0,1\}$ with $P_M(0) = 1/4$ and $P_M(1) = 3/4$, $\mathcal{K} = \{A, B\}$ with $P_K(A) = 1/4$ and $P_K(B) = 3/4$, $\mathcal{C} = \{a, b\}$ with $P_C(a)$ and $P_C(b)$ as computed below, and encryption function $E$ be defined as follows:

$$E_A(0) = a \qquad E_A(1) = b \qquad E_B(0) = b \qquad E_B(1) = a$$

The probability distributions $P_M$ and $P_K$ are independent, whereas the probability distribution $P_C$ is determined by $P_M$ and $P_K$. Due to the independence of $P_M$ and $P_K$, one can compute the probability that the plaintext message 0 is encrypted with key $A$ (yielding ciphertext $a$) according to $P_{MK}(0, A) = P_M(0) \cdot P_K(A) = \Pr[M = 0] \cdot \Pr[K = A] = 1/4 \cdot 1/4 = 1/16$. Similarly, one can compute $P_{MK}(1, A) = P_M(1) \cdot P_K(A) = \Pr[M = 1] \cdot \Pr[K = A] = 3/4 \cdot 1/4 = 3/16$ (yielding ciphertext $b$), $P_{MK}(0, B) = P_M(0) \cdot P_K(B) = \Pr[M = 0] \cdot \Pr[K = B] = 1/4 \cdot 3/4 = 3/16$ (again, yielding ciphertext $b$), as well as $P_{MK}(1, B) = P_M(1) \cdot P_K(B) = \Pr[M = 1] \cdot \Pr[K = B] = 3/4 \cdot 3/4 = 9/16$ (again, yielding ciphertext $a$). Due to the law of total probability (Theorem B.1), the ciphertext $a$ occurs with probability $P_C(a) = P_{MK}(0, A) + P_{MK}(1, B) = 1/16 + 9/16 = 10/16 = 5/8$, whereas the ciphertext $b$ occurs with probability $P_C(b) = P_{MK}(1, A) + P_{MK}(0, B) = 3/16 + 3/16 = 6/16 = 3/8$ (needless to say, $P_C(a) = 5/8$ and $P_C(b) = 3/8$ must sum up to one).

Equipped with these values, one can compute the conditional probability that the original plaintext message is 0 if ciphertext $a$ is observed. This probability $\Pr[0|a]$ equals the probability that the original plaintext message is 0 and the ciphertext is $a$ (i.e., $P_{MK}(0, A) = 1/16$) divided by the probability that the ciphertext is $a$ (i.e., $P_C(a) = 5/8$):

$$\Pr[0|a] = \frac{P_{MK}(0, A)}{P_C(a)} = \frac{1/16}{5/8} = \frac{1 \cdot 8}{16 \cdot 5} = \frac{1}{2 \cdot 5} = \frac{1}{10}$$

On the other hand, the probability that the original plaintext message is 1 if $a$ is observed (i.e., $\Pr[1|a]$) equals the probability that the original plaintext message is 1 and the ciphertext is $a$ (i.e., $P_{MK}(1, B) = 9/16$) divided by the probability that the ciphertext is $a$ (i.e., $P_C(a) = 5/8$):

$$\Pr[1|a] = \frac{P_{MK}(1,b)}{P_C(a)} = \frac{9/16}{5/8} = \frac{9 \cdot 8}{16 \cdot 5} = \frac{9}{2 \cdot 5} = \frac{9}{10}$$

The result is $9/10$. Hence, if a ciphertext $a$ is observed, then it is the encryption of 0 with a probability of $1/10$ and the encryption of 1 with a probability of $9/10$. Again, the two probabilities must sum up to one, meaning that if a ciphertext $a$ is observed, then it must be the case that the original plaintext message is either zero or one (these are the only two possible messages).

Following the same line of argumentation, one can compute the conditional probability that the original plaintext message is 0 if ciphertext $b$ is observed (i.e., $\Pr[0|b] = P_{MK}(0, B)/P_C(b) = (3/16)/(3/8) = 1/2$), and the conditional probability that the original plaintext message is 1 if ciphertext $b$ is observed (i.e., $\Pr[1|b] = P_{MK}(1, A)/P_C(b) = (3/16)/(3/8) = 1/2$). Both values are equal to $1/2$ and sum up to one.

The bottom line is that in neither of the two cases is the a posteriori probability distribution $P_{M|C}$ equal to $P_M$, and this, in turn, means that the encryption system is not perfectly secure. If one wanted to make the encryption system perfectly secure, then one would have to make the keys equally probable (i.e., $P_K(A) = P_K(B) = 1/2$).

Using the information-theoretical notion of entropy (Section C.2), one can follow the argumentation of Shannon and define a perfectly secure symmetric encryption system as captured in Definition 9.2.

**Definition 9.2 (Perfectly secure symmetric encryption system)** *A symmetric encryption system $(E, D)$ over $\mathcal{M}$, $\mathcal{C}$, and $\mathcal{K}$ is* perfectly secure *if $H(M|C) = H(M)$ for every probability distribution $P_M$.*

In his seminal work, Shannon also showed for nonrandomized symmetric encryption systems that a necessary (but usually not sufficient) condition for such a system to be perfectly secure is that the entropy of $K$ is at least as large as the entropy of $M$ (this means that the secret key must be at least as long as the total amount of plaintext that is to be transmitted). This result is formally expressed in Theorem 9.1.

**Theorem 9.1 (Shannon's Theorem)** *In a perfectly secure symmetric encryption system $H(K) \geq H(M)$.*

*Proof.*

$$\begin{aligned} H(M) = H(M|C) &\leq H(MK|C) \\ &= H(K|C) + H(M|CK) \\ &= H(K|C) \\ &\leq H(K) \end{aligned}$$

In the first line, we employ the definition of perfect secrecy, namely that $H(M) = H(M|C)$ and the fact that $H(MK|C)$ is at least $H(M|C)$. In the second line, we use the basic expansion rule for uncertainties (generalized to conditional uncertainties). In the third line, we use the fact that $H(M|CK) = 0$ for any symmetric encryption system (i.e., it is required that a plaintext can be uniquely decrypted if a ciphertext and a key are known). The inequality stated in the theorem then follows immediately.

□

A practical encryption scheme that can be shown to be perfectly secure (using Shannon's theorem) is the *one-time pad* that is usually credited to Gilbert S. Vernam [3] and Joseph O. Mauborgne. Due to a patent[10] granted to Vernam in 1919, the one-time pad is also known as the *Vernam cipher*.[11] It consists of a randomly generated and potentially infinite stream of key bits $k = k_1, k_2, k_3, \ldots$ that is shared between the sender and the recipient. To encrypt a plaintext message $m = m_1, m_2, \ldots, m_n$, the sender adds each bit $m_i$ $(1 \leq i \leq n)$ modulo 2 with a key bit $k_i$:

$$c_i = m_i \oplus k_i \text{ for } i = 1, \ldots, n$$

The ciphertext $c = c_1, c_2, c_3, \ldots, c_n$ is sent from the sender to the recipient. It is then up to the recipient to recover the plaintext by adding each ciphertext bit $c_i$ modulo 2 with the corresponding key bit $k_i$:

$$p_i = c_i \oplus k_i = (p_i \oplus k_i) \oplus k_i = p_i \oplus (k_i \oplus k_i) = p_i \oplus 0 = p_i \text{ for } i = 1, \ldots, n$$

Consequently, the plaintext is recovered by adding each ciphertext bit with the corresponding key bit. As proven below, the one-time pad provides perfect secrecy, but the proof only applies if the key is truly random and used only once. In this case, the ciphertext provides absolutely no information about the plaintext message.

10 U.S. Patent 1,310,719.

11 In 2011, it was observed by Steven M. Bellovin [4] that the one-time pad had been known almost 35 years earlier to a Californian banker named Frank Miller who published the cipher in a book entitled "Telegraphic Code to Insure Privacy and Secrecy in the Transmission of Telegrams" back in 1882. This is long before people had the tools at hand to scientifically argue about the security of the one-time pad.

**Theorem 9.2 (One-time pad)** *The one-time pad provides perfect secrecy.*

*Proof.*
Starting with Equation (9.2), one can compute the probability $\Pr[E_k(m) = c]$ for $m \in \mathcal{M}$, $c \in \mathcal{C}$, and $k \in \mathcal{K}$. This probability equals the number of keys that map $m$ to $c$ divided by the number of all possible keys:

$$\Pr[E_k(m) = c] = \frac{|\{k \in \mathcal{K} : E_k(m) = c\}|}{|\mathcal{K}|}$$

If this probability is constant, then the respective encryption system is perfectly secure. In the case of the one-time pad, there is just one (uniquely defined) key $k$ that maps a particular plaintext message $m$ to a particular ciphertext $c$ (i.e., $k = m \oplus c$), and this means that the probability is always $1/|\mathcal{K}|$. This value is constant, and hence the one-time pad provides perfect secrecy.

□

The one-time pad provides perfect security in terms of secrecy, but it neither provides integrity nor authenticity. In fact, it is possible to modify a known ciphertext so that it has a well-defined effect on the underlaying plaintext message. Assume, for example, that the one-time pad has been used to encrypt the ASCII-encoded version of the plaintext "Bob" (hence the three bytes that represent this plaintext are `0x42` for the capital letter "B," `0x6F` for the letter "o," and `0x62` for the letter "b"). Under the assumption that the adversary somehow knows or guesses that the plaintext refers to "Bob," he or she can modify the ciphertext in some predictable way. In fact, he or she can modify the ciphertext in a way that it decrypts to "Jim" (represented by the three ASCII values `0x4A`, `0x69`, and `0x6D`) instead of "Bob." He or she simply adds the two ASCII values bytewise modulo 2: `0x42` $\oplus$ `0x4A` = `0x08` for the first character, `0x6F` $\oplus$ `0x69` = `0x06` for the second character, and `0x62` $\oplus$ `0x6D` = `0x0F` for the third character. He or she then adds `0x08`, `0x06`, and `0x0F` bytewise modulo 2 to the first three characters of the ciphertext to change the underlaying plaintext message from "Bob" to "Jim."

The property of being able to modify a ciphertext in such a predictable way is called *malleability*, and, in general, *nonmalleability* is a desired property for an encryption system—be it symmetrical or asymmetrical. There are situations in which an encryption system needs to be malleable, but these situations are rather rare. The one-time pad is highly malleable because a ciphertext can be modified at will. So secrecy (or confidentiality) and integrity are different design goals, and in many applications both are important and must be provided in one way or another.

In addition to perfect security as expressed in Definition 9.2, Shannon also introduced the notion of ideal security. The idea is that an adversary does not get any information about a key from a ciphertext of arbitrary length. Alternatively speaking, no matter how much ciphertext an adversary knows, the entropy of $K$ is not decreased. This idea is captured and formally expressed in Definition 9.3, but it is not further used in this book.

**Definition 9.3 (Ideal security)** *An encryption system* $(E, D)$ *over* $\mathcal{M}$, $\mathcal{C}$, *and* $\mathcal{K}$ *is* ideally secure *if* $H(K|C^n) = H(K)$ *for all* $n \in \mathbb{N}$.

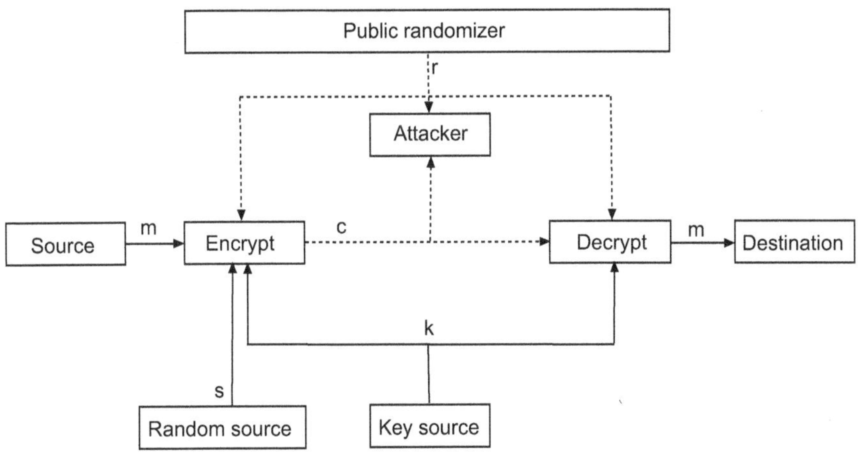


**Figure 9.3** A randomized symmetric encryption system that employs a public randomizer.

In summary, Shannon's theorem says that unless two entities initially share a secret key that is at least as long as the plaintext message to be transmitted, the adversary will always obtain some information about the plaintext message from the ciphertext. This result has caused many cryptographers to believe that perfect security (or secrecy) is impractical. This pessimism can be relativized by pointing out that Shannon's analysis assumes that, except for the secret key, the adversary has access to exactly the same information as the communicating entities and that this apparently innocent assumption is more restrictive than is generally realized. For example, Maurer showed that it is possible to develop randomized symmetric encryption systems that employ public randomizers as illustrated in Figure 9.3 to provide perfect security even if the secret keys are smaller than the plaintext messages [5]. The output of a public randomizer is assumed to be publicly accessible (also to the adversary) but impossible to modify. It can be modeled as a random variable $R$. There are basically two different ways of implementing a public randomizer: broadcasting and storing. A source (e.g., a satellite) could broadcast random data or storage devices that contain the same random data could be distributed. In the first case, it is possible to come up with a randomized symmetric encryption system that employs a public randomizer and that is perfectly secure under the sole assumption that the noise on the main channel (i.e., the channel from the source to the destination) is at least to some extent independent from the noise on the channel from the sender to the adversary. This system demonstrates that a mere difference in the signals received by the legitimate receiver and by the adversary, but not necessarily with an advantage to the legitimate receiver, is sufficient for achieving security. From Maurer's results, one may also conclude that, for cryptographic purposes, a given communication channel that is noisy is not necessarily bad. In addition, such a channel should not be turned into an error-free channel by means of error-correcting codes, but rather that cryptographic coding and error-control coding should ideally be combined.

## 9.4 COMPUTATIONALLY SECURE ENCRYPTION

Formulas (9.2) and (9.3) require the probabilities and probability distributions to be the same. This allows us to come up with an absolute notion of security (or secrecy). In practice, however, these requirements are overly restrictive, and it may be sufficient to only require that the probabilities and probability distributions are computationally indistinguishable (instead of requiring that they are the same). This allows us to come up with a weaker notion of security, and we call the respective symmetric encryption systems *computationally secure*. All currently deployed symmetric encryption systems (except the one-time pad) are not perfectly but only computationally secure. This includes all symmetric encryption systems outlined in the rest of this chapter.

In a perfectly secure encryption system, absolutely no information about a plaintext message can be feasibly extracted from a ciphertext. In a computationally secure encryption system, we deviate from this strict definition, and we want to be sure that only negligible information can be extracted (instead of no information at all). This idea led to the notion of *semantic security* as originally proposed by Shafi Goldwasser and Silvio Micali in the early 1980s.[12] Informally speaking, an encryption system is *semantically secure* if it is computationally infeasible to derive any significant information about a plaintext message from a given ciphertext. Alternatively speaking, whatever an efficient algorithm $A$ can compute about a plaintext message from a given ciphertext can also be computed by an efficient

12 As mentioned in Chapter 5, Shafi Goldwasser and Silvio Micali jointly received the Turing Award in 2012.

algorithm $A'$ that does not know the ciphertext in the first place. This must apply to every possible probability distribution for plaintext messages.

Unfortunately, it is difficult to prove the semantic security of any practically used encryption system, and hence Goldwasser and Micali showed that semantic security is equivalent to another notion of security called *ciphertext indistinguishability under CPA* (IND-CPA) [6]. This definition is more commonly used than the original definition of semantic security, mainly because it better facilitates proving the security of cryptosystems used in the field. When we say that a symmetric encryption system is computationally secure, we basically mean that it is semantically secure, and hence that it provides IND-CPA. The two notions of security are equivalent.

To understand the notion of IND-CPA it is best to start from the security game introduced in Section 1.2.2.1 (illustrated in Figure 1.2) and to fine-tune it a little bit. Let $(E, D)$ over $\mathcal{M}, \mathcal{C}$, and $\mathcal{K}$ be a symmetric encryption system for which IND-CPA has to be shown in the game-theoretic setting. This can be done as follows:

- The adversary chooses a pair of equally long plaintext messages $m_0$ and $m_1$ (i.e., $|m_0| = |m_1|$), and sends them to a challenger.
- The challenger randomly selects a key $k \in_R \mathcal{K}$ and a bit $b \in_R \{0, 1\}$, and sends $c = E_k(m_b)$ as a challenge back to the adversary.
- It is the adversary's task to decide whether $c$ is the encryption of $m_0$ (i.e., $b = 0$) or $m_1$ (i.e., $b = 1$). Obviously, he or she can always guess with a success probability of $1/2$, but the goal is to distinguish the two cases with better odds. To do so, the adversary has oracle access to the encryption function $E_k(\cdot)$, meaning that he or she can have one or several plaintext messages $m_1, m_2, \ldots, m_q$ of his or her choice be encrypted to learn the respective ciphertexts $c_i = E_k(m_i)$ for $1 \leq i \leq q$.

After $q$ oracle queries, the adversary has to choose and output $b'$ that can either be zero or one. The adversary is successful if the probability that $b' = b$ is significantly better than guessing, meaning that

$$\Pr[b' = b] = \frac{1}{2} + \epsilon(n)$$

for a function $\epsilon(n)$ that is nonnegligible. In this case, the adversary can somehow distinguish the two cases and tell them apart with a success probability that is better than guessing.

Whenever people use encryption systems in the field, they go for systems that are semantically secure and hence provide IND-CPA. This is equally true for stream ciphers and block ciphers.

## 9.5 STREAM CIPHERS

Stream ciphers have played and continue to play an important role in cryptography.[13] Remember from Section 9.1.1 that stream ciphers use internal state, and that the i-th ciphertext unit depends on the i-th plaintext unit, the secret key, and this state. Also remember that it is common to distinguish between synchronous (or additive) stream ciphers and nonsynchronous (or self-synchronizing) stream ciphers. Having the modes of operation for block ciphers in mind (Section 9.7), it is obvious that operating a block cipher in CFB mode yields a nonsynchronous (self-synchronizing) stream cipher (i.e., the next state depends on previously generated ciphertext units), whereas operating a block cipher in OFB mode yields a synchronous (additive) stream cipher (i.e., the next state does not depend on previously generated ciphertext units). Most stream ciphers in use today are synchronous (or additive); they are similar to the one-time pad (see Section 9.3) in the sense that they bitwise add modulo 2 a plaintext message and a key stream.

Let $\Sigma = \mathbb{Z}_2 = \{0, 1\}$, $\mathcal{M} = \mathcal{C} = \Sigma^*$, and $\mathcal{K} = \Sigma^n$ for some reasonably sized key length $n$. To encrypt an $l$-bit plaintext message $m = m_1 \ldots m_l \in \mathcal{M}$ using an additive stream cipher, a secret key $k \in \mathcal{K}$ must be expanded into a stream of $l$ key bits $k_1, \ldots, k_l$. The encryption function is then defined as follows:

$$E_k(m) = m_1 \oplus k_1, \ldots, m_l \oplus k_l = c_1, \ldots, c_l$$

Similarly, the decryption function is defined as follows:

$$D_k(c) = c_1 \oplus k_1, \ldots, c_l \oplus k_l = m_1, \ldots, m_l$$

The main question in the design of an additive stream cipher is how to expand $k \in \mathcal{K}$ into a potentially infinite key stream $(k_i)_{i \geq 1}$. Many designs are based on linear feedback shift registers (LFSRs). The respective LFSR-based stream ciphers are overviewed next, before we turn our attention to other stream ciphers (not based on LFSRs) that are currently more widely used in the field.

### 9.5.1 LFSR-Based Stream Ciphers

A *feedback shift register* (FSR) of length $L$ consists of that amount of storage elements (called *stages*), each capable of storing one bit and having one input and

13 The importance of stream ciphers is controversially discussed in the community. At the Cryptographer's Panel of the RSA Conference 2004, for example, Adi Shamir gave a short talk entitled "The Death of the Stream Cipher." In this talk, Shamir noticed and tried to explain why stream ciphers are losing popularity against block ciphers. On the other hand, we have also experienced a revival of stream ciphers, mainly due to their simplicity and efficiency.

one output, and a clock that controls the movement of data in the FSR. The stages are initialized with the bits $s_0, s_1, \ldots, s_{L-1}$ that collectively refer to the initial state of the FSR. During each unit of time (clock cycle), the following operations are performed:

- The content of stage 0 (i.e., $s_0$) is output as part of the output sequence.
- The content of stage $i$ (i.e., $s_i$) is moved to stage $i-1$ for each $1 \leq i \leq L-1$. If we considered the contents of the FSR as a word, then we could call this operation a shift right (denoted FSR$\hookrightarrow$ 1 in this book).
- The new content for stage $L-1$ is the feedback bit $s_j$ that is computed as $s_j = f(s_0, s_1, \ldots, s_{L-1})$ for some (feedback) function $f$.

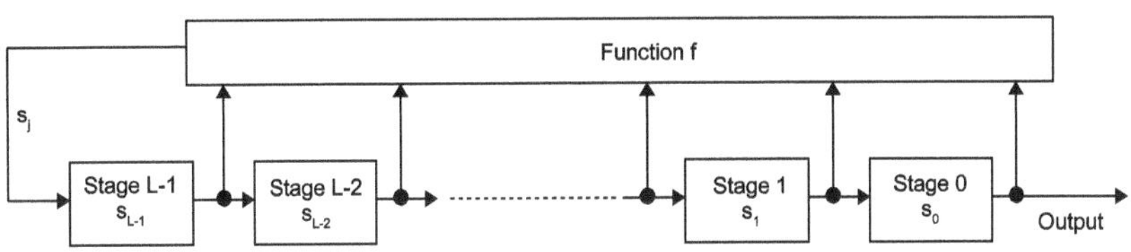


**Figure 9.4** A feedback shift register (FSR).

A respective FSR is illustrated in Figure 9.4. At every clock cycle $j$, the feedback function $f$ computes a new value $s_j$ from the stages, and this value is fed into the FSR from the left. Consequently, the contents of all stages are shifted to the right, and the content of the rightmost stage is the output of the FSR (for this clock cycle). This procedure is repeated for every clock cycle. Consequently, the FSR may be used to generate a sequence of (pseudorandom) output values. Because the length of the FSR and hence the number of possible stages is finite, the FSR represents an FSM and can be illustrated with a respective state diagram. If the register has length $L$ and there are $q$ possible stage values, then the FSR represents an FSM with $q^L - 1$ possible states, and hence the FSR has a maximal period of $q^L - 1$.[14] In mathematics, one can use Good-deBruijn graphs to argue about an FSR. Such graphs are not needed here, and hence we ignored them. Note, however, that in our case, $q$ is typically equal to two, and hence there are $2^L - 1$ possible states.

14 We have to exclude the case in which all stages comprise a zero. Otherwise, the state of the FSR does not change anymore.

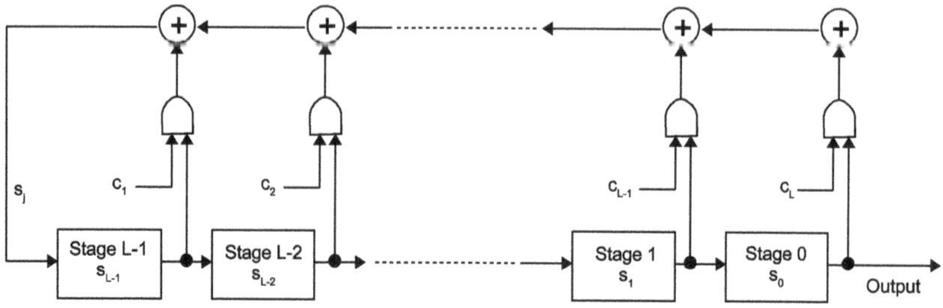


**Figure 9.5** A linear feedback shift register (LFSR).

If the feedback function $f$ is the modulo 2 sum of the contents of some stages, then the respective FSR yields an LFSR. This is illustrated in Figure 9.5. Each $c_i$ for $1 \leq i \leq L$ represents a bit that can either be 0 or 1, and it somewhat determines whether the respective stage value is taken into account in the feedback function. More specifically, $s_j$ is the modulo 2 sum of the contents of those stages $0 \leq i \leq L-1$, for which $c_{L-i} = 1$ (the closed semicircles in Figure 9.5 represent AND gates).

Using this notation, an LFSR as illustrated in Figure 9.5 can be specified as $\langle L, c(X) \rangle$, where $L$ refers to the length of the LFSR (i.e., the number of stages) and

$$c(X) = c_1X + c_2X^2 + \ldots + c_LX^L$$

refers to the *connection polynomial* that is an element of $\mathbb{F}_2[X]$. The LFSR is called *nonsingular*, if the degree of $c(X)$ is $L$, meaning that $c_L = 1$.

If $s^{(t=0)} = (s_0, s_1, \ldots, s_{L-1})$ is the initial state of a LFSR with connection polynomial $c(X)$, then the output sequence $s_L, s_{L+1}, s_{L+2}, \ldots$ is generated as

$$s_{L+t} = \sum_{i=1}^{L} c_i s_{L+t-i} = c_1 s_{L+t-1} + c_2 s_{L+t-2} + \ldots + c_{L+t} s_0$$

for $t \geq 0$. This sequence may be infinite, but (according to what has been said above) it must be cyclic after at most $2^L - 1$ steps. If, for example, $L = 4$, $c(X) = c_1X + c_2X^2 + c_3X^3 + c_4X^4$ with $(c_1, c_2, c_3, c_4) = (0,0,1,1)$, and $s^{(t=0)} = (s_0, s_1, s_2, s_3) = (1,0,1,1)$, then the resulting LFSR in its starting position is illustrated in Figure 9.6. In the first clock cycle (i.e., $t = 1$), $s_0 = 1$ is the output bit and the new stage value $s_3$ is the sum modulo 2 of 0 and 1 (i.e., 1). This means that $s^{(t=1)} = (0,1,1,1)$. Similarly, one can construct

$$\begin{aligned}
s^{(t=2)} &= (1,1,1,1) \\
s^{(t=3)} &= (1,1,1,0) \\
s^{(t=4)} &= (1,1,0,0) \\
s^{(t=5)} &= (1,0,0,0) \\
s^{(t=6)} &= (0,0,0,1) \\
s^{(t=7)} &= (0,0,1,0) \\
s^{(t=8)} &= (0,1,0,0) \\
s^{(t=9)} &= (1,0,0,1) \\
s^{(t=10)} &= (0,0,1,1) \\
s^{(t=11)} &= (0,1,1,0) \\
s^{(t=12)} &= (1,1,0,1) \\
s^{(t=13)} &= (1,0,1,0) \\
s^{(t=14)} &= (0,1,0,1) \\
s^{(t=15)} &= (1,0,1,1)
\end{aligned}$$

At this point, one can recognize that $s^{(t=15)} = s^{(t=0)}$, and hence that one period is complete and starts from scratch.

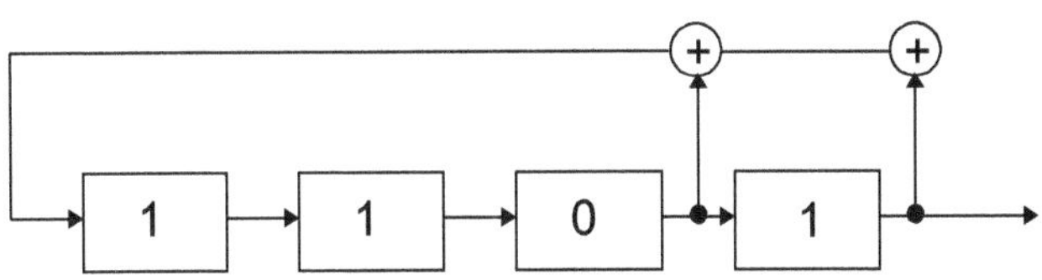


**Figure 9.6** An exemplary LFSR in its starting position.

Anyway, it can be shown that an LFSR of length $L$ has an output sequence with maximum possible period $2^L - 1$, if and only if its connection polynomial $c(X)$ has degree $L$ and is irreducible over $\mathbb{F}_2$. This means that $c(X)$ is primitive, and hence that $c(X)$ is a generator of $\mathbb{F}_{2^L}^*$, the multiplicative group of the nonzero elements of $\mathbb{F}_{2^L}$. In this sense, the exemplary LFSR from Figure 9.6 has a maximum period of $2^4 - 1 = 15$. In either case, the output sequence may be used as a key stream in a (LFSR-based) stream cipher.

Because LFSRs are well understood in theory and can be implemented very efficiently in hardware, several stream ciphers employ a single LFSR with a primitive connection polynomial. Such ciphers are susceptible to a known-plaintext attack, meaning that $2L$ plaintext-ciphertext pairs are usually sufficient to cryptanalyze and break them. A stream cipher should therefore not directly use the output of a single LFSR. In theory, people have proposed many constructions to securely use LFSRs in stream ciphers. Examples include the shrinking generator [7] and the self-shrinking generator [8] (Section 7.2), as well as the use of multiple LFSRs with irregular clocking,[15] such as A5/1 (using three LFSRs), the content scrambling system (CSS) for DVD encryption (using two LFSRs), and E0 for Bluetooth encryption (using four LFSRs). The details of these LFSR-based stream ciphers are beyond the scope of this book, but keep in mind that all of them have been cryptanalyzed and broken in the past. As such, they should no longer be used in the field.

### 9.5.2 Other Stream Ciphers

LFSR-based stream ciphers can be implemented efficiently in hardware, but they are not particularly well suited to be implemented in software. Consequently, there is room for other—preferably additive—stream ciphers optimized for implementation in software. Note that such a cipher need not depend on the availability of a computer system. For example, Bruce Schneier proposed a cipher named *Solitaire*[16] that employs a shuffled deck of (Solitaire) cards to generate a stream of pseudorandom letters (representing the key stream). On the sender's side, each letter of the plaintext message is encrypted by adding it modulo 26 to the respective letter from the key stream, and on the recipient's side each ciphertext letter is decrypted by adding it modulo 26 to the same letter from the key stream. Apart from Solitaire and a few similar ciphers, most stream ciphers in use today require computer systems to encrypt and decrypt messages. This is also the focus of this book.

In what follows, we briefly look at a historically relevant example of a non-LFSR-based stream cipher known as RC4, before we turn our attention to a more modern stream cipher known as Salsa20 and a variant known as ChaCha20. Other stream ciphers have been nominated as part of the European Network of Excellence for Cryptology (ECRYPT) Stream Cipher Project (eSTREAM[17]) that was running

15 Irregular clocking means that each LFSR has a specific position (or bit), called *clocking tap*. In each cycle, the majority bit is computed among all clocking taps. Only the LFSRs whose clocking tap agree with the majority are actually clocked. For example, A5/1 has 3 LFSRs with clocking taps 8 for a 19-bit LFSR, 10 for a 22-bit LFSR, and again 10 for a 23-bit LFSR. In this case, the majority is two, and hence at least two LFSRs are clocked, and all LFSRs are clocked if the 3 clocking taps agree on the same bit.

16 http://www.schneier.com/solitaire.html.

17 https://www.ecrypt.eu.org/stream.

from 2004 to 2008. The aim of the project was to find stream ciphers suited for fast encryption in software (profile 1) or hardware (profile 2). There were 16 finalists, and the eSTREAM profile 1 finally comprised HC-128, Rabbit, Salsa20/12, and SOSEMANUK, whereas profile 2 comprised Grain (version 1), MICKEY (version 2), and Trivium. Except for Salsa20, these stream ciphers are not addressed in this book.

#### 9.5.2.1 RC4

In 1987, Rivest proposed a stream cipher named *RC4*[18] that was quite widely deployed until a few years ago. It was, for example, used for Wired Equivalent Privacy (WEP), Wi-Fi Protected Access (WPA), and SSL/TLS protocols. The design of RC4 was originally kept as a trade secret of RSA Security. In 1994, however, the source code of an RC4 implementation was anonymously posted to the Cypherpunks mailing list, and the correctness of the posting was later confirmed by comparing its outputs to those produced by licensed implementations. Because the RC4 stream cipher is treated as a trade secret and the term RC4 is trademarked, the algorithm that was anonymously posted is sometimes referred to as *ARC4* or *ARCFOUR* (standing for "Alleged-RC4"). Hence, the terms RC4, ARC4, and ARCFOUR all refer to the same stream cipher.

In essence, RC4 is a synchronous (additive) stream cipher, meaning that a sequence of pseudorandom bytes (i.e., a key stream) is generated independently from the plaintext message or ciphertext, and this sequence is bytewise added modulo 2 to the plaintext message. The cipher takes a variable-length key $k$ that may range from 1 to 256 bytes (i.e., 2,048 bits). Its bytes are labeled $k[0], \ldots, k[255]$.

**Algorithm 9.1** The S-box initialization algorithm of RC4.

$(S, k)$

---

$j = 0$
for $i = 0$ to 255 do $S[i] = i$
for $i = 0$ to 255 do
    $j = (j + S[i] + k[i \mod |k|]) \mod 256$
    $S[i] \leftrightarrow S[j]$

---

$(S)$

To generate a key stream, RC4 employs an array $S$ of 256 bytes of state information (called S-box). The elements of the S-box are labeled $S[0], \ldots, S[255]$.

18 The acronym RC is rumored to stand for "Ron's Code." Note that RC2, RC5, and RC6 are block ciphers that are not related to RC4.

They are initialized according to Algorithm 9.1. The algorithm takes as input an empty S-box $S$ and a key $k$, and it then uses $k$ to put $S$ into some $k$-dependent state. More specifically, it initializes the 256 bytes of the state in a first for-loop with their respective index or position number:

$$\begin{aligned} S[0] &= 0 \\ S[1] &= 1 \\ &\dots \\ S[255] &= 255 \end{aligned}$$

Afterward, it pseudorandomly permutes the bytes of the S-box in a second for-loop, in which each $0 \leq j \leq 255$ is assigned $(j + S[i] + k[i \mod |k|]) \mod 256$ and the bytes at positions $i$ and $j$ are swapped.

**Algorithm 9.2** The RC4 PRG algorithm.

$(S)$

$i = (i+1) \bmod 256$
$j = (j + S[i]) \bmod 256$
$S[i] \leftrightarrow S[j]$
$k = S[(S[i] + S[j]) \bmod 256]$

$(k)$

After the S-box has been initialized according to Algorithm 9.1, $i$ and $j$ are both initialized with zero (the respective initialization steps $i \leftarrow 0$ and $j \leftarrow 0$ could be appended to Algorithm 9.1, but this is not done here). So just keep in mind that this needs to be done before the actual encryption begins.

To encrypt a plaintext message byte, the RC4 PRG algorithm outlined in Algorithm 9.2 must be executed and the resulting byte $k$ (that no longer refers to original key that is input to Algorithm 9.1) must be added modulo 2 to the respective plaintext message byte. The RC4 PRG algorithm therefore takes as input an S-box $S$ that is initialized according to Algorithm 9.1, and it generates as output a new byte of the key stream, denoted $k$ here. To generate $k$, the algorithm generates new values for $i$ and $j$: It increments $i$ modulo 256, and it adds modulo 256 the byte found at position $i$ in the S-Box to $j$. The S-box bytes at positions $i$ and $j$ are then swapped, and $k$ is taken from the S-box at position $S[i] + S[j]$ modulo 256. This algorithm is so simple that it can be implemented in only a few lines of code.

In spite of its simplicity, RC4 had been used for a very long time, until some weaknesses and statistical defects were found. Most importantly, the key stream generated by RC4 is biased, meaning that some byte sequences occur more frequently than others, and that these biases may be exploited in some attacks. For example, we know that the probability that the second byte being generated is equal to zero is $2/256 = 1/128$ (instead of $1/256$), and that the probability that a double zero byte is generated is $1/(256)^2 + 1/(256)^3$ (instead of $1/(256)^2$). Biases like these have been exploited, for example, in some published attacks against the WEP and the TLS protocols.[19] It is sometimes recommended to discard the first 256 or 512 bytes of a key stream, but it is disputable whether this is sufficient. In fact, RC4 is no longer a recommended stream cipher in most application settings for the Internet (e.g., [9]).

RC4 is an additive stream cipher that is deterministic, meaning that it always generates the same key stream if seeded with the same key. This also means that the same ciphertext is generated each time the same plaintext message is encrypted with the same key, and hence that the key should be changed periodically. Formally speaking, the encryption function can be expressed as follows:

$$c = E_k(m) = m \oplus G(k) \tag{9.4}$$

In this expression, $G$ refers to the PRG of the stream cipher, $k$ refers to the key (that represents the seed of $G$), and $m$ and $c$ refer to the plaintext message and the respective ciphertext. Many modern stream ciphers, like Salsa20 and ChaCha20, deviate from this simple model and employ an additional nonce $n$, which is a fresh and nonrepeating random value,[20] when they encrypt a plaintext message. The resulting encryption is probabilistic, meaning that a given plaintext is encrypted differently depending on the particular nonce in use. Formally, the encryption function of such a stream cipher can be expressed as follows:

$$c = E_k(m, n) = m \oplus G(k, n) \tag{9.5}$$

Again, the PRG $G$ pseudorandomly generates a bit sequence that is bitwise added modulo 2 to $m$ to form $c$. But this time, $G$ takes as input a key (seed) and a nonce. RC4 does not take a nonce as input and is therefore not of this type. However, something similar can be simulated by using a long-term key that is hashed with a nonce to construct a short-term key that is then used for only one encryption. This is always possible (not only for RC4), but makes key management more involved. The following two stream ciphers, Salsa20 and ChaCha20, use nonces by default. This makes it possible to use the same key to encrypt many and possibly very large plaintext messages.

19 The most famous of these attacks is known as RC4 NOMORE (https://www.rc4nomore.com).

20 More specifically, the random value does not repeat for a given key. Once the key is changed, the random value can be used again. So the nonrepeating value is the pair $(k, n)$ that consists of a key $k$ and a nonce $n$.

#### 9.5.2.2 Salsa20

Salsa20[21] is an additive stream cipher—or rather a family of additive stream ciphers—that follows the design principle mentioned above. It was originally developed by Dan Bernstein and submitted to the eSTREAM project in 2005. Since then, it has been widely used in many Internet security protocols and applications. Its major advantages are simplicity and efficiency both in hardware or software implementations. In a typical software implementation, for example, the throughput of Salsa20 is about five times bigger than the throughput of RC4 that is already a fast stream cipher. As mentioned before, Salsa20 uses nonces, and it is therefore less important to periodically refresh the key. As of this writing, there are no published attacks against Salsa20/20 and Salsa20/12 (i.e., the reduced-round version of Salsa20 with only 12 rounds). The best-known attack is able to break Salsa20/8 (i.e., the reduced-round version of Salsa20 with 8 rounds), but this attack is more theoretically interesting than practical.

Salsa20 operates on 64-byte (or 512-bit) blocks of data, meaning that a plaintext or ciphertext unit that is processed in one step is 64 bytes long.[22] Referring to formula (9.5), the encryption function of Salsa20 can be expressed as follows:

$$c = E_k(m,n) = Salsa20_k^{encrypt}(m,n) = m \oplus Salsa20_k^{expand}(n)$$

In this expression, $Salsa20^{encrypt}$ refers to the encryption function of Salsa20, whereas $Salsa20^{expand}$ refers to its expansion function. Both functions are keyed with $k$ (that is typically 32 bytes long) and employ an 8-byte nonce $n$.[23] For the sake of simplicity, we don't distinguish between the Salsa20 encryption and expansion functions, and we use the term Salsa20 to refer to either of them (from the context it is almost always clear whether the encryption or expansion function is referred to). In addition to the Salsa20 encryption function and expansion functions, there is also a Salsa20 hash function that takes a 64-byte argument $x$ and hashes it to a 64-byte output value $Salsa20(x)$. So the Salsa20 hash function does neither compress nor expand the argument, but it can still be thought of as being a cryptographic hash function (i.e., a function that has the same properties as a "normal" cryptographic hash function, such as pseudorandomness). We introduce the Salsa20 hash, expansion, and encryption functions in this order next.

21 https://cr.yp.to/salsa20.html.

22 In spite of this relatively large unit length, Salsa20 is still considered to be a stream cipher (and not a block cipher).

23 As explained below, $Salsa20_k^{expand}(n)$ refers to the iterated application of the Salsa20 expansion function. In each iteration, the 8-byte nonce $n$ is concatenated with an 8-byte sequence number $i$. It is iterated as many times as required until a sufficiently long key stream is generated.

### *Hash Function*

The Salsa20 hash function is word-oriented, meaning that it operates on words, referring to 4-byte or 32-bit values. It employs three basic operations on words:

- The addition modulo $2^{32}$ of two words $w_1$ and $w_2$, denoted as $w_1 + w_2$.
- The addition modulo 2 (XOR) of $w_1$ and $w_2$, denoted as $w_1 \oplus w_2$.
- The $c$-bit left rotation of word $w$, denoted as $w \stackrel{\curvearrowleft}{\hookleftarrow} c$ for some integer $c \geq 0$.[24]

For example, 0x77777777 + 0x01234567 = 0x789ABCDE, 0x01020304 $\oplus$ 0x789ABCDE = 0x7998BFDA, and 0x7998BFDA $\stackrel{\curvearrowleft}{\hookleftarrow}$ 7 = 0111 1001 1001 1000 1011 1111 1101 1010 $\stackrel{\curvearrowleft}{\hookleftarrow}$ 7 = 1100 1100 0101 1111 1110 1101 0011 1100 = 0xCC5FED3C. Mainly for performance reasons, the Salsa20 hash function only uses constant values for $c$ (in the left rotation operation) and invokes neither word multiplications nor table lookups.

The Salsa20 hash function employs the three basic operations as building blocks in the following auxiliary functions:

- Let $y$ be a 4-word value that consists of the 4 words $y_0$, $y_1$, $y_2$, and $y_3$; that is, $y = (y_0, y_1, y_2, y_3)$. This means that $y$ is $4 \cdot 32 = 128$ bits long. The quarterround function is defined as $quarterround(y) = z = (z_0, z_1, z_2, z_3)$, where

$$
\begin{aligned}
z_1 &= y_1 \oplus ((y_0 + y_3) \stackrel{\curvearrowleft}{\hookleftarrow} 7) \\
z_2 &= y_2 \oplus ((z_1 + y_0) \stackrel{\curvearrowleft}{\hookleftarrow} 9) \\
z_3 &= y_3 \oplus ((z_2 + z_1) \stackrel{\curvearrowleft}{\hookleftarrow} 13) \\
z_0 &= y_0 \oplus ((z_3 + z_2) \stackrel{\curvearrowleft}{\hookleftarrow} 18)
\end{aligned}
$$

  The quarterround function modifies the 4 words of $y$ in place; that is, $y_1$ is changed to $z_1$, $y_2$ is changed to $z_2$, $y_3$ is changed to $z_3$, and $y_0$ is finally changed to $z_0$. It is called "quarterround function," because it operates only on 4 words, whereas Salsa20 operates on 16 words (note that 4 is a quarter of 16).

24 Note that there is a subtle difference between a $c$-bit left rotation of word $w$, denoted as $w \stackrel{\curvearrowleft}{\hookleftarrow} c$ (in this book), and a $c$-bit left shift of word $w$, denoted as $w \hookleftarrow c$. While the first operator ($\stackrel{\curvearrowleft}{\hookleftarrow}$) means that the bits are rotated, meaning that the bits that fall out of the word on the left side are reinserted on the right side, this is not true for the second operator ($\hookleftarrow$). Here, zero bits are reinserted on the right side, and the bits that fall out of the word on the left side are lost. The same line of argumentation and notation apply to the $c$-bit right rotation of word $w$, denoted as $w \stackrel{\curvearrowright}{\hookrightarrow} c$, and the $c$-bit right shift of $w$, denoted as $w \hookrightarrow c$.

- Let $y$ be a 16-word value $(y_0, y_1, y_2, \ldots, y_{15})$ that can be represented as a square matrix

$$\begin{pmatrix} y_0 & y_1 & y_2 & y_3 \\ y_4 & y_5 & y_6 & y_7 \\ y_8 & y_9 & y_{10} & y_{11} \\ y_{12} & y_{13} & y_{14} & y_{15} \end{pmatrix}$$

The rowround function takes $y$ as input and modifies the rows of the matrix in parallel using the quarterround function mentioned above. More specifically, it generates a 16-word output value $z = rowround(y) = (z_0, z_1, z_2, \ldots, z_{15})$, where

$$\begin{aligned} (z_0, z_1, z_2, z_3) &= quarterround(y_0, y_1, y_2, y_3) \\ (z_5, z_6, z_7, z_4) &= quarterround(y_5, y_6, y_7, y_4) \\ (z_{10}, z_{11}, z_8, z_9) &= quarterround(y_{10}, y_{11}, y_8, y_9) \\ (z_{15}, z_{12}, z_{13}, z_{14}) &= quarterround(y_{15}, y_{12}, y_{13}, y_{14}) \end{aligned}$$

This means that each row is processed individually (and independently from the other rows), and that the four words of each row are permuted in a specific way. In fact, the words of row $i$ (for $1 \leq i \leq 4$) are rotated left for $i - 1$ positions. This means that the words of the first row are not permuted at all, the words of the second row are rotated left for one position, the words of the third row are rotated left for two positions, and the words of the fourth row are rotated left for three position before the quarterround function is applied.

- Similar to the rowround function, the columnround function takes a 16-word value $(y_0, y_1, y_2, \ldots, y_{15})$ and generates a 16-word output according to $z = columnround(y) = (z_0, z_1, z_2, \ldots, z_{15})$, where

$$\begin{aligned} (z_0, z_4, z_8, z_{12}) &= quarterround(y_0, y_4, y_8, y_{12}) \\ (z_5, z_9, z_{13}, z_1) &= quarterround(y_5, y_9, y_{13}, y_1) \\ (z_{10}, z_{14}, z_2, z_6) &= quarterround(y_{10}, y_{14}, y_2, y_6) \\ (z_{15}, z_3, z_7, z_{11}) &= quarterround(y_{15}, y_3, y_7, y_{11}) \end{aligned}$$

The columnround function is somehow the transpose of the rowround function; that is, it modifies the columns of the matrix in parallel by feeding a permutation of each column through the quarterround function.

- The rowround and columnround functions can be combined in a doubleround function. More specifically, the doubleround function is a columnround function followed by a rowround function. As such, it takes a 16-word sequence as input and outputs another 16-word sequence. If $y = (y_0, y_1, y_2, \ldots, y_{15})$ is the input, then

$$\begin{aligned} z &= (z_0, z_1, z_2, \ldots, z_{15}) \\ &= doubleround(y) \\ &= rowround(columnround(y)) \end{aligned}$$

is the respective output. This means that the doubleround function first modifies the input's columns in parallel, and then modifies the rows in parallel. This, in turn, means that each word is modified twice.

- Finally, the littleendian function encodes a word or 4-byte sequence $b = (b_0, b_1, b_2, b_3)$ in little-endian order $(b_3, b_2, b_1, b_0)$ that represents the value $b_3 \cdot 2^{24} + b_2 \cdot 2^{16} + b_1 \cdot 2^8 + b_0$. This value, in turn, is typically written in hexadecimal notation. For example, $littleendian(86, 75, 30, 9) = (9, 30, 75, 86)$ represents $9 \cdot 2^{24} + 30 \cdot 2^{16} + 75 \cdot 2^8 + 86$ that can be written as 0x091E4B56. Needless to say that the littlendian function can be inverted, so $littleendian^{-1}(\text{0x091E4B56}) = (86, 75, 30, 9)$.

Putting everything together, the Salsa20 hash function takes a 64-byte sequence $x = (x[0], x[1], \ldots, x[63])$ as input and generates another 64-byte sequence $Salsa20(x) = x + doubleround^{10}(x)$ as output. The input sequence $x$ consists of 16 words in littleendian form:

$$\begin{aligned} x_0 &= littlendian(x[0], x[1], x[2], x[3]) \\ x_1 &= littlendian(x[4], x[5], x[6], x[7]) \\ &\ldots \\ x_{15} &= littlendian(x[60], x[61], x[62], x[63]) \end{aligned}$$

If $z = (z_0, z_1, z_2, \ldots, z_{15}) = doubleround^{10}(x_0, x_1, \ldots, x_{15})$, then the output of the hash function $Salsa20(x)$ is the concatenation of the 16 words that are generated as follows:

$$\begin{aligned} &littlendian^{-1}(z_0 + x_0) \\ &littlendian^{-1}(z_1 + x_1) \\ &\ldots \\ &littlendian^{-1}(z_{15} + x_{15}) \end{aligned}$$

The 20 rounds of Salsa20 come from the fact that the doubleround function is iterated 10 times, and each iteration basically represents two rounds, one standing for the columnround function and one standing for the rowround function.

*Expansion Function*

As its name suggests, the aim of the Salsa20 expansion function is to expand a 16-byte input $n$ into a 64-byte output, using 32 or 16 bytes of keying material and 16 constant bytes. Depending on whether the keying material consists of 32 or 16 bytes, the constant bytes and the respective expansion functions are slightly different.

If the keying material consists of 32 bytes, then this material is split into two halves that represent two 16-bytes keys $k_0$ and $k_1$. In this case, the constant bytes look as follows (where each $\sigma$ value consists of four bytes that are encoded using the littleendian function):

$$\begin{aligned}\sigma_0 &= (101, 120, 112, 97) = \text{0x61707865}\\ \sigma_1 &= (110, 100, 32, 51) = \text{0x3320646E}\\ \sigma_2 &= (50, 45, 98, 121) = \text{0x79622D32}\\ \sigma_3 &= (116, 101, 32, 107) = \text{0x6B206574}\end{aligned}$$

The Salsa20 expansion function is then defined as follows:

$$Salsa20_{k_0,k_1}(n) = Salsa20(\sigma_0, k_0, \sigma_1, n, \sigma_2, k_1, \sigma_3)$$

Note that $littleendian(\sigma_0) = littleendian(101, 120, 112, 97) =$ 0x61707865, so the argument that is subject to the Salsa20 hash function starts with the four bytes 0x61, 0x70, 0x78, and 0x65.

Otherwise, for example, if the keying material consists of 16 bytes, then this material represents a single 16-byte key $k$ that is applied twice. In this case, a slightly different set of 4-byte $\tau$ constants is used

$$\begin{aligned}\tau_0 &= (101, 120, 112, 97)\\ \tau_1 &= (110, 100, 32, \underline{49})\\ \tau_2 &= (\underline{54}, 45, 98, 121)\\ \tau_3 &= (116, 101, 32, 107)\end{aligned}$$

where the two bytes that are different from the respective $\sigma$ constants are marked as underlined. In this case, the Salsa20 expansion function is defined as

$$Salsa20_k(n) = Salsa20(\tau_0, k, \tau_1, n, \tau_2, k, \tau_3).$$

In either case, $\sigma$ and $\tau$ can be seen as constants $c$, $k$ is the key, and $n$ is the argument of the Salsa20 expansion function. Hence, the input to the function can be written in a specific matrix layout:

$$\begin{pmatrix} c & k & k & k \\ k & c & n & n \\ n & n & c & k \\ k & k & k & c \end{pmatrix}$$

Obviously, this layout is somewhat arbitrary and can be changed at will. As explained later, ChaCha20 is a variant of Salsa20 that uses a different matrix layout.

*Encryption Function*

Salsa20 is an additive stream cipher, meaning that an appropriately sized key stream is generated and added modulo 2 to the plaintext message. The Salsa20 expansion function is used to generate the key stream. More specifically, let $k$ be a 32- or 16-byte key,[25] $n$ an 8-byte nonce, and $m$ an $l$-byte plaintext message that is going to be encrypted (where $0 \leq l \leq 2^{70}$). The Salsa20 encryption of $m$ with nonce $n$ under key $k$ is denoted as $Salsa20_k(m,n)$. It is computed as $m \oplus Salsa20_k(n')$, where $Salsa20_k(n')$ represents a key stream that can be up to $2^{70}$ bytes long and $n'$ is derived from $n$ by adding a counter. Hence, the key stream is iteratively constructed as follows:

$$Salsa20_k(n,0) \parallel Salsa20_k(n,1) \parallel \ldots \parallel Salsa20_k(n,2^{64}-1)$$

In each iteration, the Salsa20 expansion function is keyed with $k$ and applied to a 16-byte input that consists of the 8-byte nonce and an 8-byte counter $i$. If $i$ is written bitwise; that is, $i = (i_0, i_1, \ldots, i_7)$, then the respective counter stands for $i_0 + 2^8 i_1 + 2^{16} i_2 + \ldots + 2^{56} i_7$. Each iteration of the Salsa20 expansion function yields $64 = 2^6$ bytes, so the maximal length of the key stream that can be generated this way is $2^{64} \cdot 2^6 = 2^{64+6} = 2^{70}$ bytes. It goes without saying that only as many bytes as necessary are generated to encrypt the $l$ bytes of the message $m$. The bottom line is that the Salsa20 encryption function can be expressed as

$$c = (c[0], c[1], \ldots, c[l-1]) = (m[0], m[1], \ldots, m[l-1]) \oplus Salsa20_k(n')$$

or

$$c[i] = m[i] \oplus Salsa_k(n, \lfloor i/64 \rfloor)[i \bmod 64]$$

25 Consider the possibility of using a 16-byte key as an option. The preferred key size is 32 bytes referring to 256 bits.

for $i = 0, 1, \ldots, l-1$. Since Salsa20 is an additive stream cipher, the encryption and decryption functions are essentially the same (with $m$ and $c$ having opposite meanings).

Because the length of a nonce is controversially discussed in the community, Bernstein proposed a variant of Salsa20 that can handle longer nonces. More specifically, *XSalsa20*[26] can take nonces that are 192 bits long (instead of 64 bits) without reducing the claimed security.

#### 9.5.2.3 ChaCha20

In 2008, Bernstein proposed a modified version of the Salsa20 stream cipher named *ChaCha20* [10].[27] Again, the term refers to a family of stream ciphers that comprises ChaCha20/20 (20 rounds), ChaCha20/12 (12 rounds), and ChaCha20/8 (8 rounds). ChaCha20 is structurally identical to Salsa20, but it uses a different round function and a different matrix layout. Also, it uses a key that is always 32 bytes (256 bits) long, a nonce that is 12 bytes (96 bits) long, and a block counter that is only 4 bytes (32 bits) long. Remember that Salsa20 nonces and block counters are 8 bytes long each. Furthermore, the ChaCha20 specification also uses another notation to describe the quarterround function. Instead of using $y = (y_0, y_1, y_2, y_3)$ and $z = (z_0, z_1, z_2, z_3)$, it uses four 32-bit words $a$, $b$, $c$, and $d$. This means that

$$
\begin{aligned}
z_1 &= y_1 \oplus ((y_0 + y_3) \stackrel{\frown}{\hookleftarrow} 7) \\
z_2 &= y_2 \oplus ((z_1 + y_0) \stackrel{\frown}{\hookleftarrow} 9) \\
z_3 &= y_3 \oplus ((z_2 + z_1) \stackrel{\frown}{\hookleftarrow} 13) \\
z_0 &= y_0 \oplus ((z_3 + z_2) \stackrel{\frown}{\hookleftarrow} 18)
\end{aligned}
$$

can also be written as

$$
\begin{aligned}
b &= b \oplus ((a + d) \stackrel{\frown}{\hookleftarrow} 7) \\
c &= c \oplus ((b + a) \stackrel{\frown}{\hookleftarrow} 9) \\
d &= d \oplus ((c + b) \stackrel{\frown}{\hookleftarrow} 13) \\
a &= a \oplus ((d + c) \stackrel{\frown}{\hookleftarrow} 18)
\end{aligned}
$$

The operations performed by ChaCha20 are the same as the ones performed by Salsa20, but they are applied in a different order and each word is updated twice

26 https://cr.yp.to/snuffle/xsalsa-20081128.pdf.

27 https://cr.yp.to/chacha/chacha-20080128.pdf

instead of just once. The advantage is that the Chacha20 round function provides more diffusion than the Salsa20 round function. Also, the rotation distances are changed from 7, 9, 13, and 18 to 16, 12, 8, and 7, but this difference is less important. The ChaCha20 quarterround function updates $a$, $b$, $c$, and $d$ as follows:

$$\begin{aligned} a = a + b;\; & d = d \oplus a;\; d = d \hookleftarrow 16; \\ c = c + d;\; & b = b \oplus c;\; b = b \hookleftarrow 12; \\ a = a + b;\; & d = d \oplus a;\; d = d \hookleftarrow 8; \\ c = c + d;\; & b = b \oplus c;\; b = b \hookleftarrow 7; \end{aligned}$$

Like Salsa20, ChaCha20 is an additive stream cipher, meaning that it pseudorandomly generates a key stream that is then added modulo 2 to encrypt or decrypt a message. Hence, it is characterized by the PRG that is inherent in the design. The ChaCha20 PRG is overviewed in Algorithm 9.3. It operates on a $(4 \times 4)$-matrix $S$ of 4-byte words called the state. Hence, the state is exactly 64 bytes long. The ChaCha20 PRG takes as input a 32-byte key $k$, a 4-byte block counter $i$, and a 12-byte nonce $n$, and it generates as output a serialized version of the state. The respective bits are then added modulo 2 to the plaintext message (for encryption) or ciphertext (for decryption).

**Algorithm 9.3** The ChaCha20 PRG algorithm.

$(k, i, n)$

---

$S = \sigma \parallel k \parallel i \parallel n$
$S' = S$
for $i = 1$ to 10 do begin
  $quarterround(S'_0, S'_4, S'_8, S'_{12})$
  $quarterround(S'_1, S'_5, S'_9, S'_{13})$
  $quarterround(S'_2, S'_6, S'_{10}, S'_{14})$
  $quarterround(S'_3, S'_7, S'_{11}, S'_{15})$
  $quarterround(S'_0, S'_5, S'_{10}, S'_{15})$
  $quarterround(S'_1, S'_6, S'_{11}, S'_{12})$
  $quarterround(S'_2, S'_7, S'_8, S'_{13})$
  $quarterround(S'_3, S'_4, S'_9, S'_{14})$
end
$S = S + S'$

---

Serialized($S$)

In step one of the ChaCha20 PRG algorithm, $S$ is constructed as the concatenation of the 4 constants $\sigma_0$, $\sigma_1$, $\sigma_2$, and $\sigma_3$ that are the same as the ones defined for Salsa20, $k$, $i$, and $n$. So the 48 bytes from $k$, $i$, and $n$ are complemented with 16 constant bytes, and hence the total size of the state is 64 bytes. With regard to the matrix layout mentioned above, ChaCha20 uses the following (simplified) layout:

$$\begin{pmatrix} c & c & c & c \\ k & k & k & k \\ k & k & k & k \\ n & n & n & n \end{pmatrix}$$

The first row comprises the constants $\sigma$, the second and third rows comprise the key $k$, and the fourth row comprises the nonce $n$. To be precise, the nonce consists of a 4-byte block counter $i$ and a 12-byte value $n$ that represents the actual nonce.[28]

In step two of the ChaCha20 PRG algorithm, the state $S$ is copied to the working state $S'$. This is the value that is processed iteratively in 10 rounds. In each round, the quarterround function is applied 8 times, where the first 4 applications refer to "column rounds" and the second 4 applications refer to "diagonal rounds" (remember that Salsa20 uses column and row rounds, but no diagonal rounds). Diagonal rounds are new in the ChaCha20 design, and they stand for themselves. Ten rounds with 8 applications of the quarterround function each yield $10 \cdot 8/4 = 20$ rounds. In the end, the original content of the state $S$ is added modulo $2^{32}$ to $S'$, and the result (in serialized form) refers to the 64-byte output of the ChaCha20 PRG algorithm. Serialization, in turn, is done by subjecting the words of $S$ to the littleendian function and sequencing the resulting bytes. If, for example, the state begins with the two words 0x091E4B56 and 0xE4E7F110, then the output sequence begins with the 8 bytes 0x56, 0x4B, 0x1E, 0x09, 0x10, 0xF1, 0xE7, and 0xE4.

Similar to Salsa20, no practically relevant cryptanalytical attack against the ChaCha20 stream cipher is known to exist. It is therefore widely used on the Internet to replace RC4. Most importantly, it is often used with Bernstein's Poly1305 message authentication code (Section 10.3.3) to provide authenticated encryption.

## 9.6 BLOCK CIPHERS

As mentioned before, every practical symmetric encryption system processes plaintext messages unit by unit. In the case of a block cipher such a unit is called a *block*. Consequently, a block cipher maps plaintext message blocks of a specific length into ciphertext blocks of typically the same length; that is, $\mathcal{M} = \mathcal{C} = \Sigma^n$ for some alphabet $\Sigma$ and block length $n$ (e.g., 128 bits).

In theory, a permutation on set $S$ is just a bijective function $f : S \rightarrow S$ (Definition A.22). If we fix a block length $n$ and work with the plaintext message

28 Note that the block counter $i$ and the actual nonce $n$ sum up to 16 bytes.

and ciphertext spaces $\mathcal{M} = \mathcal{C} = \Sigma^n$, then any element $\pi$ randomly selected from $\text{Perms}[\Sigma^n]$ defines a block cipher with encryption and decryption functions ($E_\pi$ and $D_\pi$) that are defined as follows:

$$E_\pi : \begin{array}{l} \Sigma^n \longrightarrow \Sigma^n \\ w \longmapsto \pi(w) \end{array} \qquad D_\pi : \begin{array}{l} \Sigma^n \longrightarrow \Sigma^n \\ w \longmapsto \pi^{-1}(w) \end{array}$$

There are $|P(\Sigma^n)| = (\Sigma^n)!$ elements in $\text{Perms}[\Sigma^n]$, and this suggests that there are $(2^n)!$ possible permutations for $\Sigma = \{0,1\}^n$. This function grows tremendously (as illustrated in Table 9.1 for the first 10 values).

**Table 9.1**
The Growth Rate of $f(n) = (2^n)!$ for $n = 1, \ldots, 10$

| $n$ | $(2^n)!$ |
|---|---|
| 1 | $2^1! = 2! = 2$ |
| 2 | $2^2! = 4! = 24$ |
| 3 | $2^3! = 8! = 40'320$ |
| 4 | $2^4! = 16! = 20'922'789'888'000$ |
| 5 | $2^5! = 32! \approx 2.63 \cdot 10^{35}$ |
| 6 | $2^6! = 64! \approx 1.27 \cdot 10^{89}$ |
| 7 | $2^7! = 128! \approx 3.86 \cdot 10^{215}$ |
| 8 | $2^8! = 256! \approx 8.58 \cdot 10^{506}$ |
| 9 | $2^9! = 512! \approx 3.48 \cdot 10^{1166}$ |
| 10 | $2^{10}! = 1024! \approx 5.42 \cdot 10^{2639}$ |

For a typical block length $n$ of 64 bits, this suggests that there are

$$2^{64}! = 18,446,744,073,709,551,616!$$

possible permutation to choose from. This number is so huge that it requires more than $2^{69}$ bits to encode it (this is why we have to use the factorial notation in the formula given above). Consequently, if we want to specify a particular permutation from $\text{Perms}[\{0,1\}^{64}]$, then we have to introduce a numbering scheme and use a respective index (that is approximately of that size) to refer to a particular permutation. This $2^{69}$-bit number would then yield a secret key. It is doubtful whether the communicating entities would be able to manage such a long key. Instead, symmetric encryption systems are usually designed to take a reasonably long key[29] and generate a one-to-one mapping that looks random to someone who does not know the secret key. So it is reasonable to use only some possible

29 A reasonably long key has more like 69 bits than $2^{69}$ bits.

permutations of $\Sigma^n$ (from $\text{Perms}[\Sigma^n]$) as encryption and decryption functions and to use comparably short keys to refer to them.

What we basically need for a block cipher is a PRP (Section 8.1) that uses a key $k$ from a moderately sized key space $\mathcal{K}$ to define a family of bijective encryption functions $E_k : \Sigma^n \rightarrow \Sigma^n$ and a family of respective decryption function $E_k : \Sigma^n \rightarrow \Sigma^n$. To analyze the security of such a block cipher, one can study the algebraic properties of this PRP.

From a practical viewpoint, the design of symmetric encryption systems combines permutations and substitutions to generate confusion and diffusion.

- The purpose of *confusion* is to make the relation between the key and the ciphertext as complex as possible.
- The purpose of *diffusion* is to spread the influence of a single plaintext bit over many ciphertext bits. In a block cipher, diffusion propagates bit changes from one part of a block to other parts of the block.

These terms are frequently used in the literature for the design of block ciphers. Block ciphers that combine permutations and substitutions in multiple rounds (to provide a maximum level of confusion and diffusion) are sometimes called *substitution-permutation ciphers*. Many practically relevant block ciphers, including, for example, the DES, represent substitution-permutation ciphers. We overview and discuss DES and a few other block ciphers next. In the respective explanations, however, we are not as formal and formally correct as one can possibly be. Instead, we adopt some terminology and notation used in the original descriptions and specifications (to make it simpler to get into these documents).

### 9.6.1 DES

In the early 1970s, the U.S. National Bureau of Standards (NBS[30]) recognized the importance of having a standardized block cipher for commercial use. It therefore set up a competition for a standardized block cipher. Knowledge in block cipher design was not widely deployed in those days, so the only company that was able to contribute to the competition was IBM. In fact, IBM had been internally developing a block cipher called *Lucifer*, and hence IBM was able to submit Lucifer to the NBS. NBS, NSA, and IBM refined the design of Lucifer and finally standardized the result as DES in FIPS PUB 46 (that was first issued in 1977). Today, the FIPS PUBs are developed and maintained by NIST. The standard was reaffirmed in 1983, 1988, 1993, and 1999, before it was officially withdrawn in July 2004. The DES

30 In the United States, the NBS later became the National Institute of Standards and Technology (NIST).

specification that was reaffirmed in 1999, FIPS PUB 46-3 [11], is publicly and freely available on the Internet.[31] It specifies both the DES and the *Triple Data Encryption Algorithm* (TDEA) that is further addressed in Section 9.6.1.6. Contrary to the DES, the TDEA may still be used in situations that can handle moderate performance. In either case, cryptographic modules that implement FIPS 46-3 should also conform to the requirements specified in FIPS 140-1 [12]. In fact, there is an increasingly large number of DES implementations both in hardware and software that conform to the different levels specified there.

DES is a substitution-permutation cipher. It is also the major representative of a *Feistel cipher*.[32] The characteristics of a Feistel cipher are overviewed first. Then the DES encryption and decryption algorithms are described, and the security of the DES is briefly analyzed. Finally, a variant of DES (named DESX) and TDEA are overviewed, discussed, and put into perspective.

#### 9.6.1.1 Feistel Ciphers

A Feistel cipher is a block cipher with a characteristic structure (also known as a *Feistel network*). The alphabet is $\Sigma = \mathbb{Z}_2 = \{0,1\}$ and the block length is $2t$ for a reasonably sized $t \in \mathbb{N}^+$. The Feistel cipher runs in $r \in \mathbb{N}^+$ rounds, where $r$ round keys $k_1, \ldots, k_r$ are generated from $k \in \mathcal{K}$ and used on a per-round basis.

The encryption function $E_k$ starts by splitting the plaintext message block $m$ into two halves of $t$ bits each. Let $L_0$ be the left half and $R_0$ the right half of $m$; that is, $m = L_0 \parallel R_0 = (L_0, R_0)$. For $i = 1, \ldots, r$, a sequence of pairs $(L_i, R_i)$ is then recursively computed as follows:

$$(L_i, R_i) = (R_{i-1}, L_{i-1} \oplus f_{k_i}(R_{i-1})) \tag{9.6}$$

This means that $L_i = R_{i-1}$ and $R_i = L_{i-1} \oplus f_{k_i}(R_{i-1})$. For example, if $i = 1$, then $L_1$ and $R_1$ are computed as follows:

$$\begin{aligned} L_1 &= R_0 \\ R_1 &= L_0 \oplus f_{k_1}(R_0) \end{aligned}$$

Similarly, if $i = 2$, then $L_2$ and $R_2$ are computed as follows:

$$\begin{aligned} L_2 &= R_1 \\ R_2 &= L_1 \oplus f_{k_2}(R_1) \end{aligned}$$

31 http://csrc.nist.gov/publications/fips/fips46-3/fips46-3.pdf.

32 Feistel ciphers are named after the IBM researcher Horst Feistel who was involved in the original design of Lucifer and DES. Feistel lived from 1915 to 1990 and was one of the first nongovernment cryptographers.

This continues until, in the last round $r$, $L_r$ and $R_r$ are computed as follows:

$$\begin{aligned} L_r &= R_{r-1} \\ R_r &= L_{r-1} \oplus f_{k_r}(R_{r-1}) \end{aligned}$$

The pair $(L_r, R_r)$ in reverse order then represents the ciphertext block. Hence, the encryption of plaintext message $m$ using key $k$ can formally be expressed as follows:

$$E_k(m) = E_k(L_0, R_0) = (R_r, L_r)$$

The recursive formula (9.6) can also be written as follows:

$$(L_{i-1}, R_{i-1}) = (R_i \oplus f_{k_i}(L_i), L_i)$$

This means that it is possible to recursively compute $L_{i-1}$ and $R_{i-1}$ from $L_i$, $R_i$, and $k_i$, and to determine $(L_0, R_0)$ from $(R_r, L_r)$ using the round keys in reverse order (i.e., $k_r, \ldots, k_1$). Consequently, a Feistel cipher can always be decrypted using the same (encryption) algorithm and applying the round keys in reverse order. This simplifies the implementation considerably.

In theory, Michael Luby and Charles Rackoff have shown that if one builds a Feistel cipher with a round function $f$ that is a secure PRF, then three rounds are sufficient to turn the Feistel cipher into a secure PRP, and hence a secure block cipher [13]. This is a general result, and to make more precise statements about the security of a Feistel cipher, one has to look more closely at a particular round function $f$.

In addition to DES, there are many other representatives of Feistel ciphers, such as *Camellia*, which is mostly used in Japan [14]. It is possible to design and come up with iterative block ciphers that are not Feistel ciphers, but whose encryption and decryption functions (after a certain reordering or recalculation of variables) are still similar or structurally the same. One such example is the *International Data Encryption Algorithm* (IDEA) incorporated in many security products, including, for example, former versions of Pretty Good Privacy (PGP). Feistel ciphers have important applications in public key cryptography as well. For example, the optimal asymmetric encryption padding (OAEP) scheme (Section 13.3.1.4) is basically a two-round Feistel cipher. We return to the DES and its encryption and decryption functions or algorithms next.

#### 9.6.1.2 Encryption Algorithm

As mentioned before, the DES is a Feistel cipher with $t = 32$ and $r = 16$. This means that the block length of DES is 64 bits, and hence $\mathcal{M} = \mathcal{C} = \{0,1\}^{64}$, and that the DES encryption and decryption algorithms operate in 16 rounds. Furthermore, DES keys are 64-bit strings with the additional property that the last bit of each byte has odd parity. This means that the sum modulo 2 of all bits in a byte must be odd and that the parity bit is set accordingly. This can be formally expressed as follows:

$$\mathcal{K} = \{(k_1, \ldots, k_{64}) \in \{0,1\}^{64} \mid \sum_{i=1}^{8} k_{8j+i} \equiv 1 \pmod 2 \text{ for } j = 0, \ldots, 7\}$$

For example, `F1DFBC9B79573413` is a valid DES key. Its odd parity can easily be verified using the following table:

| | | | | | | | | |
|---|---|---|---|---|---|---|---|---|
| F1 | 1 | 1 | 1 | 1 | 0 | 0 | 0 | 1 |
| DF | 1 | 1 | 0 | 1 | 1 | 1 | 1 | 1 |
| BC | 1 | 0 | 1 | 1 | 1 | 1 | 0 | 0 |
| 9B | 1 | 0 | 0 | 1 | 1 | 0 | 1 | 1 |
| 79 | 0 | 1 | 1 | 1 | 1 | 0 | 0 | 1 |
| 57 | 0 | 1 | 0 | 1 | 0 | 1 | 1 | 1 |
| 34 | 0 | 0 | 1 | 1 | 0 | 1 | 0 | 0 |
| 13 | 0 | 0 | 0 | 1 | 0 | 0 | 1 | 1 |

Note that the bit specified in the last column in each row refers to the parity bit that ensures that each row has an odd number of ones. In the first row, for example, four out of seven bits are ones, and this means that the parity bit must also be set to one. Consequently, the first seven bits of a DES key byte determine the last bit, and hence the size of the resulting key space is only $2^{56}$ (instead of $2^{64}$). As mentioned earlier, the round keys derived from the DES key are the same for encryption and decryption; they are only used in reverse order. This results from the fact that DES is a Feistel cipher.

The DES encryption algorithm is specified in Algorithm 9.4 and illustrated in Figure 9.7. To encrypt a 64-bit plaintext message block $m$ using a 64-bit key $k$, the algorithm operates in three steps:

1. The initial permutation ($IP$) as illustrated in Table 9.2 is applied to $m$. If $m = m_1 m_2 m_3 \ldots m_{64} \in \mathcal{M} = \{0,1\}^{64}$, then $IP(m) = m_{58} m_{50} m_{42} \ldots m_7 \in \mathcal{M}$. This means that the 58th bit in $m$ becomes the first bit in $IP(m)$, the 50th bit becomes the second bit, and so on. The resulting $m$ is then split into two parts: $L_0$ referring to the leftmost 32 bits of $m$ (denoted $m|_{32}$), and $R_0$ referring to the rightmost 32 bits of $m$ (denoted $m|^{32}$).

2. A 16-round Feistel cipher is then applied to $L_0$ and $R_0$. The round function $f$ is illustrated in Figure 9.8 and addressed below. Remember that the swapping

**Algorithm 9.4** The DES encryption algorithm.

$(m, k)$

---

$m = IP(m)$
$L_0 = m|_{32}$
$R_0 = m|^{32}$
for $i = 1$ to 16 do
  $L_i \leftarrow R_{i-1}$
  $R_i \leftarrow L_{i-1} \oplus f_{k_i}(R_{i-1})$
$c \leftarrow IP^{-1}(R_{16}, L_{16})$

---

$(c)$

of the left and right halves is part of the structure of a Feistel cipher, and hence that $(R_{16}, L_{16})$ is the output of this step.

3. The inverse initial permutation ($IP^{-1}$) as illustrated in Table 9.3 is applied to $(R_{16}, L_{16})$ to generate the ciphertext block; that is, $c = IP^{-1}(R_{16}, L_{16})$.

**Table 9.2**
The Initial Permutation $IP$ of the DES

| | | | | | | | |
|---|---|---|---|---|---|---|---|
| 58 | 50 | 42 | 34 | 26 | 18 | 10 | 2 |
| 60 | 52 | 44 | 36 | 28 | 20 | 12 | 4 |
| 62 | 54 | 46 | 38 | 30 | 22 | 14 | 6 |
| 64 | 56 | 48 | 40 | 32 | 24 | 16 | 8 |
| 57 | 49 | 41 | 33 | 25 | 17 | 9 | 1 |
| 59 | 51 | 43 | 35 | 27 | 19 | 11 | 3 |
| 61 | 53 | 45 | 37 | 29 | 21 | 13 | 5 |
| 63 | 55 | 47 | 39 | 31 | 23 | 15 | 7 |

The DES round function $f$ operates on blocks of 32 bits and uses a 48-bit key $k_i$ in each round $i = 1, \ldots, r$, i.e., $f_{k_i} : \{0,1\}^{32} \rightarrow \{0,1\}^{32}$ for every $k_i \in \{0,1\}^{48}$. The working principle of the DES round function $f$ is illustrated in Figure 9.8. First, the 32-bit argument $R$ is expanded to 48 bits using the expansion function $E : \{0,1\}^{32} \rightarrow \{0,1\}^{48}$. As shown in Table 9.4, the expansion function basically works by doubling some input bits. If $R = r_1 r_2 \ldots r_{31} r_{32}$, then $E(R) = r_{32} r_1 \ldots r_{32} r_1$. The string $E(R)$ is added modulo 2 to the 48-bit key $k$, and the result is split into 8 blocks $B_1, \ldots, B_8$ of 6 bits each; that is, $E(R) \oplus k = B_1 B_2 B_3 B_4 B_5 B_6 B_7 B_8$ with $B_i \in \{0,1\}^6$ for $i = 1, \ldots, 8$. Next, each 6-bit block $B_i$ is transformed into a 4-bit block $C_i$ for $i = 1, \ldots, 8$ using a function $S_i : \{0,1\}^6 \longrightarrow \{0,1\}^4$ (this function is called *S-box* and explained later). For $i = 1, \ldots, 8$, we have $C_i = S_i(B_i)$,

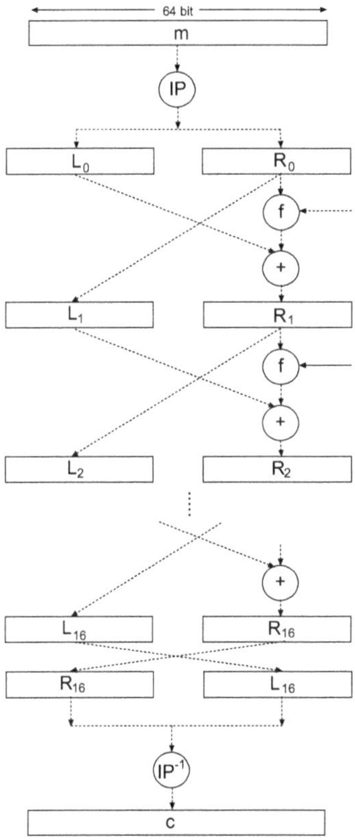


**Figure 9.7** The DES encryption algorithm.

and hence $C = C_1C_2 \ldots C_8$. Each $C_i$ for $i = 1, \ldots, 8$ is 4 bits long, so the total length of $C$ is 32 bits. It is subject to the permutation $P$, as specified in Table 9.5. If $C = c_1c_2 \ldots c_{32}$, then $P(C) = c_{16}c_7 \ldots c_{25}$. The result is $f_k(R)$, and it is the output of the round function $f$.

The eight S-boxes $S_1, \ldots, S_8$ of the DES are illustrated in Table 9.6. Each S-box can be represented by a table that consists of 4 rows and 16 columns. If $B = b_1b_2b_3b_4b_5b_6$ is input to $S_i$, then the binary string $b_1b_6 \in \{0,1\}^2$ represents a number between 0 and 3 (this number is the row index for the table), whereas the binary string $b_2b_3b_4b_5 \in \{0,1\}^4$ represents a number between 0 and 15 (this number is the column index for the table). The output of $S_i(B)$ is the number found in the table (on the row that corresponds to the row index and the column that corresponds to the column index), written in binary notation. For example, if $B = 011001$, then

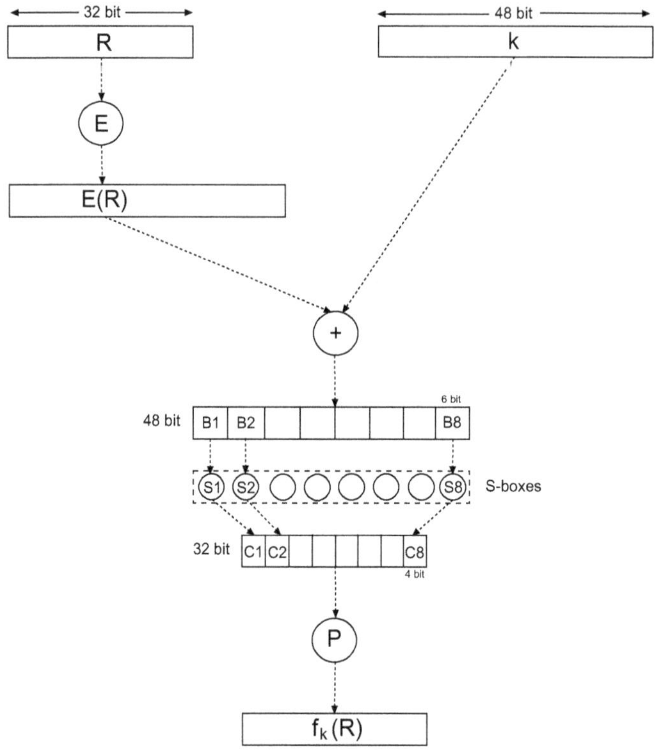


**Figure 9.8** The DES round function $f$.

the row index is $b_1 b_6 = 01 = 1$ and the column index is $b_2 b_3 b_4 b_5 = 1100 = 12$. Consequently, $S_5(011001)$ refers to the decimal number 3 that can be written as a sequence of bits (i.e., 0011). This sequence is the output of the S-box. This applies to all inputs to the respective S-boxes.

Due to the fact that the eight S-boxes $S_1, \ldots, S_8$ and the two permutations $IP$ and $IP^{-1}$ cannot be implemented efficiently (especially in software), a new version of DES known as *DES Lightweight* (DESL) has been proposed in [15]. In the design of DESL, the eight S-boxes $S_1, \ldots, S_8$ are replaced with a single S-Box $S$ and the two permutations $IP$ and $IP^{-1}$ are omitted altogether. DESL is well suited for low-computing environments, such as the ones employed by radio frequency identification (RFID) technologies, but the general phasing-out of DES also applies to DESL.

**Table 9.3**
The Inverse Initial Permutation $IP^{-1}$ of the DES

| | | | | | | | |
|---|---|---|---|---|---|---|---|
| 40 | 8 | 48 | 16 | 56 | 24 | 64 | 32 |
| 39 | 7 | 47 | 15 | 55 | 23 | 63 | 31 |
| 38 | 6 | 46 | 14 | 54 | 22 | 62 | 30 |
| 37 | 5 | 45 | 13 | 53 | 21 | 61 | 29 |
| 36 | 4 | 44 | 12 | 52 | 20 | 60 | 28 |
| 35 | 3 | 43 | 11 | 51 | 19 | 59 | 27 |
| 34 | 2 | 42 | 10 | 50 | 18 | 58 | 26 |
| 33 | 1 | 41 | 9 | 49 | 17 | 57 | 25 |

**Table 9.4**
The Expansion Function $E$ of the DES

| | | | | | | | | | | | |
|---|---|---|---|---|---|---|---|---|---|---|---|
| 32 | 1 | 2 | 3 | 4 | 5 | 4 | 5 | 6 | 7 | 8 | 9 |
| 8 | 9 | 10 | 11 | 12 | 13 | 12 | 13 | 14 | 15 | 16 | 17 |
| 16 | 17 | 18 | 19 | 20 | 21 | 20 | 21 | 22 | 23 | 24 | 25 |
| 24 | 25 | 26 | 27 | 28 | 29 | 28 | 29 | 30 | 31 | 32 | 1 |

Last but not least, we must explain the key schedule; that is, how the 16 round keys $k_1, \ldots, k_{16} \in \{0,1\}^{48}$ are generated from the DES key $k \in \{0,1\}^{64}$. We therefore define $v_i$ for $i = 1, \ldots, 16$:

$$v_i = \begin{cases} 1 & \text{if } i \in \{1,2,9,16\} \\ 2 & \text{otherwise (i.e., if } i \in \{3,4,5,6,7,8,10,11,12,13,14,15\}) \end{cases}$$

Furthermore, we use two functions called $PC1$ and $PC2$ (where PC stands for permuted choice). $PC1$ maps a 64-bit string (i.e., a DES key $k$) to two 28-bit strings $C$ and $D$ (i.e., $PC1 : \{0,1\}^{64} \rightarrow \{0,1\}^{28} \times \{0,1\}^{28}$), and $PC2$ maps two 28-bit strings to a 48-bit string (i.e., $PC2 : \{0,1\}^{28} \times \{0,1\}^{28} \rightarrow \{0,1\}^{48}$).

**Table 9.5**
The Permutation $P$ of the DES

| | | | | | | | |
|---|---|---|---|---|---|---|---|
| 16 | 7 | 10 | 21 | 29 | 12 | 28 | 17 |
| 1 | 15 | 23 | 26 | 5 | 18 | 31 | 20 |
| 2 | 8 | 24 | 14 | 32 | 27 | 3 | 9 |
| 19 | 13 | 30 | 6 | 22 | 11 | 4 | 25 |

**Table 9.6**
The S-Boxes $S_1$ to $S_8$ of the DES

| | | 0 | 1 | 2 | 3 | 4 | 5 | 6 | 7 | 8 | 9 | 10 | 11 | 12 | 13 | 14 | 15 |
|---|---|---|---|---|---|---|---|---|---|---|---|---|---|---|---|---|---|
| $S_1$ | 0 | 14 | 4 | 13 | 1 | 2 | 15 | 11 | 8 | 3 | 10 | 6 | 12 | 5 | 9 | 0 | 7 |
| | 1 | 0 | 15 | 7 | 4 | 14 | 2 | 13 | 1 | 10 | 6 | 12 | 11 | 9 | 5 | 3 | 8 |
| | 2 | 4 | 1 | 14 | 8 | 13 | 6 | 2 | 11 | 15 | 12 | 9 | 7 | 3 | 10 | 5 | 0 |
| | 3 | 15 | 12 | 8 | 2 | 4 | 9 | 1 | 7 | 5 | 11 | 3 | 14 | 10 | 0 | 6 | 13 |
| $S_2$ | 0 | 15 | 1 | 8 | 14 | 6 | 11 | 3 | 4 | 9 | 7 | 2 | 13 | 12 | 0 | 5 | 10 |
| | 1 | 3 | 13 | 4 | 7 | 15 | 2 | 8 | 14 | 12 | 0 | 1 | 10 | 6 | 9 | 11 | 5 |
| | 2 | 0 | 14 | 7 | 11 | 10 | 4 | 13 | 1 | 5 | 8 | 12 | 6 | 9 | 3 | 2 | 15 |
| | 3 | 13 | 8 | 10 | 1 | 3 | 15 | 4 | 2 | 11 | 6 | 7 | 12 | 0 | 5 | 14 | 9 |
| $S_3$ | 0 | 10 | 0 | 9 | 14 | 6 | 3 | 15 | 5 | 1 | 13 | 12 | 7 | 11 | 4 | 2 | 8 |
| | 1 | 13 | 7 | 0 | 9 | 3 | 4 | 6 | 10 | 2 | 8 | 5 | 14 | 12 | 11 | 15 | 1 |
| | 2 | 13 | 6 | 4 | 9 | 8 | 15 | 3 | 0 | 11 | 1 | 2 | 12 | 5 | 10 | 14 | 7 |
| | 3 | 1 | 10 | 13 | 0 | 6 | 9 | 8 | 7 | 4 | 15 | 14 | 3 | 11 | 5 | 2 | 12 |
| $S_4$ | 0 | 7 | 13 | 14 | 3 | 0 | 6 | 9 | 10 | 1 | 2 | 8 | 5 | 11 | 12 | 4 | 15 |
| | 1 | 13 | 8 | 11 | 5 | 6 | 15 | 0 | 3 | 4 | 7 | 2 | 12 | 1 | 10 | 14 | 9 |
| | 2 | 10 | 6 | 9 | 0 | 12 | 11 | 7 | 13 | 15 | 1 | 3 | 14 | 5 | 2 | 8 | 4 |
| | 3 | 3 | 15 | 0 | 6 | 10 | 1 | 13 | 8 | 9 | 4 | 5 | 11 | 12 | 7 | 2 | 14 |
| $S_5$ | 0 | 2 | 12 | 4 | 1 | 7 | 10 | 11 | 6 | 8 | 5 | 3 | 15 | 13 | 0 | 14 | 9 |
| | 1 | 14 | 11 | 2 | 12 | 4 | 7 | 13 | 1 | 5 | 0 | 15 | 10 | 3 | 9 | 8 | 6 |
| | 2 | 4 | 2 | 1 | 11 | 10 | 13 | 7 | 8 | 15 | 9 | 12 | 5 | 6 | 3 | 0 | 14 |
| | 3 | 11 | 8 | 12 | 7 | 1 | 14 | 2 | 13 | 6 | 15 | 0 | 9 | 10 | 4 | 5 | 3 |
| $S_6$ | 0 | 12 | 1 | 10 | 15 | 9 | 2 | 6 | 8 | 0 | 13 | 3 | 4 | 14 | 7 | 5 | 11 |
| | 1 | 10 | 15 | 4 | 2 | 7 | 12 | 9 | 5 | 6 | 1 | 13 | 14 | 0 | 11 | 3 | 8 |
| | 2 | 9 | 14 | 15 | 5 | 2 | 8 | 12 | 3 | 7 | 0 | 4 | 10 | 1 | 13 | 11 | 6 |
| | 3 | 4 | 3 | 2 | 12 | 9 | 5 | 15 | 10 | 11 | 14 | 1 | 7 | 6 | 0 | 8 | 13 |
| $S_7$ | 0 | 4 | 11 | 2 | 14 | 15 | 0 | 8 | 13 | 3 | 12 | 9 | 7 | 5 | 10 | 6 | 1 |
| | 1 | 13 | 0 | 11 | 7 | 4 | 9 | 1 | 10 | 14 | 3 | 5 | 12 | 2 | 15 | 8 | 6 |
| | 2 | 1 | 4 | 11 | 13 | 12 | 3 | 7 | 14 | 10 | 15 | 6 | 8 | 0 | 5 | 9 | 2 |
| | 3 | 6 | 11 | 13 | 8 | 1 | 4 | 10 | 7 | 9 | 5 | 0 | 15 | 14 | 2 | 3 | 12 |
| $S_8$ | 0 | 13 | 2 | 8 | 4 | 6 | 15 | 11 | 1 | 10 | 9 | 3 | 14 | 5 | 0 | 12 | 7 |
| | 1 | 1 | 15 | 13 | 8 | 10 | 3 | 7 | 4 | 12 | 5 | 6 | 11 | 0 | 14 | 9 | 2 |
| | 2 | 7 | 11 | 4 | 1 | 9 | 12 | 14 | 2 | 0 | 6 | 10 | 13 | 15 | 3 | 5 | 8 |
| | 3 | 2 | 1 | 14 | 7 | 4 | 10 | 8 | 13 | 15 | 12 | 9 | 0 | 3 | 5 | 6 | 11 |

- The function $PC1$ is illustrated in Table 9.7. The upper half of the table specifies the bits that are taken from $k$ to construct $C$, and the lower half of the table specifies the bits that are taken from $k$ to construct $D$. If $k = k_1k_2 \dots k_{64}$, then $C = k_{57}k_{49} \dots k_{36}$ and $D = k_{63}k_{55} \dots k_4$. Note that the eight parity bits $k_8, k_{16}, \dots, k_{64}$ are not considered and occur neither in $C$ nor in $D$.

- The function $PC2$ is illustrated in Table 9.8. The two 28-bit strings that are input to the function are concatenated to form a 56-bit string. If this string is $b_1b_2 \dots b_{56}$, then the function $PC2$ turns this string into $b_{14}b_{17} \dots b_{32}$. Note that only 48 bits are taken into account and that $b_9$, $b_{18}$, $b_{22}$, $b_{25}$, $b_{35}$, $b_{38}$, $b_{43}$, and $b_{54}$ are discarded.

To derive the 16 round keys $k_1, \dots, k_{16}$ from the DES key $k$, $(C_0, D_0)$ are first initialized with $PC1(k)$ according to the construction given earlier. For $i = 1, \dots 16$, the following steps are then performed:

**Table 9.7**
The Function $PC1$ of the DES

| | | | | | | |
|---|---|---|---|---|---|---|
| 57 | 49 | 41 | 33 | 25 | 17 | 9 |
| 1 | 58 | 50 | 42 | 34 | 26 | 18 |
| 10 | 2 | 59 | 51 | 43 | 35 | 27 |
| 19 | 11 | 3 | 60 | 52 | 44 | 36 |
| 63 | 55 | 47 | 39 | 31 | 23 | 15 |
| 7 | 62 | 54 | 46 | 38 | 30 | 22 |
| 14 | 6 | 61 | 53 | 45 | 37 | 29 |
| 21 | 13 | 5 | 28 | 20 | 12 | 4 |

**Table 9.8**
The Function $PC2$ of the DES

| | | | | | |
|---|---|---|---|---|---|
| 14 | 17 | 11 | 24 | 1 | 5 |
| 3 | 28 | 15 | 6 | 21 | 10 |
| 23 | 19 | 12 | 4 | 26 | 8 |
| 16 | 7 | 27 | 20 | 13 | 2 |
| 41 | 52 | 31 | 37 | 47 | 55 |
| 30 | 40 | 51 | 45 | 33 | 48 |
| 44 | 49 | 39 | 56 | 34 | 53 |
| 46 | 42 | 50 | 36 | 29 | 32 |

1. $C_i$ is set to the string that results from a cyclic shift left of $C_{i-1}$ for $v_i$ positions.
2. $D_i$ is set to the string that results from a cyclic shift left of $D_{i-1}$ for $v_i$ positions.
3. $C_i$ and $D_i$ are concatenated.
4. The round key $k_i$ is the result of applying the function $PC2$ to the result of step 3 (i.e., $k_i = PC2(C_i \parallel D_i)$).

The resulting DES key schedule calculation is summarized and illustrated in Figure 9.9.

In the relevant literature, many examples and test vectors can be found either to illustrate the working principles of the DES encryption algorithm or to verify the correct input-output behavior of a specific DES implementation.[33]

33 Test vectors for DES encryption and decryption can be found, for example, in a NIST document available at http://csrc.nist.gov/publications/nistpubs/800-17/800-17.pdf.

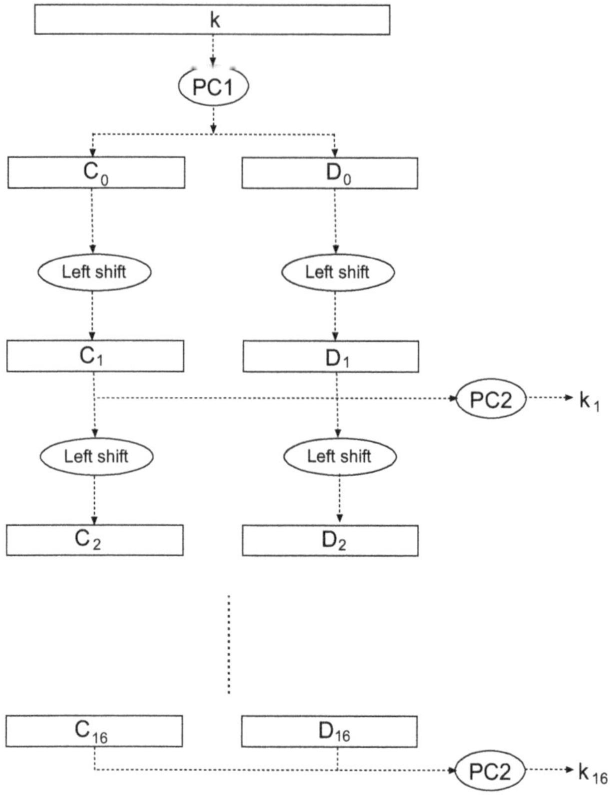


**Figure 9.9** The DES key schedule calculation.

### 9.6.1.3 Decryption Algorithm

Since DES is a Feistel cipher, the encryption and decryption algorithms are basically the same, meaning that Algorithm 9.4 can also be used for decryption. The only difference between the DES encryption and decryption algorithms is the key schedule that must be reversed in the case of decryption; that is, the DES round keys must be used in reverse order $k_{16}, \ldots, k_1$. If the key schedule is precomputed, then using the keys in reverse order is simple and straightforward. If, however, the key schedule is computed on the fly, then things are slightly more involved. In this case, the key schedule is generated as overviewed above, but cyclic shift right operations are used instead of cyclic shift left operations. Everything else remains the same.

#### 9.6.1.4 Security Considerations

Since its standardization in the 1970s, the DES has been subject to a lot of public scrutiny. For example, people found 4 weak keys and 12 semiweak keys.

- A DES key $k$ is *weak* if $DES_k(DES_k(m)) = m$ for all $m \in \mathcal{M} = \{0,1\}^{64}$, meaning that the DES encryption with $k$ is inverse to itself (i.e., if $m$ is encrypted twice with a weak key, then the result yields $m$ again).
- The DES keys $K_1$ and $K_2$ are *semiweak* if $DES_{k_1}(DES_{k_2}(m)) = m$ for all $m \in \mathcal{M} = \{0,1\}^{64}$, meaning that the DES encryptions with $k_1$ and $k_2$ are inverse to each other.

Because of their properties, weak and semiweak DES keys should not be used in practice. As there are only $16 = 2^4$ such keys, the probability of randomly generating one is only

$$\frac{2^4}{2^{56}} = 2^{-52} \approx 2.22 \cdot 10^{-16}$$

This probability is not particularly worrisome. It's certainly equally insecure to use a very small (large) key because an adversary is likely to start searching keys from the bottom (top). Consequently, there is no need to worry much about weak and semiweak keys in a given application setting.

More interestingly, several cryptanalytical attacks have been developed in an attempt to break the security of DES. Examples include differential cryptanalysis and linear cryptanalysis. Both attacks were published in the early 1990s, and are almost 30 years old.

- *Differential cryptanalysis* represents a CPA that requires $2^{47}$ chosen plaintexts to break DES [16].
- *Linear cryptanalysis* represents a known-plaintext attack that requires $2^{43}$ known plaintexts to break DES [17].

It goes without saying that both attacks require less plaintexts if one reduces the number of rounds. In either case, the amount of chosen or known plaintext is far too large to be relevant in practice. The results, however, are theoretically interesting and have provided principles and criteria for the design of secure block ciphers (people have since admitted that defending against differential cryptanalysis was one of the design goals for DES [18]). Also, all newly proposed block ciphers are routinely shown to be resistant against differential and linear cryptanalysis.

From a practical viewpoint, the major vulnerability and security problem of DES is its relatively small key length (and key space). Note that a DES key is effectively 56 bits long, and hence the key space comprises only

$$2^{56} = 72,057,594,037,927,936$$

elements. Consequently, a key search is successful after $2^{56}$ trials in the worst case and $2^{56}/2 = 2^{55}$ trials on the average.

Furthermore, the DES encryption has the *complementation property* that can be expressed as follows:

$$DES_k(m) = c \Longleftrightarrow DES_{\overline{k}}(\overline{m}) = \overline{c} \tag{9.7}$$

If one encrypts a plaintext message $m$ with a particular key $k$, then one gets ciphertext $c$. If one then encrypts the bitwise complement of the message (i.e., $\overline{m}$) with the bitwise complement of the key (i.e., $\overline{k}$), then one also gets the bitwise complement of the ciphertext (i.e., $\overline{c}$). This property can also be expressed as follows:

$$DES_k(m) = \overline{DES_{\overline{k}}(\overline{m})} \tag{9.8}$$

It can be used in a known-plaintext attack to narrow down the key space with another factor of two. If the adversary knows two plaintext-ciphertext pairs $(m, c_1)$ with $c_1 = DES_k(m)$ and $(\overline{m}, c_2)$ with $c_2 = DES_k(\overline{m})$, then he or she can compute for every key candidate $k'$ the value $c = DES_{k'}(m)$ and verify whether this value matches $c_1$ or $\overline{c_2}$:

- If $c = c_1$, then $k'$ is the correct key. This follows directly from $c = DES_{k'}(m)$, $c_1 = DES_k(m)$, and the fact that $c$ equals $c_1$.
- If $c = \overline{c_2}$, then $\overline{k'}$ is the correct key. This follows from $c = DES_{k'}(m)$, $c_2 = DES_k(\overline{m})$, and the complementation property mentioned above (note that $DES_{k'}(\overline{m}) = c_2$ means that $DES_{\overline{k'}}(m) = \overline{c_2}$).

So in every trial with key candidate $k'$, the adversary can also verify the complementary key candidate $\overline{k'}$. As mentioned earlier, this narrows down the key space with another factor of two. Hence, one can conclude that an exhaustive key search against DES is successful after $2^{54}$ trials on average.

The feasibility of an exhaustive key search was first publicly discussed by Diffie and Hellman in 1977 [19]. They estimated that a brute force machine that could find a DES key within a day would cost 20 million USD. Note that an exhaustive key search needs a lot of time but almost no memory. On the other hand, if one has a lot of memory and is willing to precompute the ciphertext $c$ for any given plaintext message $m$ and all possible keys $k$, then one can store the pairs $(c, k)$ and quickly find the correct key in a known-plaintext attack. Consequently, there is a lot of room for time-memory trade-offs (this topic was first explored by Hellman in 1980 [20]).

In either case, many people have discussed the possibility to design and actually build dedicated machines to do an exhaustive key search for DES. For example, Michael J. Wiener proposed such a design with 57,000 chips in 1993 [21]. He came to the conclusion that a 1 million USD version of such a machine would be capable of finding a DES key in 3.5 hours on the average. In 1997, he modified his estimates with a factor of 6 (i.e., a 1 million USD version of the machine would be capable of finding a DES key in 35 minutes on average and a 10,000 USD version of the machine would be capable of finding a DES key in 2.5 days on average [22]). These numbers are worrisome with regard to the security of DES against an exhaustive key search.

The first real attempts at breaking DES were encouraged by the RSA Secret-Key Challenge launched in 1997. In 1998, for example, the Electronic Frontier Foundation (EFF) won the RSA DES Challenge II-1 using a hardware-based DES search machine named *Deep Crack* [23].[34] Deep Crack cost around 200,000 USD to build and consisted of 1,536 processors, each capable of searching through 60 million keys per second. Referring to (9.1), the time to do an exhaustive key search is

$$\frac{|\mathcal{K}|t}{2p} = \frac{2^{56}}{60,000,000 \cdot 2 \cdot 1,536} = \frac{2^{55}}{60,000,000 \cdot 1,536} \approx 390,937 \text{ seconds}$$

Consequently, Deep Crack is theoretically able to recover a DES key in approximately 6,516 minutes, 109 hours, or 4.5 days. In the RSA DES Challenge II-1, Deep Crack was able to find the correct DES key within 56 hours.

More recently, a group of German researchers built a massively parallel *Cost-Optimized Parallel Code Breaker* (COPACOBANA) machine, which is optimized for running cryptanalytical algorithms and can be realized for less than 10,000 USD.[35] COPACOBANA can perform an exhaustive DES key search in less than 9 days on average.

Even more interestingly, one can spend the idle time of networked computer systems to look for DES keys and run an exhaustive key search. If enough computer systems participate in the search, then a DES key can be found without having to build a dedicated machine like Deep Crack or COPACOBANA. For example, the first DES Challenge (DESCHALL) sponsored by RSA Security (with a prize of 10,000 USD) was solved in a distributed way in 1997. In the DESCHALL project, thousands of networked computer systems were interconnected to a distributed

34 http://www.eff.org/descracker.html.

35 http://www.copacobana.org.

system and operated from January 29, 1997, until June 17, 1997 (4.5 months). The correct key was found after 17,731,502,968,143,872 elements (i.e., after having searched one-fourth of the key space). The success of the DESCHALL project had a major impact on the security of DES, because it showed that DES could be broken without dedicated hardware. Two years later, in 1999, the participants of the *Distributed.Net* project[36] broke a DES key in even less than 23 hours. More than 100,000 computer systems participated, received, and did a little part of the work. This allowed a search rate of 250 billion keys per second.

Against this background, it is obvious that the relatively small key length and the corresponding feasibility of an exhaustive key search is the most serious vulnerability and security problem of DES. There are only a few possibilities to protect a block cipher with a small key length, such as DES, against this type of attack. For example, one can frequently change keys, eliminate known plaintext, or use a complex key setup procedure (or key schedule, respectively). An interesting idea to slow down an exhaustive key search attack is due to Rivest and is known as *all-or-nothing encryption* [24]. It yields an encryption mode for block ciphers that makes sure that one must decrypt the entire ciphertext before one can determine even one plaintext message block. This means that an exhaustive key search attack against an all-or-nothing encryption is slowed down by a factor that depends on the number of ciphertext blocks. This is advantageous for the encryption of large plaintext messages. It does not help if the encrypted messages are short.

The simplest method to protect a block cipher against exhaustive key search attacks is to work with sufficiently long keys. It is commonly agreed today that modern ciphers with a key length of 128 bits and more are resistant against exhaustive key search attacks. In a 1996 paper[37] written by a group of well-known and highly respected cryptographers, it was argued that keys should be at least 75 bits long and that they should be at least 90 bits long if data must be protected adequately for the next 20 years (i.e., until 2016). Note that these numbers only provide a lower bound for the key length, and that there is no reason not to work with longer keys in the first place.[38] If, for example, a key is 128 bits long, then there are

$$2^{128} \approx 340 \cdot 10^{36}$$

possible keys. If an adversary can try out 1 million possible keys per second on a particular computer system, then an exhaustive key search takes about $10^{25}$ years. If

36 http://www.distributed.net.

37 http://www.schneier.com/paper-keylength.html.

38 It is sometimes argued that long keys slow down the encryption and decryption algorithms considerably. This argument is false. In most symmetric encryption systems, the secret key is expanded by a highly efficient key schedule algorithm, and this algorithm is largely independent from how many key bits are provided in the first place.

the same adversary has access to 1 million similar computer systems, then the attack still lasts for $10^{19}$ years. Taking into account the fact that our universe is estimated to be roughly $1.4 \cdot 10^{10}$ years old, it is quite obvious that an exhaustive key search against a 128-bit key is computationally infeasible.

In Appendix D.5, we introduce and put into perspective the notion of a quantum computer. There is an algorithm created by Lov K. Grover that requires $2^{n/2}$ steps in order to perform an exhaustive key search for an $n$-bit key cipher (with a key space of $2^n$) [25], and this algorithm can be shown to be the most efficient one [26]. In order to be resistant against such attacks, one has to double the key length, meaning that 200-bit keys correspond to 100-bit keys and 256-bit keys correspond to 128-bit keys in post-quantum secret key cryptography. This explains the standardization of 192-bit and 256-bit keys in the case of the AES (Section 9.6.2).

Coming back to DES, there are three possibilities to address (and possibly solve) the problem of its relatively small key length:

1. The DES may be modified in a way that compensates for its relatively small key length.
2. The DES may be iterated multiple times.
3. An alternative symmetric encryption system with a larger key length may be used.

The first possibility leads us to a modification of DES or the way DES is used. Rivest's all-or-nothing encryption mentioned above represents an example. Another example is known as DESX (it is addressed in the following section). The second possibility leads us to the TDEA addressed in Section 9.6.1.6. Last but not least, the third possibility leads us to the AES addressed in Section 9.6.2.

#### 9.6.1.5 DESX

In order to come up with a modification of DES that compensates for its relatively small key length, Rivest applied Merkle's key whitening technique [27] to DES and proposed DESX (sometimes also written as DES-X). DESX is practically relevant because it was the first symmetric encryption system employed by the Encrypted File System (EFS) in the Microsoft Windows 2000 operating system.

In essence, the key whitening technique consists of steps that combine the data with portions of the key (most commonly using a simple XOR) before the first round and after the last round of encryption. More specifically, the DESX construction is illustrated in Figure 9.10. In addition to the DES key $k$, the DESX construction

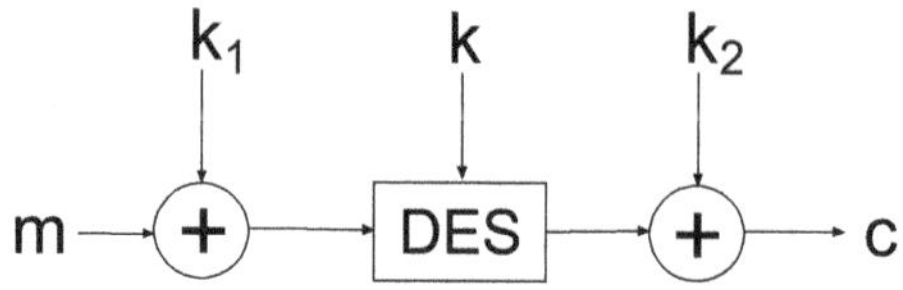


**Figure 9.10** The DESX construction.

employs two additional 64-bit keys, $k_1$ and $k_2$.[39] They are added modulo 2 to the plaintext message $m$ before and after the DES encryption takes place. Consequently, the DESX encryption of a plaintext message $m$ using keys $k$, $k_1$, and $k_2$ can be formally expressed as follows:

$$c = k_2 \oplus DES_k(m \oplus k_1)$$

Both additions modulo 2 are important, so neither $c = k_2 \oplus DES_k(m)$ nor $c = DES_k(m \oplus k_1)$ would significantly increase the resistance of DES against exhaustive key search.

DESX requires a total of $56 + 64 + 64 = 184$ bits of keying material. As such, it improves resistance against exhaustive key search considerably (e.g., [28]). In fact, it can be shown that the computational complexity of the most efficient known-plaintext attack is

$$2^{|k|+|k_1|-t}$$

where $2^t$ is the number of plaintext message-ciphertext pairs known to the adversary. In this formula, it is assumed that $k_1$ equals $k_2$, and that this value corresponds to the block length of the block cipher in use (64 bits in the case of DES). If, for example, the adversary can collect a maximum of $2^{30}$ plaintext message-ciphertext pairs, then he or she has to perform

$$2^{56+64-30} = 2^{90}$$

DES computations to distill the target key. Given today's technology, this computational task is surely out of reach. But keep in mind that key whitening in general, and DESX in particular, only protect against exhaustive key search attacks, and that they do not improve resistance against other cryptanalytical attacks such as differential or linear cryptanalysis.

39 Note that $k_1$ and $k_2$ must be different. Otherwise, the binary additions modulo 2 (i.e., XOR operations) cancel themselves out.

Note that Merkle's key whitening technique can be applied to any block cipher. If, for example, one applies it to DESL, then one ends up with a symmetric encryption system known as DESXL. It can be efficiently implemented and is not vulnerable to exhaustive key search.

#### 9.6.1.6 TDEA

As mentioned earlier, a possibility to address (or solve) the small key length problem is to iterate DES multiple times. There are two points to make:

- First, multiple iterations with the same key are not much more secure than a single encryption. This is because an adversary can also iterate the encryption functions multiple times. If, for example, DES is iterated twice (with the same key), then each step of testing a key is also twice as much work (because the adversary has to do a double encryption). A factor of two for the adversary is not considered much added security, especially because the legitimate users have their work doubled as well. Consequently, multiple iterations must always be done with different keys to improve security.
- Second, it was shown that the DES encryption functions are not closed with regard to concatenation (i.e., they do not form a group [29]). If the DES encryption functions provided a group, then there would exist a DES key $k_3$ for all pairs $(k_1, k_2)$ of DES keys, such that $DES_{k_3} = DES_{k_1} \circ DES_{k_2}$. This would be unfortunate, and the iterated use of the DES would not provide any security advantage.[40] Because the DES encryption functions do not form a group, this is not the case, and hence the adversary cannot simply perform an exhaustive key search for $k_3$ (instead of performing an exhaustive key search for $k_1$ and $k_2$).

Against this background, the first (meaningful) possibility to iterate the DES is the double encryption with two independent keys. However, it was first shown by Diffie and Hellman that double encryption is not particularly useful due to the existence of a *meet-in-the-middle attack*. Assume an adversary has a few plaintext message-ciphertext pairs $(m_i, c_i)$, where $c_i$ is derived from a double encryption of $m_i$ with $k_1$ and $k_2$, and he or she wants to find $k_1$ and $k_2$. The meet-in-the-middle attack is illustrated in Figure 9.11; it operates in the following four steps:

40 Consider the affine cipher as a counterexample. There is a single affine cipher $E_{(a_3,b_3)}$ which performs exactly the same encryption and decryption as the combination of two affine ciphers $E_{(a_1,b_1)}$ and $E_{(a_2,b_2)}$. So instead of breaking the combination, it is possible to break the single affine cipher, and hence the combination does not increase the overall security.

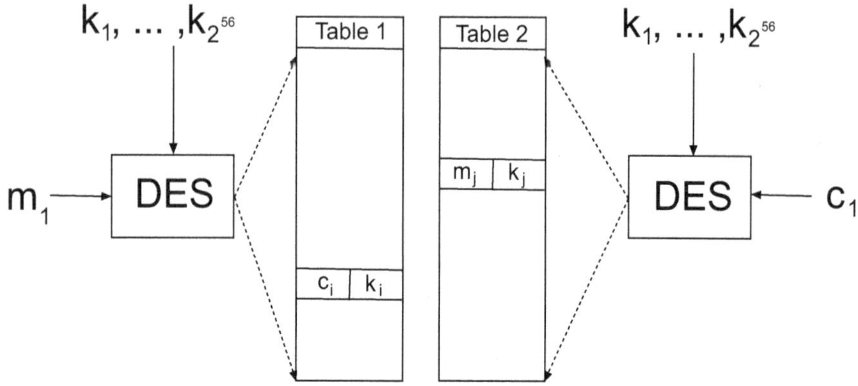


**Figure 9.11** The meet-in-the-middle attack against double DES.

1. The adversary computes a first table (i.e., Table 1) with $2^{56}$ entries. Each entry consists of a possible DES key $k_i$ and the result of applying that key to encrypt the plaintext message $m_1$. Table 1 is sorted in numerical order by the resulting ciphertexts. Hence, the entry $(c_i, k_i)$ refers to the ciphertext that results from encrypting $m_1$ with $k_i$ and $i = 1, \ldots, 2^{56}$.

2. The adversary computes a second table (i.e., Table 2) with $2^{56}$ entries. Each entry consists of a possible DES key $k_j$ and the result of applying that key to decrypt the ciphertext $c_1$. Table 2 is sorted in numerical order by the resulting plaintexts. Hence, the entry $(m_j, k_j)$ refers to the plaintext that results from decrypting $c_1$ with $k_j$ for $j = 1, \ldots, 2^{56}$.

3. The adversary searches through the sorted tables to find matching entries. Each matching entry $c_i = p_j$ yields $k_i$ as a key candidate for $k_1$ and $k_j$ as a key candidate for $k_2$ (because $k_i$ encrypts $m_1$ to a value to which $k_j$ decrypts $c_1$).

4. If there are multiple matching pairs (which there almost certainly will be),[41] the adversary tests the candidate pairs $(k_1, k_2)$ against $m_2$ and $c_2$. If multiple candidate pairs still work for $m_2$ and $c_2$, then the same test procedure is repeated for $m_3$ and $c_3$. This continues until a single candidate pair remains. Note that the correct candidate pair always works, whereas an incorrect

41 There are $2^{64}$ possible plaintext and ciphertext blocks, but only $2^{56}$ entries in each table. Consequently, each 64-bit block appears with a probability of $1/256$ in each of the tables, and of the $2^{56}$ blocks that appear in the first table, only $1/256$ of them also appear in the second table. That means that there should be $2^{48}$ entries that appear in both tables.

candidate pair will almost certainly fail to work on any particular $(m_i, c_i)$ pair.

The meet-in-the-middle attack is not particularly worrisome because it requires two tables with $2^{56}$ entries each (there are some improvements not addressed in this book). The mere existence of the attack, however, is enough reason to iterate DES three times, and to do *triple DES* (3DES) accordingly. It may be that double DES would be good enough, but because triple DES is not much harder, it is usually done in the first place. As mentioned earlier, FIPS PUB 46-3 specifies the TDEA, and this specification also conforms to ANSI X9.52.

A TDEA key consists of three keys that are collectively referred to as a key bundle (i.e., $k = (k_1, k_2, k_3)$. The TDEA encryption function works as follows:

$$c = E_{k_3}(D_{k_2}(E_{k_1}(m)))$$

Consequently, a TDEA or 3DES encryption is sometimes also referred to as EDE (encrypt-decrypt-encrypt). The reason for the second iteration of DES being a decryption (instead of an encryption) is that a 3DES implementation can then easily be turned into a single-key DES implementation by feeding all three iterations with the same key $k$. If we then compute $c = E_{k_3}(D_{k_2}(E_{k_1}(m)))$, we actually compute $c = E_k(D_k(E_k(m))) = E_k(m)$. Anyway, the TDEA decryption function works as follows:

$$m = D_{k_1}(E_{k_2}(D_{k_3}(c)))$$

FIPS PUB 46-3 specifies the following three options for the key bundle $k = (k_1, k_2, k_3)$:

- Keying option 1: $k_1$, $k_2$, and $k_3$ are independent keys.
- Keying option 2: $k_1$ and $k_2$ are independent keys and $k_3 = k_1$.
- Keying option 3: All keys are equal (i.e., $k_1 = k_2 = k_3$). As mentioned earlier, the 3DES implementation then represents a single-key DES implementation.

Again, keying option 1 is used to make 3DES implementations backward-compatible with single-key DES. Keying option 2 looks reasonable to use 3DES with an effective key length of $2 \cdot 56 = 112$ bits. Unfortunately, this is only wishful thinking, and it was shown that the effective key length of 3DES with keying option 2 is somewhere in the range of 80 bits [30]. So the remaining option to use in practice is keying option 3. But keep in mind that, due to the meet-in-the-middle attack, the effective key length of 3DES even with keying option 3 is not $3 \cdot 56 = 168$, but somewhere in the range of 112 bits.

We close the section with the remark that iterating a block cipher multiple times can be done with any block cipher and that there is nothing DES-specific about this construction. It is, however, less frequently used with other block ciphers (mainly because most of them have been designed to use longer keys in the first place).

### 9.6.2 AES

In the time between 1997 and 2000, NIST carried out an open competition to standardize a successor for the DES, called the AES. In contrast with the DES standardization effort in the mid-1970s, many parties from industry and academia participated in the AES competition. In fact, there were 15 submissions qualifying as serious AES candidates, and among these submissions, NIST selected five finalists: MARS, RC6,[42] Rijndael,[43] Serpent,[44] and Twofish.[45] On October 2, 2000, NIST decided that Rijndael would become the AES.[46] According to [32], NIST could not distinguish between the security of the finalist algorithms, and Rijndael was selected mainly because of its ease of implementation in hardware and its strong performance on nearly all platforms. The AES is officially specified in FIPS PUB 197 [33].

According to the requirements specified by NIST, the AES is a block cipher with a block length of 128 bits and a variable key length of 128, 192, or 256 bits.[47] We already mentioned why keys longer than 128 bits make sense in post-quantum (secret key) cryptography. The respective AES versions are referred to as AES-128, AES-192, and AES-256. The number of rounds depends on the key length (i.e., 10, 12, or 14 rounds). Table 9.9 summarizes the three official versions of the AES. $N_b$ refers to the block length (in number of 32-bit words), $N_k$ refers to the key length (again, in number of 32-bit words), and $N_r$ refers to the number of rounds. Note that the official versions of the AES all work with a block size of $N_b$ words, referring to $N_b \cdot 32 = 4 \cdot 32 = 128$ bits. While FIPS PUB 197 explicitly defines the allowed values for $N_b$, $N_k$, and $N_r$, future reaffirmations may include changes or additions to these values. Implementors of the AES should therefore make their implementations as flexible as possible (this is a general recommendation that does not only apply to the AES).

42 https://people.csail.mit.edu/rivest/pubs/RRSY98.pdf.

43 The Rijndael algorithm was developed and proposed by the two Belgian cryptographers, Joan Daemen and Vincent Rijmen. Its design and some background information is described in [31].

44 http://www.cl.cam.ac.uk/~rja14/serpent.html.

45 https://www.schneier.com/academic/twofish.

46 Refer to the NIST *Report on the Development of the Advanced Encryption Standard (AES)* available at http://csrc.nist.gov/archive/aes/round2/r2report.pdf.

47 Rijndael was originally designed to handle additional block sizes and key lengths (that are, however, not adopted in the current AES specification).

**Table 9.9**
The Three Official Versions of the AES

| | $N_b$ | $N_k$ | $N_r$ |
|---|---|---|---|
| AES-128 | 4 | 4 | 10 |
| AES-192 | 4 | 6 | 12 |
| AES-256 | 4 | 8 | 14 |

Before we enter a more detailed description of the AES, we have to start with a few preliminary remarks regarding the way data is processed in the encryption and decryption algorithms.

#### 9.6.2.1 Preliminary Remarks

Similar to most other symmetric encryption systems, the AES is byte oriented, meaning that the basic unit for processing is a byte (i.e., a sequence of 8 bits). Each byte represents an element of $\mathbb{F}_{2^8}$ (or $GF(2^8)$, respectively) and can be written in binary or hexadecimal notation.

- In binary notation, a byte is written as $\{b_7b_6b_5b_4b_3b_2b_1b_0\}$ with $b_i \in \{0,1\} = \mathbb{Z}_2 \cong \mathbb{F}_2$ for $i = 0, \ldots, 7$.
- In hexadecimal notation, a byte is written as 0xXY with X,Y $\in$ {0, ..., 9, A, ..., F }. Referring to the binary notation, X stands for $\{b_7b_6b_5b_4\}$ and Y stands for $\{b_3b_2b_1b_0\}$.

Alternatively, the 8 bits of a byte can also be seen and written as the coefficients of a polynomial of degree 7:

$$b_7x^7 + b_6x^6 + b_5x^5 + b_4x^4 + b_3x^3 + b_2x^2 + b_1x + b_0 = \sum_{i=0}^{7} b_ix^i$$

For example, the byte $\{10100011\}$ = 0xA3 can be written as polynomial $x^7 + x^5 + x + 1$; that is, for every bit equal to 1, the corresponding coefficient in the polynomial is set to 1.

Mathematically speaking, the AES operates in the extension field $\mathbb{F}_{2^8}$ (Section A.1.2.5). This, in turn, means that the elements of the field are polynomials over $\mathbb{F}_2$ with degree equal or smaller than seven. Using these polynomials, it is simple and straightforward to add and multiply field elements. Let us therefore have a closer look at the addition and multiplication operations.

**Addition:** If we consider bytes, then the addition is achieved by adding modulo 2 the bits in the bytes representing the two elements of $\mathbb{F}_{2^8}$. For example,

$$\begin{array}{rl} & 01010111 \\ \oplus & 10000011 \\ = & 11010100 \end{array}$$

We can also use the hexadecimal or polynomial notation to add bytes. If, for example, we use polynomials, then the addition is achieved by adding modulo 2 the coefficients for the same powers in the polynomials representing the two elements. Hence, the example given above can be written as

$$(x^6 + x^4 + x^2 + x + 1) + (x^7 + x + 1) = x^7 + x^6 + x^4 + x^2.$$

**Multiplication:** If we consider bytes in binary or hexadecimal notation, then there is no easy way to describe the multiplication operation. If, however, we consider the polynomial notation, then the multiplication operation can be easily described as a multiplication of two polynomials over $\mathbb{Z}_2$ modulo an irreducible polynomial of degree 8. In the case of the AES, the irreducible polynomial

$$f(x) = x^8 + x^4 + x^3 + x + 1$$

is standardized. The modular reduction by $f(x)$ ensures that the result is always a binary polynomial of degree less than 8, and hence that it represents a single byte. Note that the multiplication is associative and the polynomial 1—$\{00000001\}$ in binary and 0x01 in hexadecimal notation—represents the multiplicative identity element of the multiplication operation.

Referring to Appendix A.3.6, $\mathbb{Z}_2[x]_{f(x)}$ represents a field if $f(x)$ is an irreducible polynomial over $\mathbb{Z}_2$. In the case of the AES, $f(x)$ is an irreducible polynomial over $\mathbb{Z}_2$ and the degree of $f(x)$ is 8. So $\mathbb{Z}_2[x]_{f(x)}$ is a field that is isomorphic to $\mathbb{F}_{2^8}$. In cryptography, this field is also known as the *AES field*. Since it is a field, we know that every nonzero element $b(x)$ (i.e., every nonzero polynomial over $\mathbb{Z}_2$ with degree less than 8) has a multiplicative inverse element $b^{-1}(x)$. As for any field, this element can be efficiently computed with the extended Euclid algorithm (Algorithm A.2).

There are a few special cases for which the multiplication operation is simple to compute. For example, multiplying a polynomial with the polynomial 1 is trivial, since 1 represents the neutral element. Also, multiplying a polynomial with the polynomial $x$ is simple. If we start with a polynomial $b(x)$, then $b(x) \cdot x$ refers to the polynomial

$$b_7x^8 + b_6x^7 + b_5x^6 + b_4x^5 + b_3x^4 + b_2x^3 + b_1x^2 + b_0x = \sum_{i=0}^{7} b_ix^{i+1}$$

that must be reduced modulo $f(x)$. If $b_7 = 0$, then the result is already in reduced form. Otherwise, if $b_7 = 1$, then the reduction is accomplished by subtracting (i.e., adding modulo 2) the polynomial $f(x)$. It follows that multiplication by $x$ can be implemented efficiently at the byte level as a shift left and a subsequent conditional addition modulo 2 with $f(x)$. We are now ready to delve more deeply into the internals of the AES.

#### 9.6.2.2 State

Internally, the AES operates on a two-dimensional array $s$ of bytes, called the *State* (note the capital letter the term starts with). The State consists of 4 rows and $N_b$ columns. Remember from Table 9.9 that $N_b = 4$ for all official versions of the AES, so the State in all official versions of the AES has 4 columns. Each entry in the State refers to a byte $s_{r,c}$ or $s[r, c]$, where $0 \leq r < 4$ refers to the row number and $0 \leq c < 4$ refers to the column number. Note that the four bytes $s_{r,0}$, $s_{r,1}$, $s_{r,2}$, and $s_{r,3}$ ($s_{0,c}$, $s_{1,c}$, $s_{2,c}$, and$s_{3,c}$) in row $r$ (column $c$) of the State represent a 32-bit word, and hence the State can also be seen as a one-dimensional array of four 32-bit words. In fact, the views of the State as a two-dimensional 4x4 array (or matrix) of bytes or as a one-dimensional array of four 32-bit words are equivalent. We use either view in the sequel.

As illustrated in Figure 9.12, the 16 input bytes $in_0, \ldots, in_{15}$ are copied into the State at the beginning of the AES encryption or decryption process. The encryption or decryption process is then conducted on the State, and the State's final bytes are copied back to the output bytes $out_0, \ldots, out_{15}$. More formally speaking, the input array $in$ is copied into the State according to

$$s_{r,c} = in_{r+4c}$$

for $0 \leq r < 4$ and $0 \leq c < 4$ at the beginning of the encryption (or decryption) process. Similarly, the State is copied back into the output array $out$ according to

$$out_{r+4c} = s_{r,c}$$

for $0 \leq r < 4$ and $0 \leq c < 4$ at the end of the encryption (or decryption). Let us now outline the AES encryption, key expansion, and decryption algorithms, before we analyze what is currently known about the security of the AES.

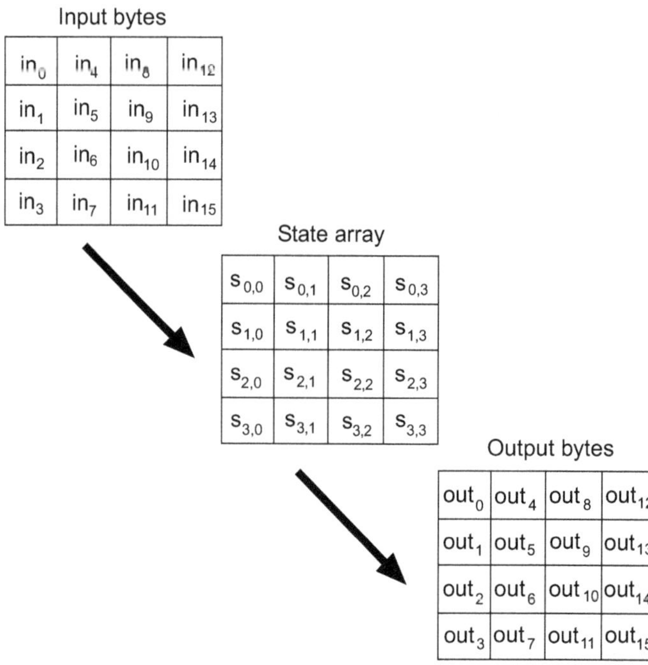


**Figure 9.12** Input bytes, State array, and output bytes of the AES.

#### 9.6.2.3 Encryption Algorithm

The AES encryption algorithm[48] is overviewed in Algorithm 9.5. The 16 input bytes $in_0, \ldots, in_{15}$ are first copied into the State $s$, before an initial AddRoundKey() transformation is applied to it. The algorithm then enters a loop that is iterated $N_r - 1$ times, where $N_r$ =10, 12, or 14 depending on the key length in use. In each iteration, a round function is applied that consists of the following four transformations:

1. In the SubBytes() transformation, the bytes of the State are substituted according to a well-defined substitution table.

2. In the ShiftRows() transformation, the rows of the State are subject to a cyclic shift (or rotation) for a row-specific number of bytes.

48 Note that the official AES specification uses the terms *cipher* and *inverse cipher* to refer to the AES encryption and decryption algorithms. This is not in line with our use of the term *cipher* in this book. In particular, we use the term to refer to a full-fledged symmetric encryption system and not only to an encryption algorithm.

**Algorithm 9.5** The AES encryption algorithm.

$(in)$

---

$s = in$
$s =$ AddRoundKey$(s, w[0, N_b - 1])$
for $r = 1$ to $(N_r - 1)$ do
    $s =$ SubBytes$(s)$
    $s =$ ShiftRows$(s)$
    $s =$ MixColumns$(s)$
    $s =$ AddRoundKey$(s, w[rN_b, (r + 1)N_b - 1])$
$s =$ SubBytes$(s)$
$s =$ ShiftRows$(s)$
$s =$ AddRoundKey$(s, w[N_rN_b, (N_r + 1)N_b - 1])$
$out = s$

---

$(out)$

3. In the MixColumns() transformation, the columns of the State are subject to a linear transformation.
4. In the AddRoundKey() transformation, a round key is added to the State (this is where the secret key and the respective key schedule come into play). Note in the notation of the algorithm that $w[i]$ refers to the i-th word in the key schedule $w$ and $w[i, j]$ refers to the $j - i + 1$ words between $w_i$ and $w_j$ in the key schedule.

Note that the SubBytes() and ShiftRows() transformations are commutative, meaning that a SubBytes() transformation immediately followed by a ShiftRows() transformation is equivalent to a ShiftRows() transformation immediately followed by a SubBytes() transformation). This is not the case for all other transformations.

In Algorithm 9.5, the loop is left after $N_r - 1$ rounds. There is a final round that is slightly different, as it does not include a MixColumns() transformation. In the end, the content of the State represents the output of the AES encryption. It is copied back into the output array. Let us now have a closer at each of the four transformations that are the ingredients of the AES encryption algorithm.

*SubBytes() Transformation*

As its name suggests, the SubBytes() transformation stands for a substitution cipher in which each byte $s_{r,c}$ of the State is replaced with another byte $s'_{r,c}$ from a substitution table (called S-box). The S-box of the AES is illustrated in Table 9.10. If, for example, the input byte is 0xXY, then 0xX refers to a particular row and 0xY refers to a particular column in the S-box and the output is the element that can be found at that particular position. For input byte 0x52 this means that the output of the S-box is the element that can be found in row 5 and column 2. This is the element 0x00, and hence the output of the S-box for the element 0x52 is 0x00; that is, S-box(0x52) = 0x00 or S-box(01010010) = 00000000.

**Table 9.10**
The S-Box of the AES Encryption Algorithm

| | 0 | 1 | 2 | 3 | 4 | 5 | 6 | 7 | 8 | 9 | A | B | C | D | E | F |
|---|---|---|---|---|---|---|---|---|---|---|---|---|---|---|---|---|
| 0 | 63 | 7C | 77 | 7B | F2 | 6B | 6F | C5 | 30 | 01 | 67 | 2B | FE | D7 | AB | 76 |
| 1 | CA | 82 | C9 | 7D | FA | 59 | 47 | F0 | AD | D4 | A2 | AF | 9C | A4 | 72 | C0 |
| 2 | B7 | FD | 93 | 26 | 36 | 3F | F7 | CC | 34 | A5 | E5 | F1 | 71 | D8 | 31 | 15 |
| 3 | 04 | C7 | 23 | C3 | 18 | 96 | 05 | 9A | 07 | 12 | 80 | E2 | EB | 27 | B2 | 75 |
| 4 | 09 | 83 | 2C | 1A | 1B | 6E | 5A | A0 | 52 | 3B | D6 | B3 | 29 | E3 | 2F | 84 |
| 5 | 53 | D1 | 00 | ED | 20 | FC | B1 | 5B | 6A | CB | BE | 39 | 4A | 4C | 58 | CF |
| 6 | D0 | EF | AA | FB | 43 | 4D | 33 | 85 | 45 | F9 | 02 | 7F | 50 | 3C | 9F | A8 |
| 7 | 51 | A3 | 40 | 8F | 92 | 9D | 38 | F5 | BC | B6 | DA | 21 | 10 | FF | F3 | D2 |
| 8 | CD | 0C | 13 | EC | 5F | 97 | 44 | 17 | C4 | A7 | 7E | 3D | 64 | 5D | 19 | 73 |
| 9 | 60 | 81 | 4F | DC | 22 | 2A | 90 | 88 | 46 | EE | B8 | 14 | DE | 5E | 0B | DB |
| A | E0 | 32 | 3A | 0A | 49 | 06 | 24 | 5C | C2 | D3 | AC | 62 | 91 | 95 | E4 | 79 |
| B | E7 | C8 | 37 | 6D | 8D | D5 | 4E | A9 | 6C | 56 | F4 | EA | 65 | 7A | AE | 08 |
| C | BA | 78 | 25 | 2E | 1C | A6 | B4 | C6 | E8 | DD | 74 | 1F | 4B | BD | 8B | 8A |
| D | 70 | 3E | B5 | 66 | 48 | 03 | F6 | 0E | 61 | 35 | 57 | B9 | 86 | C1 | 1D | 9E |
| E | E1 | F8 | 98 | 11 | 69 | D9 | 8E | 94 | 9B | 1E | 87 | E9 | CE | 55 | 28 | DF |
| F | 8C | A1 | 89 | 0D | BF | E6 | 42 | 68 | 41 | 99 | 2D | 0F | B0 | 54 | BB | 16 |

The substitution cipher defined by the AES' S-box is bijective and nonlinear. The first property means that all $2^8 = 256$ possible input elements are one-to-one mapped to unique output elements, and hence that it is possible to uniquely inverse the SubBytes() transformation. The second property means that SubBytes(s) + SubBytes(s') $\neq$ SubBytes(s + s') for two possible states $s$ and $s'$. In fact, the S-box is the only nonlinear component of the AES, and is therefore important from a security viewpoint.

Remember from DES that the S-boxes look random and mysterious, and that the underlying design principles have not been published for a long time. This is different with the AES and its S-box. This time, it is algebraically structured and follows well-defined and well-documented design principles. In fact, the AES S-box is constructed by subjecting each byte $s_{r,c}$ $(0 \leq r, c < 16)$ to a two-step mathematical transformation:

1. First, $s_{r,c}$ is mapped to its multiplicatively inverse element $b = s_{r,c}^{-1}$ in the AES field (the element 0x00 is mapped to itself).
2. Second, the byte $b = (b_7, \ldots, b_0)$ is subject to an affine transformation, meaning that $b$ is first multiplied with a constant (8x8)-bit-matrix and then

added to a constant 8-bit vector. This can be expressed as follows:

$$\begin{pmatrix} b'_0 \\ b'_1 \\ b'_2 \\ b'_3 \\ b'_4 \\ b'_5 \\ b'_6 \\ b'_7 \end{pmatrix} = \begin{pmatrix} 1 & 0 & 0 & 0 & 1 & 1 & 1 & 1 \\ 1 & 1 & 0 & 0 & 0 & 1 & 1 & 1 \\ 1 & 1 & 1 & 0 & 0 & 0 & 1 & 1 \\ 1 & 1 & 1 & 1 & 0 & 0 & 0 & 1 \\ 1 & 1 & 1 & 1 & 1 & 0 & 0 & 0 \\ 0 & 1 & 1 & 1 & 1 & 1 & 0 & 0 \\ 0 & 0 & 1 & 1 & 1 & 1 & 1 & 0 \\ 0 & 0 & 0 & 1 & 1 & 1 & 1 & 1 \end{pmatrix} \cdot \begin{pmatrix} b_0 \\ b_1 \\ b_2 \\ b_3 \\ b_4 \\ b_5 \\ b_6 \\ b_7 \end{pmatrix} + \begin{pmatrix} 1 \\ 1 \\ 0 \\ 0 \\ 0 \\ 1 \\ 1 \\ 0 \end{pmatrix}$$

The resulting byte $b'$ is then written into the S-box in row $r$ and column $c$.

Let us make an example to illustrate how the S-box is generated: For input byte 0xC3 = (11000011) we first compute the inverse element (using, for example, the extended Euclid algorithm). This element is 0xA3 = (10100011). We then subject this element to the affine transformation mentioned above. The result is 0x2E = (00101110), and this byte value can actually be found in row C (12) and column 3 of the S-box. Similarly, the element 0x00 is self-inverse; that is, for this element we only have to add the constant 8-bit vector 0x63 = (01100011). The result is 0x63, and this byte thus represents the first element of the S-box. From an implementor's viewpoint, there are two possibilities to implement the S-box: It is either precomputed and statically stored somewhere or it is generated dynamically on the fly (and not stored at all). Either possibility has advantages and disadvantages, but we are not going into the details here, mainly because this book is not an implementation manual or guideline.

Finally, we note that the nonlinerarity of the SubBytes() transformation comes from the inversion of $s_{r,c}$; that is, if the affine transformation were applied to $s_{r,c}$ instead of $s_{r,c}^{-1}$, then the SubBytes() transformation would be linear. We also note that the invertibility of the SubBytes() transformation requires that the matrix from the affine transformation is invertible (i.e., its rows and columns must be linearly independent in the AES field). This is the case here, and hence the SubBytes() transformation is invertible. This fact is required for the AES decryption algorithm and its InvSubBytes() transformation.

### *ShiftRows() Transformation*

As mentioned above and illustrated in Figure 9.13, the ShiftRows() transformation cyclically shifts (or rotates) the bytes in each row of the State. The number of bytes that are subject to a cyclic shift left is equal to the row number; that is, the bytes in row $r$ $(0 \leq r \leq 3)$ are shifted for $r$ bytes. This means that the bytes in the first row

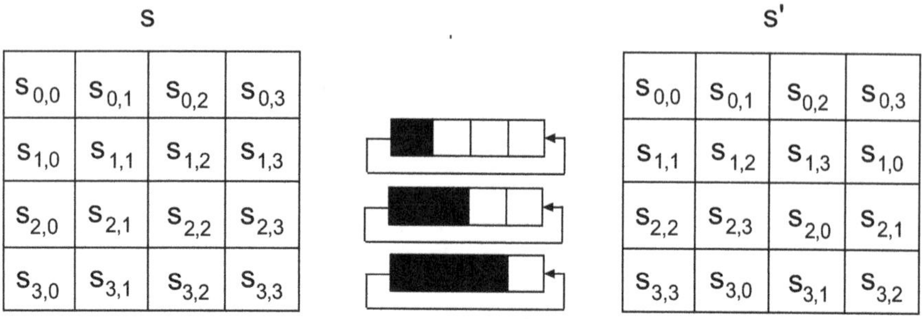


**Figure 9.13** The ShiftRows() transformation of the AES encryption algorithm.

(i.e., $r = 0$) are not shifted at all, the bytes in the second row (i.e., $r = 1$) are shifted for 1 byte, the bytes in the third row (i.e., $r = 2$) are shifted for 2 bytes, and the bytes in the fourth row (i.e., $r = 3$) are shifted for 3 bytes. Formally, for $0 \leq r < 4$ and $0 \leq c < N_b = 4$, the ShiftRows() transformation can be expressed as follows:

$$s'_{r,c} = s_{r,(c+shift(r,N_b)) \bmod N_b} \tag{9.9}$$

In this formula, the shift value $shift(r, N_b)$ only depends on the row number $r$ (because $N_b$ is always equal to 4):

$$\begin{aligned} shift(1,4) &= 1 \\ shift(2,4) &= 2 \\ shift(3,4) &= 3 \end{aligned}$$

For example, $s'_{2,1} = s_{2,(1+shift(2,4)) \bmod 4} = s_{2,(1+2) \bmod 4} = s_{2,3}$. Note that the elements of $s'$ are the same as the elements of $s$, and that only their ordering is subject to change when the State is tranformed. Also note that a cyclic shift operation can be trivially inverting by a respective cyclic shift right operation. Again, this used in the AES decryption algorithm for the InvShiftRows() transformation.

*MixColumns() Transformation*

The MixColumns() transformation operates on each column of the State individually, and—as mentioned above—it subjects each column to a linear transformation. This means that the MixColumns() transformation provides diffusion, especially if combined with the ShiftRows() transformation. In fact, it has been shown that the combination of the ShiftRows() and MixColumns() transformations makes it possible that after only three rounds every byte of the State depends on all 16 input bytes.

When the MixColumns() transformation operates on column $c$ ($0 \leq c < 4$), it considers the 4 bytes $s_{0,c}$, $s_{1,c}$, $s_{2,c}$, and $s_{3,c}$ of the State simultaneously. To keep the notation simple, we use $s_0$, $s_1$, $s_2$, and $s_3$ to refer to these bytes. They can be used as the coefficients of a polynomial $s(x)$ of degree 3:

$$s(x) = s_3x^3 + s_2x^2 + s_1x + s_0$$

Note that the coefficients are bytes and elements of $\mathbb{F}_{2^8}$. This is different from the polynomials we have seen so far (where the coefficients are bits and elements of $\mathbb{F}_2$). Also note that a column polynomial $s(x)$ can also be written as $[s_0, s_1, s_2, s_3]$ in a more compact representation.

The MixColumns() transformation operates on $s(x)$ by multiplying it with a fixed polynomial $c(x)$ of degree 3 and reducing the result modulo a polynomial of degree 4. The fixed polynomial is $c(x) = c_3x^3 + c_2x^2 + c_1x + c_0$ with $c_3$=0x03, $c_2$=0x01, $c_1$=0x01, and $c_0$=0x02, and the polynomial to reduce the product is $x^4 + 1$. Note that this polynomial need not be irreducible and is not in $\mathbb{F}_2$ (since $x^4 + 1 = (x + 1)^4$). Its sole purpose is to make sure that the multiplication of $s(x)$ and $c(x)$ yields a polynomial of degree 3 at most (so that the respective bytes can be placed in a column again). The bottom line is that the MixColumns() transformation maps $s(x)$ to $(c(x) \cdot s(x)) \mod (x^4 + 1)$. Because $c(x)$ is relatively prime to $x^4 + 1$ in $\mathbb{F}_2[x]$, the inverse polynomial $c(x)^{-1} \pmod{x^4 + 1}$ exists, and hence the MixColumns() transformation is invertible.

Because this computation is involved, it is generally simpler to compute the MixColumns() transformation for column $c$ ($0 \leq c < 4$) as follows:

$$\begin{pmatrix} s'_{0,c} \\ s'_{1,c} \\ s'_{2,c} \\ s'_{3,c} \end{pmatrix} = \begin{pmatrix} \text{0x02} & \text{0x03} & \text{0x01} & \text{0x01} \\ \text{0x01} & \text{0x02} & \text{0x03} & \text{0x01} \\ \text{0x01} & \text{0x01} & \text{0x02} & \text{0x03} \\ \text{0x03} & \text{0x01} & \text{0x01} & \text{0x02} \end{pmatrix} \cdot \begin{pmatrix} s_{0,c} \\ s_{1,c} \\ s_{2,c} \\ s_{3,c} \end{pmatrix}$$

This can also be written as follows:

$$\begin{aligned} s'_{0,c} &= (\text{0x02} \cdot s_{0,c}) \oplus (\text{0x03} \cdot s_{1,c}) \oplus s_{2,c} \oplus s_{3,c} \\ s'_{1,c} &= s_{0,c} \oplus (\text{0x02} \cdot s_{1,c}) \oplus (\text{0x03} \cdot s_{2,c}) \oplus s_{3,c} \\ s'_{2,c} &= s_{0,c} \oplus s_{1,c} \oplus (\text{0x02} \cdot s_{2,c}) \oplus (\text{0x03} \cdot s_{3,c}) \\ s'_{3,c} &= (\text{0x03} \cdot s_{0,c}) \oplus s_{1,c} \oplus s_{2,c} \oplus (\text{0x02} \cdot s_{3,c}) \end{aligned}$$

If, for example, we are given the column vector

$$\begin{pmatrix} s_{0,c} \\ s_{1,c} \\ s_{2,c} \\ s_{3,c} \end{pmatrix} = \begin{pmatrix} \text{0xd4} \\ \text{0xbf} \\ \text{0x5d} \\ \text{0x30} \end{pmatrix}$$

and we want to compute the new element $s'_{0,c}$, then we have to compute

$$\begin{aligned} s'_{0,c} &= (\text{0x02} \cdot \text{0xd4}) \oplus (\text{0x03} \cdot \text{0xbf}) \oplus (\text{0x01} \cdot \text{0x5d}) \oplus (\text{0x01} \cdot \text{0x30}) \\ s'_{0,c} &= (\text{0x02} \cdot \text{0xd4}) \oplus (\text{0x03} \cdot \text{0xbf}) \oplus \text{0x5d} \oplus \text{0x30} \end{aligned}$$

The operations $\oplus$ and $\cdot$ refer to the addition modulo 2 and the byte multiplication. The byte multiplication, in turn, is best expressed in the polynomial representation. So the first term (0x02·0xd4) refers to 0000 0010·11010100 or $x \cdot (x^7+x^6+x^4+x^2)$, and this results in $x^8+x^7+x^5+x^3$. As the degree of the polynomial is larger than 7, we have to reduce it modulo the irreducible polynomial $f(x) = x^8+x^4+x^3+x+1$. The respective division

$$(x^8 + x^7 + x^5 + x^3) : (x^8 + x^4 + x^3 + x + 1)$$

yields 1 and the remaining polynomial is $(x^7 + x^5 + x^4 + x + 1)$. This polynomial refers to the byte 1011 0011.

The second term (0x03 · 0xbf) refers to 00000011 · 10111111 or

$$(x + 1) \cdot (x^7 + x^5 + x^4 + x^3 + x^2 + x + 1).$$

This multiplication results in $x^8 + x^7 + x^6 + 1$. Again, the degree of the polynomial is larger than 7, so we reduce it modulo $x^8+x^4+x^3+x+1$. The respective division

$$(x^8 + x^7 + x^6 + 1) : (x^8 + x^4 + x^3 + x + 1)$$

yields 1, and the remaining polynomial is $(x^7 + x^6 + x^4 + x^3 + x)$ that refers to the byte 11011010.

The third and fourth terms both comprise a factor (0x01, so these terms refer to the original values 0x5d (representing 01011101) and 0x30 (representing 00110000). Finally, we add the four bytes

10110011
11011010
01011101
00110000

modulo 2 and get 0000 0100. So the resulting value for $s'_{0,c}$ is 0x04. The other values of the column vector, $s'_{1,c}$, $s'_{2,c}$, and $s'_{3,c}$, are computed similarly.

*AddRoundKey() Transformation*

In the AddRoundKey() transformation, a word of the key schedule $w$ is added modulo 2 to each column of the State. This means that

$$[s'_{0,c}, s'_{1,c}, s'_{2,c}, s'_{3,c}] = [s_{0,c}, s_{1,c}, s_{2,c}, s_{3,c}] \oplus w[rN_b + c]$$

for $0 \leq c < N_b$ and $0 \leq r \leq N_r$. Because the AddRoundKey() transformation only consists of a bitwise addition modulo 2, it represents its own inverse.

#### 9.6.2.4 Key Expansion Algorithm

The AES key expansion algorithm takes a secret key $k$ and generates a key schedule $w$ that is employed by the AddRoundKey() transformation mentioned above. The key $k$ comprises $4N_k$ bytes or $32N_k$ bits. In the byte-wise representation, $k_i$ refers to the i-th byte of $k$ ($0 \leq i < 4N_k$). The key schedule $w$ to be generated is a linear array of $N_b(N_r + 1)$ 4-byte words, meaning that it is $N_b(N_r + 1)$ words long (the algorithm requires an initial set of $N_b$ words, and each of the $N_r$ rounds requires $N_b$ additional words of key data). Again, we use $w[i]$ for $0 \leq i < N_b(N_r + 1)$ to refer to the i-th word in this array.

**Algorithm 9.6** The AES key expansion algorithm.

$(k)$

---

for $i = 0$ to $(N_k - 1)$ do
  $w[i] = [k_{4i}, k_{4i+1}, k_{4i+2}, k_{4i+3}]$
for $i = N_k$ to $(N_b(N_r + 1) - 1)$ do
  $t = w[i - 1]$
  if $(i \bmod N_k = 0)$
    then $t =$ SubWord(RotWord($t$)) $\oplus$ RCon$[i/N_k]$
    else if $(N_k > 6$ and $i \bmod N_k = 4)$
      then $t =$ SubWord($t$)
  $w[i] = w[i - N_k] \oplus t$

---

$(w)$

The AES key expansion algorithm is summarized in Algorithm 9.6 (we assume that $N_k$ is included in $k$, so we don't have to consider $N_k$ as an additional parameter). The algorithm employs a round constant word array RCon$[i]$ ($1 \leq i \leq$

$N_r$). The array consists of the values $[x^{i-1}, 0x00, 0x00, 0x00]$, with $x^{i-1}$ being an appropriate power of $x$ (computed in the field $\mathbb{F}_{2^8}$ using the reduction polynomial $f(x) = x^8 + x^4 + x^3 + x + 1$). The first 10 values of the Rcon array are summarized in Table 9.11.

**Table 9.11**
The Round Constant Word Array Rcon

| $i$ | $x^{i-1}$ | Rconf[$i$] |
|---|---|---|
| 1 | $x^{1-1} = x^0 = 1$ | 0x01000000 |
| 2 | $x^{2-1} = x^1$ | 0x02000000 |
| 3 | $x^{3-1} = x^2$ | 0x04000000 |
| 4 | $x^{4-1} = x^3$ | 0x08000000 |
| 5 | $x^{5-1} = x^4$ | 0x10000000 |
| 6 | $x^{6-1} = x^5$ | 0x20000000 |
| 7 | $x^{7-1} = x^6$ | 0x40000000 |
| 8 | $x^{8-1} = x^7$ | 0x80000000 |
| 9 | $x^{9-1} = x^8$ | 0x1b000000 |
| 10 | $x^{10-1} = x^9$ | 0x36000000 |

In addition to RCon, the algorithm also employs two auxiliary functions:

- SubWord() takes a 4-byte input word and applies the S-box of the SubBytes() transformation to each of the 4 bytes to produce an output word.
- RotWord() takes a 4-byte input word and performs a cyclic shift left (i.e., if the input word is $[a_0, a_1, a_2, a_3]$, then the output word is $[a_1, a_2, a_3, a_0]$).

The AES key expansion algorithm then works as follows: In a first loop, the first $N_k$ words of the key schedule are filled with the first $4 \cdot N_k$ bytes of the original key $k$. The rest of the key schedule is then filled in a second loop. In this loop, every word $w[i]$ is set to the sum modulo 2 of the previous word (i.e., $w[i-1]$), and the word that is located $N_k$ positions earlier, $w[i - N_k]$. For words in positions that are a multiple of $N_k$, a transformation is applied to $[w_{i-1}]$ prior to the addition modulo 2, followed by an addition modulo 2 with the round constant word RCon$[i]$. This transformation basically consists of a cyclic shift of the bytes in a word (i.e., RotWord()), followed by the application of a table lookup to all 4 bytes of the word; that is SubWord().

Finally, we note that the AES key expansion algorithm for $N_k = 8$ is slightly different than the one for $N_k = 6$ (i.e., AES-192) and $N_k = 4$ (i.e., AES-128). If $N_k = 8$ and $i - 4$ is a multiple of $N_k$, then SubWord() is applied to $w[i-1]$ prior to the addition modulo 2 operation.

#### 9.6.2.5 Decryption Algorithm

The transformations used by the AES encryption algorithm can be inverted and implemented in reverse order to produce a straightforward AES decryption algorithm. The inverse transformations used in the AES decryption algorithm are called InvSubBytes(), InvShiftRows(), and InvMixColumns(). As mentioned earlier, the AddRoundKey() transformation is its own inverse (as it only involves a bitwise addition modulo 2). Hence, the AES decryption algorithm is formally expressed in Algorithm 9.7. Its inverse transformations are briefly addressed next.

**Algorithm 9.7** The AES decryption algorithm.

$(in)$

---

$s = in$
$s =$ AddRoundKey($s$, $w[N_r N_b, (N_r+1)N_b - 1]$)
for $r = N_r - 1$ downto 1 do
  $s =$ AddRoundKey($s$, $w[rN_b, (r+1)N_b - 1]$)
  $s =$ InvMixColumns($s$)
  $s =$ InvShiftRows($s$)
  $s =$ InvSubBytes($s$)
$s =$ AddRoundKey($s$, $w[0, N_b - 1]$)
$s =$ InvShiftRows($s$)
$s =$ InvSubBytes($s$)
$out = s$

---

$(out)$

*InvMixColumns() Transformation*

As it name suggests, the InvMixColumns() transformation is the inverse of the MixColumns() transformation. Again, it operates on the State column by column, treating each column as a four-term polynomial $s(x)$ over $\mathbb{F}_{2^8}$. More specifically, the polynomial $s(x)$ is multiplied modulo $x^4 + 1$ with the polynomial

$$c^{-1}(x) = c_3'x^3 + c_2'x^2 + c_1'x + c_0'$$

where $c_3'$ =0x0b, $c_2'$ =0x0d, $c_1'$ =0x09, and $c_0'$ =0x0e. Note that the polynomial $x^4 + 1$ is the same as the one used in the MixColumns() transformation.

The InvMixColumns() transformation maps $s(x)$ to the following polynomial:

$$(c^{-1}(x) \cdot s(x)) \bmod (x^4 + 1)$$

This mapping can also be expressed as

$$\begin{pmatrix} s'_{0,c} \\ s'_{1,c} \\ s'_{2,c} \\ s'_{3,c} \end{pmatrix} = \begin{pmatrix} \text{0x0e} & \text{0x0b} & \text{0x0d} & \text{0x09} \\ \text{0x09} & \text{0x0e} & \text{0x0b} & \text{0x0d} \\ \text{0x0d} & \text{0x09} & \text{0x0e} & \text{0x0b} \\ \text{0x0b} & \text{0x0d} & \text{0x09} & \text{0x0e} \end{pmatrix} \cdot \begin{pmatrix} s_{0,c} \\ s_{1,c} \\ s_{2,c} \\ s_{3,c} \end{pmatrix}$$

or

$$\begin{aligned} s'_{0,c} &= (\text{0x0e} \cdot s_{0,c}) \oplus (\text{0x0b} \cdot s_{1,c}) \oplus (\text{0x0d} \cdot s_{2,c}) \oplus (\text{0x09} \cdot s_{3,c}) \\ s'_{1,c} &= (\text{0x09} \cdot s_{0,c}) \oplus (\text{0x0e} \cdot s_{1,c}) \oplus (\text{0x0b} \cdot s_{2,c}) \oplus (\text{0x0d} \cdot s_{3,c}) \\ s'_{2,c} &= (\text{0x0d} \cdot s_{0,c}) \oplus (\text{0x09} \cdot s_{1,c}) \oplus (\text{0x0e} \cdot s_{2,c}) \oplus (\text{0x0b} \cdot s_{3,c}) \\ s'_{3,c} &= (\text{0x0b} \cdot s_{0,c}) \oplus (\text{0x0d} \cdot s_{1,c}) \oplus (\text{0x09} \cdot s_{2,c}) \oplus (\text{0x0e} \cdot s_{3,c}) \end{aligned}$$

for $0 \leq c < 4$.

*InvShiftRows() Transformation*

The InvShiftRows() transformation is the inverse of the ShiftRows() transformation. As such, the bytes in the last three rows of the State are cyclically shifted right over $shift(r, N_b)$ bytes. Note that the $shift$ function is the same for both the encryption and decryption algorithms. Analog to (9.9), the InvShiftRows() transformation can be specified as

$$s'_{r,(c+shift(r,N_b)) \bmod N_b} = s_{r,c}$$

for $0 \leq r < 4$ and $0 \leq c < N_b = 4$. This means that the byte at position $s_{r,c}$ is shifted to the byte at position $s'_{r,(c+shift(r,N_b)) \bmod N_b}$. The resulting transformation is illustrated in Figure 9.14.

*InvSubBytes() Transformation*

The InvSubBytes() transformation is the inverse of the SubBytes() transformation. Remember that the SubBytes() transformation results from taking the multiplicative inverse of each element of $\mathbb{F}_{2^8}$ and subjecting the result to an affine transformation. These two operations are inverted in the InvSubBytes() transformation to construct the inverse S-box as illustrated in Table 9.12: First each element in $\mathbb{F}_{2^8}$ is subjected to the affine transformation that is inverse to the one used in the SubBytes() transformation, and then the multiplicative inverse element is computed in $\mathbb{F}_{2^8}$.

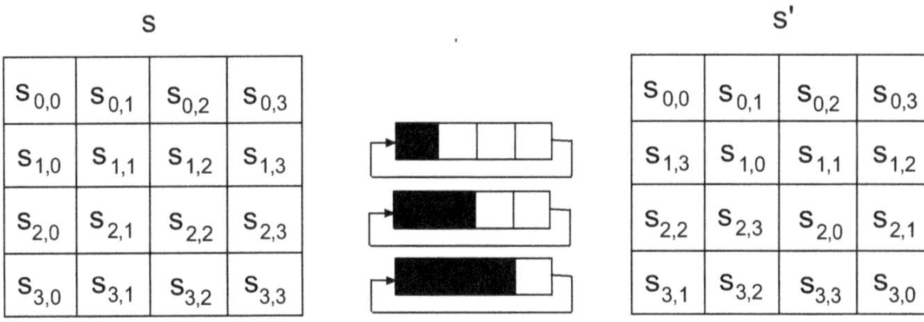


**Figure 9.14** The InvShiftRows() transformation of the AES decryption algorithm.

Using Tables 9.10 and 9.12, it is simple to verify that the two S-boxes are inverse to each other. For example, the S-box of Table 9.10 maps the element 0xa3 to 0x0a, whereas the inverse S-box of Table 9.12 maps the element 0x0a back to 0xa3.

**Table 9.12**
The Inverse S-Box of the AES Decryption Algorithm

| | 0 | 1 | 2 | 3 | 4 | 5 | 6 | 7 | 8 | 9 | A | B | C | D | E | F |
|---|---|---|---|---|---|---|---|---|---|---|---|---|---|---|---|---|
| 0 | 52 | 09 | 6A | D5 | 30 | 36 | A5 | 38 | BF | 40 | A3 | 9E | 81 | F3 | D7 | FB |
| 1 | 7C | E3 | 39 | 82 | 9B | 2F | FF | 87 | 34 | 8E | 43 | 44 | C4 | DE | E9 | CB |
| 2 | 54 | 7B | 94 | 32 | A6 | C2 | 23 | 3D | EE | 4C | 95 | 0B | 42 | FA | C3 | 4E |
| 3 | 08 | 2E | A1 | 66 | 28 | D9 | 24 | B2 | 76 | 5B | A2 | 49 | 6D | 8B | D1 | 25 |
| 4 | 72 | F8 | F6 | 64 | 86 | 68 | 98 | 16 | D4 | A4 | 5C | CC | 5D | 65 | B6 | 92 |
| 5 | 6C | 70 | 48 | 50 | FD | ED | B9 | DA | 5E | 15 | 46 | 57 | A7 | 8D | 9D | 84 |
| 6 | 90 | D8 | AB | 00 | 8C | BC | D3 | 0A | F7 | E4 | 58 | 05 | B8 | B3 | 45 | 06 |
| 7 | D0 | 2C | 1E | 8F | CA | 3F | 0F | 02 | C1 | AF | BD | 03 | 01 | 13 | 8A | 6B |
| 8 | 3A | 91 | 11 | 41 | 4F | 67 | DC | EA | 97 | F2 | CF | CE | F0 | B4 | E6 | 73 |
| 9 | 96 | AC | 74 | 22 | E7 | AD | 35 | 85 | E2 | F9 | 37 | E8 | 1C | 75 | DF | 6E |
| A | 47 | F1 | 1A | 71 | 1D | 29 | C5 | 89 | 6F | B7 | 62 | 0E | AA | 18 | BE | 1B |
| b | FC | 56 | 3E | 4B | C6 | D2 | 79 | 20 | 9A | DB | C0 | FE | 78 | CD | 5A | F4 |
| C | 1F | DD | A8 | 33 | 88 | 07 | C7 | 31 | B1 | 12 | 10 | 59 | 27 | 80 | EC | 5F |
| D | 60 | 51 | 7F | A9 | 19 | B5 | 4A | 0D | 2D | E5 | 7A | 9F | 93 | C9 | 9C | EF |
| E | A0 | E0 | 3B | 4D | AE | 2A | F5 | B0 | C8 | EB | BB | 3C | 83 | 53 | 99 | 61 |
| F | 17 | 2B | 04 | 7E | BA | 77 | D6 | 26 | E1 | 69 | 14 | 63 | 55 | 21 | 0C | 7D |

This finishes our brief exposition of the AES encryption and decryption algorithms, and we are now ready to discuss some results from its security analysis.

### 9.6.2.6 Security Analysis

In 2003, the NSA approved the AES to become a legitimate cipher to encrypt classified information:

> "The design and strength of all key lengths of the AES algorithm (i.e., 128, 192 and 256) are sufficient to protect classified information up to

> the SECRET level. TOP SECRET information will require use of either the 192 or 256 key lengths."

In consideration of the fact that the implementation represents the Achilles' heel of any cryptographic system, the NSA further required that

> "The implementation of AES in products intended to protect national security systems and/or information must be reviewed and certified by NSA prior to their acquisition and use."

This brief announcement of the NSA is remarkable because it marks the first time that the public has access to a cipher approved by NSA for the encryption of information classified as TOP SECRET (at least, if the longer-key versions of the AES are used).

Unfortunately, we don't know what the NSA currently knows about the security of the AES. Outside the NSA, however, the AES has been subject to a lot of public scrutiny—especially since its standardization. There is good and bad news:

- The good news is that the encryption system is designed in a way that lower bounds for the complexity of differential cryptanalysis, linear cryptanalysis, and some related attacks can be provided. Hence, it is not susceptible to these types of attacks.
- The bad news is that new cryptanalytical techniques have been developed (and will probably continue to be developed) that may eventually break the AES in the future. From a cryptanalyst's viewpoint, the AES is an attractive target, mainly because it is widely deployed in the field, and—maybe also—because it has a rather simple mathematical structure that may be exploitable in some yet-to-be-discovered ways.

The most common strategy to attack a block cipher is to first define a reduced-round version of it, and then trying to attack it under some specific assumptions. If, for example, an adversary can observe the operation of the cipher under different but mathematically related keys, then he or she can mount a *related-key attack*. A mathematical relationship may be that the last bits of the keys are the same, or something similar. Note that related-key attacks are theoretically interesting to explore, but that it is not generally known how to actually mount them in the field on a sufficiently large scale. In 2009, a few results were published that elaborate on the resistance of the AES against such attacks. It was shown, for example, that related-key attacks against AES-256 (AES-192) are feasible with a time complexity of $2^{119}$ ($2^{176}$). AES-128, in contrast, remains unaffected by such attacks.

If one reduces the number of rounds, then some real attacks become feasible. With a complexity of $2^{70}$, for example, one can break AES-256 up to 11 rounds. Because the full-round-version of AES-256 employs 14 rounds, there is still a security margin. This margin, however, is not particularly large, and hence people sometimes argue that future versions of the standard should also increase the number of rounds (to make the cipher more resistant against related-key attacks). This is a perfectly valid argument, and it is possible and very likely that the respective standardization bodies will take this argument into account when updating the AES specification in the future.

From a purely puristic view, a (computationally secure) encryption system is said to be *broken* if there is an attack that is more efficient than an exhaustive key search. If the efficiency gain is small, then it is very likely that the attack is infeasible to mount and hence not relevant in practice. In theory, however, the system can still be called broken. In 2011, for example, a cryptanalytical attack technique known as *biclique cryptanalysis* made press headlines because it referred to such a theoretical breakthrough. Using biclique cryptanalysis, it is possible to break the full-round AES-128 with a computational complexity of $2^{126.1}$, AES-192 with a computational complexity of $2^{189.7}$, and AES-256 with a computational complexity of $2^{254.4}$ with small memory requirements in either case. The efficiency gains against exhaustive key search are marginal, but they are still sufficient to claim that the AES has been broken (at least in theory). In practice, however, neither related-key attacks nor biclique cryptanalysis represent a serious concern for the security of AES.

The security of the AES remains an important and timely research topic. The situation is comparable to DES: Since its standardization, many people have tried to attack and eventually break the encryption system. The more time passes while nobody is finding a serious or devastating attack, the more people gain confidence in the actual security of the system. This is where we are regarding the security of the AES. Even after two decades of cryptanalytical research, nobody has found a vulnerability that can be turned into a practical attack against the AES. This has made us confident about the security of the AES.

## 9.7 MODES OF OPERATION

As its name suggests, a block cipher operates on blocks of relatively small and fixed size (typical values are 64 bits for the DES and 128 bits for the AES). Since messages may be arbitrarily long, it must be defined how a block cipher can be used to encrypt long messages. This is where modes of operation come into play. In fact, multiple modes of operation have been defined for block ciphers, where some of them actually turn a block cipher into a stream cipher (needless to say, this further obfuscates the differences between block and stream ciphers).

Historically, the most important document was FIPS PUB 81 [34] that was published by NIST in 1980, shortly after the standardization of the DES. The document specifies four modes of operation for DES, namely the block cipher modes *electronic code book* (ECB) and *cipherblock chaining* (CBC) and the stream cipher modes *output feedback* (OFB) and *cipher feedback* (CFB). In 2001, NIST released an updated document [35] that not only specifies ECB, CBC, OFB, and CFB, but also a *counter* (CTR) mode that—similar to OFB mode—turns a block cipher into a synchronous stream cipher.[49] All five modes are confidentiality modes, meaning that they protect (only) the confidentiality of messages.[50] They neither provide message authenticity nor integrity. As mentioned before, cryptographers have therefore come up with modes of operation that provide support for AE and AEAD. These modes are important and further addressed in Section 11.2.

Due to their importance in practice, NIST has recently reactivated its standardization effort for block cipher modes. In an addendum to [35],[51] for example, it has specified three variants of CBC that use a technique known as ciphertext stealing to avoid message expansion due to padding.[52] Furthermore, it has complemented [35] with the specifications of several modes of operation that are all part of the Special Publication 800-38 series:

- An authentication-only mode known as CMAC [36] (Section 10.3.1.2);
- Two AEAD modes known as counter with CBC-MAC (CCM) [37, 38] and Galois/counter mode (GCM) [39];
- Another confidentiality mode known as XTS-AES that is specifically crafted for block-oriented storage devices [40];
- A few methods for key wrapping [41] (in which cryptographic keys are protected instead of "normal" messages);

49 In some literature, the CTR mode is also known as the *integer counter mode* (ICM) or the *segmented integer counter* (SIC) mode. These terms are not used in this book.

50 In addition to [34, 35], the ANSI specified seven modes of operation for the 3DES (also known as TDEA) in X9.52-1998: The TDEA ECB (TECB), TDEA CBC (TCBC), TDEA CFB (TCFB), and TDEA OFB (TOFB) modes refer to the use of TDEA in ECB, CBC, CFB, and OFB mode. Furthermore, there are two interleaved versions of TCBC and TOFB—TDEA CBC Interleaved (TCBC-I), and one pipelined version of TCFB—TDEA CFB Pipelined (TCFB-P). Because these modes are not used in the field, they are not further addressed here (they are mentioned here only for the sake of completeness).

51 http://nvlpubs.nist.gov/nistpubs/Legacy/SP/nistspecialpublication800-38a-add.pdf

52 In some literature, such as RFCs 2040 and 3962, the respective CBC variant is called *CTS mode*, where CTS stands for "ciphertext stealing."

- A few methods for format-preserving encryption [42].[53]

In the rest of this section, we focus on the five confidentiality modes ECB, CBC, OFB, CFB, and CTR. Some modes require padding to align the plaintext message length with the block length of the block cipher, and most modes (except ECB) also require an additional input block that acts as a dummy block to kick off the encryption and decryption processes and provide some randomization for the encryption function. Such a block is usually called an *initialization vector* (IV) or a *nonce* (i.e., a random number used only once). In general, an IV (nonce) does not need to be secret, but it almost always has to unpredictable. Many cryptanalytical attacks are feasible if the adversary can somehow predict the next-to-be-used IV (nonce). In fact, the proper use of IVs (nonces) is tricky and has become a research topic of its own.

### 9.7.1 ECB

The ECB mode is the simplest and most straightforward mode of operation for a block cipher. Its working principles are illustrated in Figures 9.15 (for encryption) and 9.16 (for decryption). An arbitrarily long plaintext message $m \in \mathcal{M}$ is split into $t$ $n$-bit blocks, where $n$ represents the block length of the block cipher, and each block is then encrypted and decrypted individually. To encrypt message block $m_i$ $(1 \leq i \leq t)$, it is subject to the encryption function $E_k$; that is, $c_i = E_k(m_i)$, and to decrypt $c_i$ $(1 \leq i \leq t)$, it is subject to the respective decryption function $D_k$; that is, $m_i = D_k(c_i)$. In either case, the same key $k$ is used.

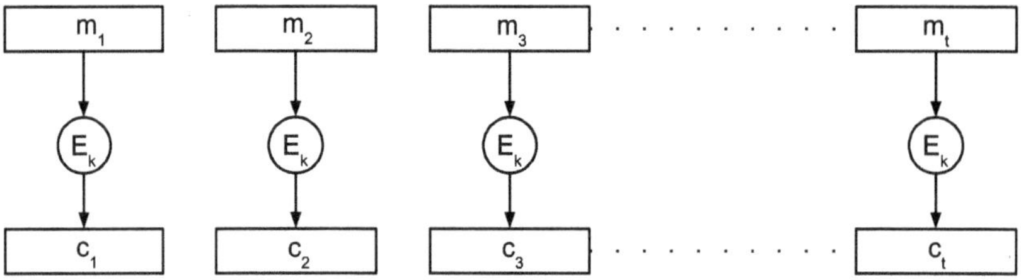


**Figure 9.15** ECB encryption.

53 In a format-preserving encryption, the output (i.e., the ciphertext) is in the same format as the input (i.e., the plaintext message), whereas the meaning of "format" varies (e.g., a 16-digit credit card number).

If the message length is not a multiple of $n$ bits (i.e., $|m| \neq k \cdot n$ for some $k \in \mathbb{N}$), then the message may need to be padded first (using either a fixed bit pattern or some random bits). This is not further addressed here.

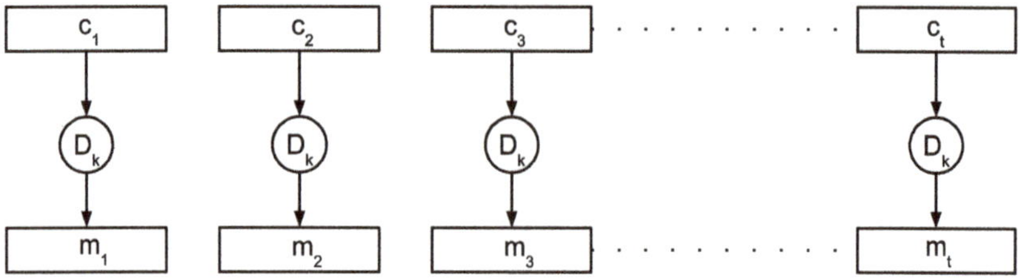


**Figure 9.16** ECB decryption.

The ECB mode has distinct properties that are advantageous or disadvantageous in a particular application setting:

- It is simple and straightforward.
- It supports parallelism (i.e., multiple blocks may be encrypted and decrypted in parallel).
- It neither provides message expansion (beyond padding) nor error propagation.
- It does not protect the sequential ordering of the message blocks, meaning that an adversary can delete or reorder particular blocks in a multiblock message. This goes hand in hand with the fact that ECB is "only" a confidentiality mode.
- As its name suggests, the mode implements a code book for plaintext message blocks, meaning that identical plaintext message blocks are always mapped to the same ciphertext blocks. This also means that—depending on the redundancy of the data—multiblock ciphertexts may reveal statistical information about the underlying plaintext messages.

Whether the last disadvantage is severe or not depends on the application in use. If a block cipher is operated in ECB mode, for example, and the plaintext message refers to a large and highly redundant image file, then large portions of this file comprise similar blocks. Using one particular key, these blocks are mapped to the same ciphertext blocks, and hence the structural properties of the image file leak. In fact, this type of statistical information is what cryptanalysts are looking for and what they usually try to exploit. This is why the ECB mode is not recommended and not used in the field. Instead, people go for modes that are more sophisticated and less susceptible to cryptanalysis.

### 9.7.2 CBC

The CBC mode of operation is often used to overcome the most important disadvantages of the ECB mode. In this mode, the encryption of a plaintext message block $m_i$ not only depends on $m_i$ and $k$, but also on the previous ciphertext block $c_{i-1}$ (or the IV in case of $m_1$). This means that the ciphertext blocks are cryptographically chained, and hence that one particular ciphertext block always depends on all previous blocks. Unless the IV and all previous blocks are the same, identical plaintext message blocks are thus mapped to different ciphertext blocks. This makes cryptanalysis more difficult. Note that the use of an IV turns the encryption function into a probabilistic one. Also note that the IV need not be kept secret, but it must be unpredictable and its integrity needs to be protected.

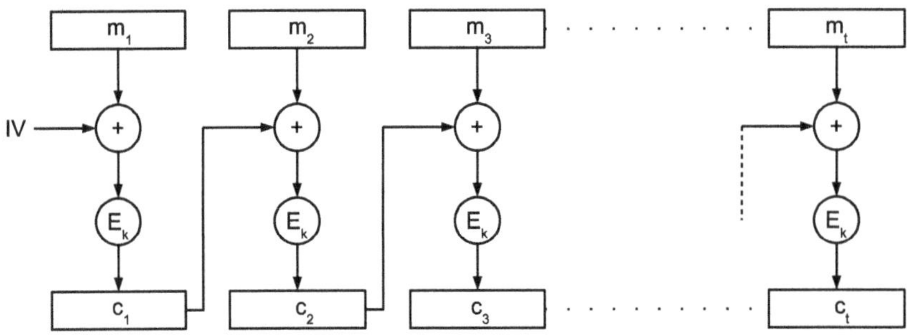


**Figure 9.17** CBC encryption.

The working principles of CBC encryption and decryption are illustrated in Figure 9.17 (for encryption) and 9.18 (for decryption). Let $m = m_1 \parallel m_2 \parallel \ldots \parallel m_t$ be a $t$-block plaintext message, where $m_t$ may be padded to be in line with the block length $n$. If one wanted to avoid padding and respective message expansion, then one would use one of the CBC variants with ciphertext stealing (Section 9.7.2.1). To encrypt $m$, $c_0$ is first initialized with an IV. All plaintext message blocks $m_i$ for $i = 1, \ldots, t$ are then added modulo 2 to $c_{i-1}$ (i.e., the previous ciphertext block), and the sum is subject to the encryption function $E_k$

to generate $c_i$:

$$\begin{aligned} c_0 &= IV \\ c_i &= E_k(m_i \oplus c_{i-1}) \text{ for } 1 \leq i \leq t \end{aligned}$$

Due to the IV, the ciphertext $c = c_1 \parallel c_2 \parallel \ldots \parallel c_t$ is one block longer than the plaintext message $m$. This means that the CBC mode is not length-preserving, and that it has a message expansion of one block (in addition to padding). The resulting $t+1$ ciphertext blocks $c_0, \ldots, c_t$ are transmitted to the decrypting device. If the decrypting device is initialized with the same IV using some out-of-band mechanism, then $c_0$ doesn't have to be transmitted. In this case, the encryption is length-preserving.

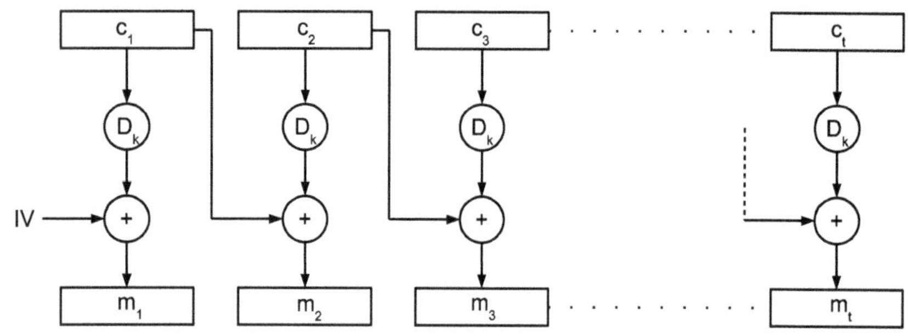


**Figure 9.18** CBC decryption.

On the receiving side, the decrypting device uses the IV to initialize $c_0$. Each $c_i$ ($i = 1, \ldots, t$) is decrypted with $k$, and the result is added modulo 2 to $c_{i-1}$:

$$\begin{aligned} c_0 &= IV \\ m_i &= D_k(c_i) \oplus c_{i-1} \text{ for } 1 \leq i \leq t \end{aligned}$$

The correctness of the decryption can be verified as follows:

$$\begin{aligned} D_k(c_i) \oplus c_{i-1} &= D_k(E_k(m_i \oplus c_{i-1})) \oplus c_{i-1} \\ &= m_i \oplus c_{i-1} \oplus c_{i-1} \\ &= m_i \end{aligned}$$

Again, the CBC mode has distinct properties that are advantageous or disadvantageous in a particular application setting:

- It is less simple and straightforward than ECB.
- It has a chaining structure and does not support parallelism.
- Due to the IV, it has a message expansion of one block (beyond padding).[54]
- Due to the chaining structure, it protects the sequential ordering of the message blocks.
- Also due to the chaining structure, some errors are propagated in this mode. If, for example, $c_i$ is transmitted with an error, then only $c_i$ and $c_{i+1}$ decrypt incorrectly, and all other ciphertext blocks (i.e., $c_1, \ldots, c_{i-1}, c_{i+2}, \ldots, c_t$) still decrypt correctly (unless there are other transmission errors).

The last item—the fact that an incorrectly transmitted ciphertext block only affects two blocks—has a useful side effect: If two entities don't share a common IV but start with different IVs, then this difference only affects the first two ciphertext blocks. From the third block onward, the difference doesn't matter anymore. This self-synchronizing property is sometimes exploited in the field.

For the sake of completeness, we address two variants of the "normal" CBC mode, namely CBC with *ciphertext stealing* (as already mentioned above) and a mode in which an error is propagated through the entire message. This mode is called *propagating CBC* (PCBC) mode.

#### 9.7.2.1 Ciphertext Stealing

Again, the purpose of ciphertext stealing is to avoid padding if the plaintext message length is not a multiple of the block length $n$, and hence to avoid a situation in which the ciphertext is longer than the plaintext message. Let $d$ be the bit length of the last plaintext message block $m_t$ (i.e., $d = |m_t|$) so that $n - d$ bits are missing in the last plaintext block to fill an entire block.

There are multiple ways to implement ciphertext stealing, and three similar variants of "normal" CBC mode are specified in the addendum mentioned above. The first variant of CBC encryption with ciphertext stealing, CBC-CS1, is illustrated

54 In some implementations, people have tried to find a shortcut here. In SSL 3.0 and TLS 1.0, for example, the last block of the previous ciphertext record was used as an IV for the encryption of the next record. This is not in line with the normal specification of CBC mode, but it saves one block. Unfortunately, this deviation from the CBC mode specification was subject to a devastating attack called *browser exploit against SSL/TLS* (BEAST) that could only be mitigated by making the IV explicit in the specification of TLS 1.1. Since this TLS version, the IV for the encryption of the next TLS record needs to be transmitted explicitly.

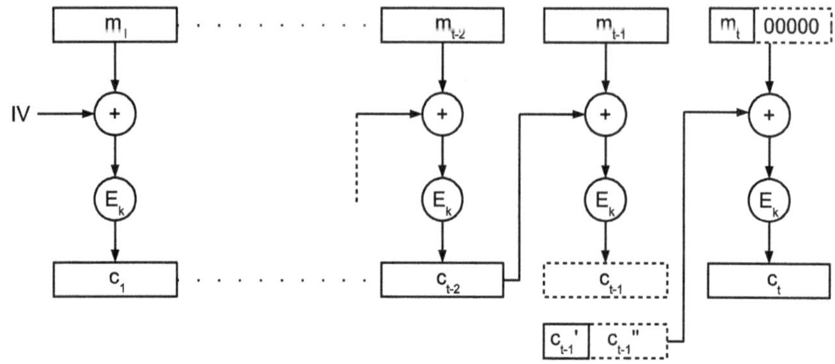


**Figure 9.19** CBC encryption with ciphertext stealing (CBC-CS1 variant).

in Figure 9.19.[55] It deviates from "normal" CBC (Figure 9.17) in the last two blocks. As usual, the last but one block $m_{t-1}$ is added modulo 2 to $c_{t-2}$, before the sum is subject to the encryption function $E_k$. The result of this encryption is $c_{t-1} = E_k(m_{t-1} \oplus c_{t-2})$, but this block is not used directly. Instead the most significant $d$ bits of $c_{t-1}$ yield $c'_{t-1}$, and the $n-d$ least significant bits of $c_{t-1}$ yield $c''_{t-1}$. While $c'_{t-1}$ is going to be part of the ciphertext $c = c_1, \ldots, c'_{t-1}, c_t$, $c'_{t-1} \parallel c''_{t-1}$ is added modulo 2 to $m_t \parallel 00\ldots0$ (where $m_t$ is padded with $n-d$ zeros to fill a block), and the sum is again subject to the encryption function $E_k$. The result of this encryption yields the final ciphertext block $c_t$:

$$c_t = E_k(m_t \parallel 00\ldots0 \oplus c'_{t-1} \parallel c''_{t-1})$$

In the end, the ciphertext $c$ consists of the blocks $c_1, c_2, \ldots, c_{t-2}, c'_{t-1}$, and $c_t$ (in this order). All blocks (except $c'_{t-1}$) are $n$ bits long, whereas $c'_{t-1}$ is only $d$ bits long. This means that $c$ is equally long as $m$, and hence there is no message expansion. The $n-d$ bits of $c''_{t-1}$ need not be part of the ciphertext because they are recovered from $c_t$ during the decryption process.

CBC decryption with ciphertext stealing turns out to be more involved. It is illustrated in Figure 9.20. It starts with the ciphertext blocks $c_1, \ldots, c_{t-2}, c'_{t-1}$, and

55 The two other variants of CBC encryption with ciphertext stealing, CBC-CS2 and CBC-CS3, work similarly but have some subtle differences regarding the order of the blocks that form the ciphertext. CBC-CS2 is used in some other literature, whereas CBC-CS3 is used, for example, in the Kerberos authentication and key distribution system.

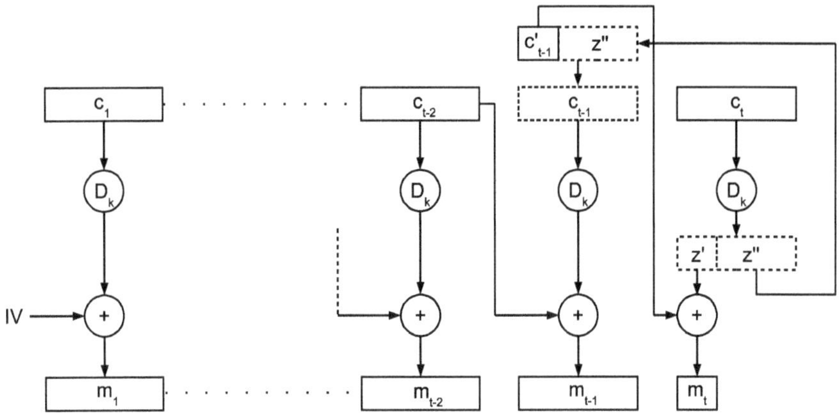


**Figure 9.20** CBC decryption with ciphertext stealing (CBC-CS1 variant).

$c_t$ (where $c'_{t-1}$ is only $d$ bits long), and recovers $m_1, \ldots, m_{t-1}, m_t$ (where $m_t$ is only $d$ bits long). CBC decryption works as usual until $c_{t-2}$ is processed and $m_{t-2}$ is decrypted accordingly. At this point in time, decryption jumps to $c_t$ and decrypts this block with $D_k$. The result is one block of plaintext $z' \parallel z''$, where $z'$ refers to the most significant $d$ bits and $z''$ refers to the least significant $n-d$ bits. Taking into account how $c_t$ is generated, $z'$ stands for $m_t \oplus c'_{t-1}$ and $z''$ stands for $00\ldots0 \oplus c''_{t-1}$. This means that $m_t$ can be recovered as $z' \oplus c'_{t-1}$. The way to recover $m_{t-1}$ is more involved: The next-to-last ciphertext block (i.e., $c'_{t-1}$ that is $d$ bits long) is concatenated with $z''$ that is $n-d$ bits long to form the ciphertext block $c_{t-1}$, and this block is then decrypted as

$$m_{t-1} = D_k(c_{t-1}) \oplus c_{t-2}$$

The resulting plaintext message $m$ consists of the $t$ blocks $m_1, m_2, \ldots, m_t$, where $m_t$ is only $d$ bits long. Again, the ciphertext stealing variants 2 and 3 (i.e., CBC-CS2 and CBC-CS3) are very similar but different in some subtle details (not addressed here).

#### 9.7.2.2 PCBC

Similar to *infinite garble extension* (IGE) as used, for example, in the Telegram E2EE messaging app [44], the design goal of PCBC was to provide an encryption mode that propagates a single bit error in a ciphertext to the entire decryption process (so that the resulting plaintext message can be rejected automatically). Remember that "normal" CBC mode only garbles two blocks of the plaintext message, whereas the prupose of IGE and PCBC is to garble all of them. The PCBC mode was originally designed for and used in Kerberos version 4. Since version 5, however, Kerberos no longer supports it, and hence it has silently sunk into oblivion (Kerberos version 5 still supports ciphertext stealing though).

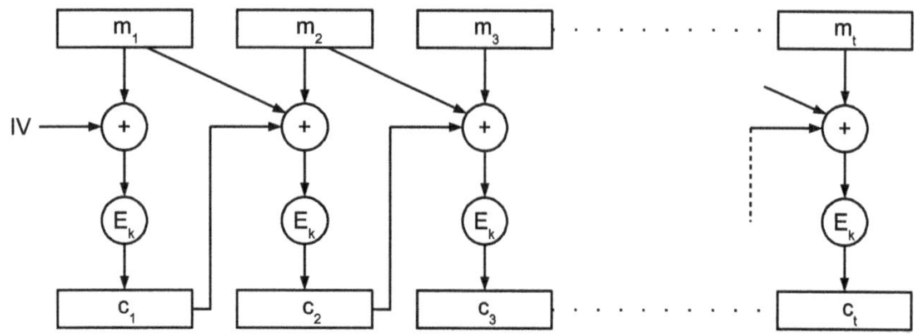


**Figure 9.21** PCBC encryption.

The PCBC encryption function slightly deviates from the "normal" CBC encryption function, because it adds not only $c_{i-1}$ modulo 2 to $m_i$ but also $m_{i-1}$ (as illustrated in Figure 9.21). The function is recursively defined as follows:

$$\begin{aligned} c_0 &= IV \\ c_i &= E_k(m_i \oplus m_{i-1} \oplus c_{i-1}) \text{ for } 1 \leq i \leq t \end{aligned}$$

Similarly, the PCBC decryption function (as illustrated in Figure 9.22) works as follows:

$$\begin{aligned} c_0 &= IV \\ m_i &= D_k(c_i) \oplus c_{i-1} \oplus m_{i-1} \text{ for } 1 \leq i \leq t \end{aligned}$$

When a block cipher is used as a true block cipher, then the CBC mode (or a variant thereof) is still in widespread use. This is about to change, because people move to AEAD modes that are the preferred choice today. If such a cipher with block length $n$ is used in CBC mode, then it is recommended to rekey the cipher

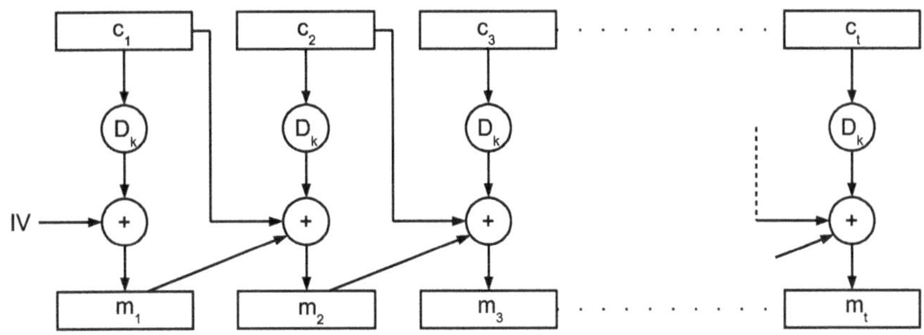


**Figure 9.22** PCBC decryption.

after at most $2^{n/2}$ blocks. This is due to the birthday paradox that suggests that the probability of finding a colliding block (i.e., a block that has already been encrypted with the same key) increases with the number of already encrypted blocks. This allows an adversary to mount specific attacks, such as Sweet32[56] against 3DES (with a block length of 64 bits) that was used in former versions of the SSL/TLS protocols. Consequently, the block length matters and this is why one usually requires modern block ciphers to have a block length of at least 128 bits.

Let us now have an informal look at the modes of operation that turn a block cipher into a stream cipher: CFB, OFB, and CTR (in this order). Having a stream cipher is particularly important for applications that don't require the transmission of large messages, such as terminal protocols that are mostly character-oriented. In any of these modes, the sending and receiving devices only use the encryption function of the underlying block cipher. This means that these modes of operation can also be used if the encryption function is replaced with a one-way function (this may be important if symmetric encryption systems are not available or their use is restricted in one way or another).

### 9.7.3 CFB

The CFB mode turns a block cipher into a stream cipher. More specifically, it uses the block cipher to generate a sequence of pseudorandom bits, and it then adds these bits modulo 2 to the plaintext bits to generate the ciphertext. The resulting stream cipher turns out be be self-synchronizing.

56 https://sweet32.info.

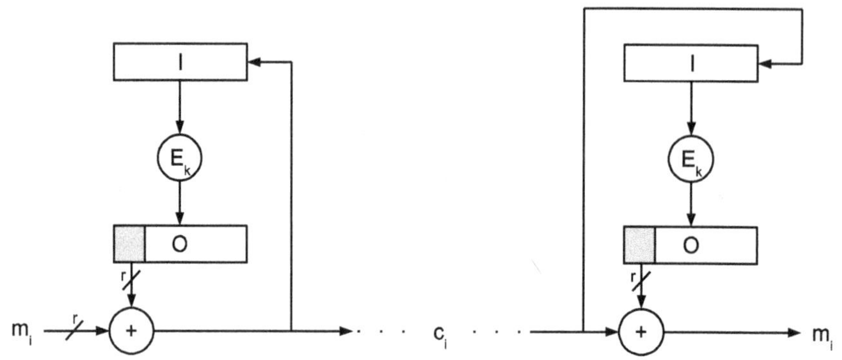


**Figure 9.23** CFB encryption and decryption.

The working principles of CFB encryption and decryption are illustrated in Figure 9.23. The encryption process is drawn on the left side, whereas the decryption device is drawn on the right side. Both processes have two registers at their disposal: an input register I and an output register O. On either side, the registers are initialized with the IV before the encryption and decryption processes start (this initialization step is not visualized in Figure 9.23). Hence, the registers I and O as well as the IV are as long as the block length $n$ of the block cipher in use (e.g., 64 bits for 3DES and 128 bits for the AES). In each step, $1 \leq r \leq n$ bits are processed (i.e., encrypted or decrypted) simultaneously. The value of $r$ is sometimes incorporated into the name of the mode (i.e., $r$-bit CFB). Typical values for $r$ are 1, 8, 64, and 128, and hence the respective modes are called 1-bit CFB, 8-bit CFB, 64-bit CFB, and 128-bit CFB (if $n$ is larger or equal than 128).

To apply $r$-bit CFB, the plaintext message $m \in \mathcal{M}$ is split into $t$ $r$-bit units $m_1, \ldots, m_t$, where each unit $m_i$ $(i = 1, \ldots, t)$ is encrypted and decrypted individually and sequentially. To encrypt $m_i$ (for $i \geq 1$), the encryption process must generate $r$ pseudorandom bits that are added modulo 2 to $m_i$. To generate these bits, the current content of I is subject to the encryption function $E_k$, and the result is fed into O. Finally, the most significant $r$ bits of O are taken as pseudorandom bits to encrypt $m_i$. The result is the ciphertext unit $c_i$ that is not only sent to the receiving process but also fed into the sending process' I register from the right side. This, in turn, changes the content of I for the encryption of the next plaintext message unit $m_{i+1}$.

On the receiving side, the decryption process uses the inverse process: It decrypts $c_i$ by adding it modulo 2 to $r$ pseudorandom bits that are generated in exactly the same way. Again, the content of I is subject to $E_k$, the result is fed into O, and the most significant $r$ bits are taken from O to serve as pseudorandom bits. The input register I is periodically updated by feeding $c_i$ into it from the right for the decryption of the next ciphertext unit.

Like CBC, CFB has a chaining structure that prohibits parallelism. There are a few other disadvantages that should be kept in mind when actually using CFB:

- The major disadvantage is performance. Note that a full encryption of $n$ bits is required to encrypt $r$ bits. The impact of this disadvantage depends on the symmetric encryption system in use and the size of $r$ relative to $n$. For example, if AES is used in the CFB mode and $r$ is 8 bits (meaning that the encryption is character-oriented), then the performance is $n/r = 128/8 = 16$ times slower than "normal" AES in ECB or CBC mode. Consequently, there is a trade-off to make to choose an optimal value for $r$, and this trade-off depends on the application.
- The size of $r$ also influences the error propagation of the encryption. Note that an incorrectly transmitted ciphertext block disturbs the decryption process until it "falls out" of the input register. Consequently, the larger $r$ is, the fewer errors are propagated in CFB mode.
- As mentioned earlier, the encryption is a simple addition modulo 2, and the block cipher is only used to generate the key stream. This generation, however, also depends on the ciphertext bits that are fed back into the input register (that's why the mode is called cipher feedback in the first place). Consequently, it is not possible to precompute the key stream.

The last point—the impossibility of precomputing the key stream—is the major difference to the OFB mode that is addressed next.

### 9.7.4 OFB

The OFB mode is similar to the CFB mode, but it represents a synchronous stream cipher. As illustrated in Figure 9.24 (and contrary to the CFB mode), the key stream is generated independently from the ciphertext blocks. This suggests that it can be precomputed and that the encryption has no error propagation. In some application contexts this is advantageous, and the OFB mode is used for exactly this reason. In other application contexts, however, the lack of error propagation is disadvantageous, and the CFB mode is used instead.

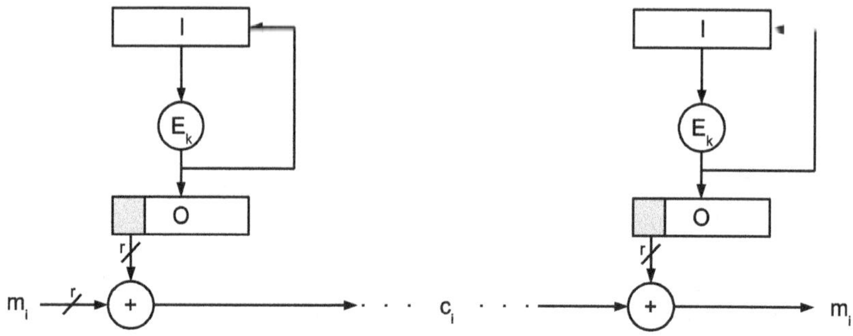


**Figure 9.24** OFB encryption and decryption.

In OFB mode, it is particularly important that the IV is unique for every message that is encrypted (with the same key $k$). Otherwise, the same key stream is generated, and this fact can easily be exploited in an attack. This must be taken into account when OFB is used in the field.

### 9.7.5 CTR

The CTR mode was introduced in [35] to complement the CFB and OFB modes. The basic idea of CTR mode is that key-stream blocks are generated by encrypting successive values of a counter. A counter, in turn, can be any function that produces a sequence that is guaranteed not to repeat for a long time. In the simplest case, the counter is just a value that is incremented by one in each step.

The working principles of CTR encryption and decryption are illustrated in Figure 9.25. The input register I is the one that is incremented successively. It starts with a value that results from the concatenation of an IV (or nonce) and a counter initialized with zero bytes. This value is then incremented in each step. The output register O yields the key stream from which $r$ bits are taken to encrypt the next $r$ plaintext unit. As usual, encryption means bitwise addition modulo 2.

The CTR mode is very similar to the OFB mode. But because there is no feedback or chaining, it has a random access property that makes it very well suited for multiprocessor machines, where blocks can be encrypted and decrypted in parallel. This is advantageous in many application settings, and hence CTR mode is often the preferred choice today.

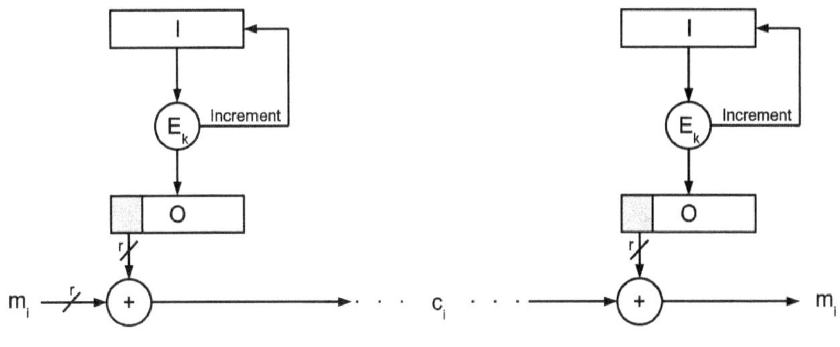


**Figure 9.25** CTR encryption and decryption.

## 9.8 FINAL REMARKS

In this chapter, we elaborated on symmetric encryption systems and had a closer look at some exemplary systems, such as the block ciphers DES (3DES) and AES, and the stream ciphers RC4, Salsa20, and ChaCha20. These systems were chosen because they are widely used in the field and sometimes even implemented in commercial off-the-shelf (COTS) products. There are many other symmetric encryption systems developed and proposed in the literature but not addressed here. Examples include the other AES finalists (i.e., the competitors of Rijndael), the symmetric encryption systems specified in ISO/IEC 18033,[57] including, for example, the 64-bit block cipher MISTY1 [45, 46] and the 128-bit block cipher Camellia [14], as well as more recent proposals, like SHACAL or FOX (sometimes also referred to as IDEA-NXT [47]).

All symmetric encryption systems in use today look somewhat similar in the sense that they all employ a mixture of more or less complex functions that are iterated multiple times (i.e., in multiple rounds) to come up with something that is inherently hard to understand and analyze. There are also details in a cryptographic design that may look mysterious or arbitrary to some extent. For example, the S-boxes of DES look arbitrary, but they are not and are well chosen to protect against

57 ISO/IEC 18033 is a multipart ISI/IEC standard that specifies symmetric encryption systems. In particular, part 3 specifies the 64-bit block ciphers TDEA, MISTY1, and CAST-128, as well as the 128-bit block ciphers AES, Camellia, and SEED, whereas part 4 specifies ways to generate a keystream as well as the dedicated keystream generators MUGI and SNOW (version 2).

differential cryptanalysis (that was published almost two decades after the specification of the DES). Against this background, one may get the impression that it is simple to design and come up with a new symmetric encryption system. Unfortunately, this is not the case, and the design of a system that is secure and efficient is a tricky endeavor. Legions of systems had been proposed, implemented, and deployed before a formerly unknown successful attack was discovered and applied to break them. Such a discovery then often brings the end to several symmetric encryption systems. For example, the discovery of differential cryptanalysis in the public brought the end to many symmetric encryption systems (and other cryptosystems), including, for example, the Fast Data Encipherment Algorithm (FEAL) and many of its variants.

Unless one enters the field of information-theoretical security, the level of security a symmetric encryption system is able to provide is inherently difficult to determine. How resistant is a symmetric encryption system against known and yet-to-be-discovered cryptanalytical attacks? This question is difficult to answer mainly because it is not possible to say what cryptanalytical attacks are known or will be discovered in the future. In this situation, it is simple to put in place and distribute rumors about possible weaknesses and vulnerabilities of encryption systems. Many of these rumors are placed for marketing reasons (rather than for security reasons). For example, there are people (typically not selling AES encryption devices) who argue that the fact that NIST has standardized the AES suggests that it contains trapdoors. There are other people (typically selling AES encryption devices) who argue that the fact that the AES has been subject to public scrutiny suggests that it does not contain a trapdoor. Who is right? Who is able to say who is right? Why would you trust any particular person or group? The bottom line is that fairly little is known about the true security of symmetric encryption systems (except for information-theoretically secure encryption systems).

## References

[1] Shannon, C.E., "A Mathematical Theory of Communication," *Bell System Technical Journal*, Vol. 27, No. 3/4, July/October 1948, pp. 379–423/623–656.

[2] Shannon, C.E., "Communication Theory of Secrecy Systems," *Bell System Technical Journal*, Vol. 28, No. 4, October 1949, pp. 656–715.

[3] Vernam, G.S., "Cipher Printing Telegraph Systems for Secret Wire and Radio Telegraphic Communications," *Journal of the American Institute for Electrical Engineers*, Vol. 55, 1926, pp. 109–115.

[4] Bellovin, S.M., "Frank Miller: Inventor of the One-Time Pad," *Cryptologia*, Vol. 35, No. 3, July 2011, pp. 203–222.

[5] Maurer, U.M., "Secret Key Agreement by Public Discussion," *IEEE Transaction on Information Theory*, Vol. 39, No. 3, 1993, pp. 733–742.

[6] Goldwasser, S., and S. Micali, "Probabilistic Encryption," *Journal of Computer and System Sciences*, Vol. 28, No. 2, April 1984, pp. 270–299.

[7] Coppersmith, D., H. Krawczyk, and Y. Mansour, "The Shrinking Generator," *Proceedings of CRYPTO '93*, Springer-Verlag, LNCS 773, 1994, pp. 22–39.

[8] Meier, W., and O. Staffelbach, "The Self-Shrinking Generator," *Proceedings of EUROCRYPT '94*, Springer-Verlag, LNCS 950, 1995, pp. 205–214.

[9] Popov, A., *Prohibiting RC4 Cipher Suites*, Standards Track RFC 7465, February 2015.

[10] Nir, Y., and A. Langley, *ChaCha20 and Poly1305 for IETF Protocols*, RFC 7539, May 2015.

[11] U.S. Department of Commerce, National Institute of Standards and Technology, *Data Encryption Standard (DES)*, FIPS PUB 46-3, October 1999.

[12] U.S. Department of Commerce, National Institute of Standards and Technology, *Security Requirements for Cryptographic Modules*, FIPS PUB 140-1, January 1994.

[13] Luby, M., and C. Rackoff, "How to Construct Pseudorandom Permutations from Pseudorandom Functions," *SIAM Journal on Computing*, Vol. 17, No. 2, 1988, pp. 373-386.

[14] Matsui, M., J. Nakajima, and S. Moriai, *A Description of the Camellia Encryption Algorithm*, Informational RFC 3713, April 2004.

[15] Poschmann, A., et al., "New Light-Weight Crypto Algorithms for RFID," *Proceedings of the IEEE International Symposium on Circuits and Systems (ISCAS 2007)*, New Orleans, LA, 2007, pp. 1843–1846.

[16] Biham, E., and A. Shamir, *Differential Cryptanalysis of DES*. Springer-Verlag, New York, 1993.

[17] Matsui, M., "Linear Cryptanalysis of DES Cipher," *Proceedings of EUROCRYPTO '93*, Springer-Verlag, New York, 1994, pp. 386–397.

[18] Coppersmith, D., "The Data Encryption Standard (DES) and Its Strength Against Attacks," *IBM Journal of Research and Development*, Vol. 38, No. 3, 1994, pp. 243–250.

[19] Diffie, W., and M.E. Hellman, "Exhaustive Cryptanalysis of the NBS Data Encryption Standard," *IEEE Computer*, Vol. 10, No. 6, 1977, pp. 74–84.

[20] Hellman, M.E., "A Cryptanalytic Time–Memory Trade-Off," *IEEE Transactions on Information Theory*, Vol. 26, No. 4, July 1980, pp. 401–406.

[21] Wiener, M.J., "Efficient DES Key Search," presented at the rump session of the CRYPTO '93 Conference and reprinted in Stallings, W. (ed.), *Practical Cryptography for Data Internetworks*, IEEE Computer Society Press, 1996, pp. 31–79.

[22] Wiener, M.J., "Efficient DES Key Search—An Update," *CryptoBytes*, Vol. 3, No. 2, Autumn 1997, pp. 6–8.

[23] Electronic Frontier Foundation (EFF), *Cracking DES: Secrets of Encryption Research, Wiretap Politics & Chip Design*. O'Reilly & Associates, Sebastopol, CA, 1998.

[24] Rivest, R.L., "All-or-Nothing Encryption and the Package Transform," *Proceedings of 4th International Workshop on Fast Software Encryption*, Springer-Verlag, LNCS 1267, 1997, pp. 210–218.

[25] Grover, L.K., "A Fast Quantum Mechanical Algorithm for Database Search," *Proceedings of the 28th ACM Symposium on Theory of Computing (STOC)*, 1996, pp. 212–219.

[26] Bennett, C.H., et al., "Strengths and Weaknesses of Quantum Computing," *SIAM Journal on Computing*, Vol. 26, Issue 5, October 1997, pp. 1510–1523.

[27] Merkle, R.C., "Fast Software Encryption Function," *Proceedings of CRYPTO '90*, Springer-Verlag, 1991, pp. 476–501.

[28] Kilian, J., and P. Rogaway, "How to Protect DES Against Exhaustive Key Search," *Proceedings of CRYPTO '96*, Springer-Verlag, 1996, pp. 252–267.

[29] Campbell, K.W., and M.J. Wiener, "DES Is Not a Group," *Proceedings of CRYPTO '92*, Springer-Verlag, 1993, pp. 512–520.

[30] Mitchell, C., "On the Security of 2-Key Triple DES," *IEEE Transactions on Information Theory*, Vol. 62, No. 11, November 2016, pp. 6260–6267.

[31] Daemen, J., and V. Rijmen, *The Design of Rijndael*. Springer-Verlag, New York, 2002.

[32] Burr, W.E., "Selecting the Advanced Encryption Standard," *IEEE Security & Privacy*, Vol. 1, No. 2, March/April 2003, pp. 43–52.

[33] U.S. Department of Commerce, National Institute of Standards and Technology, *Specification for the Advanced Encryption Standard (AES)*, FIPS PUB 197, November 2001, http://csrc.nist.gov/publications/fips/fips197/fips-197.pdf.

[34] U.S. Department of Commerce, National Institute of Standards and Technology, *DES Modes of Operation*, FIPS PUB 81, December 1980.

[35] U.S. Department of Commerce, National Institute of Standards and Technology, *Recommendation for Block Cipher Modes of Operation—Methods and Techniques*, FIPS Special Publication 800-38A, 2001.

[36] U.S. Department of Commerce, National Institute of Standards and Technology, *Recommendation for Block Cipher Modes of Operation: The CMAC Mode for Authentication*, FIPS Special Publication 800-38B, May 2005.

[37] U.S. Department of Commerce, National Institute of Standards and Technology, *Recommendation for Block Cipher Modes of Operation: The CCM Mode for Authentication and Confidentiality*, FIPS Special Publication 800-38C, May 2004.

[38] Whiting, D., R. Housley, and N. Ferguson, *Counter with CBC-MAC (CCM)*, RFC 3610, September 2003.

[39] U.S. Department of Commerce, National Institute of Standards and Technology, *Recommendation for Block Cipher Modes of Operation: Galois/Counter Mode (GCM) and GMAC*, Special Publication 800-38D, November 2007.

[40] U.S. Department of Commerce, National Institute of Standards and Technology, *Recommendation for Block Cipher Modes of Operation: The XTS-AES Mode for Confidentiality on Storage Devices*, Special Publication 800-38E, January 2010.

[41] U.S. Department of Commerce, National Institute of Standards and Technology, *Recommendation for Block Cipher Modes of Operation: Methods for Key Wrapping*, Special Publication 800-38F, December 2012.

[42] U.S. Department of Commerce, National Institute of Standards and Technology, *Recommendation for Block Cipher Modes of Operation: Methods for Key Format-Preserving Encryption*, Special Publication 800-38G, March 2016.

[43] Rogaway, P., M. Bellare, and J. Black, "OCB: A Block-Cipher Mode of Operation for Efficient Authenticated Encryption," *ACM Transactions on Information and System Security (TISSEC)*, Vol. 6, Issue 3, August 2003, pp. 365–403.

[44] Oppliger, R., *End-to-End Encrypted Messaging.* Artech House Publishers, Norwood, MA, 2020.

[45] Matsui, M., "New Block Encryption Algorithm MISTY," *Proceedings of the Fourth International Fast Software Encryption Workshop (FSE '97)*, Springer-Verlag, LNCS 1267, 1997, pp. 54–68.

[46] Otha, H., and M. Matsui, *A Description of the MISTY1 Encryption Algorithm*, RFC 2994, November 2000.

[47] Junod, P., and S. Vaudenay, "FOX: A New Family of Block Ciphers," *Proceedings of the Eleventh Annual Workshop on Selected Areas in Cryptography (SAC 2004)*, Springer-Verlag, LNCS 3357, 2004, pp. 114–129.

# Chapter 10

## Message Authentication

In this chapter, we elaborate on message authentication, MACs, and message authentication systems. More specifically, we introduce the topic in Section 10.1, overview, discuss, and put into perspective information-theoretically and—more importantly—computationally secure message authentication in Sections 10.2 and 10.3, and conclude with some final remarks in Section 10.4.

### 10.1 INTRODUCTION

In Section 2.2.4, we introduced the notion of an authentication tag, and we said that such a tag can either be a digital signature or a MAC according to Definition 2.10. There are two fundamental differences:

- Digital signatures can be used to provide nonrepudiation services, whereas MACs cannot be used for this purpose.
- A digital signature can typically be verified by everybody,[1] whereas a MAC can be verified only by somebody who knows the secret key.

Due to these differences, it must be decided in a particular application setting whether digital signatures or MACs are more appropriate to provide message authentication. There are applications that are better served with digital signatures, and there are applications that are better served with MACs. There is no single best answer on what technology to use. Digital signatures are further addressed in Chapter 14, but MACs are the topic of this chapter.

1 Note that there are also digital signature systems that limit the verifiability of the signatures to specific entities. In some literature, the respective signatures are called *undeniable*.

A MAC is basically an authentication tag that is computed and verified with a secret key. This means that the sender and the recipient(s) must share a key, and this key can then be used to compute and verify a MAC for a particular message (ideally, the same key is used to compute and verify MACs for several messages). Hence, a MAC depends on both the message that is authenticated and the secret key that only the legitimate sender and recipient(s) are supposed to know. As pointed out later in this chapter, some MAC constructions additionally require a random number that serves the prupose of a message-specific nonce.

If $\mathcal{M}$ is a message space, $\mathcal{T}$ a tag space, and $\mathcal{K}$ a key space, then—according to Definition 2.11—a *message authentication system* is a pair $(A, V)$ of families of efficiently computable functions:

- $A : \mathcal{K} \times \mathcal{M} \rightarrow \mathcal{T}$ denotes a family $\{A_k : k \in \mathcal{K}\}$ of *authentication functions* $A_k : \mathcal{M} \rightarrow \mathcal{T}$;
- $V : \mathcal{K} \times \mathcal{M} \times \mathcal{T} \rightarrow \{valid, invalid\}$ denotes a family $\{V_k : k \in \mathcal{K}\}$ of *verification functions* $V_k : \mathcal{M} \times \mathcal{T} \rightarrow \{valid, invalid\}$.

For every message $m \in \mathcal{M}$ and key $k \in \mathcal{K}$, $V_k(m,t)$ must yield $valid$ if and only if $t$ is a valid authentication tag for $m$ and $k$; that is, $t = A_k(m)$, and hence $V_k(m, A_k(m))$ yields $valid$.

The working principle of a message authentication system (using the algorithms Generate, Authenticate, and Verify) is depicted in Figure 2.8 and not repeated here. Typically, $\mathcal{M} = \{0,1\}^*$, $\mathcal{T} = \{0,1\}^{l_{tag}}$ for some tag length $l_{tag}$, and $\mathcal{K} = \{0,1\}^{l_{key}}$ for some key length $l_{key}$. It is often the case that $l_{tag} = l_{key} = 128$, meaning that the tags and keys are 128 bits long each.

To formally define a secure message authentication system, we must first define the attacks an adversary may be able to mount. In an encryption setting, the weakest possible attack is a ciphertext-only attack. Translated to message authentication, this refers to a MAC-only attack, meaning that the adversary gets to know only authentication tags (without learning the respective messages). Such an attack does not make a lot of sense, because a MAC is always transmitted together with the message. So MAC-only attacks only exist in theory and are not relevant in practice (because they do not occur). Instead, there are only two types of attacks that are relevant and need to be distinguished:

- In a *known-message attack*, the adversary gets to know one or several message-tag pairs $(m_1, t_1), (m_2, t_2), \ldots, (m_l, t_l)$ for some $l \geq 1$.
- In a *chosen-message attack* (CMA), the adversary does not only know message-tag pairs, but he or she has access to the authentication function (or the device that implements the function, respectively) and can generate

authentication tags for one or several messages $m_1, m_2, \ldots, m_l$ of his or her choice for some $l \geq 1$. For each $m_i$, the adversary gets to know the respective authentication tag $t_i$ ($1 \leq i \leq l$). In the simplest (nonadaptive) case, the adversary must choose the messages $m_1, m_2, \ldots, m_l$ before the attack begins. In an *adaptive CMA*, however, the adversary can dynamically choose messages of his or her choice while the attack is going on. Again, these attacks are more powerful than their nonadaptive counterparts.

In the second part of a security definition, one must also specify the task the adversary is required to solve in order to be successful (i.e., to break the security of the system). With decreasing severity, the following tasks can be considered:

- In a *total break*, the adversary is able to determine the secret key in use, meaning that he or she can thereafter forge a valid authentication tag for any message of his or her choice.
- In a *selective forgery*, the adversary is able to forge a valid tag for a particular—typically meaningful—message. In contrast to a total break, the adversary does not have to determine the secret key in use. Any other mechanism that allows him or her to forge a valid authentication tag is equally fine.
- In an *existential forgery*, the adversary is able to forge a valid authentication tag for an arbitrary message (that does not have to be meaningful).

Because an adversary can always guess an authentication tag that is valid, we cannot require that a MAC is truly unforgeable, but we require that the success probability of doing so is negligibly small. If the tag length is $l_{tag}$, then the probability of correctly guessing a MAC is $1/2^{l_{tag}}$, and this value is in fact negligible (since it decreases exponentially fast).

If an adversary (who can mount CMAs) is able to generate a valid message-tag pair with a success probability that is at most negligible, then we say that the MAC is *unforgeable*. More precisely, there are two notions of unforgeable MACs:

- A MAC is *weakly unforgeable* under a CMA (WUF-CMA) if it is computationally infeasible for the adversary to find a message-tag pair in which the message is "new."
- A MAC is even *strongy unforgeable* under a CMA (SUF-CMA) if it is computationally infeasible for the adversary to find a new message-tag pair.

The difference is subtle here: In the case of a MAC that is WUF-CMA, the message is "new" in the sense that the adversary has not yet seen any valid MAC for this message. This is in contrast to the case of a MAC that is SUF-CMA. Here, the message need not be "new," meaning that the adversary may have already seen some valid MACs for this message, and his or her task is to find a new one.[2] For the rest of this chapter, we say that a message authentication system is secure if the MACs it generates are SUF-CMA. This is appropriate if a key is used to authenticate multiple messages.

If a key is used to authenticate a single message, then we can go for message authentication that is information-theoretically secure. This is similar to perfectly secret symmetric encryption, such as that provided by the one-time pad (Section 9.3). The respective MACs are sometimes called *one-time MACs*. Due to the requirement of using a new key for every message, the use of information-theoretically secure message authentication systems and one-time MACs is prohibitively expensive in practice, and hence computationally secure message authentication systems and MACs that are SUF-CMA are usually the preferred choice. Information-theoretically secure and computationally secure message authentication are addressed in the following two sections.

## 10.2 INFORMATION-THEORETICALLY SECURE MESSAGE AUTHENTICATION

In this section, we address message authentication that is information-theoretically secure. As a simple and pathological example, we want to construct a message authentication system that is information-theoretically secure and can be used to authenticate messages that refer to the outcome of a coin-flipping experiment:

- H stands for head;
- T stands for tail.

This means that the message space $\mathcal{M}$ consists of the two elements H and T (i.e., $\mathcal{M} = \{\text{H,T}\}$). Furthermore, the tag space $\mathcal{T}$ is to consist of a single bit (i.e., $\mathcal{T} = \{0,1\}$), and the key space $\mathcal{K}$ is to consist of all possible 2-bit keys (i.e., $\mathcal{K} = \{0,1\}^2 = \{00,01,10,11\}$). Using these ingredients, it is possible to define an information-theoretically secure message authentication system as illustrated in Table 10.1. The rows refer to the four possible keys and the columns refer to all valid message-tag pairs. If, for example, the key is 01 and the outcome of the experiment is H, then the appropriate message-tag pair is H0, where H stands for the message and 0 stands for the tag. Similarly, if the outcome of the experiment is tail, then the appropriate message-tag pair is T1 (for the same key). So a 1-bit message is always authenticated with a 1-bit tag, and a new 2-bit key is needed that cannot be

2 Obviously, this requires the authentication function to be nondeterministic, meaning that many MACs may be valid for a particular message.

reused. This means that $2n$ bits of fresh keying material are needed to authenticate $n$ outcomes of the experiment. If, for example, we wanted to authenticate the sequence TTHTH with the five keys 00, 10, 11, 11, and 01, then the appropriate message-tag pairs would be T0, T0, H1, T1, and H0.

**Table 10.1**
An Information-Theoretically Secure Message Authentication System

| | H0 | H1 | T0 | T1 |
|---|---|---|---|---|
| 00 | H | - | T | - |
| 01 | H | - | - | T |
| 10 | - | H | T | - |
| 11 | - | H | - | T |

To prove that the message authentication system is information-theoretically secure, we must show that an adversary with infinite computing power and time has no better way to forge a MAC than to guess. Because a MAC is only one bit long, the success probability of correctly guessing it is $1/2$. If, for example, the adversary wants to forge a MAC for message H, then the MAC is 0 if the key is either 00 or 01, and 1 if it is 10 or 11. If the keys are equally probable, then the adversary has a $2/4 = 1/2$ probability of correctly guessing the MAC, and there is nothing he or she can do to improve this value. A similar line of argumentation applies if the adversary wants to modify a message-tag pair. Assume that the adversary has received H0, and he or she wants to change the message from H to T. He or she therefore has to generate a valid tag for this new message (without knowing the secret key). If the adversary has received H0, then—according to Table 10.1—the only possible keys are 00 and 01:

- If 00 is the correct key, then the message T must be authenticated with 0, and the adversary must change the message-tag pair to T0.
- If 01 is the correct key, then the message T must be authenticated with 1, and the adversary must change the message-tag pair to T1.

In either case, the adversary has a probability of $1/2$ to correctly guess the correct key and to change the message accordingly. Again, there is nothing the adversary can do to improve this value, and this means that the message authentication system is in fact information-theoretically secure.

The previous example is only to illustrate how information-theoretically secure message authentication works in principle. It is highly inefficient, so it can't be used in practice. But there are other one-time message authentication systems—typically based on polynomial evaluation—that are sufficiently efficient. Let us consider an example to illustrate this point.[3] Let $p$ be a prime that is slightly larger than the maximum value of a message block. If the block length is 128 bits, then a possible value is $p = 2^{128} + 51$ (this number is prime). The key $k$ consists of two parts, $k_1$ and $k_2$, that are both randomly chosen integers between 1 and $p - 1$. To compute a one-time MAC for message $m$, the message is first cut into $l = \lceil |m|/128 \rceil$ 128-bit blocks $m[1], m[2], \ldots, m[l]$, and these blocks then represent the coefficients of a polynomial $P_m(x)$ of degree $l$ that is defined as follows:

$$P_m(x) = m[l]x^l + m[l-1]x^{l-1} + m[l-2]x^{l-2} + \ldots + m[2]x^2 + m[1]x$$

Having this polynomial in mind, a one-time MAC (OTMAC) can be defined as the modulo $p$ sum of $P_m$ evaluated at point $k_1$ and $k_2$:

$$OTMAC_k(m) = P_m(k_1) + k_2 \pmod{p}$$

This construction yields an OTMAC that is efficient and information-theoretically secure, meaning that it leaks no information (about another MAC that could be used to authenticate another message). But it is still a one-time MAC, meaning that it can be used to authenticate a single message. If the same key were used to authenticate two or more messages, then the respective message authentication system would become insecure, meaning that an adversary would then able to construct MACs for arbitrary messages of his or her choice. Luckily, there is construction based on the universal hashing paradigm that can be used to turn an OTMAC into a MAC that can be used to authenticate multiple messages. From the names of its inventors, the resulting construction is known as a *Carter-Wegman MAC*. Carter-Wegman MACs play an increasingly important role in the field, and hence they are further addressed in Section 10.3.3.

The bottom line is that all information-theoretically secure MACs have the disadvantage that they can only be used to authenticate a single message. This is prohibitively expensive in terms of key management. So for all practical purposes, we use MAC constructions that are "only" computationally secure (this includes the case in which an OTMAC is turned into a Carter-Wegman MAC). These constructions are overviewed, discussed, and put into perspective next.

3 The example was created by Dan Boneh.

## 10.3 COMPUTATIONALLY SECURE MESSAGE AUTHENTICATION

There are basically three possibilities to design a message authentication system that is computationally secure:

- One can start with a symmetric encryption system—typically a block cipher—and derive a message authentication system that uses a symmetric encryption system.
- One can start with a cryptographic hash function and a key, and derive a message authentication system that uses a keyed hash function.
- One can start with an OTMAC and use the universal hashing paradigm to derive a Carter-Wegman MAC (as mentioned above).

All three possibilities are outlined in this section. The first two possibilities are further addressed in the multipart standard ISO/IEC 9797. While ISO/IEC 9797-1 [1] elaborates on MAC constructions that use a block cipher,[4] ISO/IEC 9797-2 [2] addresses MAC constructions that use a dedicated (cryptographic) hash function.[5] Furthermore, there are a few outdated proposals that are standardized but not widely deployed. For example, the now withdrawn *message authenticator algorithm* (MAA) was specified in ISO 8731-2 [3] (originally published in 1984). Until today, no significant structural weakness has been found in this construction. Its major problem is that it generates MACs that are only 32 bits long. This is unacceptably short for most applications in use today, and hence this construction is not further addressed here.

### 10.3.1 MACs Using A Symmetric Encryption System

If one has a symmetric encryption system—typically a block cipher, like DES or AES—at hand, then there are several possibilities to use it to authenticate messages. In the simplest case, one can employ a cryptographic hash function $h$ to compute a hash value of message $m$ and encrypt $h(m)$ with the block cipher. The respective MAC refers to $E_k(h(m))$, where $k$ is the encryption key. The length of this MAC is a multiple of the block length. If, for example, $h$ generates hash values of 128 bits and the block length is 64 bits, then two blocks need to be encrypted. If the block length is 128 bits, then only one block is needed. In addition to a symmetric encryption system, this construction requires a hash function. There are other constructions that don't have this requirement and that are more widely deployed in the field. In particular, we take a closer look at CBC-MAC, CMAC, and PMAC.

4 ISO/IEC 9797-1 specifies six such constructions.

5 ISO/IEC 9797-2 specifies three such constructions.

#### 10.3.1.1 CBC-MAC

Remember from Section 9.7.2 that CBC is a mode of operation for a block cipher that cryptographically chains all message blocks together, meaning that the last ciphertext block depends on all previous blocks. This, in turn, means that the last block may be used as a MAC for the entire message. This construction is known as *CBC residue* or *CBC-MAC*. If CBC mode is used for encryption, then a fresh IV is required for every message that is encrypted. But if CBC is used for authentication, then it is possible to always use the same IV (e.g., the zero IV). Also, the length of the CBC-MAC depends on the block cipher in use and its block length. It is common to use a block cipher that has a block length of 128 bits at least (e.g., AES). This CBC-MAC construction is adopted by many standardization organizations and bodies, such the ANSI accredited standards committee X9,[6] NIST [4],[7] and—maybe most importantly—ISO/IEC [1].

If the CBC-MAC construction uses a secure block cipher and all messages are equally long, then the construction is known to be secure [5]. The requirement that all messages are equally long is hereby essential. If this is not the case; that is, if the messages have different lengths, then an adversary who knows two message-tag pairs, $(m, t)$ and $(m', t')$, can generate a third message $m''$ whose CBC-MAC is also $t'$, meaning that $(m'', t')$ is another valid message-tag pair and hence represents an existential forgery. The message $m''$ is generated by adding modulo 2 $t$ to the first block of $m'$ and then concatenating $m$ with this modified[8] message $m'$. The resulting message $m''$ thus looks as follows:

$$m'' = m \parallel (m'_1 \oplus t) \parallel m'_2 \parallel \ldots \parallel m'_l$$

See what happens if the CBC-MAC is computed for $m''$: First, $t$ is computed as usual for $m$. When $m'$ is CBC encrypted, this value $t$ is added modulo 2 to the first block $(m'_1 \oplus t)$. So the encryption of this block refers to $E_k(m'_1 \oplus t \oplus t)$, and this means that the value $t$ cancels itself; that is, $E_k(m'_1 \oplus t \oplus t) = E_k(m'_1)$. The CBC encryption of $m''$ is thus identical to $m'$, and this, in turn, means that $t'$ is a valid tag for $m''$, and hence that $(m'', t')$ yields an existential forgery.

6 In 1986, X9 released X9.9 that provided a wholesale banking standard for authentication of financial transactions. It addressed two issues: message formatting and the particular message authentication algorithm (that was essentially DES in CBC or CFB mode). X9.9 was adopted in ISO 8730 and ISO 8731, but these standards have been withdrawn—mainly due to the insufficient security of DES.

7 Note that this standard was officially withdrawn on September 1, 2008. It used the term *data authentication algorithm* (DAA) to refer to CBC-MAC and the term *data authentication code* (DAC) to refer to a respective MAC value.

8 Note that $m'$ is modified only in its first block. All other $l-1$ blocks $m'_2, m'_3, \ldots, m'_l$ remain the same.

Put the other way round, we know that the CBC-MAC construction is insecure for messages that have different lengths (as is almost always the case). There are three possibilities to deal with this situation:

- First, one can use a different key for every possible input message length. This is not convenient, and hence this possibility is seldom used in practice.
- Second, one can prepend the length of the message to the first block. This is theoretically interesting because the respective construction can be proven secure.[9] But from a practical viewpoint, this construction has severe disadvantages, mainly because the message length may not be known when message processing begins.
- Third, one can encrypt the last block with a different key; that is, a (second) key that is independent from the key used for CBC encryption. The resulting construction is known as *encrypt-last-block* or *encrypted CBC-MAC* (ECBC-MAC).

The ECBC-MAC construction is the preferred choice used in the field. In addition, there are a few other constructions that try to turn a CBC-MAC into something that can be proven secure. The *eXtended CBC* (XCBC) [6], for example, is a construction that employs three different keys and has been used a lot in the realm of IPsec [7]. But the use of three keys is disadvantageous in practice, and hence this disadvantage has been eliminated in a pair of constructions known as *two-key CBC MAC* (TMAC[10]) and—more importantly—*one-key CBC MAC* (OMAC) [8].[11] Officially, there are two OMAC constructions (OMAC1 and OMAC2) that are both essentially the same except for a small tweak. ECBC-MAC, XCBC, TMAC, and OMAC are all variants of CBC-MAC that have distinct advantages and disadvantages. Research in this area finally culminated in a block cipher mode of operation for authentication known as *cipher-based MAC* (CMAC). CMAC is equivalent to OMAC1 and was standardized by NIST in 2005 [9]. It represents the state of the art and is addressed next.

#### 10.3.1.2 CMAC

Compared to ECBC-MAC, the CMAC construction has two advantages: It avoids the need to pad a message that is already aligned with the block length of the underlying block cipher, and it avoids the need to additionally encrypt the last

9 https://cs.uwaterloo.ca/~sgorbuno/publications/securityOfCBC.pdf.

10 http://eprint.iacr.org/2002/092.

11 http://www.nuee.nagoya-u.ac.jp/labs/tiwata/omac/omac.html.

ciphertext block after CBC encryption with a different key. To achieve this, the CMAC construction operates in two steps:

1. If $k$ is the block cipher key in use, then two equally long subkeys, $k_1$ and $k_2$, are derived from $k$.
2. The two subkeys $k_1$ and $k_2$ are used to generate a CMAC tag $t$.

**Algorithm 10.1** The CMAC subkey derivation algorithm.

$(k)$

---

$k_0 = E_k(0^b)$
if (MSB($k_0$) = 0)
    then $k_1 = k_0 \hookleftarrow 1$
    else $k_1 = (k_0 \hookleftarrow 1) \oplus R_b$
if (MSB($k_1$) = 0)
    then $k_2 = k_1 \hookleftarrow 1$
    else $k_2 = (k_1 \hookleftarrow 1) \oplus R_b$

---

$(k_1, k_2)$

The CMAC subkey derivation algorithm is outlined in Algorithm 10.1. As already mentioned above, it takes as input a key $k$, and it generates as output two equally long subkeys $k_1$ and $k_2$. First, a block of $b$ zero bits is encrypted using the block cipher and key $k$. The result is assigned to $k_0$. Depending on the most significant bit (MSB) of $k_0$, there are two cases that are distinguished when generating $k_1$:

- If the MSB of $k_0$ is equal to 0, then $k_1$ is simply assigned the bit string that results from $k_0$ after a left shift for one position (denoted $k_0 \hookleftarrow 1$). If, for example, the bit string 10101010 is subject to a left shift for 1, then the leftmost bit is discarded, all other bits are shifted to the left, and the rightmost bit is set to 0. The resulting bit string is 01010100.
- Otherwise (i.e., if the MSB of $k_0$ is not equal to 0), then $k_1$ is assigned the value that results from adding $k_0 \hookleftarrow 1$ with a constant $R_b$ modulo 2. The value of $R_b$ depends on the block length of the cipher in use (that's why $b$ is added as a subscript). For the block sizes that are relevant in practice—64, 128, and 256—these values refer to $R_{64} = 0^{59}11011$ =0x87, $R_{128} = 0^{120}10000111$ =0x87, and $R_{256}$ =0x425.

After $k_1$, $k_2$ is assigned a value in a very similar way: If the MSB of $k_1$ is equal to 0, then it is assigned $k_1 \hookleftarrow 1$. Otherwise (i.e., if the MSB of $k_1$ is not equal to 0), then it is assigned $k(k_1 \hookleftarrow 1) \oplus R_b$.

**Algorithm 10.2** The CMAC tag construction algorithm.

$(m, k_1, k_2)$

---

if $(|m'_n| = b)$
    then $m_n = m'_n \oplus k_1$
    else $m_n = (m'_n \parallel 10^j) \oplus k_2$
$c_0 = 0^b$
for $i = 1$ to $n$ do
    $c_i = E_k(c_{i-1} \oplus m_i)$
t $=$ truncate$(c_n)$

---

(t)

Once the subkeys $k_1$ and $k_2$ have been derived from $k$, the algorithm to construct the CMAC tag $t$ is simple and straightforward. It is illustrated in Algorithm 10.2. It takes as input a message $m$ and a subkey pair $(k_1, k_2)$, and it generates as output $t$. The algorithm starts by dividing $m$ into $n = \lceil |m|/b \rceil$ consecutive $b$-bit blocks:[12]

$$m = m_1 \parallel m_2 \parallel \ldots \parallel m_{n-1} \parallel m'_n$$

The first $n - 1$ blocks $m_1, m_2, \ldots, m_{n-1}$ are complete in the sense that they comprise $b$ bits, whereas the final block $m'_n$ may be incomplete. Remember that the CMAC construction tries to avoid the necessity to pad a message that is already aligned it with the block length $b$ of the underlying block cipher. This is achieved by using a special encoding of $m_n$ that uses either of the two subkeys:

- If $m'_n$ is complete (i.e., $|m'_n| = b$), then $m'_n$ is encrypted with $k_1$: $m_n = m'_n \oplus k_1$. In this case, no padding is used, and the resulting block $m_n$ is also $b$ bits long.
- Otherwise, if $m'_n$ is not complete (i.e., $|m'_n| < b$), then the final block is padded and encrypted with $k_2$: $m_n = (m'_n \parallel 10^j) \oplus k_2$ for $j = b - |m| - 1$. In this case, the padding consists of a one followed by $j$ zeros, and the resulting block $m_n$ is $b$ bits long.

Using this encoding, the message $m = m_1 \parallel m_2 \parallel \ldots \parallel m_{n-1} \parallel m_n$ is CBC encrypted with key $k$ and IV $c_0 = 0^b$. The final ciphertext block $c_n$ represents the

12 The empty message is treated as one incomplete block.

CMAC tag $t$ that my be optionally truncated to less than $b$ bits. Note, however, that it is not recommended to truncate $c_n$ to less than 64 bits. In the usual case, $t$ is sent together with the message $m$ to the recipient, and it is up to the recipient to verify the tag.

In theory, the CMAC construction can be used with any block cipher. In practice, however, it is recommended to use the construction with a block cipher that is in widespread use, such as the AES. The AES-CMAC construction is specified in [10], whereas its use in the realm of IPsec/IKE is addressed in [11, 12].

#### 10.3.1.3 PMAC

All MAC constructions mentioned so far employ a block cipher in CBC mode. This means that they operate sequentially, and that one cannot process a block until all previous blocks have been processed properly. This sequential nature may become a bottleneck in high-speed networks and respective applications. In 1995, Mihir Bellare, Roch Guérin, and Phillip Rogaway proposed an alternative design that uses a family of PRFs (instead of a block cipher that represents a family of PRPs) [13]. In theory, this means that a block cipher is not needed. In practice, however, using a block cipher is still a convenient way to build a family of PRFs, and hence a block cipher may still be used. The original Bellare-Guérin-Rogaway constructions are collectively referred to as *XOR MACs* because they make use of parallel XOR operations. From a bird's-eye perspective, an XOR MAC is computed in three steps:

1. The message $m$ is block encoded, meaning that it is encoded as a collection of $n$ blocks of equal length (i.e., $m = m_1 \parallel m_2 \parallel \ldots \parallel m_n$). Each block must include a block index (mainly to defeat block swapping attacks).
2. A PRF (from the family of PRFs) is applied to each message block individually, creating a set of PRF images.
3. The PRF images are then added modulo 2 to form the XOR MAC.

Note that there are many different choices for the block encoding of step 1 and the PRF of step 2, and that each of these choices yields a distinct XOR MAC construction. We don't delve into the details here. Instead, we point out the major advantage of any such construction, namely the possibility to execute all $n$ invocations of the PRF in parallel (even if the message blocks arrive out of order[13]). Furthermore, an XOR MAC is incremental in the sense of [14]. Whether a particular XOR MAC construction is efficient or not depends on the PRF family (or block

13 Out-of-order MAC verification is in fact a very useful property in contemporary networks, such as the Internet, because of packet losses and retransmission delays.

cipher) in use. In either case, it has been shown that the XOR MAC construction is secure if $F$ is a secure family of PRFs.

Based on the work of Bellare, Guérin, and Rogaway in the 1990s [13], John Black and Rogaway proposed a fully parallelizable MAC (PMAC) construction in 2002 [15].[14] This construction has several advantages that make it a useful choice in many practical settings. On the downside, however, the PMAC construction is patented and this may be one of the reasons why it is not so widely deployed in the field.

### 10.3.2 MACs Using Keyed Hash Functions

The idea of using cryptographic hash functions to protect the authenticity and integrity of data and program files dates back to the late 1980s [16]. In the early 1990s, people started to think more seriously about the possibility of using cryptographic hash functions (instead of symmetric encryption systems) to efficiently authenticate messages. In fact, there are a couple of arguments in favor of using cryptographic hash functions:

- There are a number of cryptographic hash functions in widespread use (refer to Chapter 6 for an overview).
- Cryptographic hash functions can be implemented efficiently in hardware or software.
- Many implementations of cryptographic hash functions are publicly and freely available.
- Cryptographic hash functions are free to use (meaning, for example, that they are not subject to patent claims and/or export controls).
- Cryptographic hash functions have well-defined security properties, such as preimage resistance and (weak or strong) collision resistance.

Some of these arguments have become obsolete (e.g., export restrictions), whereas others still apply (e.g., widespread availability and use) and will likely be applicable in the foreseeable future (e.g., efficiency).

Against this background, Li Gong and Gene Tsudik independently proposed an encryption-free message authentication based on keyed hash functions instead of a symmetric encryption system [17, 18].[15] More specifically, Tsudik proposed and discussed the following three methods to authenticate a message $m$ using a cryptographic hash function $h$ and a secret key $k$:

14 http://www.cs.ucdavis.edu/~rogaway/ocb/pmac.htm.

15 An earlier version of [18] was presented at IEEE INFOCOM '92.

- In the *secret prefix* method, $k$ is a prefix to $m$ and $h$ is applied to the composite message; that is, $MAC_k(m) = h(k \parallel m)$.
- In the *secret suffix* method, $k$ is a suffix to $m$ and $h$ is applied to the composite message; that is, $MAC_k(m) = h(m \parallel k)$.
- In the *envelope* method, there are two keys $k_1$ and $k_2$ that are a prefix and a suffix to $m$. Again, $h$ is applied to the composite message; that is, $MAC_{k_1,k_2}(m) = h(k_1 \parallel m \parallel k_2)$.

The three methods and a few variations thereof are overviewed and briefly discussed next. There are also a few other methods one could think of. For example, one could use the key $k$ as a replacement for the otherwise fixed IV in an iterated hash function, like MD5 or SHA-1. If the IV (and hence the key $k$) is $l$ bits long (according to the notation introduced in Section 6.2), then this method is essentially the same as the secret prefix method itemized above. Consequently, we don't address it separately here.

#### 10.3.2.1 Secret Prefix Method

As mentioned above, the secret prefix method consists of prepending a secret key $k \in \mathcal{K}$ to the message $m \in \mathcal{M}$ before it is hashed with $h$. This construction can be formally expressed as follows:

$$MAC_k(m) = h(k \parallel m)$$

If $m$ is an $i$-block message $m = m_1 \parallel m_2 \parallel \ldots \parallel m_i$ and $h$ is an iterated hash function, then the construction can also be expressed as follows:

$$MAC_k(m) = h(k \parallel m_1 \parallel m_2 \parallel \ldots \parallel m_i)$$

In this case, the secret prefix method is insecure. Anybody who knows a single message-MAC pair $(m, MAC_k(m))$ can selectively forge a MAC for a message $m'$ that has the known message $m$ as a prefix (i.e., $m' = m \parallel m_{i+1}$). If one considers Figure 4.2 and the way an iterated hash function $h$ is constructed (using a compression function $f$), then one easily notices that $MAC_k(m')$ can be computed from $MAC_k(m)$ as follows:

$$MAC_k(m') = f(MAC_k(m) \parallel m_{i+1})$$

Consequently, an adversary who knows $MAC_k(m)$ and $m_{i+1}$ can easily compute $MAC_k(m')$ without knowing $k$. This procedure can be repeated for an arbitrary sequence of message blocks $m_{i+2}, m_{i+3}, \ldots$ Consequently, the messages for which a MAC can be selectively forged are restricted to those having a message with a known MAC as a prefix. This restriction is not particularly strong.

Tsudik was aware of this type of *message extension* or *padding attack*, and he suggested three possibilities to defeat it:

- Only part of the hash value is taken as output (e.g., only 64 bits).
- The messages are always of fixed length.
- An explicit length field is included at the beginning of a message.

The first two possibilities are not very practical, and hence the third possibility is the one that can be used in practice. Unfortunately, almost all iterated hash functions that are used in the field follow the Merkle-Damgård construction (see Section 6.2) and encode the length of the message at the end of the message (instead of the beginning of the message). If this format were changed a little bit, then the secret prefix method could be securely used in practice. In its current form, however, the secret prefix method is too dangerous to be used in any nontrivial setting.

#### 10.3.2.2 Secret Suffix Method

Due to the message extension attack mentioned above, the secret suffix method seems to be preferable at first glance. As mentioned above, the secret suffix method consists of appending the key $k$ to the message $m$ before it is hashed with the cryptographic hash function $h$. The construction looks as follows:

$$MAC_k(m) = h(m \parallel k)$$

If $h$ is an iterated hash function, then the secret suffix method has a structural weakness.[16] Whether this weakness can be exploited mainly depends on the cryptographic hash function $h$ (or rather its compression function $f$). To understand the weakness, it is important to see that

$$\begin{aligned} MAC_k(m) &= h(m \parallel k) \\ &= h(m_1 \parallel m_2 \parallel \ldots \parallel m_i \parallel k) \\ &= f(\underbrace{f(f(\ldots f(f(m_1) \parallel m_2) \parallel \ldots) \parallel m_i)}_{h^*(m)} \parallel k) \end{aligned}$$

16 If the message is very short (i.e., there is only one iteration of the compression function), then the secret prefix method exhibits the same weakness.

In this notation, $h^*(m)$ refers to the hashed message $m$ without initialization and padding (to simplify the exposure). The point to note is that $h^*(m)$ does not depend on $k$, and hence anybody can determine this value for a message $m$ of his or her choice. This possibility can be turned into a (partly) known-message attack, in which the adversary tries to reveal the key $k$ or some partial information about it. While it is unlikely that the compression functions of currently deployed cryptographic hash functions are susceptible to this attack, other hash functions may not fare as well. Consequently, there is incentive to go for a more secure method in the first place. This is where the envelope method comes into play.

#### 10.3.2.3 Envelope Method

The envelope method combines the prefix and suffix methods. As mentioned earlier, the envelope method consists of prepending a key $k_1$ and appending another key $k_2$ to the message $m$ before it is hashed with $h$. The construction can be formally expressed as follows:

$$MAC_{k_1,k_2}(m) = h(k_1 \parallel m \parallel k_2)$$

Until the mid-1990s, people thought that this method is secure and that breaking it requires a simultaneous exhaustive key search for $k_1$ and $k_2$ (see, for example, [18] for a corresponding line of argumentation). In 1995, however, it was shown by Bart Preneel and Paul van Oorschot that this is not the case and that there are more efficient attacks against the envelope method than to do a simultaneous exhaustive key search for $k_1$ and $k_2$ [19]. Since then, the envelope method is slowly being replaced by some alternative methods.

#### 10.3.2.4 Alternative Methods

After Tsudik published his results, many cryptographers turned their interest to the problem of using keyed one-way hash functions for message authentication and finding security proofs (e.g., [20, 21]). Most importantly, Mihir Bellare, Ran Canetti, and Hugo Krawczyk developed a pair of message authentication schemes—the *nested MAC* (NMAC) and the *hashed MAC* (HMAC)—that can be proven to be secure as long as the underlying hash function is strong (in a cryptographic sense) [22]. From a practical point of view, the HMAC construction has become particularly important. It is specified in informational RFC 2104 [23] and has been adopted by many standardization bodies for Internet applications.

The HMAC construction uses the following pair of 64-byte strings:

- The string *ipad* (standing for "inner pad") consists of the byte `0x36` (i.e., `00110110`) repeated 64 times.
- The string *opad* (standing for "outer pad") consists of the byte `0x5C` (i.e., `01011100`) repeated 64 times.

The strings *ipad* and *opad* are both $64 \cdot 8 = 512$ bits long. If $h$ is a cryptographic hash function, $k$ a secret key,[17] and $m$ a message to be authenticated, then the HMAC construction can be formally expressed as follows:

$$HMAC_k(m) = h(k \oplus opad \parallel h(k \oplus ipad \parallel m))$$

This construction looks more involved than it actually is. It begins by appending zero bytes (i.e., `0x00`) to the end of $k$ to create a 64-byte or 512-bit string.[18] If, for example, $k$ is 128 bits long, then 48 zero bytes are appended. The resulting $48 \cdot 8 = 384$ bits and the 128 key bits sum up to a total of 512 bits. This key is then added modulo 2 to *ipad*, and the message $m$ is appended to this value. At this point in time, the cryptographic hash function $h$ is applied a first time to the entire data stream generated so far. The key (again, appended with zero bytes) is next added modulo 2 to *opad*, and the result of the first application of $h$ is appended to this value. To compute the final hash value, $h$ is applied a second time (note that this time the argument for the hash function is comparably short). Last but not least, the output of the HMAC construction may be truncated to a value that is shorter than $l$ (e.g., 80 or 96 bits). In fact, it has been shown that some analytical advantages result from truncating the output. In either case, $k \oplus ipad$ and $k \oplus opad$ are intermediate results of the HMAC construction that can be precomputed at the time of generation of $k$, or before its first use. This precomputation allows the HMAC construction to be implemented very efficiently. It is the state-of-the-art construction when it comes to message authentication based on cryptographic hash functions.

More recently, and in particular after the standardization of SHA-3 and KECCAK (Section 6.4.5), it has been realized that MACs using keyed hash functions are much simpler to construct if the hash function in use does not follow the Merkle-Damgård construction. In this case, the secret prefix method basically works and can be easily protected against message extension attacks. The resulting construction is known as KECCAK MAC (KMAC), and it is specified in NIST SP 800-185.[19] As of this writing, the KMAC construction is not yet used in the field, but this is subject to change.

17 The recommended minimal length of the key is $l$ bits.

18 If the key is longer than 64 bytes, then it must be truncated to the appropriate length.

19 https://nvlpubs.nist.gov/nistpubs/SpecialPublications/NIST.SP.800-185.pdf

### 10.3.3 Carter-Wegman MACs

At the end of Chapter 6, we mentioned that universal hashing as introduced in [24–26] provides an alternative design paradigm for cryptographic hash functions and message authentication systems [27, 28]. More specifically, universal hashing can be used to provide a one-time MAC, and—as mentioned earlier—a construction known as Carter-Wegman MAC can then be used to turn this one-time MAC into a MAC that can be used to authenticate multiple messages. So let us delve more deeply into the Carter-Wegman MAC construction.

A Carter-Wegman MAC has two ingredients: a one-time MAC $OTMAC_k(m)$ that maps an arbitrarily long message $m$ to an $n$-bit MAC (using a particular one-time key $k$) and a family $F$ of PRFs that map $n$-bit input strings to $n$-bit output strings. Each key $k$ can select a particular PRF $F_k$ from $F$. Furthermore, a Carter-Wegman MAC is always probabilistic, meaning that—in addition to some keying material—the authentication of a message $m$ always requires an additional $n$-bit random string $r$ that represents a *nonce*.[20] This suggests that different Carter-Wegman MACs may be generated for a particular message if the construction is invoked multiple times.

If $k_1$ and $k_2$ are two keys and $r$ is a nonce, then a Carter-Wegman MAC can be defined as follows:

$$CWMAC_{k_1,k_2}(m) = F_{k_1}(r) \oplus OTMAC_{k_2}(m)$$

The first key $k_1$ is used to select a PRF from $F$ and to map the nonce $r$ to a pseudorandom $n$-bit string, whereas the second key $k_2$ is used to one-time authenticate the message $m$. The results are added modulo 2 to form the Carter-Wegman MAC. The construction is efficient because the efficient one-time MAC is applied to the long message, whereas the less-efficient PRF is applied to the nonce that is relatively short (i.e., it is only $n$ bits long). One can formally prove that such a Carter-Wegman MAC is secure, if the two ingredients—the family of PRFs and the one-time MAC—are secure (even if the same key is used to authenticate multiple messages). In many Carter-Wegman MAC constructions, it is even possible to somehow quantify the level of security that can be achieved.

Note that the $n$-bit nonce $r$ must change with every CWMAC tag that is generated, and that the recipient needs to know which nonce was used by the sender in the first place. So some method of synchronization is required here. This can be done, for example, by explicitly sending the nonce with the message and the tag, or by agreeing upon the use of some other nonrepeating value, such as a sequence or

20 This artificial term stands for "number used only once." It need not be truly randomly generated, but it needs to be unique for the authentication of a particular message under a given key.

message number. Because the nonce need not be secret, the use of such a sequence number works perfectly fine in practice.

There are several Carter-Wegman MAC constructions that have been proposed in the past, and some of them are even being used today. The most prominent examples are UMAC, Poly1305, and GMAC.

#### 10.3.3.1 UMAC

The *universal MAC* (UMAC[21]) is a Carter-Wegman MAC construction that was proposed in 1999 [29] and is now being used in some Internet applications (e.g., [30]). Following the original design of a Carter-Wegman MAC, UMAC uses a family $F$ of PRFs and a family $H$ of universal hash functions (to serve as a OTMAC). In either case, a key is used to select a particular function from the respective families: $k_1$ to select a pseudorandom function $F_{k_1}$ from $F$ and $k_2$ to select a universal hash function $H_{k_2}$ from $H$. For each message, a fresh and unique nonce $r$ must be used in addition to $k_1$ and $k_2$. A UMAC tag is then constructed as follows:

$$UMAC_{k_1,k_2}(m) = F_{k_1}(r) \oplus H_{k_2}(m)$$

The default implementation of UMAC is based on the AES (as a family $F$ of PRFs) and a specifically crafted family $H$ of universal hash functions that is not repeated here (you may refer to the references given above). Depending on the desired security level, it can generate tags that are 32, 64, 96, or 128 bits long. In either case, the design of UMAC is optimized for 32-bit architectures, and VMAC is a closely related variant of UMAC that is optimized for 64-bit architectures.[22]

#### 10.3.3.2 Poly1305

Poly1305 is a Carter-Wegman MAC construction that was originally proposed by Daniel J. Bernstein in 2005 [31]. The construction uses a (block or stream) cipher to serve as a PRF family and polynomial evaluation to yield a OTMAC. The security properties are inherited from the cipher. In the original proposal, the AES was used as a block cipher, but more recently, the use of a stream cipher, such as Salsa20 or ChaCha20 [32], has also been suggested.

In the case of Poly1305-AES (and the 128-bit version of AES), the construction takes as input an arbitrarily long message $m$, a 128-bit nonce $r$, a 128-bit AES key $k_1$, and an additional 128-bit key $k_2$, whose possible values are restricted for

21 http://www.fastcrypto.org/umac and http://www.cs.ucdavis.edu/~rogaway/umac/.

22 http://www.fastcrypto.org/vmac.

performance reasons.[23] The output is a 128-bit authentication tag that is computed as follows:

$$\text{Poly1305-AES}_{k_1,k_2}(r,m) = \text{AES}_{k_1}(r) + P_m(k_2) \pmod{2^{128}} \qquad (10.1)$$

In this notation, $\text{AES}_{k_1}(r)$ refers to the encryption of $r$ with key $k_1$, whereas $P_m(k_2)$ refers to a polynomial defined by message $m$ (meaning that the coefficients of the polynomial are determined by $m$) and evaluated at $k_2$ in GF($2^{130}-5$).[24] To compute the 128-bit authentication tag, the two values are added modulo $2^{128}$. The details of the construction of $P_m$ from $m$ can be found in [31].

#### 10.3.3.3 GMAC

As mentioned and fully explained in Section 11.2.2, the Galois/counter mode (GCM) is specified in [33] (or [34] for the realm of IPsec/IKE) and yields a standardized mode of operation for block ciphers that provides AEAD. If operated in authentication-only mode (i.e., without encryption), GCM is also called *Galois message authentication code* (GMAC). While GCM/GMAC can be used with any block cipher that has a block length of 128 bits, it is most typically used with AES.

## 10.4 FINAL REMARKS

In this chapter, we elaborated on the possibilities to authenticate messages and to compute and verify MACs. We also focused on the notion of security (in the context of message authentication), and we overviewed and discussed several message authentication systems that are information-theoretically or computationally secure. From a practical viewpoint, computationally secure message authentication systems and respective MACs dominate the field. Most applications and standards that require message authentication in one way or another employ MACs that use keyed hash functions (this is mainly due to their efficiency). Most importantly, the HMAC construction is part of most Internet security protocols in use today, including, for example, the IPsec and SSL/TLS protocols (e.g., [35, 36]). In a typical implementation, the HMAC construction is based on an iterated cryptographic hash function,

23 The restrictions cause the actual key length of $k_2$ to be only 106 bits (instead of 128 bits). If $k_2$ consists of the 16 bytes $k_2[0], k_2[1], \ldots, k_2[15]$, then $k_2[3]$, $k_2[7]$, $k_2[11]$, and $k_2[15]$ are required to have their top four bits set to zero, and $k_2[4]$, $k_2[8]$, and $k_2[12]$ are required to have their two bottom bits set to zero. Using little-endian encoding, $k_2$ refers to the value $k_2 = k_2[0] + 2^8 k_2[1 + \ldots + 2^{120} k_2[15]$.

24 The name of the construction comes from the size of the Galois field, $2^{130} - 5$, that is also a prime number.

such as SHA-1 or a representative of the SHA-2 family. As already mentioned in Section 6.5, all of these functions have the problem that they operate strongly sequentially, meaning that they cannot be parallelized. This may lead to performance problems, especially in the realm of high-speed networks. Against this background, there is a strong incentive to go for message authentication systems and MAC constructions that can be parallelized. The XOR MAC constructions were historically the first proposals that went into this direction, but the PMAC construction is most promising today. Furthermore, when it comes to high-speed networks and provable security, then the Carter-Wegman MACs, such as UMAC, Poly1305, and GMAC, are good choices. It is probable and very likely that Carter-Wegman MACs will be more widely deployed in the future, and that some other constructions are going to be developed and proposed as well. Message authentication remains an active area of research in applied cryptography, and this is not likely going to change in the foreseeable future.

## References

[1] ISO/IEC 9797-1, *Information technology—Security techniques—Message Authentication Codes (MACs)—Part 1: Mechanisms using a block cipher*, 2011.

[2] ISO/IEC 9797-2, *Information technology—Security techniques—Message Authentication Codes (MACs)—Part 1: Mechanisms using a dedicated hash-function*, 2011.

[3] ISO 8731-2, *Banking—Approved Algorithms for Message Authentication—Part 2: Message Authenticator Algorithm*, 1992.

[4] U.S. Department of Commerce, National Institute of Standards and Technology, *Computer Data Authentication*, FIPS PUB 113, May 1985.

[5] Bellare, M., J. Kilian, and P. Rogaway, "The Security of Cipher Block Chaining," *Proceedings of CRYPTO '94*, Springer-Verlag, LNCS 839, 1994, pp. 341–358.

[6] Black, J., and P. Rogaway, "CBC MACs for Arbitrary-Length Messages: The Three-Key Constructions," *Proceedings of CRYPTO 2000*, Springer-Verlag, LNCS 1880, 2000, pp. 197–215.

[7] Frankel, S., and H. Herbert, *The AES-XCBC-MAC-96 Algorithm and Its Use With IPsec*, Standards Track RFC 3566, September 2003.

[8] Iwata, T., and K. Kurosawa, "OMAC: One-Key CBC MAC," *Proceedings of the 10th International Workshop on Fast Software Encryption (FSE 2003)*, Springer-Verlag, LNCS 2887, 2003, pp. 129–153.

[9] U.S. Department of Commerce, National Institute of Standards and Technology, *Recommendation for Block Cipher Modes of Operation: The CMAC Mode for Authentication*, FIPS Special Publication 800-38B, May 2005.

[10] Song, J.H., et al., *The AES-CMAC Algorithm*, Informational RFC 4493, June 2006.

[11] Song, J.H., R. Poovendran, and J. Lee, *The AES-CMAC-96 Algorithm and Its Use with IPsec*, Standards Track RFC 4494, June 2006.

[12] Song, J.H., et al., *The Advanced Encryption Standard-Cipher-based Message Authentication Code-Pseudo-Random Function-128 (AES-CMAC-PRF-128) Algorithm for the Internet Key Exchange Protocol (IKE)*, Standards Track RFC 4615, August 2006.

[13] Bellare, M., R. Guerin, and P. Rogaway, "XOR MACs: New Methods for Message Authentication Using Block Ciphers," *Proceedings of CRYPTO '95*, Springer-Verlag, 1995, pp. 15–28.

[14] Bellare, M., O. Goldreich, and S. Goldwasser, "Incremental Cryptography: The Case of Hashing and Signing," *Proceedings of CRYPTO '94*, Springer-Verlag, 1994, pp. 216–233.

[15] Black, J., and P. Rogaway, "A Block-Cipher Mode of Operation for Parallelizable Message Authentication," *Proceedings of EUROCRYPT 2002*, Springer-Verlag, LNCS 2332, 2002, pp. 384–397.

[16] Cohen, F., "A Cryptographic Checksum for Integrity Protection," *Computers & Security*, Vol. 6, No. 5, 1987, pp. 505–510.

[17] Gong, L., "Using One-Way Functions for Authentication," *ACM SIGCOMM Computer Communication Review*, Vol. 19, No. 5, October 1989, pp. 8–11.

[18] Tsudik, G., "Message Authentication with One-Way Hash Functions," *ACM SIGCOMM Computer Communication Review*, Vol. 22, No. 5, October 1992, pp. 29–38.

[19] Preneel, B., and P. van Oorschot, "MDx-MAC and Building Fast MACs from Hash Functions," *Proceedings of CRYPTO '95*, Springer-Verlag, LNCS 963, 1995, pp. 1–14.

[20] Kaliski, B., and M. Robshaw, "Message Authentication with MD5," *CryptoBytes*, Vol. 1, No. 1, Spring 1995, pp. 5–8.

[21] Preneel, B., and P. van Oorschot, "On the Security of Two MAC Algorithms," *Proceedings of EUROCRYPT '96*, Springer-Verlag, 1996.

[22] Bellare, M., R. Canetti, and H. Krawczyk, "Keying Hash Functions for Message Authentication," *Proceedings of CRYPTO '96*, Springer-Verlag, LNCS 1109, 1996, pp. 1–15.

[23] Krawczyk, H., M. Bellare, and R. Canetti, *HMAC: Keyed-Hashing for Message Authentication*, Informational RFC 2104, February 1997.

[24] Carter, J.L., and M.N. Wegman, "Universal Classes of Hash Functions," *Journal of Computer and System Sciences*, Vol. 18, 1979, pp. 143–154.

[25] Wegman, M.N., and J.L. Carter, "New Hash Functions and Their Use in Authentication and Set Equality," *Journal of Computer and System Sciences*, Vol. 22, 1981, pp. 265–279.

[26] Naor, M., and M. Yung, "Universal One-Way Hash Functions and Their Cryptographic Applications ," *Proceedings of the Twenty-First Annual ACM Symposium on Theory of Computing (STOC '89)*, ACM Press, pp. 33–43.

[27] Brassard, G., "On Computationally Secure Authentication Tags Requiring Short Secret Shared Keys," *Proceedings of CRYPTO '82*, 1982, pp. 79–86.

[28] Shoup, V., "On Fast and Provably Secure Message Authentication Based on Universal Hashing," *Proceedings of CRYPTO '96*, Springer-Verlag, LNCS 1109, 1996, pp. 313–328.

[29] Black, J. et al., "UMAC: Fast and Secure Message Authentication," *Proceedings of CRYPTO '99*, Springer-Verlag, LNCS 1666, 1999, pp. 216–233.

[30] Krovetz, T. (Ed.), *UMAC: Message Authentication Code using Universal Hashing*, Informational RFC 4418, March 2006.

[31] Bernstein, D.J., "The Poly1305-AES Message-Authentication Code," *Proceedings of the 12th International Workshop on Fast Software Encryption (FSE) 2005*, Springer-Verlag, LNCS 3557, 2005, pp. 32–49.

[32] Nir, Y., and A. Langley, *ChaCha20 and Poly1305 for IETF Protocols*, Informational RFC 7539, May 2015.

[33] U.S. Department of Commerce, National Institute of Standards and Technology, *Recommendation for Block Cipher Modes of Operation: Galois/Counter Mode (GCM) and GMAC*, Special Publication 800-38D, November 2007.

[34] McGrew, D., and J. Viega, *The Use of Galois Message Authentication Code (GMAC) in IPsec ESP and AH*, Standards Track Request for Comments 4543, May 2006.

[35] Oppliger, R., *Internet and Intranet Security*, 2nd edition. Artech House Publishers, Norwood, MA, 2002.

[36] Oppliger, R., *SSL and TLS: Theory and Practice*, 2nd edition. Artech House Publishers, Norwood, MA, 2016.

# Chapter 11

# Authenticated Encryption

In this chapter, we put symmetric encryption and message authentication together in what is now being called authenticated encryption (AE). We introduce the topic in Section 11.1, give some constructions in Section 11.2, and conclude with final remarks in Section 11.3. After two extensive chapters, this chapter is again a relatively short one, but this does not reduce the relevance of the topic.

## 11.1 INTRODUCTION

As mentioned in Section 2.2.5, people had used three approaches (or generic composition methods), EtM, E&M, and MtE, to combine symmetric encryption and message authentication in a particular order, before it was shown that EtM provides the best level of security [1]. Since then, cryptographic security protocols follow this method consistently and always apply message encryption prior to authentication. More specifically, people combine message encryption and authentication in AE or AE with associated data (AEAD) [2] when needed.

AE and AEAD systems are typically constructed by combining symmetric encryption that provides IND-CPA with a secure message authentication system that protects the authenticity and integrity of the message. More specifically, people sometimes distinguish whether message authentication protects the integrity of plaintext messages, abbreviated INT-PTXT, or the integrity of ciphertexts, abbreviated INT-CTXT. While INT-PTXT requires that it is computationally infeasible to generate a ciphertext that decrypts to a message that has not been encrypted before, INT-CTXT requires that it is even computationally infeasible to generate a ciphertext that has not previously been generated, independent of whether or not the underlying plaintext message is new in the sense that it has not been encrypted before. Consequently, INT-CTXT is a stronger requirement, but the difference is subtle and INT-PTXT is sometimes more intuitive.

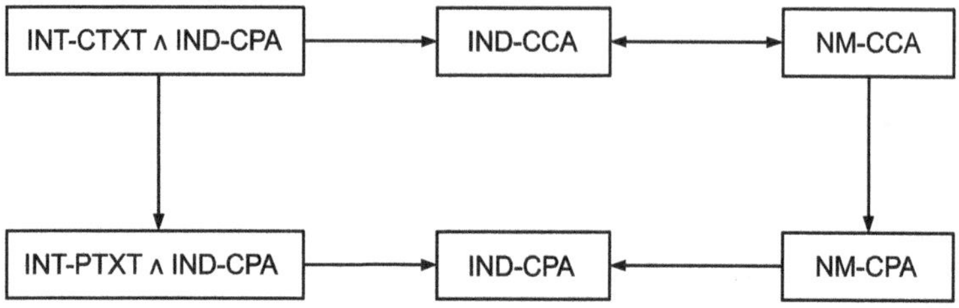


**Figure 11.1** The relations among the various notions of security for encryption (in simplified form). $A \rightarrow B$ suggests that an encryption system meeting security notion $A$ also meets notion $B$.

In Section 9.3, we already mentioned the term *malleability*, and in Chapter 13, we further delve into *nonmalleability* (NM) [3], in particular nonmalleability under CPA (NM-CPA) or under CCA (NM-CCA). As shown in [1] and illustrated in Figure 11.1 in simplified form, there are several relations among the various notions of security for (symmetric or asymmetric) encryption. First and foremost, a system that provides IND-CPA and INT-CTXT not only provides INT-PTXT (because INT-PTXT is a weaker requirement than INT-CTXT) but—more importantly—also IND-CCA. This is an important result, because IND-CCA is the preferred notion of security for every encryption system, and hence we can achieve it by requiring a system to provide IND-CPA and INT-CTXT simultaneously. The other relations are less important here, and we will revisit most of them when we address the notions of security for asymmetric encryption. In particular, we will use the result that IND-CCA and NM-CCA are equivalent a lot in the rest of the book.

## 11.2 AEAD CONSTRUCTIONS

As mentioned in Section 9.7, NIST has standardized the AEAD modes CCM [4, 5] and GCM [6]. These modes are important in the field and used in many Internet security protocols, such as TLS. With regard to encryption, CCM and GCM both use a block cipher with a block length of 128 bits (e.g., AES) operated in CTR mode. This mode is advantageous because it only requires the encryption function of the block cipher to be implemented and because respective implementations may be pipelined and parallelized. With regard to authentication, however, CCM and GCM use different MAC constructions.

In general, an AE operation has four inputs—a secret key $k$, a nonce $r$,[1] a plaintext message $m$, and some additional data $a$ that need to be authenticated (but not encrypted), and two outputs—a ciphertext $c$ and an authentication tag $t$. These two items can either represent two separate outputs or a single one that stands for both. In either case, the respective (authenticated) decryption operation takes $k$, $r$, $a$, $c$, and $t$ (or only $c$, if it also stands for $t$) as inputs and has as its single output either $m$ or a special symbol **FAIL** that states that the provided inputs are not authentic. In this case, the cryptographic operation must abort without providing any further information. Let's now have closer look at CCM and GCM.

### 11.2.1 CCM

As already mentioned, CCM uses a 128-bit block cipher (e.g., AES) operated in CTR mode for encryption and a CBC-MAC (Section 10.3.1.1) for authentication. The two operations are applied in an MtE manner, meaning that a CBC-MAC is first computed on the message and the additional data to obtain an authentication tag, and that the message and the tag are then encrypted using the block cipher in CTR mode. The resulting ciphertext stands for both the encrypted message and the authentication tag. CCM has distinct advantages and disadvantages:

- The biggest advantages are that it uses standard cryptographic primitives that are well understood, and that it can be used with a single key for both applications of the block cipher.
- Its biggest disadvantage is that it requires two applications of the block cipher for every message block. This is suboptimal to say the least.

The CCM authenticated encryption and decryption algorithms are summarized in Algorithms 11.1 and 11.2. The (authenticated) encryption algorithm takes as input a key $k$ for the block cipher, a 128-bit nonce $r$, a block of data $a$ that needs to be authenticated (but not encrypted), and a block of data $m$ that represents the plaintext message that needs to be encrypted and authenticated. The algorithm first generates a sequence $b$ of 128-bit blocks from $r$, $a$, and $m$, by formatting them according to a special formatting function named format (this function is not specified here). The resulting sequence consists of $|r|_l + |a|_l + |m|_l = 1 + |a|_l + |m|_l$ blocks $b_0, b_1, \ldots, b_{|a|_l+|m|_l}$ (note that $|x|$ refers to the bitlength of $x$, whereas $|x|_l$ refers its block length, i.e., the number of $l$-bit blocks needed to represent $x$). From these

1 Note that this nonce is sometimes also represented as initialization vector (IV). For the purpose of this explanation, it doesn't matter whether we use a nonce $r$ or an initialization vector $IV$.

**Algorithm 11.1** CCM authenticated encryption.

$(k, r, a, m)$

---

$b = \text{format}(r, a, m)$
$x_0 = E_k(b_0)$
for $i = 1$ to $(|a|_l + |m|_l)$ do $x_i = E_k(b_i \oplus x_{i-1})$
$t = \text{MSB}_{|t|}(x_{|a|_l+|m|_l})$
generate $|m|_l + 1$ counter blocks $y_0, y_1, \ldots, y_{|m|_l}$
for $i = 0$ to $|m|_l$ do $s_i = E_k(y_i)$
$s = s_1 \parallel s_2 \parallel \ldots \parallel s_{|m|_l}$
$c = (m \oplus \text{MSB}_{|m|}(s)) \parallel (t \oplus \text{MSB}_{|t|}(s_0))$

---

$(c)$

blocks, an equally long sequence of CBC encrypted blocks $x_0, x_1, \ldots, x_{|a|_l+|m|_l}$ is computed with the block cipher and key $k$ (where $x_0 = E_k(b_0)$ represents the IV). The final block is $x_{|a|_l+|m|_l}$ and the authentication tag $t$ is taken from the $|t|$ most significant bits of it. In a typical setting, $|t|$ is equal to 128 bits. After having generated $t$, the algorithm begins with the encryption part. It generates $|m|_l + 1$ counter blocks $y_0, y_1, \ldots, y_{|m|_l}$ that are ECB-encrypted to form a key stream $s_0, s_1, \ldots, s_{|m|_l}$. The first block $s_0$ is reserved to later mask $t$. All other blocks $s_1, \ldots, s_{|m|_l}$ are concatenated to form $s$. Finally, the ciphertext $c$ is generated by bitwise adding $m$ modulo 2 to the most significant $|m|$ bits of $s$, concatenated to the bitwise addition modulo 2 of $t$ and the most significant $|t|$ bits of $s_0$. Consequently, the ciphertext $c$ not only comprises the message $m$, but also the authentication tag $t$ for $m$ and $a$ in masked form. This actually turns CCM into an AEAD cipher.

The CCM (authenticated) decryption algorithm takes as input $k$, $r$, $c$, and $a$, and generates as output either $m$ or **FAIL**. The algorithm starts by verifying whether the length of the ciphertext is larger than the length of the tag (that is given in the specification). If this is not the case, then something has gone wrong and the algorithm aborts and returns **FAIL**. Otherwise (i.e., if $|c| > |t|$), then the algorithm generates the same key stream $s$ as generated by the encryption algorithm. This key stream is then used to reconstruct $m$ and $t$: $m$ is the result of a bitwise addition modulo 2 of the $|c| - |t|$ most significant bits of $c$ and the $|c| - |t|$ most significant bits of $s$, whereas $t$ is the result of a bitwise addition modulo 2 of the $|t|$ least significant bits of $c$ and the $|t|$ most significant bits of $s_0$. If at this point in time $r$, $a$, or $m$ is invalid (meaning that they are not properly formatted), then the algorithm aborts and returns **FAIL**. Otherwise, it starts verifying $t$. To do so, it constructs $b$ in the same way as the encryption algorithm has done before. The sequence of blocks from $b$ (i.e., $b_0$, $b_1, \ldots, b_{|a|_l+|m|_l}$) is CBC-encrypted to form a sequence of blocks $x_0$, $x_1, \ldots, x_{|a|_l+|m|_l}$. If the most significant $|t|$ bits of the resulting final

**Algorithm 11.2** CCM authenticated decryption.

$(k, r, c, a)$

---

if $|c| \leq |t|$ then abort and return **FAIL**
generate $|m|_l + 1$ counter blocks $y_0, y_1, \ldots, y_{|m|_l}$
for $i = 0$ to $|m|_l$ do $s_i = E_k(y_i)$
$s = s_1 \parallel s_2 \parallel \ldots \parallel s_{|m|_l}$
$m = \text{MSB}_{|c|-|t|}(c) \oplus \text{MSB}_{|c|-|t|}(s)$
$t = \text{LSB}_{|t|}(c) \oplus \text{MSB}_{|t|}(s_0)$
if $r$, $a$, or $m$ is invalid
    then abort and return **FAIL**
    else $b = \text{format}(r, a, m)$
$x_0 = E_k(b_0)$
for $i = 1$ to $(|a|_l + |m|_l)$ do $x_i = E_k(b_i \oplus x_{i-1})$
if $t \neq \text{MSB}_{|t|}(x_{|a|_l+|m|_l})$
    then return **FAIL**
    else return $m$

---

($m$ or **FAIL**)

block $x_{|a|_l+|m|_l}$ is equal to $t$, then everything is fine and $m$ is returned. Otherwise (i.e., if equality does not hold), then authentication is not guaranteed, and hence the algorithm must abort and return **FAIL** instead of $m$.

### 11.2.2 GCM

Like CCM, GCM is designed to use a 128-bit block cipher (e.g., AES) in CTR mode. But unlike CCM, the CTR mode of GCM uses a unique counter incrementation function and a message authentication construction that employs a universal hash function based on polynomial evaluation in $GF(2^{128})$. According to Section 10.3.3, this construction yields a Carter-Wegman MAC. The NIST document that specifies the GCM mode [6] also refers to an authentication-only variant of GCM called Galois message authentication code (GMAC). In short, GMAC uses GCM encryption but requires no data to be encrypted, meaning that all data are only authenticated.

$GF(2^{128})$ is an extension field of $GF(2)$. Its elements are strings of 128 bits, and its operations are addition ($\oplus$) and multiplication ($\cdot$). If $x = x_0x_1 \ldots x_{127}$ and $y = y_0y_1 \ldots y_{127}$ are two elements of $GF(2^{128})$ with $x_i$ and $y_i$ representing bits for $i = 0, 1, \ldots, 127$, then $x \oplus y$ can be implemented as bitwise addition modulo 2 and $x \cdot y$ can be implemented as polynomial multiplication modulo the distinct irreducible polynomial $f(x) = 1 + x + x^2 + x^7 + x^{128}$.

GCM employs two complementary functions: A hash function called GHASH and an encryption function called GCTR that is a variant of "normal" CTR mode encryption.

- The GHASH function is specified in Algorithm 11.3 and illustrated in Figure 11.2.[2] It takes as input a 128-bit hash subkey $h$ and $x = x_1 \parallel x_2 \parallel \ldots \parallel x_n$ that is a sequence of $n$ 128-bit blocks $x_1, x_2, \ldots, x_n$, and it generates as output a 128-bit hash value $y^{(n)}$. The function is simple and straightforward: It starts with a 128-bit block $y^{(0)}$ that is initialized with 128 zeros (written as $0^{128}$), and it then iteratively adds the next block of $x$ and multiplies the result with $h$. This is iterated $n$ times, until $y^{(n)}$ is returned as output.

**Algorithm 11.3** GHASH function used in GCM mode.

$(h, x)$

---

$y^{(0)} = 0^{128}$
for $i = 1$ to $n$ do $y^{(i)} = (y^{(i-1)} \oplus x_i) \cdot h$

---

$(y^{(n)})$

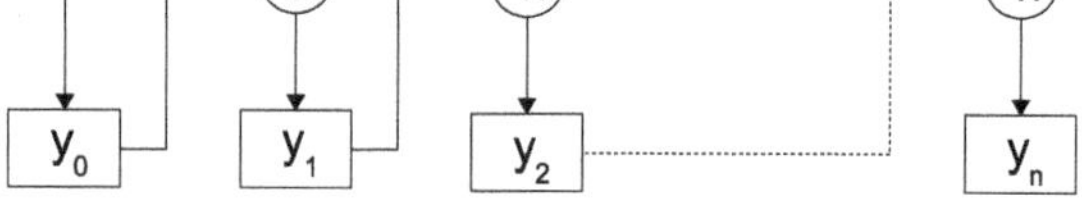


**Figure 11.2** The GHASH function.

- The GCTR encryption function is specified in Algorithm 11.4. It takes as input a key $k$, an initial counter block (ICB), and an arbitrarily long bit string $x$, and it generates as output another bit string $y$ that represents the encrypted version of $x$ using $k$ and the ICB. More specifically, $x = x_1 \parallel x_2 \parallel \ldots \parallel x_n$ is a sequence of $n$ blocks, where $x_1, x_2, \ldots, x_{n-1}$ are complete 128-bit blocks

2 While GHASH is a hash function, it is not a cryptographic one.

**Algorithm 11.4** GCTR encryption function.

$(k, \text{ICB}, x)$

---

if $x$ is empty then return empty bit string $y$
$n = \lceil |x|/128 \rceil$
$b_1 = \text{ICB}$
for $i = 2$ to $n$ do $b_i = \text{inc}_{32}(b_{i-1})$
for $i = 1$ to $n-1$ do $y_i = x_i \oplus E_k(b_i)$
$y_n = x_n \oplus \text{MSB}_{|x_n|}(E_k(b_n))$
$y = y_1 \parallel y_2 \parallel \cdots \parallel y_{n-1} \parallel y_n$

---

$(y)$

but $x_n$ does not need to be complete (i.e., $|x_n| \leq 128$). The algorithm uses a sequence of $n$ 128-bit counter blocks $b_1, b_2, \ldots, b_n$ that are encrypted and then added modulo 2 to the respective blocks of $x$. If $x_n$ is incomplete, then the respective number of $b_n$'s most significant bits are used and the remaining bits of $b_n$ are simply discarded. In the end, $y$ is compiled as the concatenation of all $n$ ciphertext blocks $y_1, y_2, \ldots, y_n$, where $y_n$ can again be incomplete. The auxiliary function $\text{inc}_s(\cdot)$ increments the least significant $s$ bits of a block and leaves the remaining $l - s$ bits unchanged. This can be formally expressed as follows:

$$\text{inc}_s(x) = \text{MSB}_{l-s}(x) \parallel [\text{int}(\text{LSB}_s(x)) + 1 \pmod{2^s}]_s$$

In GCTR and GCM, $s$ is 32 bits. This means that the first 96 bits of $x$ remain unchanged and only the last 32 bits are incremented in each step.

**Algorithm 11.5** GCM authenticated encryption.

$(k, r, m, a)$

---

$h = E_k(0^{128})$
if $|r| = 96$ then $y_0 = r \parallel 0^{31}1$
else $s = 128 \cdot \lceil |r|/128 \rceil - |r|$
$y_0 = \text{GHASH}(h, (r \parallel 0^{s+64} \parallel [|r|]_{64}))$
$c = \text{GCTR}(k, \text{inc}_{32}(y_0), m)$
$pad_a = 128 \cdot \lceil |a|/128 \rceil - |a|$
$pad_c = 128 \cdot \lceil |c|/128 \rceil - |c|$
$b = \text{GHASH}(h, (a \parallel 0^{pad_a} \parallel c \parallel 0^{pad_c} \parallel [|a|]_{64} \parallel [|c|]_{64}))$
$t = \text{MSB}_{|t|}(\text{GCTR}(k, y_0, b)$

---

$(c, t)$

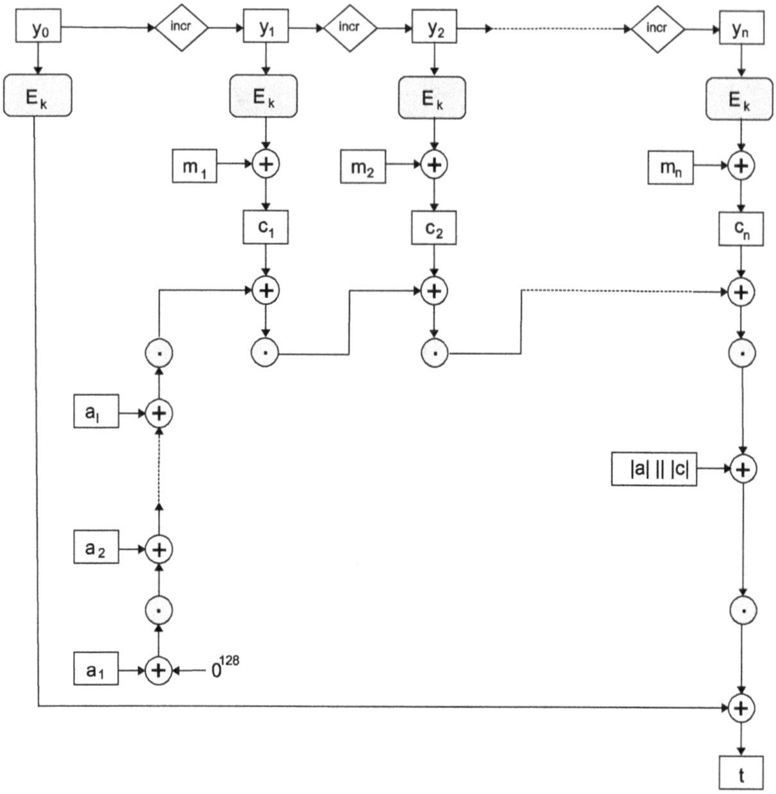


**Figure 11.3** GCM authenticated encryption.

Having prepared all the ingredients, we are now ready to delve more deeply into GCM. GCM authenticated encryption is specified in Algorithm 11.5 and partly illustrated in Figure 11.3. As is usually the case in AEAD, the algorithm takes as input a key $k$ for the block cipher in use, a variable-length nonce $r$,[3] a message $m$ that is going to be encrypted and authenticated, and some additional data $a$ that is only going to be authenticated. The message $m$ comprises $|m|$ bits that form $n$ 128-bit blocks, whereas $a$ is $|a|$ bits long and forms $l = \lceil |a|/128 \rceil$ 128-bit blocks. The output of the algorithm consists of two parts: The ciphertext $c$ that is equally long as $m$ and

3 Again, the distinction between a nonce and an IV is somehow vague. While the GCM specification uses the notion of an IV, we use the notion of a nonce. It is particularly important that the value does not repeat, and this is best characterized with the notion of a nonce.

the authentication tag $t$ that can have a variable length $|t|$. The official specification suggests $|t|$ to be 128, 120, 112, 104, or 96 bits, and for some applications even only 64 or 32 bits.

The GCM encryption algorithm first generates a subkey $h$ for the GHASH function. This value is generated by encrypting a block that consists of 128 zero bits (i.e., $0^{128}$) with the block cipher and the key $k$. It then derives a 128-bit precounter block $y_0$ from the variable-length nonce $r$. This derivation is shown in Algorithm 11.5 but is not illustrated in Figure 11.3. In the most likely case that $r$ is 96 bits long, $y_0$ is just the concatenation of $r$, 31 zero bits, and a one bit. This yields 128 bits in total. If, however, $r$ is not 96 bits long, then the construction of $y_0$ is slightly more involved. In this case, $r$ is padded some with some zero bits so that the concatenation of $r$, the zero bits, and the 64-bit length encoding of $r$ yields a string that is a multiple of 128 bits long. This string is then subject to the GHASH function with subkey $h$, so that the resulting precounter block $y_0$ is again 128 bits long. This is what Figure 11.3 starts with. The algorithm generates a sequence of counter values from $y_0$ by recursively applying the 32-bit incrementing function $\text{inc}_{32}(\cdot)$. All $y$ values except $y_0$ are then used to GCTR-encrypt the message $m$ with key $k$. This yields the ciphertext $c = c_1 \parallel c_2 \parallel \ldots \parallel c_n$. The precounter block $y_0$ is later used to encrypt the authentication tag.

To generate the authentication tag $t$, the algorithm computes the minimum numbers of zero bits, possibly none, to pad $a$ and $c$ so that the bit lengths of the respective strings are both multiples of 128 bits. The resulting values are $pad_a$ for $a$ and $pad_c$ for $c$. The algorithm then pads $a$ and $c$ with the appropriate number of zeros, so that the concatenation of $a$, $0^{pad_a}$, $c$, and $0^{pad_c}$, as well as the 64-bit length representations of $a$ and $c$ is a multiple of 128 bits long. Again, this string is subject to the GHASH function with subkey $h$. The result is $b$ and this string is input to the GCTR function—together with the key $k$ and the formerly generated precounter block $y_0$. If the tag length is $|t|$, then $t$ refers to the $|t|$ most significant bits of the output of the GCTR function. The output of the GCM authenticated encryption function is the pair that consists of $c$ and $t$.

GCM authenticated decryption works similarly, but the operations are performed in more or less reverse order. It is specified in Algorithm 11.6 and partly illustrated in Figure 11.4. The algorithm takes as input $k$ and $r$ that are the same as used for encryption, as well as $c$, $a$, and $t$, and it generates as output either $m$ or **FAIL**. First, the algorithm verifies the lengths of $r$, $c$, $a$, and $t$. If at least one of these lengths is invalid, then the algorithm aborts and returns **FAIL** (this is not explicitly mentioned in Algorithm 11.6). Next, the algorithm generates the subkey $h$ (i.e., it therefore encrypts the zero block with the block cipher and the key $k$) and the precounter block $y_0$ in the same way as before. In the next step, the message $m$ is decrypted. This step is essentially the same as in the encryption algorithm, except

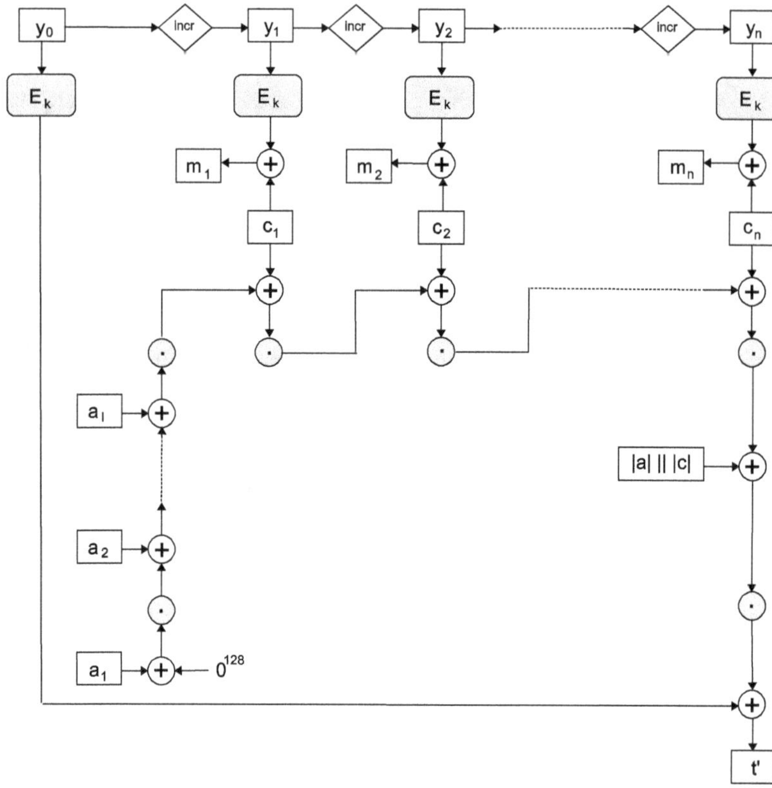


**Figure 11.4** GCM authenticated decryption.

that the $m$ and $c$ are interchanged. The computations of $pad_a$ and $pad_c$ as well as $b$ are identical. On the basis of $b$, the authentication tag $t'$ can be recomputed. Decryption is successful if and only if $t'$ equals the originally received value $t$. Otherwise, decryption fails and the algorithm signals this fact by returning **FAIL**.

It is commonly believed that the GCM mode is secure as long as a new nonce is used for every single message. If a nonce is reused, then it may become feasible to learn the authentication key (i.e., the hash subkey $h$) and to use it to forge authentication tags. Unfortunately, some Internet protocol specifications do not clearly specify how to generate nonces in a secure way. For example, the specification of the TLS protocol does not say anything about the generation of nonces for AES-GCM.

**Algorithm 11.6** GCM authenticated decryption.

$(k, r, c, a, t)$

---

verify lengths of $r$, $c$, $a$, and $t$
$h = E_k(0^{128})$
if $|r| = 96$ then $y_0 = r \,\|\, 0^{31}1$
else $s = 128 \cdot \lceil |r|/128 \rceil - |r|$
$y_0 = \text{GHASH}(h, (r \,\|\, 0^{s+64} \,\|\, [|r|]_{64}))$
$m = \text{GCTR}(k, \text{inc}_{32}(y_0), c)$
$pad_c = 128 \cdot \lceil |c|/128 \rceil - |c|$
$pad_a = 128 \cdot \lceil |a|/128 \rceil - |a|$
$b = \text{GHASH}(h, (a \,\|\, 0^{pad_a} \,\|\, c \,\|\, 0^{pad_c} \,\|\, [|a|]_{64} \,\|\, [|c|]_{64}))$
$t' = \text{MSB}_{|t|}(\text{GCTR}(k, y_0, b)$
if $t = t'$ then return $m$ else return **FAIL**

---

($m$ or **FAIL**)

Consequently, there are a few insecure implementations that reuse nonces.[4] Needless to say, these implementations are susceptible to cryptanalysis and do not provide the level of security that is otherwise anticipated with AES-GCM. To mitigate the risks of nonce reuse, people have developed and are promoting full nonce-misuse resistant AE and AEAD modes for block ciphers, such as AES-GCM-SIV[5] [7], where the acronym SIV stands for synthetic IV.[6]

## 11.3 FINAL REMARKS

In addition to CCM and GCM, there are a few other AE or AEAD modes for block ciphers, such as EAX [8] and OCB. While CCM combines CTR mode encryption with a CBC-MAC (as mentioned above), EAX combines CTR mode encryption with OMAC (Section 10.3.1.1). The result is believed to have better security properties and be less involved than CCM. But EAX is still a two-pass algorithm. So researchers have tried to develop AEAD ciphers that are less expensive and operate in a single pass. OCB is an exemplary outcome of this line of research. There are three versions of OCB in use today, ranging from OCB v1 [9] that only provides AE to OCB v3 [10] that provides full AEAD. Mainly due to some patent claims, OCB has not yet found the distribution it deserves.

4 https://eprint.iacr.org/2016/475.
5 http://eprint.iacr.org/2017/168.pdf.
6 http://cyber.biu.ac.il/aes-gcm-siv.

In spite of their importance and wide applicability in many Internet security protocols, working on AEAD ciphers and respective modes of operation for block ciphers seems to be less glamorous than working on the block ciphers themselves. This is unfortunate, because this work is key for the secure operation and use of encryption systems and block ciphers. At least some standardization bodies have initiatives in this area, such as NIST supporting AES-GCM-SIV, EAX, GCM, and OCB,[7] and ISO/IEC 19772:2009 [11] supporting OCB v2, CCM, EAX, GCM, and two additional modes for key wrapping and MtE. Outside standardization, there are a few competing proposals, such as Helix [12] and CWC [13]. These AEAD ciphers, however, are seldom used in the field, mainly because practitioners prefer standards.

More recently, people have pointed out a subtle problem that is also relevant for authenticated encryption and AE(AD) ciphers [14]: There are file formats that are not mutually exclusive in the sense that the content of a given file may be valid according to different file formats. In the case of two formats, for example, a respective file refers to a binary polyglot—it is valid for (at least) two different file formats. Two different ciphertexts may be packed into a binary polyglot. Depending on the decryption key, two different but valid plaintext messages can be recovered from the encrypted file. This works independently from the AE(AD) cipher that may otherwise be secure. It goes without saying that this may pose a serious problem and security risk in some situations. There are basically two possibilities to mitigate the risk: Either the file formats can be sharpened in a way that polyglots cannot exist, or—maybe more realistically—the AE(AD) cipher may be extended to additionally provide support for key commitment, meaning that the encryption process must also commit to the key that is being used. This should make it impossible to decrypt a given ciphertext with another key than originally anticipated. There are multiple ways to achieve this, such as adding an additional zero block prior to encryption and verifying that this block is recovered after decryption. This is conceptually similar to the quick check used in some OpenPGP implementations [15]. Anyway, it is reasonable to expect that key commitment will become relevant in authenticated encryption in the future.

## References

[1] Bellare, M., and C. Namprempre, "Authenticated Encryption: Relations among Notions and Analysis of the Generic Composition Paradigm," *Journal of Cryptology*, Vol. 21, 2008, pp. 469-491.

[2] Rogaway, P., "Authenticated-Encryption with Associated-Data.," *Proceedings of the ACM Conference on Computer and Communications Security (CCS 2002)*, ACM Press, 2002, pp. 98–107.

7 https://csrc.nist.gov/projects/block-cipher-techniques/bcm/modes-development.

[3] Dolev, D., C. Dwork, and M. Naor, "Non-Malleable Cryptography," *SIAM Journal on Computing*, Vol. 30, No. 2, 2000, pp. 391–437.

[4] U.S. Department of Commerce, National Institute of Standards and Technology, *Recommendation for Block Cipher Modes of Operation: The CCM Mode for Authentication and Confidentiality*, FIPS Special Publication 800-38C, May 2004.

[5] Whiting, D., R. Housley, and N. Ferguson, *Counter with CBC-MAC (CCM)*, RFC 3610, September 2003.

[6] U.S. Department of Commerce, National Institute of Standards and Technology, *Recommendation for Block Cipher Modes of Operation: Galois/Counter Mode (GCM) and GMAC*, Special Publication 800-38D, November 2007.

[7] Gueron, S., A. Langley, and Y. Lindell, *AES-GCM-SIV: Nonce Misuse-Resistant Authenticated Encryption*, RFC 8452, April 2019.

[8] Bellare, M., P. Rogaway, and D. Wagner, "The EAX Mode of Operation," *Proceedings of the 11th International Workshop on Fast Software Encryption (FSE 2004)*, Springer-Verlag, LNCS 3017, 2004, pp. 389–407.

[9] Rogaway, P., M. Bellare, and J. Black, "OCB: A Block-Cipher Mode of Operation for Efficient Authenticated Encryption," *ACM Transactions on Information and System Security (TISSEC)*, Vol. 6, Issue 3, August 2003, pp. 365–403.

[10] Krovetz, T., and P. Rogaway, *The OCB Authenticated-Encryption Algorithm*, RFC 7253, May 2014.

[11] ISO/IEC 19772:2009, Information technology—Security techniques—Authenticated encryption.

[12] Ferguson, N., et al., "Helix: Fast Encryption and Authentication in a Single Cryptographic Primitive," *Proceedings of the 10th International Workshop on Fast Software Encryption (FSE 2003)*, Springer-Verlag, LNCS 2887, 2003, pp. 330–346.

[13] Kohno, T., J. Viega, and D. Whiting, "CWC: A High-Performance Conventional Authenticated Encryption Mode.," *Proceedings of the 11th International Workshop on Fast Software Encryption (FSE 2004)*, Springer-Verlag, LNCS 3017, 2004, pp. 408–426.

[14] Albertini, A., et al., "How to Abuse and Fix Authenticated Encryption Without Key Commitment," Cryptology ePrint Archive, Report 2020/1456, 2020.

[15] Oppliger, R., *End-to-End Encrypted Messaging*. Artech House Publishers, Norwood, MA, 2020.

# Part III

# PUBLIC KEY CRYPTOSYSTEMS

# Chapter 12

# Key Establishment

In this chapter, we get to work in Part III and elaborate on cryptographic protocols to establish a secret key between two or more entities. More specifically, we introduce the topic in Section 12.1, outline key distribution and key agreement and respective protocols in Sections 12.2 and 12.3, address quantum cryptography in Section 12.4, and conclude with some final remarks in Section 12.5. Note that this chapter is not complete in the sense that there are many other key establishment protocols that are not addressed (see, for example, [1] for a more comprehensive overview). Also note that the problem of key establishment can also be considered for more than two entities. In this case, however, the resulting cryptographic key establishment protocols are even more involved [2], and we only briefly explore the tip of the iceberg here.

## 12.1 INTRODUCTION

In Section 2.3.1, we argued that the establishment of secret keys is a major problem and represents the Achilles' heel for the large-scale deployment of secret key cryptography, and that there are basically two approaches to resolve it:

1. The use of a KDC, such as Kerberos [3];
2. The use of a key establishment protocol.

We further made a distinction between a key distribution and a key agreement protocol—both representing key establishment protocols.

- A *key distribution protocol* can be used to securely transmit a secret key (that is generated locally or otherwise obtained) from one entity to one or several other entities.

- A *key agreement protocol* can be used by two or more entities to establish and mutually agree on a secret key. Alternatively speaking, the key is derived from information provided by all entities involved.

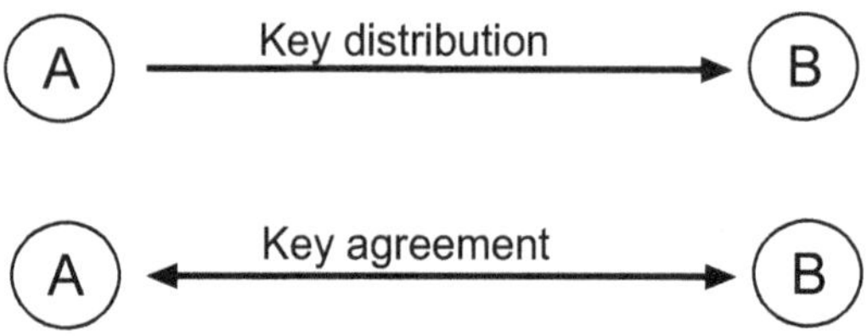


**Figure 12.1** Key distribution versus key agreement.

Figure 12.1 illustrates the notion of a key distribution as compared to a key agreement for the case of two entities. In the first case, a secret key is distributed from entity A to entity B, whereas in the second case, A and B establish and mutually agree on a secret key. So in the first case, the relationship between A and B is unidirectional, whereas the relationship is bidirectional in the second case (this is indicated by the arrows in Figure 12.1). As already mentioned in Section 2.3.1, key agreement protocols are advantageous from a security viewpoint, and hence they should be the preferred choice. The most important protocols are overviewed and briefly discussed next.

## 12.2 KEY DISTRIBUTION

With the use of asymmetric encryption, key distribution is simple and straightforward. Before we outline the respective asymmetric encryption-based key distribution protocol, we elaborate on Merkle's puzzles and Shamir's three-pass protocol. Both proposals are predecessors of what has become known as public key cryptography. As such, they are also important from a historical perspective.

### 12.2.1 Merkle's Puzzles

In 1975, Ralph C. Merkle developed and proposed an idea that is conceptually similar and closely related to public key cryptography and asymmetric encryption as it stands today [4].[1] The idea is known as *Merkle's puzzles*.

1 Note that Merkle's article appeared in the *Communications of the ACM* in 1978. It is electronically available at http://www.merkle.com/1974/PuzzlesAsPublished.pdf.

Let A and B be two entities that can communicate with each other over a public but authentic channel. A and B can then use the protocol summarized in Table 12.1 to establish a shared secret key $K_i$. The protocol takes a security parameter $n$ on either side and comprises the following three steps:

1. A generates $n$ puzzles $P_1, \ldots, P_n$ ($i = 1, \ldots, n$), randomly permutes the puzzles (using permutation $\pi$), and sends $P_{\pi(1)}, \ldots, P_{\pi(n)}$ to B. Each puzzle $P_i$ consists of an index $i$ (or a short message saying "This is puzzle $i$," respectively) and a randomly chosen secret key $k_i$, $P_i = (i, k_i)$. Solving a puzzle is feasible but requires a considerable computational effort (as explained later).

2. B randomly selects a puzzle $P_i$ from $P_{\pi(1)}, \ldots, P_{\pi(n)}$ and solves it. The solution is $(i, k_i)$, and B sends the index $i$ back to A. This transmission can be done in the clear.

3. A uses $i$ to extract the secret key $k_i$ from $P_i = (i, k_i)$, and this key can then be used to serve as a shared secret between A and B.

**Table 12.1**
Merkle's Puzzles

| A | | B |
|---|---|---|
| $(n)$ | | $(n)$ |
| Generate $P_i = (i, K_i)$ for $i = 1, \ldots, n$ Permute $P_1, \ldots, P_n$ | | |
| | $\xrightarrow{P_{\pi(1)}, \ldots, P_{\pi(n)}}$ | |
| | | Randomly select $P_i$ Solve $P_i$ |
| | $\xleftarrow{i}$ | |
| $(k_i)$ | | $(k_i)$ |

Having a closer look at the protocol, one realizes that B gets to know $k_i$ after having solved a single puzzle (i.e., $P_i$), whereas an adversary gets to know $k_i$ only after having solved all $n$ puzzles and having found the puzzle with the appropriate index $i$. On average, the adversary has to solve half of the puzzles. For sufficiently large $n$, this is computationally expensive for the adversary, and hence Merkle's puzzles can be used in theory to have two entities establish a secret key.

One possibility to generate a puzzle $P_i$ is to symmetrically encrypt $(i, k_i)$ with a key that has a fixed part and a variable part. If, for example, the variable part is 30 bits, then solving a puzzle requires $2^{30}$ tries in the worst case (or $2^{29}$ tries on average). This is the computational effort of B. If an adversary wants to compromise the key, then the computational effort is $n \cdot 2^{30}$ in the worst case (or $n/2 \cdot 2^{29} = n \cdot 2^{28}$ on the average). The computational security of Merkle's puzzles is based on the difference between $2^{30}$ (for B) and $n \cdot 2^{30}$ (for an adversary). Again, for a sufficiently large $n$, this difference may be significant.

As mentioned earlier, Merkle's puzzles are theoretically (or historically) relevant. From a more practical point of view, however, Merkle's puzzles have the problem that the amount of data that need to be transmitted from A to B is proportional to the security parameter $n$ (note that $n$ puzzles $P_{\pi(1)}, \ldots, P_{\pi(n)}$ need to be transmitted to B). This is prohibitively expensive for any reasonably sized $n$. One can play around with $n$ and the size of the variable part of the keys that are used to symmetrically encrypt the tuples $(i, k_i)$ for $i = 1, \ldots, n$, but the amount of data that need to be transmitted from A to B remains significant and prohibitively large. Furthermore, Merkle's puzzles require B to solve at least one puzzle. This is not impossible, but it may still be inconvenient for B in many situations.

### 12.2.2 Shamir's Three-Pass Protocol

Another theoretically and historically relevant key distribution protocol was proposed by Shamir in 1980. Let A and B be two entities that share no secret key initially but may have a way to encrypt and decrypt messages. $E_{k_A}$ and $E_{k_B}$ refer to A and B's encryption functions, whereas $D_{k_A}$ and $D_{k_B}$ refer to the decryption functions (the encryption function and decryption function may also be the same). In order for the encryption function and decryption function to be suitable for Shamir's three-pass protocol, they must have the property that for any plaintext message $m$, any encryption function $E_{k_1}$ with corresponding decryption function $D_{k_1}$, and any independent encryption function $E_{k_2}$,

$$D_{k_1}(E_{k_2}(E_{k_1}(m))) = E_{k_2}(m)$$

In other words, it must be possible to remove the first encryption function $E_{k_1}$ even though a second encryption function $E_{k_2}$ has been performed. This is always possible with a commutative encryption as introduced in Section 9.1. A simple three-pass protocol to secretly send a secret message, such as a session key $k$, from A to B is shown in Protocol 12.2. In this protocol, A has a secret key $k_A$ and B has another secret key $k_B$.

**Table 12.2**
Shamir's Three-Pass Protocol

| A | | B |
|---|---|---|
| $(n)$ | | $(n)$ |
| Generate $P_i = (i, K_i)$ for $i = 1, \ldots, n$ Permute $P_1, \ldots, P_n$ | | |
| | $\xrightarrow{P_{\pi(1)}, \ldots, P_{\pi(n)}}$ | |
| | | Randomly select $P_i$ Solve $P_i$ |
| | $\xleftarrow{i}$ | |
| $(K_i)$ | | $(K_i)$ |

A first randomly selects a key $k$ from the key space $\mathcal{K}$ and encrypts this key with his or her encryption function $E_{k_A}$. The resulting value

$$k_1 = E_{k_A}(k)$$

is transmitted to B. B, in turn, uses his or her encryption function $E_{k_B}$ to compute

$$k_2 \quad = \quad E_{k_B}(k_1) = E_{k_B}(E_{k_A}(k))$$

This double encrypted value is returned to A. A then uses his or her decryption function $D_{k_A}$ to decrypt $k_2$, and compute

$$\begin{aligned} k_3 &= D_{k_A}(k_2) \\ &= D_{k_A}(E_{k_B}(E_{k_A}(k))) \\ &= D_{k_A}(E_{k_A}(E_{k_B}(k))) \\ &= E_{k_B}(k) \end{aligned}$$

accordingly. This value is sent to B, and B uses his or her decryption function $D_{k_B}$ to decrypt $k_3$:

$$k = D_{k_B}(k_3) = D_{k_B}(E_{k_B}(k))$$

Both entities can now output the key $k$ and use it for symmetric encryption. But by using only symmetric encryption systems, it is not clear how to instantiate the three-pass protocol efficiently. A possibility one may think of is using additive stream ciphers, such as the one-time pad (Section 9.3). In this case, however, all encryptions cancel themselves out and the protocol gets totally insecure: Let $r_A$ be the bit sequence that A uses to compute $k_1$ and $k_3$, and $r_B$ the bit sequence that B uses to compute $k_2$. $K_1$, $k_2$, and $k_3$ can then be expressed as follows:

$$\begin{aligned} k_1 &= r_A \oplus k \\ k_2 &= r_B \oplus k_1 = r_B \oplus r_A \oplus k \\ k_3 &= r_A \oplus k_2 = r_A \oplus r_B \oplus r_A \oplus k = r_B \oplus k \end{aligned}$$

These are the values an adversary can observe in a passive (wiretapping) attack. In this case, the adversary can add $k_1$ and $k_2$ modulo 2 to retrieve $r_B$:

$$k_1 \oplus k_2 = r_A \oplus k \oplus r_B \oplus r_A \oplus k = r_B$$

This value can then be added modulo 2 to $k_3$ to determine $k$:

$$r_B \oplus k_3 = r_B \oplus r_B \oplus k = k$$

Note that this is the value that is assumed to be securely distributed. The bottom line is that a perfectly secure symmetric encryption system (i.e., one-time pad) is used, and yet the resulting key distribution protocol is totally insecure. This suggests that the use of an additive stream cipher is not reasonable to instantiate the three-pass protocol.

In his original proposal, Shamir suggested the use of modular exponentiation in $\mathbb{Z}_p^*$ instead of an additive stream cipher. The resulting three-pass protocol is known as *Shamir's three-pass protocol*—sometimes also called *Shamir's no key protocol*. It uses the mechanics of the RSA public key cryptosystem. Let A have a public key pair that consists of a public (encryption) key $e_A$ and a respective private (decryption) key $d_A$ (the capital "A" suggests that the key is long-term; i.e., not ephemeral). Both keys must be multiplicatively inverse modulo $\phi(p) = p - 1$; that is, $e_A \cdot d_A \equiv 1 \pmod{p-1}$. Let $(e_B, d_B)$ be B's public key pair that fulfills the same requirements. Shamir's three-pass protocol can then be instantiated with the following values for $k_1$, $k_2$, and $k_3$:

$$\begin{aligned} k_1 &\equiv k^{e_A} \pmod{p} \\ k_2 &\equiv (k^{e_A})^{e_B} \equiv k^{e_A e_B} \pmod{p} \\ k_3 &\equiv ((k^{e_A})^{e_B})^{d_A} \\ &\equiv ((k^{e_A})^{d_A})^{e_B} \\ &\equiv (k^{e_A d_A})^{e_B} \\ &\equiv k^{e_B} \pmod{p} \end{aligned}$$

At the end of the protocol, B can use $d_B$ to retrieve $k$:

$$k \equiv (k^{e_B})^{d_B} \equiv k^{e_B d_B} \equiv k \pmod{p}$$

In 1982, James L. Massey and Jim Omura developed and patented[2] a simple variation of Shamir's three-pass protocol that is sometimes known as the *Massey-Omura protocol*. Instead of $\mathbb{Z}_p^*$, the protocol uses modular exponentiation in an extension field $\mathbb{F}_{2^n}$ for some $n \in \mathbb{N}^+$. This allows implementations to be more efficient.

However, it is important to note that all currently known instantiations of Shamir's three-pass protocol employ modular exponentiation in one way or another. This suggests that there is no immediate advantage compared to using an asymmetric encryption system in the first place.

### 12.2.3 Asymmetric Encryption-Based Key Distribution Protocol

Asymmetric encryption-based key distribution and corresponding protocols are simple and straightforward. As illustrated in Protocol 12.3, such a protocol can be used by two entities—A and B—that share no secret key initially. B is assumed to have a public key pair of an asymmetric encryption system ($E_B$ refers to the encryption function that is keyed with $pk_B$, and $D_B$ refers to the corresponding decryption function that is keyed with $sk_B$). A randomly selects a secret key $k$ from an appropriate key space $\mathcal{K}$, encrypts it with $E_B$, and transmits $\mathsf{Encrypt}(pk_B, k)$ to B. B, in turn, uses $sk_B$ and $D_B$ to decrypt $k$; that is, $k = \mathsf{Decrypt}(sk_B, \mathsf{Encrypt}(pk_B, k))$. A and B now both share the secret key $k$ that can be used for session encryption.

**Table 12.3**
An Asymmetric Encryption-based Key Distribution Protocol

| A | | B |
|---|---|---|
| $(pk_B)$ | | $(sk_B)$ |
| $k \xleftarrow{r} \mathcal{K}$ | | |
| | $\xrightarrow{\mathsf{Encrypt}(pk_B, k)}$ | |
| | | $k = \mathsf{Decrypt}(sk_B, \mathsf{Encrypt}(pk_B, k))$ |
| $(k)$ | | $(k)$ |

2 U.S. patent 4,567,600 entitled "Method and Apparatus for Maintaining the Privacy of Digital Messages Conveyed by Public Transmission" was issued in January 1986. It expired in 2003.

Many cryptographic security protocols for the Internet make use of asymmetric encryption-based key distribution in one way or another. Examples include former versions of the SSL/TLS protocols [5] and some keying option in the Internet key exchange (IKE) protocol used in the IPsec protocol suite [6]. More recently, however, these protocols have been extended to take advantage of key agreement mechanisms.

## 12.3 KEY AGREEMENT

As mentioned in Section 1.3, Diffie and Hellman published their landmark paper entitled "New Directions in Cryptography" in 1976 [7]. The paper introduced the notion of public key cryptography and provided some evidence for its feasibility by proposing a key agreement protocol. In fact, the *Diffie-Hellman key exchange protocol*—sometimes also termed *exponential key exchange protocol*—can be used by two entities that have no prior relationship to agree on a secret key by communicating over a public but authentic channel. As such, the mere existence of the Diffie-Hellman key exchange protocol may seem paradoxical at first sight.

**Table 12.4**
Diffie-Hellman Key Exchange Protocol

| **A** | | **B** |
|---|---|---|
| $(G, g)$ | | $(G, g)$ |
| $x_a \xleftarrow{r} \mathbb{Z}_q^*$ | | $x_b \xleftarrow{r} \mathbb{Z}_q^*$ |
| $y_a = g^{x_a}$ | | $y_b = g^{x_b}$ |
| | $\xrightarrow{y_a}$ | |
| | $\xleftarrow{y_b}$ | |
| $k_{ab} = y_b^{x_a}$ | | $k_{ba} = y_a^{x_b}$ |
| $(k_{ab})$ | | $(k_{ba})$ |

The Diffie-Hellman key exchange protocol can be implemented in a cyclic group $G$ in which the DLP (Definition 5.5) is assumed to be intractable, such as the multiplicative group of a finite field. If $G$ is such a group (of order $q$) with generator $g$, then the Diffie-Hellman key exchange protocol can be formally expressed as shown in Protocol 12.4. A and B both know $G$ and $g$, and they want to agree on a shared secret key $k$. A therefore randomly selects an (ephemeral) secret exponent $x_a$ from $\mathbb{Z}_q^* = \mathbb{Z}_q \setminus \{0\} = \{1, \ldots, q-1\}$, computes the public exponent $y_a = g^{x_a}$, and sends $y_a$ to B. B does the same: It randomly selects a secret exponent $x_b$ from

$\mathbb{Z}_q^*$, computes $y_b = g^{x_b}$, and sends $y_b$ to A. A now computes

$$k_{ab} \equiv y_b^{x_a} \equiv g^{x_b x_a}$$

and B computes

$$k_{ba} \equiv y_a^{x_b} \equiv g^{x_a x_b}$$

According to the laws of exponentiation, the order of the exponents do not matter, and hence $k_{ab}$ is equal to $k_{ba}$. It is the output of the Diffie-Hellman key exchange protocol and can be used as a secret key $k$.

Let us consider a toy example to illustrate the working principles of the Diffie-Hellman key exchange protocol: From Section 5.2.1 we know that $p = 23$ is a safe prime, because $11 = (23 - 1)/2$ is prime also. This basically means that $q$ is a Sophie Germain prime and $\mathbb{Z}_{23}^* = \{1, \ldots, 22\}$ has a subgroup $G$ that consists of the 11 elements 1, 2, 3, 4, 6, 8, 9, 12, 13, 16, and 18. There are several elements that generate this group, and we take $g = 3$ here (it can be easily verified that $3^i \mod 23$ for $i = 0, 1, \ldots, 10$ generates all elements of $G$). A randomly selects $x_a = 6$, computes $y_a = 3^6 \mod 23 = 16$, and sends this value to B. B, in turn, randomly selects $x_b = 9$, computes $y_b = 3^9 \mod 23 = 18$, and sends this value to A. A now computes $y_b^{x_a} = 18^6 \mod 23 = 8$, and B computes $y_a^{x_b} = 16^9 \mod 23 = 8$. Consequently, $k = 8$ is the shared secret that may serve as a session key.

Note that an adversary eavesdropping on the communication channel between A and B knows $p$, $g$, $y_a$, and $y_b$, but does neither know $x_a$ nor $x_b$. The problem of determining $k \equiv g^{x_a x_b} \pmod{p}$ from $y_a$ and $y_b$ (without knowing $x_a$ or $x_b$) is known as the DHP (Definition 5.6). Also note that the Diffie-Hellman key exchange protocol can be transformed into a (probabilistic) asymmetric encryption system. For a plaintext message $m$ (that represents an element of the cyclic group in use), A randomly selects $x_a$, computes the common key $k_{AB}$ (using B's public key $y_b$ and following the Diffie-Hellman key exchange protocol), and combines $m$ with $k_{ab}$ to obtain the ciphertext $c$. The special case where $c = m \cdot k_{ab}$ refers to the Elgamal asymmetric encryption system that is addressed in Section 13.3.3.

If the Diffie-Hellman key exchange protocol is used natively (as outlined in Protocol 12.4), then there is a problem that is rooted in the fact that the values exchanged (i.e., $y_a$ and $y_b$) are not authenticated, meaning that the values may be modified or replaced with some other values. Assume an adversary C who is located between A and B, and who is able to actively modify messages as they are sent back and forth. Such an adversary is conventionally called a *man-in-the-middle* (MITM), and the respective attack is called a *MITM attack*. As sketched in Protocol 12.5, the Diffie-Hellman key exchange protocol is susceptible to such a MITM attack: While observing the communication between A and B, C replaces $y_a$ by $y_c$ and $y_b$ by $y_c$

(it would even be possible to use two different keys $y_c$ and $y_{c'}$ on either side of the communication channel, but this makes the attack more complex). When A receives $y_c$ (instead of $y_b$), it computes $k_{ac} = y_c^{x_a}$. On the other side, when B receives $y_c$ (instead of $y_a$), it computes $k_{bc} = y_c^{x_b}$. Contrary to a normal Diffie-Hellman key exchange, the two keys $k_{ac}$ and $k_{bc}$ are not the same, but A and B consider them to be the same. C is able to compute all keys, and to decrypt all encrypted messages accordingly. The bottom line is that A shares a key with C (i.e., $k_{ac}$) but assumes to share it with B, whereas—on the other side of the communication channel—B shares a key with C (i.e., $k_{bc}$) but assumes to share it with A. This allows C to decrypt all messages with one key and reencrypt them with the other key, making the fact that it is able to read messages that are invisible and unrecognizable to both A and B.

**Table 12.5**
A MITM Attack Against the Diffie-Hellman Key Exchange Protocol

| A | C | B |
|---|---|---|
| $(G, g)$ | | $(G, g)$ |
| $x_a \xleftarrow{r} \mathbb{Z}_q^*$ | | $x_b \xleftarrow{r} \mathbb{Z}_q^*$ |
| $y_a = g^{x_a}$ | | $y_b = g^{x_b}$ |
| | $\xrightarrow{y_a} \rightsquigarrow \xrightarrow{y_c}$ | |
| | $\xleftarrow{y_c} \rightsquigarrow \xleftarrow{y_b}$ | |
| $k_{ac} = y_c^{x_a}$ | | $k_{bc} = y_c^{x_b}$ |
| $(k_{ac})$ | | $(k_{bc})$ |

Remember that the problem that the Diffie-Hellman key exchange is susceptible to a MITM attack is rooted in the fact that the values exchanged (i.e., $y_a$ and $y_b$) are not authenticated. This means that the most obvious way to mitigate the attack is to authenticate these values. In practice, people therefore use an *authenticated Diffie-Hellman key exchange protocol* instead on an unauthenticated (native) one. In the literature, there are many proposals to authenticate the Diffie-Hellman key exchange protocol using some complementary cryptographic techniques, such as passwords, secret keys, and digital signatures with public key certificates. Many authenticated Diffie-Hellman key exchange protocols support multiple ways to provide authentication. The first proposal was the *station-to-station* (STS) protocol [8] that was proposed in the early 1990s, and that strongly influenced the IKE protocol mentioned earlier. Almost all cryptographic protocols used on the Internet today support an authenticated Diffie-Hellman key exchange in one way or another.[3]

3 One exception are the SSL/TLS protocols up to version 1.2 that also support an anonymous Diffie-Hellman key exchange (e.g., [5]).

The susceptibility of the Diffie-Hellman key exchange protocol to MITM attacks has been known since the original publication [7], and many researchers have proposed other mitigation techniques (other than to complement the Diffie-Hellman key exchange with some form of authentication). In 1984, for example, Rivest and Shamir proposed a simple technique and a respective protocol—named *interlock protocol*—that defeats some MITM attacks against a public key cryptosystem used for encryption [9]. It is outlined in Protocol 12.6. Let $(pk_a, sk_a)$ and $(pk_b, sk_b)$ be A's and B's (ephemeral) public key pairs with key length $l$. After their generation, A and B exchange the public keys and encrypt their messages with the respective public key (i.e., A encrypts $m_a$ with $pk_b$ and B encrypts $m_b$ with $pk_a$). The resulting ciphertexts $c_a$ and $c_b$ are split into two halves, and the two halves are sent in two distinct messages. This results in four messages, in which A first sends the left half of $c_a$ to B, B sends the left half of $c_b$ to A, A sends the right half of $c_a$ to B, and B sends the right half of $c_b$ to A in this order. A and B can then concatenate the two halves and decrypt the respective ciphertexts using their private keys: A decrypts $m_b$ with $sk_a$ and B decrypts $m_a$ with $sk_b$. If the two messages look reasonable, then it is unlikely that an MITM has tampered with them.

**Table 12.6**
The Interlock Protocol Used for Encryption

| **A** | | **B** |
|---|---|---|
| $(l)$ | | $(l)$ |
| $(pk_a, sk_a) = \mathsf{Generate}(1^l)$ | | $(pk_b, sk_b) = \mathsf{Generate}(1^l)$ |
| | $\xrightarrow{pk_a}$ | |
| | $\xleftarrow{pk_b}$ | |
| $c_a = \mathsf{Encrypt}(pk_b, m_a)$ | | $c_b = \mathsf{Encrypt}(pk_a, m_b)$ |
| | $\xrightarrow{\text{left half of } c_a}$ | |
| | $\xleftarrow{\text{left half of } c_b}$ | |
| | $\xrightarrow{\text{right half of } c_a}$ | |
| | $\xleftarrow{\text{right half of } c_b}$ | |
| $m_b = \mathsf{Decrypt}(sk_a, c_b)$ | | $m_a = \mathsf{Decrypt}(sk_b, c_a)$ |
| $(m_b)$ | | $(m_a)$ |

Note what happens if a MITM wants to defeat the interlock protocol. In this case, the MITM provides A with $pk_c$ (instead of $pk_b$). This means that $m_a$ gets encrypted with $pk_c$, and hence that C is, at least in principle, able to decrypt $c_a$. But C only gets the first half of $c_a$ and has to forward something meaningful; that is, something that refers to the first half of $m_a$ encrypted with $pk_b$. But C does not know

$m_a$ at this point in time, and hence it is not able to forward something meaningful. Instead, it has to forward something different that likely reveals its existence. The bottom line is that the interlock protocol does not really defeat the MITM attack; the attack can still be mounted, but it can be detected in the aftermath.

The interlock protocol as described above can be used to mitigate a MITM attack when a public key cryptosystem is used for encryption. If it is used for authentication, as suggested in [10], then some subtle attacks are feasible (e.g., [11]) and more sophisticated techniques are required. Examples include the *forced-latency protocol*,[4] Chaum's protocol,[5] and—maybe most importantly—the *encrypted key exchange* (EKE) protocol [12]. In the EKE protocol, a shared secret, such as a password, is not encrypted but used as a key to encrypt the values of a Diffie-Hellman key exchange. The original EKE protocol was later improved (e.g., [13]) and gave birth to an entire family of *authenticated key exchange* (AKE) and *password authenticated key exchange* (PAKE) methods and respective protocols.[6] Many of these protocols have been widely used and some have even been standardized.

The Diffie-Hellman key exchange and related protocols can be used in any cyclic group (other than $\mathbb{Z}_p^*$), in which the DLP is intractable, and there are basically two reasons for doing so: Either there may be groups in which the Diffie-Hellman key exchange protocol (or the modular exponentiation function) can be implemented more efficiently in hardware or software, or there may be groups in which the DLP is more difficult to solve. The two reasons are not independent from each other: If, for example, one has a group in which the DLP is more difficult to solve, then one can work with smaller key sizes (for a similar level of security). This is the major advantage of ECC introduced in Section 5.5. The ECDLP is more difficult to solve (than the DLP in $\mathbb{Z}_p^*$), and hence one can work with smaller key sizes. There are elliptic curve variants of many key agreement protocols based on the ECDLP. Examples include the elliptic curve Diffie-Hellman (ECDH) and the elliptic curve Menezes-Qu-Vanstone (ECMQV)—both of them initially became part of NSA's set of cryptographic algorithms and protocols known as Suite B (Section 18.2). ECMQV was later dropped from Suite B due to some security weaknesses.

4 The forced-latency protocol was originally proposed by Zooko Wilcox-O'Hearn in a 2003 blog entry.

5 In 2006, David Chaum filed a U.S. patent application for the protocol. In 2008, however, the U.S. Patent Office rejected most of the claims and Chaum finally abandoned the application.

6 The original EKE protocol and the augmented EKE protocol (by the same inventors) were patented in the United States (patents 5,241,599 and 5,440,635) and expired in 2011 and 2013, respectively. The expiration of these patents has simplified the situation and legal use of PAKE protocols considerably.

**Table 12.7**
ECDH Protocol

| A | | B |
|---|---|---|
| $(Curve, G, n)$ | | $(\mathbb{G}, g)$ |
| $d_a \xleftarrow{r} \mathbb{Z}_n \setminus \{0\}$ | | $d_b \xleftarrow{r} \mathbb{Z}_n \setminus \{0\}$ |
| $Q_a = d_a G$ | | $Q_b = d_b G$ |
| | $\xrightarrow{Q_a}$ | |
| | $\xleftarrow{Q_b}$ | |
| $k_{ab} = d_a Q_b$ | | $k_{ba} = d_b Q_a$ |
| $(k_{ab})$ | | $(k_{ba})$ |

Table 12.7 illustrates the ECDH protocol that is sometimes also acronymed ECDHE.[7] On either side of the protocol, $Curve$ specifies an elliptic curve $E(\mathbb{F}_q)$ over a finite field $\mathbb{F}_q$ (Section 5.5), $G$ a generator, and $n$ the order of the elliptic curve. A randomly selects a private key $d_a$ from $\mathbb{Z}_n \setminus \{0\} = \{1, \ldots, n-1\}$ and computes the public key $Q_a = d_a G$. B does the same: It randomly selects a private key $d_b$ and computes the public key $Q_b = d_b G$. A and B then exchange their public keys, so that A can compute $k_{ab} = d_a Q_b = d_a d_b G$ and B can compute $k_{ba} = d_b Q_a = d_b d_a G$. Again, the two values refer to the same point on the curve, and a session key can be derived from the two coordinates of this point. Breaking the ECDH protocol requires an adversary to solve the ECDLP. Again, this is assumed to be computationally intractable.

## 12.4 QUANTUM CRYPTOGRAPHY

Most key distribution and key agreement protocols in use today employ public key cryptographic techniques, and are based on some computational intractability assumption, such as the IFA (in the case of RSA) or the DLA (in the case of the Diffie-Hellman key exchange). The questions that arise are:

- What happens if the assumptions turn out to be wrong?
- What happens if somebody finds an efficient algorithm to solve the IFP and/or the DLP?
- What happens if somebody succeeds in building a quantum computer?

7 In this case, the final letter E stands for the fact that the Diffie-Hellman key exchange is ephemeral and short-lived.

In all of these cases, the use of public key cryptographic techniques would have to be suspended and people would have to start looking for alternatives. This is where PQC comes into play (Section 18.3). Another alternative that may become useful under these circumstances is *quantum cryptography*. This basically refers to a key establishment technology that is not based on a computational intractability assumption. Instead, it is based on the laws of quantum physics, and hence it is assumed to be secure as long as these laws apply. Consequently, any progress in solving the DLP or IFP or even the successful construction of a quantum computer is not expected to affect the security of quantum cryptography. To some extent, this is calming news, but there are also some caveats that must be considered with care and taken into account accordingly.

In this section, we provide a brief overview about quantum cryptography. More specifically, we introduce the basic principles, elaborate on a quantum key exchange protocol, and outline some historical and recent developments in the field. Note that quantum cryptography does not provide a complete replacement for public key cryptography. For example, it can neither be used to implement digital signatures nor to provide nonrepudiation services. Consequently, it is at most a fallback technology for some parts of public key cryptography.

### 12.4.1 Basic Principles

As mentioned above, quantum cryptography refers to a key establishment technology that is based on the laws of quantum physics instead of computational intractability assumptions. More specifically, it makes use of the *uncertainty principle*[8] of quantum physics to provide a secure *quantum channel*. Roughly speaking, the uncertainty principle states that certain pairs of physical quantities of an object, such as its position and velocity, cannot both be measured exactly at the same time. This theoretical result has practical implications for the exceedingly small masses of atoms and subatomic particles such as photons. In particular, it affects the way we can measure the polarization of photons.

Assume the following experiment: We send a photon to two polarization filters. The first filter is vertical, meaning that it only lets vertically polarized photons pass through, whereas the second filter is placed with some angle $t > 0$ degree. It is then reasonable to expect a photon not to pass through both filters. However, due to quantum physical effects in the world of photons, there is still a certain probability that the photon passes through both filters, and this probability depends on $t$. As $t$ increases, the probability of a vertically polarized photon passing through the second

8 The uncertainty principle was discovered by the German physicist Werner Heisenberg, and is therefore also known as the Heisenberg uncertainty principle.

filter decreases. For $t = 45^\circ$, the probability is $1/2$, and for $t = 90^\circ$, the probability reaches zero.

To measure the polarization of a photon, we can either use a *rectilinear basis* that is able to reliably distinguish between photons that are polarized with $0^\circ$ and $90^\circ$, or a *diagonal basis* that is able to reliably distinguish between photons that are polarized with $45^\circ$ and $135^\circ$. The two bases are *conjugate* in the sense that the measurement of the polarization in the first basis completely randomizes the measurement in the second basis (as in the previous experiment for $t = 45^\circ$). The bottom line is that if somebody gives a photon an initial polarization (either horizontal or vertical, but you don't know which) and you use a filter in the $45/135^\circ$ basis to measure the photon, then you cannot determine any information about the initial polarization of the photon.

These principles can be exploited to establish a quantum channel that cannot be attacked passively without detection, meaning that the fact that someone tries to eavesdrop on the channel can be detected by the communicating parties. In fact, the adversary cannot gain even partial information about the data being transmitted without altering it in a random and uncontrollable fashion that can be detected afterward. As such, the quantum channel is not secure in the sense that it cannot be attacked. But if it is attacked, then this fact can at least be detected by the communicating parties. A quantum channel can therefore be used to securely transmit information (e.g., a secret key) from one side to the other. Any attempt to tamper with the channel can be recognized by the communicating parties. Again, this fact does not depend on a computational intractability assumption; it only depends on the validity of the laws of quantum physics. Consequently, the quantum channel is provably secure in a physical sense, i.e., it is secure even against an adversary with superior technology and unlimited computational power (and even if $\mathbf{P} = \mathbf{NP}$).

### 12.4.2 Quantum Key Exchange Protocol

Based on the basic principles outlined above, Charles H. Bennett and Gilles Brassard proposed a quantum cryptography-based key exchange protocol known as *quantum key exchange* in 1984 [14]. Let A be the sender and B the receiver on a quantum channel. A may send out photons in one of four polarizations: $0^\circ$, $45^\circ$, $90^\circ$, or $135^\circ$ (we use the symbols $\longrightarrow$, $\nearrow$, $\uparrow$, and $\nwarrow$ to refer to these polarizations). On the other side of the quantum channel, B measures the polarizations of the photons it receives. According to the laws of quantum physics, B can distinguish between rectilinear polarizations (i.e., $0^\circ$ or $90^\circ$) and diagonal polarizations (i.e., $45^\circ$ or $135^\circ$), but it cannot distinguish between both types of polarization simultaneously (because the rectilinear and diagonal bases are conjugate).

**Table 12.8**
An Exemplary Execution of the Quantum Key Exchange Protocol

| | | | | | | | | | | |
|---|---|---|---|---|---|---|---|---|---|---|
| 1) | 0 | 0 | 1 | 0 | 1 | 1 | 0 | 1 | 1 | 0 |
| 2) | + | x | x | + | x | + | + | x | x | + |
| 3) | $\rightarrow$ | $\nearrow$ | $\nwarrow$ | $\rightarrow$ | $\nwarrow$ | $\uparrow$ | $\rightarrow$ | $\nwarrow$ | $\nwarrow$ | $\rightarrow$ |
| 4) | + | + | x | + | + | x | + | + | x | x |
| 5) | 0 | | 1 | 0 | *0* | *0* | 0 | *1* | 1 | *0* |
| 6) | + | | x | + | + | x | + | + | x | x |
| 7) | OK | | OK | OK | | | OK | | OK | |
| 8) | 0 | | 1 | 0 | | | 0 | | 1 | |
| 9) | | | 1 | | | | 0 | | | |
| 10) | | | OK | | | | OK | | | |
| 11) | 0 | | | 0 | | | | | 1 | |

In this situation, A and B can use Bennett and Brassard's quantum key exchange protocol to agree on a secret key. Table 12.8 summarizes an exemplary transcript of the protocol. First, A chooses a random bitstring (from which the secret key is going to be distilled) and a random sequence of polarization bases, where each basis can be either rectilinear or diagonal. In Table 12.8, this is illustrated in lines 1 and 2 (+ refers to a rectilinear basis and x refers to a diagonal basis). Each bit is then encoded with the respective polarization basis. For example, a horizontal or 45° photon can be used to represent a 0, whereas a vertical or 135° photon can be used to represent a 1. Line 3 illustrates the polarization of the photons that are actually sent from A to B. Again, the polarization can be 0° (i.e., $\rightarrow$), 45° (i.e., $\nearrow$), 90° (i.e., $\uparrow$), or 135° (i.e., $\nwarrow$). If A has chosen a rectilinear basis for a particular bit, then a zero is encoded as $\rightarrow$ and a one is encoded as $\uparrow$. If, however, A has chosen a diagonal basis, then a zero is encoded as $\nearrow$ and a one is encoded as $\nwarrow$.

If an adversary wants to measure the polarization of a photon transmitted from A to B, then he or she must decide what basis to use.

- If the measurement is made with the correct basis, then the measurement also yields the correct result.
- If, however, the measurement is made with the wrong basis, then the measurement yields the correct result with a probability of only 1/2, and—maybe more worrisome—the measurement randomizes the polarization of the photon.

Consequently, without knowing the polarization bases originally chosen by A, an adversary has only a negligible chance of guessing them and correctly measuring the polarization of all photons. More likely, he or she is going to cause errors that can be detected afterward.

As B receives the photons, it must decide what basis to use to measure a photon's polarization (note that B does not know either which bases to use for the measurements, so from the quantum channel's perspective, there is no difference between B and an adversary). So B can only randomly choose the polarization bases. In Table 12.8, the polarization bases chosen by B are illustrated in line 4. At some positions, B's choices are equal to A's choices, but at some other positions, they are not. According to line 5, B decodes each result as a either 0 or 1—depending on the outcome of the measurement. Note that not all of values need to be correct. In fact, only the ones that are measured with the correct polarization basis yield correct results. In line 5, the values that are not necessarily correct are written in italics. Also, B may miss the reception of specific photons. In line 5, for example, B has missed measuring the polarization of the second photon.

Anyway, the bit string received by B in this example is 010000110. It represents the *raw key*. A and B's task is now to find out which bits they can use (because they have used the same basis for encoding and decoding). On average, half of the raw key's bits can be used, but A and B must find out which ones. This operation is known as *sifting*: it can take place over a public channel, as long as the channel is authentic. The authenticity of the channel can be ensured physically or logically. In the second case, A and B must share a secret key that can be used to authenticate message origin with a secure MAC (Section 10.2). Needless to say, the quantum key exchange protocol then works as a method of "key expansion" rather than a method of "key generation."

A and B can use the public channel to sift the key. B therefore sends to A over the public channel the type of polarization measurements (but not the results), and A sends to B which measurements were actually of the correct type. In Table 12.8, these steps are illustrated in lines 6 and 7. The respective zeros and ones then form the *sifted key* that is typically half as long as the raw key. In the example, the sifted key is 01001.

In practice, all communication channels have a nonzero error probability. This is particularly true for quantum channels that use photons, which represent a very small amount of energy, to carry bit values. These photons may be absorbed or modified in transit. In practical implementations, the intrinsic error rate is of the order of a few percent. Consequently, a *key distillation* phase is required. It typically consists of two steps:

1. In an *error correction* step, A and B run an error correction protocol to remove all errors from the sifted key. The result is an errorless key known as the *reconciled key*. A and B can estimate from the error rate of the sifted key

and the number of parity bits revealed during error correction the amount of information an adversary could have obtained about the reconciled key.

2. In a *privacy amplification* step, A and B can reduce the amount of information to an arbitrary low level. This is achieved by compressing the reconciled key by an appropriate factor. The result is the *distilled key*.

So key distillation allows A and B to generate a secure key from the sifted one, even in the case of an imperfect quantum channel. This is not illustrated in Table 12.8.

Last but not least, it remains to be seen how A and B can decide whether the sifted, reconciled, and possibly distilled key is the same or different on either side.

- If it is the same, then A and B can be sure with a high probability that no eavesdropping has taken place on the quantum channel;
- If it is different, then the quantum channel is likely to be subject to eavesdropping.

A simple and straightforward solution is for A and B to publicly compare some of the bits on which they think they should agree. The position of these bits must be chosen randomly after the quantum transmission has been completed. Obviously, this process sacrifices the secrecy of these bits. Because the bit positions used in this comparison are only a random subset of the correctly received bits, eavesdropping on more than a few photons is likely to be detected. If all comparisons agree, A and B can conclude that the quantum channel has been free of significant eavesdropping. Therefore, most of the remaining bits can safely be used as a one-time pad for subsequent communication over the public channel. When this one-time pad is used up, the protocol can be repeated arbitrarily many times.

In line 9 of Table 12.8, B reveals 2 out of 5 bits chosen at random; that is, 1 and 0, and in line 10, A confirms these bits (if they are correct). Line 11 shows the remaining bits 0, 0, and 1; they may now serve as the shared secret key. The eavesdropping-detection subprotocol as described in lines 9 to 11 is rather wasteful because a significant proportion of the bits ($2/5$ in the example given here) is sacrificed to obtain a good probability that eavesdropping is detected even if attempted on only a few photons. There are more efficient possibilities and eavesdropping-detection subprotocols not addressed here.

### 12.4.3 Historical and Recent Developments

The field of quantum cryptography was pioneered by Stephen Wiesner in the late 1960s and early 1970s [15]. Wiesner had two applications in mind:

- Making money that is impossible to counterfeit;
- Multiplexing two or three messages in such a way that reading one destroys the other(s).

He introduced the concept of quantum conjugate coding. In the early 1980s, Bennett and Brassard took Wiesner's ideas and applied them to cryptography [14, 16]. Their most important contribution was the quantum key exchange overviewed above.

Since the early work of Bennett and Brassard, many researchers have contributed to quantum cryptography in many ways. In 1991, for example, Artur Ekert proposed an alternative quantum key exchange protocol that uses a source to send entangled photons to A and B (note that in Bennett and Brassard's quantum key exchange protocol there are photons sent from A to B, or vice versa, but there is neither a source nor entangled photons). If an adversary measures the quantum state of some photons, then he or she automatically disturbs the respective entanglements, and this, in turn, leads to the fact that a well-known inequality in quantum physics—known as Bell inequality—is no longer violated. Similar to Bennett and Brassard's quantum key exchange protocol, this reveals an eavesdropper. In addition to quantum key distribution protocols (using polarized or entangled photons), many other quantum cryptographic protocols have been developed and proposed, such as quantum protocols for oblivious transfer or bit commitment. As well, a few companies sell quantum cryptographic devices and products, such as ID Quantique[9] and MagiQ Technologies.[10]

From a practical viewpoint, the major challenge in quantum cryptography is to overcome long distances, given the fact that quantum transmissions are necessarily weak and cannot be amplified in transit (an amplifier cannot measure and retransmit the signals; otherwise it would look like an adversary from the quantum system's perspective). The first prototype implementation of the quantum key exchange of Bennett and Brassard overcame a distance of approximately 30 cm. This distance was successfully stretched. In 2002, for example, a team of researchers presented a fiber-optical quantum key distribution system and performed a key exchange over 67 km between Geneva and Lausanne [18].

In order to increase the reach and key creation rate of a quantum cryptosystem, people have tried several approaches. First of all, it is important to note that it does not appear to be realistic to improve the physical properties (e.g., transparency or attenuation) of the optical fibers. The fibers currently in use have been improved over the last decades and their quality is close to the physical limit. A more realistic possibility to increase the reach and key creation rate is to use better photon

9 https://www.idquantique.com.

10 https://www.magiqtech.com.

detectors. For example, devices based on superconducting elements that feature an essentially noise-free operation have been built and demonstrated [19]. Another approach is to replace attenuated laser pulses by single photons. There are many research and development groups working on this approach, but it is not yet ready for prime time.

Given the current situation, one can argue that experimental quantum cryptosystems will be able to generate raw keying material at rates of 10 megabit per second over a distance of 50 kilometers or 1 megabits per second over 100 kilometers within the next couple of years. This suffices for experimental use cases, but it does not suffice for commercial applications.

## 12.5 FINAL REMARKS

In this chapter, we elaborated on cryptographic protocols that two entities can use to establish a secret key. Among these protocols, key agreement protocols are particularly useful, mainly because they allow both entities to participate in the key generation process. If this is not the case (such as in the case of a key distribution protocol), then the cryptographic strength of the secret key is bounded by the quality of the entity that actually generates the key. If this entity employs a cryptographically weak PRG, then the resulting secret key is also weak. Contrary to that, all PRGs of all entities involved in a key agreement must be cryptographically weak, so that the resulting secret key is also weak.

The Diffie-Hellman key exchange protocol is omnipresent in security applications today. Whenever two entities want to establish a secret key, the Diffie-Hellman key exchange protocol can be used and provides an elegant solution. In the future, it is hoped that alternative key agreement protocols are developed and deployed. In the meantime, however, the Diffie-Hellman key exchange protocol still defines the state of the art.

In the second part of this chapter, we introduced the basic principles of quantum cryptography and elaborated on the quantum key exchange protocol. This protocol is interesting because it is unconditionally secure and does not depend on a computational intractability assumption. Instead, it depends on the laws of quantum physics. As such, the security of the quantum key exchange protocol is independent from any progress that is made in solving mathematical problems, such as the IFP or the DLP. As of this writing, quantum cryptography is not yet practical for real-world applications, but it may become useful in the future. This is particularly true if quantum computers can be built (Section D.5).

## References

[1] Boyd, C., A. Mathuria, and D. Stebila, *Protocols for Key Establishment and Authentication*, 2nd edition. Springer-Verlag, New York, 2019.

[2] Hardjono, T., and L.R. Dondeti, *Multicast and Group Security*. Artech House Publishers, Norwood, MA, 2003.

[3] Oppliger, R., *Authentication Systems for Secure Networks*. Artech House Publishers, Norwood, MA, 1996.

[4] Merkle, R., "Secure Communication over Insecure Channels," *Communications of the ACM*, Vol. 21 No. 4, April 1978, pp. 294–299.

[5] Oppliger, R., *SSL and TLS: Theory and Practice*, 2nd edition. Artech House Publishers, Norwood, MA, 2016.

[6] Frankel, S., *Demystifying the IPsec Puzzle*. Artech House Publishers, Norwoord, MA, 2001.

[7] Diffie, W., and M.E. Hellman, "New Directions in Cryptography," *IEEE Transactions on Information Theory*, Vol. 22 No.6, 1976, pp. 644–654.

[8] Diffie, W., P. van Oorshot, and M.J. Wiener, "Authentication and Authenticated Key Exchanges," *Designs, Codes and Cryptography*, Vol. 2, No. 2, 1992, pp. 107–125.

[9] Rivest, R., and A. Shamir, "How to Expose an Eavesdropper," *Communications of the ACM*, Vol. 27, No. 4, April 1984, pp. 393–395.

[10] Davies, D.W., and W.L. Price, *Security for Computer Networks*, John Wiley & Sons, New York, NY, 1989.

[11] Bellovin, S.M., and M. Merritt, "An Attack on the Interlock Protocol When Used for Authentication," *IEEE Transactions on Information Theory*, Vol. 40, No. 1, January 1994, pp. 273–275.

[12] Bellovin, S.M., and M. Merritt, "Encrypted Key Exchange: Password-Based Protocols Secure Against Dictionary Attacks," *Proceedings of the IEEE Symposium on Research in Security and Privacy*, 1992, pp. 72–84.

[13] Bellovin, S.M., and M. Merritt, "Augmented Encrypted Key Exchange: A Password-Based Protocol Secure Against Dictionary Attacks and Password File Compromise," *Proceedings of the 1st ACM Conference on Computer and Communications Security*, ACM Press, 1993, pp. 244–250.

[14] Bennett, C.H., and G. Brassard, "Quantum Cryptography: Public Key Distribution and Coin Tossing," *Proceedings of the International Conference on Computers, Systems and Signal Processing*, Bangalore, India, December 1984, pp. 175–179.

[15] Wiesner, S., "Conjugate Coding," *ACM SIGACT News*, Vol. 15, No. 1, Winter-Spring 1983, pp. 78–88, original manuscript written in 1970.

[16] Wiesner, S., "Conjugate Coding," *ACM SIGACT News*, Vol. 15, No. 1, Winter-Spring 1983, pp. 78–88, original manuscript written in 1970.

[17] Ekert, A.K., "Quantum Cryptography Based on Bell's Theorem," *Physical Review Letters*, Vol. 67, No. 6, August 1991, pp. 661–663.

[18] Stucki, D., et al. "Quantum Key Distribution over 67 km with a Plug & Play System," *New Journal of Physics*, Vol. 4, 2002, pp. 41.1–41.8.

[19] Hiskett, P.A., et al. "Long-Distance Quantum Key Distribution in Optical Fiber," *New Journal of Physics*, Vol. 8, 2006, pp. 193.1–193.7.

# Chapter 13

# Asymmetric Encryption

In this chapter, we elaborate on asymmetric encryption. More specifically, we introduce the topic in Section 13.1, address probabilistic encryption in Section 13.2, overview and discuss some asymmetric encryption systems in Sections 13.3, introduce the notions of identity-based encryption and fully homomorphic encryption in Sections 13.4 and 13.5, and conclude with some final remarks in Section 13.6.

## 13.1 INTRODUCTION

In Section 2.3.2, we introduced the idea of using a trapdoor function to come up with an encryption system that is asymmetric in nature. It is asymmetric because the encryption and decryption algorithms employ different keys (i.e., a public key $pk$ and a respective private key $sk$). We also defined an asymmetric encryption system to consist of three efficiently computable algorithms, Generate, Encrypt, and Decrypt, with two of them—Encrypt and Decrypt—being inverse to each other (Definition 2.12).

The working principle of an asymmetric encryption system is illustrated in Figure 2.9. The sender (on the left side) encrypts a plaintext message $m$ and sends the respective ciphertext $c$ to the recipient (on the right side). This is simple and straightforward, if $m$ is short, meaning that the message length is smaller than a maximum value $max$ (e.g., 2,048 bits). Otherwise, $m$ must be split into a sequence of message blocks

$$m_1, m_2, \ldots, m_n$$

each of which is shorter than $max$. Each message block $m_i$ ($i = 1, \ldots, n$) must then be encrypted individually or sequentially in a specific mode of operation.

If a message (block) $m$ is to be encrypted for recipient B, then it must be encrypted with the public key $pk_B$ of B. Hence, the encryption of $m$ can be formally expressed as follows:

$$c = \mathsf{Encrypt}(pk_B, m)$$

B knows its private key $sk_B$ and can use it to properly decrypt the original message block:

$$m = \mathsf{Decrypt}(sk_B, c)$$

For most parts of this chapter, we only consider the case in which a single message $m$ is encrypted, but keep in mind that this message can be part of a possibly much longer message. In this case, $m$ is just a message block $m_i$ in a sequence of such blocks, but the working principles are essentially the same.

Similar to a symmetric encryption system, one may wonder whether a given asymmetric encryption system is secure. First of all, we note that information-theoretic (or perfect) security does not exist in the realm of asymmetric encryption. This is because the Encrypt algorithm works with a public key (i.e., a key that is publicly known), and hence an adversary who is given a ciphertext can always use this key and mount a brute-force attack to find the appropriate plaintext message (i.e., the one that encrypts to the given ciphertext). Such an attack may take a huge amount of time, but with regard to unconditional security, this does not matter, and the fact that the attack is theoretically feasible is sufficient here. Consequently, the best one can achieve is (some possibly strong notion of) conditional or computational security.

According to Section 1.2.2, we must specify the adversary's capabilities and the task he or she is required to solve to meaningfully discuss and argue about the security of a cryptosystem. This also applies to asymmetric encryption.

- With regard to the first point (i.e., the adversary's capabilities), we introduced the notions of ciphertext-only, known-plaintext, CPA, CCA,[1] and CCA2 in Section 9.1. We reuse these terms here. Ciphertext-only and known-plaintext attacks are important and certainly attacks one always has to mitigate. Because the encryption key is public in an asymmetric encryption system, (adaptive) CPAs are also trivial to mount. This is not necessarily true for CCAs and CCA2s. Because the private key is secret, it may not be possible for an adversary to have a ciphertext of his or her choice be decrypted (unless he or she has access to a decryption device that serves as an oracle). But in 1998,

1 In the realm of asymmetric encryption, the term *chosen-ciphertext attack* was introduced in [1].

it was shown that some forms of CCAs can be mounted against cryptographic protocols like SSL [2]. This has changed the way we think about CCA(2), and how feasible it is to mount such attacks in the field. Since then, the design of asymmetric encryption systems that are secure against CCA(2) has become an important field of study.

- With regard to the second point (i.e., the task he or she is required to solve), there are again several possibilities, and these possibilities lead to different notions of security (as discussed next).

The simplest and most straightforward notion is *one-way security*, meaning that it is computationally infeasible to determine a plaintext message from a given ciphertext and public key (and hence to invert the one-way function on which the asymmetric encryption system is based). At first glance, the adversary can only mount a ciphertext-only attack in this setting, but given the fact that he or she has access to the public key, he or she can also mount a CPA—even an adaptive one. In either case, the task he or she is required to solve is to determine the plaintext message from a given ciphertext.

One-way security sounds like a reasonable requirement for asymmetric encryption, but things are more involved in practice. If, for example, an asymmetric encryption system is only one-way secure, then it may still be feasible to determine the plaintext message from a given ciphertext (e.g., if the plaintext message is of a special form or is chosen according to a special distribution). Similarly, even if the plaintext message cannot be computed efficiently, some partial information may still leak. Hence, one-way security seems to be a lower bound for the notion of security that is required, and there are other—possibly stronger—notions of security. Some of them were mentioned in Sections 9.4 and 11.1 (and illustrated in Figure 11.1).

Most importantly, the notions of semantic security and indistinguishability of ciphertext (or ciphertext indistinguishability) were formally introduced and defined by Shafi Goldwasser and Silvio Micali in the early 1980s [3] in the realm of probabilistic encryption (Section 13.2). We already introduced these notions of security for symmetric encryption systems, but they can also be used to argue about the security of asymmetric encryption systems.

The notion of semantic security is appropriate to argue informally about the security of an encryption system. But when it comes to security proofs, the notion of ciphertext indistinguishability is more adequate. In the realm of asymmetric encryption, we have to modify the IND-CPA game from Section 9.4 a little bit. Here, the challenger has a public key pair $(pk, sk)$ and the adversary is given the public key $pk$. He or she can use this key to encrypt arbitrarily many messages of his or her choice. At some point in time, the adversary has to generate two equally long plaintext messages $m_0$ and $m_1$ and send them to the challenger in random order. The challenger selects $b \in_R \{0,1\}$ and encrypts $m_b$ with $pk$; that is, $c = \text{Encrypt}(pk, m_b)$. This ciphertext $c$ is sent to the adversary, whose task is to decide whether $c$ is the encryption of $m_0$ or $m_1$. If the best he or she can do is guessing, then the encryption provides IND-CPA, and this, in turn, means that it is also semantically secure. If, in addition to mounting a CPA, the adversary has access to a decryption oracle that he or she can feed with some arbitrary ciphertexts (except, of course, the ciphertext $c$ the adversary is challenged with), then the underlying asymmetric encryption system provides IND-CCA (or IND-CCA2, if the adversary can proceed adaptively).

In spite of the fact that the notions of semantic security and ciphertext indistinguishability are similar, it is not clear that they are in fact equivalent. Intuitively, this can be seen as follows: Semantic security requires that basically nothing about the plaintext message can be computed from a ciphertext, but then, one should not be able to distinguish ciphertexts of two different plaintext messages, which is ciphertext indistinguishability. If, on the other hand, one is not able to distinguish the ciphertexts of two plaintext messages, then one cannot learn anything about a plaintext message given a ciphertext, which is semantic security. Formally, Goldwasser and Micali showed that ciphertext indistinguishability implies semantic security [3], and Micali, Charles Rackoff, and Bob Sloan later showed that semantic security implies ciphertext indistinguishability [4]. Both results are under CPA. The equivalence of semantic security and IND-CPA allows many practically relevant cryptosystems to be proven secure. But there are two remarks to make at this point:

- The encryption algorithm of an encryption system that provides IND-CPA (and hence is semantically secure) must be probabilistic. Otherwise, the adversary could simply encrypt $m_0$ and $m_1$ (with the public key $pk$ that is known to him or her), and decide which of the two ciphertexts matches $c$. Consequently, many practically relevant asymmetric encryption systems (that are deterministic in nature) cannot be semantically secure. Examples include the RSA and Rabin asymmetric encryption systems. These systems, however, can be made semantically secure by applying an appropriate padding scheme prior to encryption. We revisit this topic when we elaborate on *optimal asymmetric encryption padding* (OAEP) later in this chapter.
- Having an encryption system that provides IND-CPA (and hence is semantically secure) is particularly useful if the message space is sparse (i.e., it includes only a few messages, such as "yes" and "no" or "buy" and "sell"). This situation is typical in many practical applications and application settings.

Ciphertext indistinguishability and semantic security are the commonly accepted notions of security for (asymmetric) encryption systems. There are, however, also other notions of security. For example, a more intricate notion of security is *nonmalleability* (NM) [5]. In essence, an asymmetric encryption system is *nonmalleable* if it is computationally infeasible to modify a ciphertext in a way that it has a predictable effect on the underlying plaintext message. Alternatively speaking, an asymmetric encryption system is nonmalleable if there is no efficient algorithm that given a ciphertext $c$ can generate a different ciphertext $c'$ such that the respective plaintext messages $m$ and $m'$ are related in some known (and predictable) way. For example, when given the ciphertext of a bid in an auction, it should be computationally infeasible for an adversary to come up with a ciphertext of a smaller bid—at least not with a success probability that is greater than without being given the ciphertext.

As shown in Figure 11.1, nonmalleability under CCA (NM-CCA) is equivalent to IND-CCA, and hence the two notions of security are often used synonymously and interchangeably. Note, however, that this equivalence only holds for CCA and not for CPA. In fact, NM-CPA implies IND-CPA, but the converse in not known to be true. There are many other relationships that have been shown in theory (e.g., [6]), but we don't have to be comprehensive here. There are even other notions of security proposed in the literature, such as *plaintext awareness* briefly mentioned in Section 13.3.1.4. These notions of security are not relevant and not further addressed in this book.

Let us now address probabilistic encryption that is important in theory before we delve into symmetric encryption systems that are relevant in practice.

## 13.2 PROBABILISTIC ENCRYPTION

To argue scientifically about the security of an (asymmetric) encryption system, Goldwasser and Micali developed and proposed *probabilistic encryption* in the early 1980s [3]. The implementation they suggested was based on the QRP (Definition A.31) that is believed (but not known) to be computationally equivalent to the IFP.

To understand probabilistic encryption, it is required to understand the mathematics summarized in Appendix A.3.7. In particular, it is known that

$$x \in QR_p \Leftrightarrow \left(\frac{x}{p}\right) = 1$$

for every prime number $p$. So if we work with a prime number $p$, then the Legendre symbol of $x$ modulo $p$ is one if and only if $x$ is a quadratic residue modulo $p$. The Legendre symbol of $x$ modulo $p$ can be efficiently computed using, for example, Euler's criterion (Theorem A.11).

Things get more involved if one works with composite numbers (instead of prime numbers). If $n$ is a composite number, then

$$x \in QR_n \Rightarrow \left(\frac{x}{n}\right) = 1$$

but

$$x \in QR_n \not\Leftarrow \left(\frac{x}{n}\right) = 1$$

This means that if $x$ is a quadratic residue modulo $n$, then the Jacobi symbol of $x$ modulo $n$ must be 1, but the converse need not be true (i.e., even if the Jacobi symbol of $x$ modulo $n$ is 1, $x$ need not be a quadratic residue modulo $n$). If, however, the Jacobi symbol of $x$ modulo $n$ is $-1$, then we know that $x$ is a quadratic nonresidue modulo $n$:

$$x \in QNR_n \Leftarrow \left(\frac{x}{n}\right) = -1$$

Again referring to Appendix A.3.7, $J_n$ stands for the set of all elements $x$ of $\mathbb{Z}_n$ for which the Jacobi symbol of $x$ modulo $n$ is 1, and $\widetilde{QR_n} = J_n \setminus QR_n$ stands for the set of all pseudosquares modulo $n$. If $n = pq$, then

$$|QR_n| = |\widetilde{QR_n}| = (p-1)(q-1)/4$$

This means that half of the elements in $J_n$ are quadratic residues and the other half are pseudosquares modulo $n$. So if an arbitrary element of $J_n$ is given, it is computationally intractable to decide whether it is a quadratic residue (square) or a pseudosquare modulo $n$—unless, of course, one knows the prime factorization of $n$. Probabilistic encryption exploits this computational difficulty. Its algorithms and assessment are briefly outlined next.

### 13.2.1 Algorithms

The key generation, encryption, and decryption algorithms of probabilistic encryption are summarized in Table 13.1

#### 13.2.1.1 Key Generation Algorithm

Similar to the RSA public key cryptosystem (Section 13.3.1), the key generation algorithm Generate employed by probabilistic encryption takes as input a security parameter $l$ in unary notation, and it generates as output two $l/2$-bit primes $p$ and $q$ and a modulus $n = pq$ of respective size $l$. Furthermore, the algorithm also selects a pseudosquare $y \in \widetilde{QR_n}$. The pair $(n, y)$ then represents the public key, whereas $(p, q)$ represents the private key.

**Table 13.1**
Probabilistic Encryption System

System parameters: —

| Generate | Encrypt | Decrypt |
|---|---|---|
| $(1^l)$ | $((n, y), m)$ | $((p, q), c)$ |
| $p, q \xleftarrow{r} \mathbb{P}_{l/2}$ | for $i = 1, \ldots, w$ | for $i = 1, \ldots, w$ |
| $n = p \cdot q$ | $\quad x_i \xleftarrow{r} \mathbb{Z}_n^*$ | $\quad e_i = \left(\frac{c_i}{p}\right)$ |
| $y \xleftarrow{r} \widetilde{QR_n}$ | $\quad$ if $m_i = 1$ | $\quad$ if $e_i = 1$ |
| | $\quad\quad$ then $c_i \equiv yx_i^2 \pmod{n}$ | $\quad\quad$ then $m_i = 1$ |
| | $\quad\quad$ else $c_i \equiv x_i^2 \pmod{n}$ | $\quad\quad$ else $m_i = 0$ |
| | $c = c_1, \ldots, c_w$ | $m = m1, \ldots, m_w$ |
| $((n, y), (p, q))$ | $(c)$ | $(m)$ |

### 13.2.1.2 Encryption Algorithm

The encryption algorithm Encrypt employed by probabilistic encryption must specify how a $w$-bit plaintext message $m = m_1 m_2 \ldots m_w$ is encrypted so that only the recipient (or somebody holding the recipient's private key) is able to decrypt it. As its name suggests, the Encrypt algorithm is probabilistic. It takes as input a public key $(n, y)$ and an $w$-bit plaintext message $m$, and it generates as output the ciphertext $c$.

For every plaintext message bit $m_i$ $(i = 1, \ldots, w)$, the Encrypt algorithm chooses $x_i \in_R \mathbb{Z}_n^*$ and computes $c_i$ as follows:

$$c_i \equiv \begin{cases} x_i^2 \pmod{n} & \text{if } m_i = 0 \\ yx_i^2 \pmod{n} & \text{if } m_i = 1 \end{cases}$$

If $m_i = 0$, then $c_i \equiv x_i^2 \pmod{n}$ yields a quadratic residue modulo $n$. Otherwise, $c_i \equiv yx_i^2 \pmod{n}$ yields a pseudosquare modulo $n$. In either case, each plaintext message bit $m_i$ $(i = 1, \ldots, w)$ is encrypted with an element of $\mathbb{Z}_n^*$, and hence the resulting ciphertext $c$ is a $w$-tuple of such elements (i.e., $c = c_1, \ldots, c_w$).

#### 13.2.1.3 Decryption Algorithm

The decryption algorithm Decrypt employed by probabilistic encryption takes as input a private key $(p, q)$, and a $w$-tuple of elements of $\mathbb{Z}_n^*$ that represents a ciphertext, and it generates as output the $w$-bit plaintext message $m$. Again, the Decrypt algorithm proceeds sequentially on every ciphertext element $c_i$ ($i = 1, \ldots, w$). For $c_i$, the algorithm evaluates the Legendre symbol

$$e_i = \left(\frac{c_i}{p}\right)$$

and computes

$$m_i = \begin{cases} 1 & \text{if } e_i = 1 \\ 0 & \text{otherwise} \end{cases}$$

This means that $m_i$ is set to 1, if $c_i \in QR_n$, and otherwise it is set to 0. Remember from the Encrypt algorithm that $c_i$ is an element of $QR_n$, if and only if $m_i$ is equal to one. This is in line with the Decrypt algorithm. Finally, the plaintext message $m$ that yields the output of the Decrypt algorithm is just the bitwise concatenation of the $w$ bits $m_i$ (for $i = 1, \ldots, w$).

### 13.2.2 Assessment

Probabilistic encryption takes its security from the assumed intractability of the QRP in $\mathbb{Z}_n^*$. Under the assumption that this problem is hard, probabilistic encryption can be shown to be semantically secure [3]. In fact, probabilistic encryption was one of the first asymmetric encryption systems that was proposed with some strong security guarantee. As such, it has had a deep impact and is historically relevant, but it is not particularly useful in the field. In its original form, probabilistic encryption has a huge message expansion (because every plaintext message bit is encrypted with an element of $\mathbb{Z}_n^*$). As was first pointed out by Blum and Goldwasser [7], the message expansion can be improved considerably, and the currently known best variants of probabilistic encryption can reduce message expansion to only a constant number of bits. These variants are comparable to RSA, both in terms of performance and message expansion. But compared to RSA with appropriate padding (e.g., OAEP), there is still no advantage, and hence probabilistic encryption is not used in the field.

## 13.3 ASYMMETRIC ENCRYPTION SYSTEMS

We already mentioned in Section 1.3 that a number of public key cryptosystems were developed and proposed after the discovery of public key cryptography by Diffie and Hellman. This is particularly true for RSA, Rabin, and Elgamal, and we look at these systems here. More specifically, we address their key generation, encryption, and decryption algorithms, and we provide a security analysis and discuss improvements where appropriate. For the sake of simplicity, we assume that public keys are always published in some certified form, and we discuss the implications of this assumption at the end of this chapter and in Section 16.4.

### 13.3.1 RSA

The RSA public key cryptosystem was jointly invented by Ron Rivest, Adi Shamir, and Len Adleman at MIT in 1977. A U.S. patent application was filed on December 14, 1977, and a corresponding article was published in the February 1978 issue of *Communications of the ACM* [8].[2] On September 20, 1983, the U.S. patent 4,405,829 entitled "Cryptographic Communications System and Method" was assigned to MIT. It was one of the most important patents ever granted for an invention related to cryptography.[3] After 17 years, the patent would have expired on September 21, 2000. But on September 6, 2000 (two weeks earlier than the patent expired), the algorithm was finally released to the public domain. Recognizing the relevance of their work, Rivest, Shamir, and Adleman were granted the prestigious ACM Turing Award in 2002.

The RSA public key cryptosystem is based on the RSA family (of trapdoor permutations) introduced in Section 5.2.2. Contrary to many other public key cryptosystems, RSA yields both an asymmetric encryption system and a DSS. This basically means that the same set of algorithms can be used to encrypt and decrypt messages, as well as to digitally sign messages and verify digital signatures. The function provided only depends on the cryptographic key in use.

- If the recipient's public key is used to encrypt a message, then the RSA public key cryptosystem yields an asymmetric encryption system. In this case, the

2 The RSA public key cryptosystem was first described in Martin Gardner's column in the August 1977 issue of *Scientific American*. In this article, a 129-digit (i.e., 426-bit) number was introduced to illustrate the computational intractability of the IFP, and the security of the RSA public key cryptosystem accordingly. This number was referred to as RSA-129. In 1991, RSA Security, Inc., used it to launch an RSA Factoring Challenge. RSA-129 was successfully factored in March 1994 (see Section 5.3).

3 Note that the RSA patent was a U.S. patent, and that the RSA public key cryptosystem was not patented outside the United States.

recipient's private key must be used to decrypt the ciphertext. Ideally, this can only be done by the recipient of the message.

- If the sender's private key is used to encrypt a plaintext message (or hash value thereof), then the RSA key cryptosystem yields a DSS. In this case, the sender's public key must be used to verify the digital signature. It goes without saying that this can be done by anybody.

In this chapter, we only look at the RSA asymmetric encryption system (the RSA DSS is addressed in Section 14.2.1). The system consists of a key generation algorithm Generate, an encryption algorithm Encrypt, and a decryption algorithm Decrypt that are specified in Table 13.2.

**Table 13.2**
RSA Asymmetric Encryption System

System parameters: —

| Generate | Encrypt | Decrypt |
|---|---|---|
| $(1^l)$ | $((n,e),m)$ | $(d,c)$ |
| $p,q \xleftarrow{r} \mathbb{P}_{l/2}$<br>$n = p \cdot q$<br>select $1 < e < \phi(n)$<br>with $gcd(e,\phi(n)) = 1$<br>compute $1 < d < \phi(n)$<br>with $de \equiv 1 \pmod{\phi(n)}$ | $c \equiv m^e \pmod{n}$ | $m \equiv c^d \pmod{n}$ |
| $((n,e),d)$ | $(c)$ | $(m)$ |

#### 13.3.1.1 Key Generation Algorithm

Before the RSA asymmetric encryption system can be invoked, the key generation algorithm Generate must be used to generate a public key pair. The algorithm is probabilistic in nature. It takes as input a security parameter $1^l$ (that represents the bitlength $l$ of the RSA modulus in unary notation), and it generates as output a public key pair, where $(n,e)$ refers to the public key and $d$ refers to the private key. More specifically, the algorithm consists of two steps:

- First, it randomly selects[4] two prime numbers $p$ and $q$ of roughly the same size $l/2$ and computes the RSA modulus $n = p \cdot q$. Given the current state of the art in integer factorization (Section 5.3), a modulus size of at least 2,048 bits is usually recommended. This means that each prime must be at least 1,024 bits long.
- Second, it selects an integer $1 < e < \phi(n)$ with $gcd(e, \phi(n)) = 1$,[5] and computes another integer $1 < d < \phi(n)$ with

$$de \equiv 1 \pmod{\phi(n)}$$

using the extended Euclid algorithm (Algorithm A.2).[6] Note that $d$ then represents the multiplicative inverse of $e$ modulo $\phi(n)$.

The output of the Generate algorithm is a public key pair that consists of a public key $pk = (n, e)$ and a corresponding private key $sk = d$. The public key is mainly used for encryption, whereas the private key is mainly used for decryption. Consequently, $e$ is also referred to as the *public* or *encryption exponent*, and $d$ is referred to as the *private* or *decryption exponent.*

People are sometimes confused about the fact that one can work with fixed-size primes $p$ and $q$. They intuitively believe that the fact that both primes are, for example, 1,024 bits long can be exploited in some way to more efficiently factorize $n$. It is, however, important to note that there are $2^{1024}$ 1,024-bit numbers, and that—according to Theorem A.6—primes are dense. This makes it computationally infeasible to try all primes of that particular size to factorize $n$.

Let us consider a toy example to illustrate what is going on in the RSA Generate algorithm (and the other algorithms of the RSA system). In the first step, the algorithm selects $p = 11$ and $q = 23$, and computes $n = 11 \cdot 23 = 253$ and $\phi(253) = 10 \cdot 22 = 220$. In the second step, the RSA Generate algorithm selects $e = 3$ and uses the extended Euclid algorithm to compute $d = 147$ modulo 220. Note that $3 \cdot 147 = 441 \equiv 1 \pmod{220}$, and hence $d = 147$ is in fact the multiplicative inverse element of $e = 3$ modulo 220. Consequently, $(253, 3)$ yields the public key, and 147 yields the private key.

4 It is not possible to randomly select large primes (from the set of all prime numbers $\mathbb{P}$). Instead, large integers are randomly chosen and probabilistic primality testing algorithms are then used to decide whether these integers are prime. The respective primality testing algorithms are overviewed in Appendix A.2.4.3.

5 Note that $e$ must be odd and greater than 2 (it is not possible to set $e = 2$, because $\phi(n) = (p-1)(q-1)$ is even and $gcd(e, \phi(n)) = 1$ must hold) and that the smallest possible value for $e$ is 3. The use of $e = 3$ should be considered with care, because a corresponding implementation may be subject to a low exponent attack (Section 13.3.1.4).

6 Note that $gcd(e, \phi(n)) = 1$ suggests that an integer $d$ with $de \equiv 1 \pmod{\phi(n)}$ must exist.

#### 13.3.1.2 Encryption Algorithm

In its basic form, the RSA Encrypt algorithm is deterministic. It takes as input a public key $(n, e)$ and a plaintext message $m \in \mathbb{Z}_n$, and it generates as output the ciphertext

$$c = \text{RSA}_{n,e}(m) \equiv m^e \pmod{n}$$

To compute $c$, the RSA Encrypt algorithm must employ a modular exponentiation algorithm, such as, for example, the square-and-multiply algorithm (Algorithm A.3) that is efficient.

If we want to encrypt the plaintext message $m = 26$ in our toy example, then we compute

$$\begin{aligned} c &\equiv m^e \pmod{n} \\ &\equiv 26^3 \pmod{253} \\ &\equiv 17,576 \pmod{253} \\ &\equiv 119 \end{aligned}$$

This means that 119 is the ciphertext for the plaintext message 26, and this value is thus transmitted to the recipient(s).

At this point it is important to note that the public key $(n, e)$ is publicly known, and hence anybody can use it to compute $\text{RSA}_{n,e}(m)$ and encrypt an arbitrary plaintext message $m$. Consequently, RSA encryption provides neither data origin authentication nor data integrity, and hence complementary mechanisms must be employed and put in place to provide these security services.

#### 13.3.1.3 Decryption Algorithm

The RSA Decrypt algorithm is deterministic. It takes as input a private key $d$ and a ciphertext $c$, and it generates as output the corresponding plaintext message

$$m = \text{RSA}_{n,d}(c) \equiv c^d \pmod{n}$$

Again, a modular exponentiation algorithm must be used to compute $m$, and this computation can be done efficiently. Because the private exponent must be much larger than the public exponent (as explained below), the Decrypt algorithm typically runs slower than the Encrypt algorithm.

Before we can analyze the security of the RSA asymmetric encryption system, we must first verify its correctness. We note that

$$c^d \equiv (m^e)^d \equiv m^{ed} \pmod{n}$$

Because $e$ and $d$ are chosen in a way that

$$ed \equiv 1 \pmod{\phi(n)}$$

and $\phi(n) = (p-1)(q-1)$ hold, the following two equivalences must also be true:

$$ed \equiv 1 \pmod{p-1}$$
$$ed \equiv 1 \pmod{q-1}$$

They can be written as

$$ed \equiv k_p(p-1)+1$$
$$ed \equiv k_q(q-1)+1$$

for some $k_p, k_q \in \mathbb{Z}$. If $m$ is a multiple of $p$, then $m^{ed} \equiv 0^{ed} = 0 \equiv m \pmod{p}$. Otherwise, $m$ and $p$ are coprime (i.e., $gcd(m,p) = 1$), and hence Fermat's little theorem (Theorem A.9) applies:

$$m^{p-1} \equiv 1 \pmod{p}$$

This, in turn, means that

$$m^{ed} \equiv m^{k_p(p-1)+1} \equiv (m^{p-1})^{k_p} m \equiv 1^{k_p} m = m \pmod{p}$$

In either case, $m^{ed} \equiv m \pmod{p}$. Using the same line of argumentation one can show that $m^{ed} \equiv m \pmod{q}$ must also hold. This means that $p$ and $q$ both divide $m^{ed} - m$, and hence their product $pq = n$ must also divide $m^{ed} - m$. Consequently,

$$m^{ed} \equiv m \pmod{pq}$$

and hence

$$c^d \equiv m \pmod{n}$$

This shows that RSA decryption works and is correct.

In our toy example, the ciphertext $c = 119$ is decrypted as follows:

$$\begin{aligned} m &\equiv c^d \pmod{n} \\ &\equiv 119^{147} \pmod{253} = 26 \end{aligned}$$

If, in addition to $d$, the RSA Decrypt algorithm has access to the prime factors $p$ and $q$, then the CRT can be used to speed up decryption. Instead of directly computing $m \equiv c^d \pmod{n}$, one can compute

$$m_p \equiv c^d \pmod{p}$$

and

$$m_q \equiv c^d \pmod{q}$$

and then use the CRT to compute $m \in \mathbb{Z}_n$ with $m \equiv m_p \pmod{p}$ and $m \equiv m_q \pmod{q}$. In our toy example, we have

$$m_p \equiv c^d \pmod{p} \equiv 119^{147} \pmod{11} \equiv 4$$

and

$$m_q \equiv c^d \pmod{q} \equiv 119^{147} \pmod{23} \equiv 3$$

Using the CRT, one can compute $m \in \mathbb{Z}_{253}$ with $m \equiv 4 \pmod{11}$ and $m \equiv 3 \pmod{23}$. The resulting plaintext message is again $m = 26$.

In addition to the CRT, there are several other possibilities and techniques to speed up RSA decryption (or signature generation, respectively). Examples include batch RSA, multifactor RSA, and rebalanced RSA as overviewed, for example, in [9]. These techniques are particularly important to implement RSA on devices with limited computational power.

#### 13.3.1.4 Security Analysis

Since its discovery and publication in 1977, the security of the RSA public key cryptosystem has been subject to a lot of public scrutiny. Many people have challenged and analyzed the security of RSA (e.g., [10]). While no devastating vulnerability or weakness has been found, more than four decades of cryptanalytical research have still given us a broad insight into its security properties and have provided us with valuable guidelines for the proper implementation and use of RSA. Let us start with some observations before we turn to more specific attacks, elaborate on OAEP, and conclude with some recommendations for the proper use of RSA.

*Observations*

The first observation is that the RSA asymmetric encryption system is based on the RSA family of trapdoor permutations, and hence that its security is based on the RSAP (Definition 5.9). If the modulus $n$ is sufficiently large and the plaintext message $m$ is an integer between 0 and $n-1$, then the RSAP is believed to be computationally intractable. Note, however, that there are two caveats:

- If $n$ is small, then an adversary can try all elements of $\mathbb{Z}_n$ until the correct $m$ is found.[7]
- Similarly, if the plaintext message $m$ is known to be from a small subset of $\mathbb{Z}_n = \{0, \ldots, n-1\}$, then an adversary can try all elements from this subset until the correct $m$ is found.

If $n$ is sufficiently large and $m$ is widespread between 0 and $n-1$, then the adversary can still try to find the correct $m$ by a brute-force attack. However, such an attack has an exponential running and is therefore prohibitively expensive in terms of computational power.

The second observation goes back to Section 5.2.2, where we said that the RSAP polytime reduces to the IFP (i.e., RSAP $\leq_P$ IFP), and hence that one can invert the RSA function if one can solve the IFP, but the converse is not known to be true, meaning that it is not known whether an algorithm to solve the IFP can be constructed from an algorithm to solve the RSAP. There is some evidence that such a construction may not exist if the public exponent is very small, such as $e = 3$ or $e = 17$ [11], but for larger exponents, the question remains unanswered. This suggests that the RSAP and the IFP may not be computationally equivalent, or at least that one cannot prove such a relationship. But one can at least prove that the following problems or tasks are computationally equivalent:

- Factorize $n$;
- Compute $\phi(n)$ from $n$;
- Determine $d$ from $(n, e)$.

This means that an adversary who knows $\phi(n)$ or $d$ can also factorize $n$. It also means that there is no point in hiding the factorization of $n$ (i.e., the prime factors $p$ and $q$) from an entity that already knows $d$. If an efficient algorithm to factorize $n$, compute $\phi(n)$ or deviate $d$ from $e$ existed, then the RSA asymmetric encryption system would be totally insecure.

Even if the RSA asymmetric encryption system were secure in the sense discussed above, it could still be true that a ciphertext leaks some partial information about the underlying plaintext message. For example, it may be the case that certain

7 This basically means that he or she can compute $m'^e \pmod{n}$ for every possible plaintext message $m' \in \mathbb{Z}_n$. If the resulting value matches $c$, then he or she has found the correct plaintext message.

plaintext message bits are easy to predict from a given ciphertext. Consequently, one may ask whether the RSA asymmetric encryption system provides security to every individual bit of a plaintext message. This question can be answered in the affirmative. We already know from Section 5.1 that the LSB yields a hard-core predicate for the RSA function. In an attempt to generalize this result, it has been shown that all plaintext message bits are equally protected by the RSA function in the sense that having a nonnegligible advantage for predicting a single bit enables an adversary to invert the RSA function [12]. This property is known as the *bit security* of RSA, and it can be proven by reduction. In particular, one can show that an efficient algorithm for solving the RSAP can be constructed from an algorithm for predicting one (or more) plaintext message bit(s). Note, however, that the bit security proof of the RSA encryption system is a double-edged sword, because the security reduction used in the proof also provides a possibility to attack a leaky implementation: If an implementation of the RSA Decrypt algorithm leaks some bit(s) of a plaintext message, then this leakage can be (mis)used to solve the RSAP and decrypt a ciphertext without knowing the private key. As shown later on, this fact has been exploited in a real-world attack and has led to the development and standardization of OAEP.

Finally, the third observation is that the encryption function of the RSA asymmetric encryption system is deterministic, and hence it can neither provide IND-CPA nor can it be semantically secure. This fact has already been mentioned several times.

*Specific Attacks*

Several attacks are known and must be considered with care when it comes to an implementation of the RSA asymmetric encryption system. In addition to these attacks, an RSA implementation may always be susceptible to side-channel attacks.

*Common modulus attacks:* To avoid generating a different modulus $n = pq$ for every entity, it is tempting to work with a common modulus $n$ for multiple entities. In this case, a trusted authority (that holds the prime factors of $n$) must handle the generation and distribution of the public key pairs; that is, it must provide entity $i$ with a public key $(n, e_i)$ and a respective private key $d_i$. At first glance, this seems to work, because a ciphertext $c \equiv m^{e_i} \pmod{n}$ encrypted for entity $i$ cannot be decrypted by entity $j$, as long as entity $j$ does not know the private key $d_i$. Unfortunately, this argument is flawed, and the use of a common modulus is totally insecure. More specifically, there are common modulus attacks that can be mounted from either an insider or an outsider.

- Remember from the previous discussion that knowing the private key $d_j$ is computationally equivalent to knowing the prime factors of $n$ or $\phi(n)$. Consequently, an insider $j$ can use $d_j$ to factorize $n$, and the prime factors of $n$ can then be used to efficiently compute $d_i$ from $e_i$.
- Maybe more worrisome, even an outsider can mount a common modulus attack against a message $m$ that is encrypted with the public keys of two entities having the same modulus $n$. Let $(n, e_1)$ be the public key of the first entity and $(n, e_2)$ be the public key of the second entity. The message $m$ is then encrypted as

  $$c_1 \equiv m^{e_1} \pmod{n}$$

  for the first entity and

  $$c_2 \equiv m^{e_2} \pmod{n}$$

  for the second entity. Anybody who knows the public keys $e_1$ and $e_2$ can compute the following pair of values:

  $$\begin{aligned} t_1 &\equiv e_1^{-1} \pmod{e_2} \\ t_2 &= (t_1 e_1 - 1)/e_2 \end{aligned}$$

  From the second equation, it follows that $t_1 e_1 - 1 = e_2 t_2$, and hence $t_1 e_1 = e_2 t_2 + 1$. If an outsider sees the same message $m$ encrypted with the two public keys mentioned above (i.e., $c_1$ and $c_2$), then he or she can use $t_1$ and $t_2$ to recover $m$:

  $$\begin{aligned} c_1^{t_1} c_2^{-t_2} &\equiv m^{e_1 t_1} m^{-e_2 t_2} \pmod{n} \\ &\equiv m^{1+e_2 t_2} m^{-e_2 t_2} \pmod{n} \\ &\equiv m^{1+e_2 t_2 - e_2 t_2} \pmod{n} \\ &\equiv m^1 \pmod{n} \\ &\equiv m \end{aligned}$$

  In the first transformation, we exploit the fact that $e_1 t_1 = t_1 e_1 = e_2 t_2 + 1$, and the remaining transformations are simple and straightforward.

Due to the common modulus attacks, it is important that a modulus $n$ is never reused (i.e., used by more than one entity), and this also means that the prime numbers used to generate $n$ must be unique for a particular entity.

*Attacks that exploit the multiplicative structure of the RSA function:* There are a few attacks against the RSA public key cryptosystem that exploit the multiplicative structure (or homomorphic property) of the RSA function. If, for example, two plaintext messages $m_1$ and $m_2$ are encrypted with the same public key $(n, e)$, then

$$c_1 \equiv m_1^e \pmod{n}$$

and

$$c_2 \equiv m_2^e \pmod{n}$$

and hence the ciphertext $c$ for the plaintext message $m \equiv m_1 m_2 \pmod{n}$ can be constructed by multiplying $c_1$ and $c_2$:

$$c \equiv c_1 c_2 \pmod{n}$$

This is because $c_1 c_2 \equiv m_1^e m_2^e \equiv (m_1 m_2)^e \pmod{n}$. So $c$ can be computed without knowing $m_1$, $m_2$, or $m$.

Maybe more interestingly, an adversary who can mount a CCA can exploit the multiplicative structure of RSA to decrypt $c \equiv m^e \pmod{n}$: He or she randomly selects $r \in \mathbb{Z}_n$, computes $c' \equiv cr^e \pmod{n}$, has $c'$ (which is different from $c$) be decrypted, and derives $m$ from $rm \pmod{n}$ by using a modular division with $r$.

The multiplicative structure of the RSA function is the source of many security problems. One possibility to remedy the situation is to randomly pad the plaintext message prior to encryption. This randomizes the ciphertext and eliminates the homomorphic property. Again. we revisit this possibility when we elaborate on OAEP. The multiplicative structure of the RSA function is particularly worrisome when it is used in a DSS (Chapter 14).

*Low exponent attacks:* To improve the performance of the RSA asymmetric encryption system, one may consider the use of small (public or private) exponents. In fact, a common line of argumentation goes as follows: If one employs a small public exponent $e$, then the encryption (or signature verification) process is fast, and if one employs a small private exponent $d$, then the decryption (or signature generation) is fast. Consequently, one can make one of the two operations fast at the cost of the other. Unfortunately, this line of argumentation is oversimplistic, and there are several low-exponent attacks that must be taken into account:

- On the one hand, Michael Wiener showed in 1990 that the choice of a small private exponent $d$ can lead to a total break of the RSA asymmetric encryption system [13], and this result was later improved considerably. Given the current state of the art, the private exponent $d$ should be at least 300 bits long for a typical 1,024-bit modulus $n$ [14]. In practice, people frequently use a public exponent $e$ that is 3, 17, or $2^{16}+1 = 65{,}537$. In these cases, it is guaranteed that the corresponding private exponent $d$ is nearly as long as $n$, and hence that the attacks that exploit small private exponents do not work.

- On the other hand, it has been mentioned earlier that using a small public exponent may lead to the situation that the RSAP is simpler to solve than the IFP. Consequently, one may want to work with public exponents that are not too small. Otherwise, some well-known problems may pop up and become relevant:
  - If one uses a public exponent $e$ to encrypt a small message $m$ (i.e., $m < \sqrt[e]{n}$), then $c = m^e$, and hence $m$ can be decrypted as $m = \sqrt[e]{c}$. Note that there is no modular arithmetic involved here.
  - Maybe more interestingly, if a plaintext message $m$ is encrypted for $r \geq 2$ recipients that use a common public exponent $e$ (but different moduli $n_i$ for $i = 1, \ldots, r$), then an adversary who knows the ciphertexts $c_i \equiv m^e \pmod{n_i}$ for $i = 1, \ldots, r$ can use the CRT to compute $c$ with $c \equiv c_i \pmod{n_i}$ for $i = 1, \ldots, r$ and $0 \leq c < \prod_{i=1}^{r} n_i$. In this case, $c$ equals $m^e$, and $m$ can be efficiently computed as the e-th root of $c$. This attack is relevant only for small values of $e$. If, for example, $r = 3$, $e = 3$, $n_1 = 143$, $n_2 = 391$, $n_3 = 899$, and $m = 135$, then the adversary can solve the following system of three equivalences:

$$\begin{aligned} c_1 &\equiv m^e \pmod{n_1} \equiv 135^3 \pmod{143} = 60 \\ c_2 &\equiv m^e \pmod{n_2} \equiv 135^3 \pmod{391} = 203 \\ c_3 &\equiv m^e \pmod{n_3} \equiv 135^3 \pmod{899} = 711 \end{aligned}$$

    Using the CRT, the adversary can compute $c = 2{,}460{,}375$, and hence $m = \sqrt[e]{c} = \sqrt[3]{2,460,375} = 135$. There are many generalizations of this attack that can be found in the relevant literature; they are not further addressed here.

The bottom line is that public exponents should not be too small, and that $e = 2^{16}+1 = 65{,}537$ seems to be an appropriate choice.

*RSA-OAEP*

As mentioned above, the multiplicative structure (or homomorphic property) of the RSA function leads to a vulnerability of the RSA encryption system that can be exploited by a CCA(2). One possibility to mitigate this vulnerability is to randomly pad the plaintext message prior to encryption. This was the motivation behind the padding scheme standardized in PKCS #1 version 1.5 that was used, for example, in the SSL protocol. In 1998, this protocol was shown to be susceptible to CCA2 [2], and since then people know that CCA and CCA2 are relevant and that asymmetric encryption systems and respective padding schemes need to protect against these types of attacks.

Shortly before the attack against the SSL protocol was published in 1998, Mihir Bellare and Philip Rogaway had introduced a notion of security that they called *plaintext awareness*. They also proposed a message encoding scheme called OAEP and proved that if a (deterministic) asymmetric encryption function, such as RSA, is hard to invert without the private key, then the corresponding OAEP-based encryption system is *plaintext-aware* in the random oracle model, meaning that an adversary cannot produce a valid ciphertext without knowing the underlying plaintext [15]. RSA combined with OAEP is referred to as RSA-OAEP.

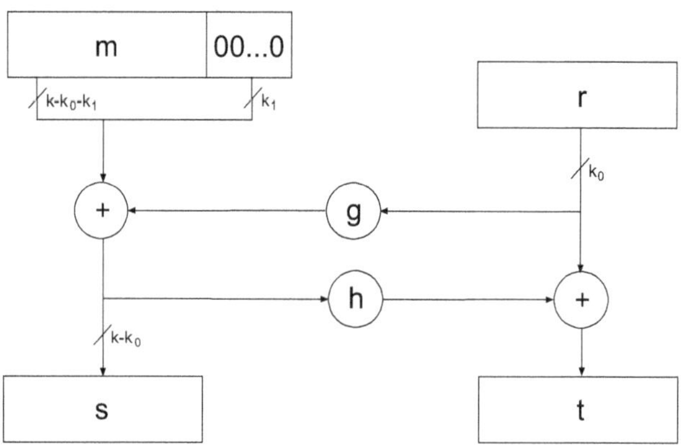


**Figure 13.1** OAEP padding as originally proposed by Bellare and Rogaway.

OAEP padding is illustrated in Figure 13.1. It has a structure that looks like a Feistel network. More specifically, it employs two cryptographic hash functions $g$ and $h$ that are used to encode a plaintext message $m$ prior to encryption. Referring to Figure 13.1, $k$ is the bitlength of the RSA modulus $n$, $k_0$ and $k_1$ are fixed integers between 0 and $k$, and $r$ is a randomly chosen $k_0$-bit string used to mask $m$. The message $m$ is $k - k_0 - k_1$ bits long and it is concatenated with $k_1$ zero bits. The resulting string is $k - k_0$ bits long. This string is added modulo 2 to $g(r)$, where $g$ hashes (or rather expands) the $k_0$ bits of $r$ to $k - k_0$ bits. The result represents $s$, and this $k - k_0$-bit string is subject to $h$ that reduces it to $k_0$ bits. This bit string is then added modulo 2 to $r$, and the result represents $t$. Hence, the output of the OAEP padding scheme comprises two bit strings—$s$ and $t$—that are concatenated to form the $k$-bit output. In summary, the OAEP padding scheme can be expressed as follows:

$$\text{OAEP}(m) = (s,t) = \underbrace{m \oplus g(r)}_{s} \parallel \underbrace{r \oplus h(m \oplus g(r))}_{t}$$

This value represents the encoded message. It is taken as input for an asymmetric encryption system, such as RSA in the case of RSA-OAEP.

In order to decrypt a ciphertext $c$ that is encrypted with RSA-OAEP, the recipient must first decrypt $c$ with the RSA Decrypt algorithm. The result represents $\text{OAEP}(m) = (s,t)$, and hence the recipient must extract $m$ and $r$ from $s$ and $t$. He or she therefore computes

$$r = t \oplus h(s) = r \oplus h(m \oplus g(r)) \oplus h(m \oplus g(r))$$

and

$$m = s \oplus g(r) = m \oplus g(r) \oplus g(r)$$

to retrieve the plaintext message $m$. Note that this string still comprises the $k_1$ zero bits that are appended to $m$. Hence, these bits need to be silently discarded.

In their 1994 paper [15], Bellare and Rogaway argued that RSA combined with OAEP yields a plaintext-aware asymmetric encryption system (i.e., an encryption system that is semantically secure against CCA2). Hence, quite naturally, OAEP was adopted in PKCS #1 version 2.0 specified in informational RFC 2437 [16]. However, it was not until 2001 that Victor Shoup showed that the security arguments provided in [15] were not formally correct, and that OAEP actually provides only semantic security against CCA [17]. Shoup also proposed an improved scheme (called OAEP+).[8] A formally correct proof of the semantic security of RSA-OAEP against CCA2 was provided in [18]. Because this security proof does not guarantee security for the key lengths typically used in practice (due to the inefficiency of the security reduction), a few alternative padding schemes have been proposed in the literature (e.g., [19]).

As of this writing, the valid version of PKCS #1 is 2.1 and it is specified in the informational RFC 3447 [20].[9] In Section 7 of this RFC, two RSA encryption schemes (RSAES)—RSAES-OAEP and RSAES-PKCS1-v1_5—are specified,

8 Instead of appending $k_1$ zero bytes to the message $m$, OAEP+ uses a third hash function $w$ (in addition to $g$ and $h$) and replaces the $k_1$ zero bytes with a hash value of $m$ and $r$ using $w$.

9 There is an updated version 2.2, but this version is not different from a security perspective.

where RSAES-OAEP refers to the OAEP-based version of RSA encryption and RSAES-PKCS1-v1_5 refers to its non-OAEP-based predecessor (due to its susceptibility to the Bleichenbacher and related CCA2, RSAES-PKCS1-v1_5 is included only for compatibility with existing applications). Contrary to the original OAEP proposal, the RSAES are specified in byte notation, meaning that the units processed are bytes (instead of bits).

RSAES-OAEP is so important in practice that we delve more deeply into the details than we usually do in this book (if you feel uncomfortable about this, then you can easily skip the rest of the paragraph and continue reading with the conclusions and recommendations regarding RSA). Let $m$ be the plaintext message to encrypt, $k$ the bytelength of the RSA modulus $n$, $h$ a cryptographic hash function with output length $|h|$, and $L$ the value of a label that may serve as an additional input ($L$ is an empty string by default). The hash function $h$ is also used in a mask generation function (MGF) that takes as input a seed and a length, and that generates as output a pseudorandomly generated bitstring of this length. There are several possibilities to instantiate such a MGF. By default, SHA-1 is used.

If $m$ is less or equal than $k - 2|h| - 2$ bytes long, then RSAES-OAEP consists of the following eight steps:

1. If $m$ is less than $k - 2|h| - 2$ bytes long, then a padding string $PS$ of $k - |m| - 2|h| - 2$ zero bytes is generated. Otherwise, the length of $PS$ is zero.

2. A data block $DB$ is generated as the concatenation of $h(L)$; that is, the hash value of the label $L$, $PS$, a byte 0x01, and the message $m$:

$$DB = h(L) \parallel PS \parallel \texttt{0x01} \parallel m$$

   The length of $DB$ is $|h| + k - |m| - 2|h| - 2 + 1 + |m| = k - |h| - 1$ bytes.

3. A random $|h|$-byte string $seed$ is generated.

4. The MGF is used to compute a $k - |h| - 1$-byte string $dbMask$; that is, $dbMask = \text{MGF}(seed, k - |h| - 1)$.

5. As its name suggests, $dbMask$ is used to mask the data block $DB$; that is, $maskedDB = DB \oplus dbMask$

6. The MGF is again used to generate an $|h|$-byte string $seedMask$ from $maskedDB$; that is, $seedMask = \text{MGF}(maskedDB, |h|)$

7. $seedMask$ is used to mask the seed; that is, $maskedSeed = seed \oplus seedMask$

8. Finally, a zero byte 0x00, the $maskedSeed$ string, and the $maskedDB$ string are concatenated to form an encoded message $EM$ that is $k$ bytes long:

$$EM = 0x00 \parallel maskedSeed \parallel maskedDB$$

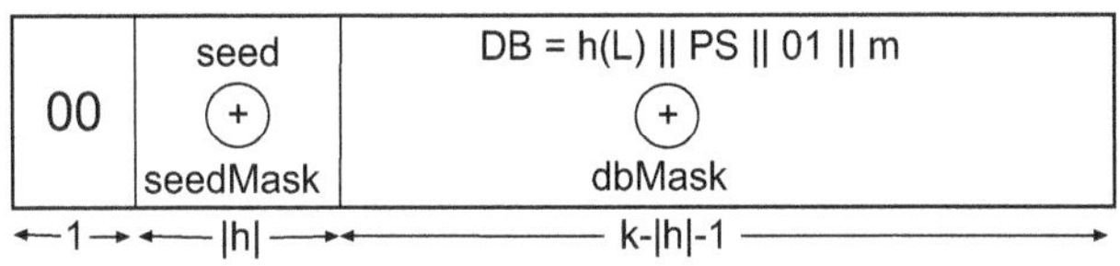


**Figure 13.2** The structure of an encoded message $EM$ according to PKCS #1 version 2.1.

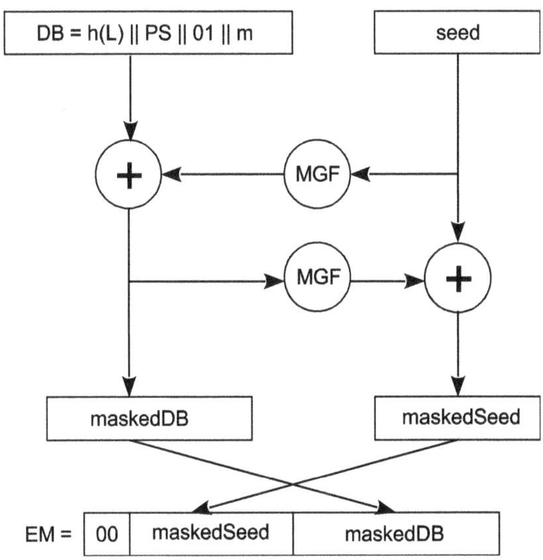


**Figure 13.3** RSAES-OAEP encoding scheme according to PKCS #1 version 2.1.

The structure of an encoded message $EM$ is illustrated in Figure 13.2. The fact that the RSAES-OAEP encoding scheme conforms to the OAEP padding scheme as illustrated in Figure 13.1 is not obvious. We therefore illustrate the RSAES-OAEP encoding scheme according to PKCS #1 version 2.1 as a variant of the originally proposed OAEP padding scheme in Figure 13.3. In this figure, MGF corresponds to the cryptographic hash functions $g$ and $h$, $DB$ corresponds to the message $m$ (padded with zeros), $seed$ corresponds to $r$, $maskedDB$ corresponds to $s$, and $maskedSeed$ corresponds to $t$ in Figure 13.1 (the compilation of $EM$ is not illustrated in Figure 13.1).

On the receiver side, the encoded message $EM$ must be separated into a single byte $Y$, an $|h|$-byte string $maskedSeed$, and an $k-|h|-1$-byte string $maskedDB$. The $seed$ can then be computed from

$$seedMask = \text{MGF}(maskedDB, |h|)$$

and

$$seed = maskedSeed \oplus seedMask$$

Similarly, $DB$ can be computed from

$$dbMask = \text{MGF}(seed, k - |h| - 1)$$

and

$$DB = maskedDB \oplus dbMask$$

The resulting string $DB$ can be separated into an $|h|$-byte string $h(L)'$, a (possibly empty) padding string $PS$, a single byte with hexadecimal value `0x01`, and a message $m$. If any of the following conditions are not fulfilled, then the decryption algorithm must stop and output an error message:

- $Y$ is not equal to zero.
- There is no byte with hexadecimal value `0x01` to separate $PS$ from $m$.
- $h(L)'$ does not equal $h(L)$.

If one (or several) of the conditions is not fulfilled, then it is important that the implementation does not leak any information about it (them). If the implementation leaks, for example, whether $Y$ is equal to zero, then it is possible to exploit this side-channel information in an attack that is able to decrypt a ciphertext. The bottom line is that side-channel attacks remain dangerous in practice, even though RSAES-OAEP mitigates CCA(2) in theory.

*Conclusions and Recommendations*

The question that is most frequently asked when it comes to an implementation or use of the RSA asymmetric encryption system is related to the size of the modulus

$n$. Obviously, $n$ should be at least as large as to make it computationally infeasible to factorize $n$. The respective algorithms are summarized in Section 5.3. As of this writing, $n$ should be at least 2,048 bits long to achieve a reasonable level of security, and people sometimes even recommend 4,096 bits for the encryption of valuable data. If one has fixed the size of the modulus, then one has implicitly also fixed the size of the prime factors $p$ and $q$ (because they should be equally long). If, for example, one wants to work with a 2,048-bit modulus $n$, then $p$ and $q$ must be about 1,024 bits long each. Unless one uses short moduli (where Pollard's P−1 algorithm can be applied), there is no need to work only with strong primes.

The most important thing to keep in mind is that—due to its deterministic encryption function—RSA is at most one-way secure (i.e., it cannot provide IND-CPA and is not semantically secure). As such, it is not on a par with the current state of the art in asymmetric encryption, and RSA-OAEP should be used instead whenever possible and appropriate.

### 13.3.2 Rabin

As mentioned earlier, the RSAP is not known to be computationally equivalent to the IFP. This means that it is theoretically possible to break the security of the RSA public key cryptosystem without solving the IFP. This possibility is worrisome, and—since the beginning of public key cryptography—people have been looking for public key cryptosystems that can be shown to be computationally equivalent to a hard problem, such as the IFP. As mentioned in Section 1.2.2, Rabin was the first researcher who proposed such a system in 1979 [21].[10] The *Rabin asymmetric encryption system* is based on the Square family of trapdoor permutations (Section 5.2.3). Its security is computationally equivalent to the IFP, meaning that there is provably no easier way to break the Rabin system than to solve the IFP and factorize a large integer $n$ accordingly.

In this section, we overview and discuss the Rabin asymmetric encryption system. The Rabin DSS is addressed in Section 14.2.3. In either case, the exposure does not follow the original publication of Rabin [21]. Instead of using the function $E_{n,b}(x) \equiv x(x+b) \pmod{n}$ for some fixed $n$ and $0 \leq b < n$, we use the Square family of trapdoor permutations. This is simpler and better reflects the basic principles of the Rabin public key cryptosystem. The Rabin asymmetric encryption system consists of the three efficient algorithms Generate, Encrypt, and Decrypt that are summarized in Table 13.3 and briefly outlined next.

10 At this time, Rabin was also working at MIT (like Rivest, Shamir, and Adleman). More specifically, he was a professor of mathematics at the Hebrew University in Jerusalem, Israel, but in 1979 he served as a visiting professor of applied mathematics at MIT.

#### 13.3.2.1 Key Generation Algorithm

The key generation algorithm Generate takes as input a security parameter $1^l$, and it generates as output a Blum integer of bitlength $l$. It first randomly selects two $l/2$-bit primes $p$ and $q$ that are both equivalent to 3 modulo 4 ($\mathbb{P}'$ refers to the set of all such primes), and it then computes the Blum integer $n = pq$.[11] In the end, $n$ represents the public key $pk$, whereas the pair $(p, q)$ represents the private key $sk$.

Let us consider a toy example to illustrate what is going on: For $p = 11$ and $q = 23$ (that are both equivalent to 3 modulo 4), the resulting Blum integer $n$ is $11 \cdot 23 = 253$. This value represents the public key; that is, $pk = 253$, whereas $(11, 23)$ represents the private key; that is, $sk = (11, 23)$.

**Table 13.3**
Rabin Asymmetric Encryption System

System parameters: —

| Generate | Encrypt | Decrypt |
|---|---|---|
| $(1^l)$ | $(n, m)$ | $((p, q), c)$ |
| $p, q \xleftarrow{r} \mathbb{P}'_{l/2}$<br>$n = p \cdot q$ | $c \equiv m^2 \pmod{n}$ | $m_1, m_2, m_3, m_4 \equiv c^{1/2} \pmod{n}$<br>Determine correct value $m_i$ |
| $(n, (p, q))$ | $(c)$ | $(m_i)$ |

#### 13.3.2.2 Encryption Algorithm

Similar to the RSA asymmetric encryption system, the Rabin system can be used to deterministically encrypt and decrypt plaintext messages that represent elements of $\mathbb{Z}_n$ (i.e., $0 \leq m < n$). The encryption algorithm Encrypt takes as input a public key $n$ and such a plaintext message $m$, and it generates as output the ciphertext $c = \text{Square}_n(m) \equiv m^2 \pmod{n}$. This can be done in just one modular squaring,[12] and hence the Encrypt algorithm is highly efficient.

11 The requirement that $p$ and $q$ are both equivalent to 3 modulo 4 (and hence $n$ is a Blum integer) is not mathematically required, but it simplifies the computation of the square roots (Appendix A.3.7).

12 By comparison, the RSA Encrypt algorithm with $e = 3$ takes one modular multiplication and one modular squaring, and for larger $e$ many more modular operations (i.e., multiplication or squaring) must be performed.

If, in our toy example, the plaintext message is $m = 158$, then the Encrypt algorithm computes $c \equiv 158^2 \pmod{253} = 170$, and the resulting ciphertext $c = 170$ is sent to the recipient.

### 13.3.2.3 Decryption Algorithm

As is usually the case for decryption, the Rabin decryption algorithm Decrypt is deterministic. It takes as input a private key $(p, q)$ and a ciphertext $c$, and it generates as output the square root of $c$ modulo $n$ that represents the plaintext message $m$. Note that the recipient can find a square root of $c$ modulo $n$ if and only if he or she knows the prime factors $p$ and $q$ of $n$. Also note that there is no single square root of $c$ modulo $n$, but that there are usually four of them. Let $m_1$, $m_2$, $m_3$, and $m_4$ be the four square roots of $c$ modulo $n$. The recipient must then decide which $m_i$ $(1 \leq i \leq 4)$ to go with (i.e., which square root represents the original plaintext message). This ambiguity is a major practical disadvantage of the Rabin asymmetric encryption system when used in the field.

Computing square roots modulo $n$ is simple if the prime factorization of $n$ is known and $n$ is a Blum integer (this is why we have required that $n$ is a Blum integer in the first place). In this case, one can first compute the square roots of $c$ modulo $p$ and modulo $q$. Let $m_p$ be the square roots of $c$ modulo $p$ and $m_q$ be the square roots of $c$ modulo $q$. This means that the following two equivalences hold:

$$\begin{aligned} m_p^2 &\equiv c \pmod{p} \\ m_q^2 &\equiv c \pmod{q} \end{aligned}$$

According to (A.5), $m_p$ and $m_q$ can be computed as follows:

$$\begin{aligned} m_p &\equiv c^{\frac{p+1}{4}} \pmod{p} \\ m_q &\equiv c^{\frac{q+1}{4}} \pmod{q} \end{aligned}$$

It can be verified that

$$\begin{aligned} m_p^2 &\equiv (c^{\frac{p+1}{4}})^2 \equiv c^{(p+1)/2} \equiv (m^2)^{\frac{(p+1)}{2}} \equiv m^{\frac{2(p+1)}{2}} \equiv m^{p+1} \equiv m^{p-1}m^2 \equiv 1m^2 \\ &\equiv m^2 \equiv c \pmod{p} \end{aligned}$$

and $m_q^2 \equiv c \pmod{q}$ following the same line of argumentation. Consequently, $\pm m_p$ are the two square roots of $c$ in $\mathbb{Z}_p$, and $\pm m_q$ are the two square roots of $c$ in $\mathbb{Z}_q$. There is a total of four possibilities to combine $\pm m_p$ and $\pm m_q$, and these possibilities result in four systems with two congruence relations each:

$$1) \quad m_1 \equiv +m_p \pmod{p}$$
$$m_1 \equiv +m_q \pmod{q}$$

$$2) \quad m_2 \equiv -m_p \pmod{p}$$
$$m_2 \equiv -m_q \pmod{q}$$

$$3) \quad m_3 \equiv +m_p \pmod{p}$$
$$m_3 \equiv -m_q \pmod{q}$$

$$4) \quad m_4 \equiv -m_p \pmod{p}$$
$$m_4 \equiv +m_q \pmod{q}$$

Each system yields a possible square root of $c$ modulo $n$, and we use $m_1, m_2, m_3$, and $m_4$ to refer to them. Note that only one solution represents the original plaintext message $m$, and that this ambiguity needs to be resolved during decryption.

The simplest way to find the four solutions $m_1, m_2, m_3$, and $m_4$ is to use the CRT to solve the system of two congruences

$$m \equiv m_p \pmod{p}$$
$$m \equiv m_q \pmod{q}$$

for $m \in \mathbb{Z}_n$. To apply (A.4), one has to compute $m_1 = n/p = q$ and $m_2 = n/q = p$, as well as $y_p \equiv m_1^{-1} \equiv q^{-1} \pmod{p}$ and $y_q \equiv m_2^{-1} \equiv p^{-1} \pmod{q}$ using, for example, the extended Euclid algorithm. Note that these computations depend only on $p$ and $q$ (i.e, the prime factors of $n$) and are independent from the plaintext message or the ciphertext, and this, in turn, means that they can be precomputed during the key generation process. Using all of these values, one can now compute $\pm r$ and $\pm s$ that refer to the four square roots of $c$ in $\mathbb{Z}_n$:

$$r = (y_p p m_q + y_q q m_p) \bmod n$$
$$-r = n - r$$
$$s = (y_p p m_q - y_q q m_p) \bmod n$$
$$-s = n - s$$

These four values represent the four possible solutions $m_1$, $m_2$, $m_3$, and $m_4$. It is now up to the recipient to decide which solution is the correct one (i.e., which solution represents the correct plaintext message).

In our toy example, the recipient gets the ciphertext $c = 170$ and wants to decrypt it with the public key $(p, q) = (11, 23)$. He or she therefore computes $y_p = -2$ and $y_q = 1$, and computes the square roots $m_p$ and $m_q$ as follows:

$$\begin{aligned} m_p &\equiv c^{(p+1)/4} \pmod{p} \equiv c^3 \pmod{11} = 4 \\ m_q &\equiv c^{(q+1)/4} \pmod{q} \equiv c^6 \pmod{23} = 3 \end{aligned}$$

Using these values, the recipient can determine $\pm r$ and $\pm s$:

$$\begin{aligned} r &= (-2 \cdot 11 \cdot 3 + 1 \cdot 23 \cdot 4) \bmod 253 = 26 \\ -r &= 253 - 26 = 227 \\ s &= (-2 \cdot 11 \cdot 3 - 1 \cdot 23 \cdot 4) \bmod 253 = 95 \\ -s &= 253 - 95 = 158 \end{aligned}$$

Consequently, the square roots of $c = 170$ modulo 253 are 26, 227, 95, and 158, and it is up to the recipient to decide that 158 yields the correct plaintext message.

An obvious drawback of the Rabin asymmetric encryption system is that the recipient must determine the correct plaintext message $m$ from the four possible values. This ambiguity in decryption can be overcome by adding redundancy to the original plaintext message prior to encryption. Then, with high probability, exactly one of the four square roots of $c$ modulo $n$ possesses the redundancy, and hence the recipient can easily determine this value. This point is further addressed below.

#### 13.3.2.4 Security Analysis

As mentioned above, the security of the Rabin asymmetric encryption system is based on the fact that breaking Rabin (in terms of computing square roots in $\mathbb{Z}_n^*$ without knowing the prime factorization of $n$) and factorizing $n$ are equally difficult—or computationally equivalent. This fact was already mentioned in Section 5.2.3 and it basically means that the Rabin asymmetric encryption system can be shown to be one-way secure.

**Theorem 13.1** *Breaking the one-way security of the Rabin asymmetric encryption system is computationally equivalent to solving the IFP.*

*Proof.* To prove the theorem, one must show (a) that somebody who can solve the IFP can also break the one-way security of the Rabin asymmetric encryption system (by inverting a trapdoor permutation), and (b) that somebody who can break the one-way security of the Rabin asymmetric encryption system can solve the IFP.

Direction (a) is obvious: Somebody who can solve the IFP for $n$ can break the one-way security of the Rabin asymmetric encryption system by first factorizing $n$ and then (mis)using the private key $(p, q)$ to decrypt any ciphertext $c$ at will.

Direction (b) is less obvious. Here, we must show that somebody who can break the one-way security of the Rabin asymmetric encryption system by decrypting a given ciphertext $c$ (without knowing the prime factorization of $n$) can factorize $n$. More specifically, an adversary who has access to a decryption oracle can find a nontrivial prime factor of $n$. Remember from Section 5.3.2 that $n$ can usually be factorized, if $x$ and $y$ with

$$x^2 \equiv y^2 \pmod{n}$$

are found. So the adversary can randomly select an element $x \in_R \mathbb{Z}_n$ (with $gcd(x, n) = 1$), pass the square $y \equiv x^2 \pmod{n}$ to the oracle, and get back one of the four square roots of $y$ modulo $n$. If $x = \pm y \pmod{n}$, then the adversary has had bad luck and must restart from scratch. Otherwise; that is, if $x \not\equiv \pm y \pmod{n}$, then he or she can determine a prime factor of $n$ as $gcd(x + y, n)$. This completes the proof.

□

The proof may be better understood if one looks at a simple example. Let $n = 253$, and hence the decryption oracle returns square roots of elements of $\mathbb{Z}_{253}$. The adversary randomly selects $x = 17$ (with $gcd(17, 253) = 1$), passes its square $x^2 \equiv 17^2 \equiv 289 \pmod{253} = 36$ to the oracle, and gets back one of the four square roots of 36 modulo 253 (i.e., 6, 17, 236, or 247). If he or she gets back 6, then he or she can recover the prime factor 23 (of 253) by computing $gcd(17+6, 253) = gcd(23, 253) = 23$. Similarly, if he or she gets back 247, then he or she can recover the other prime factor 11 (note that $23 \cdot 11 = 253$) by computing $gcd(17+247, 253) = gcd(264, 253) = 11$. In the other two cases (i.e., 17 and 236), he or she cannot recover anything useful, meaning that the success probability is $1/2$.

We mentioned earlier that one can add redundancy to the original plaintext message prior to encryption for easy identification among the four possible square roots and to simplify (or automate) the decryption process accordingly. If the Rabin asymmetric encryption system is modified this way, then its usefulness is improved significantly, but Theorem 13.1 no longer applies (because the decryption oracle always returns the square root that really represents the original plaintext message). The bottom line is that one has to make a choice: Either one goes for an encryption system that has provable security but is difficult to use in practice, or one goes for a system that is useful but cannot be proven secure—at least not in a way that is as simple as shown above. Rabin is not an (asymmetric) encryption system that can satisfy all requirements.

Last but not least, we stress another implication of the proof of Theorem 13.1: If an adversary has access to a decryption oracle (to break the one-way security of the Rabin encryption system), then he or she can also factorize $n$ and totally break the system accordingly. This makes the system susceptible to all kinds of CCAs. Again, one can employ redundancy to mitigate the attacks, and again one can then no longer prove that breaking the one-way security of the Rabin system is computationally equivalent to solving the IFP. The choice mentioned above is relevant again, and people working in the field usually prefer the more useful variants of the Rabin encryption system that use redundancy (to simplify decryption and mitigate CCAs). Unfortunately, the system as a whole is then no longer advantageous compared to RSA or Elgamal, and hence the system is hardly used in the field. This includes constructions like Rabin-OAEP.

### 13.3.3 Elgamal

As already mentioned in Section 1.3, public key cryptography started in the 1970s with the publication of [22], in which Diffie and Hellman introduced the basic idea and a key exchange protocol (Section 12.3). As its name suggests, this protocol can be used to exchange a key, but it can neither be used to encrypt and decrypt data nor to digitally sign messages and verify digital signatures accordingly. It was not until 1984 that Taher Elgamal[13] found a way to turn the Diffie-Hellman key exchange protocol into a full-fledged public key cryptosystem (i.e., a public key cryptosystem that can be used for asymmetric encryption and digital signatures [23][14]). In this section, we overview and discuss the Elgamal asymmetric encryption system. The respective DSS is addressed in Section 14.2.4.

Like the Diffie-Hellman key exchange protocol, the Elgamal public key cryptosystem (including the asymmetric encryption system) can be defined in any cyclic group $G$ in which the DLP is assumed to be intractable, such as a subgroup of $\mathbb{Z}_p^*$ (where $p$ is a safe prime) that is generated by $g \in \mathbb{Z}_p^*$ of order $q$ (Section 5.2.1) or a group of points on an elliptic curve over a finite field (Section 5.5). The Elgamal asymmetric encryption system then consists of the three efficient algorithms Generate, Encrypt, and Decrypt that are summarized in Table 13.4 and outlined next.

13 Taher Elgamal was a Ph.D. student of Hellman at Stanford University.

14 A preliminary version of [23] was presented at the CRYPTO '84 Conference.

#### 13.3.3.1 Key Generation Algorithm

The Elgamal key generation algorithm Generate has no further input other than $G$ and $g$ (that are system parameters). The algorithm randomly selects a private key $x$ from $\mathbb{Z}_q$ and generates a respective public key $y$ as $g^x$. In contrast to $x$ that is an integer between 0 and $q-1$, $y$ is an element of $G$ and need not be an integer (depending on the nature of $G$). Anyway, $(x, y)$ yields the public key pair that is the output of the Generate algorithm.

**Table 13.4**
Elgamal Asymmetric Encryption System

System parameters: $G, g$

| Generate | Encrypt | Decrypt |
|---|---|---|
| $(-)$ | $(m, y)$ | $((c_1, c_2), x)$ |
| $x \xleftarrow{r} \mathbb{Z}_q$<br>$y = g^x$ | $r \xleftarrow{r} \mathbb{Z}_q$<br>$K = y^r$<br>$c_1 = g^r$<br>$c_2 = Km$ | $K = c_1^x$<br>$m = c_2/K$ |
| $(x, y)$ | $(c_1, c_2)$ | $(m)$ |

Let us consider a toy example to illustrate the working principles of the Elgamal asymmetric encryption system. For $p = 17$, the multiplicative group $\mathbb{Z}_{17}^*$ is cyclic and $g = 7$ may serve as a generator; that is, 7 generates all $\mid \mathbb{Z}_{17}^* \mid = 16$ elements of $\mathbb{Z}_{17}^* = \{1, \ldots, 16\}$. In this setting, the Generate algorithm may randomly select a private key $x = 6$ and compute the respective public key $y \equiv 7^6 \pmod{17} \equiv 117,649 \pmod{17} = 9$. Consequently, the public key pair $(6, 9)$ is the output of the algorithm.

#### 13.3.3.2 Encryption Algorithm

Contrary to RSA, the Elgamal encryption algorithm Encryt is probabilistic, meaning that is uses a random number to encrypt a plaintext message. More specifically, it randomly selects an integer $r$ from $\mathbb{Z}_q$,[15] uses $r$ and the recipient's public key $y$ to compute $K = y^r$,[16] and uses this value to mask the message. More specifically, $r$

15 This value plays the role of the sender's private exponent in the Diffie-Hellman key exchange protocol.

16 This value plays the role of the outcome of the Diffie-Hellman key exchange protocol.

and $K$ are used to compute the following two elements of $G$:

$$\begin{aligned} c_1 &= g^r \\ c_2 &= Km \end{aligned}$$

The pair $(c_1, c_2)$ then represents the ciphertext of plaintext message $m$. This, in turn, means that the ciphertext is twice as long as the plaintext message, and hence that the encryption expands the plaintext message with a factor of two. This is a practical disadvantage of the Elgamal encryption system that severely limits its usefulness in the field.

It is important (and cannot be overestimated) that $r$ is used only once and is never reused. If $r$ is used more than once, then knowledge of a plaintext message $m_1$ enables an adversary to decrypt another plaintext message $m_2$ (for which the ciphertext is observed). Let

$$(c_1^{(1)}, c_2^{(1)}) = (g^r, Km_1)$$

and

$$(c_1^{(2)}, c_2^{(2)}) = (g^r, Km_2)$$

be the ciphertexts of $m_1$ and $m_2$ (where both messages are encrypted with the same value $r$). If $r$ is the same, then $K \equiv y^r$ is also the same. This, in turn, means that

$$\frac{c_2^{(1)}}{c_2^{(2)}} = \frac{Km_1}{Km_2} = \frac{m_1}{m_2}$$

and hence $m_2$ can be computed from $m_1$, $c_2^{(1)}$, and $c_2^{(2)}$:

$$m_2 = m_1 \frac{c_2^{(2)}}{c_2^{(1)}}$$

This is devastating, and hence a fresh and unique value $r$ is needed to encrypt a plaintext message (this requirement also applies to the Elgamal DSS).

The Elgamal Encryt algorithm requires only two modular exponentiations to encrypt a plaintext message, and hence it is efficient. The efficiency can even be improved by using precomputation. Note that $r$ and $K$ are independent from the plaintext message that is encrypted and that they can be precomputed and securely stored before they are used. This is also true for $c_1$. If $r$, $K$, and $c_1$ are precomputed, then it takes only one modular multiplication to encrypt a plaintext message. This is even more efficient than the modular exponentiation it takes to encrypt a plaintext message using the RSA asymmetric encryption system.

If, in our toy example, the plaintext message is $m = 7$, then the Elgamal Encryt algorithm may randomly select $r = 3$, compute $K \equiv 9^3 \pmod{17} = 15$, and conclude with $c_1 \equiv 7^3 \pmod{17} = 3$ and $c_2 \equiv 15 \cdot 7 \pmod{17} = 3$. Hence, the ciphertext transmitted to the recipient consists of the pair $(3, 3)$.

#### 13.3.3.3 Decryption Algorithm

The recipient of $(c_1, c_2)$ can use the Elgamal decryption algorithm Decryt to recover the original plaintext message $m$. It first recovers $K = c_1^x = (g^r)^x = (g^x)^r = y^r$, and then uses $K$ to unmask $m = c_2/K$.

In our toy example, the recipient receives $(3, 3)$ and wants to recover the plaintext message $m$. The Decrypt algorithm therefore computes $K \equiv 3^6 \pmod{17} \equiv 729 \pmod{17} = 15$ and solves $15m \equiv 3 \pmod{17}$ for $m$. The result is 7, because $15 \cdot 7 \equiv 105 \pmod{17} = 3$.

Alternatively, it is also possible to decrypt $(c_1, c_2)$ by computing $m = c_1^{-x} \cdot c_2$. In $\mathbb{Z}_p*$, this means first retrieving

$$x' = p - 1 - x$$

and then computing

$$\begin{aligned} c_1^{x'} c_2 &= g^{rx'} Km \\ &= g^{r(p-1-x)} Km \\ &= g^{r(p-1-x)} y^r m \\ &= (g^{p-1})^r (g^{-x})^r y^r m \\ &= (g^{p-1})^r (g^x)^{-r} y^r m \\ &= 1^r \cdot y^{-r} y^r m \\ &= m \end{aligned}$$

In the toy example, the recipient can decrypt the ciphertext by first retrieving $x' = 17 - 1 - 6 = 10$ and then computing $m \equiv 3^{10} \cdot 3 \pmod{17} \equiv 3^{11} \pmod{17} = 7$.

Like RSA, the Elgamal asymmetric encryption system requires a modular exponentiation to decrypt a ciphertext. But unlike RSA, there is no possibility to use the CRT to speed up the decryption algorithm.

#### 13.3.3.4 Security Analysis

The security of the Elgamal asymmetric encryption system is based on the DLA and the computational intractability of the DLP. If somebody is able to solve the DLP, then he or she can determine the private key from the public key. This totally breaks the Elgamal asymmetric encryption system.

With regard to the one-way security, Theorem 13.2 suggests that the one-way security of the Elgamal asymmetric encryption system is computationally equivalent to solving the (computational) DHP, meaning that an adversary is not able to decrypt a given ciphertext unless he or she is able to solve the DHP. Since we don't think that the adversary is able to solve the DHP, we strongly believe that the Elgamal asymmetric encryption system is (one-way) secure.

**Theorem 13.2** *Breaking the one-way security of the Elgamal asymmetric encryption system is computationally equivalent to solving the DHP.*

*Proof.* To prove the theorem, one must show (a) that somebody who can solve the DHP can break the one-way security of the Elgamal asymmetric encryption system, and (b) that somebody who can break the one-way security of the Elgamal system can solve the DHP. Let $G$ be a cyclic group with prime order $q$ and generator $g$.

For direction (a) we assume an adversary who has access to a DHP oracle $\mathcal{O}^{DHP}$. This oracle takes as input $g^a$ and $g^b$ for $a, b \in \mathbb{Z}_q$, and returns as output $g^{ab}$:

$$\mathcal{O}^{DHP}(g^a, g^b) = g^{ab}$$

An adversary can use such an oracle to decrypt a given ciphertext $(c_1, c_2)$ and retrieve the respective plaintext message $m$ accordingly: The adversary therefore invokes the oracle with the public key $y$ (representing $g^x$) and $c_1$ (representing $g^r$). The oracle responds with $\mathcal{O}^{DHP}(y, c_1) = \mathcal{O}^{DHP}(g^x, g^r) = g^{xr}$, and this value represents $K$ in the Elgamal encryption algorithm. The adversary can now retrieve $m$ by dividing $c_2$ by $K$ (i.e., $m = c_2/K$). Hence, the adversary can decrypt $(c_1, c_2)$, if he or she has access to a DHP oracle.

For direction (b) we assume an adversary who has access to an Elgamal decryption oracle $\mathcal{O}^{Elgamal}$. This oracle takes as input a public key $y$ (representing $g^x$) and a ciphertext $(c_1, c_2)$—with $c_1 = g^r$ and $c_2 = Km = g^{xr}m$—and it returns as output the respective plaintext message $m$:

$$\mathcal{O}^{Elgamal}(y, (c_1, c_2)) = m$$

An adversary can use such an oracle to solve the DHP; that is, compute $g^{xr}$ from $g^x$ and $g^r$: The adversary is challenged with $g^x$ and $g^r$. To compute $g^{xr}$, he or she randomly selects $s \in_R \mathbb{Z}_q$ and invokes the oracle with the public key $y$ and the ciphertext $(c_1, c_2)$, where $c_1 = g^r$ and $c_2 = g^s$. The oracle decrypts $(c_1, c_2)$ with the private key $x$ and sends the resulting plaintext message $m = c_1^{-x} c_2 = (g^r)^{-x} \cdot g^s = g^s / g^{-xr}$ to the adversary. The adversary computes $g^s / m = g^s \cdot g^{xr} / g^s = g^{xr}$ and sends this value back to the challenger. Since this yields a solution for the DHP the adversary has been challenged with, this finishes the proof.

□

More interestingly, the Elgamal encryption algorithm is probabilistic (i.e., nondeterministic) and the respective asymmetric encryption system can be shown to provide IND-CPA and be semantically secure under the assumption that the DDHP (Definition 5.7) is hard. This is captured in Theorem 13.3 (without a proof). Because the assumption is weaker and IND-CPA (or semantic security) is a stronger notion of security than one-way security, this result is more relevant in practice.

**Theorem 13.3** *If the DDHP is hard, then the Elgamal asymmetric encryption system provides IND-CPA and is semantically secure.*

IND-CPA (and hence semantic security) is the highest notion of security the Elgamal asymmetric encryption system can provide. The system is highly malleable and cannot provide IND-CCA. If, for example, one is given a ciphertext $(c_1, c_2)$ of some (possibly unknown) plaintext message $m$, then one can easily construct $(c_1, 2c_2)$ that represents a ciphertext for the plaintext message $2m$. The underlying reason for this malleability is the fact that the Elgamal asymmetric encryption system is multiplicatively homomorphic, meaning that the product of two ciphertexts $c_1$ and $c_2$ equals the encryption of the product of the underlying plaintext messages $m_1$ and $m_2$. Let

$$(c_1^{(1)}, c_2^{(1)}) = (g^{r_1}, y^{r_1} m_1)$$

and

$$(c_1^{(2)}, c_2^{(2)}) = (g^{r_2}, y^{r_2} m_2)$$

be the ciphertexts of $m_1$ and $m_2$. If these ciphertexts are multiplied, then the result is $(c_1, c_2)$ with

$$c_1 = c_1^{(1)} \cdot c_1^{(2)} = g^{r_1} \cdot g^{r_2} = g^{r_1 + r_2}$$

and

$$c_2 = c_2^{(1)} \cdot c_2^{(2)} = y^{r_1} m_1 \cdot y^{r_2} m_2 = y^{r_1 + r_2} m_1 m_2 = g^{x(r_1 + r_2)} m_1 m_2$$

Note that the Elgamal asymmetric encryption system is multiplicatively but not additively homomorphic. There is a variant [24, 25] that is additively homomorphic and used in applications like e-voting, but there is no variant that is multiplicatively and additively homomorphic. If there were, then we would have a solution for fully homomorphic encryption (Section 13.5).

Whether malleability and IND-CCA security are advantageous or disadvantageous depends on the application one has in mind. There are at least several possibilities and techniques that can be used to turn the Elgamal asymmetric encryption system into a variant that is nonmalleable and provides IND-CCA. One such variant, Cramer-Shoup, is described next. It is secure in the standard model. There are other variants that employ hybrid encryption, such as DHIES [26] and its ECC-counterpart ECIES.[17] They combine a Diffie-Hellman key exchange with some symmetric encryption, and it can be shown that they are secure and provide IND-CCA in the random oracle model.

### 13.3.4 Cramer-Shoup

In 1998, Ronald Cramer and Victor Shoup proposed a variant of the Elgamal asymmetric encryption system that is nonmalleable and provides IND-CCA in the standard model [27]. Like Elgamal, the Cramer-Shoup asymmetric encryption system requires a cyclic group $G$ of order $q$ in which the DDHP is computationally intractable. But unlike Elgamal, it uses two generators of $G$, $g_1$ and $g_2$, and makes use of a hash function $h$ that outputs values that can be interpreted as numbers in $\mathbb{Z}_q$.[18] Note that the Cramer-Shoup system employs a hash function, but its security proof does not depend on the assumption that this hash function is a random function (and hence the proof is not in the random oracle model).

#### 13.3.4.1 Key Generation Algorithm

$G$, $g_1$, and $g_2$ are system parameters that are known to everybody. This is why the key Cramer-Shoup generation algorithm Generate does not take an input parameter. Instead, it only randomly selects the five elements $x_1, x_2, y_1, y_2$, and $z$ from $\mathbb{Z}_q$ that collectively form the private key $sk$, and it then computes the three group elements $c$, $d$, and $e$ that collectively form the public key $pk$.

17 Provably secure elliptic curve (PSEC) encryption is another hybrid encryption technique that is similar to ECIES. Because it is mainly used to securely transmit a key, it yields a key encapsulation mechanism (KEM), and it is therefore acronymed PSEC-KEM (https://info.isl.ntt.co.jp/crypt/eng/psec/intro.html).

18 The Cramer-Shoup cryptosystem can be modified to get rid of the hash function.

#### 13.3.4.2 Encryption Algorithm

Similar to Elgamal, the Cramer-Shoup encryption algorithm Encrypt is probabilistic. In addition to the system parameters, it takes as input a plaintext message $m$ (that represents an element of $G$) and the recipient's public key $(c, d, e)$. It then randomly selects an element $r$ from $\mathbb{Z}_q$, computes $u_1 = g_1^r$, $u_2 = g_2^r$, and $v = e^r m$, computes $\alpha$ as the hash value of $u_1, u_2$, and $v$, and finally generates $w = c^r d^{r\alpha}$. The 4-tuple $(u_1, u_2, v, w)$ that consists of four elements of $G$ yields the ciphertext.

Note that the pair $(u_1, v)$ is essentially an Elgamal encryption, whereas the pair $(u_2, w)$ represents some form of an error detection code. It ensures that illegitimately constructed ciphertexts are detected and can be rejected accordingly. The element $w$ therefore acts as a "proof of legitimacy" that can afterward be verified during decryption.

**Table 13.5**
Cramer-Shoup Asymmetric Encryption System

System parameters: $G, g_1, g_2$

| Generate | Encrypt | Decrypt |
|---|---|---|
| $(-)$ | $(m, (c, d, e))$ | $((u_1, u_2, v, w),$<br>$(x_1, x_2, y_1, y_2, z))$ |
| $x_1, x_2, y_1, y_2, z \xleftarrow{r} \mathbb{Z}_q^5$<br>$c = g_1^{x_1} g_2^{x_2}$<br>$d = g_1^{y_1} g_2^{y_2}$<br>$e = g_1^z$ | $r \xleftarrow{r} \mathbb{Z}_q$<br>$u_1 = g_1^r$<br>$u_2 = g_2^r$<br>$v = e^r m$<br>$\alpha = h(u_1, u_2, v)$<br>$w = c^r d^{r\alpha}$ | $\alpha = h(u_1, u_2, v)$<br>If $w = u_1^{x_1+\alpha y_1} u_2^{x_2+\alpha y_2}$<br>then return $m = w/u_1^z$<br>else return "reject" |
| $(x_1, x_2, y_1, y_2, z)$<br>$(c, d, e)$ | $(u_1, u_2, v, w)$ | ($m$ or "reject") |

#### 13.3.4.3 Decryption Algorithm

The decryption algorithm Decrypt takes as input a ciphertext $(u_1, u_2, e, v)$ and the recipient's private key $(x_1, x_2, y_1, y_2, z)$, and it either generates as output the original plaintext message $m$ or rejects the ciphertext as being invalid. To achieve this, the algorithm first computes $\alpha$ by subjecting the first three components of the ciphertext to the hash function $h$. Using this value, the algorithm then computes $u_1^{x_1+\alpha y_1} u_2^{x_2+\alpha y_2}$ and verifies whether the resulting value matches $w$ (i.e., the fourth component of the ciphertext). To see whether this verification makes sense, we can algebraically transform $w$ to $u_1^{x_1+\alpha y_1} u_2^{x_2+\alpha y_2}$:

$$\begin{aligned} w &= c^r d^{r\alpha} \\ &= (g_1^{x_1} g_2^{x_2})^r (g_1^{y_1} g_2^{y_2})^{r\alpha} \\ &= g_1^{rx_1} g_2^{rx_2} g_1^{r\alpha y_1} g_2^{r\alpha y_2} \\ &= g_1^{rx_1} g_1^{r\alpha y_1} g_2^{rx_2} g_2^{r\alpha y_2} \\ &= u_1^{x_1} u_1^{\alpha y_1} u_2^{x_2} u_2^{\alpha y_2} \\ &= u_1^{x_1+\alpha y_1} u_2^{x_2+\alpha y_2} \end{aligned}$$

If the equation holds, then $m$ can be computed from $v$. We solve the equation $v = e^r m$ for $m$ (i.e., $m = v/h^r$). Substituting $h$ with $g_1^z$, we get $m = v/{g_1}^{zr}$, and again substituting $g_1^r$ with $u_1$, we get $m = v/{u_1}^z$. Using this formula, the recipient can actually recover the original plaintext message and hence decrypt the ciphertext. If the above-mentioned verification fails, then the algorithm returns "reject."

When the Cramer-Shoup asymmetric encryption system was proposed in the late 1990s, it was considered a major breakthrough because it was one of the first encryption systems that provided IND-CCA and is reasonably efficient (in fact, it is approximately twice as expensive as Elgamal, both in terms of computing time and the size of the ciphertexts). Unfortunately, its success in theory was not followed by a success in practice, and the Cramer-Shoup asymmetric encryption system has not been widely deployed in the field. However, it initiated a lot of research and development in CCA-secure encryption systems, and many variants of the Cramer-Shoup asymmetric encryption system have been proposed in the literature.

## 13.4 IDENTITY-BASED ENCRYPTION

In an asymmetric encryption system, every user has a public key pair, and the keys look somewhat arbitrary and random. Consequently, one faces the problem that one cannot easily attribute a given public key to a particular entity (e.g., user) and that one has to work with public key certificates. A public key certificate, in turn, is a data structure that is issued by a trusted (or trustworthy) certification authority (CA). It is digitally signed by the issuing CA, and it states that a public key really belongs to a particular entity. If there are multiple CAs in place, then one usually talks about public key infrastructures (PKIs). In general, public key certificates, CAs, and PKIs are complex topics, and their implementation has turned out to be more difficult than originally anticipated [28].

In the early 1980s, Shamir came up with an alternative idea. If one chooses a public key to uniquely identify its holder, then one no longer has to care about public key certification in the first place. Instead, a public key is then self-evident in the sense that it automatically becomes clear to whom it belongs (or at least to whom it was issued in the first place). Shamir coined the term *identity-based cryptography* to refer to this cryptographic technique. Like any other technique, identity-based cryptography has advantages and disadvantages:

- The advantages are obvious and related to the avoidance of public key certificates and respective key directory services.
- The disadvantages are less obvious. The most important ones are related to the necessity of having a unique naming scheme and the fact that a trusted authority is needed to generate public key pairs and distribute them. Hence, all entities must trust this authority not to hold illegitimate copies and misuse their private keys.

Note that in a conventional asymmetric encryption system, all entities can generate their own public key pairs. In identity-based cryptography, this cannot be the case, because the public keys must have specific values and it must not be possible for anybody (except the trusted authority) to determine the private key that belongs to a specific public key (otherwise, this person could determine all private keys in use). Consequently, in identity-based cryptography, all entities must provide their identities to the trusted authority, and the trusted authority must equip them with their respective public key pairs, using, for example, smart cards or USB tokens. Another disadvantage that may occur in practice is related to key revocation. What happens, for example, if a key pair needs to be revoked? Since the public key represents the key pair holder's identity, it is not obvious how this key pair can be replaced in a meaningful way.

In [29] Shamir introduced the notion of identity-based cryptography and proposed an identity-based digital signature system (Section 14.3). Shamir also pointed out that the development of an identity-based encryption system is more involved. Note, for example, that the RSA asymmetric encryption system cannot be easily turned into an identity-based encryption system. On the one hand, if $n$ is universal and the same for all users, then anyone who knows an encryption exponent $e$ and a respective decryption exponent $d$ can compute the factorization of $n$ and compute all private keys. On the other hand, if $n$ depends on the user's identity, then the trusted authority cannot factorize $n$ and compute the decryption exponent $d$ that belongs to an encryption exponent $e$.

It was not until 2001 that Dan Boneh and Matthew K. Franklin proposed an *identity-based encryption* (IBE) system based on bilinear maps called *pairings* on elliptic curves [30, 31]. They suggested using the IBE system as an alternative to commonly used secure messaging technologies and solutions that are based on public key certificates. In the same year, Cocks[19] proposed an IBE system based on the QRP [32]. The Cocks IBE system has a high degree of ciphertext expansion and is fairly inefficient. It is impractical for sending all but the shortest messages, such as a session key for use with a symmetric encryption system, and it is therefore not used in the field.

A comprehensive overview of IBE systems is provided in [33]. Researchers have also tried to combine conventional asymmetric encryption and IBE to overcome some disadvantages of IBE. Examples include *certificateless encryption* [34] and *certificate-based encryption* [35]. The definitions and security notions of certificateless encryption and certificate-based encryption are further addressed in [36]. In particular, there is an equivalence theorem, saying that—from a security perspective—IBE, certificateless encryption, and certificate-based encryption are equivalent, meaning that a secure certificateless or certificate-based encryption system exists if and only if a secure IBE system exists. The bottom line is that IBE (together with certificateless encryption and certificate-based encryption) is a nice idea, but it has not been able to move from theory to practice.

## 13.5 FULLY HOMOMORPHIC ENCRYPTION

In 1978, Rivest, Adleman, and Michael Leonidas Dertouzos published a paper in which they introduced the notion of *homomorphic encryption* [37]. In essence, homomorphic encryption is about encrypting data in a way that allows computations to be done only on the ciphertexts (i.e., without decryption). If, for example, $\circledcirc$ represents a computation that is applied to two plaintext messages $m_1$ and $m_2$, and $E$ refers to a homomorphic encryption function, then there is another computation $\circledast$ (that can also be the same) for which

$$E(m_1 \circledcirc m_2) = E(m_1) \circledast E(m_2)$$

holds. This means that we can compute $E(m_1 \circledcirc m_2)$ even if we only know $E(m_1)$ and $E(m_2)$; that is, without knowing $m_1$ and $m_2$. It is obvious and goes without saying that a homomorphic encryption system has many applications in the realm of outsourcing and cloud computing.

Many asymmetric encryption systems in use today are partially homomorphic. We already mentioned that (unpadded) RSA and Elgamal are multiplicatively homomorphic. In the case of RSA, for example, $E(m_1) \cdot E(m_2) \equiv m_1^e \cdot m_2^e = (m_1 m_2)^e \pmod{n} = E(m_1 \cdot m_2)$, and hence $\circledcirc$ and $\circledast$ both refer to integer multiplication modulo $n$. The same line of argumentation applies to Elgamal. Also,

19 Clifford Cocks was already mentioned in the Introduction. He was one of the GCHQ employees who discovered public key cryptography under the name NSE in the early 1970s.

we mentioned that the Paillier system is additively homomorphic. In the case of probabilistic encryption (Section 13.2), each bit $b$ is encrypted individually, and it can be shown that $E(b_1) \cdot E(b_2) = E(b_1 \oplus b_2)$. Each of these examples allows homomorphic computation of only one operation (either addition or multiplication). For three decades, it was not clear whether *fully homomorphic encryption* (FHE), in which both addition and multiplication are supported, is feasible at all. The best result supported the evaluation of an unlimited number of addition operations and at most one multiplication operation [38].

In 2009, Craig Gentry was the first who solved the problem and proposed a FHE system using lattice-based cryptography [39].[20] In spite of the fact that Gentry's proposal represents a major theoretical breakthrough, the system as originally proposed is impractical for real-world applications, mainly because the size of the ciphertext and the computation time increase sharply as one increases the security level. Hence, several researchers have tried (and are still trying) to improve the system and come up with proposals that are more practical. This is ongoing work and a lot of research and development efforts go into this area. It is sometimes argued that FHE represents the holy grail for secure cloud computing.

## 13.6 FINAL REMARKS

In this chapter, we elaborated on asymmetric encryption and what is meant by saying that an asymmetric encryption system is secure. We certainly require that such a system is one-way secure, meaning that it is not feasible to retrieve a plaintext message from a given ciphertext. All basic asymmetric encryption systems—RSA, Rabin, and Elgamal—fulfill this requirement. If a system is to provide IND-CPA and be semantically secure, then its encryption algorithm needs to be probabilistic. This is not true for basic RSA and Rabin, so we have to invoke some complementary technology, such as OAEP. RSA-OAEP and Rabin-OAEP can indeed be shown to provide IND-CPA in the random oracle model. The same is true for Elgamal that natively provides IND-CPA. The Elgamal asymmetric encryption system can even be modified to be nonmalleable and provide IND-CCA in the standard model, meaning that the security proof does not require the random oracle model. A respective construction is known as Cramer-Shoup, but it is not as widely used in the field as originally anticipated.

In addition to the asymmetric encryption systems addressed in this chapter, there are other systems that have been developed and proposed in the literature. Some of these systems have been broken and become obsolete. For example, the

20 Gentry's proposal is also addressed in his Ph.D. thesis that is electronically available at https://crypto.stanford.edu/craig/.

**NP**-complete subset sum problem has served as a basis for many public key cryptosystems. All knapsack-based public key cryptosystems proposed in the past have been broken. This also includes the Chor-Rivest knapsack cryptosystem [40, 41]. In fact, knapsack-based cryptosystems are good examples to illustrate the point that it is necessary but usually not sufficient that a public key cryptosystem is based on a mathematically hard problem. Breaking a knapsack-based public key cryptosystem is generally possible without solving the underlying subset sum problem.

There are also a few asymmetric encryption systems that have turned out to be resistant against all types of cryptanalytical attacks. For example, in 1978, Robert McEliece proposed an asymmetric encryption system back that has remained secure until today [42]. The respective McEliece asymmetric encryption system was the first to use randomization in the encryption process. Due to its inefficiency and use of prohibitively large keys, the system has never gained much acceptance in the cryptographic community. This is about to change, mainly because the McEliece asymmetric encryption system—or rather a variant proposed by Harald Niederreiter in 1986 [43]—is currently a candidate in the PQC competition (Section 18.3.1).

In Section 13.3, we said that we assume public keys to be published in some certified form. This simple and innocent assumption has huge implications in practice. How does one make sure that all entities have public keys? How does one publish them, and how does one certify them? Finally, how does one make sure that public keys can be revoked and that status information about a public key is publicly available in a timely fashion? All of these questions related to digital certificates are typically addressed (and solved) by a PKI. We already mentioned that the establishment and operation of a PKI is more involved than it looked at first glance, and we have to revisit this topic in Section 16.4.

## References

[1] Rackoff, C., and D.R. Simon, "Non-Interactive Zero-Knowledge Proof of Knowledge and Chosen Ciphertext Attack," *Proceedings of CRYPTO '91*, Springer-Verlag, LNCS 576, 1992, pp. 433–444.

[2] Bleichenbacher, D., "Chosen Ciphertext Attacks Against Protocols Based on the RSA Encryption Standard PKCS #1," *Proceedings of CRYPTO '98*, Springer-Verlag, LNCS 1462, 1998, pp. 1–12.

[3] Goldwasser, S., and S. Micali, "Probabilistic Encryption," *Journal of Computer and System Sciences*, Vol. 28, No. 2, April 1984, pp. 270–299.

[4] Micali, S., C. Rackoff, and B. Sloan, "The Notion of Security for Probabilistic Cryptosystems," *SIAM Journal on Computing*, Vol. 17, No. 2, 1988, pp. 412–426.

[5] Dolev, D., C. Dwork, and M. Naor, "Non-Malleable Cryptography," *SIAM Journal on Computing*, Vol. 30, No. 2, 2000, pp. 391–437.

[6] Bellare, M., et al., "Relations Among Notions of Security for Public-Key Encryption Schemes," *Proceedings of CRYPTO '98*, Springer-Verlag, LNCS 1462, 1998, pp. 26–45.

[7] Blum, M., and S. Goldwasser, "An Efficient Probabilistic Public Key Encryption Scheme Which Hides All Partial Information," *Proceedings of CRYPTO '84*, Springer-Verlag, 1985, pp. 289–299.

[8] Rivest, R.L., A. Shamir, and L. Adleman, "A Method for Obtaining Digital Signatures and Public-Key Cryptosystems," *Communications of the ACM*, Vol. 21, No. 2, February 1978, pp. 120–126.

[9] Boneh, D., and H. Shacham, "Fast Variants of RSA," *CryptoBytes*, Vol. 5, No. 1, 2002, pp. 1–9.

[10] Boneh, D., "Twenty Years of Attacks on the RSA Cryptosystem," *Notices of the American Mathematical Society (AMS)*, Vol. 46, No. 2, 1999, pp. 203–213.

[11] Boneh, D., and R. Venkatesan, "Breaking RSA May Not Be Equivalent to Factoring," *Proceedings of EUROCRYPT '98*, Springer-Verlag, LNCS 1403, 1998, pp. 59–71.

[12] Håstad, J., and M. Näslund, "The Security of Individual RSA Bits," *Proceedings of the 39th Annual Symposium on Foundations of Computer Science (FOCS '98)*, 1998, pp. 510–519.

[13] Wiener, M., "Cryptanalysis of Short RSA Secret Exponents," *IEEE Transactions on Information Theory*, Vol. 36, No. 3, 1990, pp. 553–558.

[14] Boneh, D., and G. Durfee, "Cryptanalysis of RSA with Private Exponent $d < N^{0.292}$," *Proceedings of EUROCRYPT '99*, Springer-Verlag, LNCS 1592, 1999, pp. 1–11.

[15] Bellare, M., and P. Rogaway, "Optimal Asymmetric Encryption," *Proceedings of EUROCRYPT '94*, Springer-Verlag, LNCS 950, 1994, pp. 92–111.

[16] Kaliski, B., and J. Staddon, *PKCS #1: RSA Cryptography Specifications Version 2.0*, Informational Request for Comments 2437, October 1998.

[17] Shoup, V., "OAEP Reconsidered," *Proceedings of CRYPT0 '01*, Springer-Verlag, LNCS 2139, 2001, pp. 239–259.

[18] Fujisaki, E., et al., "RSA-OAEP Is Secure Under the RSA Assumption," *Journal of Cryptology*, Vol. 17, No. 2, Spring 2004, pp. 81–104.

[19] Pointcheval, D., "How to Encrypt Properly with RSA," *CryptoBytes*, Vol. 5, No. 1, 2002, pp. 10–19.

[20] Jonsson, J., and B. Kaliski, *Public-Key Cryptography Standards (PKCS) #1: RSA Cryptography Specifications Version 2.1*, Informational Request for Comments 3447, February 2003.

[21] Rabin, M.O., "Digitalized Signatures and Public-Key Functions as Intractable as Factorization," MIT Laboratory for Computer Science, MIT/LCS/TR-212, 1979.

[22] Diffie, W., and M.E. Hellman, "New Directions in Cryptography," *IEEE Transactions on Information Theory*, Vol. 22, No. 6, 1976, pp. 644–654.

[23] Elgamal, T., "A Public Key Cryptosystem and a Signature Scheme Based on Discrete Logarithm," *IEEE Transactions on Information Theory*, Vol. 31, No. 4, 1985, pp. 469–472.

[24] Paillier, P., "Public-Key Cryptosystems Based on Composite Degree Residuosity Classes," *Proceedings of EUROCRYPT '99*, Springer-Verlag, LNCS 1592, 1999, pp. 223–238.

[25] Paillier, P., and D. Pointcheval, "Efficient Public-Key Cryptosystems Provably Secure Against Active Adversaries," *Proceedings of ASIACRYPT '99*, Springer-Verlag, 1999, pp. 165–179.

[26] Abdalla, M., M. Bellare, and P. Rogaway, "The Oracle Diffie-Hellman Assumptions and an Analysis of DHIES," *Proceedings of Topics in Cryptology—CT-RSA 2001*, Springer-Verlag, LNCS 2020, 2001, pp. 143–158.

[27] Cramer, R., and V. Shoup, "A Practical Public Key Cryptosystem Provably Secure Against Adaptive Chosen Ciphertext Attack," *Proceedings of CRYPTO '98*, Springer-Verlag, LNCS 1462, 1998, pp. 13–25.

[28] Lopez, J., R. Oppliger, and G. Pernul, "Why have Public Key Infrastructures Failed so Far?" *Internet Research*, Vol. 15, No. 5, 2005, pp. 544–556.

[29] Shamir, A., "Identity-Based Cryptosystems and Signatures," *Proceedings of CRYPTO '84*, Springer-Verlag, 1984, pp. 47–53.

[30] Boneh, D., and M. Franklin, "Identity-Based Encryption from the Weil Pairing," *Proceedings of CRYPTO 2001*, Springer-Verlag, 2001, pp. 213–229.

[31] Boneh, D., and M. Franklin, "Identity Based Encryption from the Weil Pairing," *SIAM Journal of Computing*, Vol. 32, No. 3, 2003, pp. 586–615.

[32] C. Cocks, "An Identity Based Encryption Scheme Based on Quadratic Residues," *Proceedings of the 8th IMA International Conference on Cryptography and Coding*, 2001, pp. 360–363.

[33] Luther, M., *Identity-Based Encryption*, Artech House Publishers, Norwood, MA, 2008.

[34] Al-Riyami, S.S., and K.G. Paterson, "Certificateless Public Key Cryptography," *Proceedings of ASIACRYPT 2003*, Springer-Verlag, 2003, pp. 452–473.

[35] Gentry, C., "Certificate-Based Encryption and the Certificate Revocation Problem," *Proceedings of EUROCRYPT 2003*, Springer-Verlag, 2003, pp. 272–293.

[36] Yum, D.H., and P.J. Lee, "Identity-Based Cryptography in Public Key Management," *Proceedings of EuroPKI 2004*, Springer-Verlag, 2004, pp. 71–84.

[37] Rivest, R.L., L. Adleman, and M.L. Dertouzos, "On Data Banks and Privacy Homomorphisms," *Proceedings of Foundations of Secure Computation*, 1978.

[38] Boneh, D., Goh, E., and K. Nissim, "Evaluating 2-DNF Formulas on Ciphertexts," *Proceedings of Theory of Cryptography (TCC)*, Springer-Verlag, LNCS 3378, 2005, pp. 325–341.

[39] Gentry, C., "Fully Homomorphic Encryption Using Ideal Lattices," *Proceedings of 41st Annual ACM Symposium on Theory of Computing (STOC)*, 2009, pp. 169–178.

[40] Chor, B., and R.L. Rivest, "A Knapsack-Type Public-Key Cryptosystem Based on Arithmetic in Finite Fields," *IEEE Transactions on Information Theory*, Vol. 34, 1988, pp. 901–909.

[41] Vaudenay, S., "Cryptanalysis of the Chor-Rivest Cryptosystem," *Journal of Cryptology*, Vol. 14, 2001, pp. 87–100.

[42] McEliece, R.J., *A Public-Key Cryptosystem Based on Algebraic Coding Theory*, Deep Space Network Progress Report 42-44, Jet Propulsion Lab., California Institute of Technology, 1978, pp. 114–116.

[43] Niederreiter, H., "Knapsack-type Cryptosystems and Algebraic Coding Theory," *Problems of Control and Information Theory/Problemy Upravlenija i Teorii Informacii*, Vol. 15, 1986, pp. 159-166.

# Chapter 14

## Digital Signatures

In this chapter, we elaborate on digital signatures and respective DSSs. More specifically, we introduce the topic in Section 14.1, outline and discuss the DSSs used in the field in Section 14.2, elaborate on identity-based signatures, one-time signatures, and a few other variants in Sections 14.3–14.5, and conclude with some final remarks in Section 14.6. Note that all books on cryptography (including the ones itemized in the Preface) address the topic, and that there even are a few books that focus entirely on digital signatures [1–4].

### 14.1 INTRODUCTION

In Section 2.3.3, we introduced, briefly discussed, and put into perspective digital signatures, and we distinguished between a DSS with appendix (Definition 2.13 and Figure 2.10) and a DSS giving message recovery (Definition 2.14 and Figure 2.11). In short, a DSS consists of three efficiently computable algorithms: Generate, Sign, and Verify in the case of a DSS with appendix or Recover in the case of a DSS giving message recovery. The Generate algorithm is to generate public key pairs (that consist of an appropriately sized signing key and a corresponding verification key), the Sign algorithm is to generate digital signatures, and the Verify or Recover algorithm is to verify the digital signatures or recover the digitally signed message respectively.

In either case, a DSS must be correct and secure:

- A DSS is *correct* if every valid signature is accepted. For DSSs with appendix, this means that $\mathsf{Verify}(pk, m, \mathsf{Sign}(sk, m))$ must return $valid$ for every public key pair $(pk, sk)$ and message $m$. Similarly, for DSSs with message recovery, this means that $\mathsf{Recover}(pk, \mathsf{Sign}(sk, m))$ must yield $m$ for every public key pair $(pk, sk)$ and message $m$.

- A DSS is *secure* if it is impossible—or rather computationally infeasible—for an adversary to forge a valid signature; that is, to generate, without knowledge of a private signing key $sk$, a signature that is considered to be valid for a public verification key $pk$ and a respective message $m$ (that may not be meaningful).

The correctness requirement is simple and straightforward, and it does not lend itself to multiple interpretations. This need not be true for the security requirement, and there are different ways to read and interpret this requirement. For example, it is possible to say that a DSS is secure if it is computationally infeasible for an adversary to generate, without knowledge of the private signing key $sk$, a digital signature for a specific message $m$. This is certainly something one would demand from a DSS. In fact, all systems overviewed and discussed in this chapter are secure in this sense. Another way to read and interpret the security requirement is that it must be computationally infeasible for an adversary to generate a valid signature for an arbitrary, randomly looking and not even meaningful message $m$. This is arguably more difficult to achieve, and not all systems overviewed and discussed in this chapter are secure in this sense. There are even more ways to read and interpret the security requirement.

To be more precise, we refer to Section 1.2.2 where we required that every security definition must specify both the adversary's capabilities and the task the adversary is required to solve in order to be successful (i.e., to break the security of the system). The terminology most frequently used in the realm of secure digital signatures was coined by Goldwasser, Micali, and Rivest in the late 1980s [5]. It has withstood the test of time and still serves as a reference today.

With regard to the adversary's capabilities, it is important to note that we are in the realm of public key cryptography, where unconditional security does not exist. Consequently, we have to make assumptions about the computational power of the adversary. The assumption most frequently made is that the adversary has computational power that is polynomially bounded (with respect to the input length of the underlying mathematical problem). Furthermore, we have to specify what type of attacks the adversary is able to mount. There are three major classes of attacks that can be distinguished here:

- In a *key-only attack*, the adversary only knows the signatory's public verification key $pk$. In particular, he or she does not know and has no access to previously signed messages and their respective signatures.

- In a *known-message attack* (KMA), the adversary knows the signatory's public verification key $pk$ and $t \geq 1$ messages $m_1, m_2, \ldots, m_t$ with their respective signatures $s_1, s_2, \ldots, s_t$. The message-signature pairs $(m_i, s_i)$ for $i = 1, 2, \ldots, t$ are known to the adversary, but they are not chosen by him or her.
- In a *chosen-message attack* (CMA), the adversary knows the signatory's public verification key $pk$ and is able to obtain digital signatures $s_1, s_2, \ldots, s_t$ for a chosen list of $t \geq 1$ messages $m_1, m_2, \ldots, m_t$. In contrast to a KMA, the adversary can choose the message-signature pairs $(m_i, s_i)$ for $i = 1, 2, \ldots, t$. There are at least three subclasses of CMAs:
  - In a *generic CMA*, the message-signature pairs chosen by the adversary are preselected (i.e., selected before the attack begins) and independent from the signatory and its public verification key $pk$ (this independence makes the attack generic).
  - In a *directed CMA*, the message-signature pairs chosen by the adversary are still preselected, but they now depend on the signatory and its public verification key $pk$ (this dependence makes the attack directed).
  - Finally, in an *adaptive CMA*, the message-signature pairs are still selected while the attack is going on, and they depend on the signatory and its public verification key $pk$. Alternatively speaking, one can say that the adversary has access to a signature generation oracle. For every message $m$ he or she provides, the oracle immediately returns a valid signature $s$.

The attacks are itemized in order of increasing severity, meaning that the adaptive CMA is the most powerful attack an adversary may be able to mount. It goes without saying that key-only attacks are always possible (because the verification keys are public). However, for all practical purposes, it is reasonable to assume that the adversary one has in mind can also mount KMAs or CMAs. While adaptive CMAs may be difficult to mount in practice, a well-designed DSS should nonetheless resist them (to maintain a security margin).

With regard to the task the adversary is required to solve, there is the possibility to compromise the signatory's private signing key $sk$ in a *total break*, or the possibility to somehow forge a valid digital signature. In fact, there are three types of forgeries that are typically distinguished:

- In a *universal forgery*, the adversary is able to forge a digital signature for every possible message. This type of forgery is very similar to a total break, but it may not be necessary to compromise $sk$.

- In a *selective forgery*, the adversary is able to forge a digital signature for a particular message (that is preselected in one way or another).
- In an *existential forgery*, the adversary is able to forge a digital signature for at least one message (that may be random-looking and meaningless).

The types of forgeries are itemized in order of decreasing difficulty and severity, meaning that the existential forgery is the least difficult to mount type of forgery, but also the least severe one. Many DSSs used natively do not protect against this type of forgery. For example, if RSA or Rabin are used natively, then each element of $\mathbb{Z}_n^*$ stands for a digital signature of a particular message (that may be random-looking and meaningless). To ensure that such a signature cannot be exploited in a particular way, one usually modifies the DSS to protect against existential forgeries.

The adversary's capabilities and the task he or she is required to solve can be combined to come up with a distinct notion of security. As usual, we combine the most powerful adversary (i.e., an adversary who can mount adaptive CMAs) with the simplest task (i.e., existentially forge a digital signature) to come up with a strong security definition: We say that a DSS is secure, if an adversary who can mount an adaptive CMA is not even able to existentially forge a signature, and we say that it is provably secure if we can prove this claim (be it in the standard model or in the random oracle model). This definition goes back to [5], and Goldwasser, Micali, and Rivest also proposed a DSS (i.e., the GMR DSS[1]) that adheres to this definition.

The GMR DSS assumes the existence of claw-free pairs of trapdoor permutations. In today's parlance, we would call a claw-free pair of trapdoor permutation collision-resistant, meaning that a pair of trapdoor permutations $f_0$ and $f_1$ over a common domain (and range) $D$ is said to be *claw-free* if it is computationally infeasible (without knowing the trapdoor information[2]) to find $x, y, z \in D$ such that

$$f_0(x) = f_1(y) = z$$

Using such a claw-free pair $(f_0, f_1)$ of trapdoor permutations, it is simple and straightforward to digitally sign an $n$-bit message $m = m_1 \ldots m_n$ (where each $m_i$ represents 0 or 1): The signatory randomly selects a fresh reference value $r \in D$ and publishes this value in some authentic form. Using this value, the signatory can generate a digital signature $s$ for $m$ as follows:

$$s = f_{m_1}^{-1}(f_{m_2}^{-1}(\ldots f_{m_n}^{-1}(r)\ldots))$$

1 The acronym GMR refers to the first letters of the respective authors' names.

2 It goes without saying that finding such a triple $x, y, z$ is trivial with the trapdoor information. In this case, one can randomly select $z \in_R D$ and compute $x = f_0^{-1}(z)$ and $y = f_1^{-1}(z)$. The resulting triple $x, y, z$ then fulfills $f_0(x) = f_1(y) = z$.

For each message bit $m_i$ ($i = 1, 2, \ldots, n$), it basically inverts either $f_0$ or $f_1$. The signatory can do this because it knows the trapdoor information, but everybody else cannot. In the end, the signature $s$ is sent along with the message $m$ to the verifier.

The verifier, in turn, uses $s$ and $m$ to compute

$$r' = f_{m_1}(f_{m_2}(\ldots f_{m_n}(s)\ldots))$$

Finally, the signature $s$ is considered to be valid for message $m$, if $r'$ equals $r$. The difficulty with this DSS is that every message requires a fresh and unique reference value $r$, and that this value must be published in some authentic form. To achieve this, some efficient authentication mechanisms, such as Merkle trees [6, 7], may be used and this where the GMR DSS distinguishes itself from some follow-up proposals (e.g, [8, 9]). Another disadvantage of the GMR DSS and most of its successors is that they are stateful in the sense that the signatory has to keep state and store information about previously signed messages.

In spite of these disadvantages, the GMR DSS and most of its successors can be proven secure in the standard model (i.e., without assuming a random oracle). This may not be true for the DSSs that follow the conventional *hash-and-sign* paradigm, where a message is hashed (using a cryptographic hash function) and the result is signed using a basic DSS, such as RSA, Rabin, or Elgamal. Although hash-and-sign DSSs are usually very efficient, it was not immediately clear how to make them provably secure.

This research question was first addressed by Bellare and Rogaway in the early 1990s using the random oracle model [10]. They realized that a hash-and-sign DSS that uses, for example, MD5, PKCS #1, and RSA, has the structural problem (or deficiency) that the set of encoded messages that are subject to the RSA function represents a sparse and highly structured subset of the domain (that is $\mathbb{Z}_n^*$ in the case of RSA). This is disadvantageous and may be exploited in cryptanalysis. In a first attempt to overcome this problem, they suggested hashing a message $m$ onto the full domain $\mathbb{Z}_n^*$ of the RSA function before signing it. Following this line of argumentation, they proposed a *full-domain-hash* (FDH) function $h_{FDH} : \{0,1\}^* \rightarrow \mathbb{Z}_n^*$ that hashes arbitrarily sized strings uniformly into $\mathbb{Z}_n^*$. The FDH signature of $m$ then refers to the signature of $h_{FDH}(m)$. Assuming that $h_{FDH}$ is ideal (i.e., it behaves like a random oracle) and RSA is a trapdoor permutation, they were able to prove the security of the FDH DSS in the random oracle model. The construction was later modified to become the *probabilistic signature scheme* (PSS) and the *probabilistic signature scheme with message recovery* (PSS-R) [11]. These are addressed in Section 14.2.2.

The PSS and PSS-R are usually used in conjunction with the RSA DSS, but similar constructions are known for Rabin and Elgamal [11, 12]. After the publication of [13], researchers started to look for hash-and-sign DSSs that can be proven secure without random oracles (such as the GMR DSS). In 1999, for example, Cramer and Shoup [14],[3] and independently also Rosario Gennaro, Shai Halevi, and Tal Rabin [15] came up with respective proposals and DSSs that can be proven secure under the strong RSA assumption (Section 5.2.2). We have a brief look at the Cramer-Shoup DSS in Section 14.2.8. Since then, several variants have been proposed that are also provably secure in the standard model (e.g., [16, 17]). Note, however, that all security proofs known so far only apply in the single-user setting, where the adversary tries to existentially forge a signature for a particular user, and that it is completely unknown how to prove security in a multiuser setting, where the adversary targets any user of his or her choice. As of this writing, we don't have the tools to scientifically argue about the security of a DSS in this setting.

## 14.2 DIGITAL SIGNATURE SYSTEMS

There are several DSSs in use today, such as RSA, PSS and PSS-R, Rabin, Elgamal, Schnorr, DSA, ECDSA, and Cramer-Shoup. The security of RSA and Rabin (with or without PSS or PSS-R) is based on the IFP, whereas the security of all other DSSs is based on the (EC)DLP. For each DSS, we specify the key generation, signature generation, and signature verification algorithms, and we provide a brief security analysis. Again, we assume that all verification keys are published in certified form. We already made this assumption in Section 13.3, and we further discuss its implications in Section 16.4.

### 14.2.1 RSA

As pointed out by Diffie and Hellman in their seminal paper [18], a family of trapdoor functions can be turned into a DSS. The RSA family represents such a family, and hence the RSA public key cryptosystem [19] can also be turned into a DSS. We first look at the case in which RSA yields a DSS with appendix. This is by far the most widely used case. The case in which RSA yields a DSS giving message recovery is discussed afterward—mainly for the sake of completeness. It is rarely used in the field.

The RSA key generation, signature generation, and signature verification are summarized in Table 14.1 and explained below.

3 Although the DSS was first proposed at the 6th ACM Conference on Computer and Communications Security in 1999, [14] was published in 2000. Also, a preliminary version of the conference paper was published as an IBM Research Report in December 1998 (not cited here).

#### 14.2.1.1 Key Generation Algorithm

The RSA key generation algorithm Generate introduced in Section 13.3.1.1 also applies for the RSA DSS. Again, it takes as input a security parameter $l$ (in unary notation), and it generates as output an appropriately sized public key pair (i.e., a public key $(n, e)$ that represents the verification key and a private key $d$ that represents the signing key).

**Table 14.1**
RSA DSS with Appendix

Domain parameters: —

| Generate | Sign | Verify |
|---|---|---|
| $(1^l)$ | $(d, m)$ | $((n, e), m, s)$ |
| $p, q \xleftarrow{r} \mathbb{P}_{l/2}$<br>$n = p \cdot q$<br>select $1 < e < \phi(n)$<br>with $gcd(e, \phi(n)) = 1$<br>compute $1 < d < \phi(n)$<br>with $de \equiv 1 \pmod{\phi(n)}$ | $s \equiv h(m)^d \pmod{n}$ | $t = h(m)$<br>$t' \equiv s^e \pmod{n}$<br>$b = (t = t')$ |
| $((n, e), d)$ | $(s)$ | $(b)$ |

Let us reconsider the toy example from Section 13.3.1.1 to illustrate the Generate algorithm. For $p = 11$, $q = 23$, $n = 253$, and $\phi(n) = (p-1)(q-1) = 10 \cdot 22 = 220$, the (public) verification key may be $(n, e) = (253, 3)$ and the (private) signing key is then $d = 147$. Note that $3 \cdot 147 = 441 \equiv 1 \pmod{220}$.

#### 14.2.1.2 Signature Generation Algorithm

The RSA signature generation algorithm Sign is deterministic. It takes as input a private signing key $d$ and a message $m$ with hash value $h(m)$ that represents an element of $\mathbb{Z}_n$ for some hash function $h$, and it generates as output the digital signature

$$s = \text{RSA}_{n,d}(h(m)) \equiv h(m)^d \pmod{n}$$

The algorithm is simple and efficient. It requires only one modular exponentiation that can be done, for example, using the square-and-multiply algorithm (Algorithm A.3). In the end, $m$ and $s$ are usually transmitted together to the recipient.

Since $h(m)$ is typically much shorter than the modulus $n$, it is usually necessary to expand $h(m)$ to the bitlength of $n$. This can be done, for example, by prepending zeros, but it can also be done by using a more sophisticated message expansion function. From a security perspective, this is the preferred case and there are many such functions to choose from. For example, PKCS #1 specifies a couple of message expansion functions for RSA. Since PKCS #1 version 1.5, the following function is used:

$$h_{\text{PKCS\#1}}(m) = \texttt{0x 00 01 FF FF...FF FF 00} \parallel hash$$

In this notation, $hash$ refers to an appropriately encoded identifier $h_{ID}$ for the hash function in use concatenated with the hash value $h(m)$; that is, $hash = h_{ID} \parallel h(m)$. This value is padded (from left to right) with a zero byte, a one byte (representing block type 1), a series of 0xFF bytes (representing 255 in decimal notation), and another zero byte. There are so many 0xFF bytes that the total bitlength of $h_{\text{PKCS\#1}}(m)$ is equal to the bitlength of $n$.

In an attempt to design and come up with a message expansion function that allows the security of a DSS to be proven in the random oracle model, Bellare and Rogaway proposed PSS and PSS-R [11]. Both schemes invoke random padding, meaning that they use random values to expand $h(m)$ to the length of $n$. The resulting message expansion functions are now part of PKCS #1 version 2.1 [20] and other security standards. They are further addressed in Section 14.2.2.

Let us assume that the signatory wants to digitally sign a message $m$ with $h(m) = 26$ in our toy example. The Sign algorithm then computes

$$s \equiv m^d \pmod{n} \equiv 26^{147} \pmod{253} = 104$$

and sends 104 together with $m$ to the recipient.

#### 14.2.1.3 Signature Verification Algorithm

The signature verification algorithm Verify is simple and straightforward. It takes as input a public verification key $(n, e)$, a message $m$, and a respective signature $s$, and it generates as output one bit $b$ of information, namely whether $s$ is a valid signature with regard to $(n, e)$ and $m$. Hence, $b$ can be viewed as either 1 or 0, or $valid$ and $invalid$.

The Verify algorithm computes a hash value $t$ from $m$; that is, $t = h(m)$, determines

$$t' = \text{RSA}_{n,e}(s) \equiv s^e \pmod{n}$$

and verifies whether $t$ is equal to $t'$. If it is, then the digital signature is valid, otherwise it is not. The rationale why signature verification works and is correct is the same as the one we saw in the case of RSA decryption.

In our toy example, the Verify algorithm computes

$$t' = \text{RSA}_{253,3}(104) \equiv 104^3 \pmod{253} = 26$$

and returns $valid$, because 26 is equal to the hash value $h(m) = 26$ we started with.

#### 14.2.1.4 DSS Giving Message Recovery

If RSA is used as a DSS giving message recovery, then almost everything remains the same, except that the message $m$ is not sent together with the signature $s$, and a Recover algorithm is used to replace the Verify algorithm mentioned above. This new algorithm only takes a public verification key $(n, e)$ and a digital signature $s$ as input, and it generates as output either the message $m$ or a notification indicating that $s$ is an invalid signature for $m$ with respect to $(n, e)$. The algorithm first computes

$$m = \text{RSA}_{n,e}(s) \equiv s^e \pmod{n}$$

and then decides whether $m$ is a valid message. Only in the positive case does it return $m$ to the verifier. The second step is more important than it looks at first glance. If every message represented a valid message, then an adversary could existentially forge an RSA signature by randomly selecting an element $s \in \mathbb{Z}_n$ and claiming that it is a valid signature for message $m \equiv s^e \pmod{n}$. Note that somebody who wanted to verify this signature would have to compute exactly this value; that is, $m \equiv s^e \pmod{n}$, and hence $s$ is indeed a valid RSA signature for $m$. If $m$ represents a meaningful message, then the signatory may be held accountable for it and cannot repudiate having signed it. There are situations in which existential forgeability represents a problem. To overcome it, one must ensure that random messages are unlikely to be meaningful, or—alternatively speaking—that the probability that a randomly chosen message is meaningful is negligible. There are basically two ways to achieve this.

- One can use a natural language to construct messages to be signed. Natural languages contain enough redundancy so that randomly chosen strings (over the alphabet in use) are likely to be meaningless.
- One can use a specific (redundancy) structure for messages to be signed. If, for example, one digitally signs $m \parallel m$ instead of $m$, then one can easily verify the structure of a message after recovery (i.e., it must then consist of

two equal halves). On the other side, it is very difficult for an adversary to find a signature that recovers a message that is structured this way. It goes without saying that more efficient redundancy structures are used in practice.

In our toy example, the RSA Recover algorithm computes

$$m = \text{RSA}_{253,3}(104) \equiv 104^3 \pmod{253} \equiv 1,124,864 \pmod{253} = 26$$

and decides whether $m$ is a valid message. If, for example, valid messages are required to be congruent to 6 modulo 20, then $m = 26$ is indeed a valid message that is returned as a result of the Recover algorithm.

#### 14.2.1.5 Security Analysis

In Section 13.3.1.4, we analyzed the security of the RSA asymmetric encryption system. Most things we said there also apply to the RSA DSS. This is particularly true for the properties of the RSA family of trapdoor permutations. If, for example, somebody is able to factorize the modulus $n$, then he or she is also able to determine the private signing key $sk$ and (universally) forge signatures at will. Consequently, the modulus $n$ must be so large that its factorization is computationally infeasible for the adversary one has in mind.

More related to the RSA DSS, the multiplicative property of the RSA function yields a severe vulnerability. If, for example, $m_1$ and $m_2$ are two messages with signatures $s_1$ and $s_2$, then

$$s = s_1 s_2 \equiv (m_1 m_2)^d \pmod{n}$$

is a valid signature for $m \equiv m_1 m_2 \pmod{n}$. In other words, if an adversary knows two valid signatures $s_1$ and $s_2$, then he or she can generate another valid signature simply by computing the product of the two signatures modulo $n$. Consequently, we reemphasize the fact that good practice in security engineering must take care of the multiplicative structure of the RSA function and mitigate respective attacks. Remember from our previous discussion that one can either require a message to have a certain (nonmultiplicative) structure or apply a well-chosen message expansion function prior to the generation of the signature.

In many applications, RSA is used as an asymmetric encryption system and a DSS. Consequently, it may be necessary to apply both the RSA Encrypt algorithm and the RSA Sign algorithm to a message $m$. The question that arises immediately is whether the order of applying the two algorithms matters. More specifically, does one have to encrypt $m$ before it is digitally signed, or does one have to digitally sign it prior to encryption? In the general case, the answer is not clear, and it matters what the purpose of the cryptographic protection really is. In many practically relevant situations, however, the second possibility is the preferred choice—mainly because people are required to see the messages they sign in the clear. Consequently, it is often recommended to use the RSA DSS to digitally sign a plaintext message and then use the RSA asymmetric encryption system to encrypt the result. In this case, one must be concerned about the relative sizes of the moduli (that can be different if the keys are different).

Assume that A wants to digitally sign and then encrypt a message $m$ for B. Also assume that $(n_A, d_A)$ is A's private signing key and $(n_B, e_B)$ is B's public encryption key. If $n_A \leq n_B$, then the application of the two algorithms is straightforward (i.e., the output of the RSA Sign algorithm is smaller than or equal to the modulus $n_B$, and hence this value can be used as input for the RSA Encrypt algorithm). If, however, $n_A > n_B$, then the output of the RSA Sign algorithm may be larger than what is allowed as input for the RSA Encrypt algorithm. Obviously, one can then split the output of the RSA Sign algorithm into two input blocks for the RSA Encrypt algorithm and encrypt each block individually, but there are situations in which this type of reblocking is not feasible. In these situations, one may consider one of the following three possibilities to avoid the reblocking problem:

- One can prescribe the form of the moduli to make sure that the reblocking problem never occurs.
- One can enforce that the operation using the smaller modulus is applied first. In this case, however, it may happen that a message is first encrypted and then digitally signed.
- One can equip each user with two public key pairs. One pair has a small modulus and is used by the RSA Sign algorithm, and the other pair has a large modulus and is used by the RSA Encrypt algorithm.

The first possibility is not recommended because it is difficult to prescribe the form of the moduli in some binding way. The second possibility is not recommended either, because conditional reordering can change the meaning of the cryptographic protection one implements. So the third possibility is often the preferred choice. But the signing key is then smaller than the encryption key, and this is unusual (to say the least). The bottom line is that one has to study the requirements of an application before one can come up with a reasonable recommendation.

Instead of applying the RSA Encrypt and Sign algorithms separately in one way or another, one can consider the use of a cryptographic technique known as *signcryption* [21]. The basic idea is to digitally sign and encrypt a message simultaneously (instead of doing the two operations separately). This is conceptually similar to AE addressed in Chapter 11, but signcryption is not as technically mature as AE.

In summary, the RSA DSS is considered to be reasonably secure. This is particularly true if the modulus $n$ is sufficiently large. In fact, $n$ must be at least large enough to make it computationally infeasible to factorize it. As said before (in the context of the RSA asymmetric encryption system), this means that $n$ should be at least 2,048 bits long. Because digital signatures are often valuable, it is often recommended to use longer moduli, such as 4,096 bits. But then the reblocking problem occurs and needs to be solved. Also, for all practical purposes, it is recommended to use RSA as a DSS with appendix. It is obvious that one then has to select an appropriate cryptographic hash function, such as a representative of the SHA-2 family. It is less obvious that one also has to select an appropriate message expansion function, such as the one specified in PKCS #1. The choice of an appropriate message expansion function is particularly important if one wants to prove the security of the resulting DSS. This is further addressed next.

### 14.2.2 PSS and PSS-R

As mentioned earlier, Bellare and Rogaway proposed the PSS and PSS-R, and they also proved the security of the two DSS in the random oracle model. While PSS yields a DSS with appendix, PSS-R yields a DSS giving message recovery.

#### 14.2.2.1 PSS

Similar to OAEP in asymmetric encryption (Section 13.3.1.4), PSS is a padding scheme that can be combined with a basic DSS, such as RSA. In this case, the resulting DSS is acronymed RSA-PSS, but the PSS can also be combined with any other DSSs, such as Rabin or Elgamal. Quite naturally, the resulting DSS are then acronymed Rabin-PSS or Elgamal-PSS, but they are not addressed here.

The RSA-PSS signature generation and verification algorithms are summarized in Table 14.2. Note that RSA-PSS uses RSA, and hence the RSA key generation algorithm Generate outlined in Table 14.1 is reused (and not repeated in Table 14.2). Also note that PSS (and hence also RSA-PSS) uses some new parameters and two hash functions (instead of only one).

- In addition to $l$ that refers to the bitlength of the RSA modulus $n$, PSS uses two additional parameters $l_0$ and $l_1$ that are numbers between 1 and $l$. Typical values are $l = 1,024$ are $l_0 = l_1 = 128$.

- PSS uses two hash functions $h$ and $g$. While $h$ is a normal hash function that compresses bit bit sequences, $g$ rather expands bit sequences. According to the terminology introduced in Section 6.4.5, $g$ refers to an XOF.

**Table 14.2**
RSA-PSS

Domain parameters: —

| Sign | Verify |
|---|---|
| $(d, m)$ | $((n, e), m, s)$ |
| $r \xleftarrow{r} \{0,1\}^{l_0}$ | $y \equiv s^e \pmod{n}$ |
| $w = h(m \parallel r)$ | break up $y$ as $b \parallel w \parallel r^* \parallel \gamma$ |
| $r^* = g_1(w) \oplus r$ | $r = r^* \oplus g_1(w)$ |
| $y = 0 \parallel w \parallel r^* \parallel g_2(w)$ | $b = (b = 0 \wedge h(m \parallel r) = w \wedge g_2(w) = \gamma)$ |
| $s \equiv y^d \pmod{n}$ | |
| $(s)$ | $(b)$ |

- The hash function $h : \{0,1\}^* \rightarrow \{0,1\}^{l_1}$ is called *compressor*. It hashes arbitrarily long bit sequences to sequences of $l_1$ bits.
- The hash function (or XOF) $g : \{0,1\}^{l_1} \rightarrow \{0,1\}^{l-l_1-1}$ is called *generator*. It hashes (or rather expands) sequences of $l_1$ bits to sequences of $l - l_1 - 1$ bits.

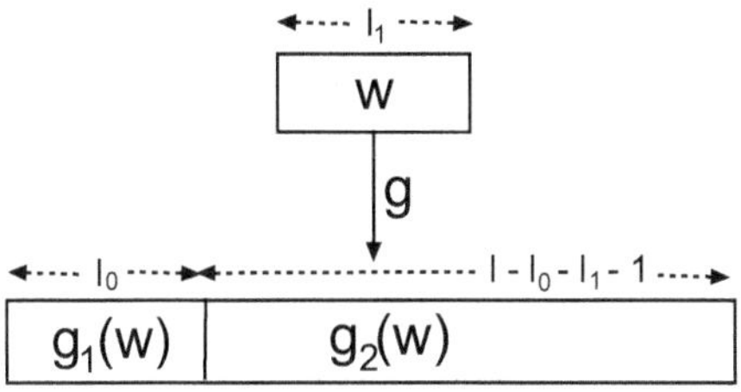


**Figure 14.1** The functions $g_1$ and $g_2$.

For the security analysis in the random oracle model, $h$ and $g$ are assumed to be ideal (i.e., they behave like random functions). For all practical purposes, however, $h$ and $g$ are implemented as cryptographic hash functions.

As illustrated in Figure 14.1, the output of the function $g$ can be subdivided into two parts, and these parts define two auxiliary functions:

- $g_1$ is the function that on input $w \in \{0,1\}^{l_1}$ returns the first $l_0$ bits of $g(w)$.
- $g_2$ is the function that on input $w \in \{0,1\}^{l_1}$ returns the remaining $l - l_0 - l_1 - 1$ bits of $g(w)$.

The RSA-PSS Sign algorithm takes as input a private signing key $d$ and a message $m$, and it generates as output a signature $s$ for $m$. As its name suggests, the PSS is probabilistic, meaning that the signature generation algorithm employs an $l_0$-bit string $r$ that is randomly selected from $\{0,1\}^{l_0}$. This bit string is appended to the message $m$, and the expression $m \parallel r$ is subject to the compressor $h$. The resulting hash value $h(m \parallel r)$ is assigned to $w$. This value, in turn, is subject to the generator $g$. The first $l_0$ bits of $g(w)$; that is, $g_1(w)$, are added modulo 2 to $r$ and the resulting bitstring is referred to $r^*$. Finally, for message preparation, a string $y$ is compiled that consists of a leading zero, $w$, $r^*$, and the right $l - l_0 - l_1 - 1$ bits of $g(w)$; that is, $g_2(w)$, in this order. This is illustrated in Figure 14.2. Note that each of the components has a fixed length, and hence $y$ can be decomposited quite easily. Last but not least, the signature $s$ is generated by putting $y$ to the power of $d$ modulo $n$. It goes without saying that this refers to a normal RSA signature. The message $m$ and the signature $s$ are both sent to the verifier.

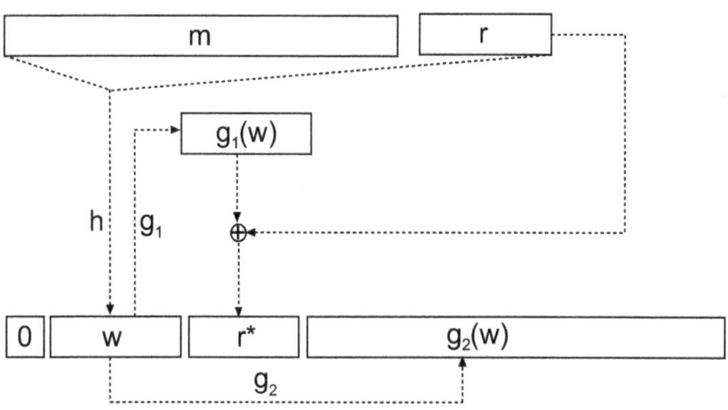


**Figure 14.2** The preparation of message $m$ for the RSA-PSS Sign algorithm.

The RSA-PSS Verify algorithm takes as input a public verification key $(n, e)$, a message $m$, and a signature $s$, and it generates as output one bit $b$ saying whether $s$ is a valid signature or not. First of all, the algorithm puts $s$ to the power of $e$ modulo

$n$. This refers to a normal RSA signature verification. The algorithm then breaks up $y$ into its four components $b$, $w$, $r^*$, and $\gamma$ according to their respective lengths (i.e., $b$ is one bit long, $w$ is $l_1$ bits long, $r^*$ is $l_0$ bits long, and $\gamma$ is $l - l_0 - l_1 - 1$ bits long). The component $r^*$ is then added modulo 2 to $g_1(w)$, and the result is assigned to $r$. Finally, the algorithm verifies whether $b = 0$, $h(m \parallel r) = w$, and $g_2(w) = \gamma$. The output bit $b$ is set to true if and only if all tests are successful.

The PSS is very efficient. The Sign and Verify algorithms both take only one application of $h$, one application of $g$, and one application of the RSA function. In the case of RSA-PSS, this is only slightly more expensive than the basic RSA DSS. RSA-PSS was therefore added in version 2.1 of PKCS #1. The corresponding encoding method is referred to as EMSA-PSS, where the acronym EMSA stands for encoding method for signature with appendix. The use of PKCS #1 version 2.1 in general, and EMSA-PSS in particular, is highly recommended and quite widely used in the field (as an alternative to ECDSA).

**Table 14.3**
RSA-PSS-R

Domain parameters: —

| Sign | Recover |
|---|---|
| $(d, m)$ | $((n, e), s)$ |
| $r \xleftarrow{r} \{0,1\}^{l_0}$<br>$w = h(m \parallel r)$<br>$r^* = g_1(w) \oplus r$<br>$m^* = g_2(w) \oplus m$<br>$y = 0 \parallel w \parallel r^* \parallel m^*$<br>$s \equiv y^d \pmod{n}$ | $y \equiv s^e \pmod{n}$<br>break up $y$ as $b \parallel w \parallel r^* \parallel m^*$<br>$r = r^* \oplus g_1(w)$<br>$m = m^* \oplus g_2(w)$<br>if $(b = 0$ and $h(m \parallel r) = w)$<br>then output $m$<br>else output *invalid* |
| $(s)$ | $(m \mid invalid)$ |

#### 14.2.2.2 PSS-R

While PSS yields a DSS with appendix, PSS-R yields a DSS giving message recovery. This means that the PSS-R Sign algorithm must fold the message $m$ into the signature $s$ in such a way that it can be recovered by the Recover algorithm afterward. When the length of $m$ is sufficiently small, then one can fold the entire message into the signature. Otherwise, the message must be cut into pieces and signed block-wise. Again, PSS-R is essentially a message padding scheme that must be combined with a basic DSS such as RSA. The resulting DSS is then acronymed RSA-PSS-R.

The RSA-PSS-R signature generation and message recovery algorithms are summarized in Table 14.3. They use the same parameters $l$, $l_0$, and $l_1$, and the same hash functions $h$ and $g$ (with the same auxiliary functions $g_1$ and $g_2$) as RSA-PSS. We assume that the messages to be signed have a maximum length $k = l - l_0 - l_1 - 1$. Suggested choices are $l = 1{,}024$, $l_0 = l_1 = 128$, and $k = 767$. This means that a 767-bit message can be folded into a single signature, and that the verifier can recover it and simultaneously check its authenticity.

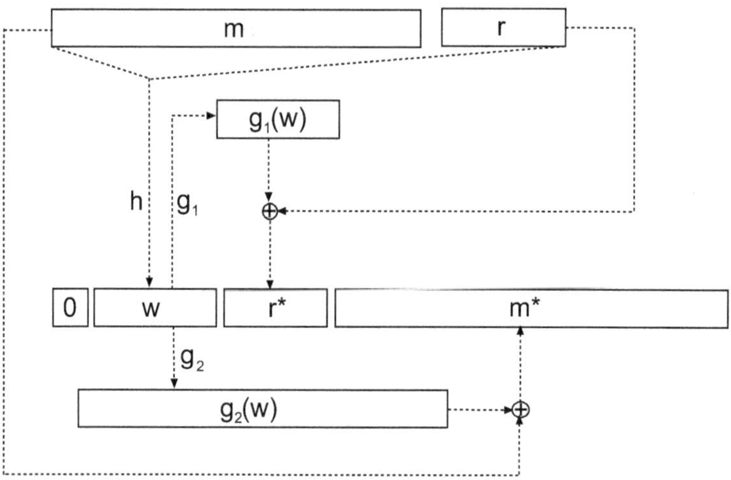


**Figure 14.3** The preparation of message $m$ for the RSA-PSS-R Sign algorithm.

The PSS-R Sign algorithm is very similar to the PSS Sign algorithm. In the RSA-PSS-R Sign algorithm, however, the last part of $y$ is $g_2(w) \oplus m$ (instead of only $g_2(w)$). This is illustrated in Figure 14.3. It means that $g_2(w)$ is used to mask the message $m$, and that in the end $m$ can be recovered from this part.

The RSA-PSS-R algorithm is also similar to the RSA-PSS Verify algorithm. The major difference is that in the RSA-PSS-R Recover algorithm the message $m$ must be recovered from $m^*$. As mentioned above, this can be achieved by adding modulo 2 $g_2(w)$ to $m^*$. Also, the output of the algorithm depends on a condition: if $b = 0$ and $h(m \parallel r) = w$, then the algorithm outputs $m$. Otherwise, it signals that the signature $s$ is invalid and the message is not recovered.

The bottom line is that RSA-PSS and RSA-PSS-R are clearly advantageous, and that they should always be used instead of native RSA. The constructions are highly efficient and yield a DSS that is provably secure in the random oracle model.

### 14.2.3 Rabin

In Section 13.3.2, we introduced the Rabin asymmetric encryption system and mentioned that it also yields a DSS [22]. Here, we briefly sketch both the DSS originally proposed by Rabin and a simplified version thereof. The key generation, signature generation, and signature verification of a simplified version of the Rabin DSS with appendix are summarized in Table 14.4 and addressed below.

#### 14.2.3.1 Key Generation Algorithm

The Rabin key generation algorithm Generate is basically the same as the one employed by the Rabin encryption system (Section 13.3.2.1). It first generates two primes $p$ and $q$, and then computes $n = pq$. In the original version of the Rabin DSS, the algorithm also selects a random value $1 < b < n$. The pair $(n, b)$ then represents the public verification key, whereas the pair $(p, q)$ represents the private signing key. In the simplified version of the Rabin DSS, $b$ is omitted and the public verification key only consists of $n$.

**Table 14.4**
Rabin DSS with Appendix (Simplified Version)

Domain parameters: —

| Generate | Sign | Verify |
|---|---|---|
| $(1^l)$ | $((p, q), m)$ | $(n, m, (U, x))$ |
| $p, q \xleftarrow{r} \mathbb{P}'_{l/2}$<br>$n = p \cdot q$ | find $U$ such that $h(m \parallel U)$ is a square modulo $n$<br>find $x$ that satisfies $x^2 \equiv h(m \parallel U) \pmod{n}$ | $b = (x^2 \equiv h(m \parallel U) \pmod{n})$ |
| $(n, (p, q))$ | $(U, x)$ | $(b)$ |

#### 14.2.3.2 Signature Generation Algorithm

The Rabin signature generation algorithm Sign is probabilistic and requires a cryptographic hash function $h$. It takes as input a private signing key $(p, q)$ and a message $m \in \mathbb{Z}_n$ to be signed, and it generates as output a respective signature. The algorithm—as originally proposed by Rabin—picks a random padding value $U$, computes $h(m \parallel U)$, and uses $(p, q)$ to compute $x$ such that

$$x(x+b) \equiv h(m \parallel U) \pmod{n}$$

Note that $b$ is part of the public key (together with $n$). If there is no solution for $x$, then a new value for $U$ must be picked. This is repeated until a solution is found (the expected number of tries is 4). The pair $(U, x)$ then represents the Rabin signature for $m$. This also means that such a signature can be twice as long as the message that is signed.

In the simplified version of the Rabin DSS with appendix, the Sign algorithm picks a random padding value $U$ and computes $h(m \parallel U)$ until $h(m \parallel U)$ turns out to be a square modulo $n$. In this case, a square root $x$ satisfying

$$x^2 \equiv h(m \parallel U) \pmod{n}$$

is guaranteed to exist and the private signing key $(p, q)$ can be used to efficiently find it. Again, the pair $(U, x)$ represents the Rabin signature for $m$.

#### 14.2.3.3 Signature Verification Algorithm

Like all signature verification algorithms, the Rabin signature verification algorithm Verify is deterministic. For message $m$ and signature $(U, x)$, it first computes $x(x+b)$ and $h(m \parallel U)$, and it then verifies whether the two values are equivalent modulo $n$. In the simplified version, the algorithm computes $x^2$ instead of $x(x+b)$, and the rest of the algorithm remains the same.

#### 14.2.3.4 Security Analysis

Recall from Section 13.3.2 that computing a square root modulo $n$ is computationally equivalent to factoring $n$. Since an adversary who can compute a square root can also forge a signature, we know that selectively forging a signature is computationally infeasible for the adversary we have in mind.

To protect against existential forgery, one has to ensure that $h(m \parallel U')$ looks random for any value $U'$ the adversary may come up with, and hence that $h(m \parallel U')$ has no structure that may be exploited in some meaningful way. This is likely to be the case if $h$ is a cryptographic hash function (that supposedly implements a random oracle). Consequently, the Rabin DSS can be shown to be secure against existential forgery in the random oracle model. But similar to the Rabin asymmetric encryption system, the Rabin DSS is not widely used in the field.

### 14.2.4 Elgamal

In Section 13.3.3, we introduced the Elgamal asymmetric encryption system and announced that the Elgamal public key cryptosystem [23] also yields a DSS. Unlike the authors of the RSA system, however, Elgamal did not file a patent application for his public key cryptosystem. Also, it uses different algorithms to encrypt and decrypt messages on the one hand, and to sign messages and verify signatures on the other hand. This is disadvantageous from an implementation viewpoint because more than one algorithm needs to be implemented. Furthermore (and similar to Rabin signatures), Elgamal signatures are twice as long as RSA signatures. Mainly for these reasons, the Elgamal DSS is not widely used in the field.

In its basic form, the Elgamal DSS is with appendix. This is also true for most variants that have been proposed in the literature (e.g., [24, 25]). But there is a variant of the Elgamal DSS created by Kaisa Nyberg and Rainer R. Rueppel that yields a DSS giving message recovery [26]. The Nyberg-Rueppel DSS is interesting and fills a niche, but it is not used in the field and therefore not addressed in this book.

Similar to the Diffie-Hellman key exchange protocol and the Elgamal asymmetric encryption system, the security of the Elgamal DSS is based on the DLA. Consequently, one needs a cyclic group in which the DLP is computationally intractable. This can be $\mathbb{Z}_p^*$ as originally proposed by Elgamal and used here, but it can also be a $q$-element subgroup of $\mathbb{Z}_p^*$ as employed in the Schnorr DSS and DSA or $E(\mathbb{F}_q)$ as employed in ECC.

**Table 14.5**
Elgamal DSS with Appendix

Domain parameters: $p, g$

| Generate | Sign | Verify |
|---|---|---|
| $(-)$ | $(x, m)$ | $(y, m, (s_1, s_2))$ |
| $x \xleftarrow{r} \{2, \ldots, p-2\}$<br>$y \equiv g^x \pmod{p}$ | $r \xleftarrow{r} \{1, \ldots, p-2\}$<br>with $gcd(r, p-1) = 1$<br>$s_1 \equiv g^r \pmod{p}$<br>$s_2 \equiv r^{-1}(h(m) - xs_1)$<br>$\pmod{p-1}$ | verify $0 < s_1 < p$<br>verify $0 < s_2 < p-1$<br>$b = (g^{h(m)} \equiv y^{s_1} s_1^{s_2} \pmod{p})$ |
| $(x, y)$ | $(s_1, s_2)$ | $(b)$ |

The key generation, signature generation, and signature verification algorithms of the Elgamal DSS with appendix are summarized in Table 14.5. There are two domain parameters, $p$ and $g$. The prime number $p$ determines the cyclic group $\mathbb{Z}_p^*$ with order $p-1$, whereas $g$ refers to a generator of this group. Domain parameters may be common for a group of entities (e.g., users). As such, they may be public and remain fixed for an extended period of time.

#### 14.2.4.1 Key Generation Algorithm

The Elgamal key generation algorithm Generate is essentially the same as the one employed by the Elgamal asymmetric encryption system (Section 13.3.3). While we used a generic cyclic group $G$ in Section 13.3.3, we use $\mathbb{Z}_p^*$ here. For every entity, the algorithm randomly selects a private signing key $x$ from $\{2, \ldots, p-2\}$ and computes the respective public verification key $y$ as $y \equiv g^x \pmod{p}$. To illustrate the Elgamal DSS, we reuse the toy example from Section 13.3.3, where $p = 17$, $g = 7$, $x = 6$, and $y = 9$.

#### 14.2.4.2 Signature Generation Algorithm

Contrary to RSA, the Elgamal signature generation algorithm Sign is probabilistic and employs a cryptographic hash function $h$. More specifically, the algorithm takes as input a private signing key $x$ and a message $m$, and it generates as output a digital signature $s$ that consists of two values $s_1$ and $s_2$ (both elements of $\mathbb{Z}_p^*$). The algorithm comprises three steps:

- First, the algorithm randomly selects a fresh $r$ from $\mathbb{Z}_{p-1} \setminus \{0\} = \{1, \ldots, p-2\}$ with $gcd(r, p-1) = 1$.[4] The requirement $gcd(r, p-1) = 1$ suggests that $r$ has a multiplicatively inverse element $r^{-1}$ in $\mathbb{Z}_p^*$, meaning that $rr^{-1} \equiv 1 \pmod{p-1}$. The inverse element can be determined with the extended Euclid algorithm (Algorithm A.2).
- Second, $r$ is used to compute $s_1 \equiv g^r \pmod{p}$ that yields the first component of the signature.
- Third, $m$ is hashed with $h$, and the result $h(m)$ is used together with $r^{-1}$, $x$, and $s_1$ to compute $s_2 \equiv r^{-1}(h(m) - xs_1) \pmod{p-1}$. If the resulting value $s_2$ is equal to zero, then the algorithm must restart with another $r$ in step one.

Note that the Sign algorithm can be made more efficient by using precomputation. In fact, it is possible to randomly select $r$ and precompute $s_1 \equiv g^r \pmod{p}$ and $r^{-1}$ modulo $p-1$. Both values do not depend on a message $m$. If one has $r$,

4 As already mentioned in Section 13.3.3.4 and further explained in Section 14.2.4.4, a value $r$ must never be used more than once. Otherwise, the system is totally insecure.

$s_1$, and $r^{-1}$, then one can digitally sign $m$ by hashing it and directly computing $s_2 \equiv r^{-1}(h(m) - xs_1) \pmod{p-1}$.

In either case, the Elgamal signature for $m$ is $s = (s_1, s_2)$. Because $m$, $s_1$, and $s_2$ are all elements of $\mathbb{Z}_p^*$, the signature is at most twice as long as $p$. Also, as mentioned earlier, the basic Elgamal DSS is with appendix, meaning that the signature $s$ must be sent along with the message $m$.

Let us revisit our toy example. If the signatory wants to sign a message $m$ with $h(m) = 6$, then the Sign algorithm may randomly select $r = 3$ (with $r^{-1} \equiv 3^{-1} \pmod{16} = 11$) and compute

$$\begin{aligned} s_1 &\equiv 7^3 \pmod{17} = 343 \bmod 17 = 3 \\ s_2 &\equiv 11(6 - 6 \cdot 3) \pmod{16} = -132 \bmod 16 = 12 \end{aligned}$$

Consequently, the Elgamal signature for $h(m) = 6$ is $s = (3, 12)$, and hence the numbers 6, 3, and 12 must all be transmitted to the verifier.

#### 14.2.4.3 Signature Verification Algorithm

Like all signature verification algorithms, the Elgamal Verify algorithm is deterministic. It takes as input a public verification key $y$, a message $m$, and an Elgamal signature $(s_1, s_2)$, and it generates as output one bit $b$ saying whether $(s_1, s_2)$ is a valid Elgamal signature for $m$ with respect to $y$. The Verify algorithm must verify $0 < s_1 < p$, $0 < s_2 < p - 1$, and

$$g^{h(m)} \equiv y^{s_1} s_1^{s_2} \pmod{p} \tag{14.1}$$

The signature is valid if and only if all verification checks succeed. Otherwise, the signature must be rejected and considered to be invalid. Note that (14.1) is correct, because

$$\begin{aligned} y^{s_1} s_1^{s_2} &\equiv g^{xs_1} g^{rr^{-1}(h(m)-xs_1)} \pmod{p} \\ &\equiv g^{xs_1} g^{h(m)-xs_1} \pmod{p} \\ &\equiv g^{xs_1} g^{-xs_1} g^{h(m)} \pmod{p} \\ &\equiv g^{h(m)} \pmod{p} \end{aligned}$$

Let us emphasize two facts that are important for the security of the Elgamal DSS: First, it is important to verify that $0 < s_1 < p$. Otherwise, it is possible to construct a new signature from a known signature [27]. Second, it is also necessary to use a cryptographic hash function $h$ and sign $h(m)$ instead of $m$ (even for short messages). Otherwise, one can existentially forge signatures.

In our toy example, the Verify algorithm must verify $0 < 3 < 17, 0 < 12 < 15$, and $7^6 \equiv 9^3 \cdot 3^{12} \pmod{17}$, which is 9 in either case. This means that $(3, 12)$ is in fact a valid Elgamal signature for $m$ with $h(m) = 6$.

#### 14.2.4.4 Security Analysis

We know from Section 13.3.3.4 that the security of the Elgamal public key cryptosystem is based on the assumed intractability of the DLP in a cyclic group. This also applies to the Elgamal DSS. If $\mathbb{Z}_p^*$ is used, then $p$ must be at least 2,048 bits long, and longer values are preferred. Furthermore, one must select $p$ so that efficient algorithms to compute discrete logarithms do not work. For example, it is necessary to select $p$ so that $p-1$ does not have only small prime factors. Otherwise, the Pohlig-Hellman algorithm [28] may be used to solve the DLP and break the Elgamal DSS accordingly. Furthermore, we assume that the function $h$ is a cryptographic hash function, and hence that it is one-way and collision-resistant.

There are other constraints that should be considered when implementing the Elgamal DSS. First, the random value $r$ that is selected at the beginning of the Sign algorithm must be kept secret. If an adversary were able to learn this value, then he or she could determine the private signing key $x$ from a message $m$ and signature $s$. To see why this is the case, we use the following sequence of equivalences:

$$\begin{aligned} s_2 &\equiv r^{-1}(h(m) - xs_1) \pmod{p} \\ rs_2 &\equiv rr^{-1}(h(m) - xs_1) \pmod{p} \\ rs_2 &\equiv h(m) - xs_1 \pmod{p} \\ xs_1 &\equiv h(m) - rs_2 \pmod{p} \\ x &\equiv (h(m) - rs_2)s_1^{-1} \pmod{p} \end{aligned}$$

Even if the adversary is not able to uniquely determine $r$ but is able to narrow down the set of possible values, he or she may still be able to mount an exhaustive search for $r$.

Second, it is necessary to use a fresh and unique value $r$ for every signature that is generated (this requirement is analog to the Elgamal asymmetric encryption system). Otherwise (i.e., if $r$ is reused), it is possible to determine the private signing key from two valid signatures: Let $s = (s_1, s_2)$ and $s' = (s_1', s_2')$ be such signatures for two distinct messages $m$ and $m'$. If $r$ is the same in either case, then $g^r \pmod{p}$ is also the same, and hence $s_1 = s_1'$. With regard to $s_2$ and $s_2'$, the following two equations hold:

$$\begin{aligned} s_2 &\equiv r^{-1}(h(m) - xs_1) \pmod{p-1} \\ s_2' &\equiv r^{-1}(h(m') - xs_1) \pmod{p-1} \end{aligned}$$

If we subtract $s_2'$ from $s_2$, we get:

$$\begin{aligned} s_2 - s_2' &\equiv r^{-1}(h(m) - xs_1) - r^{-1}(h(m') - xs_1) \pmod{p-1} \\ &\equiv r^{-1}h(m) - r^{-1}xs_1 - r^{-1}h(m') + r^{-1}xs_1 \pmod{p-1} \\ &\equiv r^{-1}h(m) - r^{-1}h(m') \pmod{p-1} \\ &\equiv r^{-1}(h(m) - h(m')) \pmod{p-1} \end{aligned}$$

If $gcd(s_2 - s_2', p-1) = 1$, then $s_2 - s_2'$ is invertible modulo $p - 1$, and this means that one can compute $r$ as follows:

$$\begin{aligned} s_2 - s_2' &\equiv r^{-1}(h(m) - h(m')) \pmod{p-1} \\ r(s_2 - s_2') &\equiv (h(m) - h(m')) \pmod{p-1} \\ r &\equiv (h(m) - h(m'))(s_2 - s_2')^{-1} \pmod{p-1} \end{aligned}$$

Given $r$, $s_2$, $s_1 = s_1'$, and $h(m)$, one can then compute the private key $x$ (as shown above). This is unfortunate, and it basically means that a fresh and unique $r$ must be chosen from the full set of all possible values for every Elgamal signature that is generated. This is a severe weakness that regularly causes problems in the field. When we address the DSA in Section 14.2.6, we also mention a possibility to deterministically select an $r$ that is fresh and unique. It goes without saying that this turns a probabilistic signature generation algorithm into a deterministic one.

### 14.2.5 Schnorr

In the late 1980s,[5] Claus-Peter Schnorr developed and patented[6] a variant of the Elgamal DSS that uses a $q$-order subgroup $G$ of $\mathbb{Z}_p^*$ with $q \mid p-1$ [29].[7] This is advantageous because the computations can be done more efficiently and the resulting signatures are shorter than the ones generated with the original Elgamal

5 While the journal version of the paper was published in 1991 [29], some preliminary versions of the paper already appeared in 1989. In fact, Schnorr gave a presentation at the rump session of EUROCRYPT '89 and officially presented the full paper at CRYPTO '89.

6 The relevant patent U.S. 4,995,082 entitled "Method for Identifying Subscribers and for Generating and Verifying Electronic Signatures in a Data Exchange System" was granted to Schnorr in 1991.

7 Remember from group theory that there is a subgroup of $\mathbb{Z}_p^*$ with $q$ elements for every prime divisor $q$ of $p-1$. Because $p$ is a large prime, $p-1$ is an even number, and hence there are at least two divisors of $p-1$: 2 and $(p-1)/2$. Consequently, there exist at least two subgroups of $\mathbb{Z}_p^*$: one with 2 elements and another one with $(p-1)/2$ elements. Whether more subgroups exist depends on $p$ and the prime factorization of $p-1$. If $q$ is a prime divisor of $p-1$, then there exists a subgroup of order $q$ (in addition to the two subgroups mentioned above). This is the group in which the modular arithmetic is done in Schnorr's DSS. In some literature, such a large prime-order subgroup of $\mathbb{Z}_p^*$ is called a *Schnorr group*.

DSS. Again, the Schnorr DSS is with appendix, but it can be turned into a DSS giving message recovery. Also, we will see that it can easily be translated in a form that is suitable for ECC.

**Table 14.6**
Schnorr DSS

Domain parameters: $p, q, g$

| Generate | Sign | Verify |
|---|---|---|
| $(-)$ | $(x, m)$ | $(y, m, (s_1, s_2))$ |
| $x \xleftarrow{r} \mathbb{Z}_q^*$ | $r \xleftarrow{r} \mathbb{Z}_q^*$ | $u = (g^{s_2} y^{-s_1}) \bmod p$ |
| $y \equiv g^x \pmod{p}$ | $r' \equiv g^r \pmod{p}$ | $v = h(u \parallel m)$ |
| | $s_1 = h(r' \parallel m)$ | $b = (v = s_1)$ |
| | $s_2 = (r + x s_1) \bmod q$ | |
| $(x, y)$ | $(s_1, s_2)$ | $(b)$ |

The key generation, signature generation, and signature verification algorithms of the Schnorr DSS are summarized in Table 14.6, where $p$ refers to a large prime number, $q$ to a smaller prime number that divides $p-1$ (i.e., $q|p-1$), and $g$ to a generator of the $q$-order subgroup $G$ of $\mathbb{Z}_p^*$. In a typical setting, $p$ was anticipated to be 1,024 bits long, whereas $q$ was anticipated to be 160 bits long. In this setting, one may wonder how a generator $g$ of $G$ can actually be found. To answer this question, one may combine the following two facts:

- First, $q|p-1$ means that $qr = p-1$ for some positive integer $r$;
- Second, $h^{p-1} \equiv 1 \pmod{p}$ according to Fermat's little theorem.

Replacing $p-1$ with $qr$ in Fermat's little theorem, it follows that $h^{rq} \equiv 1 \pmod{p}$. On the other side, we know that for every $h \in \mathbb{Z}_p^*$ either $h^r \equiv 1 \pmod{p}$ or $h^r \not\equiv 1 \pmod{p}$ must hold. In the second case, it follows that $(h^r)^q \equiv h^{rq} \equiv h^{p-1} \equiv 1 \pmod{p}$, and this means that $h^r$ is a generator of $G$. The bottom line is that every $h$ for which $h^r$ is a not equivalent to 1 modulo $p-1$ yields a generator of $G$, and hence it is simple and straightforward to find one. Like Elgamal, the Schnorr DSS employs a cryptographic hash function $h$ that generates $q$-bit values. Again, in a typical setting this is SHA-1.

#### 14.2.5.1 Key Generation Algorithm

The Schnorr key generation algorithm Generate takes no input other than the domain parameters, and it generates as output a public key pair. More specifically, it randomly selects an element $x$ from $\mathbb{Z}_q^* = \mathbb{Z}_q \setminus \{0\}$ and computes $y \equiv g^x \pmod{p}$. Similar to Elgamal, $x$ yields a private signing key, whereas $y$ yields a public verification key. While $x$ is at most as large as $q$ (or $q-1$ to be precise), $y$ is yet an element of $G$ but can be as large as $p$ (or $p-1$ to be precise).

Let us consider a toy example to illustrate the Schnorr DSS. For the two primes $p = 23$ and $q = 11$ (note that $q = 11$ divides $p-1 = 22$), the element $g = 2$ generates a Schnorr group of order 11. The Generate algorithm may randomly select a private signing key from $\mathbb{Z}_{11}^* = \{1, \ldots, 10\}$, such as $x = 5$, and compute the respective public verification key $y \equiv 2^5 \equiv 32 \pmod{23} = 9$.

#### 14.2.5.2 Signature Generation Algorithm

The Schnorr signature generation algorithm Sign is probabilistic. It takes as input a private signing key $x$ and a message $m$, and it generates as output a digital signature $s = (s_1, s_2)$ for $m$, where $s_1$ and $s_2$ are at most as large as $q$. The algorithm is very similar to Elgamal: It randomly selects an $r$ from $\mathbb{Z}_q^*$, and then computes $r' \equiv g^r \pmod{p}$, $s_1 = h(r' \parallel m)$, and $s_2 = (r + xs_1) \bmod q$. The pair $(s_1, s_2)$ yields the Schnorr signature for $m$. Alternatively, it is also possible to use the pair $(r', s_2)$ as a signature.

To digitally sign a message $m$ in our toy example, the Sign algorithm may randomly select $r = 7$ and compute $r' \equiv 2^7 \pmod{23} = 13$. If $h(r' \parallel m)$ yields 4, then $s_1 = 4$ and $s_2 = (7 + 5 \cdot 4) \mod 11 = 5$. This, in turn, suggests that the Schnorr signature for the message is $(4, 5)$, and the alternative signature is $(13, 5)$.

#### 14.2.5.3 Signature Verification Algorithm

As usual, the Schnorr signature verification algorithm Verify is deterministic. It takes as input a public verification key $y$, a message $m$, and a signature $(s_1, s_2)$, and it generates as output one bit $b$ saying whether the signature is valid or not. The algorithm computes $u = (g^{s_2} y^{-s_1}) \bmod p$ and $v = h(u \parallel m)$, and it yields $valid$ if and only if $v$ is equal to $s_1$. This verification is correct because $v = s_1$ suggests that $h(u \parallel m) = h(r' \parallel m)$ and hence $u = r'$. This equation can easily be shown to be true, because

$$\begin{aligned} u &= (g^{s_2} y^{-s_1}) \bmod p \\ &= (g^{r+xs_1} g^{-xs_1}) \bmod p \\ &= (g^r g^{xs_1} g^{-xs_1}) \bmod p \\ &= g^r \bmod p \\ &= r' \end{aligned}$$

In our toy example, the Verify algorithm computes $u = (2^5 \cdot 9^{-4}) \bmod p = (2^5 \cdot 9^7) \bmod 23 = 32 \cdot 4{,}782{,}969 \bmod 23 = 13$ and $v = h(u \parallel m) = 4$. The signature is valid because $v = 4$ equals $s_1 = 4$.

Alternatively, if the pair $(r', s_2)$ is used to represent the signature, then the signature is valid if and only if $g^{s_2}$ and $g^r y^{h(r' \| m)}$ refer to the same element in $G$. This follows from $g^r y^{h(r' \| m)} = g^r g^{xh(r' \| m)} = g^{r+xs_1}$ and $s_2 = (r + xs_1) \bmod q$. If the exponents are the same, then the resulting elements in $G$ are also the same.

#### 14.2.5.4 Security Analysis

Since the Schnorr DSS is a modified version of the Elgamal DSS, most things we said in Section 14.2.4.4 also apply. Like Elgamal, the security of the Schnorr DSS relies on the DLA and the computational intractability of the DLP in a cyclic group. Unlike Elgamal, however, the Schnorr DSS relies on the DLP in a cyclic subgroup $G \subset \mathbb{Z}_p^*$ with order $q < p - 1$. This problem can only be solved with a generic algorithm. As mentioned in Section 5.4, the best we can expect from such an algorithm is a running time that is of the order of the square root of the order of the subgroup. If, for example, the subgroup has order $2^{160}$ (as is the case here), then the best possible algorithm has a running time of order

$$\sqrt{2^{160}} = 2^{160/2} = 2^{80}$$

This is beyond the computational power of an adversary we have in mind. Consequently, it is computationally intractable to solve the DLP in a subgroup of $\mathbb{Z}_p^*$ with prime order $q$ (for sufficiently large values of $q$). From a practical perspective, the Schnorr DSS has the advantage that the signatures it generates are much shorter than the ones generated by the Elgamal DSS. Note that each component of a signature is of the order of $q$. If $q$ is 160 bits long, then a Schnorr signature is at most 320 bits long. Compare this to the $2 \cdot 1024 = 2048$ bits required for an Elgamal signature.

### 14.2.6 DSA

Based on the DSSs of Elgamal and Schnorr, NIST developed the *digital signature algorithm* (DSA) and specified a corresponding *digital signature standard* in FIPS PUB 186 [30]. Since its publication in 1994, FIPS PUB 186 has been subject to four major revisions in 1998, 2000, 2009, and 2013.[8] Note that the latest versions of FIPS PUB 186 also specify RSA and ECDSA in addition to DSA. So the digital signature standard has in fact become open in terms of supported algorithms. Also note that the original DSA was originally covered by U.S. Patent 5,231,668 entitled "Digital Signature Algorithm" assigned to David W. Kravitz, a former NSA employee, in July 1993. The patent was given to "The United States of America as represented by the Secretary of Commerce, Washington, D.C." and NIST has made the patent available worldwide without having to pay any royalty. During the second half of the 1990s, it was heavily disputed whether the DSA infringed the Schnorr patent. This dispute, however, was never brought to court, and the question has become obsolete because both patents expired a decade ago (the Schnorr patent in 2008 and the DSA patent in 2010). In what follows, we mainly focus on the DSA as it was originally specified in FIPS 186. We revisit the parameter lengths and the respective security levels as they are specified in FIPS 186-4 at the end of the section.

**Table 14.7**
DSA

Domain parameters: $p, q, g$

| Generate | Sign | Verify |
|---|---|---|
| $(-)$ | $(x, m)$ | $(y, m, (s_1, s_2))$ |
| $x \xleftarrow{r} \mathbb{Z}_q^*$ | $r \xleftarrow{r} \mathbb{Z}_q^*$ | verify $0 < s_1, s_2 < q$ |
| $y \equiv g^x \pmod{p}$ | $s_1 = (g^r \bmod p) \bmod q$ | $w = s_2^{-1} \bmod q$ |
| | $s_2 = r^{-1}(h(m) + xs_1) \bmod q$ | $u_1 = (h(m)w) \bmod q$ |
| | | $u_2 = (s_1 w) \bmod q$ |
| | | $v = (g^{u_1} y^{u_2} \bmod p) \bmod q$ |
| | | $b = (v = s_1)$ |
| $(x, y)$ | $(s_1, s_2)$ | $(b)$ |

The key generation, signature generation, and signature verification algorithms are summarized in Table 14.7. Similar to the Schnorr DSS, the DSA takes $p$, $q$, and

8 The fourth revision was made in July 2013 and led to the publication of FIPS PUB 186-4. It is electronically available at https://nvlpubs.nist.gov/nistpubs/FIPS/NIST.FIPS.186-4.

$g$ as domain parameters. Keep in mind that they can also be chosen individually for each entity, but this possibility is not considered here. While $p$ has a variable bitlength, the bitlength of $q$ is fixed.

- The prime number $p$ is $512 + 64t$ bits long (for $t \in \{0, \ldots, 8\}$);
- The prime number $q$ divides $p-1$ and is 160 bits long (i.e., $2^{159} < q < 2^{160}$).

In practice, it may be simpler to first select a 160-bit prime $q$ and then select an appropriately sized prime $p$ with the property that $q \mid p-1$. Anyway, $p$ determines the key strength, and it can be a multiple of 64 in the range between 512 and 1,024 bits.

As with the Schnorr DSS, the requirement that $q$ divides $p-1$ implies that $\mathbb{Z}_p^*$ has a subgroup of order $q$ (i.e., the subgroup has roughly $2^{160}$ elements). The value 160 is derived from the fact that the DSA (in its original form) was based on SHA-1 that generates hash values of this length. This has changed in FIPS 186-4, and the DSA now supports longer hash values and larger $q$ (as explained later). The last domain parameter is a generator $g$ that generates a $q$-order subgroup $G$ of $\mathbb{Z}_p^*$. Its construction is similar to the one employed by the Schnorr DSS.

Let us consider a toy example to illustrate the working principles of the DSA: Let $p = 23$ and $q = 11$ be again the two prime numbers that fulfill the requirement that $q = 11$ divides $p-1 = 22$. For $h = 2$, $h^{(p-1)/q} \pmod{p}$ refers to $2^{22/11} \equiv 2^2 \equiv 4 \pmod{23} = 4$ and this number is greater than 1. That means that it can be used as a generator for the subgroup $G = \{1, 2, 3, 4, 6, 8, 9, 12, 13, 16, 18\}$ of $\mathbb{Z}_{23}^*$ that has 11 elements.[9] We use this example to illustrate what is going on in the key generation, signature generation, and signature verification algorithms of the DSA.

#### 14.2.6.1 Key Generation Algorithm

The DSA key generation algorithm Generate is identical to the one employed by the Schnorr DSS. It randomly selects a private signing key $x$ from $\mathbb{Z}_q^*$ and computes the public verification key $y \equiv g^x \pmod{p}$. Again, $x$ is at most as large as $q$, whereas $y$ is at most as large as $p$.

In our toy example, the DSA Generate algorithm may randomly select $x = 3$ (that is an element of $\mathbb{Z}_{11}^* = \{1, 2, \ldots, 10\}$) and compute $y \equiv 4^3 \pmod{23} = 18$ (that is an element of $G$ and at most as large as 23). Consequently, the private signing key is $x = 3$, whereas the public verification key is $y = 18$.

9 Note that the 11 elements can be generated by computing $4^i \pmod{23}$ for $i = 1, 2, \ldots, 11$.

#### 14.2.6.2 Signature Generation Algorithm

The DSA signature generation algorithm Sign is probabilistic. It takes as input a private signing key $x$ and a message $m$, and it generates as output a DSA signature $s$ that consists of two components, $s = (s_1, s_2)$. The algorithm also requires a cryptographic hash function $h$ that generates hash values that have the same bitlength as $q$, such as SHA-1. The algorithm randomly selects a fresh $r$ from $\mathbb{Z}_q^*$, and uses this value to compute $s_1 = (g^r \bmod p) \bmod q$ and $s_2 = r^{-1}(h(m) + xs_1) \bmod q$, where $r^{-1}$ is the multiplicative inverse of $r$ modulo $q$. Neither $s_1$ nor $s_2$ must be zero. If such a case occurs, then the algorithm must restart with another $r$. Since each component of $s$ is at most 160 bits long, the total length of a DSA signature is at most 320 bits.

In our toy example, we assume a message $m$ that hashes to the value 6 (i.e., $h(m) = 6$). The DSA Sign algorithm may randomly select $r = 7$, compute $s_1 = (4^7 \bmod 23) \bmod 11 = 8$, determine $r^{-1} = 7^{-1} \bmod 11 = 8$, and compute $s_2 = 8(6 + 8 \cdot 3) \bmod 11 = 9$. Consequently, the DSA signature for the message is $(8, 9)$.

#### 14.2.6.3 Signature Verification Algorithm

The DSA signature verification algorithm Verify takes as input a public verification key $y$, a message $m$, and a signature $(s_1, s_2)$, and it generates as output one bit $b$ saying whether the signature is valid or not. The algorithm first verifies whether the components of $s$ are legitimate, meaning that they are both greater than zero and smaller than $q$ (i.e., $0 < s_1, s_2 < q$). It then computes $w = s_2^{-1} \bmod q$, $u_1 = (h(m)w) \bmod q$, $u_2 = (s_1 w) \bmod q$, and $v = (g^{u_1} y^{u_2} \bmod p) \bmod q$. Finally, the signature is valid if $v$ equals $s_1$.

The DSA is correct in the sense that the verifier always accepts valid signatures. We know that the signatory has computed $s_2 = r^{-1}(h(m) + xs_1) \bmod q$, and this means that

$$\begin{aligned} r &\equiv h(m)s_2^{-1} + xs_1 s_2^{-1} \pmod q \\ &\equiv h(m)w + xs_1 w \pmod q \end{aligned}$$

Because $g$ has order $q$, we can write

$$\begin{aligned} g^r &\equiv (g^{h(m)w + xs_1 w} \bmod p) \bmod q \\ &\equiv (g^{h(m)w} g^{xs_1 w} \bmod p) \bmod q \\ &\equiv (g^{h(m)w} y^{s_1 w} \bmod p) \bmod q \\ &\equiv (g^{u_1} y^{u_2} \bmod p) \bmod q \end{aligned}$$

The correctness then follows from $s_1 = (g^r \bmod p) \bmod q = (g^{u_1} y^{u_2} \bmod p) \bmod q = v$.

In our toy example, the DSA Verify algorithm verifies that $0 < 8, 9 < 11$ and then computes

$$
\begin{aligned}
w &= 9^{-1} \bmod 11 = 5 \\
u_1 &= (6 \cdot 5) \bmod 11 = 30 \bmod 11 = 8 \\
u_2 &= (8 \cdot 5) \bmod 11 = 40 \bmod 11 = 7 \\
v &= (4^8 18^7 \bmod 23) \bmod 11 \\
&= (65,536 \cdot 612,220,032) \bmod 23) \bmod 11 = 9
\end{aligned}
$$

This value of $v$ equals $s_1 = 9$, and hence the signature is *valid*.

#### 14.2.6.4 Security Analysis

The security analyses of the Elgamal and Schnorr DSSs also apply to the DSA. It is therefore widely believed that the DSA is secure, and that one can even increase $p$ and $q$ to further improve the security level. In Special Publication 800-57 [31], for example, NIST recommends bitlengths for $p$ that are 2,048 and 3,072, and bitlengths for $q$ that are 224 and 256 (instead of 160) to achieve a security level of 112 and 128 (instead of 80) bits. To achieve these security levels, one must replace SHA-1 with SHA-224 or SHA-256. These recommended bitlengths have also been adopted for DSA in FIPS 186-4. As mentioned above, this standard also provides support for RSA and ECDSA.

In the context of the Elgamal signature system, we already mentioned that a probabilistic signature generation algorithm can be turned into a deterministic one, by making the random value $r$ depend on the message that is signed (this mitigates the risk of reusing $r$). A respective construction for DSA and ECDSA is provided in [32]. This construction is important in environments where access to a high-quality entropy source is not available.

### 14.2.7 ECDSA

As already mentioned in Section 5.5, the acronym ECDSA refers to the elliptic curve variant of DSA. That is, instead of working in a $q$-order subgroup of $\mathbb{Z}_p^*$, one works in a group of points on an elliptic curve over a finite field $\mathbb{F}_q$, denoted $E(\mathbb{F}_q)$, where $q$ is either an odd prime or a power of 2. The history of ECDSA is also outlined in Section 5.5. Today, ECDSA is by far the most widely deployed DSS, and it has been adopted in many standards, including ANS X9.62, NIST FIPS 186 (since version

2),[10] ISO/IEC 14888[11] (and ISO/IEC 15946-1:2016 that provides the mathematical background and general techniques necessary to implement ECC and the ECDSA), IEEE 1363-2000, as well as the standards for efficient cryptography (SEC) 1 and 2. The details are subtle and beyond the scope of this book, so we only scratch the surface here.

In general, there are many possibilities to nail down a particular group that can be used for ECC in general and ECDSA in particular. The only requirement is that it is computationally intractable to solve the ECDLP (Definition 5.10) in this group. To some extent, the fact that there are so many possibilities is part of the reason why there are so many standards to choose from.

Roughly speaking, a standard for ECDSA must nail down an elliptic curve field and equation, called $Curve$, as well as a base point $G$ (on the curve) that generates a subgroup of a large prime order $n$; that is, $nG = \mathcal{O}$ (where $\mathcal{O}$ is the identity element of the group). For the sake of simplicity, we consider $Curve$, $G$, and $n$ to be domain parameters. In practice, more parameters are used,[12] but we like to keep things simple here.

**Table 14.8**
ECDSA

Domain parameters: $Curve, G, n$

| Generate | Sign | Verify |
|---|---|---|
| $(-)$ | $(d, m)$ | $(Q, m, (s_1, s_2))$ |
| $d \xleftarrow{r} \mathbb{Z}_n^*$ | $z = h(m)\mid_{len(n)}$ | verify legitimacy of $Q$ |
| $Q = dG$ | $r \xleftarrow{r} \mathbb{Z}_n^*$ | verify $0 < s_1, s_2 < n$ |
| | $(x_1, y_1) = rG$ | $z = h(m)\mid_{len(n)}$ |
| | $s_1 = x_1 \bmod n$ | $w = s_2^{-1} \bmod n$ |
| | $s_2 = r^{-1}(z + s_1 d) \bmod n$ | $u_1 = (zw) \bmod n$ |
| | | $u_2 = (s_1 w) \bmod n$ |
| | | $(x_1, y_1) = u_1 G + u_2 Q$ |
| | | $b = ((x_1, y_1) \neq \mathcal{O}) \wedge (s_1 = x_1)$ |
| $(d, Q)$ | $(s_1, s_2)$ | $(b)$ |

10 Some of the elliptic curves recommended by NIST in FIPS 186-4 are mentioned in Section 5.5.

11 The latest version of this standard is ISO/IEC 14888-3 released in 2018.

12 More specifically, domain parameters for ECDSA are of the form $(q, FR, a, b, \{seed,\}G, n, h)$, where $q$ is the order (size) of the field, $FR$ an indication for the field representation (not addressed here), $a$ and $b$ are the two elements that define the curve equation, $seed$ is an optional bit string that is present only if the curve was randomly generated in a verifiable fashion, $G$ is a base point, $n$ is the order of $G$, and $h$ is the cofactor that is equal to the order of the elliptic curve divided by $n$.

The ECDSA key generation, signature generation, and signature verification algorithms are summarized in Table 14.8 and briefly outlined below. To illustrate the working principles of ECDSA, we reuse our exemplary elliptic curve $E(\mathbb{Z}_{23})$ with its $n = 28$ elements (including $\mathcal{O}$) from Section 5.5, and we use $(3, 10)$ as a base point that generates $G$. Note that this curve is not perfectly valid because its order $n = 28$ is not prime. This means that some values may not have a multiplicatively inverse element. We circumvent this problem here by deliberately using values that have an inverse element.

#### 14.2.7.1 Key Generation Algorithm

The ECDSA key generation algorithm Generate takes no other input than the domain parameters, but it generates as output a public key pair $(d, Q)$. The private signing key $d$ is randomly selected from $\mathbb{Z}_n^*$, whereas the public verification key is computed as $Q = dG$.

In our toy example, the Generate algorithm may randomly select the private signing key $d = 3$ and compute the public verification key $Q = 3G = (19, 5)$.

#### 14.2.7.2 Signature Generation Algorithm

Like all variants of Elgamal, the ECDSA signature generation algorithm Sign is probabilistic. It takes as input a private signing key $d$ and a message $m$, and it generates as output an ECDSA signature $(s_1, s_2)$. The algorithm first computes $h(m)$, takes the $len(n)$ leftmost bits, and assigns this bitstring to $z$. The algorithm then randomly selects $r$ from $\mathbb{Z}_n^* = \{1, \ldots, n-1\}$, and adds $G$ so many times to itself. The resulting point on the curve has the coordinates $(x_1, y_1)$. From these coordinates, only $x_1$ is used to sign $m$ or $z$, respectively. In fact, $s_1$ is set to $x_1 \bmod n$, and $s_2$ is set to $r^{-1}(z + s_1 d) \bmod n$. In either of the last two steps, the resulting value $s_1$ or $s_2$ must not be equal to zero (this requirement is not indicated in Table 14.8). Otherwise, the algorithm must go back to the step in which the algorithm randomly selects $r$, and retry it with another value. In the end, the algorithm terminates with two nonzero values $s_1$ and $s_2$—they yield the ECDSA signature for $m$.

In our toy example, we want to digitally sign a message $m$, whose leftmost $len(n)$ bits of $h(m)$ is assumed to return $z = 5$. The Sign algorithm then randomly selects $r = 11$ (that has an inverse 23 modulo 28), and then computes $(18, 20) = 11 \cdot G$, $s_1 = 18$, and $s_2 = 23(5 + 18 \cdot 3) \bmod 28 = 13$. Hence, the ECDSA signature for this message is $(18, 13)$.

#### 14.2.7.3 Signature Verification Algorithm

The ECDSA signature verification algorithm Verify takes as input a public verification key $Q$, a message $m$, and a signature $(s_1, s_2)$, and it generates as output one bit $b$ saying whether the signature is valid or not. The algorithm must first verify the legitimacy of $Q$, meaning that it must verify that $Q$ is not equal to $\mathcal{O}$, $Q$ lies on the curve, and $nQ = \mathcal{O}$. Furthermore, it must verify that $0 < s_1, s_2 < n$. If everything is fine, then the algorithm reconstructs $z$ (similar to the Sign algorithm), computes $w = s_2^{-1} \bmod n$, $u_1 = (zw) \bmod n$, and $u_2 = (s_1 w) \bmod n$, and generates the point $u_1 G + u_2 Q$ with coordinates $x_1$ and $y_1$. The signature is valid if this point $(x_1, y_1)$ is not equal to $\mathcal{O}$ and $s_1 = x_1$.

To show that the signature verification is correct, we start from the point $u_1 G + u_2 Q$ and combine it with $s_2 \equiv r^{-1}(z + s_1 d) \bmod n$ or its inverse $s_2^{-1} \equiv r(z + s_1 d)^{-1} \bmod n$, respectively:

$$\begin{aligned} u_1 G + u_2 Q &= zwG + s_1 wQ \\ &= zs_2^{-1}G + s_1 s_2^{-1}Q \\ &= zs_2^{-1}G + s_1 s_2^{-1}dG \\ &= s_2^{-1}G(z + s_1 d) \\ &= r(z + s_1 d)^{-1}G(z + s_1 d) \\ &= rG \end{aligned}$$

The result is the point $(x_1, y_1)$, and $s_1$ must in fact equal $x_1$ (because it is constructed this way in the Sign algorithm).

In our toy example, we know that $Q$ is legitimate and we can easily verify that $0 < 18, 13 < 28$. Under the assumption that $z = 5$, the Verify algorithm computes $w = 13$ (13 is self-inverse modulo 28), $u_1 = 5 \cdot 13 \bmod 28 = 9$, $u_2 = 18 \cdot 13 \bmod 28 = 10$, and $(18, 20) = 9G + 10Q$. Since this point is different from $\mathcal{O}$ and its $x$-coordinate is equal to $s_1 = 18$, the signature is considered to be valid.

#### 14.2.7.4 Security Analysis

From a theoretical perspective, the ECDSA seems to fulfill the strongest security definition: It protects against existential forgery under an adaptive CMA if some well-defined conditions are met.[13] Because not all conditions are satisfied by the DSA, the security proof is not applicable, and hence the security of the ECDSA is assumed to be superior. Also, it has been shown that a variant of the ECDSA,

13 https://eprint.iacr.org/2002/026.

acronymed BLS,[14] is provably secure in the random oracle model assuming that the DHP (but not the DDHP) is hard. BLS signatures are very short (about 160 bits) and can easily be aggregated. This makes them very suitable for blockchain applications.

From a practical perspective, the ECDSA seems to have the same vulnerabilities as all Elgamal-like DSSs. In particular, it must be ensured that all signatures are generated with a fresh and unique $r$, and that an $r$ is never reused. In 2010, for example, it was shown that Sony used the same $r$ to sign software for the PlayStation 3 game console, and this reuse allowed adversaries to recover the private signing key. This attack was equally as devastating as it was embarrassing for Sony and its software developers.

**Table 14.9**
Cramer-Shoup DSS

Domain parameters: $l, l'$

| Generate | Sign | Verify |
|---|---|---|
| $(-)$ | $(sk, m)$ | $(pk, m, s)$ |
| $p, q \xleftarrow{r} \mathbb{P}^*_{l'}$ | $e \xleftarrow{r} \mathbb{P}_{l+1}$ with $e \neq e'$ | verify $e \neq e'$ |
| $n = pq$ | $y' \xleftarrow{r} QR_n$ | verify $e$ is odd |
| $f, x \xleftarrow{r} QR_n$ | solve $(y')^{e'} = x' f^{h(m)}$ for $x'$ | verify $len(e) = l + 1$ |
| $e' \xleftarrow{r} \mathbb{P}_{l+1}$ | solve $y^e = x f^{h(x')}$ for $y$ | compute $x' = (y')^{e'} f^{-h(m)}$ |
| $pk = (n, f, x, e')$ | $s = (e, y, y')$ | $b = (x = y^e f^{-h(x')})$ |
| $sk = (p, q)$ | | |
| $(pk, sk)$ | $(s)$ | $(b)$ |

### 14.2.8 Cramer-Shoup

All practical DSSs addressed so far have either no rigorous security proof or only a security proof in the random oracle model. This is different with the Cramer-Shoup DSS [14]: It is practical and can be proven secure in the standard model under the strong RSA assumption. There is also a variant that is secure in the random oracle model under the standard RSA assumption, but we leave this variant aside here. The Cramer-Shoup DSS is based on [9], but it has the big advantage that it is stateless; this makes it usable in practice.

14 The acronym BLS is taken from the names of the researchers who proposed the DSS—Dan Boneh, Ben Lynn, and Hovav Shacham.

The key generation, signature generation, and signature verification algorithms of the Cramer-Shoup DSS are summarized in Table 14.9. There are two domain parameters: $l$ and $l' > l + 1$. Reasonable choices are $l = 160$ and $l' = 512$. Also, it makes use of a collision-resistant hash function $h$ whose output values are $l$ bits long and can be interpreted as integers between 0 and $2^l - 1$. If, for example, $l = 160$, then a reasonable choice for $h$ is SHA-1. As usual, $QR_n$ denotes the subgroup of $\mathbb{Z}_n^*$ that comprises all squares (or quadratic residues) modulo $n$.

#### 14.2.8.1 Key Generation Algorithm

The Cramer-Shoup key generation algorithm Generate takes no input other than the system parameters, but generates as output a public verification key $pk$ and a private signing key $sk$. More specifically, it randomly selects two $l'$-bit safe primes $p$ and $q$ (where $\mathbb{P}^*_{l'}$ refers to the set of all safe primes with bitlength $l'$). Remember from Appendix A.2.4.4 that $p$ and $q$ are safe primes if they are of the form $p = 2p' + 1$ and $q = 2q' + 1$ for some Sophie Germain primes $p'$ and $q'$. The product $n = pq$ yields an RSA modulus. The algorithm randomly selects two quadratic residues $f$ and $x$ modulo $n$ (i.e., $f, x \in QR_n$), as well as a prime $e'$ from $\mathbb{P}_{l+1}$ (i.e., the set of all $(l + 1)$-bit primes). The public key $pk$ is the 4-tuple $(n, f, x, e')$,[15] whereas the private key $sk$ comprises the prime factors of $n$ (i.e., $p$ and $q$).

#### 14.2.8.2 Signature Generation Algorithm

The Cramer-Shoup signature generation algorithm Sign is probabilistic and requires a collision-resistant hash function $h$ (as mentioned above). It takes as input a private signing key $sk$ and a message $m$, and it generates as output a signature $s$. The algorithm randomly selects an $(l + 1)$-bit prime $e$ that is different from $e'$ and an element $y'$ from $QR_n$. It first solves the equation

$$(y')^{e'} = x' f^{h(m)}$$

for $x'$ (i.e., $x' = (y')^{e'} f^{-h(m)}$), and then uses $x'$ to solve the equation

$$y^e = x f^{h(x')}$$

for $y$ (i.e., $y = (x f^{h(x')})^{1/e}$). This can be done because the prime factorization of $n$ (i.e., $p$ and $q$) is part of the private signing key $sk$, and hence known to the signatory. We know from before that knowing the prime factorization of $n$ is computationally equivalent to be able to compute e-th roots. In the end, the signature $s$ consists of the triple $(e, y, y')$ that is sent to the verifier together with the message $m$.

15 To speed up signature generation and verification, $pk$ may comprise $f^{-1}$ instead of $f$.

#### 14.2.8.3 Signature Verification Algorithm

The Cramer-Shoup signature verification algorithm Verify takes as input a public verification key $pk$, a message $m$, and a signature $s$, and it generates as output one bit $b$ saying whether $s$ is a valid signature for $m$ with respect to $pk$. The algorithm first verifies that $e$ is not equal to $e'$, $e$ is odd, and the length of $e$ is equal to $l+1$, and it then computes $x' = (y')^{e'} f^{-h(m)}$. Finally, it verifies whether $x = y^e f^{-h(x')}$. If this is the case, then $s$ is indeed a valid signature for $m$ with respect to $pk$.

#### 14.2.8.4 Security Analysis

The security analysis can be kept very short, because we already know from the introductory remarks that the Cramer-Shoup DSS can be proven secure in the standard model under the strong RSA assumption. As usual, this is a theoretical result, and any implementation of the Cramer-Shoup DSS may still have vulnerabilities and security problems of its own.

## 14.3 IDENTITY-BASED SIGNATURES

In Section 13.4, we said that Shamir came up with the idea of identity-based cryptography in the early 1980s, and that he also proposed an identity-based DSS [33]. This DSS is relatively simple and straightforward: Let a trusted authority (that is always needed in identity-based cryptography) choose an RSA modulus $n$ (that is the product of two primes $p$ and $q$), a large number $e$ with $gcd(e, \phi(n))$, and a one-way function $f$, and publish them as domain parameters. For every user, the trusted authority then derives a public verification key $pk$ from the user's identity, and computes the private signing key $sk$ as the e-th root of $pk$ modulo $n$; that is, $sk^e \equiv pk \pmod{n}$. This key is used to digitally sign messages, whereas $pk$—together with $n$, $e$, and $f$—is used to verify digital signatures. Since the trusted authority knows the prime factorization of $n$, it can always compute $sk$ from $pk$, but for anybody else (not knowing the prime factorization of $n$) this is computationally intractable.

To digitally sign a message $m \in \mathbb{Z}_n$, the user randomly selects a value $r \in_R \mathbb{Z}_n$ and computes

$$t = r^e \bmod n$$

and

$$s = (sk \cdot r^{f(t,m)} \bmod n$$

The pair $(s, t)$ yields the identity-based signature for $m$. Note that the signatory needs $sk$ to generate the signature. To verify it, the following equivalence must hold:

$$s^e \equiv pk \cdot t^{f(t,m)} \pmod{n}$$

This is correct, because

$$\begin{aligned} s^e &\equiv (sk \cdot r^{f(t,m)})^e \pmod{n} \\ &\equiv sk^e r^{ef(t,m)} \pmod{n} \\ &\equiv pk \cdot t^{f(t,m)} \pmod{n} \end{aligned}$$

Shamir's identity-based DSS has fueled a lot of research and development in identity-based cryptography. In particular, many other identity-based DSS have been proposed, and for the last two decades we also know a few IBE systems. However, the disadvantages mentioned in Section 13.4 still apply, meaning that we don't have a unique naming scheme, that it is difficult to set up trusted authorities (to issue public key pairs), and that the key revocation problem remains unsolved. None of these problems is likely to go away anytime soon, so the potential of identity-based cryptography remains questionable (to say the least). This is also true for identity-based DSS.

## 14.4 ONE-TIME SIGNATURES

A *one-time signature system* is a DSS with the specific property that a public key pair can be used to digitally sign a single message. If the pair is reused, then it becomes feasible to forge signatures. There are advantages and disadvantages related to one-time signature systems:

- The advantages are simplicity and efficiency, meaning that one-time signature systems are particularly applicable in situations that require low computational complexity.
- The disadvantages are related to the size of the verification key(s) and the corresponding key management overhead.

To overcome the disadvantages of one-time signature systems, one-time signatures are often combined with techniques to efficiently authenticate public verification keys such as Merkle trees [6, 7].

Historically, the first one-time signature system was proposed by Michael O. Rabin in 1978. The system employed a symmetric encryption system and was too inefficient to be used in practice. In 1979, however, Leslie Lamport[16] proposed a one-time signature system that is efficient because it employs a one-way function instead of a symmetric encryption [34]. If combined with techniques to efficiently authenticate public verification keys (e.g., Merkle trees), the resulting one-time signature systems are practical.

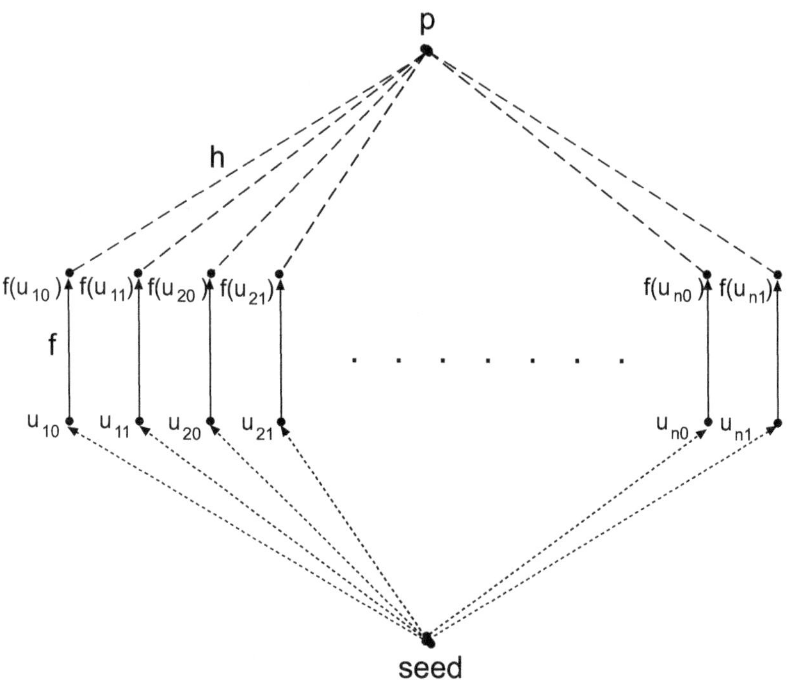


**Figure 14.4** Lamport's one-time DSS.

Let $f$ be a one-way function and $m$ the message to be signed. The bitlength of $m$ is assumed to be at most $n$ ($len(m) \leq n$), where $n$ may be 128 or 160 bits. If a message is longer than $n$ bits, it must first be shortened with a cryptographic hash function. To digitally sign $m$ with the Lamport one-time DSS, the signatory must

16 In 2013, Leslie Lamport received the ACM Turing Award for fundamental contributions to the theory and practice of distributed and concurrent systems, notably the invention of concepts such as causality and logical clocks, safety and liveness, replicated state machines, and sequential consistency.

have a private key that consists of $n$ pairs of randomly chosen preimages for $f$:

$$[u_{10}, u_{11}], [u_{20}, u_{21}], \ldots, [u_{n0}, u_{n1}]$$

Each preimage $u_{ij}$ ($i = 1, \ldots, n$ and $j = 0, 1$) may, for example, be a string of typically $n$ bits. In an efficient implementation, the $2n$ $n$-bit arguments may be generated with a properly seeded PRG. The public key then consists of the $2n$ images $f(u_{ij})$:

$$[f(u_{10}), f(u_{11})], [f(u_{20}), f(u_{21})], \ldots, [f(u_{n0}), f(u_{n1})]$$

Furthermore, in an efficient implementation, the $2n$ images $f(u_{ij})$ are hashed to a single value $p$ that represents the public key required to verify the one-time signature:

$$p = h(f(u_{10}), f(u_{11}), f(u_{20}), f(u_{21}), \ldots, f(u_{n0}), f(u_{n1}))$$

Remember that complementary techniques to efficiently authenticate verification keys are required if multiple signatures must be verified.

To digitally sign message $m$, each message bit $m_i$ ($i = 1, \ldots, n$) is individually signed using the pair $[u_{i0}, u_{i1}]$. If $m_i = 0$, then the signature bit is $u_{i0}$, and if $m_i = 1$, then the signature bit signature bit is $u_{i1}$. The Lamport one-time signature $s$ for message $m$ finally comprises all $n$ signature bits:

$$s = [u_{1m_1}, u_{2m_2}, \ldots, u_{nm_n}]$$

The signature $s$ can be verified by computing all images $f(u_{ij})$, hashing all of these values to $p'$, and comparing $p'$ with the public key $p$. The signature is valid if and only if $p' = p$.

The Lamport one-time DSS is illustrated in Figure 14.4. As mentioned earlier, a PRG and a seed are typically used to generate the $2n$ values $u_{10}, u_{11}, u_{20}, u_{21}, \ldots,$ $u_{n0}, u_{n1}$, and a cryptographic hash function $h$ is typically used to compute the public key $p$. Figure 14.5 illustrates an exemplary one-time signature for the binary message 0110. The message bit $m_1$ is digitally signed with $u_{10}$, $m_2$ with $u_{21}$, $m_3$ with $u_{31}$, and $m_4$ with $u_{40}$.

There are several possibilities to generalize and improve the efficiency of the Lamport one-time DSS. Some of these improvements are due to Merkle. Other improvements have been proposed recently to make one-time signature systems—or hash-based digital signature systems as they are sometimes called—a viable alternative for PQC (Section 18.3.2). Last but not least, we note that the Lamport one-time DSS and some variations thereof are also used in several cryptographic applications. For example, they can be used to protect against the double-spending problem in anonymous offline digital cash systems (e.g., [35]). This use is not further addressed here.

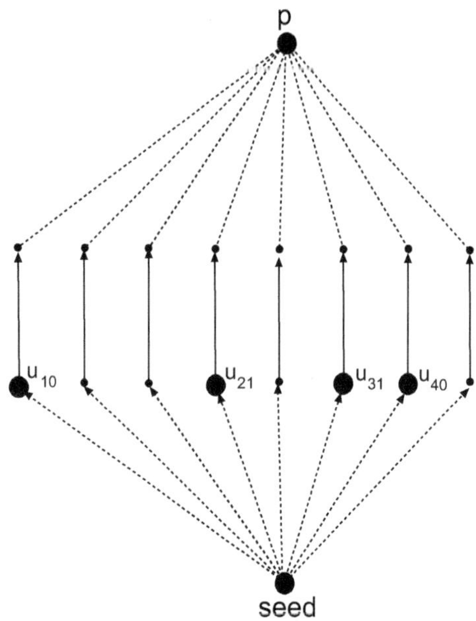


**Figure 14.5** Exemplary one-time signature.

## 14.5 VARIANTS

In practice, different use cases require different types of digital signatures, and several variants of normal digital signatures exist. In this section, we briefly overview and put into perspective blind signatures, undeniable signatures, fail-stop signatures, and group signatures. More variants are outlined and discussed in the relevant literature (e.g., [1–4]).

### 14.5.1 Blind Signatures

The idea of blind signatures was developed and originally proposed by David Chaum in the early 1980s [36, 37]. Informally speaking, a signature is *blind* if the signatory does not obtain any information about the message that is signed or the signature that is generated. Instead, the message is blinded in a way that can only be reversed by the recipient of the blind signature.

For example, the RSA DSS as introduced in Section 14.2.1 can be turned into a blind RSA DSS. Let A be a signatory with public RSA verification key $(n, e)$ and

**Table 14.10**
Protocol to Issue Blind RSA Signatures

| B | | A |
|---|---|---|
| $((n, e), m)$ | | $(d)$ |
| $r \xleftarrow{r} \mathbb{Z}_n^*$ | | |
| $t \equiv mr^e \ (bmod\ n)$ | | |
| | $\xrightarrow{t}$ | |
| | | $u \equiv t^d \equiv t^{1/e} \equiv m^d r \pmod{n}$ |
| | $\xleftarrow{u}$ | |
| $s \equiv u/r \equiv m^d r/r \equiv m^d \pmod{n}$ | | |
| $(s)$ | | |

B a recipient of a blind signature from A. The protocol summarized in Table 14.10 can then be used to have A issue a blind RSA signature $s$ for a message $m$ chosen by B. On B's side, the protocol takes as input A's public verification key $(n, e)$ and a message $m$, and it generates as output the RSA digital signature $s$ for $m$. On A's side, the protocol only takes A's private signing key $d$ as input. B randomly selects an $r$ from $Z_n^*$ and uses this random value to blind the message $m$. Blinding is performed by multiplying $m$ with $r$ to the power of $e$ modulo $n$. The resulting blinded message $t$ is transmitted to A and digitally signed there. This is done by putting $t$ to the power of $d$ modulo $n$. The resulting message $u$ is sent back to B, and B is now able to unblind the message. Unblinding is performed by dividing $u$ by $r$, or multiplying $u$ with the multiplicative inverse of $r$ modulo $n$, respectively. The inverse can be computed because $r$ is a unit (and hence invertible) in $\mathbb{Z}_n$.

At first glance, one would argue that blind signatures are not particularly useful because a signatory may want to know what it signs. Surprisingly, this is not always the case, and there are many applications for blind signatures and corresponding DSSs. Examples include anonymous digital cash and electronic voting. After Chaum published his results in the early 1980s, almost all DSS have been extended in one way or another to provide the possibility to issue blind signatures.

### 14.5.2 Undeniable Signatures

The notion of an *undeniable signature* was proposed in the late 1980s [38]. In short, an undeniable signature is a digital signature that cannot be verified with only a public verification key. Instead, it must be verified interactively, meaning that an undeniable signature can only be verified with the aid of the signatory. The signature verification algorithm is therefore replaced with a protocol that is executed between the verifier and the signatory. Because a dishonest signatory can always refuse participation in a signature verification protocol, an undeniable DSS must also comprise a disavowal protocol that can be used to prove that a given signature is a forgery.

### 14.5.3 Fail-Stop Signatures

The notion of a *fail-stop signature* was proposed in the early 1990s [39, 40]. In short, a fail-stop signature is a signature that allows the signatory to prove that a signature purportedly (but not actually) generated by itself is a forgery. This is done by showing that the underlying assumption on which the DSS is based has been compromised. After such a compromise has been shown, the DSS stops (this is why such signatures are called fail-stop in the first place). Fail-stop signatures are theoretically interesting, but practically not very relevant. Note that it is much more likely that a private signing key is compromised than the underlying assumption is compromised.

### 14.5.4 Group Signatures

The notion of a group signature was developed in the early 1990s [41]. It refers to a signature that can be generated anonymously by a member of a group on behalf of the group. For example, a group signature system can be used by an employee of a large company where it is sufficient for a verifier to know that a message was signed by an employee, but not the particular employee who signed it.

Essential to a group signature system is a group manager, who is in charge of adding group members and has the ability to reveal the original signer in the case of a dispute. In some systems the responsibilities of adding members and revoking signature anonymity are separated and given to a membership manager and revocation manager, respectively. In some other systems, the requirement of a group manager is excluded. Such a DSS is sometimes called a ring signature system. It then provides true anonymity for potential signatories.

## 14.6 FINAL REMARKS

In this chapter, we elaborated on digital signatures and DSSs, and we overviewed and discussed many examples. Note that many other DSSs—with or without specific properties—are described and discussed in the literature. There are even some DSSs that can be constructed from zero-knowledge protocols (Section 15.2), and DSSs that are qualified for PQC (Section 18.3.2 for hash-based systems and Section 18.3.3 for lattice-based systems).

It is argued (or rather hoped) that digital signatures provide the digital analog of handwritten signatures, and that they can be used to provide nonrepudiation services (i.e., services that make it impossible or useless for communicating peers to repudiate participation). Against this background, many countries and communities have put forth new legislation regarding the use of digital signatures. Examples include the European Electronic Identification and Trust Services Regulation (910/2014/EC), commonly referred to as eIDAS,[17] and the U.S. Electronic Signatures in Global and National Commerce Act, commonly referred to as E-SIGN. However, although many countries have digital signature laws in place, it is important to note that these laws have not been seriously disputed in court and that it is not clear what the legal status of digital signatures really is. The fact that digital signatures are based on mathematical formulas intuitively makes us believe that the evidence they provide is strong. This belief is seductive and often illusive (e.g., [42–44]). Contrary to handwritten signatures, digital signatures are based on many layers of hardware and software, and on each of these layers many things can go wrong. This also applies to the user of the software who may be the target of social engineering attacks.

## References

[1] Pfitzmann, B., *Digital Signature Schemes: General Framework and Fail-Stop Signatures.* Springer-Verlag, LNCS 1100, 1996.

[2] Piper, F., S. Blake-Wilson, and J. Mitchell, *Digital Signatures Security & Controls.* Information Systems Audit and Control Foundation (ISACF), 2000.

[3] Atreya, M., et al., *Digital Signatures.* RSA Press, McGraw-Hill/Osborne, Berkeley, CA, 2002.

[4] Katz, J., *Digital Signatures.* Springer-Verlag, New York, 2010.

[5] Goldwasser, S., S. Micali, and R.L. Rivest, "A Digital Signature Scheme Secure Against Adaptive Chosen-Message Attacks," *SIAM Journal of Computing*, Vol. 17, No. 2, April 1988, pp. 281–308.

[6] Merkle, R.C., "Secrecy, Authentication, and Public Key Systems," Ph.D. Thesis, Stanford University, 1979, http://www.merkle.com/papers/Thesis1979.pdf.

[7] Merkle, R.C., "A Certified Digital Signature," *Proceedings of CRYPTO '89*, Springer-Verlag, LNCS 435, 1989, pp. 218–238.

[8] Dwork, C., and M. Naor, "An Efficient Existentially Unforgeable Signature Scheme and Its Applications," *Journal of Cryptology*, Vol. 11, No. 3, 1998, pp. 187–208.

17 eIDAS took effect on July 1, 2016, and partly replaced Directive 1999/93/EC of the European Parliament and of the Council of December 13, 1999, on a Community Framework for Electronic Signatures.

[9] Cramer, R., and I. Damgård, "New Generation of Secure and Practical RSA-Based Signatures," *Proceedings of CRYPTO '96*, Springer-Verlag, LNCS 1109, 1996, pp. 173–185.

[10] Bellare, M., and P. Rogaway, "Random Oracles Are Practical: A Paradigm for Designing Efficient Protocols," *Proceedings of 1st ACM Conference on Computer and Communications Security*, ACM Press, New York, 1993, pp. 62–73.

[11] Bellare, M., and P. Rogaway, "The Exact Security of Digital Signatures—How to Sign with RSA and Rabin," *Proceedings of EUROCRYPT '96*, Springer-Verlag, LNCS 1070, 1996, pp. 399–414.

[12] Pointcheval, D., and J. Stern, "Security Proofs for Signature Schemes," *Proceedings of EUROCRYPT '96*, Springer-Verlag, LNCS 1070, 1996, pp. 387–398.

[13] Canetti, R., O. Goldreich, and S. Halevi, "The Random Oracles Methodology, Revisited," *Proceedings of 30th STOC*, ACM Press, New York, 1998, pp. 209–218.

[14] Cramer, R., and V. Shoup, "Signature Schemes Based on the Strong RSA Assumption," *ACM Transactions on Information and System Security (ACM TISSEC)*, Vol. 3, No. 3, 2000, pp. 161–185.

[15] Gennaro, R., Halevi, S., and T. Rabin, "Secure Hash-and-Sign Signatures Without the Random Oracle," *Proceedings of EUROCRYPT '99*, Springer-Verlag, LNCS 1592, 1999, pp. 123–139.

[16] Camenisch, J., and A. Lysyanskaya, "A Signature Scheme with Efficient Protocols," *Proceedings of the International Conference on Security in Communication Networks (SCN 2002)*, Springer-Verlag, LNCS 2576, 2002, pp. 268–289.

[17] Fischlin, M., "The Cramer-Shoup Strong-RSA Signature Scheme Revisited," *Proceedings of the 6th International Workshop on Theory and Practice in Public Key Cryptography*, Springer-Verlag, LNCS 2567, 2003, pp. 116–129.

[18] Diffie, W., and M.E. Hellman, "New Directions in Cryptography," *IEEE Transactions on Information Theory*, IT-22(6), 1976, pp. 644–654.

[19] Rivest, R.L., A. Shamir, and L. Adleman, "A Method for Obtaining Digital Signatures and Public-Key Cryptosystems," *Communications of the ACM*, Vol. 21, No. 2, February 1978, pp. 120–126.

[20] Jonsson, J., and B. Kaliski, *Public-Key Cryptography Standards (PKCS) #1: RSA Cryptography Specifications Version 2.1*, Request for Comments 3447, February 2003.

[21] Zheng, Y., "Digital Signcryption or How to Achieve Cost (Signature & Encryption) $\ll$ Cost (Signature) + Cost (Encryption)," *Proceedings of CRYPTO '97*, Springer-Verlag, LNCS 1294, 1997, pp. 165–179.

[22] Rabin, M.O., "Digitalized Signatures and Public-Key Functions as Intractable as Factorization," MIT Laboratory for Computer Science, MIT/LCS/TR-212, 1979.

[23] Elgamal, T., "A Public Key Cryptosystem and a Signature Scheme Based on Discrete Logarithm," *IEEE Transactions on Information Theory*, Vol. 31, No. 4, 1985, pp. 469–472.

[24] Agnew, G.B., R.C. Mullin, and S.A. Vanstone, "Improved Digital Signature Scheme Based on Discrete Exponentiation," *IEEE Electronics Letters*, Vol. 26, No. 14, July 1990, pp. 1024–1025.

[25] Horster, P., M. Michels, and H. Petersen, "Meta-ElGamal Signature Schemes," *Proceedings of 2nd ACM Conference on Computer and Communications Security*, ACM Press, New York, 1994, pp. 96–107.

[26] Nyberg, K., and R.A. Rueppel, "Message Recovery for Signature Schemes Based on the Discrete Logarithm Problem," *Designs, Codes and Cryptography*, Vol. 7, 1996, pp. 61–81.

[27] Bleichenbacher, D., "Generating ElGamal Signatures Without Knowing the Secret Key," *Proceedings of EUROCRYPT '96*, Springer-Verlag, LNCS 1070, 1996, pp. 10–18.

[28] Pohlig, S., and M.E. Hellman, "An Improved Algorithm for Computing Logarithms over $GF(p)$ and its Cryptographic Significance," *IEEE Transactions on Information Theory*, Vol. 24, No. 1, January 1978, pp. 108–110.

[29] Schnorr, C.P., "Efficient Signature Generation by Smart Cards," *Journal of Cryptology*, Vol. 4, 1991, pp. 161–174.

[30] U.S. National Institute of Standards and Technology (NIST), *Digital Signature Standard (DSS)*, FIPS PUB 186, May 1994.

[31] U.S. National Institute of Standards and Technology (NIST) Special Publication 800-57, *Recommendation for Key Management—Part 1: General (Revised)*, May 2006.

[32] Pornin, T., *Deterministic Usage of the Digital Signature Algorithm (DSA) and Elliptic Curve Digital Signature Algorithm (ECDSA)*, Informational RFC 6979, August 2013.

[33] Shamir, A., "Identity-Based Cryptosystems and Signatures," *Proceedings of CRYPTO '84*, Springer-Verlag, 1984, pp. 47–53.

[34] Lamport, L. , *Constructing Digital Signatures from a One-Way Function*, Technical Report CSL-98, SRI International, October 1979.

[35] Chaum, D., A. Fiat, and M. Naor, "Untraceable Electronic Cash," *Proceedings of CRYPTO '88*, Springer-Verlag, LNCS 403, 1988, pp. 319–327.

[36] Chaum, D., "Blind Signatures for Untraceable Payments," *Proceedings of CRYPTO '82*, Plenum Press, New York, 1983, pp. 199–203.

[37] Chaum, D., "Blind Signature System," *Proceedings of CRYPTO '83*, Plenum Press, New York, 1984, p. 153.

[38] Chaum, D., and H. van Antwerpen, "Undeniable Signatures," *Proceedings of CRYPTO '89*, Springer-Verlag, LNCS 435, 1990, pp. 212–216.

[39] Pfitzmann, B., "Fail-Stop Signatures: Principles and Applications," *Proceedings of the 8th World Conference on Computer Security, Audit and Control (COMPSEC '91)*, 1991, pp. 125–134.

[40] Pedersen, T.P., and B. Pfitzmann, "Fail-Stop Signatures," *SIAM Journal on Computing*, Vol. 26, No. 2, 1997, pp. 291–330.

[41] Chaum, D., and E. van Heyst, "Group Signatures," *Proceedings of EUROCRYPT '91*, Springer-Verlag, LNCS 547, 1991, pp. 257–265.

[42] Oppliger, R., and R. Rytz, "Digital Evidence: Dream and Reality," *IEEE Security & Privacy*, Vol. 1, No. 5, September/October 2003, pp. 44–48.

[43] Maurer, U.M., "Intrinsic Limitations of Digital Signatures and How to Cope with Them," *Proceedings of the 6th Information Security Conference (ISC '03)*, Springer-Verlag, LNCS 2851, pp. 180–192.

[44] Maurer, U.M., "New Approaches to Digital Evidence," *Proceedings of the IEEE*, Vol. 92, No. 6, June 2004, pp. 933–947.

# Chapter 15

# Zero-Knowledge Proofs of Knowledge

In this chapter, we explore the notion of an interactive proof system that may have the zero-knowledge property, and discuss its application in proofs of knowledge for entity authentication. More specifically, we introduce the topic in Section 15.1, outline some zero-knowledge authentication protocols in Section 15.2, explore non-interactive zero-knowledge in Section 15.3, and conclude with some final remarks in Section 15.4. Note that the notion of zero-knowledge is hyped today, and many things are going on. So in this chapter, we only scratch the surface and want to be sure that the latest research results can be put into proper perspective.

## 15.1 INTRODUCTION

Before we start with the main topic of this chapter—zero-knowledge proofs of knowledge—we want to elaborate a little bit on the notion of a proof. What is a proof? This simple question turns out to be difficult to answer. On a very high level of abstraction, a proof is just a method to establish truth. If we want to prove a claim (that may be any proposition one may think of), we basically mean that we want to convince somebody or even everybody that the claim is true. The details of such a proof depend on the situation, and whether one has a philosophical, legal, scientific, or mathematical perspective, but in all cases, the goal is to establish a conviction of truth.

Cryptography is about applied mathematics. so we are mainly interested in mathematical proofs here. The ultimate goal of such a proof is to derive a claimed proposition from a set of axioms, using well-defined (syntactical and semantical) derivation rules. We don't care much about who provides the proof and how it is generated in the first place, but we care a lot about its correctness and the completeness of the derivation chain. In fact, each derivation step in this chain must be comprehensible and logical. If there is a missing piece in the chain, then the entire proof must be rejected as being invalid. So the point to remember and keep in mind is that a "normal" proof in mathematics is verifier-centric, meaning that only the verifier is needed and whoever generates the proof with what computational power is more or less pointless and doesn't really matter. As a consequence of this independence from the prover, such a proof is usually transferable by default.

In the sequel, we focus entirely on decision problems that—according to Section D.2—can be expressed as a language membership problem (i.e., given a language $L \subseteq \{0,1\}^*$, decide whether a given input $x \in \{0,1\}^*$ is a member of $L$, possible answers are YES and NO). If we have a verifier $V$ that is able to check a proof $\pi$ for $x$ being a member of $L$, then we can define

$$L = \{x \mid \exists \pi : V(x, \pi) = \text{YES}\}$$

Put in other words, the language $L$ consists of all $x \in \{0,1\}^*$, for which there is a proof $\pi$ that can be verified by $V$. Following this line of argumentation, we can define a proof system for membership in $L$ as captured in Definition 15.1.

**Definition 15.1 (Proof System)** *A* proof system *for membership in L is an algorithm V, such that for all $x \in \{0,1\}^*$ the following two requirements are fulfilled:*

1. *If $x \in L$, then there exists a proof $\pi$ with $V(x, \pi) =$ YES.*
2. *If $x \notin L$, then for all proofs $\pi$ it must be the case that $V(x, \pi) =$ NO.*

While the first requirement refers to the *completeness* of the proof system, the second requirement refers to its *soundness*. A proof system is *complete* if all $x \in L$ can be proven to be so, whereas it is *sound* if no $x \notin L$ can be proven to be in $L$. Both properties are important for a proof system to be useful in practice.

So far, we have put no restriction on the efficiency of the verification algorithm $V$. This is not very realistic, and in practice we are interested in verification algorithms that are efficient, meaning that they run in polynomial time. Hence, we say that a proof system is *efficient* or is an *NP proof system*, if in addition to the requirements of Definition 15.1 (i.e., completeness and soundness), $V$ is also efficient, meaning that $V(x, \pi)$ halts after at most a polynomial number of steps (where the polynomial is taken over the length of $x$) for every $x$ and $\pi$.

At this point, it is important to note that a proof system allows us to prove language membership, but it does not allow us to prove nonmembership, meaning that $x \notin L$ cannot be proven directly. Obviously, one can verify language membership for every $x \in \{0,1\}^*$ individually and take a negative result that no $x$ is in $L$ as a naïve proof, but such a proof is exponentially large and outside the scope of an efficient or NP proof system.

This is where the work of Goldwasser and Micali—together with Charles Rackoff—comes into play [1, 2]. After the discovery of public key cryptography in the 1970s, this was the next major breakthrough in modern cryptography.[1] In fact, they added two ingredients to efficient or NP proof systems:

- Randomness and the possibility of make errors;
- Interaction.

The first point means that a proof may be erroneous, and hence that one must tolerate errors, whereas the second point means that there needs to be another entity for the verifier to interact with. This entity is called the *prover*, denoted as $P$, and there is a protocol to be executed between $P$ and $V$. The aim of this protocol is to convince $V$ about the truth of what is claimed by $P$.

Contrary to the proofs discussed so far, the work of Goldwasser, Micali, and Rackoff suggests that one can prove language membership or nonmembership simultaneously. To illustrate this point, we consider the QRP (Definition A.31) in $\mathbb{Z}_n^*$, and assume a prover $P$ that is able to solve this problem (because it knows the prime factorization of $n$ or can otherwise "magically" solve it). For any $x \in \mathbb{Z}_n^*$, $P$ can prove membership in $QR_n$ by computing a square root $w \in \mathbb{Z}_n^*$ with $w^2 \equiv x \pmod{n}$ and providing $w$ as a witness. More interestingly, $P$ can also prove nonmembership of $x$ in $QR_n$ using a protocol (executed with $V$) that operates in rounds. In each round, $V$ randomly selects a bit $b \in_r \{0,1\}$ and $y \in_R \mathbb{Z}_n^*$, computes $z \equiv x^b y^2 \pmod{n}$, and challenges $P$ with $z$. If $b = 0$, then $z = y^2 \bmod n$ and hence $z \in QR_n$. Otherwise (i.e., if $b = 1$), then $z = xy^2 \bmod n$ and hence $z \in QR_n$ if and only if $x \in QR_n$. This means that $x \notin QR_n$ implies that $z \notin QR_n$. Against this background, $P$ responds to the challenge $z$ with a bit $b'$ that is defined as follows:

$$b' = \begin{cases} 0 & \text{if } z \in QR_n \\ 1 & \text{if } z \notin QR_n \end{cases}$$

$V$ can verify $b'$ by comparing it with $b$. If the two bits are equal (i.e., $b' = b$), then $P$'s response is correct.

- If $b = 0$, then $z \equiv y^2 \pmod{n}$ implies $z \in QR_n$ (see above). This suggests that $b' = 0$, and hence $b = b'$.

1 In 1993, Goldwasser, Micali, and Rackoff—together with László Babai and Shlomo Moran—won the prestigious Gödel Prize for this work and the respective development of interactive proof systems.

- If $b = 1$, then $z \equiv xy^2 \pmod{n}$ implies $z \in QR_n$ if and only if $x \in QR_n$. This, in turn, means that non-membership of $x$ in $QR_n$ suggests nonmembership of $z$ in $QR_n$ (and hence $b = b'$).

In responding correctly in multiple rounds, $P$ can convince $V$ about the nonmembership of $x$ in $QR_n$. If the confidence level is sufficiently large, then the result can be taken as a proof.

The point to notice and keep in mind is that this type of proof is probabilistic and interactive in nature. It is conceptually similar to a factual proof we may use in daily life. Consider, for example, the situation in which somebody (acting as prover) claims that he or she is able to distinguish two balls that look the same. Another person (acting as verifier) may take the balls in the left and right hand and hide them behind his or her back, where they are either swapped or left as they are. Again showing the balls to the prover, he or she must tell whether the balls are swapped or not. If the prover is right, then the test is repeated multiple times, until the verifier is convinced that the prover is able to tell the balls apart. This is exactly the protocol sketched above in a mathematical setting. In each round, the prover can guess with a success probability of $1/2$. After two rounds, the success probability of correctly guessing is $1/4$, after three rounds it is $1/8$, and so on and so forth. After $k$ rounds, it is $1/2^k$, and hence the success probability of correctly guessing (or cheating) can be made arbitrarily small by increasing $k$.

Formally speaking, an *interactive proof system* is a pair $(P, V)$, where $P$ is an arbitrary function that refers to a prover, and $V$ is a PPT algorithm that refers to a verifier. Note that proof verification must be efficient here, whereas this is not required for $P$. If $P$ is also a PPT algorithm, then an interactive proof is called an *interactive argument*. So the distinction between an interactive proof and an interactive argument is that an argument can be generated efficiently, whereas this need not be the case for an interactive proof. In this chapter, we use the term interactive proof to sometimes also include interactive arguments. Using this terminology, we can define an *interactive proof system* as suggested in Definition 15.2.

**Definition 15.2 (Interactive Proof System)** *An* interactive proof system *for membership in L is a pair $(P, V)$ that consists of a function $P$ (representing a prover) and a PPT algorithm $V$ (representing a verifier), such that for all $x \in \{0,1\}^*$ the following two requirements are fulfilled:*

1. *If $x \in L$, then* $\Pr[(P, V)(x) = \text{YES}] \geq 2/3$.
2. *If $x \notin L$, then for all $P'$ it must hold that* $\Pr[(P', V)(x) = \text{YES}] \leq 1/3$.

Again, the first requirement refers to the completeness of the interactive proof system, whereas the second requirement refers to its soundness. The values $2/3$ and

$1/3$ used in the definition are somehow arbitrary and can technically be replaced with $1/2 + 1/p(|x|)$ (instead of $2/3$) and $1/2 - 1/p(|x|)$ (instead of $1/3$) for any polynomial $p(\cdot)$.

We now know what a proof system is and what an interactive proof system is, but we still have to explain the term *zero-knowledge*. Let us consider the situation in which there is a proof $\pi$ for $x \in QR_n$. As mentioned above, such a proof $\pi$ may provide $w$ with $w^2 \equiv x \pmod{n}$ as a witness. If the prime factorization of $n$ is unknown, then such a witness can be generated with a subexponential running time. This is not efficient, but so far we have not required the prover $P$ to be efficient (in fact we have only required the verifier $V$ to be efficient). Anyway, the witness can be verified efficiently by $V$. Having a closer look at this setting reveals the fact that $V$ learns something from $\pi$, namely a witness $w$ that stands for $x \in QR_n$. Prior to seeing $\pi$, $V$ did not know $w$ and was not able to find it (because it is limited to a PPT algorithm). This makes it obvious that $\pi$ must have leaked some information, and that $V$ was able to increase its knowledge accordingly. So this type of proof is clearly not zero-knowledge.

What we are looking for is a type of (interactive) proof that leaks no information or knowledge. Unfortunately, this property is difficult to define. One may, for example, be tempted to define it as "$V$ doesn't learn $w$," "$V$ doesn't learn any symbol of $w$," or "$V$ doesn't learn any information about $w$." All of these definitions have defects, and what we actually want to have is that $V$ doesn't learn anything at all beyond the fact that $x \in QR_n$. It goes without saying that this is a much stronger requirement.

Against this background, Goldwasser, Micali, and Rackoff proposed the *zero-knowledge* property for an interactive proof system. Informally speaking, an interactive proof system has this property, if whatever $V$ can compute when interacting with $P$ it can also compute without interaction. More formally, one can denote $V$'s view of a protocol execution with $P$ as $V(view)$. It basically includes $x$, all random values chosen by $V$, and all messages that are exchanged between $P$ and $V$. Having this notion of $V(view)$ in mind, we can say that a protocol leaks no information or knowledge, if for all possible input values $x$, $V(view)$ can be efficiently simulated without interacting with $P$. This means that there is an efficient algorithm $S$ (standing for a *simulator*) that can generate an output $S(x)$ that is indistinguishable from $V(view)$; that is, $S(x) \cong V(view)$. We can write $V(view)$ as $(P, V)(x)$ to get Definition 15.3.

**Definition 15.3 (Zero-Knowledge)** *An interactive proof system* $(P, V)$ *for* $L$ *is* zero-knowledge *if there exists a PPT algorithm* $S$*, such that for all* $x \in L$ *the relation* $S(x) \cong (P, V)(x)$ *holds.*

This simulation property or paradigm is key to zero-knowledge. Note that it allows us to define zero-knowledge without even defining what knowledge is. This is somewhat astonishing.

Depending on the notion of the $\cong$, there are various flavors of zero-knowledge. Most importantly, if $S(x) \cong (P, V)(x)$ means that $S(x)$ and $(P, V)(x)$ are computationally indistinguishable, meaning that no efficient algorithm can tell them apart. In this case, the resulting interactive proof system is *computationally zero-knowledge*, but for the purpose of this book, we often leave the word "computationally" aside. To be complete, we note that an interactive proof system has *perfect zero-knowledge* if $S(x)$ and $(P, V)(x)$ are the same, and it has *statistical zero-knowledge* if $S(x)$ and $(P, V)(x)$ are not the same, but they are statistically close, meaning that their statistical difference is negligible. We leave these subtleties aside here.

## 15.2 ZERO-KNOWLEDGE AUTHENTICATION PROTOCOLS

Because an interactive proof system yields some form of challenge-response mechanism, it can also be used for entity authentication. In this case, the zero-knowledge property of a protocol is particularly useful because it ensures that the protocol leaks no information about the (secret) authentication information in use no matter how often it is executed. In the sequel, we outline three exemplary zero-knowledge authentication protocols (that are conceptually similar and can be seen as instantiations of a single unifying protocol [3]). In all examples, we assume a mechanism that allows the verifier $V$ to learn the prover $P$'s public key in some certified and authenticated form (Section 16.4).

### 15.2.1 Fiat-Shamir

Soon after Goldwasser, Micali, and Rackoff had introduced the notion of a zero-knowledge proof, Amos Fiat and Shamir found a way to implement a zero-knowledge protocol for authentication [4]. They also found a way to use the protocol in a noninteractive setting, and the result yields a DSS (Section 15.3). Similar to the Rabin public key cryptosystem (Sections 13.3.2 and 14.2.3), the Fiat-Shamir protocol takes its security from the fact that computing square roots and factoring a modulus are computationally equivalent—that is, a square root modulo $n$ can be efficiently computed if and only if the prime factorization of $n$ is known.

Let $p$ and $q$ be two primes, and $n$ their product (i.e., $n = pq$). As usual, the prover $P$ has a private key $x \in \mathbb{Z}_n^*$ and a respective public key $y = x^2 \bmod n$. It keeps its private key $(n, x)$ secret and provides $V$ with the public key $(n, y)$. The Fiat-Shamir authentication protocol operates in rounds, where each round looks as

**Table 15.1**
A Round in the Fiat-Shamir Authentication Protocol

| P | | V |
|---|---|---|
| $(n, x)$ | | $(n, y)$ |
| $r \xleftarrow{r} \mathbb{Z}_n^*$ | | |
| $t \equiv r^2 \pmod{n}$ | $\xrightarrow{t}$ | |
| | $\xleftarrow{c}$ | $c \xleftarrow{r} \{0, 1\}$ |
| $s \equiv rx^c \pmod{n}$ | $\xrightarrow{s}$ | |
| | | $s^2 \stackrel{?}{\equiv} ty^c \pmod{n}$ |
| | | (*accept* or *reject*) |

illustrated in Table 15.1. $P$ randomly selects $r \in_R \mathbb{Z}_n^*$, computes $t \equiv r^2 \pmod{n}$, and sends this value to $V$. $V$, in turn, randomly selects a bit $c \in_R \{0, 1\}$ and uses it to challenge $P$.

- If $c = 0$, then $P$ must respond with $s = r$ and $V$ must verify $s^2 \equiv ty^0 \equiv t \pmod{n}$.
- If $c = 1$, then $P$ must respond with $s = (rx) \bmod n$ and $V$ must verify $s^2 \equiv ty \pmod{n}$.

Depending on the outcome of the verification step, the authentication is accepted or rejected. The protocol is complete, because

$$s^2 \equiv r^2(x^c)^2 \equiv t(x^2)^c \equiv ty^c \pmod{n}$$

In order to show that the system is sound, one must look at the adversary and ask what he or she can do in each round. Obviously, the adversary can randomly select a $t \in_R \mathbb{Z}_n^*$, wait for $V$ to provide a challenge $c \in_R \{0, 1\}$, and then simply guess a value for $s$. The success probability of such an attack is negligible (for a reasonably sized $n$). There are, however, more subtle attacks to consider. If, for example, the adversary was able to predict the challenge $c$, then he or she could prepare himself or herself to provide the correct response.

- If $c = 0$, then the protocol can be executed as normal; that is, the adversary can randomly select $r$ and send $t \equiv r^2 \pmod{n}$ and $s = r$ to $V$.
- If $c = 1$, then the adversary can randomly select $s \in_R \mathbb{Z}_n^*$, compute $t \equiv s^2/y \pmod{n}$, and send these values to $V$ in the appropriate protocol steps.

In either case, it is not possible for the adversary to prepare himself or herself for both cases (otherwise, he or she could also extract the private key $x$). If, for example, the adversary was able to prepare $s_0$ for $c = 0$ (i.e., $s_0 = r$), and $s_1$ for $c = 1$ (i.e., $s_1 \equiv rx \pmod{n}$), then he or she could extract $P$'s private key $x = s_1/s_0$.

Because the adversary can predict the challenge $c$ with a probability of $1/2$ and prepare himself or herself accordingly, the cheating probability is also $1/2$ in each round. This suggests that the protocol must be executed in multiple rounds (until an acceptably small cheating probability is achieved). If, for example, the protocol is repeated $k$ times (where $k$ is a security parameter), then the cheating probability is $1/2^k$. This value decreases exponentially and can be made arbitrarily small.

The Fiat-Shamir authentication protocol has the zero-knowledge property, because a dishonest verifier $V'$ can use an efficient program $S$ to simulate the protocol and compute transcripts and respective triples $(t, c, s)$ that are indistinguishable from real triples. Let us consider a toy example to illustrate this point: If $p = 3$, $q = 5$, $n = pq = 15$, $x = 7$, $y \equiv 7^2 \pmod{15} = 4$ and $y^{-1} \bmod 15 = 4$ (note that $4 \cdot 4 = 16 \equiv 1 \pmod{15}$), then $S$ can assume $c = 0$, randomly select $r = 2$, and compute $t \equiv 2^2 \pmod{15} = 4$ and $s = 2$. This yields a first triple $(4, 0, 2)$ that is computationally indistinguishable from a real protocol transcript. It is valid, because $2^2 \equiv 4 \pmod{15}$. Next, $S$ can assume $c = 1$, randomly select $s = 3$, and compute $t \equiv 3^2 \cdot 4 \pmod{15} = 6$ to generate a second triple $(6, 1, 3)$. Again, it is valid because $3^2 \equiv 6 \cdot 4 \pmod{15}$. Next, $S$ can assume $c = 1$, randomly select $s = 7$, and compute $t \equiv 7^2 \cdot 4 \pmod{15} = 1$. The result is a third triple $(1, 1, 7)$ that is valid, because $7^2 \equiv 1 \cdot 4 \pmod{15}$. Finally, $S$ can assume $c = 0$, randomly select $r = 8$, and compute $t \equiv 8^2 (\bmod 15) = 4$ and $s = 8$. This yields a fourth triple $(4, 0, 8)$ that is valid because $8^2 \equiv 4 \pmod{15}$. Needless to say, we can have $S$ generate as many triples as we like, and there is no need for $S$ to interact with $P$ in the first place. In fact, the property that valid triples can be generated without any interaction with $P$ makes the protocol zero-knowledge.

The Fiat-Shamir protocol is conceptually simple, but it is not very efficient (because the success probability for an adversary is divided only by two in each round). Consequently, there are several variants of the Fiat-Shamir protocol that speed things up using some form of parallelization.

- As already suggested in [4], one can execute $k$ rounds in parallel by replacing $t$, $c$, and $s$ with respective vectors of $k$ values each. This variant is very efficient, but it has the disadvantage that it is not known how to construct an efficient simulator, and hence it cannot be shown to be zero-knowledge in a strict sense.

- Another variant that supports some form of parallelization was proposed in [5]. The idea is that $P$ has $z$ public key pairs $(x_1, y_1), \ldots, (x_z, y_z)$ instead of only one. As before, the first message is $t = r^2 \bmod n$, but the second message now consists of $z$ challenge bits $c_1, \ldots, c_z$. This also means that $P$ must respond with $s \equiv r \sum_{i=1}^{z} x_i^{c_i} \pmod{n}$. In some sense, all $z$ challenges are mixed into one single response value. Instead of $1/2$, the cheating probability in this variant is $1/2^z$, and this means that the protocol can be executed with fewer rounds.

Other variants are described in the literature. The authentication protocol addressed next can also be seen as a generalization of the Fiat-Shamir protocol.

### 15.2.2 Guillou-Quisquater

Only two years after the publication of Fiat and Shamir, Louis C. Guillou and Jean-Jacques Quisquater proposed another variant that is more efficient [6]. Instead of working with squares and binary challenges, this protocol works with e-th powers (where $e$ is prime) and challenges between 0 and $e - 1$ (instead of 0 or 1). The security of the resulting protocol is based on the RSA problem; that is, computing e-th roots modulo $n$ without knowing the prime factorization of $n$ or $\phi(n)$.

**Table 15.2**

A Round in the Guillou-Quisquater Authentication Protocol

| P | | V |
|---|---|---|
| $(n, x)$ | | $(n, y)$ |
| $r \xleftarrow{r} \mathbb{Z}_n^*$ | | |
| $t \equiv r^e \pmod{n}$ | $\xrightarrow{t}$ | |
| | $\xleftarrow{c}$ | $c \xleftarrow{r} \{0, \ldots, e-1\}$ |
| $s \equiv rx^c \pmod{n}$ | $\xrightarrow{s}$ | |
| | | $s^e \stackrel{?}{\equiv} ty^c \pmod{n}$ |
| | | $(accept$ or $reject)$ |

Again, the Guillou-Quisquater authentication protocol operates in rounds, where each round looks as illustrated in Table 15.2. In each round, $P$ proves that it knows the private key $x$ that refers to the public key $y \equiv x^e \pmod{n}$. $P$ therefore randomly selects an $r \in_R \mathbb{Z}_n^*$, computes $t \equiv r^e \pmod{n}$, and sends this value as a commitment to $V$. $V$, in turn, randomly selects a number between 0 and $e - 1$ and challenges $P$ with this value. $P$ computes $s \equiv rx^c \pmod{n}$ and sends this response to $V$. Finally, $V$ must verify $s^e \equiv ty^c \pmod{n}$, and the result indicates whether the round is accepted or rejected.

The protocol is complete, because

$$s^e \equiv (rx^c)^e \equiv r^e(x^e)^c \equiv ty^c \pmod{n}$$

It is sound because an attacker can either guess $s$ or prepare himself or herself for one challenge $c$ before committing to $t$. In the first case, the success probability (i.e., the probability to correctly guess $s$) is negligible for any reasonably sized $n$. In the second case, the adversary guesses $c$, randomly selects $s$, computes $t \equiv s^e y^{-c} \pmod{n}$, and uses this value as a commitment. If $V$ really used $c$ as a challenge, then the adversary could respond with $s$. $V$, in turn, would verify $ty^c \equiv s^e y^{-c} y^c \equiv s^e \pmod{n}$ and accept the proof as valid. Of course, if the guess was wrong and $V$ provided another challenge, then the adversary could only try to correctly guess $s$. Again, this can done with only a negligible success probability.

One may wonder whether an adversary can prepare himself or herself for different challenges to improve his or her success probability. Let us assume that an adversary can prepare himself or herself for two challenges $c_1$ and $c_2$ and sends a respective commitment $t$ to $V$. In this case, the following pair of equivalences must hold:

$$s_1^e \equiv ty^{c_1} \pmod{n}$$
$$s_2^e \equiv ty^{c_2} \pmod{n}$$

Dividing the two equivalences yields $(s_1/s_2)^e \equiv y^{c_1-c_2} \equiv (x^{c_1-c_2})^e \pmod{n}$, and hence $s_1/s_2 \equiv x^{c_1-c_2} \pmod{n}$. Because $gcd(e, c_1 - c_2) = 1$ (note that $e$ is prime), the adversary can use the Euclid extended algorithm to compute $u$ and $v$ with $u(c_1 - c_2) + ve = 1$. He or she can then compute $P$'s private key $x$ as follows:

$$(s_1/s_2)^u y^v \equiv x^{u(c_1-c_2)} x^{ve} \equiv x \pmod{n}$$

Consequently, if an adversary was able to prepare himself or herself for at least two challenges, then he or she could extract e-th roots without knowing the group order $\phi(n)$. Against this background, it is reasonable to assume that the adversary can prepare himself or herself for one challenge at most, and hence the success probability for cheating is $1/e$ per round. If $e$ is sufficiently large, then success probability for cheating in $k$ rounds is $1/e^k$. Again, this probability can be made arbitrarily small by increasing $k$.

The Fiat-Shamir and Guillou-Quisquater protocols are based on the difficulty of computing square or e-th roots in $\mathbb{Z}_n^*$ with unknown prime factorization and order. There are also protocols that are based on the DLP. In such a protocol, $P$ is to prove that it knows the discrete logarithm of a public key.

### 15.2.3 Schnorr

In Section 14.2.5, we introduced a DLP-based DSS that is due to Schnorr. In [7, 8], Schnorr also proposed a DLP-based authentication protocol that is in line with the Fiat-Shamir and Guillou-Quisquater protocols. It is assumed that a large prime $p$ and a generator $g$ of $\mathbb{Z}_p^*$ are known (they can be either system parameters or part of the public key pairs), and that $P$ has a private key $x$ and a respective public key $y \equiv g^x \pmod{p}$.

**Table 15.3**
A Round in the Schnorr Authentication Protocol

| P | | V |
|---|---|---|
| $(p, g, x)$ | | $(p, g, y)$ |
| $r \xleftarrow{r} \mathbb{Z}_p^*$ | | |
| $t \equiv g^r \pmod{p}$ | $\xrightarrow{t}$ | |
| | $\xleftarrow{c}$ | $c \xleftarrow{r} \{0, \ldots, 2^k - 1\}$ |
| $s = r + cx \pmod{p-1}$ | $\xrightarrow{s}$ | |
| | | $g^s \stackrel{?}{\equiv} ty^c \pmod{p}$ |
| | | (*accept* or *reject*) |

As usual, the Schnorr authentication protocol operates in multiple rounds, where a round is illustrated in Table 15.3. $P$ randomly selects $r \in_R \mathbb{Z}_p^*$, computes $t \equiv g^r \pmod{p}$, and sends this value to $V$. $V$, in turn, randomly selects a $k$-bit value $c$ (where $k$ is again a security parameter) and challenges $P$ with this value. Finally, $P$ must respond with $s \equiv r + cx \pmod{p-1}$, and $V$ must verify $g^s \equiv ty^c \pmod{p}$.

The protocol is complete, because

$$g^s \equiv g^r g^{cx} \equiv ty^c \pmod{p}$$

It is sound because the adversary can either guess or prepare himself or herself for one challenge. If the adversary was able to prepare himself or herself for two challenges $c_1$ and $c_2$, then he or she could determine $P$'s private key $x$. From $g^{s_1} \equiv ty^{c_1} \pmod{p}$ and $g^{s_2} \equiv ty^{c_2} \pmod{p}$, it follows that $g^{s_1 - s_2} \equiv y^{c_1 - c_2} \equiv g^{x \cdot (c_1 - c_2)} \pmod{p}$, and hence $x \equiv (s_1 - s_2) \cdot (c_1 - c_2)^{-1} \pmod{p-1}$.

## 15.3 NONINTERACTIVE ZERO-KNOWLEDGE

The major advantage of a zero-knowledge proof or protocol is due to the fact that one can mathematically show that the verifier learns nothing more than the correctness of the statement that is proven. This is particularly useful in the realm of authentication protocols. The major disadvantage, however, is that a zero-knowledge proof or protocol is interactive by default, meaning that messages need to be sent back and forth (between the prover and the verifier). There are many application settings in which this level of interaction is neither possible nor welcome. Consequently, people have been looking for possibilities to prove statements in zero-knowledge without requiring any form of interaction. This leads to the notion of a noninteractive zero-knowledge proof, in which a single message is sent from the prover to the verifier. In some sense, the resulting proofs can be seen as the probabilistic analog of a conventional (noninteractive) proof.

After having introduced the notion of zero-knowledge in general, and the Fiat-Shamir authentication protocol in particular, it became clear that the latter can be turned into a DSS (that is also noninteractive by default). The trick [4] is to replace the challenge $c$ that is randomly selected by the verifier with a hash value $c = h(m,t)$ that takes into account the message $m$ and the commitment $t$. The pair $(t,s)$ then yields a digital signature for $m$. The resulting Fiat-Shamir DSS is summarized in Table 15.4. It speaks for itself.

**Table 15.4**
Fiat-Shamir DSS

System parameters: —

| Generate | Sign | Verify |
|---|---|---|
| $(1^l)$ | $((n,x),m)$ | $((n,y),m,(t,s))$ |
| $p,q \xleftarrow{r} \mathbb{P}_{l/2}$ | $r \xleftarrow{r} \mathbb{Z}_n^*$ | $b = (s^2 \equiv ty^c \pmod{n})$ |
| $n = p \cdot q$ | $t \equiv r^2 \pmod{n}$ | |
| $x \xleftarrow{r} \mathbb{Z}_n^*$ | $c = h(m,t)$ | |
| $y \equiv x^2 \pmod{n}$ | $s \equiv rx^c \pmod{n}$ | |
| $((n,x),(n,y))$ | $(t,s)$ | $(b)$ |

The Fiat-Shamir DSS is secure against CMA as long as $h$ is a random oracle (meaning that the proof is in the random oracle model). Obviously, the construction does not require interaction and is noninteractive. Because the zero-knowledge property still applies, it provides a possibility to implement noninteractive zero-knowledge. The construction is known as the *Fiat-Shamir heuristic*—it is generally applicable and can be used to turn an interactive proof of knowledge into a DSS. Consequently, similar DSS can be constructed from the Guillou-Quisquater protocol, and—as outlined in Section 14.2.5 and Table 14.6—from the Schnorr protocol.

More generally, noninteractive zero-knowledge proofs are zero-knowledge proofs in which no interaction is necessary between the prover and the verifier. This means that a single message is sent from the prover to the verifier, and this structure matches the needs of many cryptographic protocols and has therefore many use cases and applications. In 1988, Blum, Paul Feldman, and Micali showed that a common reference string generated by a trusted party and accessible to both the prover and the verifier is sufficient to achieve computational zero-knowledge without interaction [9].[2] Their model is commonly called the *common reference string* model, and it is used in many noninteractive zero-knowledge proofs.

In the early 1990s, Oded Goldreich and Yair Oren proved that noninteractive zero-knowledge is impossible to achieve in the standard model; that is, without any further assumption, like the common reference string model or the random oracle model [11], and in the early 2000s, Goldwasser and Yael Tauman Kalai even published an instance of the Fiat-Shamir protocol for which any concrete hash function yields an insecure DSS [12]. At first glance, these results seem to contradict [9, 10] and [4], but this is not the case. Note that the impossibility results neither hold in the common reference string model (used in [9, 10]) nor in the random oracle model (used in [4]). But noninteractive zero-knowledge seems to draw a clear line between what can be achieved in the standard model and what can be achieved in more powerful models, like the common reference string or random oracle model. Without such a model, noninteractive zero-knowledge remains impossible to achieve. Needless to say, people have been exploring other models for noninteractive zero-knowledge. This is actually an ongoing research effort.

## 15.4 FINAL REMARKS

The notion of zero-knowledge was invented in the 1980s. For the first three decades, it was a theoretically stimulating research topic but was not used in the field. Even the zero-knowledge authentication protocols that looked very promising in the beginning (especially for smartcard implementations) were not widely deployed. This is unfortunate, because the zero-knowledge property of a proof or protocol is advantageous from a security viewpoint. It means, for example, that an authentication

2 There is also a journal version of the paper that appeared in 1991 [10].

protocol can be executed arbitrarily many times without having to change the underlying authentication information (since the protocol execution does not leak any information or knowledge about it). The biggest problem one faces with these types of protocols is the high degree of interaction, which means that many messages must be sent back and forth between the prover and the verifier. This limits the performance one can achieve in practice.

Against this background, zero-knowledge has experienced a strong revival in its noninteractive form in the past decade. As a matter of fact, people have been using the Fiat-Shamir heuristic to come up with the notion of a SNARK, an acronym standing for a succinct noninteractive argument of knowledge, or even a zk-SNARK, standing for a zero-knowledge SNARK [13]. Zk-SNARKs are currently used, for example, in a blockchain-based crypto currency called Zcash.[3] They have initiated an entirely new line of research that seeks to make noninteractive zero-knowledge proofs or arguments as efficient as possible (since they need to be stored and processed in blockchains). Examples of such techniques include Bulletproofs[4] and zk-STARKs.[5] Whenever one needs to be assured that some parties behave honestly in a cryptographic protocols, noninteractive zero-knowledge proofs or arguments may provide a viable solution.

## References


[1] Goldwasser, S., S. Micali, and C. Rackoff, "The Knowledge Complexity of Interactive Proof Systems," *Proceedings of the seventeenth annual ACM Symposium on Theory of Computing (STOC '85)*, ACM Press, 1985, pp. 291-304.

[2] Goldwasser, S., S. Micali, and C. Rackoff, "The Knowledge Complexity of Interactive Proof Systems," *SIAM Journal of Computing*, Vol. 18, No. 1, 1989, pp. 186–208.

[3] Maurer, U.M., "Unifying Zero-Knowledge Proofs of Knowledge," *Proceedings of AFRICACRYPT 2009*, Springer, LNCS 5580, 2009, pp. 272–286.

[4] Fiat, A., and A. Shamir, "How to Prove Yourself: Practical Solutions to Identification and Signature Problems," *Proceedings of CRYPTO '86*, Springer, LNCS 263, 1987, pp. 186–194.

[5] Feige, U., and A. Shamir, "Zero Knowledge Proofs of Knowledge in Two Rounds," *Proceedings of CRYPTO '89*, Springer-Verlag, LNCS 435, 1989, pp. 526–544.

[6] Guillou, L.C., and J.J. Quisquater, "A Practical Zero-Knowledge Protocol Fitted to Security Microprocessor Minimizing Both Transmission and Memory," *Proceedings of EUROCRYPT '88*, Springer-Verlag, LNCS 330, 1988, pp. 123–128.

[7] Schnorr, C.P., "Efficient Identification and Signatures for Smart Cards," *Proceedings of CRYPTO '89*, Springer-Verlag, 1989, pp. 239–251.


3 https://z.cash.

4 http://web.stanford.edu/~buenz/pubs/bulletproofs.pdf.

5 https://eprint.iacr.org/2018/046.pdf.

[8] Schnorr, C.P., "Efficient Signature Generation by Smart Cards," *Journal of Cryptology*, Vol. 4, 1991, pp. 161–174.

[9] Blum, M., P. Feldman, and S. Micali, "Non-Interactive Zero-Knowledge and Its Applications," *Proceedings of the Twentieth Annual ACM Symposium on Theory of Computing (STOC '88)*, ACM Press, pp. 103–112.

[10] Blum, M., et al., "Noninteractive Zero-Knowledge," *SIAM Journal of Computing*, Vol. 20, No. 6, 1991, pp. 1084–1118.

[11] Goldreich, O., and Y. Oren, "Definitions and Properties of Zero-Knowledge Proof Systems," *Journal of Cryptology*, Vol. 7, No. 1, 1994, pp. 1–32.

[12] Goldwasser, S., and Y.T. Kalai, "On the (In)security of the Fiat-Shamir Paradigm," *Proceedings of the 44th Annual IEEE Symposium on Foundations of Computer Science (FOCS '03)*, IEEE, 2003, pp. pp. 102–113.

[13] Bitansky, N., et al., "From Extractable Collision Resistance to Succinct Non-Interactive Arguments of Knowledge, and Back Again," *Proceedings of the 3rd Innovations in Theoretical Computer Science Conference (ITCS 2012)*, ACM Press, New York, 2012, pp. 326–349.

# Part IV

# CONCLUSIONS

# Chapter 16

# Key Management

In this chapter, we discuss some aspects related to key management. More specifically, we introduce the topic and elaborate on the life cycle of a cryptographic key in Section 16.1, address secret sharing, key recovery, and certificate management (also known as PKI) in Sections 16.2–16.4, and conclude with some final remarks in Section 16.5. This chapter briefly addresses all relevant topics related to key management, but it does not delve into the details (a more thorough and comprehensive treatment of these topics is beyond the scope of this book).

## 16.1 INTRODUCTION

According to [1], the term *key management* refers to "the process of handling and controlling cryptographic keys and related material (such as initialization values) during their life cycle in a cryptographic system, including ordering, generating, distributing, storing, loading, escrowing, archiving, auditing, and destroying the material." This process is so complex that it represents the Achilles' heel of almost all systems that employs cryptography and cryptographic techniques in one way or another (we already made this point in Section 2.3.1). There are (at least) two conclusions to draw:

- First, if one is in charge of designing a security system, then one has to get the key management process right. Otherwise, there is no use in employing cryptography and cryptographic techniques in the first place.
- Second, if one is in charge of breaking a security system, then one should also start with the key management process. Most successful attacks against cryptographic systems are exploits of vulnerabilities or weaknesses related to key management.

Consequently, the key management process is by far the most important part of a security system that employs cryptography and cryptographic techniques. This is equally true for the cryptographer who designs the system and the cryptanalyst who may want to break it. Because the key management process is so comprehensive and complex, there is usually no single standard to refer to. Instead, there are many standards that address specific questions and problems related to key management. Some of these standards are overviewed, discussed, and put into perspective in [2].

According the definition given above, the life cycle of a cryptographic key includes many tasks, including "ordering, generating, distributing, storing, loading, escrowing, archiving, auditing, and destroying" keying material. The more important tasks, namely key generation, distribution, storage, and destruction are briefly addressed next, whereas key escrow is discussed in Section 16.3.

### 16.1.1 Key Generation

Unless one is in the realm of unkeyed cryptosystems, the use of a cryptographic system always requires the generation of cryptographic keys and related material (e.g., IVs) in one way or another. The generation of this material, in turn, requires the usc of a random generator as addressed in Chapter 3. Either the random generator is used directly to generate the cryptographic keys or—more preferably—the random bit generator is used to seed a PRG (Chapter 7). In either case, it is important to know and properly understand the possible realizations and implementations of random generators and PRGs.

### 16.1.2 Key Distribution

Ideally, the cryptographic keys are used where they are generated, and hence the distribution of the cryptographic keys does not represent a problem. In all other cases, however, the distribution of the keys is a problem and must be considered with care. In fact, it must be ensured that the keys cannot be attacked passively or actively during their distribution. This is an important and challenging engineering task. Some key establishment techniques have been discussed in Chapter 12. Many other key distribution protocols and systems have been developed, proposed, implemented, and partly deployed in the field (e.g., [3]). Again, there are many subtle details that must be considered and addressed with care.

### 16.1.3 Key Storage

Unless a cryptographic key is ephemeral (short-lived), it is typically used for a long period of time (i.e., between its generation and destruction). In this entire period, the key must be securely stored, meaning that it must be stored in a way that cannot be attacked passively or actively. Again, this is an important and challenging engineering task where many things can go wrong. Compared to the key distribution problem, the key storage problem is theoretically and practically even more challenging. One reason is that the storage of a key can only be considered in the context of a specific operating system. So the key storage and operating system security problems are not independent from each other, and the first problem depends on the second. Unfortunately, we know that the security of currently deployed operating systems is not particularly good, and hence there are many low-level details that must be considered when one wants to provide a (secure) solution for key storage.

If there is no single place to store a key, then one may consider using a secret sharing scheme as addressed in the following section to store the key in a decentralized and distributed way. As of this writing, however, these schemes are not as widely deployed as one would expect considering their theoretical importance and usefulness.

### 16.1.4 Key Destruction

At the end of its life cycle, a cryptographic key may be archived and must be destructed. Due to its electronic nature, the destruction of a key is not as simple as it might seem at first glance. There are two reasons:

- First, it is technically difficult to destroy data that is stored electronically. In practice, one usually requires to wipe the memory location by overwriting it multiple times with random bit patterns.
- Second, there may be (many) temporary copies of the key held in memory. This means that a memory dump with a subsequent analysis may reveal all keys that have been used recently.

Note that the feasibility of recovering electronically stored data was, for example, demonstrated by the cold boot attack mentioned in Section 1.2.2.2. Also note that the question whether and to what extent keys can be securely destroyed mainly depends on the operating system in use.

## 16.2 SECRET SHARING

As already mentioned, there are situations in which it may be useful to split a secret value (e.g., a cryptographic key) into multiple parts and have different parties hold and manage these parts. If, for example, one wants to have $n$ parties share a secret value $s$, then one can randomly choose $n-1$ values $s_1, \ldots, s_{n-1}$, compute

$$s_n = s \oplus s_1 \oplus \ldots \oplus s_{n-1}$$

and distribute $s_1, \ldots, s_n$ to the $n$ parties. The secret value $s$ can then only be recovered if all $n$ parties contribute their parts. Consequently, such a *secret splitting system* requires that all parties are available and reliable, and that they all behave honestly in one way or another. If only one party is not available, loses its part, or refuses to contribute it, then the secret value $s$ can no longer be recovered. Needless to say, this is a major drawback of a secret splitting system that severely limits its usefulness in the field.

In 1979, Shamir [4] and George Blakley [5] independently came up with the notion of *secret sharing* as a viable alternative to secret splitting. In a *secret sharing system*, it is generally not required that all parties are available and reliable, and that they all behave honestly. Instead, the reconstruction of the secret value $s$ requires only the parts of a well-defined subset of all parts (in this case, the parts are called shares). More specifically, a secret sharing system allows an entity, called the dealer, to share a secret value $s$ among a set $P$ of $n$ players; that is, $P = \{P_1, \ldots, P_n\}$, such that only a qualified subset of $P$ can reconstruct $s$ from their shares. It is usually required that all nonqualified subsets of the players get absolutely no information about $s$ (as mentioned later, the secret sharing system is then called perfect). The secret and the shares are usually elements of the same domain, most often a finite field.

Formally, the set of qualified subsets is a subset of the power set $2^P$ and is called the *access structure* $\Gamma$ of the secret sharing system. If, for example, $\Gamma = \{\{P_1, \ldots, P_n\}\}$ (i.e., only all players are qualified), then the secret sharing system is a secret splitting system as described earlier. More generally, a *k-out-of-n secret sharing scheme* can be defined as suggested in Definition 16.1.

**Definition 16.1 (K-out-of-n secret sharing system)** *A* k-out-of-n secret sharing system *is a secret sharing system in which the access structure is*

$$\Gamma = \{M \subseteq 2^P : |M| \geq k\}$$

A k-out-of-n secret sharing system is *perfect* if $k-1$ players who collaborate (and pool their shares) are not able to recover $s$ or retrieve any information about $s$. As mentioned above, there are two proposals of systems that fulfill this requirement.

### 16.2.1 Shamir's System

Shamir's k-out-of-n secret sharing system [4] is based on polynomial interpolation. More specifically, the system employs the fact that that a polynomial $f(x)$ of degree $k-1$ (over a field) can be uniquely interpolated from $k$ points. This means that a polynomial of degree 1 can be interpolated from 2 points, a polynomial of degree 2 can be interpolated from 3 points, and so on. The corresponding interpolation algorithm has been around for a long time. It is usually attributed to Lagrange. Let

$$f(x) = r_0 + r_1 x + \ldots + r_{k-1} x^{k-1} = \sum_{i=0}^{k-1} r_i x^i \tag{16.1}$$

be a polynomial of degree $k-1$ that passes through the $k$ points

$$\begin{array}{c} (x_1, f(x_1) = y_1) \\ (x_2, f(x_2) = y_2) \\ \ldots \\ (x_k, f(x_k) = y_k) \end{array}$$

The Lagrange interpolating polynomial $P(x)$ is then given by

$$P(x) = \sum_{i=1}^{k} P_i(x)$$

where

$$P_i(x) = y_i \prod_{j=1;j\neq i}^{k} \frac{x - x_j}{x_i - x_j}$$

Written explicitly,

$$\begin{aligned} P(x) &= P_1(x) + P_2(x) + \ldots + P_k(x) \\ &= y_1 \frac{(x-x_2)(x-x_3)\cdots(x-x_k)}{(x_1-x_2)(x_1-x_3)\cdots(x_1-x_k)} \\ &\quad + y_2 \frac{(x-x_1)(x-x_3)\cdots(x-x_k)}{(x_2-x_1)(x_2-x_3)\cdots(x_2-x_k)} \\ &\quad + \ldots \\ &\quad + y_k \frac{(x-x_1)(x-x_2)\cdots(x-x_{k-1})}{(x_k-x_1)(x_k-x_2)\cdots(x_k-x_{k-1})} \end{aligned}$$

In Shamir's k-out-of-n secret sharing system, the secret (to be shared) represents the coefficient $r_0$. The dealer randomly selects $k-1$ coefficients $r_1, \ldots, r_{k-1}$ to define a polynomial according to (16.1). For every player $P_i$, the dealer then assigns a fixed nonzero field element $x_i$ and computes $y_i = f(x_i)$. The pair $(x_i, f(x_i))$ then yields $P_i$'s share.

Anybody who is given $k$ shares can compute the secret $r_0$ by evaluating the Lagrange interpolating polynomial at point zero; that is, $s = r_0 = P(0)$. Anybody who is given fewer than $k$ shares cannot compute the secret. More precisely, anybody who is given fewer than $k$ shares does not obtain any (partial) information about the secret. This means that Shamir's k-out-of-n secret sharing system is perfect.

### 16.2.2 Blakley's System

Independent from Shamir's work, Blakley proposed a secret sharing system that is geometric in nature [5]. In Blakley's system, the secret is a point in a $k$-dimensional space. The $n$ shares are constructed with each share defining an affine hyperplane in this space; an affine hyperplane, in turn, is the set of solutions $x = (x_1, \ldots, x_k)$ to an equation of the form

$$a_1x_1 + \ldots + a_kx_k = b$$

By finding the intersection of any $k$ of these planes, the secret (i.e., the point of intersection) can be obtained. Note that this system is not perfect, as a person with a share of the secret knows the secret is a point on his or her hyperplane (this is more than a person without a share). Nevertheless, it is fair to say that the system can be modified to also achieve perfect security (e.g., [6]).

### 16.2.3 Verifiable Secret Sharing

K-out-of-n secret sharing systems are interesting from a theoretical viewpoint. This is particularly true if the systems are perfect. From a more practical viewpoint, however, there are at least two problems that must be addressed and considered with care:

- If a malicious player is not honest and provides a wrong share, then the secret that is reconstructed is also wrong.
- If the dealer is malicious or untrusted, then the players may want to have a guarantee that they can in fact put together the correct secret.

A *verifiable secret sharing system* can be used to overcome these problems. For example, Shamir's k-out-of-n secret sharing system can be made verifiable by having the dealer make commitments to the coefficients of the polynomial $f(x)$ and providing the players with help-shares they can use to verify shares. We don't delve deeper into verifiable secret sharing in this book. Note, however, that verifiable secret sharing systems play a central role in many cryptographic systems and applications, such as electronic cash or electronic voting.

### 16.2.4 Visual Cryptography

In 1994, Shamir and Moni Naor developed and proposed a visual variant of secret sharing that is commonly referred to as visual cryptography [7]. In short, visual cryptography can be used to share a secret picture among $n$ participants. The picture is divided into $n$ transparencies (that play the role of the shares) such that if any $k$ transparencies are placed together, the picture becomes visible, but if fewer than $k$ transparencies are placed together, nothing can be seen at all. Such a visual cryptosystem can be constructed by viewing the secret picture as a set of black and white pixels and handling each pixel individually. Most visual cryptosystems proposed in the literature are perfectly secure and can be implemented easily without any cryptographic computation. A further improvement allows each transparency to represent an innocent-looking picture, such as a picture of a landscape or a picture of a flower), thus concealing the fact that secret sharing is taking place. Consequently, steganographic techniques are sometimes used in combination with visual cryptography.

## 16.3 KEY RECOVERY

If one uses cryptographic techniques for data encryption, then one may also be concerned about the fact that (encryption and decryption) keys get lost. What happens, for example, if all data of a company are securely encrypted and the decryption key is lost? How can the company recover its data? The same questions occur if only the data of a specific employee are encrypted. What happens if the corresponding decryption key gets lost? What happens if the employee himself or herself gets lost? It is obvious that a professional use of cryptography and cryptographic techniques for data encryption must take into account a way to recover cryptographic keys.

According to [1], the term *key recovery* refers to "a process for learning the value of a cryptographic key that was previously used to perform some cryptographic operation" [1]. Alternatively, one may also use the term to refer to "techniques that provide an intentional, alternate (i.e., secondary) means to access the key used for data confidentiality service." There are basically two classes of key recovery techniques:

- *Key escrow* is "a key recovery technique for storing knowledge of a cryptographic key or parts thereof in the custody of one or more third parties called escrow agents, so that the key can be recovered and used in specified circumstances" [1]. In this context, escrow agents are sometimes also called trusted third parties (TTPs).
- *Key encapsulation* is "a key recovery technique for storing knowledge of a cryptographic key by encrypting it with another key and ensuring that only certain third parties called *recovery agents* can perform the decryption operation to retrieve the stored key. Key encapsulation typically allows direct retrieval of the secret key used to provide data confidentiality" [1]. Key encapsulation has sometimes been used in security protocols that don't have key recovery as their primary goal (e.g., [8–10]).

The basic principles of key escrow and key encapsulation are illustrated in Figure 16.1. In key escrow, the cryptographically protected data is sent from A to B, whereas the key recovery data is sent to a TTP. In key encapsulation, either data is sent directly from A to B. Another way to look at things is to say that key escrow refers to out-band key recovery, whereas key encapsulation refers to in-band key recovery. These terms, however, are less frequently used in the literature.

Key recovery in general, and key escrow in particular, became hotly debated research topics in the mid-1990s [11], and the discussion was even more intensified when the U.S. government published the *Escrowed Encryption Standard* (EES) [12] and released an implementation of it in the so-called Clipper chip. The EES was basically a secret splitting system with two governmental bodies acting as escrow agents. This was the major problem of the EES. People were concerned about the possibility of having the government illegitimately decrypting their communications—notably without any restriction in time. Also, it was argued that key escrow on transmitted data is neither necessary nor particularly useful (because either end of the communication can provide the data in unencrypted form). The controversy about the EES and the Clipper chip suddenly came to an end when it was shown that the original design of the EES was flawed [13].[1] The flaw was an authentication field that was too short to provide protection against a brute-force attack.

In 1997, a group of recognized cryptographers wrote a highly influential paper about the risks related to key recovery, key escrow, and TTP encryption [15]. Today, the U.S. export controls are relaxed, but state-controlled cryptography remains an issue. Most of the same group of cryptographers therefore wrote a follow-up paper

1 You may also refer to [14] for the entire story about the EES, the Clipper chip, and the crypto debate.

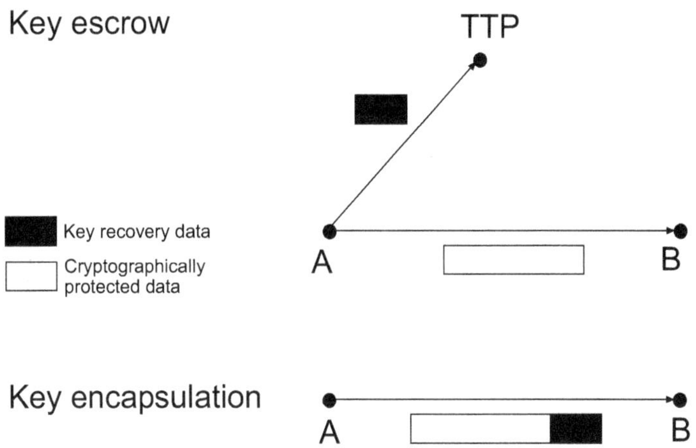


**Figure 16.1** Key escrow and key encapsulation.

in 2015 [16]. Both papers can be used as starting point to discuss the crypto wars that have been going on for decades.

## 16.4 CERTIFICATE MANAGEMENT

Most cryptographic technologies and protocols in use today employ public key cryptography and public key certificates. The management of these certificates is an involved topic that is briefly addressed here. We introduce the topic in Section 16.4.1, elaborate on X.509 certificates and OpenPGP certificates in Sections 16.4.2 and 16.4.3, and briefly address the state of the art in Section 16.4.4.

### 16.4.1 Introduction

According to [1], the term *certificate* refers to "a document that attests to the truth of something or the ownership of something." This definition is fairly broad and applies to many subject areas, not necessarily related to cryptography or even public key cryptography. In this particular area, the term certificate was coined and first used by Loren M. Kohnfelder [17] to refer to a digitally signed record holding a name and a public key. As such, it was positioned as a replacement for a public file[2] that had been used before. A respective certificate is to attest to the legitimate ownership of a public key and to attribute the key to a particular entity, such as a person, a hardware device, or anything else. Quite naturally, such a certificate is called a *public key certificate*. Such certificates are used by many cryptographic security technologies and protocols in use today in one way or another. Again referring to [1], a public key certificate is a special case of a certificate, namely one "that binds a system entity's identity to a public key value, and possibly to additional data items." As such, it is a digitally signed data structure that attests to the true ownership of a particular public key.

More generally (but still in accordance with [1]), a certificate can not only be used to attest to the legitimate ownership of a public key (as in the case of a public key certificate), but also to attest to the truth of some arbitrary property that could be attributed to the certificate owner. This more general class of certificates is commonly referred to as *attribute certificates*. The major difference between a public key and an attribute certificate is that the former includes a public key (i.e., the public key that is certified) whereas the latter includes a list of attributes (i.e., the attributes that are certified). In either case, the certificates are issued (and possibly revoked) by authorities that are recognized and trusted by a community of users.

- In the case of public key certificates, the authorities in charge are called *certification authorities* (CAs[3]) or—more related to digital signature legislation—*certification service providers* (CSPs);
- In the case of attribute certificates, the authorities in charge are called *attribute authorities* (AAs).

It goes without saying that a CA and an AA may be the same organization. As soon as attribute certificates start to take off, it is possible and very likely that CAs will also try to establish themselves as AAs. It also goes without saying that a CA can have one or several *registration authorities* (RAs)—sometimes also called *local registration authorities* or *local registration agents* (LRAs). The functions an RA carries out vary from case to case, but they typically include the registration and authentication of the entities (typically human users) that want to become certificate owners. In addition, the RA may also be involved in tasks like token distribution, certificate revocation reporting, key generation, and key archival. In fact, a CA can delegate some of its tasks (apart from certificate signing) to an RA. Consequently,

2 A public file was just a flat file that included the public keys and names of the key owners in any particular order (e.g., sorted alphabetically with regard to the names of the key owners). The entire file could be digitally signed if needed.

3 In the past, CAs were often called trusted third parties (TTPs). This is particularly true for CAs that are operated by government bodies.

RAs are optional components that are transparent to the users. Also, the certificates that are generated by the CAs may be made available in online directories and certificate repositories.

While the notion of a CA is well defined and sufficiently precise, the notion of a *public key infrastructure* (PKI) is more vague. According to [1], a PKI is "a system of CAs that perform some set of certificate management, archive management, key management, and token management functions for a community of users," that employ public key cryptography (as one may be tempted to add here). Another way to look at a PKI is as an infrastructure that can be used to issue, validate, and revoke public keys and public key certificates. Hence, a PKI comprises a set of agreed-upon standards, CAs, structures among multiple CAs, methods to discover and validate certification paths, operational and management protocols, interoperable tools, and supporting legislation.

In the past, PKIs have experienced a great deal of hype, and many companies and organizations have started to provide certification services on a commercial basis. Unfortunately (and for the reasons discussed in [18]), most of these service providers have failed to become commercially successful. In fact, the PKI business has turned out to be particularly difficult to make a living from, and there are only a few CAs that are self-feeding. Most CAs that are still in business also have other sources of revenue.

Many standardization bodies are working in the field of public key certificates and the management thereof. Most importantly, the Telecommunication Standardization Sector of the International Telecommunication Union (ITU-T) has released and is periodically updating a recommendation that is commonly referred to as ITU-T X.509 [19], or X.509 in short. The respective certificates are addressed in Section 16.4.2. Meanwhile, ITU-T X.509 has also been adopted by many other standardization bodies, including the International Organization for Standardization (ISO) and the International Electrotechnical Committee (IEC) Joint Technical Committee 1 (JTC1) [20]. Furthermore, a few other standardization bodies also work in the field of profiling ITU-T X.509 for specific application environments.[4] In 1995, for example, the IETF recognized the importance of public key certificates for Internet security, and chartered an IETF Public-Key Infrastructure X.509 (PKIX[5]) WG to develop Internet standards for an X.509-based PKI. The PKIX WG initiated and stimulated a lot of standardization and profiling activities within the IETF, and was closely aligned with the activities of the ITU-T. In spite of the practical importance of the specifications of the IETF PKIX WG, we do not delve deeper into the details

4 To profile ITU-T X.509—or any general standard or recommendation—basically means to fix the details with regard to a specific application environment. The result is a profile that elaborates on how to use and deploy ITU-T X.509 in the environment.

5 http://www.ietf.org/html.charters/pkix-charter.html.

in this book (as this is a topic for a book on its own). The IETF PKIX WG was concluded in 2013, almost 20 years after it was chartered.[6]

As mentioned before, a public key certificate comprises at least the following three main pieces of information:

- A public key;
- Some naming information;
- One or more digital signatures.

The *public key* is the raison d'être for the public key certificate, meaning that the certificate only exists to certify the public key in the first place. The public key, in turn, can be from any public key cryptosystem, like RSA, Elgamal, Diffie-Hellman, DSA, or anything else. The format (and hence also the size) of the public key depends on the system in use.

The *naming information* is used to identify the owner of the public key and public key certificate. If the owner is a user, then the naming information typically consists of at least the user's first name and surname—also known as the family name. In the past, there has been some discussions about the namespace that can be used here. For example, the ITU-T recommendation X.500 introduced the notion of a *distinguished name* (DN) that can be used to identify entities, such as public key certificate owners, in a globally unique namespace. However, since then, X.500 DNs have not really taken off, at least not in the realm of naming persons. In this realm, the availability and appropriateness of globally unique namespaces have been challenged in the research community (e.g., [21]). In fact, the Simple Distributed Security Infrastructure (SDSI) initiative and architecture [22] has started from the argument that a globally unique namespace is not appropriate for the global Internet, and that logically linked local namespaces are simpler and therefore more likely to be deployed (this point is further explored in [23]). As such, work on SDSI inspired the establishment of a Simple Public Key Infrastructure (SPKI) WG within the IETF Security Area. The WG was chartered in 1997 to produce a certificate infrastructure and operating procedure to meet the needs of the Internet community for trust management in a way that was as easy, simple, and extensible as possible. This was partly in contrast (and in competition) to the IETF PKIX WG. The IETF SPKI WG published a pair of experimental RFCs [24, 25] before its activities were abandoned in 2001.[7] Consequently, the SDSI and SPKI initiatives have turned out to be dead ends for the Internet as a whole. They barely play a role in today's discussions about

6 To be precise, the IETF PKIX WG was chartered on October 26, 1995, and it was concluded on October 31, 2013. It was therefore active for slightly more than 18 years.

7 The WG was formally concluded in February 2001, only four years after it was chartered.

the management of public key certificates. But the underlying argument that globally unique namespaces are not easily available remains valid.

Last but not least, the *digital signature(s)* is (are) used to attest to the fact that the other two pieces of information (i.e., the public key and the naming information) belong together. The digital signature(s) turn(s) the public key certificate into a data structure that is useful in practice, mainly because it can be verified by anybody who knows the signatory's (i.e., CA's) public key. These keys are normally distributed with particular software, be it at the operating system or application software level.

As of this writing, there are two types of public key certificates that are practically relevant and in use: *X.509* and *OpenPGP certificates.* While their aims and scope are somewhat similar, they use different certificate formats and trust models. A *trust model*, in turn, refers to the set of rules that a system or application uses to decide whether a certificate is valid. In the direct trust model, for example, a user trusts a public key certificate only because he or she knows where it came from and considers this entity to be trustworthy. In addition to the direct trust model, there is a hierarchical trust model, as employed, for example, by ITU-T X.509, and a cumulative trust model, as employed, for example, by OpenPGP. These trust models can also be called *centralized* and *distributed.* It then becomes clear that there is hardly anything in between. Hence, coming up with alternatives to the direct, hierarchical, and cumulative trust models is somewhat challenging.

### 16.4.2 X.509 Certificates

As mentioned before (and as their name suggests), X.509 certificates conform to the ITU-T recommendation X.509 [19] first published in 1988 as part of the X.500 directory series of recommendations. It specifies both a certificate format and a certificate distribution scheme (while the specification language used was ASN.1). The original X.509 certificate format has gone through two major revisions:

- In 1993, the X.509 version 1 (X.509 v1) format was extended to incorporate two new fields, resulting in the X.509 version 2 (X.509 v2) format.
- In 1996, the X.509 v2 format was revised to allow for additional extension fields. This was in response to the attempt to deploy certificates on the global Internet. The resulting X.509 version 3 (X.509 v3) specification has since then been reaffirmed every couple of years.

When people today refer to X.509 certificates, they essentially refer to X.509 v3 certificates (and the version denominator is often left aside in the acronym). Let us now have a closer look at the X.509 certificate format and the hierarchical trust model it is based on.

#### 16.4.2.1 Certificate Format

With regard to the use of X.509 certificates, the profiling activities within the IETF PKIX WG are particularly important. Among the many RFC documents produced by this WG, RFC 5280 [26] is the most relevant one (with some RFC documents that yield some updates on particular topics. Without delving into the details of the respective ASN.1 specification for X.509 certificates, we note that an X.509 certificate is a data structure that basically consists of the following fields (remember that any additional extension fields are possible):[8]

- *Version:* This field is used to specify the X.509 version in use (i.e., version 1, 2, or 3).
- *Serial number:* This field is used to specify a serial number for the certificate. The serial number is a unique integer value assigned by the (certificate) issuer. The pair consisting of the issuer and the serial number must be unique—otherwise, it would not be possible to uniquely identify an X.509 certificate.
- *Algorithm ID:* This field is used to specify the object identifier (OID) of the algorithm that is used to digitally sign the certificate. For example, the OID 1.2.840.113549.1.1.5 refers to `sha1RSA`, which stands for the combined use of SHA-1 with RSA encryption. Many other OIDs are specified in respective standards.
- *Issuer:* This field is used to name the issuer. As such, it comprises the DN of the CA that issues (and digitally signs) the certificate.
- *Validity:* This field is used to specify a validity period for the certificate. The period, in turn, is defined by two dates, namely a start date (i.e., Not Before) and an expiration date (i.e., Not After).
- *Subject:* This field is used to name the subject (i.e., the owner of the certificate, typically using a DN).
- *Subject Public Key Info:* This field is used to specify the public key (together with the algorithm) that is certified.

8 From an educational viewpoint, it is best to compare the field descriptions with the contents of real certificates. If you run a Windows operating system, then you may look at some certificates by running the certificate snap-in for the management console (just enter certmgr on a command line interpreter). The window that pops up summarizes all certificates that are available at the operating system level.

- *Issuer Unique Identifier:* This field can be used to specify some optional information related to the issuer of the certificate (only in X.509 versions 2 and 3).
- *Subject Unique Identifier:* This field can be used to specify some optional information related to the subject (only in X.509 versions 2 and 3). This field typically comprises some alternative naming information, such as an e-mail address or a DNS entry.
- *Extensions:* This field can be used to specify some optional extensions that may be critical or not (only in X.509 version 3). While critical extensions need to be considered by all applications that employ the certificate, noncritical extensions are truly optional and can be considered at will. With regard to secure messaging on the Internet, the most important extensions are "Key Usage" and "Basic Constraints."

  - The *key usage extension* uses a bit mask to define the purpose of the certificate (i.e., whether it is used for normal digital signatures (0), legally binding signatures providing nonrepudiation (1), key encryption (2), data encryption (3), key agreement (4), digital signatures for certificates (5) or certificate revocation lists (CRLs) addressed below (6), encryption only (7), or decryption only (8)). The numbers in parentheses refer to the respective bit positions in the mask.
  - The *basic constraints extension* identifies whether the subject of the certificate is a CA and the maximum depth of valid certification paths that include this certificate. This extension should not appear in a leaf (or end entity) certificate.

  Furthermore, there is an *Extended Key Usage* extension that can be used to indicate one or more purposes for which the certified public key may be used, in addition to or in place of the basic purposes indicated in the key usage extension field.

The last three fields make X.509v3 certificates very flexible, but also very difficult to deploy in an interoperable manner. Anyway, the certificate must come along with a digital signature that conforms to the digital signature algorithm specified in the Algorithm ID field.

A distinguishing feature of an X.509 certificate is that there is one single piece of naming information, namely the content of the subject field, that is bound to a public key, and that there is one single signature that vouches for this binding. This is different in the case of an OpenPGP certificate. In such a certificate, there can be multiple pieces of naming information bound to a particular public key, and there can even be multiple signatures that vouch for this binding.

#### 16.4.2.2 Hierarchical Trust Model

X.509 certificates are based on the hierarchical trust model that is built on a hierarchy of (commonly) trusted CAs. As illustrated in Figure 16.2, such a hierarchy consists of a set of *root CAs* that form up the top level and that must be trusted by default. The respective certificates are self-signed, meaning that the issuer and subject fields refer to the same entity (typically an organization). Note that from a theoretical point of view, a self-signed certificate is not particularly useful. Anybody can claim something and issue a certificate for this claim. Consequently, a self-signed certificate basically says: "Here is my public key, trust me." There is no argument that speaks in favor of this claim. However, to bootstrap hierarchical trust, one or several root CAs with self-signed certificates are unavoidable (because the hierarchy is finite and must have a top level).

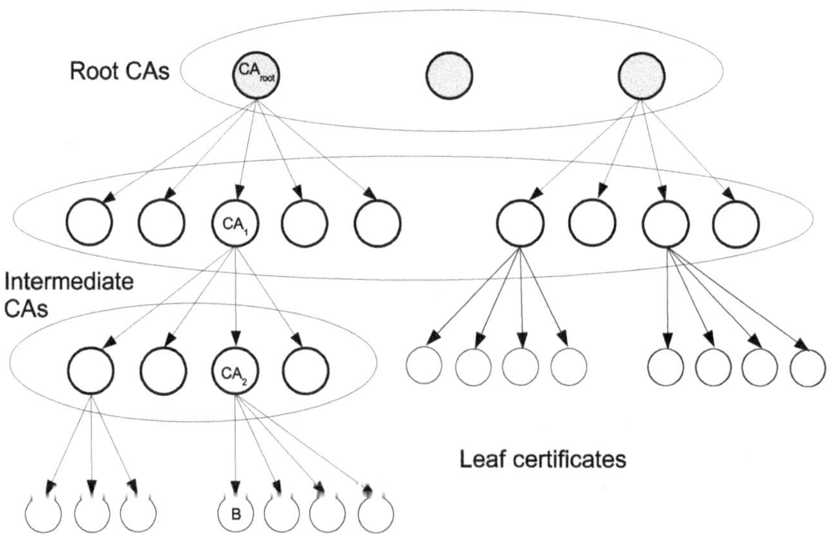


**Figure 16.2** A hierarchy of trusted root and intermediate CAs that issue leaf certificates.

In Figure 16.2, the set of root CAs consists of only three CAs (the three shadowed CAs at the top of the figure). In reality, we are talking about several dozens of root CAs that come preconfigured in a client software—be it an operating system or application software. Each root CA may issue certificates for other CAs that are called *intermediate CAs*. The intermediate CAs may form up multiple layers in the hierarchy. At the bottom of the hierarchy, the intermediate CAs may issue certificates for end users or other entities, such as Web servers. These certificates are called *leaf certificates* and they cannot be used to issue other certificates. This, by the way, is controlled by the basic constraints extension mentioned earlier. In a typical setting, a commercial CSP operates a CA that represents a trusted root CA and several subordinate CAs that may represent intermediate CAs. Note, however, that it is up to the client software to make a distinction between these types of CAs—either type is considered to be trustworthy.

Equipped with one or several root CAs and respective root certificates, a user may try to find a *certification path*—or *certification chain*—from one of the root certificates to a leaf certificate. Formally speaking, a certification path or chain is defined in a tree or wood of CAs (root CAs and intermediate CAs), and refers to a sequence of one or more certificates that leads from a trusted root certificate to a leaf certificate. Each certificate certifies the public key of its successor. Finally, the leaf certificate is typically issued for a person or end system. Let us assume that $CA_{root}$ is a root certificate and $B$ is an entity for which a certificate must be verified. In this case, a certification path or chain with $n$ intermediate CAs (i.e., $CA_1, CA_2, \ldots,$ $CA_n$) may look as follows:

$$
\begin{aligned}
&CA_{root} \ll CA_1 \gg \\
&CA_1 \ll CA_2 \gg \\
&CA_2 \ll CA_3 \gg \\
&\ldots \\
&CA_{n-1} \ll CA_n \gg \\
&CA_n \ll B \gg
\end{aligned}
$$

In Figure 16.2, a certification path with 2 intermediate CAs is illustrated. The path consists of $CA_{root} \ll CA_1 \gg$, $CA_1 \ll CA_2 \gg$, and $CA_2 \ll B \gg$. If a client supports intermediate CAs, then it may be sufficient to find a sequence of certificates that lead from a trusted intermediate CA's certificate to the leaf certificate. This may shorten certification chains considerably. In our example, it may be the case that $CA_2$ represents a (trusted) intermediate CA. In this case, the leaf certificate $CA_2 \ll B \gg$ would be sufficient to verify the legitimacy of B's public key.

The simplest model one may think of is a certification hierarchy representing a tree with a single root CA. In practice, however, more general structures are possible, using multiple root CAs, intermediate CAs, and CAs that issue cross certificates. In such a general structure, a certification path may not be unique and multiple certification paths may exist. In such a situation, it is required to have authentication metrics in place that allow one to handle multiple certification paths. The design and analysis of such metrics is an interesting and challenging research topic not further addressed in this book (you may refer to [27] for a respective introduction and overview).

As mentioned above, each X.509 certificate has a validity period, meaning that it is well-defined when the certificate is supposedly valid. However, in spite of this information, it may still be possible that a certificate needs to be revoked ahead of time. For example, it may be the case that a user's private key gets compromised or a CA goes out of business. For situations like these, it is necessary to address certificate revocation in one way or another. The simplest way is to have the CA periodically issue a *certificate revocation list* (CRL). A CRL is basically a blacklist that enumerates all certificates (by their serial numbers) that have been revoked so far or since the issuance of the last CRL in the case of a delta CRL. In either case, CRLs can be tremendously large and impractical to handle. Due to the CRLs' practical disadvantages, the trend goes to retrieving online status information about the validity of a certificate. The protocol of choice to retrieve this information is the Online Certificate Status Protocol (OCSP) [28] that has problems of its own. There are a few alternative or complementary technologies, such as Google's *Certificate Transparency*[9] or technologies that employ DNS, such as DNS Certification Authority Authorization (CAA) or DNS-based Authentication of Named Entities (DANE). The bottom line is that certificate revocation remains a challenging issue (e.g., [29]), and that many application clients that employ public key certificates either do not care about it or handle it incompletely or even improperly.

In spite of the fact that we characterize the trust model employed by ITU-T X.509 as being hierarchical, it is not so in a strict sense. The possibility to define cross-certificates, as well as forward and reverse certificates, enables the construction of a mesh (rather than a hierarchy). This means that something similar to PGP's web of trust can also be established using X.509. The misunderstanding partly occurs because the X.509 trust model is mapped to the directory information tree (DIT), which is hierarchical in nature (each DN represents a leaf in the DIT). Hence, the hierarchical structure is a result of the naming scheme rather than the certificate format. This should be kept in mind when arguing about trust models.

9 https://www.certificate-transparency.org.

### 16.4.3 OpenPGP Certificates

We already mentioned that an OpenPGP certificate is similar to an X.509 certificate, but that it uses a different format. The most important difference is that an OpenPGP certificate may have multiple pieces of naming information (user IDs) and multiple signatures that vouch for them. Hence, an OpenPGP certificate is inherently more general and flexible than an X.509 certificate. Also, OpenPGP employs e-mail addresses (instead of DNs) as primary naming information.

Let us first look at the OpenPGP certificate format before we more thoroughly address the cumulative trust model that is used in the realm of OpenPGP and OpenPGP certificates.

#### 16.4.3.1 Certificate Format

Like an X.509 certificate, an OpenPGP certificate is a data structure that binds some naming information to a public key.

- The naming information consists of one or several user IDs, where each user ID includes a user name and an e-mail address put in angle brackets (< and >). The e-mail address basically makes the user ID unique. An exemplary user ID is Rolf Oppliger <rolf.oppliger@esecurity.ch>.
- The public key is the key that is certified by the certificate. It is a binary string that is complemented by a fingerprint, a key identifier (key ID), an algorithm name (i.e., RSA, Diffie-Hellman, or DSA), and a respective key length. A fingerprint represents an SHA-1 hash value of the public key (and some auxiliary data), whereas the key ID refers to the least significant 64 (or 32) bits of the fingerprint.

In addition to the naming information and public key, an OpenPGP certificate may also comprise many other fields (depending on the implementation). The following fields are commonly used:

- *Version number:* This field is used to identify the version of OpenPGP. The current version is 4. Version 3 is deprecated.
- *Creation and expiration dates:* These fields determine the validity period (or lifetime) of the public key and certificate. In fact, it is valid from the creation date to the expiration date. In many cases, the expiration date is not specified, meaning that the respective certificate does not expire by default. Again, this is a difference between X.509 and OpenPGP certificates. While X.509

certificates typically expire after a few years, OpenPGP certificates typically don't expire at all (unless an expiration date is specified).

- *Self-signature:* This field is used to hold a self-signature for the certificate. As its name suggests, a self-signature is generated by the certificate owner using the private key that corresponds to the public key associated with the certificate. Note that X.509 certificates normally do not include self-signatures—except for root CA certificates.
- *Preferred encryption algorithm:* This field is used to identify the encryption algorithm of choice for the certificate owner.

One may think of an OpenPGP certificate as a public key with one or more labels attached to it. For example, several user IDs may be attached to it. Also, one or several photographs may be attached to an OpenPGP certificate to simplify visual authentication. Note that this is a feature that is not known to exist in the realm of X.509 certificates. Also note that the use of photographs in certificates is controversially discussed within the security community. While some people argue that it simplifies user authentication, others argue that it is dangerous because certificates that come along with a photograph only look trustworthy (whereas in fact they may not be trustworthy at all, or at least not more trustworthy than any certificate without a photograph). Hence, there are implementations that support the attachment of photographs, and there are implementations that don't. In either case, it is possible to bring in arguments that speak in favor of the respective choice. Therefore, it is a matter of taste whether one wants to use photographs or not.

#### 16.4.3.2 Cumulative Trust Model

The hierarchical trust model of X.509 starts from central CAs that are assumed to be commonly trusted. Contrary to that, the cumulative trust model negates the existence of such CAs, and starts from the assumption that there is no central CA that is trusted by everybody. Instead, every user must decide for himself or herself who he or she is willing to trust. If a user trusts another user, then this other user may act as an *introducer* to him or her, meaning that any PGP certificate signed by him or her will be accepted by the user. It goes without saying that different users may have different introducers they trust and start from

In practice, things are more involved, mainly because there is no unique notion of trust and trust can come in different flavors (or degrees, respectively). PGP, for example, originally distinguished between *marginal* and *full* trust, and this distinction has been adapted by most OpenPGP implementations. The resulting trust model is cumulative in the sense that more than one introducer can vouch for the validity and trustworthiness of a particular certificate. The respective signatures are accumulated in the certificate, and the more people sign a certificate, the more likely it is going to be trusted (and hence accepted) by a third party. The resulting certification and trust infrastructure is distributed and called a *web of trust*.

In practice, the implementation and deployment of a web of trust is more involved than it may look at first sight. For example, certificate revocation is particularly challenging in a web of trust, mainly because there is no central authority that can be contacted in this matter. The cumulative trust model and the web of trust are seldom used in the field and have turned out be dead ends to some extent.

### 16.4.4 State of the Art

Since public key certificates represent the Achilles' heel of public key cryptography, the management of these certificates represents an important and practically relevant topic. A user who wants to send a confidential and cryptographically protected message to a recipient must have access to this recipient's public key. A valid certificate is one way to achieve this. Similarly, the recipient must have access to a valid certificate for the sender's public key if he or she wants to verify the signature of that message. If certificates can be faked, then any form of active attack becomes feasible and difficult to mitigate.

While the PKI industry has been partly successful in deploying server-side certificates, the client-side deployment of certificates has remained poor. This is equally true for hardware and software certificates.

- Hardware certificates refer to hardware devices or tokens that comprise public key pairs. Examples include smartcards or USB tokens. The relevant standards are PKCS #11 and PKCS #15. The question of whether the public key pairs should be generated inside or outside the hardware device or token is controversially discussed within the community.
  - In the first case, it can be ensured that no private key can leak the device or token, but the quality of the random number generator may be poor;
  - In the second case, the quality of the random number generator can be controlled, but it may be possible to export the keying material from the device or token (because the respective import function must be supported by default).
- Software certificates do not require hardware. Instead, the public key pairs are entirely stored in memory—hopefully in some encrypted form (while not being used).

It goes without saying that software certificates are generally more vulnerable and simpler to attack than hardware certificates. Using hardware certificates, one can reasonably argue that extracting private keying material is technically difficult. This is not true for software certificates. Here, the respective commands (to extract private keys) can be disabled by default, but it is very difficult to technically avoid an adversary who may find a way to extract a private key anyway. The bottom line is that for high-secure environments, hardware certificates are advantageous and should be the preferred choice (this applies to X.509 and OpenPGP certificates). However, the deployment of hardware certificates is more involved and expensive, and we hardly see such certificates deployed and used in the field.

Another problem that appears is that there are not many publicly available directories that can be used to retrieve user certificates. The main reason for this lack of directories is that organizations hesitate to make their information publicly available, mainly because they are afraid of people misusing it for spam and targeted headhunting. Hence, they keep this information internal, and this severely restricts its usefulness. Inside an organization, the situation is simpler, because there are usually possibilities to roll out user certificates at moderate costs.

## 16.5 FINAL REMARKS

In this chapter, we elaborated on key management (i.e., the process of handling and controlling cryptographic keys and related material during their life cycle in a cryptographic system). Key management is a very complex process, and it does not come as a surprise that it is the Achilles' heel of almost every system that employs cryptography and cryptographic techniques. The key life cycle includes many important phases, and we had a closer look at key generation, distribution, storage, and destruction.

If there are keys that are so valuable that there is no single entity that is trustworthy enough to serve as a key repository, then one may look into secret splitting schemes or—more importantly—secret sharing systems. In fact, secret sharing systems are likely to be more widely deployed in future systems. The same is true for key recovery. If data encryption techniques are implemented and widely used, then mechanisms and services for key recovery are valuable and in many situations unavoidable (especially in the business world). Following this line of argumentation, the first products that implement and make use of key recovery features already appeared on the marketplace a few years ago. For example, the commercial versions of PGP have support key recovery on a voluntary basis. This trend is likely to continue in the future. Last but not least, we briefly elaborated on digital certificates and PKIs. This is a very difficult topic, both from a theoretical and practical point of view.

## References


[1] Shirey, R., *Internet Security Glossary, Version 2*, RFC 4949 (FYI 36), August 2007.

[2] Dent, A.W., and C.J. Mitchell, *User's Guide to Cryptography and Standards*. Artech House Publishers, Norwood, MA, 2004.

[1] Boyd, C., A. Mathuria, and D. Stebila, *Protocols for Key Establishment and Authentication*, 2nd edition. Springer-Verlag, New York, 2019.

[4] Shamir, A., "How to Share a Secret," *Communications of the ACM*, Vol. 22, 1979, pp. 612–613.

[5] Blakley, G.R., "Safeguarding Cryptographic Keys," *AFIPS Conference Proceedings*, Vol. 48, 1979, pp. 313–317.

[6] Simmons, G.J., (ed.), *Contemporary Cryptology—The Science of Information Integrity*. IEEE Press, New York, 1992.

[7] Naor, M., and A. Shamir, "Visual Cryptography," *Proceedings of EUROCRYPT '94*, Springer-Verlag, 1994, pp. 1–12.

[8] Ioannidis, J., and M. Blaze, "The Architecture and Implementation of Network-Layer Security Under UNIX," *Proceedings of USENIX UNIX Security Symposium IV*, October 1993, pp. 29–39.

[9] Caronni, G., et al., "SKIP—Securing the Internet," *Proceedings of WET ICE '96*, Workshops on Enabling Technologies: Infrastructure for Collaborative Enterprises, June 1996, pp. 62–67.

[10] Oppliger, R., *Internet and Intranet Security*, 2nd edition. Artech House Publishers, Norwood, MA, 2002.

[11] Denning, D.E., and D.K. Branstad., "A Taxonomy for Key Escrow Encryption Systems," *Communications of the ACM*, Vol. 39, No. 3, March 1996, pp. 34–40.

[12] U.S. Department of Commerce, National Institute of Standards and Technology, *Escrowed Encryption Standard*, FIPS PUB 185, February 1994.

[13] Blaze, M., "Protocol Failure in the Escrowed Encryption Standard," *Proceedings of the 2nd ACM Conference on Computer and Communications Security*, Fairfax, VA, November 1994, pp. 59–67.

[14] Hoffman, L.J., (ed.), *Building in Big Brother: The Cryptographic Policy Debate*. Springer-Verlag, New York, 1995.

[15] Abelson, H., et al., "The Risks of Key Recovery, Key Escrow, and Trusted Third-Party Encryption," May 1997, https://www.schneier.com/academic/paperfiles/paper-key-escrow.pdf.

[16] Abelson, H., et al., "Keys Under Doormats: Mandating insecurity by requiring government access to all data and communications," July 2015, https://dspace.mit.edu/bitstream/handle/1721.1/97690/MIT-CSAIL-TR-2015-026.pdf.

[17] Kohnfelder, L.M., "Towards a Practical Public-Key Cryptosystem," Massachusetts Institute of Technology (MIT), Cambridge, MA, May 1978, http://groups.csail.mit.edu/cis/theses/kohnfelder-bs.pdf.

[18] Lopez, J., Oppliger, R., and G. Pernul, "Why Have Public Key Infrastructures Failed So Far?" *Internet Research*, Vol. 15, No. 5, 2005, pp. 544–556.

[19] ITU-T X.509, *Information technology–Open systems interconnection–The Directory: Public-key and attribute certificate frameworks*, 2012.

[20] ISO/IEC 9594-8, *Information technology–Open systems interconnection–The Directory: Public-key and attribute certificate frameworks*, 2001.

[21] Ellison, C., "Establishing Identity Without Certification Authorities," *Proceedings of the 6th USENIX Security Symposium*, 1996, pp. 67–76, http://static.usenix.org/publications/library/proceedings/sec96/ellison.html.

[22] Rivest, R.L., and B. Lampson, "SDSI—A Simple Distributed Security Infrastructure," September 1996, http://people.csail.mit.edu/rivest/sdsi10.html.

[23] Abadi, M., "On SDSI's Linked Local Name Spaces," *Journal of Computer Security*, Vol. 6, No. 1–2, September 1998, pp. 3–21.

[24] Ellison, C., "SPKI Requirements," RFC 2692, September 1999.

[25] Ellison, C., et al., "SPKI Certificate Theory," RFC 2693, September 1999.

[26] Cooper, D., et al., "Internet X.509 Public Key Infrastructure Certificate and Certificate Revocation List (CRL) Profile," RFC 5280, May 2008.

[27] Reiter, M.K., and S.G. Stubblebine, "Authentication Metric Analysis and Design," *ACM Transactions on Information and System Security*, Vol. 2, No. 2, May 1999, pp. 138–158.

[28] Myers, M., et al., "X.509 Internet Public Key Infrastructure Online Certificate Status Protocol—OCSP," RFC 2560, June 1999.

[29] Oppliger, R., "Certification Authorities under Attack: A Plea for Certificate Legitimation," *IEEE Internet Computing*, Vol. 18, No. 1, January/February 2014, pp. 40–47.

# Chapter 17

# Summary

In this book, we overviewed, discussed, and put into perspective many cryptographic systems in use today. In doing so, we made a distinction between unkeyed, secret key, and public key cryptosystems. We also noted that this distinction is somewhat arbitrary and that other classification schemes may work equally well. Because we think that the scheme is still useful and appropriate—especially for didactic purposes—we reuse it in this chapter to summarize the situation.

## 17.1 UNKEYED CRYPTOSYSTEMS

Unkeyed cryptosystems play a fundamental role in cryptography, and they are heavily used as building blocks in more advanced cryptographic systems and applications. In Part I of the book, we had a closer look at the representatives of unkeyed cryptosystems in terms of random generators, random functions, one-way functions, and cryptographic hash functions.

- Randomness is deeply intertwined with cryptography, and most cryptographic systems and applications employ random bits (or random numbers, respectively) in one way or another. Consequently, random generators play a fundamental and enabling role in cryptography. We saw that there are various types of hardware-based and software-based random generators, and that it is important to test the statistical randomness properties of the output of such a generator before it is used in the field. Many random generators have statistical deficiencies that are surprisingly simple to find and exploit, and the use of such generators has thus led to many cryptographic systems and applications that have failed to provide security in the past. In fact, the history of cryptography is full of such examples.

- In contrast to a random generator, a random function is not characterized by its output. Instead, it is a function that is randomly chosen from a huge set of possibilities. Many cryptographic systems try to simulate the characteristics of a random function, making it necessary for an adversary to go through all possibilities to break the security. It is therefore important to properly understand the notion of a random function and to be able to apply it in security proofs. Unlike many other cryptographic systems, a random function is a purely theoretical construct that is not meant to be implemented in practice. So we are not going to see entirely new and ingenious random functions be proposed in the literature or promoted on the market. This is inherently different from all other types of cryptosystems addressed in this book.
- One-way functions (and trapdoor functions) are at the core of modern cryptography. This may come as a surprise, especially if one considers the fact that no function has been shown to be one way in a mathematically strong sense, and that even the existence of one-way functions has not been proven so far. In fact, there are only a few candidate one-way functions (i.e., functions that are conjectured to be one way) in use today. Examples include the discrete exponentiation function, the RSA function, and the modular square function. The fact that it is currently not known how to efficiently invert these functions gives us a good feeling when we use these functions in public key cryptography and respective systems and applications. Unfortunately, we don't know how justified the feeling really is. If somebody found an algorithm to efficiently invert a candidate one-way function, then many deployed systems and applications would become totally insecure and useless. This is also what PQC is all about: If somebody were able to build a sufficiently large quantum computer[1] and use it to solve the RSAP or DLP efficiently, then again many deployed systems and applications would become insecure and useless, and we would need alternatives that remain secure.
- In many cryptographic systems and applications, cryptographic hash functions (i.e., hash functions that are one way and collision resistant) are used and play a fundamental role. This is particularly true for digital signatures with appendix and corresponding DSSs. If one can make the idealized assumption that a cryptographic hash function behaves like a random function, then one is often able to prove security properties for cryptographic systems that one is not able to prove without making this assumption (the corresponding proofs are then valid in the random oracle model). In spite of their fundamental role in cryptography, there are not many practically relevant cryptographic hash

1 Refer to Section D.5 to learn more about quantum computers.

functions to choose from. In fact, most cryptographic hash functions in use today follow the Merkle-Damgård construction (i.e., they iteratively apply a compression function to subsequent blocks of a message), and—maybe also surprisingly—there are only a few alternative designs.

For each of these representatives, it is important to define and properly understand the notion of security (i.e., what is meant by saying that such a cryptosystem is secure). In short, a random generator is considered to be secure if its output fulfills a well-defined set of statistical randomness tests; a random function is considered to be secure if it is randomly chosen from a really huge set of functions; a one-way function is considered to be secure if it is not known how to invert it efficiently; and finally a cryptographic hash function is considered to be secure if it is one way and collision-resistant. These security properties have been more precisely defined in the respective chapters of Part I of the book.

## 17.2 SECRET KEY CRYPTOSYSTEMS

Secret key cryptosystems, and in particular symmetric encryption systems, are the cryptographic systems one usually thinks about when people talk about cryptography. Some of these systems have a long tradition and have been used for a long period of time (e.g., to protect the secrecy of messages). In Part II of this book, we had a closer look at the representatives of secret key cryptosystems in terms of PRGs, PRFs, symmetric encryption, message authenticated, and AE.

- As the prefix "pseudo" suggests, a PRG tries to simulate a random generator in the sense that it its output is very similar and hence indistinguishable from the output of a random generator. In contrast to a random generator, a PRG has a relatively short input, called a seed, that is stretched into a potentially very long sequence of (pseudorandom) bits. Whenever random bits are needed, it is usually efficient to use a random generator to generate a seed, and then use a PRG to stretch this value into a sequence of pseudorandom bits that is as long as needed. As such, there are many applications and use cases for PRGs in the field.
- Contrary to PRGs, PRFs do not generate an output that meets specific (randomness) requirements. Instead, a PRF tries to model the input-output behavior of a random function. From a theoretical perspective, many other cryptographic systems can be seen as a PRF (or a PRP, respectively). Most importantly, a cryptographic hash function can be seen as PRF, and a block cipher can be seen as a PRP. Also, PRGs and PRFs are closely related to each other

in the sense that a PRF can be used to construct a PRG, and—vice versa—a PRG can be used to construct a PRF.

- Symmetric encryption has a long and thrilling history, and there are many attempts to design and come up with symmetric encryption systems that are secure. From a high level of abstraction, one usually distinguishes between block and stream ciphers that have distinct properties that make them appropriate for distinct applications and use cases. However, if one has a block cipher, then one can use a particular mode of operation to turn it into a stream cipher. The opposite direction is not possible, meaning that there is no mode of operation that can turn a stream cipher into a block cipher (and stream ciphers do not have modes of operation in the first place). In practice, there are many symmetric encryption systems that have been proposed and standardized by different key players and organizations.
- While symmetric encryption is to protect the confidentiality of messages, message authentication is to protect their authenticity and integrity—using a secret key (this distinguishes a MAC from a digital signature that is generated and verified with a public key pair). Message authentication is usually very efficient (compared to digital signatures), but it cannot be used to provide nonrepudiation. This is because both the sender and the recipient share a secret key that is needed to generate and verify a MAC. Note, however, that nonrepudiation is not always needed, and that in some application settings it is not even a desired property. The example that immediately comes to mind is off-the-record messaging.
- Finally, symmetric encryption and message authentication is currently typically combined in what is known as AE(AD). This type of encryption is resistant to many attacks and clearly represents the state of the art in cryptography, meaning that whenever data needs to be protected cryptographically, people should consider its use. In fact, there is hardly any reason not to use and take advantage of AE with or without additional data.

Again, for each of these representatives, it is important to define and properly understand the notion of security (i.e., what is meant by saying that such a cryptosystem is secure). In short, a PRG is considered to be secure if its output is (computationally) indistinguishable from the output of a true random generator, whereas a PRF is considered to be secure if it (computationally) indistinguishable from a random function. The security discussion of symmetric encryption and message authentication is more subtle because it has to distinguish between unconditional and conditional security. It was shown by Shannon in the late 1940s that a symmetric encryption system can only be unconditionally secure (and hence provide perfect secrecy), if the key is at least as long as the plaintext message. The one-time pad is the reference example here. Unfortunately, the key length requirement of an unconditionally secure symmetric encryption system is almost always prohibitively expensive in practice, so that most symmetric encryption systems in use today are "only" conditionally secure. As such, they can be broken theoretically by mounting an exhaustive key search. Consequently, it is important to make the key space so large that an exhaustive key search is not feasible. This is certainly the case if the key has a size of 100 bits or more. The same line of argumentation and distinction between unconditional and conditional security also apply to message authentication and respective MACs. All such constructions used in the field are "only" conditionally secure. Anyway, combining conditionally secure symmetric encryption with conditionally secure message authentication leads to AE(AD). Again, the notions of security for all of these representatives of secret key cryptosystems were more precisely defined in the respective chapters of Part II of the book.

## 17.3 PUBLIC KEY CRYPTOSYSTEMS

Public key cryptosystems have been developed since the late 1970s and are typically associated with modern cryptography. In fact, digital signatures and key establishment were the two major driving forces behind the invention and development of public key cryptography in general, and public key cryptosystems in particular. In Part III of this book, we had a closer look at the representatives of public key cryptosystems in terms of key establishment, asymmetric encryption, digital signatures, and zero-knowledge proofs of knowledge.

- Whenever a secret key cryptosystem is used, a respective key shared between the participating entities needs to be established in one way or another. This is where key establishment comes into play. There are many protocols for key establishment, but the Diffie-Hellman key exchange protocol is by far the most important one. This is astonishing because the protocol was the first public key cryptosystem ever published in the open literature, and in spite of its age, it is still in widespread use on the Internet.
- Asymmetric encryption serves the same purpose as symmetric encryption, but—due to its inefficiency compared to symmetric encryption—it is mostly used to protect the confidentiality of small messages that are transmitted, such as authentication information or secret keys (i.e., keys from a symmetric encryption system). Mathematically speaking, an asymmetric encryption system is based on one or several one-way functions, and its security therefore depends on the assumed intractability of these functions. Unless one employs

IBE, the use of an asymmetric encryption system requires the availability of digital certificates and a respective PKI. This is a topic of its own that is only very briefly addressed in this book.

- Many public key cryptosystems can be used as an asymmetric encryption system or a DSS. In fact, the possibility to digitally sign electronic documents and verify digital signatures is very powerful, and it is often argued that it is a prerequisite for the successful deployment of electronic commerce. This line of argumentation may be a little bit exaggerated, but digital signatures and DSSs clearly play a pivotal role in the provision of nonrepudiation services.
- Finally, zero-knowledge proofs of knowledge allow one to proof knowledge of something without leaking information about it. Although the idea may seem paradoxical at first glance, there are still many applications of such protocols. In the realm of entity authentication, for example, it allows one to proof knowledge of a secret (e.g., password) without leaking information about it. Normally, zero-knowledge proofs are highly interactive, but there are also variants that are noninteractive. These variants have interesting applications in the field, such as in the realm of blockchain and DLT.

Once again, for each of these representatives, it is important to define and properly understand the notion of security (i.e., what is meant by saying that such a cryptosystem is secure). In short, a key establishment protocol is considered to be secure if somebody knowing the transcript of a protocol execution is not able to determine the key that is established. Similar to symmetric encryption, asymmetric encryption may be attacked in many different ways. Most importantly, CPAs are trivial to mount because the public keys are publicly available by definition. Consequently, an asymmetric encryption system must protect against different attacks, including CPAs and CCAs, and there are many notions of security that can be distinguished. A similar line of argumentation and subtle security discussion applies to digital signatures and respective DSSs. Finally, a proof of knowledge—or more generally—a protocol is zero-knowledge, if a valid-looking transcript can be generated (or simulated) without interaction. Again, the notions of security for all of these representatives of public key cryptosystems were more precisely defined in the respective chapters of Part III of the book.

## 17.4 FINAL REMARKS

In practice, unkeyed, secret key, and public key cryptosystems are often combined to complement each other. For example, we saw that a random generator can be used to seed a PRG and that symmetric and asymmetric encryption are usually combined in hybrid encryption. In fact, public key cryptosystems are often used for authentication and key distribution, whereas secret key cryptosystems are often used for bulk data encryption and message authentication (if performance is an issue). Consequently, real applications typically combine all types of cryptosystems (including unkeyed cryptosystems) to come up with a design that can be implemented in an efficient and secure way.

It is sometimes argued that public key cryptography is inherently more secure than secret key cryptography. This argument is flawed; there are secure and insecure public key and secret key cryptosystems. If one has to decide what cryptosystem to use, then one has to look at the requirements from an application's point of view. If, for example, it is required that data can be authenticated efficiently, then a MAC is usually a good choice. If, however, it is required that the sender cannot later repudiate having sent a particular message, then a DSS is the more preferred choice. Consequently, there is no single best cryptosystem to be used for all purposes and applications. Instead, it is important to understand the working principles, advantages, and disadvantages, as well as the shortcomings and limitations of all practically relevant and deployed cryptosystems, and to design and implement a security architecture that is appropriate for the particular purpose and application one has in mind. This is not a simple task and should be dealt with professionally. There is usually much more to say than "data must be encrypted with a 128-bit key."

# Chapter 18

## Outlook

> *It would appear that we have reached the limits of what is possible to achieve with computer technology, although one should be careful with such statements, as they tend to sound pretty silly in five years.*
>
> — John von Neumann

After having overviewed, discussed, and put into perspective the state of the art in cryptography, it may be worthwhile to elaborate on possible or likely trends and developments in the future. In spite of John von Neumann's epigraph, we try to provide an outlook that goes beyond the next five years. Note, however, that the outlook is highly subjective and based on the author's own assessment, and that other people working in the field may think differently and come up with different conclusions and predications.

Before we begin, we want to stress the fact that cryptography has become mature and established itself as a self-standing field of study and area of scientific research (we relativize this point toward the end of the chapter, but at the moment we start from here). As such, an increasingly large number of universities provide courses and degrees on cryptography and information security, and we experience a significant level of diversification and specialization in cryptographic research. In the past, we have seen cryptographers who were able to talk about all aspects related to cryptography. Today, this is no longer the case, and there are cryptographers who are specialized in integer factorization algorithms and algorithms to solve the DLP, cryptographers who are specialized in stream ciphers, cryptographers who are specialized in block ciphers, cryptographers who are specialized in modes of operation for block ciphers, cryptographers who are specialized in RSA, cryptographers who are specialized in ECC, and so on. There are plenty of cryptographic fields of study that are each populated with a distinct research community of its own. This development (and trend toward specialization) goes hand in hand with the maturity level of a particular science, and it sometimes makes it difficult to still see the forest for the trees. In this book, we have tried to act as a counterbalance.

In the rest of this chapter, we provide an overview from a theoretical and practical viewpoint in Sections 18.1 and 18.2, outline the state of the art in PQC in Section 18.3, and complete the book with some closing remarks in Section 18.4.

## 18.1 THEORETICAL VIEWPOINT

From a theoretical viewpoint, the central theme in cryptography is provability: How can one define security, and how can one prove that a given cryptographic system meets this definition? Starting from Shannon's notion of perfect secrecy that is applicable to symmetric encryption only, many researchers have defined and come up with different notions of security that are not only applicable to symmetric encryption, but to many other types of cryptographic systems as well. Most of these notions are introduced, discussed, and put into perspective in this book—at least at an informal level. For some of these notions we know that they are equivalent or relate to each other in a specific way, whereas for other notions the cryptographic research community is still trying to figure out the details. Being able to scientifically argue about different notions of security related to different types of cryptographic systems is an important skill in today's cryptographic research community.

In security discussions, people often prefer cryptosystems that are provably secure. We introduced the notion of provable security in Chapter 1, and we discussed its applicability to different types of cryptosystems in many subsequent chapters. One usually assumes that a particular (mathematical) problem is intractable, and one then shows that a cryptographic system is secure (according to a particular notion of security) as long as this intractability assumption holds. Put in other words: If somebody is able to break the system, then he or she is also able to solve the problem, and this is not likely to be the case. Also, it is sometimes assumed that a cryptographic hash function behaves like a random function (in addition to the intractability assumption of the underlying mathematical problem), and one is then able to show that a cryptographic system is provably secure in the random oracle model.

The bottom line (and fact to keep in mind) is that it has not been possible to provide an absolute proof for the security of a cryptographic system. We are only able to prove the security (properties) of a cryptographic system if we make assumptions. Some of these assumptions are implicit (and appear too trivial to be mentioned in the first place). For example, when we talk about encryption systems, we often make the implicit assumption that telepathy does not exist or does not work (otherwise, encrypting data does not make a lot of sense). Similarly, we assume that randomness exists (otherwise, secret keys cannot exist in principle). Other assumptions are less obvious. As mentioned above, we often work with intractability assumptions when we prove the security of a cryptographic system. These intractability assumptions are often related to a specific adversary and assumptions about his or her capabilities and computational power. For example, if we assume that an adversary is an illiterate (i.e., he or she cannot read and write), then it is fairly trivial to come up with a secure encryption system.[1] More realistic assumptions are related to the computing power, available time, and available memory. Last but not least, we often make assumptions about the correct behavior of system entities and human users. These assumptions are particularly difficult to make, and many cryptographic security protocols can be broken if an adversary does not play by the rules (we revisit this theme in Section 18.4).

According to [1], all assumptions that are made implicitly and explicitly must be taken into account and considered with care when one considers cryptography and security proofs. It is particularly important to note:

- That every security proof for a cryptographic system is only relative to certain assumptions;
- That assumptions should be made explicit;
- That assumptions should always be as weak as possible.

A major goal in cryptographic research remains to reduce the necessary assumptions to a set of realistic assumptions while preserving the practicality of the systems. This is particularly true for computational intractability assumptions.

The more one enters the field of cryptographic protocols (as opposed to cryptographic algorithms), the more formal methods are used to scientifically argue about the security of these protocols. In this area, the question how to properly model the real world and come up with a appropriate notion of security is less clear than it is with cryptographic algorithms. Hence, the use of formal methods in the design of cryptographic protocols is an important and timely research topic.

## 18.2 PRACTICAL VIEWPOINT

From a practical viewpoint, the use of cryptography boils down to standards and profiles. There are simply too many and too complex cryptographic systems (i.e.,

1 This is why the Caesar cipher was secure. It was used in a time when most people were illiterate.

cryptographic algorithms and protocols) and modes of operation to choose from. Anybody not actively working in the field is likely to be overtaxed. The DES is a success story mainly because its promoters (i.e., NIST) realized the need for a standardized symmetric encryption system in the 1970s. In the late 1990s, NIST repeated (and improved) the success story with the AES, and more recently with SHA-3 and PQC (Section 18.3).

On February 16, 2005, the NSA announced Suite B—a set of cryptographic algorithms as an interoperable cryptographic base for both unclassified information and most classified information. Initially, Suite B comprised the following cryptographic algorithms and protocols:

- Symmetric encryption systems: AES-128 and AES-256
- Cryptographic hash functions: SHA-256 and SHA-384
- Key agreement protocols: ECDH and ECMQV
- Digital signature system: ECDSA

As briefly mentioned in Section 12.3, MQV is an authenticated version of the Diffie-Hellman key agreement protocol, and ECMQV is the elliptic curve version thereof. Prior to its incorporation in Suite B, the NSA had licensed Certicom's patents on ECMQV. However, the security of ECMQV had been discussed controversially, so the protocol was finally dropped from Suite B. Hence, ECDH is currently the only key agreement protocol that is still part of Suite B.

The key agreement protocol (i.e., ECDH) and the digital signature system (i.e., ECDSA) employ elliptic curves with 256- and 384-bit prime moduli. Elliptic curves over 256-bit prime moduli, SHA-256, and AES-128 are sufficient for protecting classified information up to the secret level. Elliptic curves over 384-bit prime moduli, SHA-384, and AES-256 are sufficient for the protection of top secret information. In this case, however, the implementation of the algorithms must also be evaluated and certified. This requirement takes into account that the implementation of an algorithm is at least as important as the algorithm itself—at least from a security perspective.

In addition to Suite B, the NSA has specified another set of cryptographic algorithms known as Suite A. This suite comprises algorithms named ACCORDION, BATON, MEDLEY, SHILLELAGH, and WALBURN. These algorithms are unpublished and intended for highly sensitive communication and critical authentication systems. Without knowing the details, it is very difficult if not impossible to make meaningful statements about the security of these algorithms.

Outside the United States, several other (national and international) standardization bodies, forces, and groups are working on cryptography. Examples include

ANSI, the IEEE, the IETF, and the W3C. Unfortunately, many of these bodies have problems of their own, and hence the current state of international standardization is not particularly good. This is worrisome but must be addressed elsewhere. In the meantime, industry-sponsored standardization activities, like the PKCS, are important to fill the gap. These activities have come up with complementary standards for cryptographic systems and their use, such as HMAC for message authentication (Section 10.3.2), OAEP for asymmetric encryption (Section 13.3.1.4), PSS and PSS-R for digital signatures (Section 14.2.2), and many more. Again, a comprehensive overview about the standards that are relevant in practice is provided in [2].

The more we can prove about the security properties of standardized cryptographic systems, the better the odds that they are successful and get widely deployed. The best we can hope is that the complexity of cryptographic systems is hidden in a reference implementation and library, such as cryptlib, Bouncy Castle,[2] OpenSSL[3] or its fork LibreSSL,[4] NaCl (pronounced "salt"),[5] and many more. Ideally, a cryptographic library provides a standardized application programming interface (API), such as Microsoft's CryptoAPI. This makes it possible to easily replace one cryptographic library with another, and hence to provide cryptographic agility.

## 18.3 PQC

Throughout the book, we have mentioned several times that there are new cryptographic techniques being explored in PQC (i.e., under the security assumption that the adversary has a quantum computer). The aim of this section is to provide a respective overview, and to round up and complete the book in this regard. In particular, we briefly address code-based, hash-based, lattice-based, isogeny-based, and multivariate-based cryptography and respective cryptosystems (in this order). Note that PQC is a moving target, and that there are many things going on concurrently. Consequently, we can only overview the tip of the iceberg, and this section is not meant to be comprehensive at all.

As already mentioned in Section 1.3 and touched upon above, there is a NIST competition going on to evaluate, and standardize one or more post-quantum public key cryptographic algorithms or systems. The competition started in 2017 with 69 valid submissions. The first round lasted until January 2019, during which NIST selected 26 algorithms to move forward to the second round.[6] The second

2 https://www.bouncycastle.org.
3 https://www.openssl.org.
4 https://www.libressl.org.
5 https://nacl.cr.yp.to.
6 https://csrc.nist.gov/publications/detail/nistir/8240/final.

**Table 18.1**
The NIST Competition Round 3 Finalists

| Algorithm | Type | Category |
|---|---|---|
| Classic McEliece | Encryption | Code-based |
| CRYSTALS-KYBER | Encryption | Lattice-based |
| NTRU | Encryption | Lattice-based |
| SABER | Encryption | Lattice-based |
| CRYSTALS-DILITHIUM | Signature | Lattice-based |
| FALCON | Signature | Lattice-based |
| Rainbow | Signature | Multivariate-based |

round lasted until July 2020, during which NIST selected 7 algorithms to move forward to the third round.[7] As summarized in Table 18.1, the third-round finalist public-key encryption and key-establishment algorithms are classic McEliece (code-based), CRYSTALS-KYBER, NTRU, and SABER (lattice-based), whereas the finalists for digital signatures are CRYSTALS-DILITHIUM and FALCON (lattice-based), as well as Rainbow (multivariate-based). In addition, 8 alternate candidate algorithms also advanced to the third round: The 5 algorithms for encryption and key establishment are FrodoKEM and NTRU Prime (lattice-based), BIKE and HQC (code-based), and SIKE (isogeny-based), whereas the 3 algorithms for signatures are SPHINCS+ (hash-based), GeMSS (multivariate-based), and Picnic (a new construction based on zero-knowledge proofs). The alternate candidates are summarized in Table 18.2; they are also considered for standardization, although this is unlikely to occur in reality.

**Table 18.2**
The NIST Competition Round 3 Alternate Candidates

| Algorithm | Type | Category |
|---|---|---|
| FrodoKEM | Encryption | Lattice-based |
| NTRU Prime | Encryption | Lattice-based |
| BIKE | Encryption | Code-based |
| HQC | Encryption | Code-based |
| SIKE | Encryption | Isogeny-based |
| SPHINCS+ | Signature | Hash-based |
| GeMSS | Signature | Multivariate-based |
| Picnic | Signature | — |

7 https://csrc.nist.gov/publications/detail/nistir/8309/final.

All NIST competition round 3 finalists and alternate candidates are outlined and compared in an ENISA report.[8] Given the fact that the NIST competition is still ongoing (at least as of this writing), it is not obvious what post-quantum cryptosystems are going to succeed and prevail in the long term. This also means that it may be an appropriate strategy to go for hybrid implementations that use a mixture of different systems—either pre-quantum or post-quantum ones, or preferably both. Again, the keyword is agility, meaning that it should be possible to invoke and play around with different cryptographic algorithms in a given implementation. Except for the fact that this increases the complexity and the amount of code, there is nothing wrong with this strategy. Its advantages outweigh its disadvantages by far.

### 18.3.1 Code-based Cryptosystems

As already mentioned in Section 13.6, McEliece proposed an asymmetric encryption system based on error-correcting codes and the **NP**-hardness of decoding general linear codes in 1978 [3]. More specifically, the private key is an error-correcting code for which an efficient decoding algorithm is known (typically a binary Goppa code), whereas the public key is derived from the private key by disguising it as a general linear code. Each of these codes needs to be represented by a large matrix, making the public and private keys relatively large (this is still the major practical disadvantage of code-based cryptosystems). There are many variants of the McEliece encryption system using different types of codes. Most have been broken or proven to be less secure than the originally proposed system. Strictly speaking, classic McEliece refers to a variant that was proposed by Harald Niederreiter in 1986 [4]. This variant also yields a DSS, but classic McEliece is only used for encryption in the realm of the NIST competition. As mentioned above, it is a third-round finalist public-key encryption and key-establishment algorithm. Furthermore, two other code-based encryption and key establishment systems, BIKE and HQC, have been nominated as alternate candidates. They both use special codes in order to reduce the size of the public key.

### 18.3.2 Hash-based Cryptosystems

As introduced in Section 14.4, one-time signature systems—or hash-based signature systems as they are called in the context of PQC—yield a good starting point to develop signature systems that are resistant to quantum computers (mainly because they only use cryptographic hash functions). In fact, there are several such systems that have been proposed to sign long messages or multiple messages. They fall into

8 https://www.enisa.europa.eu/publications/post-quantum-cryptography-current-state-and-quantum-mitigation.

two categories: Stateless and stateful signature systems. While the former work as normal signatures, the latter are more involved to use, mainly because the signatory needs to keep track of some state information, such as the number of signatures generated using a particular key. This severely limits the usefulness and applicability of such signatures. Examples of such signature systems that have been standardized (outside the NIST competition) include the eXtended Merkle Signature Scheme (XMSS) [5] and the Leighton-Micali signature (LMS) system [6]. Furthermore, SPHINCS+, an updated version of SPHINCS,[9] is a stateless hash-based signature system that has made it into the third round of the NIST competition as an alternate candidate. Due to the nature of hash-based cryptosystems, they cannot be used for encryption and key establishment.

### 18.3.3 Lattice-based Cryptosystems

In mathematics, the term *lattice* is ambiguous and used for many different meanings. For the purpose of this, however, we are mainly interested in point lattices that refer to discrete subgroups of $\mathbb{R}^n$ under addition. More specifically, let $\mathbb{R}^n$ be the $n$-dimensional real Euclidean space and $\{b_1, b_2, \ldots, b_n\}$ a set of linearly independent vectors of $\mathbb{R}^n$. A lattice $\mathcal{L}$ in $\mathbb{R}^n$ then refers to the set of all integer linear combinations of these vectors; that is,

$$\mathcal{L}(b_1, b_2, \ldots, b_n) = \left\{ \sum_{i=1}^{n} x_i b_i : x_i \in \mathbb{Z} \right\}$$

Such a lattice is closed under addition and inverses, and every point has a neighborhood in which it is the only lattice point. It can be visualized as a regularly spaced array of points. This is particularly simple in a plane, but gets more involved in a higher-dimensional space. Mathematical descriptions of lattices look similar to codes, but the entries in the vectors are large numbers instead of zeros and ones. In either case, there may be error vectors added to these elements. There are several problems that can be specified in a point lattice, such as the shortest vector problem (SVP), the closest vector problem (CVP), learning with errors (LWE), and many more. Every lattice-based cryptosystem is typically based on such a problem.

The first lattice-based cryptosystems were developed in the 1990s and published in 1998 [7, 8]. The latter is known as the NTRU public key encryption system, where NTRU stands for "Nth degree Truncated polynomial Ring Units," and it has made its way into the third round of the NIST competition. In addition to NTRU, CRYSTALS-KYBER and SABER are also lattice-based finalists for encryption and key establishment, whereas CRYSTALS-DILITHIUM and FALCON

9 https://sphincs.org.

are lattice-based finalists for signatures. Furthermore, NTRU Prime (a variant of NTRU) and FrodoKEM are lattice-based alternate candidates for encryption and key establishment, whereas there are no lattice-based alternate candidates for signatures. A key establishment algorithm that has seen some usage in the field (mainly by Google and a few other software companies) but is not considered as a candidate for the NIST competition is called A New Hope.[10] In the past few years, lattice-based cryptography has become a hot topic in cryptographic research.

### 18.3.4 Isogeny-based Cryptosystems

An isogeny is a nonconstant map between elliptic curves that can be written as a fraction of polynomials and is compatible with the point addition on both curves, meaning that the sum of two points on the first curve is equal to the sum of the point images, when computed on the second curve. The isogeny problem is to find such an isogeny between two elliptic curves over finite fields. It was originally proposed to design new hash functions [9], but it has also been used as a basis for isogeny-based cryptosystems. One such system called supersingular isogeny key encapsulation (SIKE[11]) was submitted to the NIST competition and is in the third round as an alternate candidate. In addition to SIKE, however, there has been some recent interest in isogeny-based cryptography and respective cryptosystems. Examples include supersingular isogeny DiffieHellman key exchange (SIDH) and commutative SIDH (CSIDH[12] pronounced "seaside") that look promising but were developed after the launch of the NIST competition.

### 18.3.5 Multivariate-based Cryptosystems

As its name suggests, multivariate-based cryptosystems are based on the computational hardness of finding a solution for a system of multivariate quadratic equations over a finite field. It is possible to come up with cryptosystems from such systems using uniformly random coefficients. The resulting systems are assumed to be very secure, but not so efficient. The more efficient systems use coefficients that appear to be random, but are constructed deterministically, such as using a trapdoor (such systems are sometimes called oil-and-vinegar systems). Thanks to the structure of such a system, is possible to find solutions more efficiently (than in the case of using random coefficients).

As of this writing, the best currently available multivariate-based cryptosystems are still not very efficient and come along with very large public keys and

10 https://eprint.iacr.org/2015/1092.

11 https://sike.org.

12 https://csidh.isogeny.org.

long decryption times. In the case of DSSs, the situation is better, and there are at least two multivariate-based DSSs that have made it into the third round of the NIST competition: Rainbow as a finalist and GeMSS as an alternate candidate. The signatures they generate are very short (66 bytes in the case of Rainbow and 258 bits in the case of GeMSS), but the public keys are relatively large (158 kilobytes in the case of Rainbow and more than double in the case of GeMSS).

## 18.4 CLOSING REMARKS

In theory, we have many statements and proofs related to cryptography, and this might suggest that cryptography is a mature science. We already made this point in Section 18.1. In practice, however, this is only partly true, and the maturity level of cryptography as a science to protect information (or data that encodes information, respectively) is not particularly high, meaning that it is not so clear whether and to what extent cryptography can really help protecting data. If we want to authenticate and/or encrypt data, then we usually have many cryptographic algorithms to choose from. But all of these algorithms require some keying material that needs to be protected adequately. What this basically means is that we have reduced the data protection problem to a key management problem, and this line of reasoning applies to almost everything that can be done cryptographically. It always boils down to some keying material that needs to be protected, and this protection almost always represents the Achilles' heel of the system under consideration.

Against this background, you may remember the serious discussion from the Preface regarding the relationship between science and magic. We quoted Clarke saying that "any sufficiently advanced technology is indistinguishable from magic," and we mentioned a talk given by Massey in which he provocatively asked whether cryptography is science or magic. Massey was referring to public key cryptography, but we may open the scope of the question here, and ask—more broadly—whether the protection cryptography can provide is real or only illusive.

If a cryptography-savvy person encrypts data in a particular way, is that data really protected or does it only seem to be so? This question may sound unreal and absurd, but it may become more clear if one considers the line of action of an illusionist. Such a person usually has many tricks at his or her disposal, so that the observers believe what they see is real, whereas in fact it is unreal and results from illusion. Every piece of legerdemain performed by the illusionist is well-prepared, combines several tricks (in sometimes ingenious ways), and uses distraction to confuse the observers. The better the illusionist, the more he or she is able to lull the observers into believing something that is not real. To improve the performance, he or she may even have some observers be part of the show and let them scrutinize something, for example, check whether a box doesn't have double floor or whether some chain is made of solid metal.

The point I want to make at the end of the book is that telling *cryptography* apart from *cryptographic illusion*—let's call it *cryptollusion*—is sometimes difficult if not impossible. Like an illusionist who can use many tricks and distractions to confuse the observers, somebody implementing and making use of cryptography does not have to play by the rules and may cheat at will. No trick is impossible, and it is perfectly fine to invoke any trick at any step of the hardware and software development processes—including the entire supply chain. In the most extreme case this may even include standardization—some may remember the backdoor built into the Dual_EC_DRBG standard (Section 5.5). There are less obvious tricks (to build in a backdoor) that look like "normal" software bugs, and are thus indistinguishable from them. Examples include Heartbleed and Apple's goto fail bug. In most of these cases, it is impossible to tell whether a particular software bug is the result of a programming error or has been built in on purpose. In some sense, it may act as a double floor in a software product.

A similar situation refers to the question of whether it is possible to construct an AE-ciphertext (e.g., AES-GCM) that can be decrypted to two distinct plaintext messages. Naïvely speaking, this should not be possible, because an AE cipher ideally behaves like a PRP, for which is is impossible to find a collision. But in Section 11.3, we already answered the question in the affirmative way, at least for some file formats and if the AE cipher is without key commitment. The trick is to use some file formats' features to hide multiple ciphertexts in a particular file (that then decrypt to different plaintext messages if triggered with the respective keys). In the illusionist's world, this is like performing magic and taking a pigeon out of the top hat, whereas in reality the pigeon was already hidden in the top hat in the first place.

Consequently, a key question is as follows: If I am given an implementation of a fancy cryptographic product, how can I be sure that it really works as expected and is secure as claimed? Or alternatively speaking: Is the claimed security real or only illusive? The person promoting the product is certainly going to provide all kinds of arguments to convince me. This may also include mathematical proofs, as well as security evaluation reports and respective certificates of all kinds. But I may still have my doubts—for very legitimate reasons. The devil is in the details, and it is generally simple to obfuscate them. Hence, human skepticism is a valuable characteristic in most situations in daily life; clearly, it is also valuable or even inevitable in cryptography. Take every security argument or proof you see with the grain of salt it deserves.

## References

[1] Maurer, U.M., *Cryptography 2000 ±10*, Springer-Verlag, New York, LNCS 2000, 2000, pp. 63–85.

[2] Dent, A.W., and C.J. Mitchell, *User's Guide to Cryptography and Standards*. Artech House Publishers, Norwood, MA, 2004.

[3] McEliece, R.J., *A Public-Key Cryptosystem Based on Algebraic Coding Theory*, Deep Space Network Progress Report 42-44, Jet Propulsion Lab., California Institute of Technology, 1978, pp. 114–116.

[4] Niederreiter, H., "Knapsack-type Cryptosystems and Algebraic Coding Theory," *Problems of Control and Information Theory*, Vol. 15, No. 2, 1986, pp. 159-166.

[5] Huelsing, A., *XMSS: eXtended Merkle Signature Scheme*, RFC 8391, May 2018.

[6] McGrew, D., Curcio, M, and S. Fluhrer, *Leighton-Micali Hash-Based Signatures*, RFC 8554, April 2019.

[7] Ajtai, M., "The Shortest Vector Problem in L2 is NP-hard for Randomized Reductions (Extended Abstract)," *Proceeding of the 30th Annual ACM Symposium on Theory of Computing*, ACM Press, May 1998, pp. 10–19.

[8] Hoffstein, J., Pipher, J., and J.H. Silverman, "NTRU: A Ring-based Public Key Cryptosystem," *Proceeding of the International Algorithmic Number Theory Symposium (ANTS 1998)*, Springer-Verlag, LNCS 1423, 1998, pp. 267–288.

[9] Charles, D.X., Lauter, K.E., and E.Z. Goren, "Cryptographic Hash Functions from Expander Graphs," *Journal of Cryptology*, Vol. 22, No. 1, January 2009, pp. 93–113.

# Appendix A

# Discrete Mathematics

In this appendix, we summarize some aspects of discrete mathematics that are relevant for the topic of this book. More specifically, we introduce algebraic basics in Section A.1, and we elaborate on integer and modular arithmetic in Sections A.2 and A.3. Note that this appendix is kept relatively compact, and that many facts are stated without a proof. There are many books on discrete mathematics or algebra that contain the missing proofs, put the facts into perspective, and provide more background information (e.g., [1–6]).[1]

## A.1 ALGEBRAIC BASICS

The term *algebra* refers to the mathematical field of study that deals with sets of elements (e.g., sets of numbers) and operations on these elements.[2] The operations must satisfy some rules that are stated as *axioms*. These axioms are defined abstractly, but most of them are motivated by existing mathematical structures, such as the set of integers with the addition and multiplication operations.

### A.1.1 Preliminary Remarks

Let $S$ be a nonempty set and $*$ a binary operation on the elements of this set.[3] For example, $S$ may be one of the following sets of numbers that are frequently used in mathematics.

1 [6] is electronically available at http://www.shoup.net/ntb.

2 For the purpose of this book, we assume familiarity with set theory at a basic level.

3 The choice of the symbol $*$ is arbitrary. The operations most frequently used in algebra are addition (denoted as $+$) and multiplication (denoted as $\cdot$).

- The set $\mathbb{N} := \{0, 1, 2, \ldots\}$ of *natural numbers* (also known as *nonnegative integers*). In some literature, the term $\mathbb{N}^+$ is used to refer to $\mathbb{N}$ without zero; that is, $\mathbb{N}^+ = \mathbb{N} \setminus \{0\}$.
- The set $\mathbb{Z} := \{\ldots, -2, -1, 0, 1, 2, \ldots\}$ of *integer numbers*, or *integers* in short. In addition to the natural numbers, this set also comprises the negative numbers.
- The set $\mathbb{Q}$ of *rational numbers*. Roughly speaking, a rational number is a number that can be written as a ratio of two integers. More specifically, a number is rational if it can be written as a fraction where the numerator and denominator are integers and the denominator is not equal to zero. This can be formally expressed as follows:

$$\mathbb{Q} := \left\{\frac{a}{b} \mid a, b \in \mathbb{Z} \text{ and } b \neq 0\right\}$$

- The set $\mathbb{R}$ of *real numbers*. Each real number can be represented by a converging infinite sequence of rational numbers (i.e., the limit of the sequence refers to the real number). There are two subsets within the set of real numbers: The algebraic numbers and the transcendental numbers. Roughly speaking, an *algebraic number* is a real number that is the root of a polynomial equation with integer coefficients, whereas a *transcendental number* is a real number that is not the root of a polynomial equation with integer coefficients. Examples of transcendental numbers are $\pi$ and $e$. Real numbers are the most general and most frequently used mathematical objects to model real-world phenomena. A real number that is not rational is called *irrational*, and hence the set of irrational numbers is $\mathbb{R} \setminus \mathbb{Q}$. In some literature, the term $\mathbb{R}^+$ is used to refer to the set of real numbers that are nonnegative.
- The set $\mathbb{C}$ of *complex numbers*. Each complex number can be specified by a pair $(a, b)$ of real numbers, and hence $\mathbb{C}$ can be defined as follows:

$$\mathbb{C} := \{a + bi \mid a, b \in \mathbb{R} \text{ and } i = \sqrt{-1}\}$$

  In this notation, $a$ refers to the *real part* and $b$ refers to the *imaginary part* of the complex number $(a, b)$ or $a + bi$. Note that the second part can also be written as a multiple of $i = \sqrt{-1}$, meaning that the imaginary part of $a + bi$ is written as $b$ (instead of $bi$). Complex numbers are not used in this book.

To formally define a binary operation, we have to introduce the notion of a function. As illustrated in Figure A.1, a *function* $f : X \rightarrow Y$ is a mapping from a *domain* $X$ to a *codomain* $Y$ that assigns to every $x \in X$ a unique $f(x) \in Y$. The

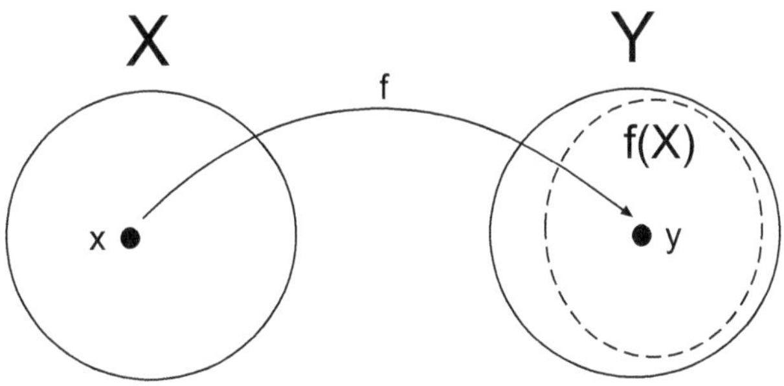


**Figure A.1** A function $f : X \rightarrow Y$.

*range* of $f$ is the subset of values from $Y$ that are reached by the function in one way or another. This may be the entire codomain or only a subset of it (i.e., $f(X) \subseteq Y$). In Figure A.1, the range $f(X)$ is the subset of $Y$ that is drawn with a dotted line.

A function $f : X \rightarrow Y$ may be injective and/or surjective:

- It is *injective* (or *one to one*) if for all $x_1, x_2 \in X$ it holds that $x_1 \neq x_2 \Rightarrow f(x_1) \neq f(x_2)$; that is, if two preimages are different, then the corresponding images must also be different. Equivalently, $f(x_1) = f(x_2) \Rightarrow x_1 = x_2$; that is, if two images are equal, then the corresponding preimages must also be equal.
- It is *surjective* (or *onto*) if for all $y \in Y$ there is an $x \in X$ with $y = f(x)$, meaning that $f(X) = Y$; that is, the codomain and the range are equal.

A function that is injective and surjective is called *bijective*. If $f : X \rightarrow Y$ is a bijective function, then it has an inverse function $f^{-1} : Y \rightarrow X$ with $f^{-1} \circ f = id$. In this notation, $f^{-1} \circ f$ refers to the composite mapping $f^{-1} \circ f = f^{-1}f : X \rightarrow X$ with $(f^{-1} \circ f)(x) = f^{-1}f(x) = x$, and $id$ refers to the identity map.

Let $F$ be a set of functions $f_k : X \rightarrow Y$ that take a key $k$ from a key space $\mathcal{K}$ as an additional input parameter, then we call $F$ a *function family* or *family of functions*. It is defined as

$$F := \{f_k : X \rightarrow Y \mid k \in \mathcal{K}\}$$

where $X$ and $Y$ are the domain and codomain of each function $f_k$. For every $k \in K$, the function $f_k : X \rightarrow Y$ is defined as $f_k(x) = f(k, x)$ and yields an instance of $F$. In this book, we prefer the term family when referring to a family of functions, whereas other authors use the terms collection or ensemble to mean the same thing.

If one has a function family $F$ and requires a particular function $f_k$ from $F$, then one must sample a $k \in \mathcal{K}$ according to some probability distribution. If the distribution is uniform (.e., each element occurs with the same probability), then we say that $k$ is sampled *uniformly*, and we even say that it is sampled *uniformly at random* if $k$ is randomly selected from all uniformly distributed possibilities. We write

$$k \xleftarrow{r} \mathcal{K}$$

to denote that $k$ is sampled uniformly at random from $K$, and

$$f \xleftarrow{r} F$$

to denote that $f$ is sampled uniformly at random from $F$. This can be translated into the sequence

$$k \xleftarrow{r} \mathcal{K}; f \longleftarrow f_k$$

In other words, $f$ refers to the function $f_k$, where $k$ is a key that is sampled uniformly at random from $\mathcal{K}$. This terminology is frequently used in cryptography.

$\text{Funcs}[X, Y]$ refers to the set or family of all functions that map elements of the domain $X$ to elements of the codomain $Y$, whereas $\text{Perms}[X]$ refers to $\text{Funcs}[X, Y]$, where $X = Y$ and all functions yield permutations. Permutations and families of permutations are further addressed in Section A.1.4.

The fact that $*$ is a binary operation on $S$ means that it defines a function from $S \times S$ into $S$. For $a, b \in S$, the use of $*$ can be expressed as follows:

$$\begin{aligned} * : S \times S &\longrightarrow S \\ (a, b) &\longmapsto a * b \end{aligned}$$

This expression suggests that two arbitrary elements $a, b \in S$ are mapped to a new element $a * b \in S$. In this setting, the operation $*$ may have specific properties. In algebra, we are mainly interested in commutative and associative operations as captured in Definitions A.1 and A.2. In what follows, the symbol $\forall$ stands for the quantifier "for all" and the symbol $\exists$ stands for the quantifier "there exists."

**Definition A.1 (Commutative operation)** *A binary operation $*$ is* commutative *if $\forall\, a, b \in S$ it holds that $a * b = b * a$.*

**Definition A.2 (Associative operation)** *A binary operation $*$ is* associative *if $\forall\, a$, $b$, $c \in S$ it holds that $a * (b * c) = (a * b) * c$.*

Let $S$ be a set and $*$ a binary operation on $S$. The operation $*$ may have (left and right) identity elements according to Definitions A.3 to A.5.

**Definition A.3 (Left identity element)** *An element $e \in S$ is a* left identity element *if $\forall\, a \in S$ it holds that $e * a = a$.*

**Definition A.4 (Right identity element)** *An element $e \in S$ is a* right identity element *if $\forall\, a \in S$ it holds that $a * e = a$.*

**Definition A.5 (Identity element)** *An element $e \in S$ is a* identity element *(or a* neutral element*) if it is both a left identity element and a right identity element (i.e., $\forall\, a \in S$ it holds that $e * a = a * e = a$).*

Note that the operation $*$ does not need to be commutative in these definitions. For example, the identity matrix is the identity element of the matrix multiplication, but this operation is not commutative. Also note that an identity element does not need to exist in all cases, but if it exists, it must be unique. This can easily be shown by assuming that $e_1$ and $e_2$ are both identity elements. It then follows that $e_1 = e_1 * e_2 = e_2$, and hence that $e_1 = e_2$.

If an identity element $e \in S$ for the binary operation $*$ exists, then some elements of $S$ may be invertible and have inverse elements. Again, one has to distinguish whether an element is left invertible or right invertible, or whether it is both left and right invertible. We then call this element two-sided invertible, and the respective inverse element two-sided inverse or inverse in short. We only consider this case in Definition A.6.

**Definition A.6 (Inverse element)** *Let $S$ be a set, $*$ a binary operation with identity element $e$, and $a$ an element from $S$. If there exists an element $b \in S$ with $a * b = b * a = e$, then $a$ is* invertible *and $b$ is the* inverse (element) *of $a$.*

Note that not all elements in a given set must be invertible and have an inverse element with respect to the operation under consideration. As we will see later, the question whether all elements in a given set are invertible is the distinguishing fact between a group and a monoid, as well as between a field and a ring. Groups, monoids, fields, and rings are algebraic structures that are introduced next.

### A.1.2 Algebraic Structures

An *algebraic structure*[4] consists of a nonempty set $S$ with one or more binary operations. For the sake of simplicity, we sometimes omit the operation(s) and use

4 In some literature, an algebraic structure is also called an *algebraic system*.

$S$ to denote the entire structure. This is not precise, but convenient. Let us now overview and briefly discuss the algebraic structures that are most frequently used in algebra. Among these structures, groups, rings, and finite fields are particularly important in cryptography.

#### A.1.2.1 Semigroups

The simplest algebraic structure is a semigroup as captured in Definition A.7.

**Definition A.7 (Semigroup)** *A* semigroup *is an algebraic structure $\langle S, * \rangle$ that consists of a nonempty set $S$ and an associative binary operation $*$. The semigroup must be closed; that is, for all $a, b \in S$, $a * b$ must yield an element in $S$.*

Note that this definition does not require a semigroup to have an identity element. For example, the set of even integers (i.e., $\{\ldots, -4, -2, 0, 2, 4, \ldots\}$) with the multiplication operation is a semigroup without an identity element.[5]

#### A.1.2.2 Monoids

According to Definition A.8, a monoid is a semigroup with the additional property (or requirement) that it must have an identity element.

**Definition A.8 (Monoid)** *A* monoid *is a semigroup $\langle S, * \rangle$ that has an identity element $e \in S$ with respect to $*$.*

Examples are $\langle \mathbb{N}, \cdot \rangle$, $\langle \mathbb{Z}, \cdot \rangle$, $\langle \mathbb{Q}, \cdot \rangle$, and $\langle \mathbb{R}, \cdot \rangle$ with the identity element 1. Also, the set of even integers with the addition operation and the identity element 0, as well as the set of all binary sequences of nonnegative and finite length with the string concatenation operation and the empty string representing the identity element, are all monoids.

#### A.1.2.3 Groups

According to Definition A.9, a group is a monoid in which every element is invertible and has an inverse.

**Definition A.9 (Group)** *A* group *is a monoid $\langle S, * \rangle$ in which every element $a \in S$ has an inverse element in $S$; that is, every element $a \in S$ is invertible.*

5 The identity element with respect to multiplication would be 1 (which is not even).

Because $\langle S, *\rangle$ is a group and the operation $*$ is associative, one can easily show that the inverse element of an element must be unique (see above).

Formally speaking, a group can be defined as an algebraic structure $\langle S, *\rangle$ that satisfies the following four axioms:

1. *Closure axiom:* $\forall\, a, b \in S:\ a * b \in S$
2. *Associativity axiom:* $\forall\, a, b, c \in S:\ a * (b * c) = (a * b) * c$
3. *Identity axiom:* $\exists$ a unique identity element $e \in S$ such that $\forall\, a \in S : a * e = e * a = a$
4. *Inverse axiom:* $\forall\, a \in S$ $\exists$ a unique inverse element $a^{-1} \in S$ such that $a * a^{-1} = a^{-1} * a = e$

The operations most frequently used in groups are addition $(+)$ and multiplication $(\cdot)$, and the respective groups are called *additive* and *multiplicative*. For multiplicative groups, the symbol $\cdot$ is often omitted, and $a \cdot b$ is written as $ab$. For additive and multiplicative groups, the identity elements are usually denoted as 0 and 1, whereas the inverse elements of element $a$ are usually denoted as $-a$ and $a^{-1}$. Consequently, a multiplicative group is assumed in the fourth axiom itemized above.

Commutative operations are relevant in practice, and the notion of a commutative group is captured in Definition A.10.

**Definition A.10 (Commutative group)** *A group* $\langle S, *\rangle$ *is* commutative *if the operation* $*$ *is commutative; that is,* $a * b = b * a, \forall\, a, b \in S$.

Commutative groups are also called *Abelian*. Otherwise, if the group is not commutative or Abelian, then it is called *noncommutative* or *non-Abelian*. For example, $\langle \mathbb{Z}, +\rangle$, $\langle \mathbb{Q}, +\rangle$, and $\langle \mathbb{R}, +\rangle$ are commutative groups with the identity element 0. The inverse element of $a$ is $-a$. Similarly, $\langle \mathbb{Q} \setminus \{0\}, \cdot\rangle$ and $\langle \mathbb{R} \setminus \{0\}, \cdot\rangle$ are commutative groups with the identity element 1. Furthermore, the set of real-valued $n \times n$ matrices is a commutative group with respect to matrix addition, whereas the subset of nonsingular (i.e., invertible) matrices is a noncommutative group with respect to matrix multiplication.

*Finite Groups*

Depending on the number of elements, a group can be finite or infinite. The notion of a finite group is captured in Definition A.11. Such groups play a pivotal role in (public key) cryptography.

**Definition A.11 (Finite group)** *A group* $\langle S, * \rangle$ *is* finite *if it contains only finitely many elements.*

The order of a finite group $\langle S, * \rangle$ is the number of elements and refers to the cardinality of $S$, denoted $|S|$. Hence, another way to define a finite group is to say that $\langle S, * \rangle$ is finite if $|S| < \infty$. For example, the set of permutations of $n$ elements is very large, but still finite (it has $n!$ elements). It is a noncommutative group with respect to the composition of permutations (as further addressed in Section A.1.4). Also, $\langle \mathbb{Z}_n, + \rangle$ and $\langle \mathbb{Z}_n^*, \cdot \rangle$ are finite groups that have many cryptographic applications. As explained later in this chapter, $\mathbb{Z}_n$ consists of the $n$ integers between 0 and $n-1$, whereas $\mathbb{Z}_n^*$ consists of the $\phi(n)$ integers between 1 and $n-1$ that have no common divisor with $n$ that is greater than 1.[6] In this context, $\phi$ refers to Euler's totient function that is relevant in cryptography and introduced in Section A.2.6.

If $\langle S, * \rangle$ is a group, then for every element $a \in S$ and every positive integer $i \in \mathbb{N}$, $a^i \in S$ refers to the element

$$\underbrace{a * a * \ldots * a}_{i \text{ times}}$$

Due to the closure axiom, this element must again be an element in $S$. Note that we use $a^i$ only as a shorthand representation for this element, and that the operation between the group element $a$ and the integer $i$ is not the group operation (instead $a^i$ stands for applying the group operation $i$ times to $a$). For additive groups, $a^i$ is written as $i \cdot a$, or $ia$ in short. Again, $i \cdot a$ only represents the resulting group element and $\cdot$ is not the group operation. This subtlety is important to keep in mind.

*Cyclic Groups*

If $\langle S, * \rangle$ is a finite group with identity element $e$, then the order of an element $a \in S$, denoted $ord(a)$, is the least positive integer $n$ such that $a^n$ equals the identity element $e$:[7]

$$\underbrace{a * a * \ldots * a}_{n \text{ times}} = e$$

6 Note that the star in $\mathbb{Z}_n^*$ does not represent a binary operation here.

7 Note that such an integer $n$ always exists in a finite group: Consider the list of elements $a^0, a^1, a^2, a^3, \ldots, a^m$, where $m$ is the number of elements of the group. Since this list contains $m + 1$ elements, due to the pigeonhole principle, there must be 2 elements that are equal. In other words, there exist integers $i$ and $j$ with $i \neq i$ and $i > j$ such that $a^i = a^j$. Since we are in a group, every element has an inverse. This also applies to $a^j$, and we can define $a^{-j} = (a^j)^{-1}$. Using this notation, we have $a^i a^{-j} = a^j (a^j)^{-1}$, and hence $a^{i-j} = e$. This, in turn, means that $i - j$ is a possible candidate for $n$.

Alternatively speaking, the order of an element $a \in S$ in a multiplicatively written group is defined as $ord(a) := min\{n \geq 1 \mid a^n = e\}$. If there exists an element $a \in S$ such that the elements

$$\begin{array}{c} a \\ a * a \\ a * a * a \\ \ldots \\ \underbrace{a * a * \ldots * a}_{n \text{ times}} \end{array}$$

are all different and represent the elements of $S$, then the group $\langle S, * \rangle$ is called *cyclic* and $a$ is called a *generator* of the group or a *primitive root* of the group. If $a$ generates the group (in the sense that $a$ is a generator of it), then one sometimes writes $S = \langle a \rangle$.

If a finite group is cyclic and $n$ is the order of that group, then there are typically several generators. In fact, there are $\phi(n)$ generators[8] that are similar and can be used interchangeably. For example, $\langle \mathbb{Z}_n, + \rangle$ is a cyclic group with $\phi(n)$ generators. The most obvious generator is 1. This basically means that every element of $\mathbb{Z}_n = \{0, 1, 2, 3, \ldots, n-1\}$ can be generated by adding 1 modulo $n$ a certain number of times:

$$\begin{array}{rcl} 0 & = & \overbrace{1 + 1 + \ldots + 1}^{n \text{ times}} \\ 1 & = & 1 \\ 2 & = & 1 + 1 \\ 3 & = & 1 + 1 + 1 \\ & \ldots & \\ n-1 & = & \underbrace{1 + 1 + \ldots + 1}_{n-1 \text{ times}} \end{array}$$

As illustrated in Figure A.2, $\langle \mathbb{Z}_7^*, \cdot \rangle$ is another cyclic group with generator 3, i.e., $\langle \mathbb{Z}_7^*, \cdot \rangle = \langle 3 \rangle$.[9] This means that every element of $\mathbb{Z}_7^* = \{1, 2, \ldots, 6\}$ can be

8 $\phi$ is Euler's totient function that is introduced and further addressed in Section A.2.6.

9 Because $\phi(6) = 2$, there are 2 generators (the other generator would be 5, i.e., $\langle \mathbb{Z}_7^*, \cdot \rangle = \langle 5 \rangle$).

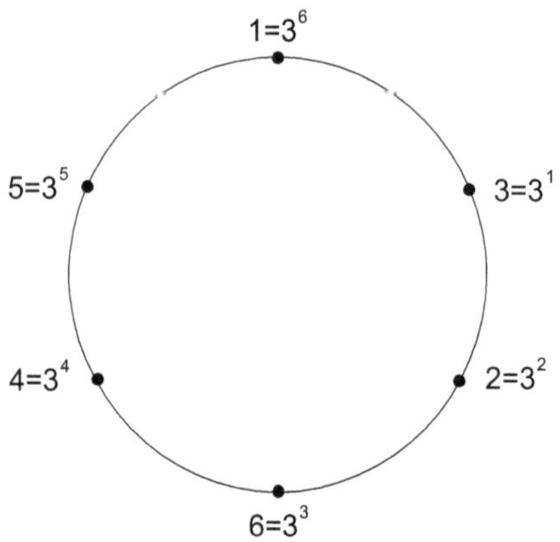


**Figure A.2** The cyclic group $\langle \mathbb{Z}_7^*, \cdot \rangle$.

represented by 3 to the power of an integer modulo 7:

$$\begin{aligned} 3^1 \pmod 7 &= 3 \\ 3^2 \pmod 7 &= 2 \\ 3^3 \pmod 7 &= 6 \\ 3^4 \pmod 7 &= 4 \\ 3^5 \pmod 7 &= 5 \\ 3^6 \pmod 7 &= 1 \end{aligned}$$

The same is true for $\mathbb{Z}_{17}^*$ with the generator 7.[10] It generates all $\mid \mathbb{Z}_{17}^* \mid = 16$ elements of $\mathbb{Z}_{17}^*$):

$$\begin{aligned} 7^1 \pmod{17} &= 7 & 7^9 \pmod{17} &= 10 \\ 7^2 \pmod{17} &= 15 & 7^{10} \pmod{17} &= 2 \\ 7^3 \pmod{17} &= 3 & 7^{11} \pmod{17} &= 14 \\ 7^4 \pmod{17} &= 4 & 7^{12} \pmod{17} &= 13 \\ 7^5 \pmod{17} &= 11 & 7^{13} \pmod{17} &= 6 \end{aligned}$$

10 Because $\phi(16) = 8$, there are 8 generators (the other generators would be 3, 5, 6, 10, 11, 12, and 14).

$$7^6 \pmod{17} = 9 \qquad 7^{14} \pmod{17} = 8$$
$$7^7 \pmod{17} = 12 \qquad 7^{15} \pmod{17} = 5$$
$$7^8 \pmod{17} = 16 \qquad 7^{16} \pmod{17} = 1$$

Note that a cyclic group must be finite, but the converse need not be true. In fact, there are finite groups that are not cyclic, and hence do not have a generator. Also note that all cyclic groups must be Abelian. Again, the converse need not be true, and there are Abelian groups that are not cyclic.

*Subgroups*

When we elaborate on groups and their basic properties, it is sometimes useful to consider subgroups. The notion of a subgroup is captured in Definition A.12.

**Definition A.12 (Subgroup)** *A subset $H$ of a group $G$ is a* subgroup *of $G$, denoted $H \subseteq G$, if it is closed under the operation of $G$ and forms a group on its own.*

For example, the integers are a subgroup of both the rational and the real numbers (with respect to the addition operation). Furthermore, $\{0, 2, 4\}$ is a subgroup of $\langle \mathbb{Z}_6, + \rangle$ with regard to addition modulo 6, and $\{0\}$ and $\{1\}$ are (trivial) subgroups of every additive and multiplicative group (if the group contains numbers). We note that $\{e\}$ is a trivial subgroup of every group (with the neutral element $e$).

An important class of subgroups of a finite group are those that are generated by an element $a$, denoted as $\langle a \rangle := \{a^j \mid 0 \leq j \in \mathbb{N}\}$. The subgroup $\langle a \rangle$ has $ord(a)$ elements and can be used to build cosets according to Definitions A.13–A.15.

**Definition A.13 (Left coset)** *Let $G$ be a group and $H \subseteq G$ a subset of $G$. $\forall\, a \in G$, the sets $a * H := \{a * h \mid h \in H\}$ are called* left cosets *of $H$.*

**Definition A.14 (Right coset)** *Let $G$ be a group and $H \subseteq G$ a subset of $G$. $\forall\, a \in G$, the sets $H * a := \{h * a \mid h \in H\}$ are called* right cosets *of $H$.*

**Definition A.15 (Coset)** *Let $G$ be a (commutative) group and $H \subseteq G$ a subset of $G$. $\forall\, a \in G$, the sets $a * H = H * a$ are called* cosets *of $H$.*

In either case, the $|G|$ elements of $G$ are partitioned into $|G|/|H|$ distinct subsets that represent the cosets. If, for example, $G = \langle \mathbb{Z}_6, + \rangle$ and $H = \{0, 2, 4\}$, then the elements of $G = \{0, 1, 2, 3, 4, 5\}$ can be partitioned into the following two left cosets of $H$:

$$1 + H = 3 + H = 5 + H = \{1, 3, 5\}$$
$$0 + H = 2 + H = 4 + H = \{0, 2, 4\}$$

The notion of a coset is important to prove the following Theorem A.1 that is due to Lagrange.[11]

**Theorem A.1 (Lagrange's Theorem)** *If $H$ is a subgroup of $G$, then $|H| \mid |G|$ (i.e., the order of $H$ divides the order of $G$).*

*Proof.* If $H = G$, then $|H| \mid |G|$ holds trivially. Consequently, we only consider the case in which $H \subset G$. For any $a \in G \backslash H$, the coset $a * H$ is a subset of $G$. The following can be shown:

i) For any $a \neq a'$, if $a \notin a' * H$ then $(a * H) \cap (a' * H) = \emptyset$.

ii) $|a * H| = |H|$.

For (i), suppose there exists a $b \in (a * H) \cap (a' * H)$. Then there exist $c, c' \in H$ such that $a * c = b = a' * c'$. Applying various group axioms, we have $a = a * e = a * (c * c^{-1}) = b * c^{-1} = (a' * c') * c^{-1} = a' * (c' * c^{-1}) \in a' * H$. This contradicts our assumption (that $a \notin a' * H$).

For (ii), $|a * H| \leq |H|$ holds trivially (by the definition of a coset). Suppose that the inequality is rigorous. This is only possible if there are $b, c \in H$ with $b \neq c$ and $a * b = a * c$. Applying the inverse element of $a$ on either side of the equation, we get $b = c$, contradicting to $b \neq c$.

In summary, $G$ is partitioned by $H$ and the family of its mutually disjoint cosets, each has the size $|H|$, and hence $|H| \mid |G|$. This proves the theorem.

□

*Quotient Groups*

Let $\langle G, * \rangle$ be a (commutative) group with the identity element $e$, and $H \subseteq G$ a subgroup of $G$. The *quotient group* of $G$ modulo $H$, denoted $G/H$, then refers to the group that consists of the cosets of $H$ (i.e., the sets $a * H$ with $a \in G$) that represent the "normal" group elements and $e * H$ that represents the identity element. If, for example, $G$ refers to the integers (i.e., $G = \mathbb{Z}$) and $H$ refers to the subgroup of $\mathbb{Z}$ that consists of all multiples of a positive integer $n \in \mathbb{N}^+$ (i.e., $H = n\mathbb{Z} = \{0, \pm n, \pm 2n, \ldots\}$), then the quotient group of $\mathbb{Z}$ modulo $n\mathbb{Z}$ is written as

$$\mathbb{Z}/n\mathbb{Z} = \{x + n\mathbb{Z} \mid x \in \mathbb{Z}\}$$

11 Joseph Louis Lagrange was a French mathematician who lived from 1736 to 1813.

or $\mathbb{Z}_n$ in short. It has the following $n$ elements:

$$\begin{array}{rcl} 0 & + & n\mathbb{Z} \\ 1 & + & n\mathbb{Z} \\ 2 & + & n\mathbb{Z} \\ & \cdots & \\ n-1 & + & n\mathbb{Z} \end{array}$$

This quotient group is frequently used in cryptography (for large values of $n$).

So far, we have only looked into algebraic structures that have a single operation. There are at least two algebraic structures that comprise two operations, rings and fields. They are addressed next.

#### A.1.2.4 Rings

The simpler algebraic structure that comprises two operations is the ring that is formally introduced in Definition A.16.

**Definition A.16 (Ring)** *A* ring *is an algebraic structure $\langle S, *_1, *_2 \rangle$ with a set $S$ and two associative binary operations $*_1$ and $*_2$ that fulfill the following requirements:*

1. *$\langle S, *_1 \rangle$ is a commutative group with identity element $e_1$.*
2. *$\langle S, *_2 \rangle$ is a monoid with identity element $e_2$.*
3. *The operation $*_2$ is distributive over the operation $*_1$. This means that the following two distributive laws must hold $\forall\, a, b, c \in S$:*

$$\begin{array}{rcl} a *_2 (b *_1 c) & = & (a *_2 b) *_1 (a *_2 c) \\ (b *_1 c) *_2 a & = & (b *_2 a) *_1 (c *_2 a) \end{array}$$

According to the first requirement, the operation $*_1$ must be commutative. This is not the case for the operation $*_2$, and hence the ring is called *commutative* (*noncommutative*) if $*_2$ is (not) commutative.

For example, $\langle \mathbb{Z}, +, \cdot \rangle$, $\langle \mathbb{Z}_n, +, \cdot \rangle$,[12] $\langle \mathbb{Q}, +, \cdot \rangle$, and $\langle \mathbb{R}, +, \cdot \rangle$ are commutative rings. Also, the set of real-valued $n \times n$ matrices form a ring with the zero matrix as neutral element of the addition and the identity matrix as the neutral element of the multiplication. Contrary to the previous examples, however, this ring is noncommutative.

12 If $n$ is a prime, then $\langle \mathbb{Z}_n, +, \cdot \rangle$ is a field.

#### A.1.2.5 Fields

If we have a ring $\langle S, *_1, *_2 \rangle$ and require $\langle S \setminus \{e_1\}, *_2 \rangle$ to be a group (instead of a monoid), then the resulting algebraic structure is a *field*. This is captured in Definition A.17.

**Definition A.17 (Field)** *A ring $\langle S, *_1, *_2 \rangle$ in which $\langle S \setminus \{e_1\}, *_2 \rangle$ is a group is a* field.

Another way of saying that $\langle S \setminus \{e_1\}, *_2 \rangle$ is a group is that every nonidentity element (with respect to $*_1$) has an inverse element (with respect to $*_2$).

A field $\langle S, *_1, *_2 \rangle$ is *finite* if it contains a finite number of elements (i.e., $|S| < \infty$). All finite fields with $n$ elements can be shown to be structurally equivalent or isomorphic. Consequently, it suffices to consider and examine only one finite field with $n$ elements to basically understand all of them. This field is called a *Galois field*,[13] and it is denoted as $\mathbb{F}_n$ or $GF(n)$. For every prime number $p$, there is a *prime field* with $p$ elements, $\mathbb{F}_p$ or $GF(p)$, and an *extension field*, $\mathbb{F}_{p^n}$ or $GF(p^n)$ for $1 < n \in \mathbb{N}$.

- The elements of the prime field $\mathbb{F}_p$ can be represented by the $p$ integers ranging from 0 up to $p - 1$. The addition and multiplication are defined as usual, but all results are always reduced modulo $p$. The simplest prime field is $\mathbb{F}_2$ that comprises only the two elements 0 and 1. It is frequently used in computer science.

- The elements of the extension field $\mathbb{F}_{p^n}$ can be represented by polynomials of degree $n - 1$ with coefficients from $\mathbb{F}_p$. The addition and multiplication are defined over polynomials, where the resulting polynomial is always reduced modulo an irreducible polynomial (that needs to be fixed). Hence, in an extension field, the fixed and irreducible polynomial plays a similar role as the prime in a prime field; that is, it ensures that the result of adding and multiplying two elements is again an element of the field. In cryptography, a widely used extension field is the AES field $\mathbb{F}_{2^8}$ (Section 9.6.2).

When we discussed the notion of a group, we said that every group can have subgroups. Similarly, every field can have subfields, where the notion of a subfield is captured in Definition A.18.

**Definition A.18 (Subfield)** *If $F$ is a field, then a subset $H$ of $F$ is a* subfield *of $F$ if it closed under the (same) operations and also forms a field.*

13 The term was chosen in honor of Evariste Galois, a French mathematician who lived from 1811 to 1832.

Using the notion of a subfield, one can also say that a prime field is a field that contains no proper subfield. For example, $\mathbb{Q}$ is a prime field that is infinite (i.e., it has infinitely many elements). Since $\mathbb{Q}$ is a proper subfield of $\mathbb{R}$, $\mathbb{R}$ cannot be a prime field either.

### A.1.3 Homomorphisms

Let $\langle A, \circ \rangle$ and $\langle B, \bullet \rangle$ be two algebraic structures of the same type, such as a group.[14] According to Definition A.19, a homomorphism is a structure-preserving mapping of $A$ into $B$.

**Definition A.19 (Homomorphism)** *A mapping $f : A \rightarrow B$ is a* homomorphism *of $A$ into $B$, if for every pair $x, y \in A$ it holds that $f(x \circ y) = f(x) \bullet f(y)$.*

This means that combining $x$ and $y$ with the operator $\circ$ in $A$ and subjecting the result to $f$ must yield the same result as subjecting $x$ and $y$ to $f$ individually and then combining the result with $\bullet$ in $B$. It goes without saying that equality does not hold in the general case (only if $f$ is homomorphic). In the realm of cryptography, for example, we say that an encryption system is additively homomorphic if the encryption of the sum of two plaintext messages is the same as the sum of the respective ciphertexts. Similarly, we say that it is multiplicatively homomorphic if the encryption of the product of two plaintext messages is the same as the product of the respective ciphertexts. There are many asymmetric encryption systems that are either additively or multiplicatively homomorphic, but there is no practical system that satisfies both requirements. Finding such a system is the big challenge of fully homomorphic encryption (Section 13.5).

If the algebraic structure is a group, then the homomorphism is a *group homomorphism*. If it is a ring, then it is a *ring homomorphism*. In this case, however, the mapping must preserve the ring addition, the ring multiplication, and the multiplicative identity.

There are two special cases of homomorphisms: Isomorphisms and automorphisms (as captured in Definitions A.20 and A.21).

**Definition A.20 (Isomorphism)** *A homomorphism $f : A \rightarrow B$ is an* isomorphism *if and only if it is bijective (injective and surjective). In this case, $A$ and $B$ are structurally identical. We say that $A$ and $B$ are* isomorphic *and we write $A \cong B$.*

**Definition A.21 (Automorphism)** *An isomorphism $f : A \rightarrow A$ is an* automorphism.

14 For the sake of simplicity, we assume algebraic structures with only one operation each.

Another way of saying that two algebraic structures are isomorphic is to say that they are structurally equivalent. It is known that all cyclic groups with order $n$ are isomorphic to $\langle \mathbb{Z}_n, + \rangle$, that $\langle \mathbb{Z}_p^*, \cdot \rangle$ is a cyclic group for every prime number $p$, and that this group is isomorphic to $\langle \mathbb{Z}_{p-1}, + \rangle$. This, in turn, means that the function $f(x) = g^x \bmod p$, where $g$ is a generator of $\mathbb{Z}_p^*$, defines an isomorphism between $\langle \mathbb{Z}_{p-1}, + \rangle$ and $\langle \mathbb{Z}_p^*, \cdot \rangle$, and hence that $g^{x+y} \equiv g^x \cdot g^y \pmod{p}$.

### A.1.4 Permutations

Permutations are important building blocks for symmetric encryption systems in general, and block ciphers in particular. The notion of a permutation is captured in Definition A.22.

**Definition A.22 (Permutation)** *Let $S$ be a set. A map $f : S \rightarrow S$ is a* permutation *if it is bijective (i.e., injective and surjective). The* set of all permutations *of $S$ is denoted* $\text{Perms}[S]$.

If $S = \{1, 2, 3, 4, 5\}$, then an example of a permutation can be expressed as follows:

$$\begin{pmatrix} 1 & 2 & 3 & 4 & 5 \\ 5 & 3 & 4 & 2 & 1 \end{pmatrix}$$

This permutation maps every element in the first row to the respective element in the second row; that is, 1 is mapped to 5, 2 is mapped to 3, 3 is mapped to 4, 4 is mapped to 2, and 5 is mapped to 1. Using this notation, it is possible to specify any permutation of a finite set $S$. If $S$ has $n$ elements, then $\text{Perms}[S]$ comprises $n! = 1 \cdot 2 \cdot \ldots \cdot n$ elements and $\langle \text{Perms}[S], \circ \rangle$ yields a noncommutative group for $n \geq 3$ (where $\circ$ represents the composition operator, meaning that for $A, B \in \text{Perms}[S]$, $A \circ B$ refers to the permutation that results by applying $B$ and $A$ in this order).

Let $S = \{0, 1\}^n$ be the set of all binary strings of length $n$. A *bit permutation* is defined as a permutation of the bit positions of $S$. To specify a bit permutation $f$, one selects $\pi \in \text{Perms}[S]$ and sets

$$\begin{array}{rcl} f : \{0,1\}^n & \longrightarrow & \{0,1\}^n \\ b_0 \ldots b_{n-1} & \longmapsto & b_{\pi(0)} \ldots b_{\pi(n-1)} \end{array}$$

Every bit permutation can be described in this way, and hence there are $n!$ possible bit permutations for binary strings of length $n$. If the length of a binary string is fixed, then the respective string is also called a *word*. A word is typically 32 or 64 bits long.

There are some bit permutations on words that are frequently used in cryptography, especially when it comes to hash functions and stream ciphers. Examples include left and right rotation that stand for circular shifts in the respective direction. If $w$ is a word and $c \in \mathbb{N}$ is a positive integer, typically in the range between 1 and $|w| - 1$, then $w \overset{\frown}{\hookleftarrow} c$ refers to the $c$-bit left rotation (circular shift) of $w$, and $w \overset{\frown}{\hookrightarrow} c$ to the $c$-bit right rotation of $w$. In this notation $w \hookleftarrow c$ ($w \hookrightarrow c$) refers to the $c$-bit left (right) shift of $w$.

## A.2 INTEGER ARITHMETIC

According to Carl Friedrich Gauss,[15] "mathematics is the queen of sciences and number theory is the queen of mathematics." Number theory is also called integer arithmetic. As such, it elaborates on the ring $\langle \mathbb{Z}, +, \cdot \rangle$ and its basic properties. It is an important and fundamental mathematical topic that has had (and continues to have) a deep impact on all natural sciences. One fascinating aspect of number theory is that many of its problems, such as the integer factorization problem (Section 5.2.2), can be easily understood even by nonmathematicians, but still remain hard to solve. This is in sharp contrast to many other areas of mathematics, where problems cannot easily be understood by nonexperts.

In this section, we look at the aspects of integer arithmetic or number theory that are relevant for the topic of this book. More specifically, we address integer division, common divisors and multiples, Euclidean algorithms, prime numbers, factorization, and Euler's totient function in this order.

### A.2.1 Integer Division

In an algebraic structure with the multiplication operation, one usually divides two elements $a$ and $b$ by multiplying the first with the multiplicatively inverse element of the second:

$$\frac{a}{b} = ab^{-1}$$

This construction requires $b$ to have an inverse element. This is always the case in a group (or a field). If, however, the algebraic structure is only a monoid (or a ring), then there are elements that have no inverse, and hence it may not be possible to divide one element by another. This is a major difference between a group and a monoid, or between a field and a ring.

15 Carl Friedrich Gauss was a German mathematician who lived from 1777 to 1855.

In $\langle \mathbb{Z}, \cdot \rangle$ (group) or $\langle \mathbb{Z}, +, \cdot \rangle$ (field) we say that element $a$ *divides* element $b$, denoted $a|b$, if there exists another element $c$ in $\mathbb{Z}$ such that $b = ac$. Alternatively speaking, $a$ is a *divisor* of $b$ and $b$ is a *multiple* of $a$. For example, $2|6$ because $6 = 2 \cdot 3$, but 3 does not divide 2. Also, 1 divides every integer and the largest divisor of $a$ is $|a|$. Furthermore, every integer $a \in \mathbb{Z}$ divides 0 (because $c = 0$ satisfies $0 = a \cdot 0$), and hence 0 has no largest divisor. Theorem A.2 enumerates some basic laws that are used to compute with divisors.

**Theorem A.2** *For all $a, b, c, d, e \in \mathbb{Z}$, the following rules apply:*

1. *If $a|b$ and $b|c$, then $a|c$.*
2. *If $a|b$, then $ac|bc$ for all $c$.*
3. *If $c|a$ and $c|b$, then $c|da + eb$ for all $d$ and $e$.*
4. *If $a|b$ and $b \neq 0$, then $|a| \leq |b|$.*
5. *If $a|b$ and $b|a$, then $|a| = |b|$.*

*Proofs.*

1. If $a|b$ and $b|c$, then there exist $f, g \in \mathbb{Z}$ with $b = af$ and $c = bg$. Consequently, we can write $c = bg = (af)g = a(fg)$ to express $c$ as a multiple of $a$. The claim (i.e., $a|c$) follows directly from this equation.
2. If $a|b$, then there exists $f \in \mathbb{Z}$ with $b = af$. Consequently, we can write $bc = (af)c = f(ac)$ to express $bc$ as a multiple of $ac$. The claim (i.e., $ac|bc$) follows directly from this equation.
3. If $c|a$ and $c|b$, then there exist $f, g \in \mathbb{Z}$ with $a = fc$ and $b = gc$. Consequently, we can write $da + eb = dfc + egc = (df + eg)c$ to express $da + eb$ as a multiple of $c$. The claim (i.e., $c|da + eb$) follows directly from this equation.
4. If $a|b$ and $b \neq 0$, then there exists $0 \neq f \in \mathbb{Z}$ with $b = af$. Consequently, $|b| = |af| \geq |a|$ and the claim (i.e., $|a| \leq |b|$) follows immediately.
5. Let us assume that $a|b$ and $b|a$. If $a = 0$ then $b = 0$, and vice versa. If $a, b \neq 0$, then it follows from Proof 4 that $|a| \leq |b|$ and $|b| \leq |a|$, and hence $|b| = |a|$.

□

Theorem A.3 is called the division theorem and it is attributed to Euclid.[16]) The theorem states that it is always possible to divide an integer with another integer.

16 Euclid, also known as Euclid of Alexandria, was a Greek mathematician who was born around 300 B.C. He is best known for his book *Elements*, which is the most influential textbook in the history of mathematics.

**Theorem A.3 (Division theorem)** *For all $a, d \in \mathbb{Z}$ with $d \neq 0$, there exist unique $q, r \in \mathbb{Z}$ such that $a = qd + r$ and $0 \leq r < |d|$.*

In this setting, $a$ is called the *dividend*, $d$ is called the *divisor*, $q$ is called the *quotient*, and $r$ is called the *remainder*. In some literature, the remainder $r$ is written as $R_d(n)$ and we sometimes use this notation in this book as well.

For example, $R_7(16) = 2$ (because $16 = 2 \cdot 7 + \underline{2}$), $R_7(-16) = 5$ (because $-16 = -3 \cdot 7 + \underline{5}$), and $R_{25}(104) = 4$ (because $104 = 4 \cdot 25 + \underline{4}$). In all three examples, the remainder is underlined. Obviously, $R_d(n) = 0$ means that $d$ divides $n$ with remainder zero, and hence $d$ is a divisor of $n$. Furthermore, $R_d(n + i \cdot d)$ is equal to $R_d(n)$ for all $i \in \mathbb{Z}$, and this suggests that $R_7(1) = R_7(8) = R_7(15) = R_7(22) = R_7(29) = \ldots = 1$.

### A.2.2 Common Divisors and Multiples

Two integers can have many common divisors, but only one of them can be the largest one. Quite naturally, this divisor is called the *greatest common divisor* (gcd). It is formally introduced in Definition A.23.

**Definition A.23 (Common divisors and greatest common divisor)** *For $a, b \in \mathbb{Z}$, $c \in \mathbb{Z}$ is a* common divisor *of $a$ and $b$ if $c|a$ and $c|b$. Furthermore, $c$ is the* greatest common divisor, *denoted $gcd(a, b)$, if it is the largest integer that divides $a$ and $b$.*

Another possibility to define the greatest common divisor of $a$ and $b$ is to say that $c = gcd(a, b)$ if any common divisor of $a$ and $b$ also divides $c$. It holds that $gcd(0, 0) = 0$, and $gcd(a, 0) = |a|$ for all $a \in \mathbb{Z} \setminus \{0\}$. If $a, b \in \mathbb{Z} \setminus \{0\}$, then $1 \leq gcd(a, b) \leq min\{|a|, |b|\}$ and $gcd(a, b) = gcd(\pm|a|, \pm|b|)$. Consequently, the greatest common divisor of two integers can never be negative—even if one or both arguments are negative. Furthermore, two integers $a, b \in \mathbb{Z}$ are *relatively prime* or *coprime* if their greatest common divisor is one (i.e., $gcd(a, b) = 1$).

Similar to the (greatest) common divisor, it is possible to define the (least) common multiple of two integers. This is captured in Definition A.24.

**Definition A.24 (Common multiples and least common multiple)** *For $a, b \in \mathbb{Z}$, $c \in \mathbb{Z}$ is a* common multiple *of $a$ and $b$ if $a|c$ and $b|c$. Furthermore, $c$ is the* least common multiple, *denoted as $lcm(a, b)$, if it is the smallest integer that is divided by $a$ and $b$.*

Another possibility to define the least common multiple of $a$ and $b$ is to say that $c = lcm(a, b)$, if $c$ divides all common multiples of $a$ and $b$.

The gcd and lcm operators can be generalized to more than two arguments in a simple and straightforward way: $gcd(a_1, \ldots, a_k)$ is the largest integer that divides all $a_i$ $(i = 1, \ldots, k)$ and $lcm(a_1, \ldots, a_k)$ is the smallest integer that is divided by all $a_i$ $(i = 1, \ldots, k)$.

### A.2.3 Euclidean Algorithms

If the prime factorization of two integers $a$ and $b$ is known, then it is easy to compute the greatest common divisor (Appendix A.2.5). Otherwise, one can still use the *Euclidean algorithm* (or *Euclid's algorithm*) to compute it.[17] According to Theorem A.3, $a, b \in \mathbb{Z}$ with $a \geq b$ and $b \neq 0$ can be written as

$$a = bq + r$$

for some quotient $q \in \mathbb{Z}$ and remainder $r \in \mathbb{Z}$ (standing for $a \bmod b$) with $0 \leq r < b$. Since $gcd(a, b)$ divides both $a$ and $b$, $gcd(a, b)$ must also divide $r$. This follows from the equation stated above. Consequently, $gcd(a, b)$ equals $gcd(b, r)$, and hence

$$gcd(a, b) = gcd(b, r) = gcd(b, a \bmod b) = gcd(b, R_b(a))$$

This equation can be recursively applied to compute $gcd(a, b)$. For example, for $a = 100$ and $b = 35$, $gcd(100, 35)$ can be computed as follows:

$$\begin{aligned} gcd(100, 35) &= gcd(35, R_{35}(100)) = gcd(35, 30) \\ &= gcd(30, R_{30}(35)) = gcd(30, 5) \\ &= gcd(5, R_5(30)) = gcd(5, 0) \\ &= 5 \end{aligned}$$

The gcd of 100 and 35 thus equals 5.

This way of computing $gcd(a, b)$ is at the core of the Euclidean algorithm. Let us consider the following series of $k$ equations:

$$\begin{aligned} a &= bq_1 + r_1 \\ b &= r_1 q_2 + r_2 \\ r_1 &= r_2 q_3 + r_3 \\ &\ldots \\ r_{k-3} &= r_{k-2} q_{k-1} + r_{k-1} \\ r_{k-2} &= r_{k-1} q_k + r_k \end{aligned}$$

17 The Euclidean algorithm is one of the oldest algorithms known; it appeared in Euclid's *Elements* around 300 B.C.

All quotients $q_1, \ldots, q_k$ and remainders $r_1, \ldots, r_k$ are integers. Only $r_k$ is equal to zero; all other values $q_1, q_2, \ldots, q_k, r_1, r_2, \ldots, r_{k-1}$ are nonzero. If $r_k$ reaches zero, then the last equation implies that $r_{k-1}$ divides $r_{k-2}$ (i.e., $r_{k-1} \mid r_{k-2}$). The second-to-last equation implies that it also divides $r_{k-3}$. This line of argumentation can be continued until the first equation, and hence $r_{k-1}$ also divides $a$ and $b$. None of the other remainders $r_{k-2}, r_{k-3}, \ldots, r_1$ has this property.[18] The bottom line is that $r_{k-1}$ is a common divisor of $a$ and $b$, and one can show that it is the greatest one; that is, $r_{k-1} = gcd(a, b)$.

**Algorithm A.1** The Euclidean algorithm.

```
(a, b)
--------------------
a = |a|
b = |b|
while b ≠ 0 do
    t = a
    a = b
    b = t mod b
--------------------
(a)
```

The resulting Euclidean algorithm is illustrated in Algorithm A.1. It takes as input two integers $a, b \in \mathbb{Z}$ with $|a| \geq |b|$ and $b \neq 0$,[19] and it computes as output $a$ that yields the greatest common divisor of $a$ and $b$. First, the algorithm replaces $a$ and $b$ with their absolute values (note that this does not change the value of the greatest common divisor). Then the previously mentioned rule that $gcd(a, b) = gcd(b, a \bmod b)$ is applied recursively until $b$ reaches the value of zero. At this point, $a$ yields the greatest common divisor and is returned as output of the algorithm. In the example given above, $a = 100$ and $b = 35$. This means that the while-loop is iterated three times: In the first iteration, $t \leftarrow 100$, $a \leftarrow 35$, and $b \leftarrow 30$; in the second iteration, $t \leftarrow 35$, $a \leftarrow 30$, and $b \leftarrow 5$, and in the third iteration, $t \leftarrow 30$, $a \leftarrow 5$, and $b \leftarrow 0$. Now $b$ has reached the value of zero and the while-loop terminates. The algorithm returns the value $a = 5$ that stands for the greatest common divisor of 100 and 35.

The Euclidean algorithm explained so far can be used to compute the greatest common divisor of two integers $a$ and $b$. During its execution, all intermediate results (i.e., all quotients $q_1, \ldots, q_k$ and remainders $r_1, \ldots, r_k$) are discarded. This makes the algorithm simple to implement. If, however, one does not throw away

18 That's why they are called remainders in the first place (not divisors). Only $r_{k-1}$ is a divisor in the last equation.

19 Strictly speaking, the condition $|a| \geq |b|$ is not necessary. If $b > a > 0$, then $a$ and $b$ can be swapped after the first step.

all intermediate results but accumulates them during the execution of the algorithm, then one obtains more information than only the greatest common divisor. In this case, the *extended Euclidean algorithm* can be used to compute two integers $x$ and $y$ that satisfy *Bézout's identity* (also known as *Bézout's lemma*) captured in Theorem A.4 (without proof).

**Theorem A.4 (Bézout's identity)** *For every pair of integers $a, b \in \mathbb{Z}$, there exists another pair of integers $x, y \in \mathbb{Z}$ such that $xa + yb = gcd(a, b)$. More generally, all integers of the form $xa + yb$ are multiples of $gcd(a, b)$.*

The first equation of the previously mentioned series of $k$ equations (i.e., $a = bq_1 + r_1$) can be resolved for $r_1$:

$$\begin{aligned} r_1 &= a - bq_1 \\ &= a + b(-q_1) \end{aligned}$$

If one multiplies either side of the equation with $q_2$, then one gets

$$aq_2 + b(-q_1q_2) = r_1q_2$$

Using this expression in the second equation of the series, one gets

$$\begin{aligned} b &= r_1q_2 + r_2 \\ &= aq_2 + b(-q_1q_2) + r_2 \end{aligned}$$

and hence

$$\begin{aligned} b - aq_2 - b(-q_1q_2) &= \\ a(-q_2) + b(1 + q_1q_2) &= r_2 \end{aligned}$$

A similar construction can be used to express each $r_i$ for $i = 3, 4, \ldots, k$ as a linear combination of $a$ and $b$:

$$ax_i + by_i = r_i$$

where $x_i$ and $y_i$ represent integers. So far, $x_1 = 1$, $y_1 = -q_1$, $x_2 = -q_2$, and $y_2 = 1 + q_1q_2$. The sequence of $r_i$ ($i = 1, \ldots, k$) terminates with $r_k = 0$, and $r_{k-1}$ yields $gcd(a, b)$. As we can write

$$\begin{aligned} gcd(a, b) &= r_{k-1} \\ &= ax_{k-1} + by_{k-1} \end{aligned}$$

we must determine $x_{k-1}$ and $y_{k-1}$ to compute $gcd(a,b)$. The extended Euclidean algorithm provides a way to accumulate the intermediate quotients to determine $x_{k-1}$ and $y_{k-1}$. As specified in Algorithm A.2, it takes as input two integers $a, b \in \mathbb{Z}$ with $|a| \geq |b|$ and $b \neq 0$, and it computes as output the two integers $x = x_{k-1}$ and $y = y_{k-1}$ that are in line with Bézout's identity $xa + yb = gcd(a,b)$.

**Algorithm A.2** The extended Euclidean algorithm.

```
(a, b)
------------------------------
i = 0
r_{-1} = a
r_0 = b
x_{-1} = 1
y_{-1} = 0
x_0 = 0
y_0 = 1
while (r_i = ax_i + by_i ≠ 0) do
    q = r_{i-1} div r_i
    x_{i+1} = x_{i-1} - qx_i
    y_{i+1} = y_{i-1} - qy_i
    i = i + 1
x = x_{i-1}
y = y_{i-1}
------------------------------
(x, y)
```

In our example, the extended Euclidean algorithm can be used to determine $x$ and $y$ that satisfy $gcd(a,b) = gcd(100,35) = 5 = x \cdot 100 + y \cdot 35$. After the initialization phase of the algorithm, we come to the first incarnation of the while-loop with $i = 0$. We compute

$$r_0 = ax_0 + by_0 = 100 \cdot 0 + 35 \cdot 1 = 35$$

Because this value is not equal to 0, we enter the loop. The variable $q$ is set to $r_{-1}$ div $r_0$. In our example, this integer division yields 100 div 35 = 2. Using $q = 2$, we compute the following pair of values:

$$\begin{aligned} x_1 &= x_{-1} - qx_0 = 1 - 2 \cdot 0 = 1 \\ y_1 &= y_{-1} - qy_0 = 0 - 2 \cdot 1 = -2 \end{aligned}$$

After having incremented $i$ with 1, we have $i = 1$ and come back to the second incarnation of the while-loop. We compute

$$r_1 = ax_1 + by_1 = 100 \cdot 1 + 35 \cdot (-2) = 100 - 70 = 30$$

Because this value is again not equal to 0, we enter the loop. This time, the variable $q$ is set to $r_0$ div $r_1 = 35$ div 30 = 1. Using $q = 1$, we then compute the following pair of values:

$$\begin{aligned} x_2 &= x_0 - qx_1 = 0 - 1 = -1 \\ y_2 &= y_0 - qy_1 = 1 - (1 \cdot (-2)) = 1 + 2 = 3 \end{aligned}$$

After having incremented $i$ with 1, we have $i = 2$ and come back to the third incarnation of the while-loop. We compute

$$r_2 = ax_2 + by_2 = 100 \cdot (-1) + 35 \cdot 3 = -100 + 105 = 5$$

Because this value is not equal to 0, we enter the loop. This time, the variable $q$ is set to $r_1$ div $r_2 = 30$ div 5 = 6. Using $q = 6$, we compute the following pair of values:

$$\begin{aligned} x_3 &= x_1 - qx_2 = 1 + 6 = 7 \\ y_3 &= y_1 - qy_2 = -2 - 6 \cdot 3 = -20 \end{aligned}$$

Finally, we increment $i$ and come back to the fourth incarnation of the while-loop with $i = 3$. We now compute

$$r_3 = ax_3 + by_3 = 100 \cdot 7 + (-20) \cdot 35$$

and realize that this value equals 0. Consequently, we don't reenter the while-loop, but return $(x, y) = (x_2, y_2) = (-1, 3)$ as the result of the algorithm. It can easily be verified that this result is correct, because

$$gcd(100, 35) = 5 = -1 \cdot 100 + 3 \cdot 35$$

### A.2.4 Prime Numbers

Prime numbers (or primes) as formally introduced in Definition A.25 are frequently used in mathematics and even more so in public key cryptography.[20]

**Definition A.25 (Prime number)** *A natural number* $1 < n \in \mathbb{N}$ *is called a* prime number *(or* prime*) if it is divisible only by* 1 *and itself.*

20 The first recorded definition of a prime was again given by Euclid. There is even some evidence that the concept of primality was known earlier to Aristotle and Pythagoras.

Contrary to that, a natural number $1 < n \in \mathbb{N}$ that is not prime is called *composite* (note that 1 is neither prime nor composite). In this book, the set of all prime numbers is denoted as $\mathbb{P}$. This set is infinitely large (i.e., $|\mathbb{P}| = \infty$), and the first 8 elements are $2, 3, 5, 7, 11, 13, 17$, and $19$.

Suppose you want to find the set of all primes that are less or equal than a certain threshold $n$ (e.g., $n = 20$). In the third century B.C., Eratosthenes proposed an algorithm to systematically find these primes, and this algorithm introduced the notion of a *sieve*. The sieve starts by writing down the set of all natural numbers between 2 and $n$. In our example, this looks as follows:

$$\{2, 3, 4, 5, 6, 7, 8, 9, 10, 11, 12, 13, 14, 15, 16, 17, 18, 19, 20\}$$

Next, all numbers bigger than 2 (i.e., the smallest prime) which are multiples of 2 are removed from the set (this means that all even numbers are removed). The following set remains:

$$\{2, 3, 5, 7, 9, 11, 13, 15, 17, 19\}$$

This step is repeated for every prime number that is less or equal than $\sqrt{n}$. In our example, $\sqrt{20} \approx 4.472$, and this means that the step must be repeated only for the prime number 3. The following set remains:

$$\{2, 3, 5, 7, 11, 13, 17, 19\}$$

What is left is the set of prime numbers less than 20. In this example, the cardinality of the prime number set is 8. In general, it is measured by the prime counting function $\pi(n)$ that is introduced next.

### A.2.4.1 Prime Counting Function

The *prime counting function* $\pi(n)$ counts the number of primes that are less or equal to $n \in \mathbb{N}$:

$$\pi(n) := |\{p \in \mathbb{P} \mid p \leq n\}|$$

The following table illustrates the first couple of values of the prime counting function $\pi(n)$. Note that the function grows monotonically.

| $n$ | 2 | 3 | 4 | 5 | 6 | 7 | 8 | 9 | 10 | 11 | 12 | 13 | 14 | ... |
|---|---|---|---|---|---|---|---|---|---|---|---|---|---|---|
| $\pi(n)$ | 1 | 2 | 2 | 3 | 3 | 4 | 4 | 4 | 4 | 5 | 5 | 6 | 6 | ... |

In public key cryptography, one often uses very large prime numbers. Consequently, one may ask whether there are arbitrarily sized prime numbers. As stated in Theorem A.5, this question can be answered in the affirmative.

**Theorem A.5** *There are infinitely many primes.*

*Proof.* Assume that there are finitely many primes $p_1, \ldots, p_n$. Consider the number $m = p_1 \cdots p_n + 1$. Because $m$ is bigger than any prime, it must be composite, and hence it must be divisible by some prime. We note, however, that $m$ is not divisible by $p_1$, as when we divide $m$ by $p_1$ we get the quotient $p_2 \cdots p_n$ and a remainder of 1. Similarly, $m$ is not divisible by any $p_i$ for $i = 2, \ldots, n$. Consequently, we get a contradiction and hence the assumption (i.e., there are finitely many primes) must be wrong. This proves the theorem.

□

Although there are infinitely many primes, it may still be the case that they are sparse and that finding a large prime is prohibitively difficult. Consequently, a somewhat related question asks for the density of the prime numbers: How likely does an interval of a given size comprise a prime number? We can use the prime density theorem addressed next to answer this and similar questions.

#### A.2.4.2 Prime Density Theorem

Theorem A.6 is called the *prime density theorem*. It states that arbitrarily sized prime numbers do in fact exist, and that finding them is not difficult—even for very large numbers. We give the theorem without a proof.

**Theorem A.6 (Prime density theorem)**

$$\lim_{n \to \infty} \frac{\pi(n) \ln(n)}{n} = 1$$

More precisely, it is known that

$$\pi(n) \geq \frac{n}{\ln(n)}$$

for $2 < n \in \mathbb{N}$ and that

$$\pi(n) \leq 1.25506 \frac{n}{\ln(n)}$$

for $17 \leq n \in \mathbb{N}$. This means that $\pi(n) \approx n/\ln(n)$ is indeed a very good approximation for almost all $n \in \mathbb{N}$.

The prime density theorem suggests that for sufficiently large $n$ the value $\pi(n)$ is about $n/\ln(n)$. This implies that roughly every $\ln(n)^{th}$ number of the size of $n$ is prime, and hence that the probability of $n$ being prime is approximately $1/\ln(n)$—or $2/\ln(n)$ if we only consider odd numbers $n$. In the case of RSA with a 2,048-bit modulus $n$, for example, the respective prime numbers $p$ and $q$ have a length of about 1,024 bits. The probability that a random odd number is a prime is roughly

$$\frac{2}{\ln(2^{1024})} = \frac{2}{1024 \cdot \ln(2)} \approx \frac{2}{710} = \frac{1}{355}$$

There are several open conjectures on prime numbers. For example, it is conjectured that there exist infinitely many twin primes; that is, primes $p$ for which $p+2$ is also prime, or that every even number is the sum of two primes.

#### A.2.4.3 Generating Large Primes

In cryptography, one often needs large primes, and there are basically two approaches to generate them:

- One can construct provable primes (e.g., [7]).
- One can randomly choose large odd integers and test them for primality (or compositeness, respectively).

In practice, one usually prefers the second approach, meaning that one randomly chooses large odd integers and subjects them to primality (or compositeness) testing. If a number turns out to be composite, then it is discarded and the next odd integer is taken into consideration.[21] This is repeated until sufficiently many primes are found. The prime density theorem suggests that this happens within a reasonable amount of time. To follow this approach, however, an algorithm to solve the primality decision problem captured in Definition A.26 is needed.

**Definition A.26 (Primality decision problem)** *Given a positive integer $n \in \mathbb{N}$, decide whether $n$ is prime (i.e., $n \in \mathbb{P}$) or composite (i.e., $n \notin \mathbb{P}$).*

The simplest algorithm to solve the primality decision problem is *trial division*: Test all primes between 2 and $\sqrt{n}$ to see if any of them divides $n$. If such a prime exists, then $n$ is not prime; otherwise, it is. Note that it is sufficient to find only one prime that divides $n$. If such a prime is found, then the algorithm can abort and need not go through all other primes.

21 This simple approach is often used in libraries, but one can also throw a new random number that is not divided by a given list of small primes. Also, there are more sophisticated methods to construct the next odd integer (not addressed here).

Trial division is simple and straightforward, but it is also highly inefficient. It requires a list of all prime numbers between 2 and $\sqrt{n}$, and it must test them all in the worst case (i.e., if $n$ happens to be prime). According to Theorem A.6, this means that the algorithm must perform

$$\frac{\sqrt{n}}{\ln \sqrt{n}}$$

trial divisions to verify that $n$ is prime. If, for example, $n$ is larger than $10^{75}$, this means

$$\frac{\sqrt{10^{75}}}{\ln \sqrt{10^{75}}} > 3.66 \cdot 10^{35}$$

trial divisions, and this is clearly beyond the computational capabilities of everybody. The bottom line is that trial division has a running time that is exponential in the input length, and such algorithms are computationally intractable to execute even for moderately sized numbers.

It is obvious that the primality decision problem can be solved if the IFP (Definition 5.8) can be solved. For a long time, however, it was not known whether the primality decision problem is simpler to solve than the IFP. Since the publication of [8], however, we know that the primality decision problem is in the complexity class $\mathbf{P}$, meaning that there are deterministic algorithms to solve the primality decision problem in polynomial time. This is not known to be true for the IFP (i.e., it is unknown whether the IFP is in $\mathbf{P}$ or $\mathbf{NP}$). A polynomial-time algorithm is theoretically better than an exponential-time one, but the degree of the polynomial matters and may still be too large to be used in the field. This applies to the algorithm proposed in [8] and some of its successors.

In practice, one prefers nondeterministic or probabilistic algorithms[22] that also solve the primality decision problem in polynomial time, but for which the degree of the polynomial is significantly smaller than the one proposed in [8]. In such an algorithm, one usually applies several primality tests to reveal the fact that the integer in question is composite. If this is the case, then one can be absolutely sure that the integer is composite. Otherwise, one assumes that the integer is prime, and one can increase the level of assurance by running more tests. Integers that are not known to be prime, but have passed several tests in the affirmative, are sometimes called *probable primes*.

In the sequel, we outline some primality tests that are used in the field (i.e., the Fermat, Solovay-Strassen, and Rabin-Miller test). Each test makes use of one or

22 Some of these algorithms could be converted into deterministic ones if one knew that the extended Riemann hypothesis is true. Most mathematicians believe that this is the case.

several randomly chosen auxiliary numbers $1 < a < n$. If such a number $a$ provides evidence for $n$ to be prime (composite), then it is called a *witness* for the primality (compositeness) of $n$. A problem is that some $a$ may be *false witnesses* (or *liars*), meaning that they provide evidence that $n$ is prime (composite), whereas in reality it is not. The respective $n$ is then sometimes called *pseudoprime* with respect to the test in use. It goes without saying that a primality test is good if it has not too many liars and pseudoprimes.

### *Fermat Test*

As its name suggests, the Fermat test goes back to Pierre de Fermat[23] and Fermat's little theorem (Theorem A.9). In short, the theorem states that for every prime $p$ and number $a$ not divisible by $p$, the equivalence $a^{p-1} \equiv 1 \pmod{p}$ must hold. Consequently, one can test the primality—or rather the compositeness—of $n$ by randomly choosing an $a$ (not divisible by $n$) and computing $a^{n-1} \pmod{n}$. If this value is not equal to 1, then $n$ is definitively not a prime (and we have found a witness for the compositeness of $n$). Unfortunately, the converse is not true; that is, finding an $a$ for which $a^{n-1} \equiv 1 \pmod{n}$ does not imply that $n$ is prime.[24] In fact, there is an entire class of integers $n$ that are composite but for which $a^{n-1} \equiv 1 \pmod{n}$ is true for every $a < n$ with $gcd(a, n) = 1$, meaning that all such $a$ are liars. These numbers are called *Carmichael numbers*,[25] and one can show that there exist infinitely of them. Every Carmichael number yields a pseudoprime for the Fermat test, and the existence of (infinitely many) Carmichael numbers is certainly one of the main reasons why the Fermat test is not so widely used in the field.

### *Solovay-Strassen Test*

The Solovay-Strassen test is another probabilistic compositeness testing algorithm developed and proposed by Robert M. Solovay and Volker Strassen in 1977 [9]. Similar to the Fermat test, it proves the compositeness of $n$ with certainty, but not its primality. The test is based on quadratic residuosity ( Appendix A.3.7) in general, and Euler's criterion (Theorem A.11) in particular. This criterion suggests that for every prime number $n$ and $1 \leq a \leq n-1$ with $gcd(a, n) = 1$, the Jacobi symbol $(a|n)$ and the value $a^{(n-1)/2} \pmod{n}$ must be the same; that is, 1 for $a \in QR_n$ and $-1$ for $a \in QNR_n$. If these two values are not the same, then $n$ must be composite.

23 Pierre de Fermat was a French mathematician who lived from 1607 to 1665.

24 For this reason, the Fermat test is sometimes also referred to a compositeness test.

25 It can be shown that a Carmichael number must be odd, square free, and divisible by at least three prime numbers. The first three Carmichael numbers are $561 = 3 \cdot 11 \cdot 17$, $1105 = 5 \cdot 13 \cdot 17$, and $1729 = 7 \cdot 13 \cdot 19$.

As further explained in Appendix A.3.7, the Jacobi symbol $(a|n)$ can be computed if the prime factorization of $n$ is known, but it can also be computed if the prime factorization of $n$ is unknown (as is the case here).

The Solovay-Strassen test thus follows the following rationale: If for a given $n$ one can find an $a$ between 1 and $n-1$ with $gcd(a,n) = 1$ for which $(a|n) \neq a^{(n-1)/2} \pmod{n}$, then one can conclude that $n$ is composite (and $a$ then represents a witness for this fact). The algorithm randomly chooses several values for $a$ and verifies whether $(a|n)$ and $a^{n-1/2} \pmod{n}$ are the same. If they are, then the level of assurance that $n$ is prime increases. If they are not, then an *Euler witness* for the compositeness of $n$ is found and the algorithm terminates.

If, for example, one wants to apply the Solovay-Strassen test to $n = 13 \cdot 17 = 221$ (that is composite and not prime), then one can randomly select $a = 47$ (note that $gcd(47, 221) = 1$) and compute $(47|221) = -1$ and $47^{(221-1)/2} \pmod{221} = -1$. This suggests that 221 is either prime, or 47 is an Euler liar for 221. The algorithm continues with $a = 2$ (again, note that $gcd(2, 221) = 1$) and computes $(2|221) = -1$ and $2^{(221-1)/2} \pmod{221} = 30$. This suggests that 2 is an Euler witness for the compositeness of 221, and hence that 221 is definitively composite and not prime.

It can be shown that for all $n$, half of the numbers $a$ between 1 and $n-1$ with $gcd(a,n) = 1$ are Euler witnesses, meaning that they can truly test the compositeness of $n$. This means that in every round (and for every $a$), there is a probability of $1/2$ that the two values match and $1/2$ that they don't match. So if the test is executed $k$ times and the two values are the same in every execution, then $n$ is assumed to be prime with a probability of $1 - 2^{-k}$. By increasing $k$, this probability can be made as close to 1 as needed. Most importantly, there are no classes of numbers (like the Carmichael numbers in the case of the Fermat test) that can pass the Solovay-Strassen test for more than the half possible values a without being prime. It goes without saying that this is a big advantage when used in the field.

*Miller-Rabin Test*

The Miller-Rabin test is yet another probabilistic compositeness testing algorithm that was developed and proposed by Gary L. Miller and Michael O. Rabin in the 1970s.[26] Its original version, created by Miller [10], is deterministic and relies on the unproven generalized Riemann hypothesis. Rabin later turned the deterministic algorithm into a probabilistic one [11]. This version is usually meant when people refer to the Miller-Rabin test.

26 It is rumored that the test was coinvented and also used by the U.S. mathematician John L.Selfridge, who lived from 1927 to 2010.

Like the Fermat test, the Miller-Rabin test is also based on Fermat's little theorem (Theorem A.9), but it is more stringent in the sense that it is able to detect liars the Fermat test is not able to detect. The Miller-Rabin test employs the fact that the unity 1 has only trivial square roots; that is, 1 and $-1$, in $\mathbb{Z}_n^*$ if $n$ is prime,[27] and that this is not the case if $n$ is composite, meaning that 1 may then also have nontrivial square roots. So if a nontrivial square root of 1 is found in $\mathbb{Z}_n^*$, then $n$ must be composite; otherwise, $n$ can still be prime.

Before we start explaining the Miller-Rabin test, we observe that for every odd integer $n > 2$, $n-1$ must be even and can be written as $2^s r$, where $s \geq 1$ and $r$ are integers and $r$ must be odd (i.e., 2 does not divide $r$). More specifically, the remainder $r$ results from repeatedly factoring out 2 from $n-1$. If, for example, $n = 89$ (which is prime), then $n-1 = 88$ and this number can be written as $2^3 \cdot 11$; that is, $s = 3$ and $r = 11$. If $n = 105$ (which is not prime because $105 = 3 \cdot 5 \cdot 7$), then $n-1 = 104$ and this number can be written as $2^3 \cdot 13$; that is, $s = 3$ and $r = 13$. All computations are done modulo $n$.

The following computations are done modulo $n$. One can compute $a^{n-1}$ for any $1 \leq a \leq n-1$ by first computing $a^r$ and then squaring the result $s$ times. In our first example with $n = 89$, this suggests that one can compute $a^{88}$ by first computing $a^r = a^{11}$ and then squaring the result $s = 3$ times (this yields $a^{11}$, $a^{22}$, $a^{44}$, and $a^{88}$). This can be done for arbitrary values of $a$. For $a = 3, a = 5, a = 11$, and $a = 2$, for example, this may look as follows (all values are modulo 89):

| $a$ | $a^{11}$ | $a^{22}$ | $a^{44}$ | $a^{88}$ |
|---|---|---|---|---|
| 3 | 37 | 34 | -1 | 1 |
| 5 | 55 | -1 | 1 | 1 |
| 11 | -1 | 1 | 1 | 1 |
| 2 | 1 | 1 | 1 | 1 |

Note that -1 actually refers to the element $88 \in \mathbb{Z}_{89}$. It is used in this tabular representation only to emphasize the fact that the square root (and hence the entry to the left) of every 1 must either be 1 or -1. This requirement is fulfilled here because 89 is prime. As a side remark we note that the Fermat test only verifies whether the entry in the last column is 1 for every value of $a$. It goes without saying that this is fulfilled here, and hence the Fermat test would also come to the conclusion that 89 is probably prime. But the Miller-Rabin test does more in verifying that in each row a 1 cannot be preceded by something other than -1 or 1.

27 To show that there are no other (i.e., nontrivial) square roots of 1, one supposes that $x$ is a square root of 1 $(\text{mod } p)$. This means that $x^2 \equiv 1 \pmod{p}$, and hence $x^2 - 1 = (x+1)(x-1) \equiv 0 \pmod{p}$. Therefore $p \mid (x+1)(x-1)$. Since $p$ is prime, we must have $p \mid (x+1)$ or $p \mid (x-1)$. It means that $x = -1$ or $+1 \pmod{p}$.

To see what this may look like, we revisit the second example given above, namely $n = 105$ (that is not prime). In this case, $n - 1 = 104 = 2^3 \cdot 13$, and hence $s = 3$ and $r = 13$. To compute $a^{104}$, one first computes $a^r = a^{13}$ and then squares the result $s = 3$ times (to yield $a^{26}$, $a^{52}$, and $a^{104}$). For $a = 2$ and $a = 8$, the tabular representation may look as follows (all values are modulo 105):

| $a$ | $a^{13}$ | $a^{26}$ | $a^{52}$ | $a^{104}$ |
|---|---|---|---|---|
| 2 | 2 | 4 | 16 | 46 |
| 8 | 8 | 64 | $\underline{1}$ | 1 |

The point to stress is that the square root of $8^{52} = 1$ (underlined above) is 64 and neither 1 nor -1. This means that we have found a square root of $8^{52} = 1$ (i.e., $8^{26}$) that is not 1 or -1, and hence 105 cannot be prime. This fact is not found by the Fermat test (since $8^{104} = 1$ holds). The bottom line is that $a = 8$ is a witness for the compositeness of $n = 105$ in the Miller-Rabin test but a liar in the Fermat test. Needless to say, the Fermat test would still find the fact that 105 is not prime by verifying $2^{104} \neq 1$.

Informally speaking, the Miller-Rabin test starts with a tabular representation as given above and adds rows for different values of $a$. In each row, the test verifies whether the row consists of only ones (like the row for $a = 2$ in the first example) or each 1 is preceded by either -1 or 1. If any other value is found to precede a 1 (like in the row for $a = 8$ in the second example), then the test terminates with the firm statement that $n$ is composite.

More formally, the Miller-Rabin test tries to find an $2 \leq a \leq n - 2$ with

$$a^r \not\equiv 1 \pmod{n}$$

and

$$a^{2^j r} \not\equiv -1 \pmod{n}$$

for all $0 \leq j \leq s - 1$. If such an $a$ is found, then it yields a witness for the compositeness of $n$. Otherwise, it is probable (but not absolutely sure) that $n$ is prime. More specifically, it can be shown that a composite number passes the Miller-Rabin test for at most $1/4$ of the possible values of $a$. If, for example, $n = 221 = 13 \cdot 17, n - 1 = 220 = 2^2 \cdot 55$ (i.e., $s = 2$ and $r = 55$, and $a = 174$), then $a^{55} = 47$, $a^{110} = -1$, and $a^{220} = 1$. It then looks as if 174 speaks in favor of the primality of 221. But this is a fallacy, and 174 is actually a Miller-Rabin liar (and 221 is a composite number).[28] Fortunately, Miller-Rabin liars are relatively rare, and one

28 The other liars are 21, 47, and 200 in this example.

can show that the probability that the Miller-Rabin test passes $k$ tests for a composite number $n$ is at most $1/4^k$. This means that the error probability of the Miller-Rabin test decreases exponentially fast—even faster than the Solovay-Strassen test.

In spite of the fact that the Miller-Rabin test looks involved, its execution is actually very simple, and hence the Miller-Rabin test is the primality (or compositeness) testing algorithm of choice by almost all practitioners working in the field.

#### A.2.4.4 Safe and Strong Primes

In cryptography, one often uses primes with specific properties. For example, a prime number $p$ is called *safe*, if it is equal to $2p' + 1$ for some other prime number $p'$ (i.e., $p = 2p' + 1$ with $p, p' \in \mathbb{P}$).[29] This other prime number $p'$ is then called a *Sophie Germain prime*. This means that a prime number $p'$ is a Sophie Germain prime if $p = 2p' + 1$ is also prime (and it is then called safe). For example, 23 is a Sophie Germain prime because it is prime and $2 \cdot 23 + 1 = 47$ is also prime. The first 8 safe primes are

$$\begin{aligned}
5 &= 2 \cdot 2 + 1 \\
7 &= 2 \cdot 3 + 1 \\
11 &= 2 \cdot 5 + 1 \\
23 &= 2 \cdot 11 + 1 \\
47 &= 2 \cdot 23 + 1 \\
59 &= 2 \cdot 29 + 1 \\
83 &= 2 \cdot 41 + 1 \\
107 &= 2 \cdot 53 + 1
\end{aligned}$$

They are generated with the 8 Sophie Germain primes $2, 3, 5, 11, 23, 29, 41$, and $53$. The set of all Sophie Germain primes is indeed a distinct subset of $\mathbb{P}$, and there are prime numbers $p' \in \mathbb{P}$ for which $p = 2p' + 1$ is not prime, such as $15 = 2 \cdot 7 + 1$ for $p' = 7$ or $27 = 2 \cdot 13 + 1$ for $p' = 13$. In fact, we know that for every $p' \in \mathbb{P}$, $p = 2p' + 1$ is either a safe prime or it is not prime.

Another distinct property of some primes is their strength. In cryptography,[30] a large prime $p$ is said to be *strong* if the following conditions are satisfied:

- $p - 1$ has large prime factors (i.e., $p = a_1 q_1 + 1$ for some integer $a_1$ and large prime $q_1$).

29 Following a similar line of argumentation, $p$ is called *quasi-safe*, if it is equal to $2p' + 1$ for some prime power $p'$ (instead of a prime). The notion of a quasi-safe prime is not further used in this book.

30 Note that the notion of a strong prime is defined in a different way in number theory.

- $q_1 - 1$ has large prime factors (i.e., $q_1 = a_2 q_2 + 1$ for some integer $a_2$ and large prime $q_2$).
- $p+1$ has large prime factors (i.e., $p = a_3 q_3 - 1$ for some integer $a_3$ and large prime $q_3$).

Strong primes are popular in public key cryptography mainly because it is assumed that they better resist algorithms to factorize large integers and compute discrete logarithms. In the case of RSA, for example, it is sometimes recommended that the modulus $n$ should be chosen as the product of two strong primes. This defeats Pollard's $p-1$ algorithm (Section 5.3.1.2), but it does not defeat other integer factorization algorithms that may even be more powerful. Similarly, in the case of cryptosystems based on the DLP, it is sometimes required that $p$ is strong because this requirement ensures that $p-1$ has at least one large prime factor, and this, in turn, defeats the Pohlig-Hellman algorithm [12]. But this requirement it already ensured if $p$ is safe. The bottom line is that the importance of strong primes is discussed controversially in the community (e.g., [13]).

### A.2.5 Factorization

It is well known that a prime $p$ that divides the product $ab$ of two natural numbers $a, b \in \mathbb{N}$ divides at least one of the two factors (i.e., $a$ or $b$). To prove this fact, one assumes that $p$ divides $ab$ but not $a$ and one then shows that $p$ must divide $b$. The fact that $p$ is prime implies that $gcd(a,p) = 1$, and this, in turn, means that—according to Bézout's identity (Theorem A.4)—there exist $x, y \in \mathbb{N}$ with $1 = ax + py$. This equation can be multiplied with $b$ to get $b = abx + pby$. Because $p$ divides $(ab)x$ and $pby$ on the right side of the equation, $p$ must also divide $b$ on the left side of the equation. This result can be generalized to more than two factors: If $p$ divides a product

$$\prod_{i=1}^{k} q_i$$

of $k$ prime factors $q_i$, then $p$ must be equal to one of the prime factors $q_1, \ldots, q_k$. This brings us to one of the fundamental theorems of integer arithmetic, namely that every natural number has a unique prime factorization. This theorem was proven by Gauss in 1801, and it is captured in Theorem A.7 (without a proof).

**Theorem A.7 (Unique factorization)** *Every natural number $n \in \mathbb{N}$ can be factored uniquely (up to a permutation of the prime factors):*

$$n = \prod_{p \in \mathbb{P}} p^{e_p(n)}$$

In this formula, $e_p(n)$ refers to the exponent of prime $p$ in the factorization of $n$. For almost all $p \in \mathbb{P}$ this value is zero, meaning that it is greater than zero only for finitely many primes $p$. Using this notation, the greatest common divisor and least common multiple are defined as follows:

$$gcd(a,b) = \prod_{p \in \mathbb{P}} p^{min(e_p(a),e_p(b))}$$

$$lcm(a,b) = \prod_{p \in \mathbb{P}} p^{max(e_p(a),e_p(b))}$$

The algorithms we learn in school to compute greatest common divisors and least common multiples are directly derived from these equations. Note, however, that these algorithms can only be used if the prime factorizations of $a$ and $b$ are known.

In the realm of integer factorization algorithms, the notion of a smooth integer is sometimes used. Informally speaking, an integer is *smooth* if it is the product of only small prime factors. More specifically, we must say what a "small prime factor" is, and hence one has to define smoothness with respect to a bound $B$. This is captured in Definition A.27.

**Definition A.27 (B-smooth integer)** *Let $B$ be an integer. An integer $n$ is* B-smooth *if every prime factor of $n$ is less than $B$.*

For example, the integer $n = 4^3 \cdot 5^{2345} \cdot 17^2$ is 18-smooth (because 17 is the largest prime factor of $n$ and is less than 18). If we know that an integer is B-smooth, then we know something about the internal structure of the integer, and this knowledge may be exploited in an integer factorization algorithm.

### A.2.6 Euler's Totient Function

As its name suggests, Euler's totient function was proposed by Leonhard Euler[31] as a function that counts the numbers that are smaller than $n \in \mathbb{N}$ and have no other common divisor with $n$ other than 1 (i.e., they are coprime with $n$). More formally, $\phi : \mathbb{N} \rightarrow \mathbb{N}$ is defined by $\phi(1) = 1$ and $\phi(n) = |\mathbb{Z}_n^*|$ for $n \geq 2$. Note that the

31 Leonhard Euler was a Swiss mathematician who lived from 1707 to 1783.

notation is not fixed, and that Euler's totient function is sometimes also denoted as $\varphi$ (instead of $\phi$). Also note that an equivalent definition of the function is as follows:

$$\phi(n) = |\{a \in \{0, \ldots, n-1\} \mid gcd(a, n) = 1\}|$$

In either case, Euler's totient function has the following properties:

- If $p$ is prime, then every number smaller than $p$ is coprime with $p$. Consequently, $\phi(p) = p - 1$ for every prime number $p$.
- If $p$ is prime and $1 \leq k \in \mathbb{Z}$, then $\phi(p^k) = p^k - p^k/p$. This is because every p-th number between 1 and $p^k$ is not coprime with $p^k$ (because $p$ is a common divisor of $p^i$ for $i = 1, \ldots, k-1$ and $p^k$) and we have to subtract $p^k/p$ from $p^k$ accordingly. Note that $p^k - p^k/p = p^k - p^{k-1} = p^{k-1}(p-1)$, and hence $\phi(p^k) = (p-1)p^{k-1}$. This is the equation that is usually found in textbooks.
- If $n$ is the product of two distinct primes $p$ and $q$; that is, $n = pq$, then $\phi(n) = \phi(p)\phi(q) = (p-1)(q-1)$. This is because the numbers $0, p, 2p, \ldots, (q-1)p, q, 2q, \ldots, (p-1)q$ are not coprime with $n$, and there are $1 + (q-1) + (p-1) = p + q - 1$ of these numbers (they are all different from each other if $p \neq q$). Consequently, $\phi(n) = pq - (p+q-1) = pq - p - q + 1 = (p-1)(q-1)$.

  The bottom line is that for any integer $n$ with prime factorization

$$n = \prod_i q_i^{k_i}$$

$\phi(n)$ can be computed as follows:

$$\phi(n) = \prod_i (q_i - 1)q_i^{k_i - 1}$$

Using this formula, for example, one can easily derive $\phi(45)$ from the prime factorization of $45 = 3^2 \cdot 5$: $\phi(45) = (3-1) \cdot 3^{2-1} \cdot (5-1) \cdot 5^{1-1} = 2 \cdot 3^1 \cdot 4 \cdot 5^0 = 2 \cdot 3 \cdot 4 \cdot 1 = 24$.

But if the prime factorization of $n$ is unknown, then it is difficult to compute $\phi(n)$. In fact, one can show that for $n$ being a product of two primes $p$ and $q$; that is, $n = pq$, computing $\phi(n)$ is as hard as finding the prime factorization of $n$. This means that if one can compute $\phi(n)$, then one can also factorize $n$. To prove this fact, one starts with $\phi(pq) = (p-1)(q-1) = pq - (p+q) + 1 = n - (p+q) + 1$, and derives equation (A.1) from it:

$$p + q = n - \phi(pq) + 1 \quad \text{(A.1)}$$

On the other hand, one knows that $(p-q)^2 = p^2-2pq+q^2 = p^2+2pq+q^2-4pq = (p+q)^2 - 4pq = (p+q)^2 - 4n$. Computing the square root on either side of this equation, one gets $p - q = \sqrt{(p+q)^2 - 4n}$. In this equation, one can substitute $p+q$ with the right side of equation (A.1). The result is equation (A.2):

$$p - q = \sqrt{(n - \phi(pq) + 1)^2 - 4n} \tag{A.2}$$

By adding (A.1) and (A.2), one gets the following formula to compute $(p+q) + (p-q) = p+q+p-q = 2p$:

$$2p = n - \phi(pq) + 1 + \sqrt{(n - \phi(pq) + 1)^2 - 4n}$$

The fact that only $n$ and $\phi(pq) = \phi(n)$ appear on the right side of the equation implies that one can compute $2p$ (and hence $p$) if one knows $n$ and $\phi(pq)$, and this, in turn, means that one can factorize $n$.

## A.3 MODULAR ARITHMETIC

Modular arithmetic elaborates on the ring[32] $\langle \mathbb{Z}_n, +, \cdot \rangle$ that consists of a complete residue system modulo $n$ (denoted as $\mathbb{Z}_n$) and two operations ($+$ and $\cdot$). In this setting, $+$ refers to the addition modulo $n$, and $\cdot$ refers to the multiplication modulo $n$. In this section, we only look at the aspects of modular arithmetic that are relevant for cryptography.

### A.3.1 Modular Congruence

Two integers are said to be *congruent* modulo a natural number if they represent the same value when computed modulo this number. This notion of modular congruence is formally expressed in Definition A.28.

**Definition A.28** *Let $a, b \in \mathbb{Z}$ and $n \in \mathbb{N}$. The element $a$ is* congruent to $b$ modulo $n$, *denoted $a \equiv b \pmod{n}$ if $n$ divides $a - b$; that is, $n \mid (a - b)$.*

For example, $7 \equiv 12 \pmod{5}$, $4 \equiv -1 \pmod{5}$, $12 \equiv 0 \pmod{2}$, and $-2 \equiv 19 \pmod{21}$.

It can be shown that congruence modulo $n$ defines an *equivalence relation* over $\mathbb{Z}$ (i.e., a relation that is reflexive, symmetric, and transitive).

- *Reflexivity*: For all $n \in \mathbb{N}$ and $a \in \mathbb{Z}$: $a \equiv a \pmod{n}$;

32 If $n$ is prime, then $\langle \mathbb{Z}_n, +, \cdot \rangle$ is actually a field.

- *Symmetry*: For all $n \in \mathbb{N}$ and $a, b \in \mathbb{Z}$: If $a \equiv b \pmod{n}$, then $b \equiv a \pmod{n}$;
- *Transitivity*: For all $n \in \mathbb{N}$ and $a, b, c \in \mathbb{Z}$: If $a \equiv b \pmod{n}$ and $b \equiv c \pmod{n}$, then $a \equiv c \pmod{n}$.

An equivalence relation over a set partitions the set into *equivalence classes*. In $\mathbb{Z}$, the equivalence classes are called *residue classes*. Every $a \in \mathbb{Z}$ is congruent modulo $n$ to some $b \in \{0, \ldots, n-1\}$, and hence $R_n(a)$ refers to a particular residue class that consists of all $x \in \mathbb{Z}$ that are congruent to $a$ modulo $n$. This can be formally expressed as follows:

$$R_n(a) := \{x \in \mathbb{Z} \mid a \equiv x \pmod{n}\}$$

In some literature, $\overline{a}$ or $a + n\mathbb{Z}$ are used to refer to $R_n(a)$. Furthermore, one employs the term residue to actually refer to a residue class. For example, the residue (class) of 0 modulo 2 refers to the set of all even integers, whereas the residue (class) of 1 modulo 2 refers to the set of all odd integers. Similarly, the residue classes modulo 4 are defined as follows:

$$\begin{aligned}
\overline{0} &= 0 + 4\mathbb{Z} = R_4(0) = \{0, 0 \pm 4, 0 \pm 2 \cdot 4, \ldots\} = \{0, -4, 4, -8, 8, \ldots\} \\
\overline{1} &= 1 + 4\mathbb{Z} = R_4(1) = \{1, 1 \pm 4, 1 \pm 2 \cdot 4, \ldots\} = \{1, -3, 5, -7, 9, \ldots\} \\
\overline{2} &= 2 + 4\mathbb{Z} = R_4(2) = \{2, 2 \pm 4, 2 \pm 2 \cdot 4, \ldots\} = \{2, -2, 6, -6, 10, \ldots\} \\
\overline{3} &= 3 + 4\mathbb{Z} = R_4(3) = \{3, 3 \pm 4, 3 \pm 2 \cdot 4, \ldots\} = \{3, -1, 7, -5, 11, \ldots\}
\end{aligned}$$

As mentioned in Section A.1.2.3, $\mathbb{Z}_n$ is sometimes used in short to refer to $\mathbb{Z}/n\mathbb{Z}$, and $\mathbb{Z}/n\mathbb{Z}$ stands for the quotient group of $\mathbb{Z}$ modulo $n\mathbb{Z}$. It consists of all residue classes modulo $n$ (there are $n$ such classes, because all residues $0, 1, \ldots, n-1$ can occur in the division with $n$). In fact, the set $\{0, \ldots, n-1\}$ is called a *complete residue system modulo* $n$. Other elements could also be used to get a complete residue system modulo $n$, but the ones mentioned above are the smallest ones.

The modulo $n$ operator defines a mapping $f : \mathbb{Z} \longrightarrow \mathbb{Z}_n$, and this mapping represents a homomorphism from $\mathbb{Z}$ onto $\mathbb{Z}_n$. This means that we can add and multiply residues (or residue classes, respectively) similar to integers. The corresponding rules are as follows:

$$\begin{aligned}
R_n(a + b) &= R_n(R_n(a) + R_n(b)) \\
R_n(a \cdot b) &= R_n(R_n(a) \cdot R_n(b))
\end{aligned}$$

Consequently, intermediate results of a modular computation can be reduced (i.e., computed modulo $n$) at any time without changing the result. The following examples illustrate this point:

$$\begin{aligned} R_7(12+18) &= R_7(R_7(12)+R_7(18)) = R_7(5+4) = R_7(9) = 2 \\ R_7(12\cdot 18) &= R_7(R_7(12)\cdot R_7(18)) = R_7(5\cdot 4) = R_7(20) = 6 \\ R_7(8^{37}+9^4) &= R_7(1^{37}+2^4) = R_7(1+16) = R_7(17) = 3 \end{aligned}$$

The fact that $\langle \mathbb{Z}_n, +, \cdot \rangle$ is (only) a ring implies that not all elements have inverse elements with regard to multiplication. For example, the multiplicative inverse element of 3 modulo 10 is 7 (because $3 \cdot 7 = 21 \equiv 1 \pmod{10}$), but the inverse of 4 modulo 10 does not exist.[33] One can show that $a$ has a multiplicative inverse modulo $n$ if and only if $gcd(a,n) = 1$, meaning that $a$ and $n$ must be coprime. In this case, the multiplicative inverse modulo $n$ can be computed using the extended Euclidean algorithm (Algorithm 3.2). If one replaces $b$ with $n$ in Bézout's identity (Theorem A.4), then one gets

$$xa + yn = gcd(a,n) = 1 \tag{A.3}$$

This is equivalent to $xa = 1 - yn$. Because all multiples of $n$ are equivalent to 0 modulo $n$, it follows that the right side of the equation is 1. This, in turn, suggests that $x$ is the multiplicative inverse of $a$ modulo $n$. Contrary to that, if $gcd(a,n) = k > 1$, then one can show that $a$ doesn't have a multiplicative inverse modulo $n$.

In the literature, $\mathbb{Z}_n^*$ is used to refer to the subset of $\mathbb{Z}_n$ that comprises all elements that are coprime with $n$, meaning all elements of $\mathbb{Z}_n$ that are invertible. More formally,

$$\mathbb{Z}_n^* := \{x \in \mathbb{Z}_n \mid gcd(x,n) = 1\}$$

Using this notation, $\langle \mathbb{Z}_n^*, \cdot \rangle$ is a commutative group that is sometimes abbreviated with $\mathbb{Z}_n^*$. If $n$ is prime, then $\mathbb{Z}_n^* = \mathbb{Z}_n \setminus \{0\}$ and $|\mathbb{Z}_n^*| = n - 1$.

### A.3.2 Modular Exponentiation

In cryptography, a frequently used computation is modular exponentiation. If, for example, we want to compute $a^b \pmod{n}$ for some $a \in \mathbb{Z}_n^*$ and $b \in \mathbb{N}$, then the simplest algorithm is to iteratively multiply ($a$ modulo $n$) $b$ times. If $b = 23$, then the

33 Note that 4 and 10 have a common factor 2, and that $4 \cdot a$ always contains a factor 2 and therefore cannot be 1 (for all $a \in \mathbb{Z}_n$).

following sequence of equations yields the result with 22 modular multiplications:

$$
\begin{aligned}
R_n(a^2) &= R_n(a \cdot a) \\
R_n(a^3) &= R_n(a \cdot R_n(a^2)) \\
R_n(a^4) &= R_n(a \cdot R_n(a^3)) \\
&\ldots \\
R_n(a^{23}) &= R_n(a \cdot R_n(a^{22}))
\end{aligned}
$$

This can be simplified considerably, and the following sequence of equations also yields the correct result but requires only seven modular multiplications:

$$
\begin{aligned}
R_n(a^2) &= R_n(a \cdot a) \\
R_n(a^4) &= R_n(R_n(a^2) \cdot R_n(a^2)) \\
R_n(a^5) &= R_n(a \cdot R_n(a^4)) \\
R_n(a^{10}) &= R_n(R_n(a^5) \cdot R_n(a^5)) \\
R_n(a^{11}) &= R_n(a \cdot R_n(a^{10})) \\
R_n(a^{22}) &= R_n(R_n(a^{11}) \cdot R_n(a^{11})) \\
R_n(a^{23}) &= R_n(a \cdot R_n(a^{22}))
\end{aligned}
$$

This method can be generalized and the resulting *square-and-multiply algorithm* as captured in Algorithm A.3 works for modular exponentiation in any (multiplicative) group. Let $\langle G, \cdot \rangle$ be such a group, $a$ an element of this group, and $b$ a natural number (i.e., $b \in \mathbb{N}$). If we want to compute the element $a^b$ of $G$, then we must have a binary representation of the exponent $b$ (i.e., $b = b_{k-1} \ldots b_1 b_0$) and process this bitstring from one end to the other. More specifically, we process the exponent from the most significant bit (i.e., $b_{k-1}$) to the least significant bit (i.e., $b_0$). The other direction is also possible and works similarly. In Algorithm A.3, the variable $s$ is used to accumulate the result. It is initially set to 1, and the exponent is processed from $b_{k-1}$ to $b_0$. For each exponent bit, the value $s$ is squared. If the bit is equal to one, then the result is multiplied with $a$. Otherwise, nothing is done. After iterating this step for every bit of the exponent, the algorithm returns $s$ that yields the element $a^b$ of $G$.

If we consider the group $\mathbb{Z}_{41}^*$ and want to compute $7^{22}$ in this group (note that 7 is a generator of $\mathbb{Z}_{41}^*$), then we write the exponent in binary notation; that is, $b = (22)_{10} = (10110)_2$, and set $s$ to one. According to Algorithm 3.3, the

**Algorithm A.3** The square-and-multiply algorithm.

$(a \in G, b \in \mathbb{N})$

---

$s = 1$
for $i = k - 1$ down to 0 do
    $s = s \cdot s$
    if $b_i = 1$ then $s = s \cdot a$

---

$(s)$

computation works as follows:

$$\begin{aligned}
7^{(1)_2} &= 1^2 \cdot 7 \equiv 7 \pmod{41} \\
7^{(10)_2} &= 7^2 \equiv 8 \pmod{41} \\
7^{(101)_2} &= 8^2 \cdot 7 \equiv 38 \pmod{41} \\
7^{(1011)_2} &= 38^2 \cdot 7 \equiv 22 \pmod{41} \\
7^{(10110)_2} &= 22^2 \equiv 33 \pmod{41}
\end{aligned}$$

In the first iteration, $s$ is squared and multiplied with 7 modulo 41. The result is 7. In the second iteration, this value is squared modulo 41. The result is 8. In the third iteration, this value is squared and multiplied with 7 modulo 41. The result is 38. In the fourth iteration, this value is squared and multiplied with 7 modulo 41. The result is 22. Finally, in the fifth and last iteration, this value is squared modulo 41. The result is 33. Consequently, $7^{22} \pmod{41}$ yields 33.

### A.3.3 Chinese Remainder Theorem

The *Chinese remainder theorem* (CRT) captured in Theorem A.8 suggests that—under certain conditions—a system of $k$ congruences $x \equiv a_i \pmod{n_i}$ for $i = 1, \dots k$ has a unique solution that fulfills all congruences simultaneously. The theorem is given without a proof here.

**Theorem A.8 (Chinese remainder theorem)** *Let*

$$\begin{aligned}
x &\equiv a_1 \pmod{n_1} \\
x &\equiv a_2 \pmod{n_2} \\
&\dots \\
x &\equiv a_k \pmod{n_k}
\end{aligned}$$

*be a system of k congruences with pairwise coprime moduli* $n_1, \dots, n_k$. *The system has a unique and efficiently computable solution* $x$ *in* $\mathbb{Z}_n$ *with* $n = \prod_{i=1}^{k} n_i$.

The fact that the solution is *unique* in $\mathbb{Z}_n$ means that all other solutions are not elements of $\mathbb{Z}_n$, meaning that they are outside of $\{0, 1, \ldots, n-1\}$. Formally, the set of all solutions is equal to the set of integers $y$ that are equivalent to $x$ modulo $n$; that is, $y \equiv x \pmod{n}$.

To solve the system of $k$ congruences, one sets $m_i = n/n_i$ and $y_i = m_i^{-1} \pmod{n_i}$ for $i = 1, \ldots, k$, meaning that $y_i$ is the multiplicative inverse of $m_i$ modulo $n_i$. Because all moduli are assumed to be pairwise coprime, $y_i$ is well defined for all $i = 1, \ldots, k$. The solution $x$ can then be computed as follows:

$$x \equiv \sum_{i=1}^{k} a_i m_i y_i \pmod{n} \tag{A.4}$$

For example, consider the following system of $k = 3$ congruences:

$$\begin{aligned} x &\equiv 5 \pmod{7} \\ x &\equiv 3 \pmod{11} \\ x &\equiv 11 \pmod{13} \end{aligned}$$

In this example, $n_1 = 7$, $n_2 = 11$, $n_3 = 13$ (note that these integers are pairwise coprime), and $n = 7 \cdot 11 \cdot 13 = 1001$. Furthermore, $a_1 = 5$, $a_2 = 3$, and $a_3 = 11$. To determine the solution $x$ in $\mathbb{Z}_{1001}$, one must first compute

$$\begin{aligned} m_1 &= 1001/7 = 143 \\ m_2 &= 1001/11 = 91 \\ m_3 &= 1001/13 = 77 \end{aligned}$$

and then

$$\begin{aligned} y_1 &\equiv 143^{-1} \pmod{7} = 5 \\ y_2 &\equiv 91^{-1} \pmod{11} = 4 \\ y_3 &\equiv 77^{-1} \pmod{13} = 12 \end{aligned}$$

Afterward, the solution $x$ can be computed as follows:

$$\begin{aligned} x &\equiv \sum_{i=1}^{k} a_i m_i y_i \pmod{n} \\ &\equiv a_1 m_1 y_1 + a_2 m_2 y_2 + a_3 m_3 y_3 \pmod{n} \\ &\equiv 5 \cdot 143 \cdot 5 + 3 \cdot 91 \cdot 4 + 11 \cdot 77 \cdot 12 \pmod{1001} \\ &\equiv 3575 + 1092 + 10164 \pmod{1001} \\ &\equiv 14,831 \pmod{1001} \\ &= 817 \end{aligned}$$

Consequently, $x = 817$ is the solution in $\mathbb{Z}_{1001}$, and $\{i \in \mathbb{Z} \mid 817 + i \cdot 1001\}$ is the set of all solutions in $\mathbb{Z}$.

### A.3.4 Fermat's Little Theorem

For every prime number $p \in \mathbb{P}$, $\langle \mathbb{Z}_p, +, \cdot \rangle$ is a field and $\langle \mathbb{Z}_p^*, \cdot \rangle$ is its multiplicative group with every element being invertible. As suggested by its name, Theorem A.9 was created by Pierre de Fermat.[34]

**Theorem A.9 (Fermat's little theorem)** *If p is a prime number and a an element of $\mathbb{Z}_p^*$, then $a^{p-1}$ must be congruent to 1 modulo p; that is, $a^{p-1} \equiv 1 \pmod{p}$.*

Because $\phi(p) = p - 1$ for every prime number $p$, Fermat's little theorem directly follows from Euler's theorem (see below). It has many applications in cryptography. It can, for example, be used to find the multiplicative inverse of $a$ modulo $p$. If we divide the equivalence $a^{p-1} \equiv 1 \pmod{p}$ by $a$ on either side, we get

$$a^{(p-1)-1} \equiv a^{p-2} \equiv a^{-1} \pmod{p}$$

This means that one can find the multiplicative inverse element of $a$ modulo $p$ by computing $a^{p-2} \pmod{p}$.[35] For example, if $p = 7$, then the multiplicative inverse of 2 modulo 7 can be computed as follows: $2^{-1} \equiv 2^{7-2} \equiv 2^5 \equiv 32 \pmod{7} = 4$. This is obviously correct, because $2 \cdot 4 = 8$ and 8 modulo 7 is equal to 1. Another application of Fermat's little theorem was already mentioned in Section A.2.4.3 in the realm of primality testing.

34 We already came across Pierre de Fermat in Section A.2.4.3.

35 This method is generally slower than the extended Euclidean algorithm, but is sometimes used when an implementation for modular exponentiation is already available. In addition, it does not work for computation modulo a large integer $n$, since it requires knowing the value of $\phi(n)$.

### A.3.5 Euler's Theorem

Fermat's little theorem goes back to the 17th century. In the 18th century, Leonhard Euler[36] generalized it to the case where $p$—or rather $n$—is not prime. This means that $\langle \mathbb{Z}_n, +, \cdot \rangle$ is a ring with the additive group $\langle \mathbb{Z}_n, + \rangle$ (with the neutral element 0) and the monoid $\langle \mathbb{Z}_n, \cdot \rangle$ (with the neutral element 1). If one restricts the set of numbers to $\mathbb{Z}_n^* = \{a \in \mathbb{Z}_n \mid gcd(a, n) = 1\}$ (i.e., the set of elements of $\mathbb{Z}_n$ that have inverse elements), then $\langle \mathbb{Z}_n^*, \cdot \rangle$ is a multiplicative group and inverse elements exist for all elements of this group. Its order can be computed with Euler's totient function (Appendix A.2.6); that is, $|\mathbb{Z}_n^*| = \phi(n)$. For example, $\phi(45) = 3 \cdot (3-1)^{2-1} \cdot (5-1) = 24$, and this value equals $|\mathbb{Z}_{45}^*| = |\{1, 2, 4, 7, 8, 11, 13, 14, 16, 17, 19, 22, 23, 26, 28, 29, 31, 32, 34, 37, 38, 41, 43, 44\}| = 24$.

In essence, Euler's theorem captured in Theorem A.10 says that any element $a$ in $\mathbb{Z}_n^*$ is equivalent to 1 modulo $n$ if it is multiplied $\phi(n)$ times.

**Theorem A.10 (Euler's theorem)** *For all $a, n \in \mathbb{N}$ with $gcd(a, n) = 1$ it must hold that $a^{\phi(n)} \equiv 1 \pmod{n}$.*

*Proof.* Let $G$ be a finite multiplicative group with $|G|$ elements. As a corollary of Lagrange's theorem (Theorem A.1), any element $a \in G$ raised to the $|G|$-th power yields 1 (the identity element of the multiplication). Here, $G$ is $\mathbb{Z}_n^*$ and $|G|$ is $\phi(n)$. Euler's theorem $a^{\phi(n)} \equiv 1 \pmod{n}$ thus follows directly from the corollary.

□

Because $\phi(n) = n - 1$ if $n$ is a prime, Fermat's little theorem directly follows from Euler's theorem.

### A.3.6 Finite Fields Modulo Irreducible Polynomials

As mentioned several times so far, finite fields play a pivotal role in cryptography today. In Section A.1.2.5, we mentioned that there is a prime field $\mathbb{F}_p$ (or $GF(p)$) for every $p \in \mathbb{P}$, and that there are also extension fields $\mathbb{F}_{p^n}$ (or $GF(p^n)$) for every positive integer $n \in \mathbb{N}^+$. Each element of such an extension field $\mathbb{F}_{p^n}$ can be seen as a polynomial of degree $n-1$ with coefficients from $\mathbb{F}_p$. The notion of a polynomial is introduced in Definition A.29.

**Definition A.29 (Polynomial)** *Let $A$ be an algebraic structure with addition and multiplication (e.g., a ring or a field). A function $p(x)$ is a* polynomial *in $x$ over $A$ if*

36 We already came across Leonhard Euler in Section A.2.6.

*it has the form*

$$p(x) = \sum_{i=0}^{n} a_i x^i = a_0 + a_1 x + a_2 x^2 + \ldots + a_n x^n$$

*where $n$ is a positive integer; that is, the degree of $p(x)$, denoted $deg(p)$, the coefficients $a_i$ $(0 \leq i \leq n)$ are elements in A, and $x$ is a variable not necessarily belonging to A.*

The set of all polynomials over $A$ is denoted as $A[x]$. Together with the addition and multiplication of polynomials, $A[x]$ forms a ring, and this means that one can add and multiply polynomials as if they were integers. Euclid's division theorem (Theorem A.3) applies; that is, if $f, g \in A[x]$ with $g \neq 0$, then $f = gq + r$ for $q, r \in A[x]$ and $deg(r) < deg(g)$, and hence one can also apply the Euclidean algorithms in $A[x]$. In this case, $r$ is the remainder of $f$ divided by $g$, denoted $r \equiv f \pmod g$. The set of all remainders of all polynomials in $A[x]$ modulo polynomial $g$ is denoted by $A[x]_g$.

To distinguish whether $A[x]_g$ is a ring or field, one needs to know whether $g$ is reducible or not. Roughly speaking, a nonconstant polynomial $f \in A[x]$ with $deg(f) > 1$ is *reducible* over $A$ if it can be factored into the product of two nonconstant polynomials $g, h \in A[x]$ with $deg(g) > 1$ and $deg(h) > 1$. Otherwise (i.e., if it cannot be factored into the product of two such polynomials), it is *irreducible* over $A$. Note that the reducibility of a polynomial depends on the algebraic structure $A$ over which the polynomial is defined (i.e., a polynomial can be reducible over one algebraic structure and irreducible over another).

If $\mathbb{F}$ is a field with $p$ elements and $f$ is a polynomial over $\mathbb{F}$ with degree $n > 0$, then $\mathbb{F}[x]_f$ is known to be a ring. If $f$ is irreducible over $\mathbb{F}$, then $\mathbb{F}[x]_f$ is even a field. Consequently, there is a field—or rather an extension field—$\mathbb{F}_{p^n}$ (or $GF(p^n)$) for every prime $p$ and positive integer $n$ (as already mentioned in Section A.1.2.5). All such fields are isomorphic, and we can use any polynomial $f$ with degree $n$ that is irreducible over $\mathbb{F}$; that is, $\mathbb{F}[x]_f$, to refer to $\mathbb{F}_{p^n}$ or $GF(p^n)$.

In a simple example, we may consider $GF(2^3)$ and the irreducible polynomial $x^3 + x + 1$ (another irreducible polynomial would be $x^3 + x^2 + 1$). $GF(2^3)_{(x^3+x+1)}$ (or $GF(2^3)_{(x^3+x^2+1)}$) yields an extension field in which one can add and multiply. The elements are 3-bit strings that represent polynomial (e.g., 101 for $x^2 + 1$). Addition is equivalent to the XOR of like terms, such as $(x + 1) + x = 1$. Multiplication can be implemented using polynomials. The respective multiplication table is shown in Table A.1. If, for example, we consider the underlined entry $x^2$, then this elements results from multiplying the polynomial $x^2 + 1$ (referring to 101) and the polynomial $x + 1$ (referring to 011). The resulting polynomial is

**Table A.1**
Multiplication Table in $GF(2^3)_{(x^3+x+1)}$

| | 000 | 001 | 010 | 011 |
|---|---|---|---|---|
| 000 | 0 | 0 | 0 | 0 |
| 001 | 0 | 1 | $x$ | $x+1$ |
| 010 | 0 | $x$ | $x^2$ | $x^2+x$ |
| 011 | 0 | $x+1$ | $x^2+x$ | $x^2+1$ |
| 100 | 0 | $x^2$ | $x+1$ | $x^2+x+1$ |
| 101 | 0 | $x^2+1$ | 1 | $\underline{x^2}$ |
| 110 | 0 | $x^2+x$ | $x^2+x+1$ | 1 |
| 111 | 0 | $x^2+x+1$ | $x^2+1$ | $x$ |
| | **100** | **101** | **110** | **111** |
| 000 | 0 | 0 | 0 | 0 |
| 001 | $x^2$ | $x^2+1$ | $x^2+x$ | $x^2+x+1$ |
| 010 | $x+1$ | 1 | $x^2+x+1$ | $x^2+1$ |
| 011 | $x^2+x+1$ | $x^2$ | 1 | $x$ |
| 100 | $x^2+x$ | $x$ | $x^2+1$ | 1 |
| 101 | $x$ | $x^2+x+1$ | $x+1$ | $x^2+x$ |
| 110 | $x^2+1$ | $x+1$ | $x$ | $x^2$ |
| 111 | 1 | $x^2+x$ | $x^2$ | $x+1$ |

$(x^3+x^2+x+1)$, and this is $x^2$ if taken modulo $(x^3+x+1)$. Instead of polynomials, the table could also be filled with the respective 3-bit strings, so $x^2$ could also be written as 100.

In a more realistic setting (at least for cryptographic applications), we may consider $GF(2^8)$ and the irreducible polynomial $x^8+x^4+x^3+x+1$. This yields the field $GF(2^8)_{(x^8+x^4+x^3+x+1)}$ in which every element of can be written as

$$b_7x^7 + b_6x^6 + b_5x^5 + b_4x^4 + b_3x^3 + b_2x^2 + b_1x + b_0$$

with binary coefficients $b_i \in \mathbb{Z}_2$ (for $0 \leq i \leq 7$). Again, every element of this field an be viewed as a sequence of 8 bits (i.e., one byte) or as a polynomial in $GF(2^8)_{(x^8+x^4+x^3+x+1)}$. For example, the byte 11010111 stands for the polynomial $x^7+x^6+x^4+x^2+x+1$. This field is used, for example, for the AES (Section 9.6.2).

### A.3.7 Quadratic Residuosity

In integer arithmetic, an $x \in \mathbb{Z}$ is a *square* if there is a $y \in \mathbb{Z}$ such that $x = y^2$. If such a $y$ exists, then it is called a *square root* of $x$. For example, 25 is a square with square root 5, whereas 20 is not a square. Similarly, all negative numbers are not squares in $\mathbb{Z}$ (because there is no integer whose square can be negative). If an integer $x$ is a square, then it has precisely two square roots in $\mathbb{Z}$ (i.e., $y$ and $-y$), and these values can be computed efficiently from $x$—even if $x$ is very large.

In modular arithmetic, things are similar but more involved. Instead of $\mathbb{Z}$, we are now working in $\mathbb{Z}_n$ and squares are called *quadratic residues* while everything else remains the same. The notions of quadratic residues and square roots are introduced in Definition A.30.

**Definition A.30 (Quadratic residue and square root)** *An element $x \in \mathbb{Z}_n$ is a* quadratic residue *modulo $n$ if there exists an element $y \in \mathbb{Z}_n$ such that $x = y^2 \pmod{n}$. If such a $y$ exists, then it is a* square root *of $x$ modulo $n$.*

The set of quadratic residues in $\mathbb{Z}_n^*$, denoted $QR_n$, is formally defined as follows:

$$QR_n := \{x \in \mathbb{Z}_n^* \mid \exists y \in \mathbb{Z}_n^* : y^2 \equiv x \pmod{n}\}$$

$QR_n$ is a multiplicative subgroup of $\mathbb{Z}_n^*$: If $x_1, x_2 \in QR_n$ with square roots $y_1$ and $y_2$, then the square root of $x_1x_2$ is $y_1y_2$ (because $(y_1y_2)^2 \equiv y_1^2y_2^2 \equiv x_1x_2 \pmod{n}$) and the square root of $x_1^{-1}$ is $y_1^{-1}$ (because $(y_1^{-1})^2 \equiv (y_1^2)^{-1} \equiv x_1^{-1} \pmod{n}$).

It is obvious that every element in $\mathbb{Z}_n^*$ is either a quadratic residue or not. In the second case, it is called a *quadratic nonresidue*. Consequently, the set of all quadratic nonresidues in $\mathbb{Z}_n^*$ is the complement of $QR_n$ with respect to $\mathbb{Z}_n^*$ (i.e., $QNR_n = \mathbb{Z}_n^* \setminus QR_n$). To further discuss the properties of quadratic residues and nonresidues, it is useful to distinguish whether $n$ is prime or composite. We explore these cases in separate sections next.

#### A.3.7.1 $QR_n$ When $n$ Is Prime

If $n$ is prime, then we use $p$ to refer to it. In this case, $\langle \mathbb{Z}_p, +, \cdot \rangle$ is a field and $\langle \mathbb{Z}_p^*, \cdot \rangle$ with $\mathbb{Z}_p^* = \mathbb{Z}_p \setminus \{0\}$ its multiplicative group. In this group, quadratic residuosity is relatively simple and not too different from integer arithmetic. For example, it can be shown that every $x \in QR_p$ has exactly two square roots modulo $p$; that is, $y$ and $-y = p - y$.[37] But some things are still fundamentally different from integer arithmetic. For example, in integer arithmetic squares are sparse, and they get sparser and sparser for larger $n$ (i.e., there are only about $\sqrt{n}$ perfect squares in the interval $[1, n]$). Contrary to that, half of the elements of $\mathbb{Z}_p^*$ are quadratic residues and hence

37 Note that $y$ and $p - y$ are always distinct if $p$ is a prime greater than 2.

elements of $QR_p$. In fact, the following equation holds for every prime $p > 2$:

$$|QR_p| = \frac{p-1}{2}$$

For example, in $\mathbb{Z}_7^*$ the elements $\{1, 2, 3, 4, 5, 6\}$ can be set to the power of 2 to figure out the quadratic residues:

| $x$ | 1 | 2 | 3 | 4 | 5 | 6 |
|---|---|---|---|---|---|---|
| $x^2$ | 1 | 4 | 2 | 2 | 4 | 1 |

Note that 1 and $7 - 1 = 6$ are mapped to 1, 2 and $7 - 2 = 5$ are mapped to 4, and 3 and $7 - 3 = 4$ are mapped to 2, so 1, 2, and 4 are the quadratic residues modulo 7; that is, $QR_7 = \{1, 2, 4\}$ and $QNR_7 = \mathbb{Z}_7^* \setminus QR_7 = \{3, 5, 6\}$.[38] Using the same line of argumentation, one can easily show that

$$QR_{19} = \{1, 4, 5, 6, 7, 9, 11, 16, 17\}$$

and

$$QNR_{19} = \mathbb{Z}_{19}^* \setminus QR_{19} = \{2, 3, 8, 10, 12, 13, 14, 15, 18\}$$

for $p = 19$. Consequently, for every prime $p > 2$, $\mathbb{Z}_p^*$ is partitioned into two equally sized subsets $QR_p$ and $QNR_p$; either subset comprises $(p - 1)/2$ elements.

We know from Fermat's little theorem that $x^{p-1} \equiv 1 \pmod{p}$ for every prime $p$ and $x \in \mathbb{Z}_p^*$. We also know that every element $x \in \mathbb{Z}_p^*$ has only two square roots, and that this also applies to 1 (with the two square roots 1 and $-1$). Combining these two facts suggests that the square root of $x^{p-1}$ (i.e., $x^{(p-1)/2}$) can either be 1 or -1. These two possibilities lead to Euler's criterion (Theorem A.11) that can be used to efficiently decide whether an $x \in \mathbb{Z}_p^*$ is a quadratic residue modulo $p$ (i.e., $x \in QR_p$) or a quadratic nonresidue modulo $p$ (i.e., $x \in QNR_p$).

**Theorem A.11 (Euler's criterion)** *Let $p \in \mathbb{P}$ be a prime number. For $x \in \mathbb{Z}_p^*$, $x \in QR_p$ if and only if*

$$x^{\frac{p-1}{2}} \equiv 1 \pmod{p}$$

38 Even though it is known that half of the elements in $\mathbb{Z}_p^*$ are quadratic nonresidues modulo $p$, there is no deterministic polynomial-time algorithm known for finding one. A randomized algorithm for finding a quadratic nonresidue is to simply select random integers $a \in \mathbb{Z}_p^*$ until one is found (using, for example, Euler's criterion). The expected number of iterations before a nonresidue is found is 2, and hence the algorithm is highly efficient and runs in polynomial time.

If Euler's criterion is not met, then

$$x^{\frac{p-1}{2}} \equiv -1 \pmod{p}$$

and hence $x$ is a quadratic nonresidue modulo $p$ (i.e., $x \in QNR_p$). The result of Euler's criterion is captured by the *Legendre symbol*[39] of $x$ modulo $p$ that is defined as follows:

$$\left(\frac{x}{p}\right) = \begin{cases} 0 & \text{if } x \equiv 0 \pmod{p} \\ 1 & \text{if } x \in QR_p \\ -1 & \text{if } x \in QNR_p \end{cases}$$

In some literature, the Legendre symbol is also written as $L(x,p)$ or $(x|p)$. For $p > 2$, it can be computed using Euler's criterion:

$$\left(\frac{x}{p}\right) \equiv x^{\frac{p-1}{2}} \pmod{p}$$

The following facts follow directly from the definition of the Legendre symbol:

$$\begin{aligned} \left(\frac{1}{p}\right) &\equiv 1^{\frac{p-1}{2}} \pmod{p} = 1 \\ \left(\frac{-1}{p}\right) &\equiv (-1)^{\frac{p-1}{2}} \text{ for every prime } p \\ \left(\frac{x^2}{p}\right) &= 1 \text{ for every } x \in \mathbb{Z}_p^* \\ \left(\frac{x}{p}\right) &= \left(\frac{y}{p}\right) \text{ for } x \equiv y \pmod{p} \end{aligned}$$

Furthermore, the Legendre symbol is multiplicative, meaning that

$$\left(\frac{xy}{p}\right) = \left(\frac{x}{p}\right) \cdot \left(\frac{y}{p}\right)$$

39 The Legendre symbol is named after Adrien-Marie Legendre, a French mathematician who lived from 1752 to 1833.

This fact directly results from Euler's criterion:

$$\begin{aligned}\left(\frac{xy}{p}\right) &\equiv (xy)^{\frac{p-1}{2}} \pmod p \\ &\equiv x^{\frac{p-1}{2}} y^{\frac{p-1}{2}} \pmod p \\ &\equiv x^{\frac{p-1}{2}} \pmod p \; y^{\frac{p-1}{2}} \pmod p \\ &= \left(\frac{x}{p}\right) \cdot \left(\frac{y}{p}\right)\end{aligned}$$

For example,

$$\left(\frac{6}{7}\right) = \left(\frac{2}{7}\right) \cdot \left(\frac{3}{7}\right) = 1 \cdot (-1) = -1$$

This result suggests that 6 is a quadratic nonresidue modulo 7 (i.e., $6 \in QNR_7$).

Note that Euler's criterion is not constructive in the sense that it provides an algorithm to compute the square roots of $x \in QR_p$. Only if $p \equiv 3 \pmod 4$ is there a simple formula to compute the square root $y$ of $x \in QR_p$:

$$y = x^{\frac{p+1}{4}} \pmod p \tag{A.5}$$

If $p \equiv 3 \pmod 4$, then $p + 1$ is a multiple of 4, and hence $\frac{p+1}{4}$ is an integer. In this case, it can be easily verified that $y^2 \equiv x \pmod p$:

$$\begin{aligned}y^2 &\equiv (x^{\frac{p+1}{4}})^2 \pmod p \\ &\equiv x^{\frac{p+1}{2}} \pmod p \\ &\equiv x^{\frac{p-1}{2}} \cdot x^{\frac{2}{2}} \pmod p \\ &\equiv x^{\frac{p-1}{2}} \cdot x \pmod p \\ &\equiv 1 \cdot x \\ &\equiv x \pmod p\end{aligned}$$

If $p \not\equiv 3 \pmod 4$, then the situation is more involved. Nevertheless, there is still an efficient probabilistic algorithm to compute the square roots of $x \in QR_p$. This algorithm is not addressed here.

### A.3.7.2 $QR_n$ When $n$ Is Composite

If $n$ is prime, then we have just seen that the question whether $x \in \mathbb{Z}_n^*$ is a quadratic residue can be answered easily (using Euler's criterion). The same question cannot be answered easily if $n$ is composite. The respective question leads to the *quadratic residuosity problem* (QRP) captured in Definition A.31.[40] The QRP is assumed to be hard, and it is the mathematical basis for quite a few cryptographic systems, such as probabilistic encryption (Section 13.2).

**Definition A.31 (QRP)** *Let $n \in \mathbb{N}$ be a composite positive integer and $x \in \mathbb{Z}_n^*$. The QRP is to decide whether $x$ is a quadratic residue modulo $n$ (i.e., $x \in QR_n$) or not (i.e., $x \in QNR_n$).*

We already know that computing square roots in $\mathbb{Z}_n^*$ and factoring $n$ are computationally equivalent. So if we can factorize $n$, then we can also compute square roots in $\mathbb{Z}_n^*$ and solve the QRP accordingly. But if we cannot factorize $n$, then it is not known how to solve the QRP in an efficient way. From Section A.3.3 we know that there are functions modulo $n$ that are simpler to compute modulo the prime factors of $n$. For example, one can show that if $n = pq$, then $x \in QR_n$ if and only if $x \in QR_p$ and $x \in QR_q$, and that every $x \in QR_n$ has four square roots in $\mathbb{Z}_n^*$. This can be generalized with the Jacobi symbol modulo $n$.[41] If $n \geq 3$ is an odd integer with prime factorization $n = p_1^{e_1} p_2^{e_2} \ldots p_k^{e_k}$ and $x \in \mathbb{N}$, then the Jacobi symbol $\left(\frac{x}{n}\right)$ is defined as follows:

$$J(x,n) = \left(\frac{x}{n}\right) = \left(\frac{x}{p_1}\right)^{e_1} \left(\frac{x}{p_2}\right)^{e_2} \cdots \left(\frac{x}{p_k}\right)^{e_k} = \prod_{i=1}^{k} \left(\frac{x}{p_i}\right)^{e_i}$$

In the above-mentioned special case where $n$ is the product of two primes $p$ and $q$, $J(x, n)$ can be computed as follows:

$$J(x,n) = \left(\frac{x}{n}\right) = \left(\frac{x}{p}\right)\left(\frac{x}{q}\right)$$

Like the Legendre symbol, the value of $J(x, n)$ is 0, 1, or $-1$. But unlike the Legendre symbol, the statements $J(x, n) = 1$ and $x \in QR_n$ are not equivalent. From $x \in QR_n$ it follows that $J(x, n) = 1$, but from $J(x, n) = 1$ it does not follow that $x \in QR_n$. To see why this is the case, consider the case in which $n$ is the product of two primes $p$ and $q$. There are two cases that lead to $J(x, n) = 1$: Either $x \in QR_p$ and $x \in QR_q$ or $x \notin QR_p$ and $x \notin QR_q$ (meaning that $x \in QNR_p$ and $x \in QNR_q$). Among these two cases, $x \in QR_n$ applies only in the first case. In

40 The QRP is a well-known problem in number theory and is one of the four main algorithmic problems discussed by Gauss in his *Disquisitiones Arithmeticae*.

41 The Jacobi symbol is named after Carl Gustav Jacob Jacobi, a German mathematician who lived from 1804 to 1851.

the more general case $J(x, n) = 1$ only suggests that all factors $L(x, p_i)$ multiplied together result in 1. It does not suggest that $x$ is a quadratic residue modulo $n$. This point is further addressed below.

If the prime factorization of $n$ is known, then it is simple and straightforward to compute $J(x, n)$ using the formula given above. But even if the prime factorization of $n$ is not known, is it still possible to compute $J(x, n)$ using an efficient algorithm that exploits some computational laws that apply to the Jacobi symbol.[42] For example, from $gcd(x, n) \neq 1$ it follows that $J(x, n) = 0$ (because the Legendre symbol of $x$ modulo this divisor of $n$ is zero and this value is multiplied into $J(x, n)$). For $x \equiv y \pmod{n}$, it also follows that $J(x, n) = J(y, n)$. Next, the Jacobi symbol is multiplicative in both the numerator and the denominator:

$$\left(\frac{xy}{n}\right) = \left(\frac{x}{n}\right) \cdot \left(\frac{y}{n}\right)$$

$$\left(\frac{x}{mn}\right) = \left(\frac{x}{m}\right) \cdot \left(\frac{x}{n}\right)$$

Consequently, for every $x$ the Jacobi symbol $J(x^2, n)$ must be equal to one (because $J(x^2, n) = J(x, n) \cdot J(x, n) = 1$ if $gcd(x, n) = 1$).

Furthermore, Gauss' law of quadratic reciprocity is heavily used to compute the Jacobi symbol. If $m, n \geq 3$ are two odd integers, then

$$\left(\frac{m}{n}\right)\left(\frac{n}{m}\right) = (-1)^{\frac{(n-1)}{2} \cdot \frac{(m-1)}{2}}$$

There are two extension laws:

$$\left(\frac{-1}{n}\right) = (-1)^{\frac{(n-1)}{2}} = \begin{cases} 1 & \text{if } n \equiv 1 \pmod{4} \\ -1 & \text{if } n \equiv 3 \pmod{4} \end{cases}$$

$$\left(\frac{2}{n}\right) = (-1)^{\frac{(n^2-1)}{8}} = \begin{cases} 1 & \text{if } n \equiv \pm 1 \pmod{8} \\ -1 & \text{if } n \equiv \pm 3 \pmod{8} \end{cases}$$

It follows that $J(m, n) = J(n, m)$ for $n \equiv 1 \pmod{4}$ or $m \equiv 1 \pmod{4}$, and $J(m, n) = -J(n, m)$ for $n \equiv m \equiv 3 \pmod{4}$.

Putting everything together, the (efficient) algorithm to compute $J(x, n)$ for $x, n \in \mathbb{Z}$ and odd $n \geq 3$ works as follows:

42 The algorithm has a running time complexity of $O((\ln n)^3)$.

- Determine $x'$ with $0 \leq x' < n$ and $x \equiv x' \pmod{n}$. This means that $x$ is reduced modulo $n$. It follows that $J(x,n) = J(x',n)$. If $x' = 0$ or $x' = 1$, then the algorithm is done and can return the result.
- Otherwise, $x'$ is written as $x' = 2^v y$ with $y$ being odd. If $x'$ is already odd, then $y = x'$ and $v = 0$. It then follows that

$$\left(\frac{x'}{n}\right) = \left(\frac{2}{n}\right)^v \left(\frac{y}{n}\right)$$

  and $J(2,n) = \pm 1$ can be computed with the second extension law. If $y = 1$, then the algorithm can abort and return the result.
- Otherwise, Gauss' law of quadratic reciprocity can be applied:

$$\left(\frac{y}{n}\right) = (-1)^{\frac{(y-1)}{2} \cdot \frac{(n-1)}{2}} \left(\frac{n}{y}\right)$$

  This equation even holds if $gcd(y,n) \neq 1$, because either side is then equal to zero. To compute $J(n,y)$, one can recursively apply the same algorithm and start with step one.

Because the moduli in the Jacobi symbol decrease in each iteration, the algorithm terminates within a finite number of steps.

If, for example, one has to compute $J(740, 211)$, then one can apply the algorithm to compute the result 1:

$$\begin{aligned} \left(\frac{740}{211}\right) &= \left(\frac{107}{211}\right) = -\left(\frac{211}{107}\right) = -\left(\frac{104}{107}\right) = -\left(\frac{2}{107}\right)^3 \left(\frac{13}{107}\right) \\ &= \left(\frac{13}{107}\right) = \left(\frac{107}{13}\right) = \left(\frac{3}{13}\right) = \left(\frac{13}{3}\right) = \left(\frac{1}{3}\right) = \left(\frac{3}{1}\right) = 1 \end{aligned}$$

As mentioned earlier, the fact that the Jacobi symbol of $x$ modulo $n$ is equal to 1 does not imply that $x$ is a quadratic residue modulo $n$ (i.e., $x \in QR_n$). Let $J_n$ be the set of all elements of $\mathbb{Z}_n^*$ with Jacobi symbol 1:

$$J_n = \{x \in \mathbb{Z}_n^* \mid \left(\frac{x}{n}\right) = 1\}$$

We know that all elements from $QR_n$ have a Jacobi symbol 1, and hence $QR_n \subset J_n$. But there are elements $x \in \mathbb{Z}_n^*$ that also have a Jacobi symbol 1 but are quadratic nonresidues; they are called *pseudosquares* modulo $n$. The set of all pseudosquares modulo $n$ is denoted as $\widetilde{QR_n}$. It comprises the elements of $J_n$ minus the elements of $QR_n$: $\widetilde{QR_n} = J_n \setminus QR_n$.

Let $n = pq$ be the product of two primes. Then $\mathbb{Z}_n^*$ has $\phi(n) = (p-1)(q-1)$ elements, and these elements can be partitioned into two equally large sets. One half of the elements (i.e., $J_n$) has Jacobi symbol 1, and the other half of the elements has Jacobi symbol $-1$. $J_n$ can be further partitioned into two equally large sets (i.e., $QR_n$ and $\widetilde{QR_n}$) with $|QR_n| = |\widetilde{QR_n}| = (p-1)(q-1)/4$. For example, if $p = 3$ and $q = 7$, then $n = 3 \cdot 7 = 21$, $\mathbb{Z}_{21}^* = \{1, 2, 4, 5, 8, 10, 11, 13, 16, 17, 19, 20\}$, and $\phi(21) = 2 \cdot 6 = 12$. If we compute the squares of all elements $x$ of $\mathbb{Z}_{21}^*$, then we get the following table:

| $x$ | 1 | 2 | 4 | 5 | 8 | 10 | 11 | 13 | 16 | 17 | 19 | 20 |
|---|---|---|---|---|---|---|---|---|---|---|---|---|
| $x^2$ | 1 | 4 | 16 | 4 | 1 | 16 | 16 | 1 | 4 | 16 | 4 | 1 |

Note that the only elements that appear in the second row are 1, 4, and 16. These elements form $QR_{21}$ (i.e., the set of all quadratic residues modulo 21). But $J_{21}$ has six elements and these elements can be partitioned into $QR_{21}$ and $\widetilde{QR_{21}}$:

$$
\begin{aligned}
J_{21} &= \{1, 4, 5, 16, 17, 20\} \\
QR_{21} &= \{1, 4, 16\} \\
\widetilde{QR_{21}} &= \{5, 17, 20\}
\end{aligned}
$$

This means that 5, 17, and 20 are pseudosquares modulo 21. This example is illustrated in Figure A.3. All elements of $\mathbb{Z}_{21}^*$ appear in the circle. Each quarter of the square comprises three elements. The upper right quarter represents $QR_{21}$ and the lower left quarter represents $\widetilde{QR_{21}}$. $QNR_{21}$ comprises all elements of $\mathbb{Z}_{21}^*$ that are not elements of $QR_{21}$; that is, $QNR_{21} = \mathbb{Z}_{21}^* \setminus QR_{21} = \{2, 5, 8, 10, 11, 13, 17, 19, 20\}$.

Consequently, in this example the QRP is to decide whether a particular element of $J_{21}$ is an element of $QR_{21}$ or $\widetilde{QR_{21}}$. It goes without saying that the problem can be solved if the prime factorization of 21 is known. It is not known how to solve it efficiently without knowing the prime factorization of 21 though.

### A.3.8 Blum Integers

Some public key cryptosystems require Blum integers as formally introduced in Definition A.32.

**Definition A.32 (Blum integer)** *A composite number $n \in \mathbb{Z}$ is a* Blum integer *if $n = pq$ with $p, q \in \mathbb{P}$, $p \neq q$, and $p \equiv q \equiv 3 \pmod 4$.*

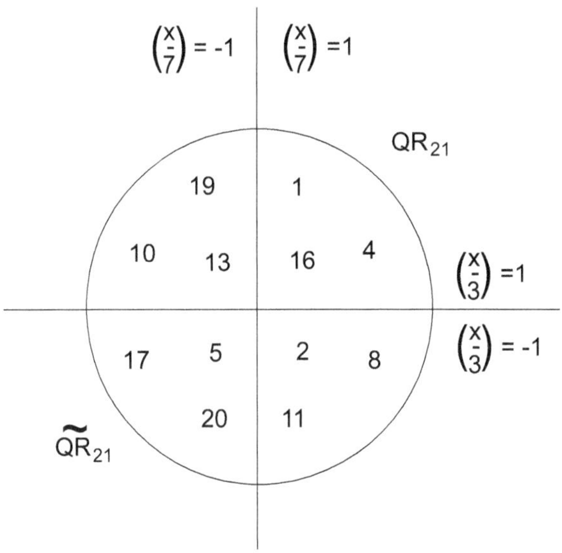


**Figure A.3** The elements of $\mathbb{Z}_{21}^*$, $QR_{21}$, and $\widetilde{QR_{21}}$.

This means that a Blum integer is the product of two distinct primes that are both equivalent to 3 modulo 4. In some literature, such primes (that are equivalent to 3 modulo 4) are called *Blum primes*, but this term is not commonly used.

For every positive integer $n \in \mathbb{N}^+$, $QR_n$ comprises the elements of $\mathbb{Z}_n^*$ that are quadratic residues (and hence have square roots modulo $n$). If $n$ is a Blum integer, then every $x \in QR_n$ has four square roots modulo $n$, of which one is again an element of $QR_n$. This unique square root of $x$ is called the *principal square root* of $x$ modulo $n$.

If we revisit the example given above, then it is easy to see that $n = 21 = 3 \cdot 7$ is a Blum integer (because $3 \equiv 7 \equiv 3 \pmod 4$). This means that every $x \in QR_n$ has 4 four square roots. This follows immediately from the table given above. 1 has the square roots $\underline{1}$, 8, 13, and 20; 4 has the square roots 2, 5, $\underline{16}$, and 19; 16 has the square roots $\underline{4}$, 10, 11, and 17. In all three cases, the respective principal square root is underlined.

The function $f : \mathbb{Z}_n^* \rightarrow \mathbb{Z}_n^*$ with $f(x) = x^2 \pmod n$ is conjectured to be one way or—more precisely—a trapdoor function (with the prime factorization of $n$ representing the trapdoor). If we restrict the domain and range to $QR_n$ (i.e., the set of quadratic residues modulo $n$), then $f$ yields a trapdoor permutation $QR_n \rightarrow QR_n$ with $f(x) = x^2 \pmod n$. Without knowing the prime factors of $n$ (i.e., $p$ and $q$), it is not known how to compute the inverse function $f^{-1}$. But knowing them, one can efficiently compute it as follows:

$$f^{-1}(x) = x^{((p-1)(q-1)+4)/8} \pmod{n}$$

In the example given above, one can use this formula to verify $f^{-1}(1) = 1$, $f^{-1}(4) = 16$, and $f^{-1}(16) = 4$. The restriction to $QR_n$ is often used in cryptography, especially if one wants to use the modular square function as a permutation.

## References


[1] Koblitz, N.I., *A Course in Number Theory and Cryptography*, 2nd edition. Springer-Verlag, New York, 1994.

[2] Koblitz, N.I., *Algebraic Aspects of Cryptography*. Springer-Verlag, New York, 1998.

[3] Rosen, K.H., *Discrete Mathematics and Its Applications*, 7th edition. McGraw-Hill Education, New York, 2011.

[4] Johnsonbaugh, R., *Discrete Mathematics*, 8th edition. Pearson, 2017.

[5] Dossey, J.A., et al., *Discrete Mathematics*, 5th edition. Pearson, 2017.

[6] Shoup, V., *A Computational Introduction to Number Theory and Algebra*, 2nd edition. Cambridge University Press, Cambridge, U.K., 2008.

[7] Maurer, U.M., "Fast Generation of Prime Numbers and Secure Public-Key Cryptographic Parameters," *Journal of Cryptology*, Vol. 8, No. 3, 1995, pp. 123–155.

[8] Agrawal, M., Kayal, N., and N. Saxena, "PRIMES Is in P," *Annals of Mathematics*, Vol. 160, No. 2, 2004, pp. 781–793.

[9] Solovay, R.M., and V. Strassen, "A Fast Monte-Carlo Test for Primality," *SIAM Journal on Computing*, Vol. 6, No. 1, 1977, pp. 84–85.

[10] Miller, G.L., "Riemann's Hypothesis and Tests for Primality," *Journal of Computer and System Sciences*, Vol. 13, No. 3, 1976, pp. 300–317.

[11] Rabin, M.O., "Probabilistic Algorithm for Testing Primality," *Journal of Number Theory*, Vol. 12, No. 1, 1980, pp. 128–138.

[12] Pohlig, S., and M.E. Hellman, "An Improved Algorithm for Computing Logarithms over GF(p)," *IEEE Transactions on Information Theory*, Vol. 24, No. 1, January 1978, pp. 106–110.

[13] Rivest, R.L., and R.D. Silverman, "Are 'Strong' Primes Needed for RSA," Cryptology ePrint Archive: Report 2001/007.

# Appendix B

# Probability Theory

Probability theory plays a central role in information theory and cryptography. In fact, the ultimate goal of a cryptographer is to make the probability that an attack against the security of a cryptographic system succeeds equal to zero. In reality, this goal is too ambitious (because security is not absolute), and it is usually sufficient to require the probability to be negligible. Probability theory provides the formalism required for this kind of reasoning. In this appendix, we summarize the basic principles of (discrete) probability theory. More specifically, we introduce some basic terms and concepts in Section B.1, and elaborate on random variables and their use in Section B.2. The entire appendix is intentionally kept short; further information can be found in any textbook on probability theory (e.g., [1–7] in chronological order).

## B.1 BASIC TERMS AND CONCEPTS

The notion of a discrete probability space as formally introduced in Definition B.1 is key for probability theory and its applications.

**Definition B.1 (Discrete probability space)** *A* discrete probability space *consists of a finite or countably infinite set* $\Omega$ *of elementary elements,*[1] *called the* sample space, *and a probability measure or distribution* $\Pr : \Omega \longrightarrow \mathbb{R}^+$ *with* $\sum_{\omega \in \Omega} \Pr[\omega] = 1$.[2]

If we run a (discrete) random experiment in such a probability space, then every elementary event of the sample space yields a possible outcome of the

1 In some literature, elementary events are also called *simple events* or *indecomposable events*, but these terms are not used in this book.

2 Alternative notations for the probability measure $\Pr[\cdot]$ are $P(\cdot)$, $P[\cdot]$, and $\text{Prob}[\cdot]$.

experiment. The probability measure or distribution $\Pr[\cdot]$ assigns a nonnegative real value to every elementary event $\omega \in \Omega$, such that all (probability) values sum up to one (this already suggests that all probability values must be larger or equal than zero and smaller or equal to one). There is no general and universally valid requirement on how to assign probability values. In fact, it is often the case that many elementary events of $\Omega$ occur with probability zero. If all $|\Omega|$ possible values occur with the same probability (i.e., $\Pr[\omega] = 1/|\Omega|$ for all $\omega \in \Omega$), then the probability distribution is called *uniform*. Uniform probability distributions are frequently used in probability theory and its applications.

As mentioned in Definition B.1, sample spaces are assumed to be finite or countably infinite for the purpose of this book (things get more involved if this assumption is not made). The term *discrete probability theory* is sometimes used to refer to the restriction of probability theory to finite or countably infinite sample spaces. In this book, however, we only focus on discrete probability theory, and hence the terms probability theory and discrete probability theory are used synonymously and interchangeably. Furthermore, we say a "finite" sample space when we actually refer to a "finite or countably infinite" sample space.

For example, flipping a coin can be understood as a random experiment taking place in a discrete probability space. The sample space consists of $head$ and $tail$ (or 0 and 1 if these binary values are used to encode *head* and *tail*) and the probability measure assigns $1/2$ to either $head$ or $tail$ (i.e., $\Pr[head] = \Pr[tail] = 1/2$). The resulting probability distribution is uniform. If the coin is flipped five times, then the sample space is $\{head, tail\}^5$ (or $\{0,1\}^5$, respectively) and the probability measure assigns $1/2^5 = 1/32$ to every possible outcome of the experiment. Similarly, rolling a dice can be understood as a random experiment taking place in a discrete probability space. In this case, the sample space is $\{1, \ldots, 6\}$ and the probability measure assigns $1/6$ to every possible outcome of the experiment (i.e., $\Pr[1] = \ldots = \Pr[6] = 1/6$). If the dice is rolled $n$ times (or $n$ dice are rolled simultaneously), then the sample space is $\{1, \ldots, 6\}^n$ and the probability measure assigns $1/6^n$ to every possible outcome of the experiment. In either case, the probability distribution is uniform if the coins are unbiased and fair.

Instead of looking at elementary events of a sample space, one may also look at sets of such elements. In fact, an *event* refers to a subset $\mathcal{A} \subseteq \Omega$, and its probability equals the sum of the probabilities of its elementary events. This can be formally expressed in the following way:

$$\Pr[\mathcal{A}] = \sum_{\omega \in \mathcal{A}} \Pr[\omega]$$

$\Pr[\Omega]$ is conventionally set to 1, and $\Pr[\emptyset]$ is set to 0. Furthermore, one frequently needs the complement of an event $\mathcal{A}$, denoted as $\overline{\mathcal{A}}$. It consists of all elements of $\Omega$

that are not elements of $\mathcal{A}$, and its probability can be computed as

$$\Pr[\overline{\mathcal{A}}] = \sum_{\omega \in \Omega \setminus \mathcal{A}} \Pr[\omega]$$

If we know $\Pr[\mathcal{A}]$, then we can easily compute $\Pr[\overline{\mathcal{A}}] = 1 - \Pr[\mathcal{A}]$, because $\Pr[\mathcal{A}]$ and $\Pr[\overline{\mathcal{A}}]$ must sum up to 1.

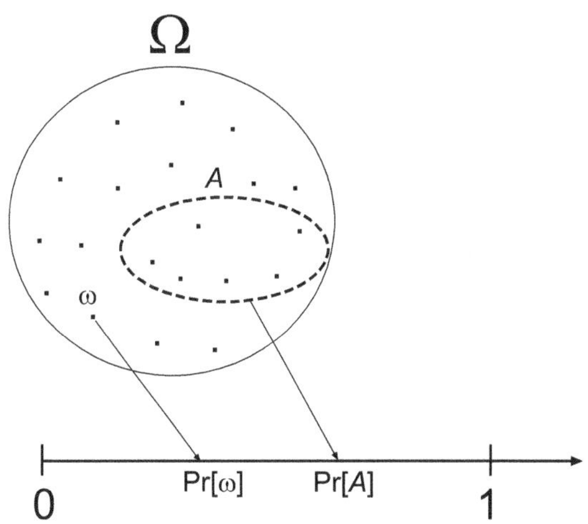


**Figure B.1** A discrete probability space.

A discrete probability space is illustrated in Figure B.1. There is a sample space $\Omega$ and a probability measure $\Pr[\cdot]$ that assign a value between 0 and 1 to every elementary event $\omega \in \Omega$ or event $\mathcal{A} \subseteq \Omega$.

If, for example, we want to compute the probability of the event that, when flipping five coins, we get three heads, then the sample space is $\Omega = \{1, 0\}^5$ and the probability distribution is uniform. This basically means that every element $\omega \in \Omega$ occurs with the same probability (i.e., $\Pr[\omega] = 1/2^5 = 1/32$). Let $\mathcal{A}$ be the subset of $\Omega = \{1, 0\}^5$ that contains bitstrings with exactly three ones and let us ask for the probability $\Pr[\mathcal{A}]$. It can easily be shown that $\mathcal{A}$ consists of the following 10 elements:

| | |
|---|---|
| 00111 | 10110 |
| 01011 | 10101 |
| 01101 | 11001 |
| 01110 | 11010 |
| 10011 | 11100 |

Consequently, Pr$[\mathcal{A}] = 10/32 = 5/16$. The example can be generalized to $n$ flips with a biased coin. If the coin flips are independent and the probability that each coin turns out heads is $0 \leq p \leq 1$, then the sample space is $\{1,0\}^n$ and the probability for a specific event $\omega$ in this space is

$$\Pr[\omega] = p^k(1-p)^{n-k}$$

where $k$ is the number of ones in $\omega$. In the example given earlier, we had $p = 1-p = 1/2$, and the corresponding distribution over $\{1,0\}^n$ was uniform. If $p=1$ ($p=0$), then $0^n$ ($1^n$) has probability 1 and all other elements have probability 0. Consequently, the interesting cases occur when $p$ is greater than 0 but smaller than 1; that is, $p \in (0,1)$. This brings up the notion of a *binominal distribution*. If we have such a distribution with parameter $p$ and ask for the probability of the event $\mathcal{A}_k$ that we get a string with $k$ ones, then this probability can be computed as follows:

$$\Pr[\mathcal{A}_k] = \left(\begin{array}{c} n \\ k \end{array}\right) p^k(1-p)^{n-k}$$

In this formula, $\left(\begin{array}{c} n \\ k \end{array}\right)$ is read "$n$ choose $k$" and can be computed as follows:

$$\left(\begin{array}{c} n \\ k \end{array}\right) = \frac{n!}{k!(n-k)!}$$

Here, $n!$ refers to the factorial of integer $n$. It is recursively defined with $0! = 1$ and $n! = (n-1)!n$.

More generally, if we have two events $\mathcal{A}, B \subseteq \Omega$, then the probability of the *union event* $\mathcal{A} \cup \mathcal{B}$ is computed as follows:

$$\Pr[\mathcal{A} \cup \mathcal{B}] = \Pr[\mathcal{A}] + \Pr[\mathcal{B}] - \Pr[\mathcal{A} \cap \mathcal{B}]$$

Consequently, $\Pr[\mathcal{A} \cup \mathcal{B}] \leq \Pr[\mathcal{A}] + \Pr[\mathcal{B}]$ and $\Pr[\mathcal{A} \cup \mathcal{B}] = \Pr[\mathcal{A}] + \Pr[\mathcal{B}]$ if and only if $\mathcal{A} \cap \mathcal{B} = \emptyset$. The former inequality is known as the *union bound*.

Similarly, we may be interested in the *joint event* $\mathcal{A} \cap \mathcal{B}$. Its probability is computed as follows:

$$\Pr[\mathcal{A} \cap \mathcal{B}] = \Pr[\mathcal{A}] + \Pr[\mathcal{B}] - \Pr[\mathcal{A} \cup \mathcal{B}]$$

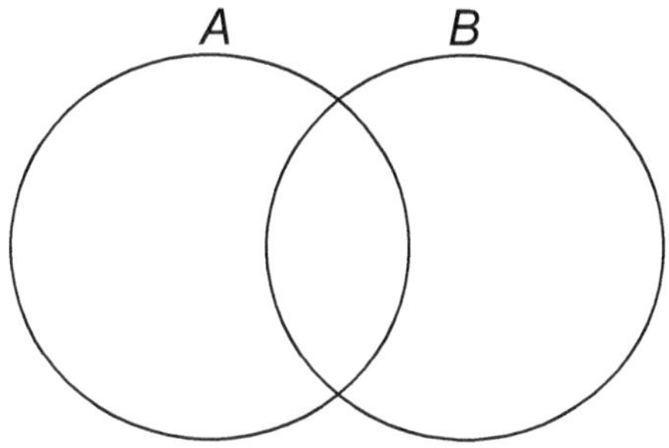


**Figure B.2** A Venn diagram with two events.

Venn diagrams can be used to illustrate the relationship of events. A Venn diagram is made up of two or more overlapping circles (each circle representing an event). For example, Figure B.2 shows a Venn diagram for two events $\mathcal{A}$ and $\mathcal{B}$. The intersection of the two circles represents $\mathcal{A} \cap \mathcal{B}$, whereas the union represents $\mathcal{A} \cup \mathcal{B}$.

The two events $\mathcal{A}$ and $\mathcal{B}$ are *independent* if $\Pr[\mathcal{A} \cap \mathcal{B}] = \Pr[\mathcal{A}] \cdot \Pr[\mathcal{B}]$, meaning that the probability of one event does not influence the probability of the other.

This notion of independence can be generalized to more than two events. In this case, it must be distinguished whether the events are pairwise or mutually independent. Let $\mathcal{A}_1, \ldots, \mathcal{A}_n \subseteq \Omega$ be $n$ events in a sample space $\Omega$:

- $\mathcal{A}_1, \ldots, \mathcal{A}_n$ are *pairwise independent* if for every $i, j \in \{1, \ldots, n\}$ with $i \neq j$ it holds that $\Pr[\mathcal{A}_i \cap \mathcal{A}_j] = \Pr[\mathcal{A}_i] \cdot \Pr[\mathcal{A}_j]$.
- $\mathcal{A}_1, \ldots, \mathcal{A}_n$ are *mutually independent* if for every subset of indices $I \subseteq \{1, 2, \ldots, n\}$ with $I \neq \emptyset$ it holds that

$$\Pr[\bigcap_{i \in I} \mathcal{A}_i] = \prod_{i \in I} \Pr[\mathcal{A}_i]$$

Sometimes it is necessary to compute the probability of an elementary event $\omega$ given that an event $\mathcal{A}$ with $\Pr[\mathcal{A}] > 0$ holds. The resulting *conditional probability*, denoted $\Pr[\omega|\mathcal{A}]$, can be computed as follows:

$$\Pr[\omega|\mathcal{A}] = \begin{cases} \frac{\Pr[\omega]}{\Pr[\mathcal{A}]} & \text{if } \omega \in \mathcal{A} \\ 0 & \text{otherwise} \end{cases}$$

If $\omega \in \mathcal{A}$, then $\Pr[\omega|\mathcal{A}]$ must have a value that is proportional to $\Pr[\omega]$, and the factor of proportionality is $1/\Pr[\mathcal{A}]$ (so that all probabilities sum up to 1). Otherwise (i.e., if $\omega \notin \mathcal{A}$), it is impossible that $\omega$ holds, and hence $\Pr[\omega|\mathcal{A}]$ must be equal to 0—independent from the probability of $\mathcal{A}$.

The definition of $\Pr[\omega|\mathcal{A}]$ can be generalized to arbitrary events. In fact, if $\mathcal{A}$ and $\mathcal{B}$ are two events, then the probability of event $\mathcal{B}$ given that event $\mathcal{A}$ holds is the sum of the probabilities of all elementary events $\omega \in \mathcal{B}$ given that $\mathcal{A}$ holds. This can be formally expressed as follows:

$$\Pr[\mathcal{B}|\mathcal{A}] = \sum_{\omega \in \mathcal{B}} \Pr[\omega|\mathcal{A}]$$

In the literature, $\Pr[\mathcal{B}|\mathcal{A}]$ is sometimes also defined as follows:

$$\Pr[\mathcal{B}|\mathcal{A}] = \frac{\Pr[\mathcal{A} \cap \mathcal{B}]}{\Pr[\mathcal{A}]}$$

Consequently, if two events $\mathcal{A}$ and $\mathcal{B}$ are independent and $\Pr[\mathcal{A}] > 0$, then $\Pr[\mathcal{B}|\mathcal{A}] = \Pr[\mathcal{A} \cap \mathcal{B}]/\Pr[\mathcal{A}] = \Pr[\mathcal{A}] \cdot \Pr[\mathcal{B}]/\Pr[\mathcal{A}] = \Pr[\mathcal{B}]$ . This means that whether $\mathcal{B}$ holds or not is not influenced by the knowledge that $\mathcal{A}$ holds. This also applies in the other direction; that is, if $\Pr[\mathcal{B}] > 0$, then $\Pr[\mathcal{A}|\mathcal{B}] = \Pr[\mathcal{B} \cap \mathcal{A}]/\Pr[\mathcal{B}] = \Pr[\mathcal{B}] \cdot \Pr[\mathcal{A}]/\Pr[\mathcal{B}] = \Pr[\mathcal{A}]$. In summary, if two events are independent, then whether one holds or not is not influenced by the knowledge that the other holds or not, and vice versa.

Because $\Pr[\mathcal{A} \cap \mathcal{B}] = \Pr[\mathcal{B} \cap \mathcal{A}]$ one can solve either equation for this term and equalize the results. This yields *Bayes' theorem* that is frequently used in probability theory:

$$\Pr[\mathcal{A}|\mathcal{B}] = \frac{\Pr[\mathcal{B} \cap \mathcal{A}]}{\Pr[\mathcal{B}]} \tag{B.1}$$

and put $\Pr[\mathcal{B}|\mathcal{A}]$ and $\Pr[\mathcal{A}|\mathcal{B}]$ into perspective. In this case, the formula

$$\Pr[\mathcal{A}|\mathcal{B}] = \frac{\Pr[\mathcal{A}]\Pr[\mathcal{B}|\mathcal{A}]}{\Pr[\mathcal{B}]}$$

Furthermore, one can also formally express a *law of total probability* as captured in Theorem B.1.

**Theorem B.1 (Law of total probability)** *If the events* $\mathcal{B}_1, \ldots, \mathcal{B}_n$ *form a partition of the sample space (i.e.,* $\cup_{i=1}^n \mathcal{B}_i = \Omega$ *and* $\mathcal{B}_i \cap \mathcal{B}_j = \emptyset$ *for all* $i \neq i$*), then*

$$\Pr[\mathcal{A}] = \sum_{i=1}^{n} \Pr[\mathcal{A}|\mathcal{B}_i] \cdot \Pr[\mathcal{B}_i]$$

*must hold for every event* $\mathcal{A} \subseteq \Omega$.

*Proof.* The event $\mathcal{A}$ can be written as

$$\mathcal{A} = \mathcal{A} \cap \Omega = \bigcup_{i=1}^{n} (\mathcal{A} \cap \mathcal{B}_i)$$

where $(\mathcal{A} \cap \mathcal{B}_i)$ and $(\mathcal{A} \cap \mathcal{B}_j)$ are disjoint (and hence mutually exclusive) for $i \neq j$. If one computes the probability on either side of the equation, then one gets $\Pr[\mathcal{A}]$ on the left side and—according to (B.1)— $\sum_{i=1}^{n} \Pr[\mathcal{A}|\mathcal{B}_i] \cdot \Pr[\mathcal{B}_i]$ on the right side. This finishes the proof.

□

The law of total probability is useful and frequently employed to compute the probability of an event $\mathcal{A}$, which is conditional given some other mutually exclusive events, such as an event $\mathcal{B}$ and its complement $\overline{\mathcal{B}}$.

## B.2 RANDOM VARIABLES

If we have a discrete probability space and run a random experiment, then we might be interested in some values that depend on the outcome of the experiment (rather than the outcome itself). If, for example, we roll two dice, then we may be interested in their sum. Similarly, if we run a randomized algorithm, then we may be interested in its output or running time. This is where the notion of a random variable as formally introduced in Definition B.2 comes into play.

**Definition B.2 (Random variable)** *Let* $(\Omega, \Pr)$ *be a discrete probability space with sample space* $\Omega$ *and probability measure* $\Pr[\cdot]$. *Any function* $X : \Omega \rightarrow \mathcal{X}$ *from the sample space to a measurable set* $\mathcal{X}$ *is a* random variable, *where* $\mathcal{X}$ *is the* range *of the random variable* $X$.

Consequently, a random variable is a function that on input an arbitrary element of the sample space of a discrete probability space outputs an element of the range. In a typical setting, $\mathcal{X}$ is either a subset of the real numbers (i.e., $\mathcal{X} \subseteq \mathbb{R}$) or a subset of the binary strings of a specific length $n$ (i.e., $\mathcal{X} \subseteq \{0,1\}^n$).[3]

3 In some literature, a random variable is defined as a function $X : \Omega \rightarrow \mathbb{R}$.

If $x$ is in the range of $X$ (i.e., $x \in \mathcal{X}$), then the expression $(X = x)$ refers to the event $\{\omega \in \Omega \mid X(\omega) = x\}$, and hence Pr$[X = x]$ is defined and something potentially interesting to compute. If only one random variable $X$ is considered, then Pr$[X = x]$ is sometimes also written as Pr$[x]$.

If, for example, we roll two fair dice, then the sample space is $\Omega = \{1, 2, \ldots, 6\}^2$ and the probability distribution is uniform (i.e., Pr$[\omega_1, \omega_2] = 1/6^2 = 1/36$ for every $(\omega_1, \omega_2) \in \Omega$). Let $X$ be the random variable that associates $\omega_1 + \omega_2$ to every $(\omega_1, \omega_2) \in \Omega$. Then the range of the random variable $X$ is $\mathcal{X} = \{2, 3, \ldots, 12\}$. For every element of the range, we can compute the probability that $X$ takes this value. In fact, by counting the number of elements in every possible event, we can easily determine the following probabilities:

$$\begin{aligned}
\Pr[X = 2] &= 1/36 \\
\Pr[X = 3] &= 2/36 \\
\Pr[X = 4] &= 3/36 \\
\Pr[X = 5] &= 4/36 \\
\Pr[X = 6] &= 5/36 \\
\Pr[X = 7] &= 6/36
\end{aligned}$$

The remaining probabilities (i.e., Pr$[X = 8], \ldots,$ Pr$[X = 12]$) can be computed by observing that Pr$[X = x] =$ Pr$[X = 14 - x]$. Consequently, we have

$$\begin{aligned}
\Pr[X = 8] &= \Pr[X = 14 - 8] = \Pr[X = 6] = 5/36 \\
\Pr[X = 9] &= \Pr[X = 14 - 9] = \Pr[X = 5] = 4/36 \\
\Pr[X = 10] &= \Pr[X = 14 - 10] = \Pr[X = 4] = 3/36 \\
\Pr[X = 11] &= \Pr[X = 14 - 11] = \Pr[X = 3] = 2/36 \\
\Pr[X = 12] &= \Pr[X = 14 - 12] = \Pr[X = 2] = 1/36
\end{aligned}$$

It can easily be verified that all probabilities sum up to one:

$$\frac{1}{36} + \frac{2}{36} + \frac{3}{36} + \frac{4}{36} + \frac{5}{36} + \frac{6}{36} + \frac{5}{36} + \frac{4}{36} + \frac{3}{36} + \frac{2}{36} + \frac{1}{36} = \frac{36}{36} = 1$$

We next look at some probability distributions of random variables.

### B.2.1 Probability Distributions

If $X : \Omega \rightarrow \mathcal{X}$ is a random variable with sample space $\Omega$ and range $\mathcal{X}$, then the *probability distribution* of $X$, denoted $P_X$, is a mapping from $\mathcal{X}$ to $\mathbb{R}^+$. It is formally defined as follows:

$$\begin{aligned} P_X : \mathcal{X} &\longrightarrow \mathbb{R}^+ \\ x &\longmapsto P_X(x) = P(X = x) = \sum_{\omega \in \Omega : X(\omega) = x} \Pr[\omega] \end{aligned}$$

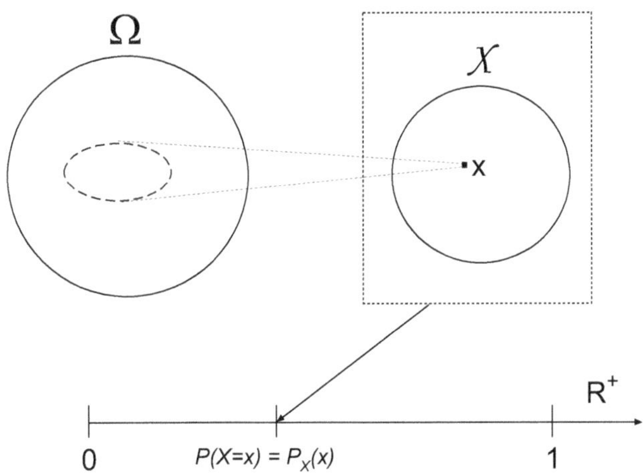


**Figure B.3** The probability distribution of a random variable $X$.

The probability distribution of a random variable $X$ is illustrated in Figure B.3. Some events from the sample space $\Omega$ (on the left side) may be mapped to $x \in \mathcal{X}$ (on the right side), and the probability that $x$ occurs as a map is $P(X = x) = P_X(x)$.

It is possible to define more than one random variable for a discrete random experiment. If, for example, $X$ and $Y$ are two random variables with ranges $\mathcal{X}$ and $\mathcal{Y}$, then $P(X = x, Y = y)$ refers to the probability that $X$ takes on the value $x \in \mathcal{X}$ and $Y$ takes on the value $y \in \mathcal{Y}$. Consequently, the *joint probability distribution* of $X$ and $Y$, denoted $P_{XY}$, is a mapping from $\mathcal{X} \times \mathcal{Y}$ to $\mathbb{R}^+$. It is formally defined as follows:

$$\begin{aligned} P_{XY} : \mathcal{X} \times \mathcal{Y} &\longrightarrow \mathbb{R}^+ \\ (x, y) &\longmapsto P_{XY}(x, y) = \\ & \quad\; P(X = x, Y = y) = \\ & \quad\; \sum_{\omega \in \Omega : X(\omega) = x; Y(\omega) = y} \Pr[\omega] \end{aligned}$$

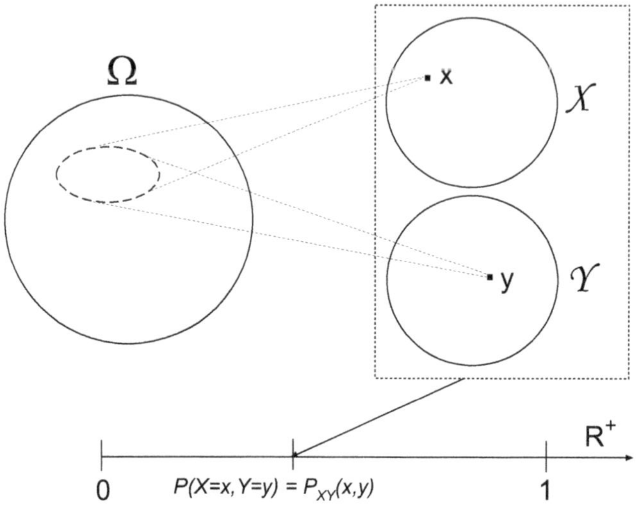


**Figure B.4** The joint probability distribution of two random variables $X$ and $Y$.

The joint probability distribution of two random variables $X$ and $Y$ is illustrated in Figure B.4. Some events from the sample space $\Omega$ (on the left side) may be mapped to $x \in \mathcal{X}$ and $y \in \mathcal{Y}$ (on the right side), and the probability that $x$ and $y$ occur as maps is $P(X = x, Y = y) = P_{XY}(x, y)$.

Similarly, for $n$ random variables $X_1, \ldots, X_n$ (with ranges $\mathcal{X}_1, \ldots, \mathcal{X}_n$), one can compute the probability that $X_i$ takes on the value $x_i \in \mathcal{X}_i$ for $i = 1, \ldots, n$. In fact, the *joint probability distribution* of $X_1, \ldots, X_n$, denoted $P_{X_1 \ldots X_n}$, is a mapping from $\mathcal{X}_1 \times \ldots \times \mathcal{X}_n$ to $\mathbb{R}^+$ that is formally defined as follows:

$$\begin{array}{rcl} P_{X_1 \ldots X_n} : \mathcal{X}_1 \times \ldots \times \mathcal{X}_n & \longrightarrow & \mathbb{R}^+ \\ (x_1, \ldots, x_n) & \longmapsto & P_{X_1 \ldots X_n}(x_1, \ldots, x_n) = \\ & & P(X_1 = x_1, \ldots, X_n = x_n) = \\ & & \displaystyle\sum_{\omega \subset \Omega : X_1(\omega) = x_1; \ldots; X_n(\omega) = x_n} \Pr[\omega] \end{array}$$

The joint probability distribution of $n$ random variables $X_1, \ldots, X_n$ is illustrated in Figure B.5. Some events from the sample space $\Omega$ (on the left side) may be mapped to $x_1 \in \mathcal{X}_1$, $\ldots$, $x_n \in \mathcal{X}_n$ (on the right side), and the probability that $x_1, \ldots, x_n$ occur as maps is $P(X_1 = x_1, \ldots, X_n = x_n) = P_{X_1 \ldots X_n}(x_1, \ldots, x_n)$.

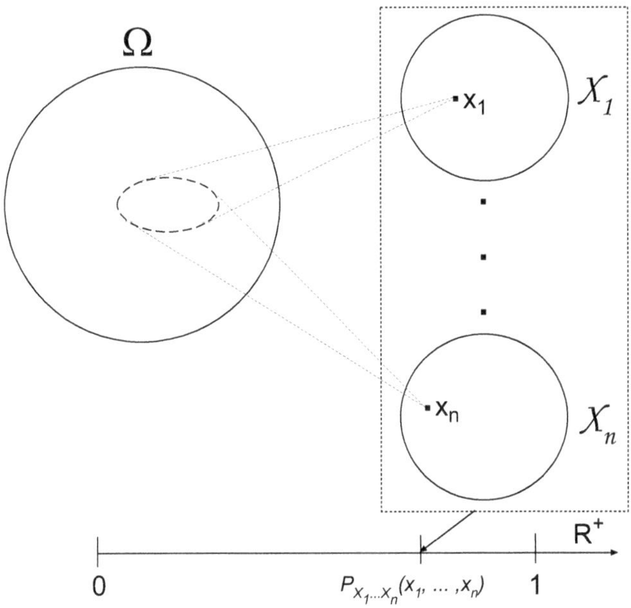


**Figure B.5** The joint probability distribution of $n$ random variables $X_1, \ldots, X_n$.

### B.2.2 Marginal Distributions

If $X$ and $Y$ are two random variables with joint probability distribution $P_{XY}$, then the two *marginal distributions* $P_X$ and $P_Y$ are defined as follows:

$$P_X(x) = \sum_{y \in \mathcal{Y}} P_{XY}(x, y)$$

$$P_Y(y) = \sum_{x \in \mathcal{X}} P_{XY}(x, y)$$

Again, the notion of a marginal distribution can be generalized to more than two random variables. If $X_1, \ldots, X_n$ are $n$ random variables with ranges $\mathcal{X}_1, \ldots, \mathcal{X}_n$ and joint probability distribution $P_{X_1 \ldots X_n}$, then for any $m < n$ the marginal distribution $P_{X_1 \ldots X_m}$ is defined as follows:

$$P_{X_1 \ldots X_m}(x_1, \ldots, x_m) = \sum_{(x_{m+1}, \ldots, x_n) \in \mathcal{X}_{m+1} \ldots \mathcal{X}_n} P_{X_1 \ldots X_n}(x_1, \ldots, x_n)$$

### B.2.3 Conditional Probability Distributions

Let $(\Omega, \Pr)$ be a discrete probability space and $\mathcal{A}$ an event with $\Pr[\mathcal{A}] > 0$. If $X$ is a random variable in that space, then the *conditional probability distribution* $P_{X|\mathcal{A}}$ of $X$ given that event $\mathcal{A}$ holds is defined as follows:

$$P_{X|\mathcal{A}}(x) = \Pr[X = x|\mathcal{A}]$$

Note that $P_{X|\mathcal{A}}$ is a regular probability distribution and hence that the probabilities $P_{X|\mathcal{A}}(x)$ must sum up to one:

$$\sum_{x \in \mathcal{X}} P_{X|\mathcal{A}}(x) = 1$$

If the conditioning event involves another random variable $Y$ defined on the same sample space $\Omega$, then the *conditional probability distribution* of $X$ given that $Y$ takes on a value $y$ is defined as

$$P_{X|Y=y}(x) = \frac{P_{XY}(x,y)}{P_Y(y)}$$

whenever $P_Y(y) > 0$. More specifically, the conditional probability distribution $P_{X|Y}$ of $X$ given $Y$ is a two-argument function that is defined as follows:

$$\begin{aligned} P_{X|Y} : \mathcal{X} \times \mathcal{Y} &\longrightarrow \mathbb{R}^+ \\ (x,y) &\longmapsto P_{X|Y}(x,y) = P(X = x|Y = y) = \\ &\quad \frac{P(X = x, Y = y)}{P(Y = y)} = \frac{P_{XY}(x,y)}{P_Y(y)} \end{aligned}$$

Note that the two-argument function $P_{X|Y}(\cdot,\cdot)$ is not a probability distribution on $\mathcal{X} \times \mathcal{Y}$, but that for every $y \in \mathcal{Y}$, the one-argument function $P_{X|Y}(\cdot, y)$ is a probability distribution, meaning that $\sum_{x \in \mathcal{X}} P_{X|Y}(x,y)$ must sum up to 1 for every $y$ with $P_Y(y) > 0$. In fact, $P_{X|Y}(x,y)$ is defined only for values with $P(Y = y) = P_Y(y) \neq 0$.

### B.2.4 Expectation

The expectation of a random variable gives some information about its order of magnitude, meaning that if the expectation is small (large), then large (small) values are unlikely to occur. More formally, let $X : \Omega \rightarrow \mathcal{X}$ be a random variable and $\mathcal{X}$

be a finite subset of the real numbers (i.e., $\mathcal{X} \subset \mathbb{R}$). Then the *expectation* or *mean* of $X$, denoted $E[X]$, is defined as

$$E[X] = \sum_{x \in \mathcal{X}} \Pr[X = x] \cdot x = \sum_{x \in \mathcal{X}} P_X(x) \cdot x \tag{B.2}$$

The expectation of a random variable is best understood in terms of betting. Consider the situation in which a player can win one USD with a probability of $2/3$, lose two USD with a probability of $1/6$, or end in a draw with a probability of $1/6$. This can be modeled with a discrete probability space that consists of a sample space $\Omega = \{W, L, D\}$ (where $W$ stands for "win," $L$ stands for "lose," and $D$ stands for "draw") and a probability measure that assigns $\Pr[W] = 2/3$ and $\Pr[L] = \Pr[D] = 1/6$. Furthermore, the random variable $X$ is used to specify wins and losses; that is, $X(W) = 1$, $X(L) = 2$, and $X(D) = 0$, and one may be interested in the expectation of $X$. Referring to (B.2), this value can be computed as follows:

$$E[X] = \frac{1}{6} \cdot (-2) + \frac{1}{6} \cdot 0 + \frac{2}{3} \cdot 1 = \frac{1}{3}$$

Consequently, if one plays the game, then one can expect to win one third of a dollar on the average (i.e., the game is advantageous for the player).

Another typical application of a random variable's expectation refers to the running time of a randomized algorithm. Remember from Section 1.2 that a randomized algorithm depends on internal random values and that a complete analysis of the algorithm would be a specification of the running time of the algorithm for every possible sequence of random values. This is clearly impractical, and one may analyze the expected running time of the algorithm instead. This refers to a single value that still provides some useful information about the typical and likely temporal behavior of the algorithm.

The expectation of a random variable $X$ is linear, meaning that $E[aX] = aE[X]$ for all $a \in \mathbb{R}$, and for multiple random variables $X_1, X_2, \ldots, X_n$ over the same sample $E[X_1+X_2+\ldots+X_n] = E[X_1]+E[X_2]+\ldots+E[X_n]$. If, for example, we wanted to compute the expected number of heads when flipping a coin $n$ times, then this computation would be involved without making use of the linearity of expectations. Making use of the linearity, however, this computation becomes simple and straightforward: Let $X$ be the sum of $n$ random variables $X_1, X_2, \ldots, X_n$, where $X_i$ is 1 if the i-th coin flip is 1 ($X_i$ is 0 otherwise). Then $E[X_i] = \frac{1}{2} \cdot 0 + \frac{1}{2} \cdot 1 = \frac{1}{2}$ and $E[X] = E[X_1+X_2+\ldots+X_n] = E[X_1]+E[X_2]+\ldots+E[X_n] = n \cdot \frac{1}{2} = \frac{n}{2}$. This result is intuitive and meets our expectations.

More generally, if $f$ is a real-valued function whose domain includes $\mathcal{X}$, then $f(X)$ is a real-valued random variable with an expected value that can be computed as follows:

$$E[f(X)] = \sum_{x \in \mathcal{X}} f(x) P_X(x)$$

More specifically, if $f$ is a convex function, then *Jensen's inequality* holds:

$$E[f(X)] \geq f(E[X])$$

Many inequalities used in information theory can be derived from Jensen's inequality.

Last but not least, the *conditional expected value* $E[X|\mathcal{A}]$ of a random variable $X$ given event $\mathcal{A}$ is defined as

$$E[X|\mathcal{A}] = \sum_{x \in \mathcal{X}} P_{X|\mathcal{A}}(x) \cdot x$$

and can be computed accordingly.

### B.2.5 Independence of Random Variables

Let $X$ and $Y$ be two random variables over the same sample space $\Omega$. $X$ and $Y$ are *independent* if for all $x \in \mathcal{X}$ and $y \in \mathcal{Y}$, the events $(X = x)$ and $(Y = y)$ are independent. This is formally expressed in Definition B.3.

**Definition B.3 (Independent random variables)** *Two random variables $X$ and $Y$ are* statistically independent *(or* independent *in short) if and only if $P_{XY}(x, y) = P_X(x) \cdot P_Y(y)$ for all $x \in \mathcal{X}$ and $y \in \mathcal{Y}$.*

This definition suggests that the joint probability distribution of two independent random variables $X$ and $Y$ is equal to the product of their marginal distributions.

If two random variables $X$ and $Y$ are independent, then the conditional probability distribution $P_{X|Y}$ of $X$ given $Y$ is

$$P_{X|Y}(x, y) = \frac{P_{XY}(x, y)}{P_Y(y)} = \frac{P_X(x) P_Y(y)}{P_Y(y)} = P_X(x)$$

for all $x \in \mathcal{X}$ and $y \in \mathcal{Y}$ with $P_Y(y) \neq 0$. This basically means that knowing the value of one random variable does not tell anything about the distribution of the other (and vice versa).

If $X$ and $Y$ are independent random variables, then $E[XY] = E[X] \cdot E[Y]$. For more than two random variables, there are two notions of independence: Pairwise independence that requires two arbitrarily chosen random variables to be independent (Definition B.4), and mutual independence that requires all random variables to be independent (Definition B.5). In either case, let $X_1, \ldots, X_n$ be $n$ random variables over the same sample space $\Omega$.

**Definition B.4 (Pairwise independent random variables)** *$X_1, \ldots, X_n$ are pairwise independent if for every $i, j \in \{1, 2, \ldots, n\}$ with $i \neq j$, it holds that the two random variables $X_i$ and $X_j$ are independent; that is, $P_{X_iX_j}(x_i, x_j) = P_{X_i}(x_i) \cdot P_{X_j}(x_j)$.*

**Definition B.5 (Mutually independent random variables)** *$X_1, \ldots, X_n$ are mutually independent if for every subset of indices $I \subseteq \{1, 2, \ldots, n\}$ with $I \neq \emptyset$, it holds that*

$$P_{X_{I_1} \ldots X_{I_m}}(x_{I_1}, \ldots, x_{I_m}) = P_{X_{I_1}}(x_{I_1}) \cdot \ldots \cdot P_{X_{I_m}}(x_{I_m}) = \prod_{i=1}^{m} P_{X_{I_i}}(x_{I_i})$$

Note that the notion of mutual independence is stronger than the notion of pairwise independence. In fact, a collection of random variables that is mutually independent is also pairwise independent, whereas the converse need not always be true (i.e., a collection of random variables can be pairwise independent without being mutually independent). For example, consider the situation in which two coins are tossed. The random variable $X$ refers to the result of the first coin, the random variable $Y$ refers to the result of the second coin, and the random variable $Z$ refers to the addition modulo 2 of the results of the two coins. Obviously, all random variables have values of either 0 or 1. Then $X$, $Y$, and $Z$ are pairwise independent, but they are not mutually independent (because the value of $Z$ is entirely determined by the values of $X$ and $Y$).

Similar to the case with two random variables, one can show that if $n$ random variables $X_1, X_2, \ldots, X_n$ are mutually independent, then

$$E[X_1 \cdot X_2 \cdot \ldots \cdot X_n] = E[X_1] \cdot E[X_2] \cdot \ldots \cdot E[X_n]$$

### B.2.6 Markov's Inequality

Markov's inequality provided in Theorem B.2 (without a proof) puts into perspective the expectation of a random variable $X$ and the probability that its value is larger than a real-valued threshold $k$.

**Theorem B.2 (Markov's inequality)** *If $X$ is a nonnegative random variable, then*

$$\Pr[X \geq k] \leq \frac{E[X]}{k}$$

*holds for every $k \in \mathbb{R}$.*

For example, if $E[X] = 10$, then

$$\Pr[X \geq 1,000,000] \leq \frac{10}{1,000,000} = \frac{1}{100,000}$$

This means that it is very unlikely that the value of $X$ is greater than or equal to 1,000,000 if its expectation is 10. This result is strongly supported by intuition.

Sometimes the order of magnitude given by Markov's inequality is extremely bad, but the bound is as strong as possible if the only information available about $X$ is its expectation. For example, suppose that $X$ counts the number of heads in a sequence of $n$ coin flips, i.e., $\Omega = \{0,1\}^n$ with uniformly distributed elements. If $X$ is the number of ones in the string, then $E[X] = n/2$. In this example, Markov's inequality provides the following upper bound for $\Pr[X \geq n]$:

$$\Pr[X \geq n] \leq \frac{E[X]}{n} = \frac{n/2}{n} = \frac{1}{2}$$

Obviously, the correct value is $2^{-n}$, and the result provided by Markov's inequality is totally off (it does not even depend on $n$). On the other hand, if $n$ coins are flipped and the flips are glued together (so that the only possible outcomes are $n$ heads or $n$ tails, both with probability $1/2$), then $X$ counts the number of heads and $E[X] = n/2$. In this case, the inequality is tight, and $\Pr[X \geq n]$ is in fact $1/2$.

The bottom line is that Markov's inequality is useful because it applies to every nonnegative random variable with known expectation. According to the examples given above, the inequality is accurate when applied to a random variable that typically deviates a lot from its expectation, and it is bad when applied to a random variable that is concentrated around its expectation. In the latter case, more powerful methods are required to achieve more accurate estimations. Most of these methods make use of the variance and standard deviation introduced next.

### B.2.7 Variance and Standard Deviation

For a random variable $X$, one may consider the complementary random variable

$$X' = |X - E[X]|$$

to provide some information about the likelihood of $X$ deviating from its expectation. More specifically, if $X'$ is expected to be small, then $X$ is not likely to deviate a lot from its expectation. Unfortunately, $X'$ is not easier to analyze than $X$, and hence $X'$ is not particularly useful to consider as a random variable.

As a more viable alternative, one may consider the complementary random variable

$$X'' = (X - E[X])^2$$

Again, if the expectation of $X''$ is small, then $X$ is typically close to its expectation. In fact, the expectation of the random variable $X''$ turns out to be a useful measure. It is called the *variance* of $X$, denoted $Var[X]$, and it is formally defined as follows:

$$Var[X] = E[X''] = E[(X - E[X])^2] = \sum_{x \in \mathcal{X}} P_X(x) \cdot (x - E[X])^2$$

Alternatively, the variance of $X$ can also be expressed as follows:

$$\begin{aligned} Var[X] &= E[(X - E[X])^2] \\ &= E[X^2 - 2XE[X] + (E[X])^2] \\ &= E[X^2] - 2E[XE[X]] + (E[X])^2 \\ &= E[X^2] - 2E[X]E[X] + (E[X])^2 \\ &= E[X^2] - 2(E[X])^2 + (E[X])^2 \\ &= E[X^2] - (E[X])^2 \end{aligned}$$

For example, let $X$ be a random variable that is equal to zero with probability $1/2$ and 1 with probability $1/2$. Then $E[X] = \frac{1}{2} \cdot 0 + \frac{1}{2} \cdot 1 = \frac{1}{2}$, $X = X^2$ (because $0 = 0^2$ and $1 = 1^2$), and

$$Var[X] = E[X^2] - (E[X])^2 = \frac{1}{2} - \frac{1}{4} = \frac{1}{4}$$

The variance of a random variable is useful because it is often easy to compute, but it still gives rise to sometimes strong estimations on the probability that a random variable deviates a lot from its expectation.

The value $\sigma[X] = \sqrt{Var[X]}$ is called the *standard deviation* of $X$. In general, one may expect the value of a random variable $X$ to be in the interval $E[X] \pm \sigma[X]$.

If $X$ is a random variable, then

$$Var[aX + b] = a^2 Var[X]$$

for every $a, b \in \mathbb{R}$. Similarly, if $X_1, \ldots, X_n$ are pairwise statistically independent random variables over the same sample space, then

$$Var[X_1 + \ldots + X_n] = Var[X_1] + \ldots + Var[X_n]$$

For example, let $X$ be again the random variable that counts the number of heads in a sequence of $n$ independent coin flips (i.e., $E[X] = n/2$). Computing the variance according to the definition given above seems difficult. If, however, we view the random variable $X$ as the sum $X_1 + \ldots + X_n$ (where all $X_i$ are mutually independent random variables such that for each $i$, $X_i$ takes the value 1 with probability $1/2$ and the value zero with probability $1/2$), then $Var[X_i] = \frac{1}{4}$, and hence $Var[X] = n \cdot \frac{1}{4} = \frac{n}{4}$.

### B.2.8 Chebyshev's Inequality

Chebyshev's inequality specified in Theorem B.3 can be used to provide an upper bound for the probability that a random variable $X$ deviates from its expectation more than a real-valued threshold $k \in \mathbb{R}$. To make use of Chebyshev's inequality, the variance of $X$ must be known.

**Theorem B.3 (Chebyshev's inequality)** *If $X$ is a random variable, then*

$$\Pr[|X - E[X]| \geq k] \leq \frac{Var[X]}{k^2}$$

*holds for every $k \in \mathbb{R}$.*

*Proof.*

$$\begin{aligned} \Pr[|X - E[X]| \geq k] &= \Pr[(X - E[X])^2 \geq k^2] \\ &\leq \frac{E[(X - E[X])^2]}{k^2} \\ &= \frac{Var[X]}{k^2} \end{aligned}$$

In the first step, the argument of the probability function is squared on either side of the relation (this does not change the probability value). In the second step, Markov's inequality is applied (for $X - E[X]$).

□

Let us test Chebyshev's inequality on the example given above. $X$ is a random variable defined over the sample space $\Omega = \{0,1\}^n$, Pr is the uniform distribution, and $X$ counts the number of ones in the elementary event. If we want to compute $\Pr[X \geq n]$ using $Var[X] = n/4$, then we get

$$\Pr[X \geq n] \leq \Pr[|X - E[X]| \geq n/2] \leq \frac{1}{n}$$

Obviously, this result is much better than the one we get from Markov's inequality. It linearly decreases with $n$, but it is still far apart from the correct value $2^{-n}$.

Using the standard deviation (instead of the variance) and setting $k = c \cdot \sigma[X]$, Chebyshev's inequality can also be expressed as follows:

$$\Pr[|X - E[X]| \geq c \cdot \sigma[X]] \leq \frac{Var[X]}{c^2(\sigma[X])^2} = \frac{(\sigma[X])^2}{c^2(\sigma[X])^2} = \frac{1}{c^2}$$

## References

[1] Chung, K.L., *A Course in Probability Theory*, 3rd edition. Academic Press, Cambridge, MA, 2000.

[2] Jaynes, E.T., *Probability Theory: The Logic of Science*. Cambridge University Press, Cambridge, U.K., 2003.

[3] Klenke, A., *Probability Theory: A Comprehensive Course*. Springer, Berlin, Germany, 2007.

[4] Ross, S.M., *Introduction to Probability and Statistics for Engineers and Scientists*, 5th edition. Academic Press, Cambridge, MA, 2014.

[5] Ross, S.M., *A First Course in Probability*, 9th edition. Pearson, New York, 2015.

[6] Ghahramani, S., *Fundamentals of Probability: With Stochastic Processes*, 4th edition. CRC Press, Boca Raton, FL, 2019.

[7] Blitzstein, J.K., and J. Hwang, *Introduction to Probability*, 2nd edition. CRC Press, Boca Raton, FL, 2019.

# Appendix C

# Information Theory

As mentioned in Section 1.3, Claude E. Shannon developed a mathematical theory of communication [1] and a related communication theory of secrecy systems [2] that started a branch of research commonly referred to as *information theory*. Information theory has had—and continues to have—a deep impact on cryptography, especially when it comes to unconditionally or perfectly secure encryption systems.

In this appendix, we summarize the basic principles and results from information theory as far as they are relevant for cryptography. More specifically, we introduce the topic in Section C.1, elaborate on the entropy to measure the uncertainty of information in Section C.2, address the redundancy of languages in Section C.3, and focus on the key equivocation and unicity distance in Section C.4. Again, this appendix is intentionally kept short, and further information can be found in any textbook on information theory (e.g., [3–5]).

## C.1 INTRODUCTION

Information theory is concerned with the analysis of a *communication system* that has traditionally been represented by a block diagram as illustrated in Figure C.1. The aim of the communication system is to communicate or transfer information (i.e., messages) from a source (on the left side) to a destination (on the right side). The following entities are involved in one way or another:

- The *source* is a person or machine that generates the messages to be communicated or transferred.
- The *encoder* associates with each message an object that is suitable for transmission over the channel. In digital communications, the object is typically

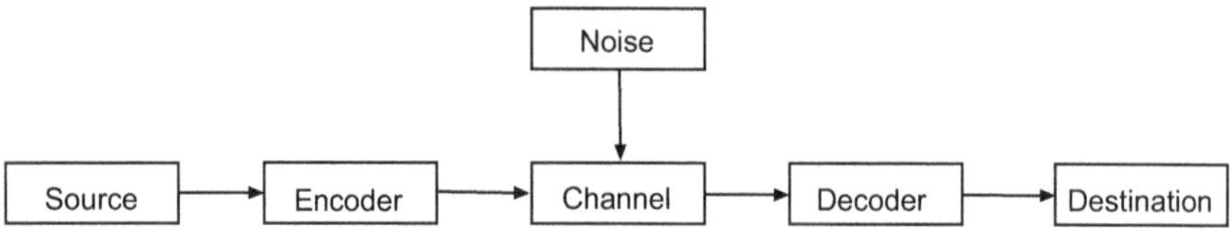


**Figure C.1** A communication system.

a sequence of bits. In analog communication, however, the object can be a signal represented by a continuous waveform.

- The *channel* is the medium over which the objects prepared by the encoder are actually communicated or transferred.
- The channel may be subject to *noise*. This noise, in turn, may cause some objects to be modified or disturbed.
- The *decoder* operates on the output of the channel and attempts to associate a message with each object it receives from the channel.
- Similar to the source, the *destination* can be a person or machine. In either case, it receives the messages that are communicated or transferred.

**Table C.1**
The Entities of a Communication System with Their Input and Output Parameters

| Entity | Input | Output |
|---|---|---|
| Source | | Message |
| Encoder | Message | (Input) object |
| Channel | (Input) object | (Output) object |
| Decoder | (Output) object | Message |
| Destination | Message | |

The entities of a communication system with their input and output parameters are summarized in Table C.1. Note that the objects mentioned earlier are divided into (input) objects that are input to the channel and (output) objects that are output to the channel.

The ultimate goal of information theory is to provide mathematically precise answers to practically relevant questions in information processing, such as how one can optimally (i.e., most efficiently) compress and transmit information or information-encoding data. Against this background, information theory can only be applied if the question can be modeled by stochastic phenomena.

One of the major results is the *fundamental theorem of information theory*. It basically says that it is possible to transmit information through a noisy channel at any rate less than channel capacity with an arbitrarily small error probability. There are many terms (e.g., "information," "noisy channel," "transmission rate," and "channel capacity") that need to be clarified before the theorem can be applied in some meaningful way. Nevertheless, an initial example may provide some preliminary ideas about the theorem.

Imagine a source of information that generates a sequence of bits at the rate of 1 bit per second. The bits 0 and 1 occur equally likely and are generated independently from each other. Suppose that the bits are communicated over a noisy channel. The nature of the noisy channel is unimportant, except that the probability that a particular bit is received in error is $1/4$ and that the channel acts on successive inputs independently. The statistical properties of the channel are illustrated in Figure C.2. We further assume that bits can be transmitted over the channel at a rate not to exceed 1 bit per second.

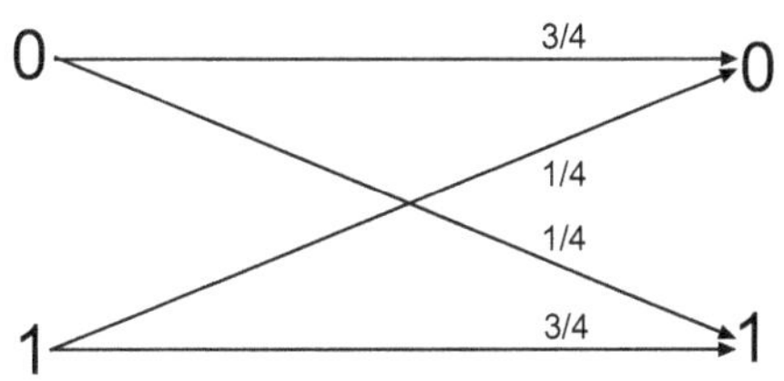


**Figure C.2** The statistical properties of a noisy channel.

If an error probability of $1/4$ is too high for a specific application, then one must find ways of improving the reliability of the channel. One way that immediately comes to mind is transmitting each source bit over the noisy channel more than once (typically an odd number of times). For example, if the source generated a zero, then one could transmit a sequence of three zeros, and if the source generated a one, then one could send a sequence of three ones. At the destination, one receives a sequence of 3 bits for each source bit. Consequently, one faces the problem of how to properly decode each sequence (i.e., make a decision, for each sequence received, as to the identity of the source bit). A reasonable way to decide is by means of a majority selector, meaning that there is a rule that if more ones than zeros are received, then the sequence is decoded as a one, and if more zeros than ones are received, then the sequence is decoded as a zero. For example, if the source generated a one, then a sequence of three ones would be transmitted over the noisy channel. If the first and third bits were received incorrectly, then the received sequence would be 010 and the decoder would incorrectly decide that a zero was transmitted.

In this example, one may calculate the probability that a given source bit is received in error. It is the probability that at least 2 of a sequence of 3 bits are received incorrectly, where the probability of a given bit being incorrect is $1/4$ and the bits are transmitted independently. The corresponding error probability $\Pr[error]$ (i.e., the probability of incorrectly receiving $\geq 2$ bits) may be computed as follows:

$$\Pr[error] = \binom{3}{2}\left(\frac{1}{4}\right)^2\frac{3}{4} + \left(\frac{1}{4}\right)^3 = \frac{10}{64}$$

Obviously, $10/64 < 1/4$, and the error probability is reduced considerably. There is, however, a price to pay for this reduction: the sequence to be transmitted is three times as long as the original one. This means that if one wants to synchronize the source with the channel, then one must slow down the rate of the source to $1/3$ bit per second (while keeping the channel rate fixed at 1 bit per second).

This procedure can be generalized. Let $\beta < 1/2$ be the error probability for each bit and each bit be represented by a sequence of $2n+1$ bits.[1] Hence, the effective transmission rate of the source is reduced to $1/(2n+1)$ bits per second. In either case, a majority selector is used at the receiving end. The probability $\Pr[error]$ of incorrectly decoding a given sequence of $2n+1$ bits is equal to the probability of having $n+1$ or more bits in error. This probability can be computed as follows:

$$\Pr[error] = \sum_{k=n+1}^{2n+1} \binom{2n+1}{k} \beta^k (1-\beta)^{2n+1-k}$$

It can be shown that $\lim_{n\to\infty} \Pr[error] = 0$, meaning that the probability of incorrectly decoding a given sequence of $2n+1$ bits can be made arbitrarily small for sufficiently large $n$. In other words, one can reduce the error probability to an arbitrarily small value at the expense of decreasing the effective transmission rate toward zero.

The essence of the fundamental theorem of information theory is that in order to achieve arbitrarily high reliability, it is not necessary to reduce the transmission rate to zero, but only to a number called the *channel capacity*. The means by which this is achieved is called *coding*, and the process of coding involves an *encoder*, as illustrated in Figure C.1. The encoder assigns to each of a specified group of source signals (e.g., bits) a sequence of symbols called a *code word* suitable for transmission over the noisy channel. In the example given above, we have seen a very primitive

1 This basically means that each source bit is represented by a bit sequence of odd length.

form of coding (i.e., the source bit 0 is assigned a sequence of zeros, whereas the source bit 1 is assigned a sequence of ones). In either case, the code words are transmitted over the noisy channel and received by a decoder, which attempts to determine the original source signals. In general, to achieve reliability without sacrificing speed of transmission in digital communications, code words must not be assigned to single bits or bytes but instead to longer bit blocks. In other words, the encoder waits for the source to generate a block of bits of a specified length and then assigns a code word to the entire block. The decoder, in turn, examines the received sequence and makes a decision as to the identity of the original source bits. In practice, encoding and decoding are much more involved than this simple example may suggest.

In order to make these ideas more concrete, we need a mathematical measure for the information conveyed by a message, or—more generally—a measure of information. This is where the notion of entropy as addressed next comes into play.

## C.2 ENTROPY

Let $(\Omega, \Pr)$ be a discrete probability space and $X : \Omega \rightarrow \mathcal{X}$ a random variable with range $\mathcal{X} = \{1, 2, \ldots, 5\}$ and uniformly distributed elements. If we have no prior knowledge about $X$ and try to guess the correct value of $X$, then we have a probability of $1/|\mathcal{X}| = 1/5$ of being correct. If, however, we have some prior knowledge and know, for example, that $1 \leq X \leq 2$, then we have a higher probability of correctly guessing $X$ (i.e., $1/2$ in this case). In other words, there is less uncertainty about the second situation, and knowing that $1 \leq X \leq 2$ has in fact reduced the uncertainty about the value of $X$. It thus appears that if we could pin down the notion of uncertainty, we would also be able to measure precisely the transfer of information.

Suppose that a random experiment involves the observation of a random variable $X$, and let $X$ take on a finite number of possible values $x_1, \ldots x_n$. The probability that $X$ takes on $x_i$ $(i = 1, \ldots, n)$ is $\Pr[X = x_i] = P_X(x_i)$ and is abbreviated as $p_i$ (note that all $p_i \geq 0$ and $\sum_{i=1}^{n} p_i = 1$). Our goal is to come up with a measure for the uncertainty associated with $X$. To achieve this goal, we construct the following two functions:

1. We define the function $h$ on the interval $[0, 1]$. The value $h(p)$ can be interpreted as the uncertainty associated with an event that occurs with probability $p$. If the event $(X = x_i)$ has probability $p_i$, then we say that $h(p_i)$ is the uncertainty associated with the event $(X = x_i)$ or the uncertainty removed (or information conveyed) by revealing that $X$ has taken on value $x_i$.

2. We define the function $H_n$ for $n \in \mathbb{N}$ probability values $p_1, \ldots, p_n$. The value $H_n([p_1, \ldots, p_n])$ represents the average uncertainty associated with the events $(X = x_i)$ for $i = 1, \ldots, n$ (or the average uncertainty removed by revealing $X$, respectively). More specifically, we require that

$$H_n([p_1, \ldots, p_n]) = \sum_{i=1}^{n} p_i h(p_i)$$

In this book, we write $H([p_1, \ldots, p_n])$ instead of $H_n([p_1, \ldots, p_n])$ most of the time.

The function $h$ is only used to introduce the function $H$. The function $H$ is then used to measure the uncertainty of a probability distribution or a random variable. In fact, $H(X)$ is called the *entropy* of the random variable $X$, and it measures the average uncertainty of an observer about the value taken on by $X$. The entropy plays a pivotal role in data compression. In fact, it can be shown that an optimal data compression technique can compress the output of an information source arbitrarily close to its entropy, but that error-free compression below this value is not possible.

In the literature, the function $H$ is usually introduced by first setting up requirements (or axioms), and then showing that the only function satisfying these requirements is

$$H([p_1, \ldots, p_n]) = -C \sum_{i:1 \leq i \leq n; p_i > 0} p_i \log p_i$$

where $C$ is an arbitrary positive number, and the logarithm base is any number greater than one. In this case, we have

$$h(p_i) = \log \frac{1}{p_i} = -\log p_i$$

and $h(p_i)$ measures the unexpectedness of an event with probability $p_i$. The units of $H$ are usually called bits; thus the units are chosen so that there is one bit of uncertainty associated with the toss of an unbiased coin. Unless otherwise specified, we assume $C = 1$ and take logarithms to the base 2.

At this point it is important to note that the average uncertainty of a random variable $X$; that is, $H(X)$, does not depend on the values the random variable assumes, or on anything else related to $\mathcal{X}$ except the probabilities associated with all values. That is why we said earlier that the entropy is defined for a random variable or a probability distribution. If we want to express the entropy of a random variable

$X$, then we can use the following formula:

$$H(X) = -\sum_{x \in \mathcal{X}: P_X(x) \neq 0} P_X(x) \log_2 P_X(x) \tag{C.1}$$

Alternatively speaking, $H(X) = E[-\log_2 P_X(X)] = E[g(X)]$ with $g(\cdot) = -\log_2 P_X(\cdot)$.

There are some intuitive properties of the entropy (as a measure of uncertainty). For example, if we add some values to a random variable that are impossible (i.e., their probability is zero), then the entropy does not change. This property can be formally expressed as follows:

$$H([p_1, \ldots, p_n]) = H([p_1, \ldots, p_n, 0])$$

Furthermore, a situation involving a number of alternatives is most uncertain if all possibilities are equally likely. This basically means that

$$0 \leq H([p_1, \ldots, p_n]) \leq \log_2 n$$

with equality on the left side if and only if one value occurs with probability one (and all other values occur with probability zero), and with equality on the right side if and only if all values are equally likely (i.e., $p_i = 1/n$). Similarly, we have

$$0 \leq H(X) \leq \log_2 |\mathcal{X}|$$

with the same conditions for equality on either side as mentioned earlier. In particular, if $X$ is uniformly distributed, then we have $H(X) = \log_2 |\mathcal{X}|$.

If we increase the number of alternatives, then we also increase the entropy of the corresponding probability distribution. This property can be formally expressed as follows:

$$H\left(\left[\frac{1}{n}, \ldots, \frac{1}{n}\right]\right) < H\left(\left[\frac{1}{n+1}, \ldots, \frac{1}{n+1}\right]\right)$$

If $p = \sum_{i=1}^{k} p_i$ and $q = \sum_{i=1}^{l} q_i$, then the following equation holds and can be used:

$$\begin{aligned} H([p_1, \ldots, p_k, q_1, \ldots, q_l]) = H([p, q]) &+ pH\left(\left[\frac{p_1}{p}, \ldots, \frac{p_k}{p}\right]\right) \\ &+ qH\left(\left[\frac{q_1}{q}, \ldots, \frac{q_l}{q}\right]\right) \end{aligned}$$

We now turn to the problem of characterizing the uncertainty associated with more than one random variable (associated with the same discrete probability space or random experiment). This is where the notion of a joint entropy comes into play.

### C.2.1 Joint Entropy

First of all, it is important to note that a vector of random variables (associated with the same discrete probability space or random experiment) can always be viewed as a single random variable. If, for example, we have two random variables $X$ and $Y$ with $n$ and $m$ possible outcomes, then $X$ and $Y$ have joint probability $P_{XY}(x_i, y_j) = \Pr[X = x_i, Y = y_j] = p(x_i, y_j) = p_{ij}$ for $i = 1, \ldots, n$ and $j = 1, \ldots, m$. The resulting experiment has a total of $nm$ possible outcomes, and the outcome $(X = x_i, Y = y_j)$ has probability $p_{ij} = p(x_i, y_j)$.

Against this background, the *joint entropy* (or joint uncertainty) of $X$ and $Y$, denoted as $H(XY)$, is defined as follows:

$$H(XY) = -\sum_{i=1}^{n}\sum_{j=1}^{m} p(x_i, y_j) \log_2 p(x_i, y_j)$$

More formally, $H(XY)$ can be expressed as follows:

$$H(XY) = -\sum_{(x,y)} P_{XY}(x, y) \log_2 P_{XY}(x, y) \tag{C.2}$$

On the right side of (C.2), the index of the sum goes through all possible pairs $(x, y)$ with $x \in \mathcal{X}$ and $y \in \mathcal{Y}$, or—equivalently—all $(x_i, y_j)$ for $i = 1, \ldots, n$ and $j = 1, \ldots, m$.

Equation (C.2) can be generalized to the joint entropy of more than two random variables. In fact, the joint entropy of $n$ random variables $X_1, X_2, \ldots, X_n$ can be expressed as follows:

$$H(X_1 \cdots X_n) = -\sum_{(x_1,\ldots,x_n)} P_{X_1 \cdots X_n}(x_1, \ldots, x_n) \log_2 P_{X_1 \cdots X_n}(x_1, \ldots, x_n)$$

In this equation, $P_{X_1 \cdots X_n}$ refers to the joint probability distribution of $X_1, \ldots, X_n$. Consequently, the joint entropy of $X_1, \ldots, X_n$ equals the entropy of the joint probability distribution $P_{X_1 \cdots X_n}$:

$$H(X_1 \cdots X_n) = H(P_{X_1 \cdots X_n})$$

There is a relation regarding the joint entropy of $n$ random variables $X_1, \ldots, X_n$ and their individual entropies. In fact, it can be shown that

$$H(X_1 \cdots X_n) \leq H(X_1) + \ldots + H(X_n)$$

with equality if and only if $X_1, \ldots, X_n$ are mutually independent.

### C.2.2 Conditional Entropy

Equation (C.1) also covers the case where the probability distribution is conditioned on an event $\mathcal{A}$ with $\Pr[\mathcal{A}] > 0$. Consequently,

$$\begin{aligned} H(X|\mathcal{A}) &= H(P_{X|\mathcal{A}}) \\ &= - \sum_{x \in \mathcal{X}: P_{X|\mathcal{A}}(x) \neq 0} P_{X|\mathcal{A}}(x) \log_2 P_{X|\mathcal{A}}(x) \end{aligned}$$

Remember from Section B.2.3 that $P_{X|\mathcal{A}}$ is a regular probability distribution.

Let $X$ and $Y$ be two random variables. If we know the event $Y = y$, then we can replace $\mathcal{A}$ with $Y = y$ and rewrite the formula given above:

$$\begin{aligned} H(X|Y = y) &= H(P_{X|Y=y}) \\ &= - \sum_{x \in \mathcal{X}: P_{X|Y=y}(x) \neq 0} P_{X|Y=y}(x) \log_2 P_{X|Y=y}(x) \end{aligned}$$

Using the conditional entropy $H(X|Y = y)$, we can define the conditional entropy of the random variable $X$ when given the random variable $Y$ as the weighted average of the conditional uncertainties of $X$ given that $Y = y$:

$$\begin{aligned} H(X|Y) &= \sum_y P_Y(y) H(X|Y = y) \\ &= - \sum_y P_Y(y) \sum_x P_{X|Y=y}(x) \log_2 P_{X|Y=y}(x) \\ &= - \sum_y \sum_x P_Y(y) \frac{P_{XY}(x,y)}{P_Y(y)} \log_2 P_{X|Y}(x,y) \\ &= - \sum_{(x,y)} P_{XY}(x,y) \log_2 P_{X|Y}(x,y) \end{aligned}$$

In this series of equations, the indices of the sums are written in a simplified way. In fact, $x$ stands for $x \in \mathcal{X} : P_{X|Y=y}(x) \neq 0$, $y$ stands for $y \in \mathcal{Y} : P_Y(y) \neq 0$, and—similar to (C.2)—$(x, y)$ stands for all possible pairs $(x, y)$ with $x \in \mathcal{X}$ and $y \in \mathcal{Y}$ or all $(x_i, y_j)$ for $i = 1, \ldots, n$ and $j = 1, \ldots, m$.

Note that in contrast to the previously introduced entropies, such as $H(X) = H(P_X)$, $H(XY) = H(P_{XY})$, or $H(X|Y = y) = H(P_{X|Y=y})$, the entropy $H(X|Y)$ is not the entropy of a specific probability distribution, but rather the expectation of the entropies $H(X|Y = y)$. It can be shown that

$$0 \leq H(X|Y) \leq H(X)$$

with equality on the left if and only if $X$ is uniquely determined by $Y$ and with equality on the right if and only if $X$ and $Y$ are (statistically) independent. More precisely, it can be shown that

$$H(XY) = H(X) + H(Y|X) = H(Y) + H(X|Y)$$

(i.e., the joint entropy of $X$ and $Y$ is equal to the entropy of $X$ plus the entropy of $Y$ given $X$, or the entropy of $Y$ plus the entropy of $X$ given $Y$). This equation is sometimes referred to as *chain rule* and can be used repeatedly to expand $H(X_1 \cdots X_n)$ as

$$\begin{aligned} H(X_1 \cdots X_n) &= H(X_1) + H(X_2|X_1) + \ldots + H(X_n|X_1 \cdots X_{n-1}) \\ &= \sum_{i=1}^{n} H(X_i|X_1 \cdots X_{i-1}) \end{aligned}$$

Note that the order in which variables are extracted is arbitrary. For example, if we have three random variables $X$, $Y$, and $Z$, we can compute their joint entropy as follows:

$$\begin{aligned} H(XYZ) &= H(X) + H(Y|X) + H(Z|XY) \\ &= H(X) + H(Z|X) + H(Y|XZ) \\ &= H(Y) + H(X|Y) + H(Z|XY) \\ &= H(Y) + H(Z|Y) + H(X|YZ) \\ &= H(Z) + H(X|Z) + H(Y|XZ) \\ &= H(Z) + H(Y|Z) + H(X|YZ) \end{aligned}$$

Similarly, we can compute the joint entropy of $X_1 \cdots X_n$ given $Y$ as follows:

$$\begin{aligned} H(X_1 \cdots X_n|Y) &= H(X_1|Y) + H(X_2|X_1Y) + \ldots + H(X_n|X_1 \cdots X_{n-1}Y) \\ &= \sum_{i=1}^{n} H(X_i|X_1 \cdots X_{i-1}Y) \end{aligned}$$

### C.2.3 Mutual Information

The *mutual information* $I(X;Y)$ between two random variables $X$ and $Y$ is defined as the amount of information by which the entropy (uncertainty) of $X$ is reduced by learning $Y$. This can be formally expressed as follows:

$$I(X;Y) = H(X) - H(X|Y)$$

The mutual information is symmetric in the sense that $I(X;Y) = H(X) - H(X|Y) = H(Y) - H(Y|X) = I(Y;X)$.

The conditional mutual information between $X$ and $Y$, given the random variable $Z$, is defined as follows:

$$I(X;Y|Z) = H(X|Z) - H(X|YZ)$$

We have $I(X;Y|Z) = 0$ if and only if $X$ and $Y$ are statistically independent when given $Z$. Furthermore, the conditional mutual information between $X$ and $Y$ is also symmetric, meaning that $I(X;Y|Z) = I(Y;X|Z)$.

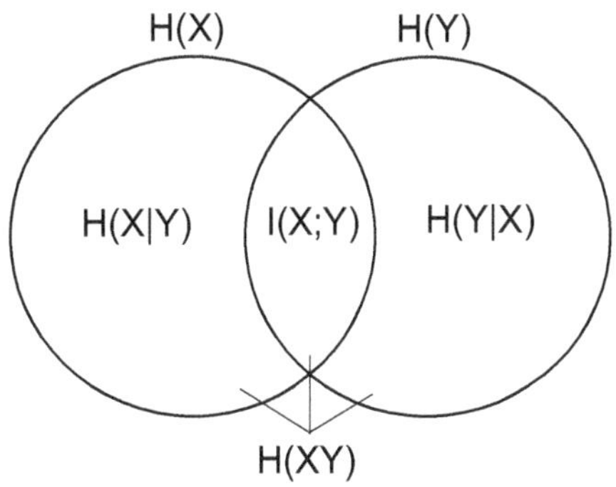


**Figure C.3** A Venn diagram graphically representing information-theoretic quantities related to two random variables.

Let $X$ and $Y$ be two random variables. Then the information-theoretic quantities $H(X)$, $H(Y)$, $H(XY)$, $H(X|Y)$, $H(Y|X)$, and $I(X;Y)$ can be graphically represented by a Venn diagram, as shown in Figure C.3.

## C.3 REDUNDANCY

If $L$ is a natural language with alphabet $\Sigma$, then one may be interested in the entropy per letter, denoted by $H_L$. In the case of the English language, $\Sigma = \{A, B, \ldots, Z\}$ and $|\Sigma| = 26$. If every letter occurred with the same probability and was independent from the other letters, then the entropy per letter would be

$$\log_2 26 \approx 4.70$$

This value represents the *absolute rate* of the language $L$ and is an upper bound for $H_L$ (i.e., $H_L \leq 4.70$). The actual value of $H_L$, however, is smaller, because one must consider the fact that letters are typically not uniformly distributed, that they occur with frequencies (that depend on the language), and that they are also not independent from each other. If $X$ is a random variable that refers to the letters of the English language (with their specific probabilities), then $H(X)$ is an upper bound for $H_L$:

$$H_L \leq H(X) \approx 4.14$$

Hence, instead of 4.7 bits of information per letter, we have around 4.14 bits of information per letter if we take into account the (statistical) letter frequencies of the English language. However, this is still an overestimate, because the letters are not independent. For example, in the English language a Q is always followed by a U, and the bigram TH is likely to occur frequently. So one would suspect that a better statistic for the amount of entropy per letter could be obtained by looking at the distribution of bigrams (instead of letters). If $X^2$ denotes the random variable of bigrams in the English language, then we can refine the upper bound for $H_L$:

$$H_L \leq \frac{H(X^2)}{2} \approx 3.56$$

This can be continued with trigrams and—more generally—$n$-grams. In the most general case, the entropy of the language $L$ is defined as follows:

$$H_L = \lim_{n \to \infty} \frac{H(X^n)}{n}$$

The exact value of $H_L$ is hard to determine. All statistical investigations show that

$$1.0 \leq H_L \leq 1.5$$

for the English language. So each letter in an English text gives at most 1.5 bits of information. This implies that the English language (like all natural languages) contains a high degree of redundancy. The *redundancy* of language $L$, denoted by $R_L$, is defined as follows:

$$R_L = 1 - \frac{H_L}{|\Sigma|}$$

In the case of the English language, we have $H_L \approx 1.25$ and $|\Sigma| = \log_2 26 \approx 4.7$. So the redundancy of the English language is

$$R_L \approx 1 - \frac{1.25}{4.7} \approx 0.75$$

This suggests that we are theoretically able to losslessly compress an English text to one-fourth its size. This means that a 10-MB file can be compressed to 2.5 MB. Note that redundancy in a natural language occurs because there are known and frequently appearing letter sequences and that these letter sequences are the major starting point for cryptanalysis.

## C.4 KEY EQUIVOCATION AND UNICITY DISTANCE

In addition to the notion of redundancy, Shannon introduced and formalized a couple of other concepts that can be used to analyze the security of deterministic (symmetric) encryption systems. Let $M^n$ and $C^n$ be random variables that denote the first $n$ plaintext message and ciphertext bits, and $K$ be a random variable that denotes the key that is in use. An interesting question one may ask is how much information about $K$ is leaked as $n$ increases. This brings us to the notion of the key equivocation formally introduced in Definition C.1.

**Definition C.1 (Key equivocation)** *The* key equivocation *is the function* $H(K|C^n)$ *(i.e., the entropy of the key as a function of the number of observed ciphertext bits).*

We generally assume that the plaintext and the key are statistically independent, meaning that $H(M|K) = H(M)$. We can show that

$$H(K|C^n) = H(K) + H(M^n) - H(C^n)$$

for a deterministic cipher. This is because

$$\begin{aligned} H(K) + H(M^n) &= H(KM^n) \\ &= H(KM^nC^n) \\ &= H(KC^n) \\ &= H(C^n) + H(K|C^n) \end{aligned}$$

In the first line, we exploit the fact that $K$ and $M^n$ are statistically independent. In the second and third line, we exploit the fact that $H(C^n|KM^n) = 0$ and $H(M^n|KC^n) = 0$.

We make the realistic assumption that the entropy of the plaintext grows approximately proportional to its length. That is,

$$H(M^n) \approx (1 - R_L)n$$

where $R_L$ is the redundancy of the plaintext language.

Against this background, it is interesting to analyze the key equivocation when $n$ grows. For every $n$, there are possible keys (and it is hoped that the size of the set of possible keys decreases as $n$ increases). More specifically, there is one correct key and a set of spurious keys (a spurious key is defined as a possible but not correct key). The most interesting question is how large $n$ must be in order to be theoretically able to uniquely determine the key. This is where the notion of the unicity distance as introduced in Definition C.2 comes into play.

**Definition C.2 (Unicity distance)** *The* unicity distance $n_u$ *is the approximate value of $n$ for which the key is uniquely determined by the ciphertext (i.e., $H(K|C^n) \approx 0$).*

In other words, the unicity distance $n_u$ is the minimum value for $n$ so that the expected number of spurious keys equals zero. This is the average amount of ciphertext that is needed before an adversary can determine the correct key (again, assuming the adversary has infinite computing power). The unicity distance can be approximately determined as follows:

$$n_u \approx \frac{H(K)}{R_L}$$

If $n \geq n_u$ ciphertext bits are given, it is theoretically possible to uniquely determine the key. For many practically relevant ciphers, $n_u$ is surprisingly small.

## References

[1] Shannon, C.E., "A Mathematical Theory of Communication," *Bell System Technical Journal*, Vol. 27, No. 3/4, July/October 1948, pp. 379–423/623–656.

[2] Shannon, C.E., "Communication Theory of Secrecy Systems," *Bell System Technical Journal*, Vol. 28, No. 4, October 1949, pp. 656–715.

[3] Ash, R.B., *Information Theory*. Dover Publications, 2012.

[4] Cover, T.M., and J.A. Thomas, *Elements of Information Theory*, 2nd edition. John Wiley & Sons, Hoboken, NJ, 2006.

[5] McEliece, R., *Theory of Information and Coding*, 2nd edition. Cambridge University Press, Cambridge, U.K., 2002.

[6] Shannon, C.E., "Prediction and Entropy of Printed English," *Bell System Technical Journal*, Vol. 30, January 1951, pp. 50–64.

# Appendix D

## Complexity Theory

As its name suggests, complexity theory is the mathematical theory that allows us to scientifically argue about the complexity of computational tasks. In cryptography, it is used to argue about the computational security of cryptosystems. In this appendix, we summarize the fundamentals of complexity theory as far as they are relevant for cryptography. More specifically, we start with some preliminary remarks in Section D.1, introduce the topic in Section D.2, overview an asymptotic order notation in Section D.3, elaborate on efficient computations in Section D.4, address computational models in Section D.5, focus on complexity classes in Section D.6, and conclude with some final remarks in Section D.7. More information is available in textbooks about complexity theory (e.g., [1–3]).

### D.1 PRELIMINARY REMARKS

In theoretical computer science, one often uses a nonempty set of *characters* or *symbols* that is referred to as an *alphabet* and denoted as $\Sigma$. For example, the following alphabet comprises all capital letters used in the English language:

$$\Sigma = \{A, B, C, \ldots, Z\}$$

The length of the alphabet $\Sigma$ corresponds to the number of elements (i.e., $|\Sigma|$). Here, the length of the alphabet is $|\{A, B, C, \ldots, Z\}| = 26$.

Another alphabet frequently used in computer science is the American Standard Code for Information Interchange (ASCII) character set. As illustrated in Table D.1, the ASCII character set assigns a value between 0 and 0xFF (in hexadecimal notation) or 127 (in decimal notation) to each character or symbol. This means that only 7 bits of each byte are used for this purpose ($2^7 = 128$ possible values). There is also an extended ASCII character set with $2^8 = 256$ characters or symbols that uses all 8 bits of each byte.

**Table D.1**
7-Bit ASCII Character Set with Hexadecimal Values

| | 0x00 | 0x10 | 0x20 | 0x30 | 0x40 | 0x50 | 0x60 | 0x70 |
|---|---|---|---|---|---|---|---|---|
| +0 | NUL | DLE | | 0 | @ | P | ` | p |
| +1 | SOH | DC1 | ! | 1 | A | Q | a | q |
| +2 | STX | DC2 | " | 2 | B | R | b | r |
| +3 | ETX | DC3 | # | 3 | C | S | c | s |
| +4 | EOT | DC4 | $ | 4 | D | T | d | t |
| +5 | ENQ | NAK | % | 5 | E | U | e | u |
| +6 | ACK | SYN | & | 6 | F | V | f | v |
| +7 | BEL | ETB | ' | 7 | G | W | g | w |
| +8 | BS | CAN | ( | 8 | H | X | h | x |
| +9 | HT | EM | ) | 9 | I | Y | i | y |
| +A | LF | SUB | * | : | J | Z | j | z |
| +B | VT | ESC | + | ; | K | [ | k | { |
| +C | FF | FS | , | < | L | \ | l | \| |
| +D | CR | GS | - | = | M | ] | m | } |
| +E | SO | RS | . | > | N | ^ | n | ~ |
| +F | SI | US | / | ? | O | _ | o | DEL |

Instead of using letters or ASCII characters, computer systems usually operate on strings of *binary digits* (*bits*). Consequently, the alphabet most frequently used in computer science is $\Sigma = \{0, 1\}$ and its length is $|\{0, 1\}| = 2$.

If an alphabet $\Sigma$ is finite (which is almost always the case), then its length is less than infinity (i.e., $|\Sigma| = n < \infty$). In this case, the $n$ elements of $\Sigma$ can also be associated with the $n$ elements (residue classes) of $\mathbb{Z}_n = \{0, 1, \ldots, n-1\}$. Consequently, it is possible to work in $\mathbb{Z}_n$ instead of any character set with $n$ elements. This simplifies things considerably, because we can work in mathematical structures that we know and are familiar with.

Let $\Sigma$ be an alphabet. The term *word* (or *string*) over $\Sigma$ refers to a finite sequence of characters or symbols from $\Sigma$, including, for example, the empty word $\varepsilon$. The length of a word $w$ over $\Sigma$, denoted $|w|$, corresponds to the number of characters. The empty word has length zero (i.e., $|\varepsilon| = 0$). The set of all words over $\Sigma$ (again, including the empty word) is referred to as $\Sigma^*$. For every $n \in \mathbb{N}$, $\Sigma^n$ refers to the set of all words of length $n$ over $\Sigma$. For example, $\{0, 1\}^n$ denotes the set of all $n$-bit sequences, whereas $\{0, 1\}^*$ denotes the set of all binary words of arbitrary length.

This can be formally expressed as follows:

$$\{0,1\}^* = \bigcup_{n \in \mathbb{N}} \{0,1\}^n$$

Using the binary alphabet, a positive integer $n \in \mathbb{N}$ can always be encoded as a binary word $b_{l-1} \ldots b_1 b_0 \in \{0,1\}^l$ of some length $l$:

$$n = \sum_{i=0}^{l-1} b_i 2^i$$

In complexity theory, a positive integer $n \in \mathbb{N}$ is sometimes also encoded using the *unary representation*. This encoding looks as follows:

$$n = 1^n = \underbrace{11 \cdots 1}_{n \text{ times}}$$

The relevant operation for words is (string) concatenation, denoted $\|$. If $v, w \in \Sigma^*$, then $v \parallel w$ results from concatenating $v$ and $w$. The empty word $\varepsilon$ is the neutral element of the concatenation operation, hence $v \parallel \varepsilon = \varepsilon \parallel v = v$ for every $v \in \Sigma^*$. It can be shown that $\langle \Sigma^*, \| \rangle$ yields a monoid (Section A.1.2.2), but this fact is not used in this book.

## D.2 INTRODUCTION

*Complexity theory* is a central field of study in theoretical computer science. According to [4], the

> main goal of complexity theory is to provide mechanisms for classifying computational problems according to the resources needed to solve them. The classification should not depend on a particular computational model, but rather should measure the intrinsic difficulty of the problem. The computational may include time, storage space, random bits, number of processors, etc., but typically the main focus is time, and sometimes space.

The important points are: (1) that the computational problems should be classified according to the resources needed to solve them, and (2) that this classification should be independent from a particular computational model. Hence, complexity theory is different from benchmark testing as used in the trade press to compare the computational power of different computer systems, products, and models. Instead, complexity theory is used to determine the computational resources (e.g., time, space, and randomness) needed to compute a particular function or solve a particular problem. The computational resources, in turn, can be determined exactly or at approximately specifying lower and upper bounds.[1] Alternatively, one can also consider the effects of limiting computational resources on the class of functions (problems) that can be computed (solved) in the first place.

In complexity theory, there are many functions and problems to consider. For example, for a positive integer $n \in \mathbb{N}$, one may look at the problem of deciding whether $n$ is prime (or composite). This problem is a decision problem and it is solved by providing a binary answer; that is, YES or NO.[2] An instance of this problem would be whether $n = 81$ is prime (which is arguably wrong, because $81 = 9 \cdot 9$). Consequently, a problem refers to a well-defined and compactly described (possibly very large) class of instances characterized by some input and output parameters. Examples include deciding primality (as mentioned above), factoring integers, or deciding graph isomorphisms. Against this background, it does not make a lot of sense to define the computational difficulty or complexity of a problem instance. There is always a trivial algorithm to solve the instance, namely the algorithm that simply outputs the correct solution. Consequently, the computational difficulty or complexity of a problem must always refer to a class of instances. This is important to properly understand complexity theory and its results.

We mentioned above that results from complexity theory should be largely independent from a particular computational model (refer to Section D.5 for an overview about the various computational models in use today). Nevertheless, one must still have a model in mind when one works in theoretical computer science and complexity theory. The computational model of choice is the Turing machine (Section D.5).[3] Looking at Turing machines is sufficient, because there is a famous thesis in theoretical computer science—called *Church's thesis*[4]—that most results from complexity theory hold regardless of the computational model in use (as long as it is "reasonable" in one way or another). More specifically, the thesis says that

1 Proving an upper bound is comparably simple. It suffices to give an algorithm together with an analysis of its computational complexity. Proving a lower bound is much more involved, because one must prove the nonexistence of an algorithm that is more efficient than the one one has in mind. Consequently, it does not come as a big surprise that no significant lower bound has been proven for the computational complexity of any practically relevant problem.

2 Today, we know that there is a deterministic polynomial-time algorithm to solve this problem (see Section A.2.4.3). Consequently, this problem is known to be in $\mathbf{P}$ (the notion of the complexity class $\mathbf{P}$ is introduced later in this appendix).

3 Turing machines are named after Alan M. Turing, who lived from 1912 to 1954.

4 The thesis is named after Alonzo Church, who lived from 1903 to 1995.

any physical computing device can be simulated by a Turing machine (in a number of steps that is polynomial in the resources used by the computing device).

For the rest of this chapter, we associate Turing machine $M$ with the function $f_M$ it computes, and we sometimes write $M(x)$ instead of $f_M(x)$ for an input $x$. As mentioned earlier, there are several things a Turing machine $M$ can do, including, for example, computing a function, solving a search problem, or making a decision.

- A Turing machine $M$ may compute a function $f : \{0,1\}^* \rightarrow \{0,1\}^*$. In this case, $M$ is given an input $x \in \{0,1\}^*$, and it computes as output $M(x) = f_M(x) = y \in \{0,1\}^*$.
- A Turing machine $M$ may solve a search problem that is defined as a binary relation $S$ (i.e., $S \subseteq \{0,1\}^* \times \{0,1\}^*$). The relation specifies input and output pairs $(x,y)$ that belong to $S$ (i.e., $(x,y) \in S$). In this case, $M$ is given input $x \in \{0,1\}^*$, and it computes as output a solution $M(x) = f_M(x) = y' \in \{0,1\}^*$ with $(x,y') \in S$. If there is more than one solution for the search problem, then any $y'$ is fine.
- Finally, a Turing machine $M$ may solve a decision problem (i.e., a problem that can be posed as a YES-NO question for some input value). Formally, one can define a language $L \subseteq \{0,1\}^*$ (over the binary alphabet) and have $M$ decide whether an input $x \in \{0,1\}^*$ is a member of $L$. If it is, then $M$ outputs YES (i.e., $M(x) = f_M(x) = 1$); otherwise it outputs NO (i.e., $M(x) = f_M(x) = 0$). Note that $L$ can be anything, such as all binary encoded prime numbers. In this case, the decision problem whether a number is prime can be seen as a membership problem for that particular language.

Due to their expressive power, complexity theory mainly focuses on decision problems. This is not too restrictive, because all computational problems can be phrased as decision problems in such a way that an efficient algorithm for the decision problem yields an efficient algorithm for the computational problem, and vice versa. This also applies to search problems.

## D.3 ASYMPTOTIC ORDER NOTATION

In complexity theory, one is mainly interested in the asymptotic behavior of the complexity of a computation as a function of its input size[5] or some other parameter(s). This is where asymptotic analysis and order notation come into play. In the following, we only consider functions that are defined for positive integers and take

5 The input size is the length of the (binary) word that is needed to represent the input (for a well-defined representation method).

on real values that are positive for some $n \geq n_0$. Let $f : \mathbb{N} \rightarrow \mathbb{R}^+$ and $g : \mathbb{N} \rightarrow \mathbb{R}^+$ be such functions. The following asymptotic bounds are used in complexity theory.

**Upper bound:** If there exist a positive constant $c$ and a positive integer $n_0$ such that $0 \leq f(n) \leq cg(n)$ for all $n \geq n_0$, then we write

$$f(n) = O(g(n))$$

If $g(n)$ is constant (and hence independent from $n$), then we write $f(n) = O(1)$. Note that this applies even if the constant is very large, such as $2^{128}$. Furthermore, if $0 \leq f(n) < cg(n)$ for all $n \geq n_0$, then we write

$$f(n) = o(g(n))$$

In this case, $f(n)$ is strictly smaller than $cg(n)$, whereas in the previous case, $f(n)$ can also be equal to $cg(n)$. In either case, the function $g$ yields an *asymptotic upper bound* for $f$.

**Lower bound:** If there exist a positive constant $c$ and a positive integer $n_0$ such that $0 \leq cg(n) \leq f(n)$ for all $n \geq n_0$, then we write

$$f(n) = \Omega(g(n))$$

In this case, the function $g$ yields an *asymptotic lower bound* for $f$.

**Tight bound:** If there exist positive constants $c_1$ and $c_2$, and a positive integer $n_0$ such that $c_1g(n) \leq f(n) \leq c_2g(n)$ for all $n \geq n_0$, then we write

$$f(n) = \Theta(g(n))$$

In this case, the function $g$ yields both an asymptotic lower bound and an asymptotic upper bound for $f$. This means that $g$ yields an *asymptotic tight bound* for $f$.

Intuitively, $f(n) = O(g(n))$ means that $f(n)$ doesn't grow asymptotically faster than $g(n)$ within a constant multiple, whereas $f(n) = \Omega(g(n))$ means that $f(n)$ grows asymptotically at least as fast as $g(n)$ within a constant multiple. For example, if $f(n) = 2n^2 + n + 1$, then $2n^2 + n + 1 \leq 4n^2$ for all $n \geq 1$, and hence $f(n) = O(n^2)$. Similarly, $2n^2 + n + 1 \geq 2n^2$ for all $n \geq 1$, and hence $f(n) = \Omega(n^2)$. Consequently, $f(n) = \Theta(n^2)$. In practice, the asymptotic upper bound (i.e., the big-$O$-notation) is most frequently used.

In complexity-theoretic discussions and considerations, one often uses functions that are polynomials according to Definition A.29. Polynomials are useful because they have the property that they are closed under addition, multiplication, and composition. This basically means that one can add, multiply, and compose two (or even more) polynomials, and that the result again yields a polynomial.

To characterize functions that result in very small values, one uses the notion of a negligible function as captured in Definition D.1.

**Definition D.1 (Negligible function)** *A function $f : \mathbb{N} \rightarrow \mathbb{R}$ is* negligible *if and only if for every $c \in \mathbb{N}$ there exists an integer $n_0 \in \mathbb{N}$ such that for all integers $n \geq n_0$ the relation $|f(n)| < n^{-c}$ holds.*

This basically means that $f(n)$ diminishes to zero faster than the reciprocal of any polynomial. The canonical example of a negligible function is $f(n) = x^{-n}$ for any $x \geq 2$. Other examples include $f(x) = c^{-\sqrt{n}}$, $f(n) = n^{-\log n}$, and even $f(n) = (\log n)^{-\log n}$. Note that negligible functions are closed under addition and multiplication with a polynomial; that is, if $f(\cdot)$ and $g(\cdot)$ are negligible functions and $p(\cdot)$ is a polynomial, then the functions $f(\cdot) + g(\cdot)$ and $p(\cdot)f(\cdot)$ are also negligible. Using the asymptotic order notation, we can say that $f(n) = o(n^{-c})$ for every constant $c \in \mathbb{N}$.

If $f(n)$ is not negligible, then it can be called *nonnegligible*. A function that is nonnegligible does not satisfy Definition D.1, and hence there exists an integer $c \in \mathbb{N}$ such that for all $n_0 \in \mathbb{N}$ there exists at least one integer $n \geq n_0$ that satisfies $|f(n)| \geq n^{-c}$. If, for example, $f(n) = n^{-1000}$, then this function is clearly nonnegligible: If $c = 1001$, then for all $n_0 \in \mathbb{N}$ all $n \geq n_0$ satisfy $|n^{-1000}| \geq n^{-1001}$.

The notion of negligibility divides the set of all functions into two distinct subsets: Functions that are negligible and functions that are nonnegligible. Furthermore, the term *noticeable* as captured in Definition D.2 refers to some functions that are nonnegligible.

**Definition D.2 (Noticeable function)** *A function $f : \mathbb{N} \rightarrow \mathbb{R}$ is* noticeable *if and only if there exist integers $c \in \mathbb{N}$ and $n_0 \in \mathbb{N}$ such that for all $n \geq n_0$ the relation $|f(n)| \geq n^{-c}$ holds.*

While the definitions of noticeable and nonnegligible functions look very similar, it is important to note that a nonnegligible function is not necessarily a noticeable function. A nonnegligible function only needs to have one particular integer $n \geq n_0$ for which $f(n) \geq n^{-c}$, whereas a noticeable function must satisfy this property for every integer $n \geq n_0$. This difference is subtle and makes it difficult to correctly use the terms.

Lat but not least, we note that there are functions that are neither negligible nor noticeable. For example, if we take a negligible function and a noticeable function and interleave them (for odd and even integer arguments), then the resulting function is nonnegligible and nonnoticeable. As an example we consider the function

$$f(n) = \begin{cases} 2^{-n} & \text{if } x \text{ is even} \\ n^{-3} & \text{if } x \text{ is odd} \end{cases}$$

This function cannot be negligible, because for every odd integer it yields a result that is only polynomially small, but it cannot be noticeable either, because for every even integer it yields a result that is exponentially small.

## D.4 EFFICIENT COMPUTATIONS

In practice, one is often interested in finding the most efficient (i.e., fastest running) algorithm to compute a function or solve a problem. Hence, the notion of efficiency is closely related to the *running time* of an algorithm (i.e., the number of primitive operations or "steps" it takes to process an input of a particular size). Often a step is just a bit operation, but it can also be something more comprehensive, such as a comparison, a machine instruction, a machine clock cycle, a modular multiplication, or anything else. If the input is a string, then the size of the input is its length. If it is an integer $n$, then its size is the logarithm of $n$. Since the logarithms of $n$ for different bases only differ by a constant factor, it does not really matter what basis one takes; that is, one can either take $\log n$ (base 10), $\ln n$ (base $e$), or any other basis—the asymptotic order notation remains unaffected by that.

The running time of an algorithm (expressed as a function of the input size) can either be measured in the *worst case* or *average case*. In the worst case, the running time represents an upper bound, meaning that the algorithm is guaranteed to terminate in this amount of time. In the average case, one can only expect the algorithm to terminate in this time. In complexity theory, one is mainly interested in the worst-case running time of an algorithm. As captured in Definition D.3, this is used to distinguish between polynomial-time and super-polynomial-time algorithms.

**Definition D.3 (Polynomial-time algorithm)** *An algorithm is called* polynomial-time *if its worst-case running time is polynomial in the input size. Otherwise (i.e., if the running time cannot be bounded by such a polynomial) it is called* super-polynomial-time.

Hence, the worst-case running time of a polynomial-time algorithm is of the form $O((\ln n)^c)$, where $\ln n$ is the input size and $c$ is some fixed (typically small) integer. The most important examples of super-polynomial-time algorithms are exponential-time algorithms (i.e., algorithms that run in time exponential in the input size). The worst-case running time of such an algorithm is of the form $O(e^{\ln n})$, but the base $e$ can also be replaced by any other real number greater than one.

Complexity theory considers polynomial-time algorithms (be they deterministic or probabilistic) as being efficient and super-polynomial-time algorithms as being inefficient. This is particularly true if the polynomials are of small degrees (e.g., $c \leq 10$).[6] In practice, however, the distinction between efficient and inefficient may be fuzzy, because a super-polynomial-time algorithm may be practical for the input size of interest, whereas a polynomial-time algorithm may be completely impractical (if, for example, the degree of the polynomial is very large).

Interestingly, there are functions that are exponential in nature, but whose exponent grows very slowly. Take an algorithm with a worst-case running time of the form $O(c^{\ln \ln n})$ as an example. Instead of the input size $\ln n$, the exponent refers to the logarithm of this value. Needless to say, this exponent grows very slowly compared to the exponent of a "normal" exponential function. Such functions are sometimes called *subexponential* because their worst-case running time is less than exponential. It goes without saying that there exist infinitely many such functions. To classify and characterize them, people often use the *L-notation* that was introduced in [5]. It yields another asymptotic notation (somewhat similar in spirit to the big-O notation).

According to its name, the L-notation employs a function $L_n$ to characterize the asymptotic behavior of an algorithm's worst-case running time for inputs of size $n$. It is defined as follows:

$$L_n[u, v] = e^{v(\ln(n))^u(\ln(\ln(n)))^{1-u}}$$

The function looks involved, so let's dissect it here. The function takes two real-valued parameters $u$ and $v$ between zero and one (i.e., $u, v \in \mathbb{R}$ and $u, v \in [0, 1]$). Roughly speaking, $u$ and $v$ stand for the efficiency of an algorithm: The smaller the values, the more efficient the respective algorithms are. The parameter $u$ is the important one, because it controls whether the running time is polynomial (if $u = 0$) or exponential (if $u = 1$) in $\ln n$ :

- If $u = 0$, then $L_n[0, v] = e^{v(\ln(n))^0(\ln(\ln(n)))^1} = e^{v(\ln(\ln(n)))} = (\ln(n))^v$ (because $e^{\ln(\ln(n))} = \ln(n)$ follows from the definition of the logarithm

6 Consider a polynomial-time algorithm with a complexity equal to $O((log_2(n))^c)$. If you double the input size (from $n$ to $2n$), then the algorithm will have a complexity that is equal to $O((log_2(n) + 1)^c) = O((log_2(n))^c + c((log_2(n))^{c-1} + \ldots$ If $log_2(n)$ is much larger than $c$, then the algorithm must essentially perform $c((log_2(n))^{c-1}$ additional steps. For example, if $c = 1$, then one has to perform only one additional step. If $c = 2$, then this is only $2(log_2(n))$ additional steps.

function). The result $(\ln(n))^v$ is polynomial in $\ln n$, where $v$ represents the degree of the polynomial.

- If $u = 1$, then $L_n[1, v] = e^{v(\ln(n))^1(\ln(\ln(n)))^0} = e^{v(\ln(n))} = e^{(\ln(n))v}$. The result $e^{\ln(n)\cdot v}$ is exponential in $\ln n$, where $v$ represents a multiplicative factor.

In either case, the parameter $v$ is less important than $u$. To be formally even more precise, $v$ is sometimes written as $v + o(1)$, where $o(1)$ is a function that asymptotically approaches zero. This formalism is not used here.

The bottom line is that $L_n$ is polynomial (in the length of the input) for $u = 0$ and exponential for $u = 1$, and that the interesting cases occur between these two extreme values. If $0 < u < 1$, then $L_n[u, v]$ is neither polynomial nor exponential; it is then called *subexponential.* A subexponential-time algorithm runs asymptotically slower than a polynomial-time algorithm, but it runs asymptotically faster than an exponential-time algorithm. There are subexponential-time algorithms to solve many computational problems that are relevant in cryptography, including the IFP and the DLP. In the running-time analyses of integer factoring algorithms and algorithms to compute discrete logarithms, for example, $u$ is typically either $1/2$ or $1/3$:

$$\begin{aligned} L_n[1/2, v] &= e^{v(\ln n)^{1/2}(\ln\ln n)^{1/2}} = e^{v\sqrt{\ln n}\sqrt{\ln\ln n}} = e^{v\sqrt{\ln n \ln\ln n}} \\ L_n[1/3, v] &= e^{v(\ln n)^{1/3}(\ln\ln n)^{2/3}} \end{aligned}$$

The second case occurs with the most efficient algorithm to factorize integers and compute discrete logarithms (see Chapter 5), although there are some recent improvements in computing discrete logarithms in some particular groups.

## D.5 COMPUTATIONAL MODELS

In theoretical computer science, one usually considers the following models to do a computation:

- Boolean circuits;
- Turing machines;
- Random access machines.[7]

7 A random access machine is similar to a Turing machine. The major distinguishing feature is that it provides access to arbitrary (i.e., randomly chosen) memory cells. As such, the random access machine even more closely represents computer systems as they exist today.

As mentioned earlier, Church's thesis claims that all three computational models are equivalent, meaning that if a function (problem) is computable (solvable) in one model, then it is also computable (solvable) in the other models. It also means that the computational complexities are equal—maybe up to a polynomial transformation. For example, simulating a random access machine using a Turing machine generally squares the number of steps. Consequently, from a theoretical point of view, it doesn't matter which model is used. As mentioned earlier, the most frequently used model is the Turing machine, which is a primitive but sufficiently general computational model.

In spite of the fact that it carries the word "machine" in its name, a Turing machine can be thought of as a computer program (software) rather than an actual computer or machine (hardware). As such, it is computer-independent and can be implemented on different computing devices. In short, a Turing machine consists of a (finite-state) control unit and one (or several) tape(s), each equipped with a tapehead (i.e., a read/write head). Each tape is marked off into (memory) cells that can be filled with at most one symbol from a given alphabet. The tapehead is able to read and/or write exactly one cell, namely the one that is located directly below it. Hence, the operations of the Turing machine are limited to reading and writing symbols on the tapes and moving along the tapes to the left or to the right. As such, it represents a *finite state machine* (FSM). This basically means that the machine has a finite number of states and is in exactly one of these states at any given point in time.

The Turing machine solves a problem (instance) by having a tapehead scanning a finite input string that is placed sequentially in the leftmost cells of one tape (i.e., the input tape). Each symbol occupies one cell and the remaining cells to the right on that tape are blank. The scanning starts from the leftmost cell while the machine is in a designated *initial state*. At any time, only one tapehead of the Turing machine is accessing its tape. A step of access made by a tapehead on its tape is called a *move*. If the machine starts from an initial state, makes one move after another, completes scanning the input string, eventually causes a satisfaction of a terminating condition and thereby terminates, then the machine is said to recognize the input. Otherwise, the machine has no move to make at some point, and hence the machine halts without recognizing the input. An input that is recognized by a Turing machine is called an *instance* in a recognizable language.

Upon termination, the number of moves that a Turing machine $M$ has taken to recognize the input is said to be the *running time* or *time complexity* of $M$. It is denoted as $T_M$. It goes without saying that $T_M$ can be expressed as a function $T_M(n) : \mathbb{N} \rightarrow \mathbb{N}$ where $n$ is the length or size of the input (i.e., the number of symbols that represent the input string when $M$ is in the initial state). It is always the case that $T_M(n) \geq n$, because the machine must at least read the input (typically encoded using the unary representation). In addition to the time requirement, $M$ may also have a space requirement $S_M$ that refers to the number of tape cells that the tapeheads of $M$ have visited. The quantity $S_M$ can also be expressed as a function $S_M(n) : \mathbb{N} \rightarrow \mathbb{N}$ and is said to be the *space complexity* of $M$.

A Turing machine is called polynomial-time if its worst-case running time is polynomial in the input size. For all practical purposes, polynomial-time Turing machines can perform computations that can also be carried out on today's computer systems within reasonable amounts of time. Such machines work in a deterministic way, meaning that they repeatedly execute one (or several) deterministic step(s). This need not be the case, and there are at least two alternative types of Turing machines:

- *Nondeterministic Turing machine:* This is a polynomial-time Turing machine that works in a nondeterministic way.
- *Probabilistic Turing machine:* This is a polynomial-time Turing machine that works in a probabilistic way.

A nondeterministic Turing machine is a purely theoretical construct, meaning that it is generally not possible to build such a machine. As its name suggests, it works in a nondeterministic way and is able to solve a computational problem if such a solution exists.

In contrast to a nondeterministic Turing machine, a probabilistic Turing machine can be built. It works in a probabilistic (and nondeterministic) way. Similar to a deterministic Turing machine, a probabilistic Turing machine may have a plurality of tapes. One of these tapes is called a *random tape* and contains uniformly distributed random symbols. During the scanning of an input instance, the machine interacts with the random tape, picks up a random string, and then proceeds like a deterministic Turing machine. The random string is called the *random input* to the probabilistic Turing machine. With the involvement of the random input, the recognition of an instance by a probabilistic Turing machine is no longer a deterministic function of the instance, but is associated with a random variable (i.e., a function of the Turing machine's random input). This random variable typically assigns error probabilities to the event of recognizing the problem instance (this is explored further later). Remarkably, there are many problems for which probabilistic Turing machines can be constructed that are more efficient, both in terms of time and space, than the best-known deterministic counterparts. Consequently, probabilistic Turing machines are an important field of study in complexity theory and also have many applications in cryptography.

Last but not least, it is important to note that there are computational models that are not equivalent to the models itemized above. Examples include quantum computers and DNA computers.

- A *quantum computer* is a computational device that makes use of quantum mechanical principles to solve a computational problem. A conventional computer operates on bits that represent either 0 or 1. In contrast, a quantum computer operates on *quantum bits* (*qubits*) that represent vectors in the two-dimensional Hilbert space. More specifically, a qubit is a linear combination or superposition of $|0\rangle$ and $|1\rangle$—with $|0\rangle$ and $|1\rangle$ representing an orthonormal basis. A qubit can be written in terms of function $\psi = \alpha|0\rangle + \beta|1\rangle$ with $\alpha, \beta \in \mathbb{C}$ and $|\alpha|^2 + |\beta|^2 = 1$. It is a fundamental law of quantum mechanics that once one measures the state of a qubit $\psi$, one either gets $|0\rangle$ or $|1\rangle$ as a result. More precisely, one measures $|0\rangle$ with probability $|\alpha|^2$ and $|1\rangle$ with probability $|\beta|^2$. The fundamental difference between bits and qubits is that qubits may be in states between $|0\rangle$ and $|1\rangle$. Only by measuring the state of a qubit, one can get one of the states $|0\rangle$ or $|1\rangle$. A quantum register of length $n$ is built of $n$ qubits $|q_k\rangle$ with $k = 1, \ldots, n$. Each $|q_k\rangle$ is of the form $\alpha_k|0\rangle + \beta_k|1\rangle$. Due to superposition, a quantum register may be in all of the $2^n$ possible states at the same time. A quantum computer may exploit this fact and make use of such a quantum register to solve particular problem instances. In 1994, Peter W. Shor proposed randomized polynomial-time algorithms for factoring integers and computing discrete logarithms on a quantum computer [6, 7]. Also, as mentioned in Section 9.6.1.4, Grover's algorithm can be used on a quantum computer to reduce the steps required to perform an exhaustive key search for an $n$-bit cipher from $2^n$ to $2^{n/2}$ [8, 9]. In spite of the fact that the reduction is significant, the resulting algorithm is still not running in polynomial time.[8] In 2001, Shor's algorithm was used to factor the integer 15 [10], but it is still unknown whether a quantum computer of useful size will ever be built. To factorize an integer $n$, a quantum register of length $\ln n$ is required. For the typical length of an RSA modulus, this translated to a few thousand qubits. As of this writing, people are able to build quantum computers with registers of 50–70 qubits. So there is still a long way to go until quantum computers are ready to implement Shor's algorithm against RSA moduli with 2048 bits or even more. Also, no polynomial-time algorithm for solving any **NP**-complete problem[9] on a quantum computer has been found so far.

8 To make things worse, the operations needed in Grover's algorithm are inherently sequential, meaning that they cannot be parallelized.

9 Refer to Section D.6 and Definition D.10 for the notion of an **NP**-complete problem.

- A *DNA computer* is a computational device that makes use of molecular biology to solve computational problems. More specifically, molecules of deoxyribonucleic acid (DNA) are used to encode problem instances, and standard protocols and enzymes are used to perform the steps of the corresponding computations. In 1994, Leonard M. Adleman[10] demonstrated the feasibility of using a small DNA computer to solve an (arguably small) instance of the directed Hamiltonian path problem, which is known to be $\mathbf{NP}$-complete [11]. Further information about the DNA computing can be found in [12, 13] or any newer book on the topic (if there are any).

It is neither presently known how to build a quantum or DNA computer of a sufficiently large size, nor is it even known to be possible at all. Nevertheless, should either quantum computers or DNA computers ever become feasible and practical, they would have a tremendous impact on theoretical computer science in general and cryptography in particular. In fact, many cryptographic systems that are computationally secure today would become totally insecure and worthless. This is particularly true for many public key cryptosystems, and this is why people are looking into PQC (Section 18.3).

## D.6 COMPLEXITY CLASSES

In the previous section, we introduced deterministic, nondeterministic, and probabilistic polynomial-time Turing machines. These machines can be used to define complexity classes. In short, deterministic polynomial-time Turing machines can be used to define the complexity class $\mathbf{P}$, nondeterministic polynomial-time Turing machines can be used to define the complexity class $\mathbf{NP}$ (or $\mathbf{coNP}$, respectively), and probabilistic polynomial-time Turing machines can be used to define the complexity class $\mathbf{PP}$ and some subclasses thereof. All of these classes are overviewed, discussed, and put into perspective next. For the sake of simplicity, we only focus on Turing machines and decision problems. Note, however, that it is also possible to formally define the previously mentioned complexity classes on the basis of algorithms (instead of Turing machines) and for problems other than decision problems.

If we want to make precise statements about the computational difficulty of a problem, then we use reductions. In short, a *reduction* is an algorithm for transforming one problem into another problem (to show that the second problem is at least as difficult to solve as the first problem). More specifically, problem 1 is reducible to problem 2 if an algorithm for solving problem 2 can be used as a subroutine (in this context sometimes also called an oracle) to solve problem 1. If this

10 Leonard M. Adleman is a coinventor of the RSA public key cryptosystem.

is true, then solving problem 1 cannot be substantially harder than solving problem 2 (where "harder" means having a higher computational difficulty). Following this line of argumentation, Definition D.4 captures the notion of a polynomial (or polynomial-time) reduction for two decision problems.

**Definition D.4 (Polynomial-time reduction)** *Let $D_1, D_2 \subseteq \{0,1\}^*$ be two decision problems. $D_1$* polytime reduces *to $D_2$, denoted $D_1 \leq_P D_2$ if there is an algorithm that can solve $D_1$ in polynomial time when using an algorithm for solving $D_2$ as a subroutine (oracle).*

If $D_1 \leq_P D_2$, then $D_1$ is not harder to solve than $D_2$, or—alternatively speaking—$D_2$ is at least as difficult to solve than $D_1$. Polynomial-time reductions are transitive, meaning that if $D_1 \leq_P D_2$ and $D_2 \leq_P D_3$, then $D_1 \leq_P D_3$. If both $D_1 \leq_P D_2$ and $D_2 \leq_P D_1$, then $D_1$ and $D_2$ are *computationally equivalent*, denoted $D_1 \equiv_P D_2$.

### D.6.1 Complexity Class P

Informally speaking, the complexity class **P** (polynomial-time) refers to the class of decision problems that can be solved by a deterministic Turing machine in polynomial time. This is formally expressed in Definition D.5.

**Definition D.5 (Complexity class P)** *The complexity class* **P** *refers to the class of decision problems $D \subseteq \{0,1\}^*$ that are solvable in polynomial time by a deterministic Turing machine; that is, there exists a deterministic polynomial-time Turing machine $M$ with $M(x) = 1$ if and only if $x \in D$.*

In practice, **P** is by far the most widely used complexity class. It comprises all problems that can currently be solved efficiently, using standard computing facilities.

### D.6.2 Complexity Classes NP and coNP

Similar to **P**, one can define the complexity class **NP** (nondeterministic polynomial-time) to refer to the class of decision problems that can be solved by a nondeterministic Turing machine in polynomial time. This is formally expressed in Definition D.6.

**Definition D.6 (Complexity class NP)** *The complexity class* **NP** *refers to the class of decision problems $D \subseteq \{0,1\}^*$ that are solvable in polynomial time by a nondeterministic Turing machine; that is, there exists a nondeterministic polynomial-time Turing machine $M$ with $M(x) = 1$ if and only if $x \in D$.*

As mentioned earlier, nondeterministic Turing machines are purely theoretical constructs, and it is not currently known how to build one. If a solution exists, then a nondeterministic Turing machine somehow finds it and its validity can be verified efficiently. This idea is captured in Definition D.7 that yields another possibility to define **NP**.

**Definition D.7 (Complexity class NP)** *The complexity class* **NP** *refers to the class of decision problems* $D \subseteq \{0,1\}^*$ *for which a YES answer can be verified in polynomial time given some extra information, called a* certificate *or* witness.

It must be emphasized that if a decision problem is in **NP**, then it may not be the case that a certificate for a YES answer can be obtained easily. What is asserted is only that such a certificate exists, and, if known, can be used to efficiently verify a YES answer. For example, the problem of deciding whether a positive integer $n$ is composite (i.e., whether there exist integers $1 < p_1, p_2, \ldots, p_k \in \mathbb{N}$ such that $n = p_1 p_2 \ldots p_k$) belongs to **NP**. This is because if $n$ is composite, then this fact can be verified in polynomial time if one is given a divisor $a$ of $n$. In this case, the certificate is a divisor $p_i$ for $1 \leq i \leq k$.

It is not clear whether the existence of an efficient verification algorithm for YES answers also implies the existence of an efficient verification algorithm for NO answers. As an example, remember the work of Goldwasser, Micali, and Rackoff (Section 15.1 on how to prove language membership and nonmembership simultaneously. In general, this need not be the case and there is room for a complementary complexity class **coNP** as captured in Definition D.8 (that follows the line of argumentation from Definition D.7).

**Definition D.8 (Complexity class coNP)** *The complexity class* **coNP** *refers to the class of decision problems* $D \subseteq \{0,1\}^*$ *for which a NO answer can be verified in polynomial time given some extra information, called a* certificate *or* witness.

It is conjectured that **coNP**$\neq$ **NP**. Note, however, that this is only a conjecture, and that nobody has been able to prove it (or the converse) so far.

**NP** (**coNP**) refers to the class of decision problems for which a YES (NO) answer can be verified in polynomial time given an appropriate certificate or witness. Contrary to that, **P** consists of the class of decision problems for which an answer can be found in polynomial time. It is obvious that **P** $\subseteq$ **NP** and **P** $\subseteq$ **coNP**. But we don't know whether the existence of an efficient verification algorithm for decision problems (be it for YES or NO answers) also implies the ability to efficiently provide an answer for such a problem. This question can be phrased in the single most important open question in theoretical computer science and complexity theory, namely whether

$$\mathbf{NP} = \mathbf{P} \text{ or } \mathbf{NP} \neq \mathbf{P}$$

If this question were answered in the affirmative (i.e., $\mathbf{NP} = \mathbf{P}$), then every problem (function) in $\mathbf{NP}$ would theoretically solvable (computable) in polynomial time. It is, however, widely believed that the opposite (i.e., $\mathbf{P} \neq \mathbf{NP}$) is true, meaning that $\mathbf{P} \subset \mathbf{NP}$. This belief is also supported by our intuition that solving a problem is usually more involved than verifying a solution. Empirical evidence toward the conjectured inequality is given by the fact that literally thousands of problems in $\mathbf{NP}$, coming from a wide variety of mathematical and scientific disciplines, are not known to be solvable in polynomial time (in spite of extensive research attempts aimed at providing efficient algorithms to solve them).

If $\mathbf{P} = \mathbf{NP}$ were true, then there would be no computationally secure cryptosystem in a mathematically strong sense. Nevertheless, there would still be cryptographic systems that are computationally secure for all practical purposes, provided that the complexity ratio between using the system and breaking it is a polynomial of sufficiently high degree. Also, all unconditionally (i.e., information-theoretically) secure cryptographic systems would remain unaffected by $\mathbf{P} = \mathbf{NP}$.

In the literature, problems are often called $\mathbf{NP}$-hard or $\mathbf{NP}$-complete, so let us briefly explain what these terms means. According to Definition D.9, a decision problem $D$ is $\mathbf{NP}$-hard, if every decision problem in $\mathbf{NP}$ polytime reduces to it.

**Definition D.9 (NP-hard problem)** *A decision problem $D \subseteq \{0,1\}^*$ is $\mathbf{NP}$-hard if $D_1 \leq_P D$ for every decision problem $D_1 \in \mathbf{NP}$.*

If, in addition to be $\mathbf{NP}$-hard, $D$ itself is also in $\mathbf{NP}$, then $D$ is $\mathbf{NP}$-complete. This is captured in Definition D.10.

**Definition D.10 (NP-complete problem)** *A decision problem $D \subseteq \{0,1\}^*$ is $\mathbf{NP}$-complete if $D \in \mathbf{NP}$ and $D_1 \leq_P D$ for every decision problem $D_1 \in \mathbf{NP}$.*

Note that this definition can't be used to show that a decision problem $D$ is $\mathbf{NP}$-complete. This is because it is difficult to show the second condition for every $D_1 \in \mathbf{NP}$. If, however, we already know that a specific decision problem $D_1$ is $\mathbf{NP}$-complete, then we can prove the $\mathbf{NP}$-completeness of $D$ by showing that it is in $\mathbf{NP}$ and that it polytime reduces to $D_1$. More specifically, the following three steps can be used to prove that a decision problem $D$ is $\mathbf{NP}$-complete:

1. Prove that $D \in \mathbf{NP}$.
2. Select a decision problem $D_1$ that is known to be $\mathbf{NP}$-complete.
3. Prove that $D_1 \leq_P D$.

Consequently, $\mathbf{NP}$-complete (decision) problems are universal in the sense that providing a polynomial-time algorithm for solving one of them immediately implies polynomial-time algorithms for solving all of them. More specifically, if there exists a single **NP**-complete decision problem that can be shown to be in **P**, then **P** = **NP** follows immediately. Similarly, if there exists a single **NP**-complete decision problem that can be shown in **coNP**, then **coNP** = **NP** follows immediately. Such a result would be extremely surprising, and a proof that a problem is **NP**-complete generally provides strong evidence for its computational intractability.

At first glance, it may be surprising that **NP**-complete problems exist in the first place. But there are literally thousands of problems known to be **NP**-complete (e.g., [1]), coming from a wide range of mathematical and scientific disciplines and fields of study. For example, deciding whether a Boolean formula is satisfiable and deciding whether a directed graph has a Hamiltonian cycle are both **NP**-complete decision problems. Furthermore, the subset sum problem is **NP**-complete and has been used as a basis for many public key cryptosystems in the past (i.e., knapsack-based cryptosystems). The subset sum problem is that given a set of positive integers $\{a_1, a_2, \ldots, a_n\}$ and a positive integer $s$, determine whether or not there is a subset of the $a_i$ that sum to $s$.

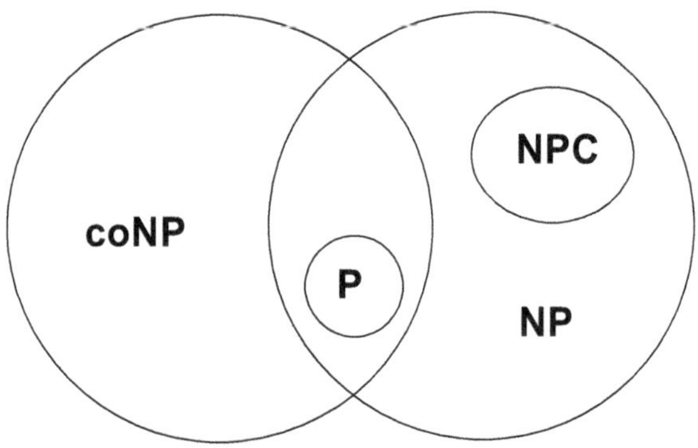


**Figure D.1** The conjectured relationship between **P**, **NP**, **coNP**, and **NPC**.

The class of all **NP**-complete (decision) problems is sometimes also denoted by **NPC**. Figure D.1 illustrates the conjectured relationship between the complexity classes **P**, **NP**, **coNP**, and **NPC**. Again, we know that **P** $\subseteq$ **NP** and **P** $\subseteq$ **coNP**, as well as **NPC** $\subseteq$ **NP**. We do not know, however, whether **P** = **NP**, **P** = **coNP**, or **P** = **NP** $\cap$ **coNP**. Most experts believe that the answers to the last three questions is NO (but keep in mind that these are conjectures that have not been proven so far).

Referring to Definitions D.9 and D.10, it is easy to see that an **NP**-complete problem must always be **NP**-hard, but that the converse need not be true (i.e., an

$\mathbf{NP}$-hard problem need not be $\mathbf{NP}$-complete). In fact, there are (decision) problems that are $\mathbf{NP}$-hard but not $\mathbf{NP}$-complete. For example, the *halting problem* (i.e., the problem to decide whether a given program with a given input will halt or run forever) is $\mathbf{NP}$-hard but not $\mathbf{NP}$-complete. On one hand, one can show that there exists an $\mathbf{NP}$-complete problem (e.g., the satisfiability problem) that polytime reduces to the halting problem. On the other hand, one can show that the halting problem is not in $\mathbf{NP}$ (because all problems in $\mathbf{NP}$ must be decidable but the halting problem is not). Also, finding a satisfying assignment for a Boolean formula or finding a Hamiltonian cycle in a directed graph are $\mathbf{NP}$-hard problems. We can also revisit the subset sum problem mentioned above. Given positive integers $\{a_1, a_2, \ldots, a_n\}$ and a positive integer $s$, a computational version of the subset sum problem would ask for a subset of the $a_i$ that sum up to $s$, provided that such a subset exists. This problem can also be shown to be $\mathbf{NP}$-hard.

The complexity classes $\mathbf{P}$, $\mathbf{NP}$, $\mathbf{coNP}$, and $\mathbf{NPC}$ are defined with deterministic and nondeterministic Turing machines that run in polynomial time. If one considers probabilistic Turing machines, then new complexity classes pop up.

### D.6.3 Complexity Class $\mathbf{PP}$ and Its Subclasses

If one replaces the deterministic Turing machine from Definition D.5 with a probabilistic one, then one enters the realms of some new complexity classes. The most general class is $\mathbf{PP}$ (probabilistic polynomial-time) as captured in Definition D.11.

**Definition D.11 (Complexity class $\mathbf{PP}$)** *The complexity class* $\mathbf{PP}$ *refers to the class of decision problems* $D \subseteq \{0,1\}^*$ *that are solvable in polynomial time by a probabilistic Turing machine.*

If a decision problem is in $\mathbf{PP}$, then there must be an algorithm that is allowed to flip coins and make random decisions and that is guaranteed to run in polynomial time. If the answer is YES, then the algorithm answers YES with a probability greater than $1/2$. If the answer is NO, then the algorithm answers YES with a probability less or equal than $1/2$. These requirements lead the machine to be better than guessing. In more practical terms, $\mathbf{PP}$ is the class of problems that can be solved to any fixed degree of accuracy by running a randomized, polynomial-time algorithm a sufficient (but bounded) number of times.

$\mathbf{PP}$ contains $\mathbf{NP}$ and many subclasses with distinct error probabilities. For example, the suclass of $\mathbf{PP}$ that comprises all decision problems $D \subseteq \{0,1\}^*$ for which a probabilistic polynomial-time Turing machine $M$ exists that always outputs correct results is called zero-sided-error probabilistic polynomial time and is denoted as $\mathbf{ZPP}$. It is captured in Definition D.12.

**Definition D.12 (Complexity class ZPP)** **ZPP** *is the subclass of* **PP** *that comprises all decision problems* $D \subseteq \{0,1\}^*$ *for which there exists a probabilistic polynomial-time Turing machine* $M$ *such that for every input* $x \in \{0,1\}^*$

$$\Pr[M \text{ outputs YES } \mid x \in D] = 1$$

*and*

$$\Pr[M \text{ outputs YES } \mid x \notin D] = 0$$

The fact that $M$ is always correct means that the errors are *zero-sided.* In the more general case, one has to consider *one-sided* and *two-sided* errors, and this brings in the complexity classes **RP** and **BPP** as captured in Definitions D.13 and D.14.

**Definition D.13 (Complexity class RP)** **RP** *is the subclass of* **PP** *that comprises all decision problems* $D \subseteq \{0,1\}^*$ *for which there exists a probabilistic polynomial-time Turing machine* $M$ *such that for every input* $x \in \{0,1\}^*$

$$\Pr[M \text{ outputs YES } \mid x \in D] > 1/2$$

*and*

$$\Pr[M \text{ outputs YES } \mid x \notin D] = 0$$

This definition says that a YES answer is always correct, whereas a NO answer may be wrong (i.e., it may be the case that $x \in D$ and yet $M$ outputs NO). Note that the fraction $1/2$ in the definition is somehow arbitrary, and that **RP** contains exactly the same problems if $1/2$ is replaced by any other value $\epsilon$ between $1/2$ and 1 (i.e., $\epsilon \in (1/2, 1)$). Also note that the class **coRP** is defined in an analog way with YES replaced with NO. This means that **coRP** comprises all decision problems for which there exists a probabilistic polynomial-time Turing machine $M$ whose NO answers are always correct.

Next, we take a look at Turing machines that have the YES and NO answers wrong. This leads to the complexity class **BPP**, which stands for bounded-error probabilistic polynomial time.

**Definition D.14 (Complexity class BPP)** **BPP** *is the subclass of* **PP** *that comprises all decision problems* $D \subseteq \{0,1\}^*$ *for which a probabilistic polynomial-time Turing machine* $M$ *exists such that for every input* $x \in \{0,1\}^*$

$$\Pr[M \text{ outputs YES } \mid x \in D] \geq \epsilon$$

*and*

$$\Pr[M \text{ outputs YES } \mid x \notin D] \leq \delta$$

*with* $\epsilon \in (1/2, 1)$ *and* $\delta \in (0, 1/2)$.

Note that we must require that $\epsilon \neq 1$ and $\delta \neq 0$. Otherwise, the subclass $\mathbf{BPP}$ degenerates to either $\mathbf{ZPP}$ or $\mathbf{RP}$.

The complexity class $\mathbf{P}$ and the various subclasses of $\mathbf{PP}$ can be ordered as follows:

$$\mathbf{P} \subseteq \mathbf{ZPP} \subseteq \mathbf{coRP} \subseteq \mathbf{BPP} \subseteq \mathbf{PP}$$

The challenging question is whether the inclusions are strict or not. In either case, algorithms that can solve problems from any of these complexity classes (not only $\mathbf{P}$) are called *efficient*, and the problems themselves are called *tractable* (in this class). Problems that are not tractable in this sense are called *intractable*. However, keep in mind that polynomials can have vastly different degrees, and hence algorithms that solve tractable problems can still have very different complexities (regarding time and/or space). Therefore, an efficient algorithm for solving a tractable problem need not be efficient in practice, and people sometimes use the term *practically efficient* to refer to polynomial-time algorithms with polynomials of sufficiently small degrees. These algorithms can be realistically executed on contemporary computing machinery.

## D.7 FINAL REMARKS

Complexity theory is a useful tool to argue about the computational complexity of a particular problem. From a bird's eye perspective, it allows us to distinguish efficient algorithms (to solve the problem) and inefficient ones. In short, they are efficient if they run in polynomial time, and they are inefficient otherwise. This distinction is rather coarse, and there may be efficient algorithms that are yet efficient, but whose polynomials have such a large degree that they are not practically efficient. So one has to be cautious whenever one applies complexity-theoretic arguments. There are two additional subtleties to keep in mind and consider with care:

- First, complexity theory deals with the worst-case complexity of problems, meaning that there may still be instances (of a particular problem) that are easy to solve. Worst-case complexity-theoretic reasoning does not exclude this possibility.

- Second, it may be that finding the exact solution for a problem is difficult (in the sense of complexity theory), but finding an approximation of it is relatively simple. If such an approximation is sufficient, then the complexity-theoretic difficulty of finding the exact solution is not particularly meaningful. Again, complexity-theoretic reasoning does not exclude this possibility.

In light of these subtleties, one may add that complexity theory is not only a useful tool, but also an imperfect one. The discussion about its adequateness for cryptography was intensified with the launch of quantum and DNA computers (at least at a conceptual level) and the question whether the Turing machine is still the right computational model. The entire field is a moving target, and it will be interesting to see how it will evolve in the future.

## References

[1] Garey, M.R., and D.S. Johnson, *Computers and Intractability: A Guide to the Theory of NP-Completeness.* W. H. Freeman & Co., New York, 1979.

[2] Arora, S., *Computational Complexity: A Modern Approach.* Cambridge University Press, New York, 2009.

[3] Hopcroft, J.E., R. Motwani, and J.D. Ullman, *Introduction to Automata Theory, Languages, and Computation,* 3rd edition. Pearson, 2008.

[4] Menezes, A., P. van Oorschot, and S. Vanstone, *Handbook of Applied Cryptography.* CRC Press, Boca Raton, FL, 1996.

[5] Buhler, J., H. Lenstra, and C. Pomerance, "Factoring Integers with the Number Field Sieve," *The Development of the Number Field Sieve,* Springer-Verlag, New York, LNCS 1554, 1993, pp. 50–94.

[6] Shor, P.W., "Algorithms for Quantum Computation: Discrete Logarithms and Factoring," *Proceedings of the IEEE 35th Annual Symposium on Foundations of Computer Science (FOCS),* Santa Fe, NM, November 1994, pp. 124–134.

[7] Shor, P.W., "Polynomial-Time Algorithms for Prime Factorization and Discrete Logarithms on a Quantum Computer," *SIAM Journal of Computing,* October 1997, pp. 1484–1509.

[8] Grover, L.K., "A Fast Quantum Mechanical Algorithm for Database Search," *Proceedings of the 28th ACM Symposium on Theory of Computing (STOC),* 1996, pp. 212–219.

[9] Bennett, C.H., et al., "Strengths and Weaknesses of Quantum Computing," *SIAM Journal on Computing,* Vol. 26, Issue 5, October 1997, pp. 1510–1523.

[10] Vandersypen, L.M.K., et al., "Experimental Realization of Shor's Quantum Factoring Algorithm Using Nuclear Magnetic Resonance," *Nature,* Vol. 414, 2001, pp. 883–887.

[11] Adleman, L.M., "Molecular Computation of Solutions to Combinatorial Problems," *Science,* Vol. 266, November 1994, pp. 1021–1024.

[12] Lipton, R.J., "DNA Solution of Hard Computational Problems," *Science*, Vol. 268, April 1995, pp. 542–545.

[13] Păun, G., G. Rozenberg, and A. Salomaa, *DNA Computing: New Computing Paradigms*. Springer-Verlag, New York, 2006.

# List of Symbols

| | |
|---|---|
| $\forall$ | quantifier "for all" |
| $\exists$ | quantifier "there exists" |
| $\square$ | end of proof |
| $=$ | equality |
| | assignment (operator) |
| $\equiv$ | congruence relation |
| $id$ | identity map |
| $\sum$ | sum |
| $\prod$ | product |
| $\infty$ | infinity |
| $\mathcal{O}$ | point at infinity (in ECC) |
| $S$ | set |
| $\lvert S \rvert$ | cardinality of $S$ (i.e., the number of elements in $S$) |
| $2^S$ | power set of $S$ (i.e., the set of all subsets of $S$) |
| $\emptyset$ | empty set |
| $x \in S$ | $x$ is an element of $S$ |
| $x \notin S$ | $x$ is not an element of $S$ |
| $x \in_R S$ | $x$ is a random (i.e., randomly chosen) element of $S$ |
| $x \in (a, b)$ | $x$ is an element from the open interval $(a, b)$ (i.e., $a < x < b$) |
| $x \in [a, b]$ | $x$ is an element from the closed interval $[a, b]$ (i.e., $a \leq x \leq b$) |
| $A \cup B$ | union of sets $A$ and $B$ |
| $A \cap B$ | intersection of sets $A$ and $B$ |
| $A \setminus B$ | difference of sets $A$ and $B$ |
| $A \subseteq B$ | set $A$ is a subset of set $B$ (or $B$ is a superset of $A$) |
| $\mathbb{N}$ | natural numbers |
| $\mathbb{N}^+$ | positive natural numbers (i.e., $\mathbb{N}^+ = \mathbb{N} \setminus \{0\}$) |

| | |
|---|---|
| $\mathbb{Z}$ | integer numbers (i.e., integers) |
| $\mathbb{Z}^+$ | positive integers |
| $\mathbb{Z}^-$ | negative integers |
| $\mathbb{Z}_n$ | integers modulo $n$ |
| $\mathbb{Z}_n^*$ | multiplicative group of integers modulo $n$ |
| $L(x,p)$ or $(x\|p)$ | Legendre symbol of $x$ modulo $p$ (where $p$ is prime) |
| $J(x,n)$ or $(x\|n)$ | Jacobi symbol of $x$ modulo $n$ (where $n$ is a composite number) |
| $J_n$ | elements of $\mathbb{Z}_n^*$ with Jacobi symbol 1 |
| $QR_n$ | set of quadratic residues modulo $n$ |
| $QNR_n$ | set of quadratic nonresidues modulo $n$ |
| $\widetilde{QR_n}$ | set of pseudosquares modulo $n$ |
| $\mathbb{Q}$ | rational numbers |
| $\mathbb{R}$ | real numbers |
| $\mathbb{R}^+$ | positive real numbers |
| $\mathbb{R}^-$ | negative real numbers |
| $\pi$ | transcendental number that represents the ratio of the circumference of a perfect circle to its diameter ($\pi = 3.14159\ldots$) |
| $e$ | transcendental number that represents the base of the natural logarithm ($e = 2.71828\ldots$) |
| $\mathbb{C}$ | complex numbers |
| $i$ | $\sqrt{-1}$ |
| $\mathbb{F}_q$ or $GF(q)$ | finite field or Galois field with $q$ elements (i.e., $\|\mathbb{F}_q\| = q$) |
| $E(\mathbb{F}_q)$ | elliptic curve over $\mathbb{F}_q$ |
| $\mathbb{P}$ | set of all primes |
| $\mathbb{P}_l$ | set of all $l$-bit primes |
| $\mathbb{P}^*$ | set of all safe primes |
| $*$ | binary operation |
| $+$ | addition |
| $-$ | subtraction |
| $\cdot$ | multiplication |
| $/$ | division |
| $x \xleftarrow{P_X} X$ | $x$ is sampled from $X$ according to probability distribution $P_X$ (sometimes only written as $x \leftarrow X$) |
| $x \xleftarrow{r} X$ | $x$ is sampled uniformly at random from $X$ (i.e., $x \in_R X$) |
| $\neg X$ | bitwise negation of the Boolean variable $X$ (NOT) |
| $X \wedge Y$ | bitwise and of the Boolean variables $X$ and $Y$ (AND) |
| $X \vee Y$ | bitwise or of the Boolean variables $X$ and $Y$ (OR) |
| $X \oplus Y$ | bitwise exclusive of the Boolean variables $X$ and $Y$ (XOR) |
| $a = b$ | $a$ is equal to $b$ |

| | |
|---|---|
| $a < b$ | $a$ is smaller than $b$ |
| $a \ll b$ | $a$ is much smaller than $b$ |
| $a > b$ | $a$ is greater than $b$ |
| $a \gg b$ | $a$ is much greater than $b$ |
| $\lvert x \rvert$ | absolute (nonngeative) value of $x$ |
| $len(x)$ | bitlength of $x$ (i.e., $\lceil \log_2 x \rceil$) |
| $\lfloor x \rfloor$ | greatest integer less than or equal to $x$ (i.e., floor of $x$) |
| $\lceil x \rceil$ | smallest integer greater than or equal to $x$ (i.e., ceiling of $x$) |
| div | integer division |
| mod | modulo operator |
| $log_b$ | logarithm to the basis of $b$ |
| $log$ | logarithm to the basis of 10 ($b = 10$) |
| $ln$ | logarithm to the basis of $e$ (logarithm naturalis) |
| $a \mid b$ | integer $a$ divides integer $b$ |
| $a \nmid b$ | integer $a$ does not divide integer $b$ |
| $gcd(a_1, \ldots, a_k)$ | greatest common divisor of integers $a_1, \ldots, a_k$ |
| $lcm(a_1, \ldots, a_k)$ | least common multiple of integers $a_1, \ldots, a_k$ |
| $n!$ | factorial of integer $n$ ($0! = 1$) |
| $\pi(n)$ | prime counting function of integer $n$ |
| $\phi(n)$ or $\varphi(n)$ | Euler's totient function of integer $n$ |
| $L$ | formal language |
| $\Sigma$ | alphabet |
| $\Sigma_{in}$ | input alphabet |
| $\Sigma_{out}$ | output alphabet |
| $\{0, 1\}$ | binary alphabet |
| $w$ | word (i.e., a string over an alphabet) |
| $\lvert w \rvert$ | length of word $w$ |
| $w\rvert_c$ | $c$ leftmost bits of $w$ |
| $w\rvert^c$ | $c$ rightmost bits of $w$ |
| $\varepsilon$ | empty word |
| $\Sigma^n$ | set of all words of length $n$ over alphabet $\Sigma$ |
| $\Sigma^*$ | set of all words over alphabet $\Sigma$ |
| $\|$ or $\circ$ | string concatenation |
| $w \overset{\frown}{\hookleftarrow} c$ | $c$-bit left rotation (circular left shift) of word $w$ |
| $w \hookleftarrow c$ | $c$-bit left *shift* of word $w$ |
| $w \overset{\frown}{\hookrightarrow} c$ | $c$-bit right rotation (circular right shift) of word $w$ |
| $w \hookrightarrow c$ | $c$-bit right shift of word $w$ |
| $f$ | function |
| $X \rightarrow Y$ | mapping from domain $X$ to codomain $Y$ |
| $f(X) \subseteq Y$ | range of $f$ |

| | |
|---|---|
| $f^{-1}$ | inverse function |
| $F$ | function family |
| $f_k$ | instance of function family $F$ |
| $\text{Funcs}[X,Y]$ | family of all functions of $X$ to $Y$ |
| $\text{Perms}[X]$ | family of all permutations on $X$ |
| $h$ | hash function |
| $H$ | hash function family |
| $p(x)$ | polynomial in $x$ |
| $deg(p)$ | degree of polynomial $p$ |
| $A[x]$ | set of all polynomials over $A$ |
| $A[x]_g$ | set of all polynomials in $A[x]$ modulo polynomial $g$ |
| $L_n[a,c]$ | running time function |
| $ord(x)$ | order of group element $x$ |
| $ord_n(x)$ | order of $x$ modulo $n$ |
| $\Omega$ | finite or countably infinite set representing a sample space |
| $\omega$ | elementary event |
| $\mathcal{A}$ | event |
| $\overline{\mathcal{A}}$ | complement of event $\mathcal{A}$ |
| Pr | probability measure |
| $\Pr[\mathcal{A}]$ | probability of event $A$ |
| $\Pr[\omega\|\mathcal{A}]$ | conditional probability of $\omega$ given that $\mathcal{A}$ holds |
| $\Pr[\mathcal{A}\|\mathcal{B}]$ | conditional probability of $\mathcal{A}$ given that $\mathcal{B}$ holds |
| $X$ | random variable |
| $P_X$ | probability distribution of $X$ |
| $P_{XY}$ | joint probability distribution of $X$ and $Y$ |
| $P_{X_1 \ldots X_n}$ | joint probability distribution of $X_1, \ldots, X_n$ |
| $P_{X\|\mathcal{A}}$ | conditional probability distribution of $X$ given that $\mathcal{A}$ holds |
| $P_{X\|Y}$ | conditional probability distribution of $X$ given that $Y$ holds |
| $E[X]$ | expectation (or mean) of $X$ |
| $E[X\|\mathcal{A}]$ | conditional expected value of $X$ given that $\mathcal{A}$ holds |
| $Var[X]$ | variance of $X$ |
| $\sigma[X]$ | standard deviation of $X$ |
| $H(X)$ | entropy of $X$ |
| $H(XY)$ | joint entropy of $X$ and $Y$ |
| $H(X\|Y=y)$ | conditional entropy of $X$ when $Y=y$ |
| $H(X\|Y)$ | conditional entropy of $X$ when given $Y$ |
| $I(X;Y)$ | mutual information between $X$ and $Y$ |
| $H_L$ | entropy of language $L$ |
| $R_L$ | redundancy of language $L$ |
| $n_u$ | unicity distance |

| | |
|---|---|
| $M$ | Turing machine |
| $S_M$ | space complexity of Turing machine $M$ |
| $T_M$ | time complexity of Turing machine $M$ |
| $\mathbf{P}$ | polynomial-time" complexity class |
| $\mathbf{NP}$, $\mathbf{coNP}$ | nondeterministic polynomial-time complexity classes |
| $\mathbf{PP}$ | probabilistic polynomial-time complexity class |
| $\mathbf{ZPP}$ | zero-sided error probabilistic polynomial-time complexity class |
| $\mathbf{RP}$ | one-sided error probabilistic polynomial-time complexity class |
| $\mathbf{BPP}$ | bounded-error probabilistic polynomial-time complexity class |
| $\mathcal{M}$ | (plaintext) message space |
| $\mathcal{C}$ | ciphertext space |
| $\mathcal{K}$ | key space |
| $k$ | secret key (i.e., $k \in \mathcal{K}$) |
| $E$ | family $\{E_k : k \in \mathcal{K}\}$ of encryption functions $E_k : \mathcal{M} \rightarrow \mathcal{C}$ |
| $D$ | family $\{D_k : K \in \mathcal{K}\}$ of decryption functions $D_k : \mathcal{C} \rightarrow \mathcal{M}$ |
| $\mathcal{T}$ | authentication tag space |
| $t$ | authentication tag (i.e., $t \in \mathcal{T}$) |
| $A$ | family $\{A_k : k \in \mathcal{K}\}$ of authentication functions $A_k : \mathcal{M} \rightarrow \mathcal{T}$ |
| $V$ | family $\{V_k : K \in \mathcal{K}\}$ of verification functions $V_k : \mathcal{M} \times \mathcal{T}$ $\rightarrow \{valid, invalid\}$ |
| $(pk, sk)$ | public key pair |
| $s$ | digital signature |
| $\Gamma$ | access structure (employed by a secret sharing scheme) |

# Abbreviations and Acronyms

| | |
|---|---|
| 2FA | two-factor authentication |
| AA | attribute authority |
| ACM | Association for Computing Machinery |
| AE | authenticated encryption |
| AEAD | authenticated encryption with associated data |
| AES | advanced encryption standard |
| AKE | authenticated key exchange |
| AI | artificial intelligence |
| ANS | American National Standard |
| ANSI | American National Standards Institute |
| API | application programming interface |
| APT | advanced persistent threat |
| ASCII | American Standard Code for Information Interchange |
| ASN.1 | abstract syntax notation one |
| ATM | automatic teller machine |
| BBS | Blum, Blum, Shub |
| BCP | best current practice |
| BEAST | browser exploit against SSL/TLS |
| BER | basic encoding rules |
| BIS | Bureau of Industry and Security |
| bit | binary digit |
| BLS | Boneh, Lynn, Shacham |
| BXA | Bureau of Export Administration |

| | |
|---|---|
| CA | certification authority |
| CBC | cipherblock chaining |
| CCA | chosen-ciphertext attack |
| CCA2 | adaptive CCA |
| CCM | counter with CBC-MAC |
| CFB | cipher feedback |
| CLUSIS | Association for the Security of Information Services |
| CMA | chosen-message attack |
| CMAC | cipher-based MAC |
| COCOM | Coordinating Committee for Multilateral Export Controls |
| COTS | commercial off-the-shelf |
| CPA | chosen-plaintext attack |
| CRA | Chinese remainder algorithm |
| CRL | certificate revocation list |
| CRT | Chinese remainder theorem, cathode-ray tube |
| CSIDH | commutative SIDH |
| cSHAKE | customizable SHAKE |
| CSP | certification service provider |
| CSS | Central Security Service |
| | content scrambling system |
| CTR | counter |
| CTS | ciphertext stealing |
| CVP | closest vector problem |
| | |
| DAA | data authentication algorithm |
| DAC | data authentication code |
| DBPA | differential branch prediction analysis |
| DDHP | Decisional Diffie-Hellman problem |
| DEA | data encryption algorithm |
| DER | distinguished encoding rules |
| DES | data encryption standard |
| DESCHALL | DES challenge |
| DESL | DES lightweight |
| DH | Diffie-Hellman |
| DHIES | Diffie-Hellman integrated encryption scheme |
| DHP | Diffie-Hellman problem |
| DIT | directory information tree |
| DLA | discrete logarithm assumption |
| DLP | discrete logarithm problem |
| DLT | distributed ledger technology |

| | |
|---|---|
| DN | distinguished name |
| DNA | deoxyribonucleic acid |
| DNS | domain name system |
| DoC | Department of Commerce |
| DRAM | dynamic random access memory |
| DSA | digital signature algorithm |
| DSS | digital signature system |
| DVD | digital versatile disc |
| | |
| E2EE | end-to-end encrypted |
| EAR | Export Administration Regulations |
| ECB | electronic code book |
| ECBC-MAC | encrypt-last-block or encrypted CBC-MAC |
| ECC | elliptic curve cryptography |
| ECDH | elliptic curve DH |
| ECDLP | elliptic curve DLP |
| ECDSA | elliptic curve digital signature algorithm |
| ECIES | elliptic curve integrated encryption scheme |
| ECM | elliptic curve method |
| ECMQV | elliptic curve MQV |
| ECRYPT | European Network of Excellence for Cryptology |
| ECSTR | Efficient and Compact Subgroup Trace Representation |
| EES | escrowed encryption standard |
| EFF | Electronic Frontier Foundation |
| EFS | encrypted file system |
| EKE | encrypted key exchange |
| EMSA | encoding method for signature with appendix |
| EMSEC | emanations security, emission security |
| ENISA | European Union Agency for Cybersecurity |
| EtM | Encrypt-then-MAC |
| E&M | Encrypt-and-MAC |
| | |
| FDH | full-domain-hash |
| FEAL | fast data encipherment algorithm |
| FIPS | Federal Information Processing Standards |
| FSM | finite state machine |
| FSR | feedback shift register |
| | |
| gcd | greatest common divisor |
| GCHQ | Government Communications Headquarters |

| | |
|---|---|
| GCM | Galois/counter mode |
| GDH | gap Diffie-Hellman |
| GMAC | Galois message authentication code |
| GMR | Goldwasser, Micali, Rivest |
| GNFS | general NFS |
| GPS | global positioning system |
| GSM | Groupe Speciale Mobile |
| | |
| HFE | hidden field equations |
| HKDF | HMAC-based extract-and-expand key derivation function |
| HMAC | hashed MAC |
| | |
| IACR | International Association for Cryptologic Research |
| IBE | identity-based encryption |
| IBM | International Business Machines |
| ICB | initial counter block |
| ICM | integer counter mode |
| | index calculus method |
| ICSI | International Computer Sciencc Institute |
| IDEA | international data encryption algorithm |
| IEC | International Electrotechnical Committee |
| IEEE | Institute of Electrical and Electronics Engineers |
| IETF | Internet Engineering Task Force |
| IFA | integer factoring assumption |
| IFIP | International Federation for Information Processing |
| IFP | integer factoring problem |
| IGE | infinite garble extension |
| IKE | Internet key exchange |
| IKM | input keying material |
| IND | indistinguishability |
| IND-CCA | indistinguishability under CCA |
| IND-CCA2 | indistinguishability under adaptive CCA |
| IND-CPA | indistinguishability under CPA |
| IP | Internet Protocol, initial permutation |
| IPsec | IP security |
| ISO | International Organization for Standardization |
| ISOC | Internet Society |
| ISP | Internet service provider |
| IST | Information Societies Technology |
| IT | information technology |

| | |
|---|---|
| ITU | International Telecommunication Union |
| ITU-T | ITU Telecommunication Standardization Sector |
| IV | initialization vector |
| JTC1 | Joint Technical Committee 1 |
| KDC | key distribution center |
| KDF | key derivation function |
| KEM | key encapsulation mechanism |
| KMA | known-message attack |
| KMAC | KECCAK MAC |
| KMOV | Koyama, Maurer, Okamoto, Vanstone |
| KW | Key wrap (using AES) |
| KWP | Key wrap with padding (using AES) |
| LCG | linear congruential generator |
| LFSR | linear feedback shift register |
| LMS | Leighton-Micali signature |
| LNCS | Lecture Notes in Computer Science |
| LRA | local registration agent |
| LSB | least significant bit |
| LWE | learning with errors |
| MAA | message authenticator algorithm |
| MAC | message authentication code |
| MD | message digest |
| MFA | multifactor authentication |
| MGF | mask generation function |
| MIC | message integrity code |
| MIME | multipurpose Internet mail extensions |
| MIT | Massachusetts Institute of Technology |
| MQV | Menezes, Qu, Vanstone |
| MPAA | Motion Picture Association of America |
| MPC | multiparty computation |
| MSB | most significant bit |
| MtE | MAC-then-Encrypt |
| NATO | North Atlantic Treaty Organization |
| NBS | National Bureau of Standards |
| NCSC | National Cyber Security Centre |

| | |
|---|---|
| NESSIE | New European Schemes for Signatures, Integrity and Encryption |
| NFS | number field sieve |
| NIST | National Institute of Standards and Technology |
| NM | nonmalleability |
| NM-CCA | nonmalleability under CCA |
| NM-CPA | nonmalleability under CPA |
| NMAC | nested MAC |
| NSA | National Security Agency |
| NSE | nonsecret encryption |
| NTRU | Nth degree Truncated polynomial Ring Units |
| | |
| OAEP | optimal asymmetric encryption padding |
| OCB | offset codebook |
| OCSP | online certificate status protocol |
| OFB | output feedback |
| OID | object identifier |
| OKM | output keying material |
| OMAC | one-key CBC MAC |
| OPIE | one-time passwords in everything |
| OTMAC | one-time MAC |
| | |
| PAKE | password-authenticated key exchange |
| PBKDF | password-based key derivation function |
| PC | permuted choice |
| | personal computer |
| PCBC | propagating CBC |
| PER | packet encoding rules |
| PGP | Pretty Good Privacy |
| PHC | Password Hashing Competition |
| PIN | personal identification number |
| PKCS | public key cryptography standard |
| PKI | public key infrastructure |
| PKIX | PKI X.509 |
| PMAC | parallelizable MAC |
| PPT | probabilistic polynomial-time |
| PQC | post-quantum cryptography |
| PRBG | pseudorandom bit generator |
| PRF | pseudorandom function |
| PRG | pseudorandom generator |
| PRP | pseudorandom permutation |

| | |
|---|---|
| PSEC | provably secure elliptic curve |
| PSS | probabilistic signature scheme |
| PSS-R | probabilistic signature scheme with message recovery |
| | |
| QCG | quadratic congruential generator |
| QRP | quadratic residuosity problem |
| QS | quadratic sieve |
| qubit | quantum bit |
| | |
| RA | registration authority |
| RC | Ron's Code |
| RFC | request for comments |
| RFID | radio frequency identification |
| RSA | Rivest, Shamir, Adleman |
| RSAES | RSA encryption scheme |
| RSAP | RSA problem |
| | |
| SDSI | simple distributed security infrastructure |
| SEC | standards for efficient cryptography |
| SECG | Standards for Efficient Cryptography Group |
| SHA | secure hash algorithm |
| SHAKE | secure hash algorithm with KECCAK |
| SHS | secure hash standard |
| SIC | segmented integer counter |
| SIDH | supersingular isogeny DiffieHellman key exchange |
| SIKE | supersingular isogeny key encapsulation |
| SIV | synthetic IV |
| SNARK | succinct non-interactive argument of knowledge |
| SNFS | special NFS |
| SoK | systematization of knowledge |
| SP | special publication |
| SPKI | simple PKI |
| SSH | secure shell |
| SSL | secure sockets layer |
| STS | station-to-station |
| SUF | strongly unforgeable |
| SVP | shortest vector problem |
| | |
| TAN | transaction authentication number |
| TCBC | triple data encryption algorithm cipherblock chaining |

| | |
|---|---|
| TCBC-I | triple data encryption algorithm cipherblock chaining interleaved |
| TCFB | triple data encryption algorithm cipher feedback |
| TCFB-P | triple data encryption algorithm cipher feedback pipelined |
| TDEA | triple data encryption algorithm |
| TECB | triple data encryption algorithm electronic code book |
| TKW | Key wrap (using TDEA) |
| TLS | transport layer security |
| TMAC | two-key CBC MAC |
| TOFB | triple data encryption algorithm output feedback |
| TOFB-I | triple data encryption algorithm output feedback interleaved |
| TTP | trusted third party |
| TWINKLE | The Weizmann Institute Key-Locating Engine |
| TWIRL | The Weizmann Institute Relation Locator |
| | |
| UC | University of California |
| UMAC | universal MAC |
| URL | uniform resource locator |
| US | United States |
| USB | universal serial bus |
| USD | US dollar |
| | |
| W3C | World Wide Web Consortium |
| WEP | wired equivalent privacy |
| WG | working group |
| WPA | Wi-Fi protected access |
| WUF | weakly unforgeable |
| WWW | World Wide Web |
| | |
| X3DH | eXtended Triple Diffie-Hellman |
| XCBC | eXtended CBC |
| XMSS | eXtended Merkle Signature Scheme |
| XOF | extendable-output function |
| XTR (ECSTR) | Efficient and Compact Subgroup Trace Representation |
| | |
| YASD | Yet Another Sieving Device |
| | |
| zk-SNARK | zero-knowledge SNARK |

# About the Author

Rolf Oppliger[1] received an M.Sc. and a Ph.D. in computer science from the University of Berne, Switzerland, in 1991 and 1993, respectively. After spending a year as a postdoctoral researcher at the International Computer Science Institute (ICSI[2]) of UC Berkeley, he joined the federal authorities of the Swiss Confederation—nowadays called the National Cyber Security Centre (NCSC[3])—in 1995 and continued his research and teaching activities at several universities in Switzerland and Germany. In 1999, he received the venia legendi for computer science from the University of Zurich, Switzerland, where he still serves as an adjunct professor. Also in 1999, he founded eSECURITY Technologies Rolf Oppliger[4] to provide scientific and state-of-the-art consulting, education, and engineering services related to information security and began serving as the editor of Artech House's Information Security and Privacy Series. Dr. Oppliger has published numerous papers, articles, and books, holds a few patents, regularly serves as a program committee member of internationally recognized conferences and workshops, and is a member of the editorial board of some prestigious periodicals in the field. He is a senior member of the Association for Computing Machinery (ACM), the Institute of Electrical and Electronics Engineers (IEEE) and its Computer Society, as well as a member of the IEEE Computer Society and the International Association for Cryptologic Research (IACR). Besides, he has also served as the vice-chair of the International Federation for Information Processing (IFIP) Technical Committee 11 (TC11) Working Group 4 (WG4) on network security. His full curriculum vitae is available online.[5]

1 https://rolf-oppliger.ch and https://rolf-oppliger.com.
2 https://www.icsi.berkeley.edu.
3 https://www.ncsc.admin.ch.
4 https://www.esecurity.ch.
5 https://www.esecurity.ch/Flyers/cv.pdf.

# Index

## Recent Titles in the Artech House Computer Security Series

Rolf Oppliger, Series Editor

*Bluetooth Security,* Christian Gehrmann, Joakim Persson, and Ben Smeets

*Computer Forensics and Privacy,* Michael A. Caloyannides

*Computer and Intrusion Forensics,* George Mohay, et al.

*Contemporary Cryptography, Second Edition,* Rolf Oppliger

*Cryptography 101: From Theory to Practice,* Rolf Oppliger

*Cryptography for Security and Privacy in Cloud Computing,* Stefan Rass and Daniel Slamanig

*Defense and Detection Strategies Against Internet Worms,* Jose Nazario

*Demystifying the IPsec Puzzle,* Sheila Frankel

*Developing Secure Distributed Systems with CORBA,* Ulrich Lang and Rudolf Schreiner

*Electric Payment Systems for E-Commerce, Second Edition,* Donal O'Mahony, Michael Peirce, and Hitesh Tewari

*Engineering Safe and Secure Software Systems,* C. Warren Axelrod

*Evaluating Agile Software Development: Methods for Your Organization,* Alan S. Koch

*Implementing Electronic Card Payment Systems,* Cristian Radu

*Implementing the ISO/IEC 27001 Information Security Management System Standard,* Edward Humphreys

*Implementing Security for ATM Networks,* Thomas Tarman and Edward Witzke

*Information Hiding,* Stefan Katzenbeisser and Fabien Petitcolas, editors

*Internet and Intranet Security, Second Edition*, Rolf Oppliger

*Introduction to Identity-Based Encryption,* Luther Martin

*Java Card for E-Payment Applications*, Vesna Hassler, Martin Manninger, Mikail Gordeev, and Christoph Müller

*Multicast and Group Security,* Thomas Hardjono and Lakshminath R. Dondeti

*Non-repudiation in Electronic Commerce,* Jianying Zhou

*Outsourcing Information Security,* C. Warren Axelrod

*The Penetration Tester's Guide to Web Applications*, Serge Borso

*Privacy Protection and Computer Forensics, Second Edition,* Michael A. Caloyannides

*Role-Based Access Control, Second Edition*, David F. Ferraiolo, D. Richard Kuhn, and Ramaswamy Chandramouli

*Secure Messaging with PGP and S/MIME,* Rolf Oppliger

*Securing Information and Communications Systems: Principles, Technologies and Applications,* Javier Lopez, Steven Furnell, Sokratis Katsikas, and Ahmed Patel

*Security Fundamentals for E-Commerce,* Vesna Hassler

*Security Technologies for the World Wide Web, Second Edition,* Rolf Oppliger

*Techniques and Applications of Digital Watermarking and Content Protection,* Michael Arnold, Martin Schmucker, and Stephen D. Wolthusen

*User's Guide to Cryptography and Standards*, Alexander W. Dent and Chris J. Mitchell

For further information on these and other Artech House titles, including previously considered out-of-print books now available through our In-Print-Forever® (IPF®) program, contact:

Artech House
685 Canton Street
Norwood, MA 02062
Phone: 781-769-9750
Fax: 781-769-6334
e-mail: artech@artechhouse.com

Artech House
16 Sussex Street
London SW1V 4RW UK
Phone: +44 (0)20 7596-8750
Fax: +44 (0)20 7630-0166
e-mail: artech-uk@artechhouse.com

Find us on the World Wide Web at: www.artechhouse.com